PERMAFROST

Fourth International Conference

FINAL PROCEEDINGS

July 17-22, 1983

Organized by
University of Alaska and
National Academy of Sciences

NATIONAL ACADEMY PRESS
Washington, D.C. 1984

NATIONAL ACADEMY PRESS 2101 Constitution Avenue NW Washington, DC 20418

The National Research Council was established by the National Academy of Sciences in 1916 to associate the broad community of science and technology with the Academy's purposes of furthering knowledge and of advising the federal government. The Council operates in accordance with general policies determined by the Academy under the authority of its congressional charter of 1863, which establishes the Academy as a private, nonprofit, self-governing membership corporation. The Council has become the principal operating agency of both the National Academy of Sciences and the National Academy of Engineering in the conduct of their services to the government, the public, and the scientific and engineering communities. It is administered jointly by both Academies and the Institute of Medicine. The National Academy of Engineering and the Institute of Medicine were established in 1964 and 1970, respectively, under the charter of the National Academy of Sciences.

Library of Congress Catalog Card Number 85-60050

International Standard Book Number 0-309-03533-3
Permafrost: Fourth International Conference, Final Proceedings

Printed in the United States of America

National Academy Press

The National Academy Press was created by the National Academy of Sciences to publish the reports issued by the Academy and by the National Academy of Engineering, the Institute of Medicine, and the National Research Council, all operating under the charter granted to the National Academy of Sciences by the Congress of the United States.

Preface

Perennially frozen ground, or permafrost, underlies an estimated 20 percent of the land surface of the earth. It affects many human activities, causing unique problems in agriculture, mining, water supply, sewage disposal, and construction of airfields, roads, railroads, urban areas, and oil and gas pipelines. Therefore, understanding of its distribution and behavior is essential.

Although the existence of permafrost has been known to the inhabitants of Siberia for centuries, not until 1836 did scientists of the Western world take seriously the accounts of thick frozen ground existing under the forests and tundra of northern Eurasia. In that year, Alexander Theodor von Middendorf measured temperatures to a depth of approximately 107 m in permafrost in the Shargin Shaft, an unsuccessful well dug in Yakutsk for the governor of the Russian-Alaskan Trading Company. It was estimated that the permafrost there was 215 m thick. For over a century since then, scientists and engineers in Siberia have been actively studying permafrost and applying the results of their research in the development of the region. Similarly, prospectors and explorers have been aware of permafrost in northern North America for many years, but not until World War II were systematic studies undertaken by scientists and engineers in the United States and Canada.

As a result of the explosive increase in research into the scientific and engineering aspects of frozen ground since the late 1940s in Canada, the United States, the USSR, and, more recently, the People's Republic of China, and Japan, among other countries, it became apparent that scientists and engineers working in the field needed to exchange information on an international level. The First International Conference on Permafrost was therefore held in the United States at Purdue University in 1963. This relatively small conference was extremely successful and yielded a publication that is still used throughout the world. In 1973 approximately 400 participants attended the Second International Conference on Permafrost in Yakutsk, Siberia, USSR. By that time it had become apparent that a conference should be held every 5 years or so, to bring together scientists and engineers to hear and discuss the latest developments in their fields. Thus in 1978 Canada hosted the Third International Conference on Permafrost in Edmonton, Alberta, including field trips to northern Canada. Approximately 450 participants from 14 nations attended, and Chinese scientists were present for the first time. The published proceedings of all three of these conferences are available (see p. 413).

In Edmonton it was decided that the United States would host the fourth conference, and a formal offer was made by the University of Alaska. Subsequently, the Fourth International Conference on Permafrost was held at the University of Alaska at Fairbanks, July 17-22, 1983. It was organized and cosponsored by committees of the Polar Research Board of the National Academy of Sciences and the State of Alaska. Local and extended field trips to various parts of the state and northwestern Canada, to examine permafrost features, were made an integral part of the conference.

Approximately 900 people participated in the numerous activities, and 350 papers and poster displays from 25 countries were presented at the conference. Many engineering and scientific disciplines were represented, including civil and mechanical engineering, soil mechanics, glacial and periglacial geology, geophysics, marine science, climatology, soil science, hydrology, and ecology. The formal program consisted of panel discussions followed by paper and poster presentations. The panels considered the following themes: pipelines, climatic change and geothermal regime, deep foundations and embankments, permafrost terrain and environmental protection, frost heave and ice segregation, and subsea permafrost.

A total of 276 contributed papers were published in the first volume of the proceedings. Reports of panel and plenary sessions, additional contributed papers and abstracts, summaries of field trips, and lists of participants are included in this second volume.

The U.S. Organizing Committee is indebted to the many sponsors for their financial support, to the technical and professional organizations which participated in the program, and to the local Fairbanks organizers for their efforts, which resulted in a highly successful meeting.

Acknowledgments

We wish to acknowledge all those who contributed their time and resources to the success of the conference. The names of all contributing sponsors are listed on p. vi.

Dixie R. Brown, Executive Director, Office of Development, University of Alaska, and Jay Barton, President, University of Alaska Statewide System, organized the private sector fund raising. The conference itself was organized by the National Academy of Sciences under a U.S. Organizing Committee of the Polar Research Board and its Committee on Permafrost. Members of the U.S. Organizing Committee are listed on p. vii. W. Timothy Hushen, Muriel A. Dodd, Ruth Siple, Barbara Valentino, and Louis DeGoes of the Polar Research Board provided invaluable administrative assistance in arranging the conference and processing manuscripts for publication. The Paper Review and Publication Committee (see Appendix C), chaired by Robert D. Miller, reviewed more than 300 papers and abstracts. Literally hundreds of reviewers assisted the Committee; their names are also listed in Appendix C. Louis DeGoes served as secretary to both the U.S. Organizing Committee and the Paper Review and Publication Committee.

Editing of the first proceedings volume was under the direction of Robert C. Rooney of the National Academy of Sciences. Roseanne R. Price and Chris McShane provided editorial assistance for the 276 papers published in the first volume. The staff of the National Academy Press, under the direction of William M. Burns, expeditiously processed more than 2000 pages of camera-ready manuscript for the two-volume compendium. Throughout the publication of both volumes, the editorial, graphics and word processing staffs at the Cold Regions Research and Engineering Laboratory assisted in many ways, including editing and preparation of the final camera-ready copy of much of the present volume. Individuals involved included Maria Bergstad and Stephen L. Bowen, editing; Donna B. Harp, Barbara A. Jewell and Nancy G. Richardson, word processing; Edward M. Perkins and Matthew H. Pacillo, graphics; and Barbara A. Gaudette, Sandra J. Smith and Donna R. Murphy, composition and paste-up. Dianne M. Nelson of CRREL provided secretarial support for many aspects of conference organization. The camera-ready copy of the Soviet papers was prepared by Mary Ann Barr. Michael Hammond, interpreter, assisted in editing some of the Soviet contributed papers.

The actual organization of the conference was accomplished in Alaska by local organizing committees. The Fairbanks activities were under the supervision of Gunter E. Weller, Vice Chairman of the U.S. Organizing Committee. Dr. Weller's leadership and foresight were major contributing factors in the success of the conference. William D. Harrison and William M. Sackinger were the chairmen of the program committee and the local arrangements committee, respectively. They were ably assisted by Kathryn W. Coffer, Cheryl E. Howard and other support staff of the Geophysical Institute of the University of Alaska, including Jim Coccia, photographer, and the editorial staff of the Northern Engineer under Carla Helfferich. Numerous local volunteers, who cannot all be listed here, helped with the preparations. S. Russell Stearns, President, American Society of Civil Engineers, presented the keynote address at the conference banquet.

The Department of Conferences and Institutes of the University of Alaska provided much of the organizational support for the actual conference. This included all details on pre-registration and housing arrangements, actual conference registration, and support for many aspects of the social and technical activities that took place during the week. These efforts were under the direction of John R. Hickey and Nancy Bachner; Elaine Walker, Gary Walker and Sharon Oien played major roles. Personnel of the Housing Office, including Eric A. Jozwiak and Mark Van Rhyn, were most helpful in accommodating the hundreds of participants and their families. Timothy L. Jensen and Ronald W. Monteleone of the ARA Food Service did an excellent job in preparing and serving refreshments for the opening session and the barbecue.

Several other activities deserve special attention. Editing and publication of the field guidebooks were performed by the Alaska Division of Geological and Geophysical Surveys, Fairbanks. The ADGGS editorial and support staff includes Cheri L. Daniels and Frank L. Larson, Publication Specialists; Karen S. Pearson and Ann C. Schell, Cartographers; Marlys J. Stroebele and Roberta A. Mann, Clerk Typists; and student interns Duncan R. Hickmott, Barbara A. Kelly, Terry M. Owen, Kathy M. Smoyer and David A. Vogel.

A special organizing committee was established in Anchorage to provide for local arrangements there. This committee was chaired by Wayne C. Hanson of Dames and Moore, and was ably assisted by numerous volunteers, including some of the staff of the University of Alaska's Arctic Environmental Information and Data Center. The World Affairs Council hosted a reception prior to the railroad field trip for the Chinese delegation and others. Finally, the chairman's office at Arizona State University was ably assisted throughout the course of the pre- and post-conference activities by Frances S. Lunsford, Alice Spears, and Joan F. Bahamonde, who provided secretarial support for many aspects of the conference and the publication effort. In addition, James T. Bales, Jr., and Glenn R. Bruck were most helpful in preparation of field maps for the extended field trips; and Susan M. Selkirk, Artist-in-Residence of the Museum of Geology, provided artistic and cartographic support for several publications.

The Executive Committee of the U.S. Organizing Committee (Troy L. Péwé, Chairman, Gunter E. Weller, Vice-Chairman, and Jerry Brown) and Louis DeGoes were responsible for overseeing and organizing the conference and its publications. Brown and Péwé were responsible for organizing and supervising the preparation of this volume for publication. We again thank all those who contributed so much to this conference and its publications.

Troy L. Péwé, Chairman
U.S. Organizing Committee
Fourth International Conference on Permafrost

Jerry Brown, Chairman
Committee on Permafrost
Polar Research Board

Contributing Sponsors

The following sources contributed financial support to the Fourth International Conference on Permafrost.

Federal Agencies

Army Research Office
Bureau of Land Management
Cold Regions Research and Engineering Laboratory
Forest Service
Minerals Management Service
National Aeronautics and Space Administration
National Oceanic & Atmospheric Administration
National Science Foundation
Office of the Federal Inspector (ANGTS)
Office of Naval Research
U.S. Department of Energy
U.S. Geological Survey

State of Alaska

Alaska Council on Science and Technology
Department of Transportation and Public Facilities
State Legislature

Industry and Private

SOHIO Alaska Petroleum Company
Atlantic Richfield Foundation
Exxon Production Research and Exxon Company USA
John Kiely

Bechtel Petroleum, Inc., and Bechtel Power Corporation
Chevron USA, Inc.
Alyeska Pipeline Service Company
Mapco/North Pole Refining
Alaska Construction Industry Advancement Program
Dresser Foundation
British Petroleum, Ltd., Inc.
Morrison-Knudsen Company, Inc.
Kodiak-Nabors (Anglo-Energy)
Marathon Oil Foundation
Northwest Pipeline Service Company
Shell Oil Company
Union Oil of California Foundation
Cameron Iron Works, Inc.
Alaska Industrial Resources, Inc.
Phillips Petroleum
3-M Company
Alfred Jurzykowski Foundation, Inc.
National Geographic Society
Law Engineering Testing
R & M Consultants
Anaconda Copper Company
Brown and Root (Halliburton)
CONOCO, Inc.
Dames and Moore
Ebasco Services, Inc.
Getty Oil Company
Harza Engineering
Nerco Minerals (Resource Associates of Alaska)
Rocky Mountain Energy
Schlumberger Well Services
Tesoro Petroleum Corporation
Conrad Topp Engineering
Tryck, Nyman, Hayes
Alaska International Constructors
Doyon Ltd.
Shannon and Wilson
R.A. Kreig & Associates, Inc.
DOWL Engineers
Fryer, Pressley, Elliott
Peter Kiewit and Sons
William J. King
Stutzman Engineering Associates
Unwen Scheben Korynta Huettl
Arctic Slope Consulting Engineers

U.S. Organizing Committee

Troy L. Péwé (Chairman), Arizona State University
Gunter E. Weller (Vice-Chairman), University of Alaska
Amos J. Alter, Technical Council on Cold Regions Engineering, ASCE
Jay Barton, University of Alaska
Jerry Brown, Cold Regions Research and Engineering Laboratory
Oscar J. Ferrians, Jr., U.S. Geological Survey
Hans O. Jahns, Exxon Production Research Company
John R. Kiely, Bechtel Corporation
Arthur H. Lachenbruch, U.S. Geological Survey
Robert D. Miller, Cornell University
Richard D. Reger, Alaska Division of Geological and Geophysical Surveys
A. Lincoln Washburn, University of Washington
James H. Zumberge, University of Southern California
Louis DeGoes, Fourth International Conference on Permafrost

Program Chairman: William D. Harrison, University of Alaska
Paper Review and Publication Chairman: Robert D. Miller, Cornell University
Field Trips Chairman: Oscar J. Ferrians, Jr., U.S. Geological Survey
Local Arrangements Chairman: William M. Sackinger, University of Alaska

Participating Technical Organizations

Alaska Academy of Engineering and Science
Alaska Council on Science and Technology
Alaska Oil and Gas Association
American Association for the Advancement of Science (Arctic Division)
American Geophysical Union (Committee on Space Hydrology)
American Society of Civil Engineers (Technical Council on Cold Regions Engineering)
American Society of Mechanical Engineers (Low Temperature Heat Transfer Committee)
American Meteorological Society (Committee on Polar Meteorology)
Cold Regions Research and Engineering Laboratory
Comité Arctique International
International Geographical Union (Periglacial Commission)
U.S. Man and the Biosphere Program (MAB, Arctic Directorate)

Contents

Contributed Soviet Papers 195

PERMAFROST

Fourth International Conference

FINAL PROCEEDINGS

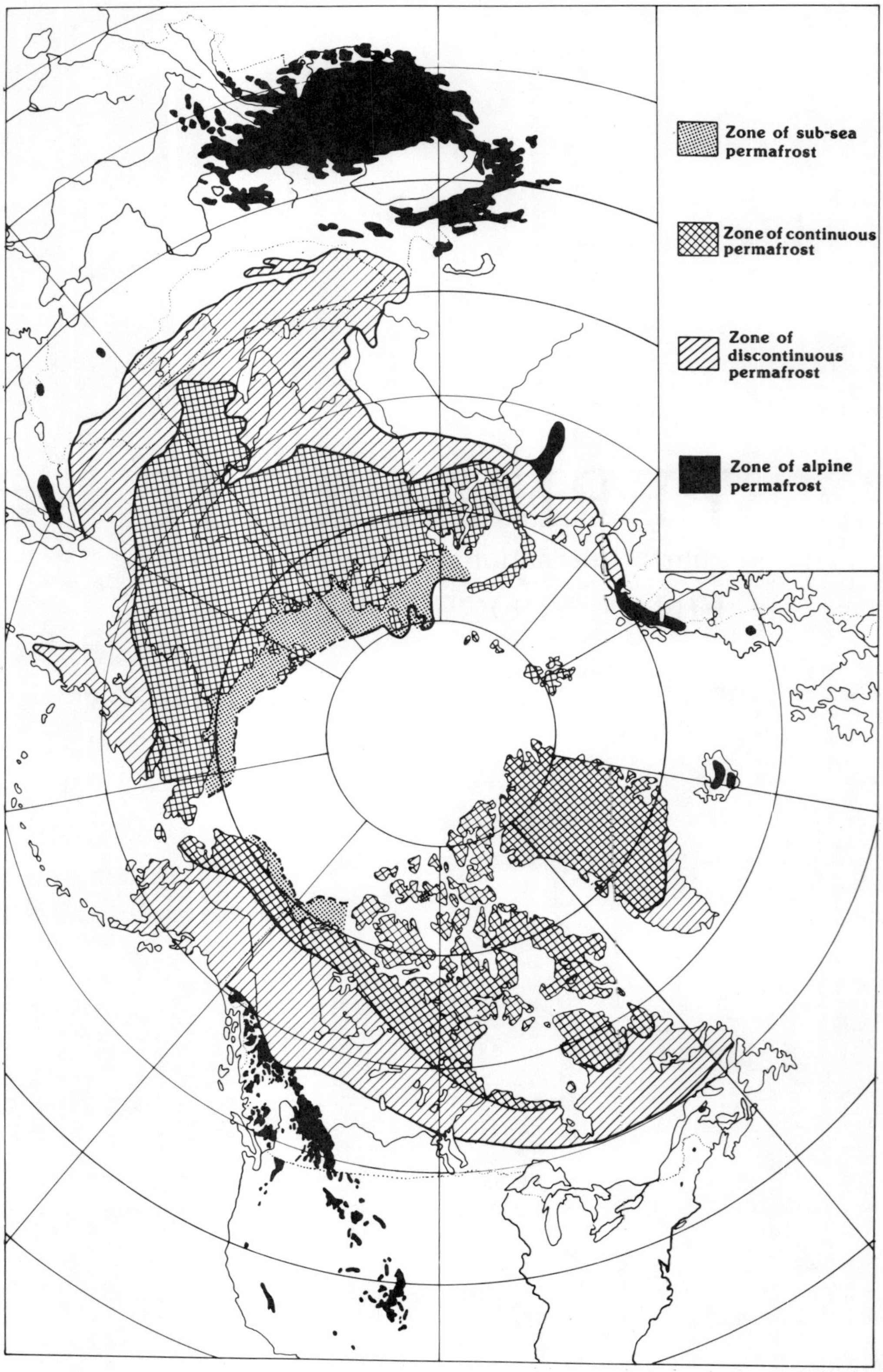

Frontispiece. Distribution of permafrost in the Northern Hemisphere. Modified from Péwe, T.L., 1983, The Periglacial Environment in North America During Wisconsin Time, In The Late Pleistocene, Stephen C. Porter, ed., University of Minnesota Press, Minneapolis, Chap. 9, p. 157-189.

Opening Plenary Session
Monday, July 18, 1983

TROY L. PÉWÉ - Ladies and gentlemen: Welcome to the formal sessions of the Fourth International Conference on Permafrost. I am Troy Péwé, Chairman of the Organizing Committee for the Conference. We, the Organizing Committee, subcommittees, staff, and many others, are pleased to have you here and invite you to participate in and attend the panel, paper, and poster sessions, the local field trips, the extended field trips, and social events. Today, I will first introduce our front-table dignitaries and then offer a few words about permafrost research history. This will be followed by words of welcome from Alaskan and international participants.

It is my pleasure to introduce the persons at the front table. At my far right [not visible in photograph] is Dr. Jerry Brown of the Cold Regions Research and Engineering Laboratory, who is representing the Polar Research Board of the United States National Academy of Sciences and is the Chairman of the Permafrost Committee of the Polar Research Board. He is also a member of our Organizing Committee. Next to Dr. Brown is Mr. Li Yusheng, Vice Principal of the Chinese Academy of Railway Sciences of the Ministry of Railways from Beijing, China. He is the co-leader of the delegation from the People's Republic of China. Next is Dr. Jay Barton, President of the University of Alaska Statewide System and a member of the Organizing Committee. On my immediate right is Academician P.I. Melnikov of the Academy of Sciences of the USSR, Director of the Permafrost Institute of Yakutsk, Siberia, and the head of the delegation from the USSR. On my immediate left [not visible in photograph] is Professor Shi Yafeng, Director of the Lanzhou Institute of Glaciology and Cryopedology, and Deputy Chief, Division of Earth Sciences, Academia Sinica. He is the co-leader of the delegation from the People's Republic of China. Next is Mr. Daniel A. Casey, Commissioner of the Department of Transportation and Public Facilities of the State of Alaska, representing the Governor of Alaska. We are pleased to have Dr. Hugh French of the University of Ottawa, and Chairman of the Permafrost Subcommittee of the National Research Council of Canada. He is representing the Canadian delegation. Next to Dr. French is Mr. Bill B. Allen, Mayor of the Fairbanks North Star Borough, Alaska. On my far left is Dr. J. Ross Mackay, Professor at the University of British Columbia, and former Chairman of the Ground Ice Division of the International Commission on Snow and Ice, International Union of Geodesy and Geophysics.

Scientists and engineers of the United States, and especially of Alaska, have waited many years for this occasion, a time when the experts in the field of the study of frozen ground from all over the world would be gathered here in central Alaska, an area of perennially frozen ground. Although the first invitation to meet here was issued only 10 years ago, at the Second International Conference on Permafrost in Yakutsk, Siberia, earlier workers dealing with frozen ground in Alaska, 70 and 80 years ago -- the gold miners -- must have been thinking that miners in other countries surely had easier ways to thaw the ground to obtain the gold than to use a little wood-burning steam boiler that was available at the time.

Those miners were not the first ones to think about permafrost in Alaska. The best-documented early account of permafrost is from a visit in 1816 by the German explorer Lt. Otto von Kotzebue, to what is now called Kotzebue Sound in western Alaska. He described the fossils in the frozen ground at a place later termed "Elephant Point" by the Englishman Lt. Beechey in 1836, who also examined the permafrost cropping out in the sea cliffs. Other early reports of scientists seeking remains of Pleistocene vertebrates in the frozen ground also describe the permafrost phenomena.

But it was the gold miners in central and western Alaska, from the end of the 19th century until the 1940's, who were the main sophisticated and unsophisticated students of the study of permafrost in Alaska, but only as a by-product of extracting the gold from the ground.

Only 10 km from here, many a miner became all too familiar with the small scale problems of frozen ground as he painfully thawed a shaft down to bedrock. Later, large-scale mining operations were invented to remove the gold from the frozen ground, creating probably the largest single artificial exposure of permafrost in North America, the Gold Hill Cut, which was 1.5 km long and 70 m deep. It was from such exposures that basic information on the origin and age of permafrost and associated ground ice was obtained.

But it was during and after World War II that the critical problems were encountered on a large scale--problems dealing with permafrost in the construction of airfields, roads, and buildings in this perennially frozen terrain of North America. It was in 1944 that the first book in English on construction problems in permafrost terrain appeared. It was compiled from the Soviet literature by Dr. Simon W. Muller.

With the rapid growth of permafrost knowledge in the late 40's, 50's and early 60's, investigators from the United States, Canada, and Siberia felt more of a need to exchange information on an international scale, and a small international conference on permafrost, with mainly the United States, Canada, and the USSR participating, was held in 1963, 20 years ago, at Purdue University. This has become known as the First International Conference on Permafrost. It was followed by the Second International Conference on Permafrost 10 years later in Yakutsk, Siberia, under the guidance of Academician P.I. Melnikov.

In the last 10 years, studies on the science and engineering of frozen ground literally exploded in all areas, especially in two new major fields, one dealing with the production and transportation of oil and gas in permafrost regions, and the other with the study of permafrost under the sea, permafrost of the Arctic Continental Shelf. These efforts, especially construction of the Alaska pipeline, have pushed the term "permafrost" and knowledge of its general importance into the minds of the general public, especially when it affects the price of the gasoline that man must put in his automobiles.

The Third Permafrost Conference was held in Edmonton in 1978, under the direction of the late Dr. R.J.E. Brown. It hosted 500 people, and 150 papers were presented. Here, the western world first learned of the existence and wide distribution of permafrost in the People's Republic of China, and of the research being undertaken on frozen ground there. Starting with the Third Conference, details for the 5 years of planning for the Fourth International Conference began to take shape.

Today, the Fourth International Conference on Permafrost, in the heart of permafrost country in the United States, is finally underway. I have been told that, so far, 800 people have registered and that 350 contributions are listed on the program.

To welcome us here today, as the representative of Bill Sheffield, Governor of the State of Alaska, is his Commissioner of Transportation and Public Facilities, Mr. Daniel A. Casey.

DANIEL A. CASEY - Thank you very much, Dr. Péwé. Ladies and gentlemen, and distinguished foreign visitors: First, let me say that I hope you have time to get out and do a little playing and not only working and studying. I know that such is your primary reason for being here, but Fairbanks is one of the best places to be in Alaska in the summer. I hope that you have taken the time to schedule some fishing, sightseeing, or a leisurely trip outside the state at the end of your conference.

On behalf of Governor Sheffield, it gives me great pleasure to extend a welcome to the scientists and engineers from all over the world who have assembled here for this conference. As Commissioner of the Alaska Department of Transportation and Public Facilities, I can say my department deals daily with the broad range of permafrost-related problems, which affect highways, bridges, airports, and shortly, I imagine, railroads, building foundations and communication-supporting structures. It is also for this reason that the department, on behalf of the State of Alaska, is a co-sponsor of this conference.

Because permafrost underlies more than four-fifths of our state, we are continuously affected by the engineering challenges presented to us by these phenomena, and we must rely on the other polar countries to help us solve these challenges. Many of these problems will be evident to those of you who take the local field trips or who have had an opportunity to go to Prudhoe Bay or down the Richardson Highway. The problems are unique and fascinating; and so are some of the solutions. They sometimes prove an embarrassment to the engineers of our department, as the traveling public assumes that we have ready solutions in hand. I am proud to note that several employees of the Department of Transportation will be presenting some of these solutions in papers during this conference.

Just prior to joining the Governor's cabinet earlier this year, it was my pleasure to head the Atlantic Richfield Company's construction program at Prudhoe Bay. Over the six years that I was associated with that effort, my understanding of permafrost grew tremendously. The project, which many of you have had or will have an opportunity to see on the North Slope, began in the classic phases that a project goes through, but compounded by the fact that it was in the Far North. The first question that we were challenged with back in 1974-1975 in the design phases was: Would it work? If we put those big modules on pilings, would they sink into the ground or would they stay up? And, of course, some of the more subtle questions in terms of long-term creep and road stability and road clearing were foremost in our minds.

As everyone knows, those challenges were met, just as the challenges of road construction and building construction throughout the state have been met. But I share with you one personal perspective, and that is that the challenge today in permafrost engineering is not what will work, but what will work most efficiently. How can we do the same things that we have now become accustomed to doing in the permafrost regions, but do them for fewer dollars so that we can get further along in the construction of our facilities and so they can last a longer time.

Let me close with a final comment. It has been stated in the scientific community that a global warming trend may be at hand. It would be useful, I think, for us practitioners if you discussed your outlook on the global warming trend. And if Alaska is to be faced in the foreseeable future with such a warming trend, as well as the rest of the polar-rim countries, we need to know the significance of that trend and the timing of it. As a consequence of this threat, our highways, airports, bridges, buildings, and pipelines that are built on permafrost may face an increasing rate of deterioration as their foundations, potentially, slowly give way. We need to know from you how serious this future threat is, and how, if at all, we should plan for that possibility in today's designs.

Again, on behalf of the Governor of Alaska, and the Alaska Department of Transportation and Public Facilities, welcome to the United States and our state.

TROY L. PÉWÉ - The next speaker might be described as the host of the conference, inasmuch as he is Dr. Jay Barton, President of the University of Alaska Statewide System. Dr. Barton is not only in a position to welcome us to the University, but has been an integral part of the Organizing Committee, involved with the production of the con-

ference on local and international scales. Ladies and gentlemen, the President of the University of Alaska, Dr. Jay Barton.

JAY BARTON - Dr. Péwé, distinguished guests, fellow scientists, ladies and gentlemen: It is with honor and pride that I welcome you on behalf of the Board of Regents of the University of Alaska to the Fourth International Conference on Permafrost. It is a truly significant occasion for the University of Alaska and one that is quite appropriate to its mission. You know, some years ago, Clark Kerr, in a facetious moment of whimsy, said that a modern university is really little more than a collection of conflicting constituencies held together mostly by a common heating system and a common shared grievance over parking. But it is on an occasion like this, I think, that we recognize the university for what it truly is, a community of learners, young and old, each learning from the other. This is, in Jacques Barzun's words, the house of intellect.

Scientists are fond, perhaps, of being pictured as solitary thinkers. You all saw the Sigma Xi report the other day that showed a little cartoon of a scientist in deep thought wearing a T-shirt that said, "I think, therefore I am paid." But, actually, I think we are very gregarious people, because it is in our interactions with each other that we apply the error-correcting mechanism that is intrinsic in our scientific method. It is through the free exchange of ideas and knowledge at conferences like this, bringing together scientists from throughout the world, that we deepen our understanding of nature and that we come close to achieving some measure of scientific truth. It is through the free exchange of ideas and knowledge at conferences like this, which bring together scientists from throughout the world, that we come close to achieving true peace.

I hope that you enjoy your sessions in the next week. Welcome to the house of intellect, to the University of Alaska.

TROY L. PÉWÉ - The University of Alaska and the area where we'll be investigating permafrost on local field trips, as well as the City of Fairbanks, falls in an area called the Fairbanks North Star Borough. This is a vital area in central Alaska, and here to welcome you for the Borough is the mayor, Mr. Bill B. Allen.

BILL B. ALLEN - Ladies and gentlemen, and honored guests visiting from foreign countries: Welcome to our part of the world. Fairbanks North Star Borough is certainly proud to have the University of Alaska at Fairbanks as a site for your conference. A tremendous amount of energy has been put into this conference by the organizers and contributors. Fairbanks, the Golden Heart of Alaska, is honored that those efforts have culminated in our community with this conference. Thank you for coming to our home, and I hope that you will make it your home in the next week. I have seen from your program that many of you will be participating in some of our local activities while you are here. I encourage you to take advantage of various attractions in our fine community during the next five days. As noted in your schedule, our Golden Heart City is celebrating its annual Golden Days, which commemorate those pioneers who came north at the turn of the century, seeking their fortunes in gold. And although some found gold, more found permafrost. So enjoy the hospitality of our home as we re-create and remember those who struggled in this land, and who founded our community.

To give those of you who are visiting us some perspective of our community, the local government is responsible for an area larger than Northern Ireland. We have a population of some 60,000 people and an abundance of geological features. To put that in the context of this conference, we have more permafrost per capita than anywhere else in the United States, and I might add that this is one reason why this conference should be held here in Alaska. We certainly hope that we have an opportunity to host you again. We have numerous examples of permafrost-related phenomena for your field trips, especially prepared for you. Also, we have nice weather for you, by proclamation of the mayor. Strange that we do have those powers, but we do. We have plenty of thermokarst pits, pingos, and frost heaves. I would suggest that we have more frost heaves per road mile than anywhere else in the United States. Just ask Mr. Casey

about his budget for maintenance. Or, just drive down any one of our roads a year after it has been constructed and you will see that this is an excellent laboratory for gathering information on some of the permafrost problems which are faced by urban centers.

As a practical matter, our community is interested in promoting research with which you are involved. We are glad to have you observe the typical permafrost features of this area, because we are searching for solutions to the problems that they cause. That was obvious to me on my way here this morning. I live out on Farmer's Loop Road; I hit a permafrost bump that was not there a week ago, and I almost broke my neck. I hope that you will do something about those types of things, and help us with our problems in that area. With situations like this all around us, it is easy to see that the people of Alaska are interested in the research that you are doing. I support using local government resources in finding answers that are essential if we are to understand and live in harmony with the permafrost that permeates our existence in so many obvious ways. And as a local government representative, I will work to encourage those at the state, national and international levels to make the necessary financial and moral commitments to help you find answers to the questions that you will be addressing this week. They are questions that impact the people who live in this region of the Borough and who must continually cope with the idiosyncrasies of permafrost.

We have been anxiously watching the progress through the United States Congress of the Arctic Research Policy Bill, which is sponsored by our Senators Stevens and Murkowski. This conference helps reinforce the needs which legislation fulfills for the development in the United States of a national policy on arctic research. We are one of the few nations with territories in the North which is still without such a policy. I am excited about the potential positive impact as a result of the passage of this legislation. Not only will the proposed legislation establish a national arctic policy, it will simultaneously provide for a clearing house to facilitate better coordination and reduce duplication in research. It also proposes $25 million annually in research funds. This focused effort is essential if the United States accepts its role in the responsibilities as a partner nation of arctic communities.

The U.S. should provide specific support for basic research, which is essential in the development and preservation of community environments in this region. Unique problems pose interesting challenges to those who wish to live, work, or play in this part of the world. The true potential of our community, our state, and all regions of the Arctic and Subarctic cannot be reached until our leaders recognize the importance of the type of work which this conference most aptly represents. We must all do what we can to promote the education of those who will be making those decisions which have an impact on the course of arctic research.

In closing, let me thank you again for the contributions you have made in permafrost research, which have helped many people of the world to cope with the environment they live in. We are glad you have come to our community. We look forward to the conference and papers which will be offered this week and to supporting your efforts in the future. Thank you.

Dr. Péwé, I have a Proclamation, if I could take a minute to read it [see next page].

TROY L. PÉWÉ - Thank you, Mr. Allen, for the invitation to inspect your permafrost. I am sure that in the next few days we will be examining exposures of frozen ground here and the effects of frozen ground on life in the North.

In this opening session, it is our privilege to have some information presented on the nature of the latest research and related activities going on in the countries of the world, especially in the major areas where permafrost is being studied. First, I would like to call on Academician Melnikov, who has long been the leading figure in the study of perennially frozen ground in the USSR and the chief delegate of the USSR to all the International Permafrost Conferences. He was host for the Second Conference in Yakutsk in 1973. Academician Melnikov.

P.I. MELNIKOV - Respected colleagues: On behalf of the Soviet delegation, I would like to greet all the participants of this outstanding geocryological forum who have assembled here in the wondrous State of Alaska to bear witness to the

Office of the Mayor

Fairbanks North Star Borough

Proclamation

WHEREAS, the Fourth International Conference on Permafrost is being hosted by the University of Alaska in Fairbanks from July 18 through 22, 1983, and organized by the State of Alaska and the National Academy of Sciences; and

WHEREAS, this International Conference on Permafrost convenes every five years or so to bring together scientific, engineering, and user-communities to discuss the state-of-the-art in these respective fields; and

WHEREAS, this International Conference on Permafrost will bring together over one thousand participants representing over 25 nations to exchange information on permafrost; and

WHEREAS, the selection of Fairbanks as the location for this conference brings great pride to our community and state as we are humbled at the honor of hosting those who are at the forefront of such important and relevant research; and

NOW THEREFORE, I, B.B. Allen, Mayor of the Fairbanks North Star Borough, by the authority vested in me, and in an effort to create a greater awareness of the critical activities occurring in relation to the Fourth International Conference on Permafrost, do hereby proclaim the week of July 18 through 22, 1983 as:

BOROUGH RECOGNIZATION WEEK FOR THE
4TH INTERNATIONAL CONFERENCE ON PERMAFROST

IN WITNESS WHEREOF, I have hereunto set my hand this 18th day of July, 1983.

MAYOR B.B. ALLEN

Attest:

Borough Clerk

achievements made in fundamental and applied geocryology since the third conference.

It has been five years since the Third International Conference on Permafrost was held in Canada. We fondly remember the organizers of that conference who did their best to make the sessions and field trips as interesting as possible. We regret that two prominent Canadian geocryologists, Dr. R.J.E. Brown and V. Poppe, who did much to develop this science in Canada and to further scientific contacts with foreign geocryologists, have passed away.

In the course of the past five years, fundamental and applied geocryological studies have progressed together with the industrial and economic development of northern permafrost regions. The principal results of these investigations are set forth in the reports. Due to considerable scientific and practical effort, we have been able to increase the efficiency of the development of northern territories which are rich in natural resources, including precious minerals.

Exploration and extraction of these minerals, however, has proven to be both an expensive and difficult undertaking. Science is now faced with the task of developing new, more effective research methods, which include satellite and geophysical surveys. We must increase both the depth and the reliability of our data. We must also create more precise instruments and maximize automation of research observations and data processing. It has become necessary to perfect methods for interpreting satellite imagery and to equip these satellites with instruments capable of determining not only the presence of permafrost, but also its thickness, ice content, discontinuity and the depth of seasonal thawing. It is also desirable to monitor the thermal-physical characteristics. These data will, to a great extent, facilitate the study of vast, almost inaccessible territories lying within the permafrost zone. We must learn more about permafrost evolution, composition, texture and conditions for distribution.

The perennially frozen ground in a major portion of the extreme north contains considerable ground ice underlying the thick, thermal-insulating peat and moss cover. Disturbance of this cover results in the formation of thermokarsts and thaw basins. It has therefore become necessary to define strict regulations governing economic development, land use, construction and environmental protection in these territories.

In order to successfully develop permafrost regions, we should extend the training of geocryologists. Both Moscow State University and Yakutsk State University have departments of geocryology and cryology. There are also special geocryological academic councils which monitor doctoral and candidate theses in the technical, geological, and geographical sciences. Geocryologists worldwide should make a greater effort to develop this science, while maintaining close scientific contact with their colleagues in other countries.

We wish all of you every success in your scientific work and in the activities in this conference.

TROY L. PÉWÉ - Our next speaker, representing the delegation of the People's Republic of China, is Professor Shi Yafeng, who is Director of the Lanzhou Institute of Glaciology and Cryopedology of the Academia Sinica. He has been one of the organizers of permafrost research in China for many years.

SHI YAFENG - Mr. Chairman, ladies and gentlemen: First of all, on behalf of the delegation of the People's Republic of China, I would like to express our heartfelt thanks to the organizers from the National Academy of Sciences, the State of Alaska, and also the University of Alaska, and especially to Professor Troy L. Péwé, Professor Gunter Weller, Dr. Jerry Brown, and all the members of the U.S. Organizing Committee for the excellent and persistent work required to prepare such a large conference. Please accept the sincere thanks of the Chinese delegation. This conference is really well organized, and it will greatly promote research and development in the world.

As we all know, permafrost programs have an important effect on the economic development of cold regions. China is an ancient civilized country with a long history. About 2,000 years ago, seasonal freezing processes in China had already been decoded in the ancient Chinese classic, the Book of Leaves, as follows: In the first events of winter, earth and water begin to freeze; in the

second, ice becomes hard, and earth cracks; in the third, freezing reaches its climax. And in the events of spring, thawing is promoted by an easterly breeze.

Permafrost in China is estimated to underlie about 2 million square kilometers, second only to the USSR and Canada. If we include ground that is seasonally frozen to a depth greater than 1/2 meter, the frozen ground occupies two-thirds of the entire country. Such a wide area of frozen soil exerts a great influence on China's economy, but it also expedites the study of permafrost in China. Since the founding of the People's Republic of China in 1949, 2000 kilometers of railroads, several thousand kilometers of highways, and many other types of construction, including coal mining, irrigation canals, oil pipelines, industry, and civil facilities, have been built in permafrost areas. All these construction projects encountered many difficulties; therefore, it was necessary for several study groups on frozen ground to be successively established in the Department of Railways, the Department of Water Conservancy, the Department of Forestry, and the Department of Construction. In the 1960's, a special permafrost research institute was also established within the Chinese Academy of Sciences, Academia Sinica, and also in the Chinese Academy of Railway Sciences. During the 10 years of disturbance of the so-called Cultural Revolution, in 1966-1976, permafrost research in China was inactive.

In 1978 the First National Conference on Permafrost was held. Only 68 reports on permafrost were presented. Three years later at the Second National Conference on Permafrost, held in 1981, 185 papers were presented. Now the study of permafrost in China includes permafrost, physical and mechanical processes of frozen ground, perfecting and testing techniques, and engineering construction problems in frozen ground. Permafrost research in China is still young, and we need to learn much from other countries, so that we can attain a higher level for the benefit of China's modernization. We welcome all types of cooperation, including the exchange of data and publications, and the exchange of scholars and graduate students to do such cooperation can be developed further. The Chinese delegation also supports the proposed International Permafrost Association, so that international cooperation in research on permafrost can more effectively be realized.

The Chinese delegation wishes to express warmest congratulations on the successful opening of the Fourth International Conference on Permafrost. We wish all the members of this conference a healthy and happy time.

TROY L. PÉWÉ - Thank you, Professor Shi. Our next speaker is Dr. Hugh French, who chairs the delegation from Canada at this meeting.

HUGH FRENCH - Mr. Chairman, ladies and gentlemen: On behalf of the Canadian delegation, I wish to state that we are very pleased to be here at the opening of the Fourth International Conference on Permafrost in Fairbanks, Alaska. As many of you may remember, Canada was the host of the Third International Conference on Permafrost, held in Edmonton in 1978. As such, we are very well aware of the tremendous amount of planning, time, and energy which must have gone into the preparation of this particular conference. We are sure that it will be successful, and we look forward to learning new things, renewing old acquaintances, and making new friends.

Permafrost is particularly important to Canada, since it underlies approximately 50% of our country. The permafrost regions are vast and extend within 1200 km of the North Pole. They include extensive areas of polar desert, of tundra, and of boreal forest, with a wide diversity of geological and topographic conditions. Some of these areas are rich in renewable and non-renewable resources. Their development, to the mutual benefit of northern indigenous people and of the Canadians, presents numerous challenges, which are technological, scientific, and social in nature. In Canada, we are anxious to meet these challenges, and to learn through the experiences of others.

For these reasons, Mr. Chairman, Canada is especially glad to participate in this Fourth International Conference on Permafrost.

I would now like to take this opportunity to briefly highlight some of the recent events which have occurred in Canada since the Third International Conference in 1978.

First of all, as has already been mentioned, the death of Roger J.E. Brown, under whose leadership the Third International Conference was organized, and who was a major pioneer of permafrost science in Canada, was a great loss. He was well known both nationally and internationally, and will be difficult to replace.

Let me turn now to the current state of permafrost studies in Canada. An ever-increasing number of individuals, agencies, and organizations are becoming involved in permafrost problems in Canada. At the same time, these problems are becoming more complex and multi-disciplinary in nature. One of our strengths, perhaps, in Canada, is in the private sector where, in addition to major industrial and commercial concerns, a number of engineering and consulting companies now possess considerable experience with permafrost-related problems. Drilling technology in the search for oil and gas in the Far North is an important area of current concern. Problems are encountered and compounded by the occurrence of substantial areas of offshore permafrost in the Beaufort Sea. Gas hydrates are now recognized as important, but their nature, occurrence, and distribution are still little known in Canada. The question of pipelining in permafrost is also of current concern to us. The proposed gas pipeline from Alaska through the Yukon, and the possibility of transporting gas from the high Arctic islands to southern markets, are central to our concerns. Finally, there is the broad area of permafrost engineering; in Canada we are continually seeking to improve our engineering design and construction techniques in the permafrost zones. Linked with this is the realization that permafrost science, involving the origin, distribution, and nature of permafrost, and related fields such as permafrost hydrology, are essential to sound engineering design.

The Government of Canada, through its various agencies and departments, is very active in permafrost research. The Department of Energy, Mines, and Resources, the Department of Indian and Northern Affairs, the Department of the Environment, and the National Research Council of Canada are all currently undertaking major programs designed to increase our understanding of both permafrost science and engineering in Canada. Other departments, such as the Department of Transport and Public Works, and the various provincial and territorial agencies are active at the practical level, providing the infrastructure for resource development in northern Canada.

A number of universities are developing expertise in permafrost-related problems. There is also the strong possibility that a major cold regions engineering research center supported by the National Research Council of Canada will be established in the next few years in western Canada.

Many of the recent advances in permafrost research in Canada were summarized at the Fourth Canadian Conference on Permafrost, held in Calgary in March of 1981. This conference saw the presentation of over 100 scientific papers and brought together over 250 scientists, including a number from outside of Canada. Ten sessions were organized at that meeting, dealing with such topics as laboratory testing of frozen soils, permafrost hydrology, climate and permafrost, gas hydrates, and engineering applications. Now, just two years later, many of these same people are here in Fairbanks and are presenting more of their findings. They are evidence of the extremely rapid pace of current permafrost research in Canada.

To conclude these brief remarks, Mr. Chairman, I would like to restate our pleasure to be here in Alaska at the opening of the Fourth International Conference on Permafrost. Many of us have seen, or will see, during the next few days on the field excursions, some of the progress that has been made here in Alaska, much of it at the University of Alaska in Fairbanks, towards the solving of permafrost problems. It is very impressive and we look forward to learning much more in the next few days. Thank you very much.

TROY L. PÉWÉ - Thank you, Dr. French. I think it would be well at this time for the U.S. Organizing Committee to acknowledge the great amount of help it received from the Canadian colleagues. Their recent experience in organizing an international conference was most helpful to us. The effort that they put forth in working with our program and in running one of the major field trips, as well as in the review of many of the contributed papers, is greatly appreciated. We wish to particularly announce our thanks to the Canadian group. Our next speaker is Dr. J. Ross Mackay, leading permafrost researcher and holder of many honors. He is a long-time friend of the late Dr.

Roger J.E. Brown of Canada, who has been mentioned two or three times already. Dr. Brown participated in the first three congresses and it is known that he was, of course, the chairman of the third conference. At this time: Dr. Mackay.

J. ROSS MACKAY - Mr. Chairman, ladies and gentlemen: I have been asked to say a few words in remembrance of the late Dr. R.J.E. Brown. As many of you know, he died in 1980 after a long and courageous struggle against cancer. It was just five years ago last week that Canada hosted the Third International Conference on Permafrost at Edmonton, Alberta. Roger was the perfect choice for chairman of the Canadian organizing committee. Much of the success of the conference was due to Roger's unselfish and untiring efforts. He was as much at home with permafrost scientists from the United States and abroad as he was with his fellow Canadians. He traveled in the Soviet Union and the People's Republic of China, always with the objective of fostering friendships and international collaboration. Roger had the cheerful and infectious smile, a genuine interest in people, and a happy facility for making friendships. He had looked forward with great anticipation to being able to attend this Fourth International Conference on Permafrost which is beginning so auspiciously today. In tribute to Dr. Brown, the proceedings of the Fourth Canadian Permafrost Conference were published last year in a memorial volume. The Canadian Geotechnical Society has established a Roger J.E. Brown Memorial Award, which will be presented for outstanding contributions to permafrost at the next and future international conferences on permafrost.

TROY L. PÉWÉ - Thank you, Dr. Mackay. Our next speaker is Dr. Jerry Brown, who is Chairman of the Committee on Permafrost of the Polar Research Board of the United States National Academy of Sciences.

JERRY BROWN - Mr. Chairman, ladies and gentlemen: On behalf of the U.S. National Academy of Sciences, and its president, Dr. Frank Press, it is my pleasure to welcome the participants to the Fourth International Conference on Permafrost. I would like to thank our hosts here in Alaska and at the university for providing this opportunity to meet in Fairbanks. Dr. Gunter Weller, Vice Chairman of the U.S. Organizing Committee, deserves particular thanks for undertaking the local arrangements. Special acknowledgment is appropriate to our Canadian colleagues who have provided considerable assistance in the review process for the contributed papers and in undertaking the post-conference field trip to the Mackenzie River Valley.

In the remaining few minutes of this opening session, I would like to briefly discuss the status of permafrost research in the United States. Scientific and engineering investigations on frozen ground are conducted by many institutions and organizations, including universities, federal and state organizations, and privately financed industrial organizations. Numerous professional organizations are involved in organizing meetings, seminars, and conferences on the subject of permafrost and freezing and thawing of soils. Notable are the American Society of Civil Engineers and the American Society of Mechanical Engineers. The geographic and disciplinary diversity of these U.S. activities is characterized by the approximately 100 U.S. papers which will be presented during the next five days. A special conference bibliography produced by the World Data Center for Glaciology in Boulder, Colorado, identifies many other U.S. publications which have become available in the last five years.

Current investigations are underway in three major regions: (1) offshore or subsea permafrost (particularly in the Beaufort Sea), (2) land-based permafrost areas in Alaska, and (3) in the alpine regions of the contiguous United States. In the past five years, major efforts have been directed to the continental shelf in association with petroleum exploration. There the occurrence and distribution of ice-bonded permafrost has been identified by both drilling and geophysical methods off Prudhoe Bay and in Harrison Bay.

On land in the Arctic, the temperature and depth distribution of permafrost have been further defined using boreholes drilled for oil exploration. The bottom of ice-bearing permafrost has been delineated through the examination of borehole data from the arctic coast to the Brooks Range. The distribution of near-surface massive ice has been further characterized. The occur-

rence of gas hydrates in both land and subsea permafrost is being examined. Considerable attention has been directed to concerns regarding potential changes in permafrost terrain due to human activities and carbon-dioxide-induced climatic warming.

In the Subarctic, or in the region of discontinuous permafrost, investigations continue on construction and maintenance techniques to mitigate adverse effects of permafrost degradation and frost heave. The influence of permafrost on surface and groundwater supplies continues to be investigated.

In alpine regions, the general distribution of permafrost has been delineated and several site-specific studies are underway. In the Antarctic, permafrost research is generally conducted in relation to periglacial and geologic investigations.

In the past year, the U.S. Committee on Permafrost undertook and completed an assessment of future permafrost research requirements for the remaining years of the twentieth century. Included in this study are assessments on subjects related to the chemistry and physics of frozen ground, subsea permafrost, geophysical techniques, soil mechanical behavior, engineering and environmental considerations, and hydrology. A series of recommendations include the need for additional research on ground ice, its detection and origin, monitoring of performance of existing facilities on permafrost, monitoring of active layer and ground temperatures, and continued international cooperation. A second study on ice segregation and frost heaving is also to be published. It is obvious that there remains a need for additional basic and applied research on permafrost and deep seasonally frozen ground. This conference will help to identify what research should be conducted in the future.

In closing, I wish you a successful conference. It is gratifying to me to see such a large attendance from so many countries. Thank you.

TROY L. PÉWÉ - Thank you, Dr. Brown. One of the most pleasing aspects of the opening session is to be able to acknowledge the great supporting staff that worked virtually for years to produce this conference. The unfortunate aspect is that the group is so large that all the supporting workers cannot be mentioned here, but their names will appear in the final conference publications.

The Organizing Committee of the conference has worked for five years on this effort, and I would especially like to mention the tremendous contributions of Jerry Brown, Gunter Weller and Lou DeGoes, who were active in all aspects of the conference. Also, Richard Reger for overseeing the publication of the guidebooks, Robert Miller for chairing the review of manuscripts for the proceedings, Oscar Ferrians for guiding the field trip details, and especially Jay Barton for his support in many national and local details. Thanks are also extended to Hans Jahns, John Kiely, Art Lachenbruch, Link Washburn and Jim Zumberge.

Special mention must be made of Will Harrison and Bill Sackinger, of the University of Alaska, along with Gunter Weller, for arranging the program and the host of local details. Also, special thanks to Dixie Brown, for organizing the private contributions. It is well known that this conference is possible only with the tireless support of the office staffs of these individuals, and our sincere thanks goes to the small army of secretaries, clerks, and aides who have been working on this conference for so long here in Alaska and in the contiguous states. Special thanks are extended to the Department of Conferences and Institutes of the University of Alaska.

Financial support for the conference has come from various sources. Monetary contributions have been received from the State of Alaska, Federal agencies, and the private sector. A list of these benefactors is presented in several of the conference publications (see page 413). In addition, valuable in-kind support has been received from a host of institutions, especially from the Cold Regions Research and Engineering Laboratory and the Geophysical Institute of the University of Alaska.

Before concluding, I call your attention to the banner on the table in front of me. To those who are not aware, the circles you see there are logos that were used in this and the previous conferences. The banner originated with the Canadian delegation five years ago and was passed on to the U.S. for this conference.

We again welcome you and appreciate your coming; have a successful conference. This concludes the opening session. Thank you.

PROGRAM

TIME	MONDAY 18 July	TUESDAY 19 July	WEDNESDAY 20 July	THURSDAY 21 July	FRIDAY 22 July
8:30 to 10:00 a.m.	Opening Plenary Session	Panel Session: Environmental Protection of Permafrost Terrain	Panel Session: Deep Foundations and Embankments	Panel Session: Frost Heave and Ice Segregation	Panel Session: Subsea Permafrost
10:00–10:30 a.m.	Coffee Break				
	Panel Session: Pipelines in Northern Regions	Paper Sessions: •Thermodynamics •Mechanics of frozen soils •Mountain and plateau permafrost •Effects of man-made disturbance •Planetary permafrost Poster Session	Paper Sessions: •Roads and railways—Thermal aspects •Foundations •Frost mounds and other periglacial features •Effect of man-made and natural disturbance •Invited Soviet Session Poster Session	Paper Sessions: •Frost heave •Embankments, roads and railways •Patterned ground •Hydrology •Geophysics Poster Session	Paper Sessions: •Excavations, mining and municipal •Cold climate rock weathering •Groundwater in permafrost •Subsea permafrost •Ecology of natural systems
12:00 noon 12:30 p.m.	LUNCH 12:00–1:30	LUNCH 12:30–2:00 p.m.			
1:30 p.m. 2:00 p.m.	•Thermal design •Pipelines •Ice and soil wedges •Invited Chinese Session	Local Field Trips	Panel Session: Climate Change and Geothermal Regime	Local Field Trips	Closing Plenary Session
3:30–4:00 p.m.	Coffee Break		Coffee Break		
4:00 to 6:00 p.m.	•Thermal analysis •Mechanics of frozen soils •Pleistocene permafrost conditions •Invited Chinese Session •Remote sensing		•Foundations •Ground ice and solifluction •Hydrology •Invited Soviet Session •Climate		Walking Tour of Local Permafrost Features
		BARBEQUE		BANQUET	

See Appendix A for descriptions of local field trips. See Appendix B for titles of presentations.

Reports of Panel Sessions

During the early planning for the Fourth International Conference on Permafrost, there was a consensus that certain topics of high current interest would be highlighted, and in as interesting a fashion as possible. The latter requirement seemed to rule out the presentation of invited review papers by individuals as at past conferences, but at the same time the tremendous value of the written versions of these reviews was recognized. This led to the idea of replacing single authors with panels of international experts on each of the topics to be highlighted. Each panel was scheduled for a plenary session during which no other activities were to take place. In these sessions, each panel reviewed its topic and, if time allowed, provided a discussion with audience participation. It was felt from the beginning that whatever the outcome of the panels at the conference, their most important task, just as with invited speakers of past conferences, would be the production of written reviews of each of the topics highlighted.

These reports follow, but not in the order of the program presentation. They are based on the reviews of subfields supplied by each panel member to his panel chairman. The degree to which these components have been integrated by the chairmen varies, but in most cases they remain relatively unmodified despite editorial changes. As such, they are not of uniform length, style, or format. On the other hand, together with the introduction of the panel chairman, they bring a depth of perspective and diversity to a topic that would be difficult for a single author to duplicate. Recommendations for future research and engineering investigations are considered the opinions of the panel members and do not necessarily reflect the opinion of the publisher.

A great deal of credit is due to the panel chairmen and to the panel members. Each panel chairman selected and organized his own panel, a task made particularly challenging by its international composition and by the uncertainty as to whether all the panel members could attend the conference. Some of the originally designated chairmen and panelists were indeed unable to attend; particular credit is due to the replacements who assumed responsibility on relatively short notice. Both the Soviet and the Chinese panelists submitted their contributions in English, which greatly aided the panel chairmen, all North Americans, in organizing and presenting the panel reviews.

William D. Harrison, Chairman
Program Committee
University of Alaska

Deep Foundations and Embankments

PANEL MEMBERS

Norbert R. Morgenstern (Chairman), University of Alberta, Edmonton, Canada

S.S. Vyalov, Research Institute of Foundations and Underground Structures, Moscow, USSR

Ding Jinkang, Northwest Institute, Academy of Railway Sciences, Lanzhou, People's Republic of China

David C. Esch, Alaska Department of Transportation and Public Facilities, Fairbanks, Alaska, USA

Francis H. Sayles, Cold Regions Research and Engineering Laboratory, Hanover, New Hampshire, USA

Branko Ladanyi, Ecole Polytechnique, Montreal, Canada

INTRODUCTION

N.R. Morgenstern

This panel presentation combines two of the most important aspects of permafrost engineering -- deep foundations and embankments. The underlying theme of the presentation is to review current practice, identify deficiencies, and draw attention to the major problems that need further attention. The validation of current practice from performance records is emphasized wherever possible.

The presentations concerned with deep foundations consist of three regional reports. The report on North American practice discusses developments related to pile installation methods, the behavior of piles under load, the improvement of bearing capacity using special types of piles, and pile design. It concludes that we now have a relatively good idea of how a pile, installed by a given method, will behave under an axial load. Much less is known, however, about the effect of seasonal changes of permafrost temperature on pile behavior. The need for more studies of thermopiles is emphasized.

Current Soviet practice in pile design and construction is reviewed. Emphasis is given to some of the special problems that arise in weaker soils and to methods for improving and validating pile foundations in frozen ground.

In China, deep foundations in permafrost have been tested and used since the early 1960s. The types employed are described. To provide a basis for rational design, studies have focused on the bearing characteristics of single reinforced piles, the freezeback rate of soil around a pile, concrete for use under negative temperatures, the resistance of foundations to frost heave, and methods for protecting pile foundations from frost-heaving. As a result of experience various problems have been identified, and they are discussed.

A distinction is made between road/railway embankments on permafrost and water-retaining embankments. The review of current practice for the design of road and railway embankments distinguishes between the preliminary investigation phase, the design stage, the construction stage, and the maintenance and operation stage. A variety of research needs for each of these stages is identified. Examples include the need for remote detection of subsurface conditions that affect embankment performance and the need to quantify the thermal effects of alternative embankment sideslope surfaces, surface vegetation covers, and the role of snow-cover in embankment performance. Several case histories of embankment performance are discussed.

Both Soviet and North American experience with water-retaining embankments in permafrost is summarized. The indications are that the design, construction, and maintenance of safe, economical earth dams and the preservation of natural slopes in the Arctic and Subarctic are strongly dependent upon such considerations as structural stability, seepage control, handling of materials, erosion control, and the environmental effect of the impoundment. Extensive development of both analytical and construction techniques is required to address the attendant problems.

PLACING OF DEEP PILE FOUNDATIONS IN PERMAFROST IN THE USSR

S.S. Vyalov*

Research Institute of Foundations and Underground Structures
Moscow, USSR

In the Soviet Union, data on the design and placement of pile foundations acquired and verified through research and practical experience are published in the Building Codes and Regulations (BC&R) "Foundations in Permafrost, 1977" and "Manual for the Designing of Foundations in Permafrost, 1980."

TYPES OF PILES AND PILE-DRIVING TECHNIQUES

The following types of piles are in use: reinforced concrete, metal, wood, composite, encased, and pillar piles.

The techniques of driving piles are as follows:

(a) At sites where no thawing of permafrost is employed, use is made of frozen-in (floating) piles driven with the aid of steam. Suitable alternatives are to lower piles into boreholes and then fix them in place with soil mortar poured into the holes, and piles driven into small-diameter pilot boreholes.

(b) At sites where the thawing of permafrost is used, the recourse is to strutted piles and deep piles bearing upon ground that displays low compressibility.

PRINCIPLES OF PILE DESIGN

Piles of all types are designed to ensure a requisite bearing power (strength). Those used in frozen plastic soil and ice-rich ground are designed with due regard for the deformation (settlement) of the soil.

In proportioning frozen-in piles to obtain adequate bearing power, the long-term ultimate shear strength along the side surface of the pile (the adfreeze strength) is taken into account as well as the long-term ultimate strength of the soil resisting the end bearing of the piling. Recommended design values of the strength characteristics are given in the BC&R. The settlement of soil is assumed to be nonlinear.

Computer-aided solutions obtained by using the finite element method have been reduced to simple formulas that permit calculations to be carried out with the aid of alignment charts. In addition, a formula that incorporates an equation of creep for settlement and an equation of long-term strength has been derived that paves the way to designing in terms of both strength and deformation with due allowance for time.

Horizontal loadings are measured, taking into account the various restraints on the pile in the permafrost layer (depending on the depth of seasonal thawing and the permafrost temperature).

The criteria for designing laterally supported piles for thawing soils include the strength of low-compressible underlying rock that resists indentation and the strength of the pile material with due allowance for its buckling. Extra loading due to negative friction is also taken into account.

Strength of Freezing of a Pile to the Soil

The physical essence of the strength of freezing has been examined, noting that a film of ice forms at the interface between the soil and the pile. The shear strength of frozen soil is thought of as a sum of the freezing forces at the interface and the frictional forces arising from the radial compaction of the soil. An equation of long-term shear strength of frozen soils has been derived. A comparison of the shear strength of the soil mortar at the interface with the pile after freezing vs the shear strength of the frozen soil mortar at the interface with the surrounding frozen soil has revealed that in the former case the shear strength is lower. A method of designing piles for weak soils (saline, ice-rich, clay) has been developed based on the assumption that the strength of the soil mortar at the side surface of the pile and that at the perimeter of the borehole are the same.

Design values of the long-term ultimate strength of several kinds of frozen soils varying with temperature have been determined at the surfaces of reinforced concrete, metal, and wood piles; similar values of the strength of soil and soil mortar at the periphery of the borehole have also been determined.

An equation of long-term strength for varying temperature and loading has been derived, based on the kinetic theory of strength.

Improving the Efficiency of Pile Foundations

An increase in pile capacity and a decrease in costs may be achieved by

- resorting to composite wood/concrete or wood/metal piles;
- dumping crushed rock into pilot boreholes or using concrete pipes with bulb-shaped bases;
- pouring soil mortar of an optimum composition into boreholes;
- artificially chilling high-temperature permafrost, including the use of thermopiles;
- ensuring the setting of molded-in-place concrete piles at the permafrost interface;

*Not present; summary presented by N.R. Morgenstern.

- streamlining pile design so that variable temperatures and loading are taken into account.

Allowance for Variable Temperature and Loading

The existing method of structural analysis relies on the least favorable conditions, piles being designed on the assumption that the permafrost temperature is high and the loading gives rise to high forces and moments. In fact, the permafrost temperature varies with time, depending on the air temperature and the chilling effect of basements, channels, and thermopiles. Pile loadings also vary with time, being low during the period of construction and susceptible to transient forces, in particular those of climatic origin (snow, wind, soil heaving). All in all, maximum loading is never timed to occur simultaneously with the soil's minimum bearing capacity. For example, loads due to snow, wind, and soil heaving are at the maximum in autumn and winter, whereas the soil has a minimum bearing capacity in summer. This fact is of particular importance in designing light engineering structures (pipelines, flyovers, power lines) and buildings of light-weight material where transient loads account for a significant fraction of the total loading sustained.

Some design methods consider simultaneously loading, soil temperatures, the long-term strength of soil and soil settlement, all varying with time.

Piles for Use in Specific Permafrost Conditions

Methods have been developed for designing and driving piles in ice-rich ground, subterranean ice, saline permafrost, and thawing ground. At the sites where permafrost-thawing techniques are used, preference is commonly given to staggered piles of considerable length and to molded-in-place piles. Floating piles find application in conjunction with local pre-thawing of permafrost, which is then compacted as the pile is driven into place.

PILE TESTING PROCEDURES

Laboratory tests based on the kinetic theory of strength have revealed the influence of the size of the pile model on the adfreeze strength. A method of allowing for this scale factor has been developed.

Field tests of piles are performed at all large construction sites, and a standard procedure has been established in this connection.

A novel accelerated test technique has been developed that relies on a "dynamometric" apparatus as a means of applying load to the pile. An initial stress set in the gauge decreases with the sinking of the pile into the soil. The stress recorded when no further settlement of the pile is evident is adopted as the long-term ultimate stress.

IMPROVING DURABILITY OF REINFORCED CONCRETE PILES

The extreme conditions of temperature and humidity in the Arctic, where temperature variations and recurrent seasonal freezing and thawing of the surface layer cause the high moisture content of concrete and freeze and thaw water in the voids of the concrete, have an adverse effect on reinforced concrete piles. Means are available to prevent the low-temperature destruction of concrete that extend the service life of reinforced concrete piles.

RESEARCH PROGRAM

The following problems require further attention:

- Improving the techniques of pile driving in permafrost (especially coarse fragmental permafrost);
- Producing piles of advanced construction;
- Streamlining the techniques of emplacing molded piles;
- Widening the field of application of piles with pre-thawing of permafrost;
- Improving the durability of reinforced concrete piles;
- Streamlining methods for computing the bearing power of piles (compiling new tables, constructing new alignment charts, etc.);
- Developing a single methodology of pile testing to be recommended for use in all countries;
- Collecting a wealth of experimental data on the strength of freezing and the bearing power of piles under various conditions with an eye to providing design criteria for practical use on analyzing these data (achievable if researchers from all countries pool their efforts).

STUDY AND PRACTICE ON DEEP FOUNDATIONS IN PERMAFROST AREAS OF CHINA

Ding Jinkang

Northwest Institute
Lanzhou, People's Republic of China

In China permafrost occupies an area of about 2,150,000 km^2, covering almost 22% of the territory of the country. The high mountainous regions of the Qinghai-Xizang Plateau, the Tian Shan Mountains, the Great Xing'an Mountains, the Qilian Mountains, and the Altai Mountains are the primary regions where permafrost exists. Among these, permafrost in the Qinghai-Xizang Plateau is the most highly developed, with the highest elevation, the most wide-spread area, the deepest layer, and the lowest temperatures, as compared to the low-latitude regions of the world. The Plateau has an area of about 1,490,000 km^2, which is almost 70% of the total permafrost area of China.

Since liberation, with the exploitation and utilization of the forest resources of the Great Xing'an Mountains and the development of such borderlands as Qinghai and Xizang, the Ya-Lin, Nen-Lin, and Chao-Wu railroads have been constructed in the permafrost area of the Great Xing'an Mountains, and highways have spread throughout the forests of the region. The highway linking Ge'rmu in Qinghai Province and Lhasa in Xizang has been constructed, crossing a permafrost belt of more than 600 km. To develop the Reshui and Jiangcang coal mines in the permafrost region of the Qilian Mountains, railroads and highways were built and mining and residential areas were established. In addition, an oil pipeline was constructed from Ge'rmu to Lhasa in the 1970s. All these projects have accelerated the study and development in China of foundations in permafrost areas.

In early construction projects, conventional methods used in temperate regions for surveying and designing ground foundations were used due to lack of knowledge about permafrost. As a result, deterioration and even destruction frequently occurred during operation. The most serious problems occurred in building construction; some buildings had to be built over and over again due to frost-heaving and thawing settlement. We were forced to consider carefully and thoroughly the specific features of foundations in permafrost areas and to inquire into suitable forms of foundations.

In the early 1960s, a series of tests were made on piles, which at that time were used as the building foundations in permafrost areas. Explosion-expanded pile foundations (Figure 1) were first tested in the permafrost region of the Great Xing'an Mountains (TSDI, 1976a,b) on a 1.4-km water supply pipeline built in 1967 and an inhabited house. The pipeline was an overhead one supported by 340 explosion-expanded piles, each 40 cm in diameter and buried to a depth of 2.5 m. Under operation, it was found that the frost table under the gravel soil was relatively deeper and the burial depth of the explosion-expanded piles was a

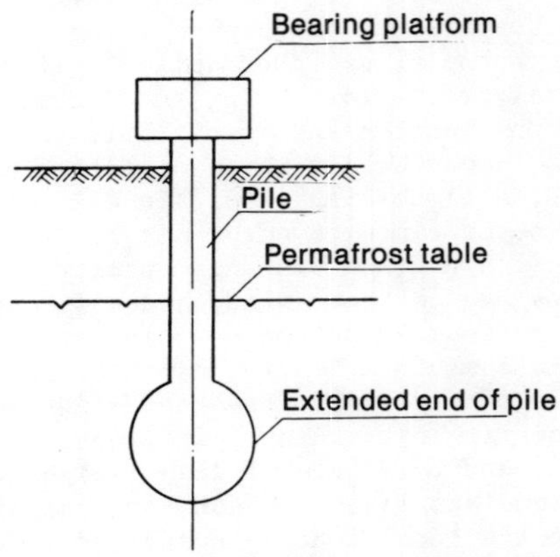

FIGURE 1 Explosion-expanded pile foundation.

little shallower; this caused serious frost heaving and led to the deformation of the pipeline. Within the sector of clayey soil, however, where piles were buried in permafrost to depths of 1-3 m, no frost heaving took place. The foundation of the tested house consisted of seven 25-cm-diameter piles buried to a depth of 2 m in permafrost. The wide end of each pile was 60 cm in diameter. No significant frost-heaving was observed. Despite deficiencies and incompleteness, the tests did provide valuable experience in designing and constructing explosion-expanding pile foundations.

Further experiments were made on explosion-expanding pile foundations for buildings, bridges, and culverts built in the Great Xing'an Mountains and on the Qinghai-Xizang Plateau in the early 1970s (TSDI, 1971; Ma, 1981). The dimensions and burial depth of piles as well as the construction technology used were changed and improved. For instance, 30-cm-diameter explosion-expanded piles, buried at depths of 4-5 m with an 80-cm-diameter widened end were used in house construction, and piles of 35 cm diameter buried to 4-5.5 m were adopted in culvert engineering. The construction method was to drill first and then use explosions to shape borings and the widened ends to the specifications. All houses and culverts built during this time have operated normally so far.

Inserted and poured piles were used in highway engineering in permafrost areas (Figures 2 and 3) in the middle of the 1960s. The diameter and burial depth of poured piles ranged from 50-125 cm and 8-10 m respectively. For inserted piles, hollow reinforced concrete piles with a 70-cm diameter and a 5-cm shell were used. After the pile was inserted in the borehole, concrete was poured into the hollow center, and clay mortar was poured

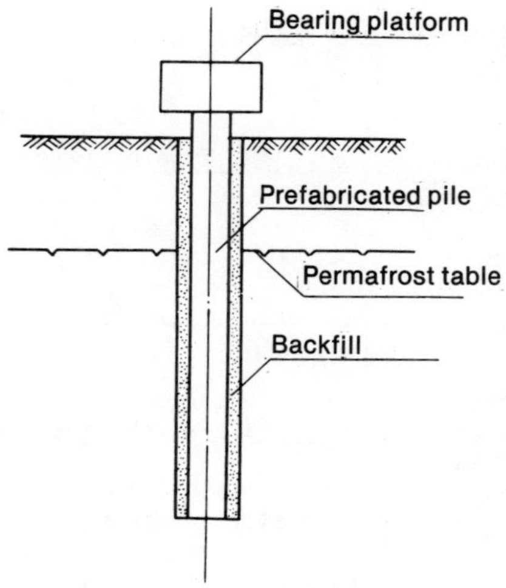

FIGURE 2 Inserted pile foundation.

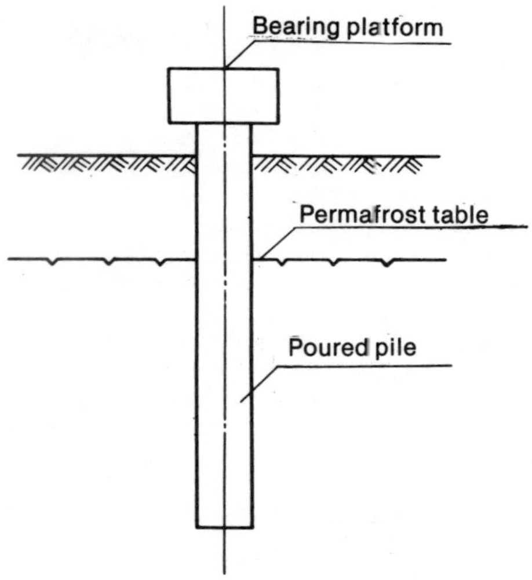

FIGURE 3 Poured pile foundation.

around the pile. The burial depth was 8-10 m. During this period, these piles were used in about 50 large and middle-sized bridges. No obvious deformation has been found since they were completed, and they are still in regular operation.

The successful use of bored piles in highway bridge engineering was followed by their trial application to house building in the early 1970s. The loading characteristics of various kinds of bored piles were studied to acquire data for designing foundations with them. Two sectors in the Qinghai-Xizang Plateau with differing features were chosen as the test sites. Vertical and horizontal loading tests of inserted, driven, and poured piles were made (NICARS, 1977). Based on the results, a test house with an inserted-pile foundation and an engineering production building with a driven-pile foundation were constructed on the Plateau (FSDI, 1978). The former was the boiler room and kitchen of the Frost Research Station and the latter was the water tower, boiler room and pump house, and electrical machinery room of a pump station. Hollow reinforced concrete piles were used at both sites; they had a diameter of 40 cm, were buried to a depth of 6 m, and ventilation was applied to keep the ground frozen. The building was completed in 1977 and no obvious deformation has yet been discovered.

Meanwhile, two test houses with poured-pile foundations -- the residences of the Frost Research Station -- were designed by the Jingtao Frost Station, located in the permafrost region of the Great Xing'an Mountains (He and Xiao, 1981). A raised foundation was not used in this case, using the thawing disc of ground to keep the base temperature inside the buildings as constant as possible. Because of the required maximum computed thawing depth of the disc and the soil bearing capacity, the designed pile diameter and burial depth were 26 cm and 7.5 m respectively. The two buildings were completed in 1976, and observations and tests verify that the strength of the concrete and the bearing capacity of the piles all meet the design specifications.

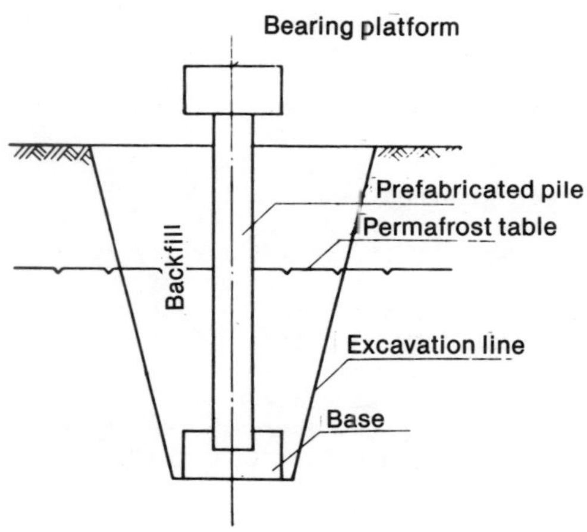

FIGURE 4 Buried pile foundation.

While bored-pile foundations were tested and put into use, experiments on other types of deep foundations were started in various permafrost areas. For instance, buried-pile foundations were tested in house construction in the permafrost region of the Great Xing'an Mountains (Qiqiha'r Railway Admin., 1975; Wang, 1982). The pile foundation was composed of prefabricated, reinforced concrete pillars (or rock pillars with mortar), ring-shaped base (or widened base), and bearing platform (Figure 4). The construction was carried out in sequence as follows:

1) Excavate the foundation pit.
2) Bury the base.
3) Put the piles vertically on the base.
4) Refill the pit.
5) Pour in concrete to cast the bearing platform.

The tests were conducted at two separate sites; at one a raised foundation was constructed to provide natural ventilation to prevent the ground from thawing. The section of pile was 20 x

20 cm, that of the base was 110 × 110 cm, and the burial depth was 3.35 m. At the other site, the foundation was not raised, i.e. ground soil was permitted to thaw. The section of pile was 25 × 25 cm, that of the base was 75 × 75 cm, and the burial depth was 5.5 m. The two houses were completed in 1974 and 1976, respectively. Observations showed that the maximum thawing depth of the former was 2.1 m, while that of the latter was 5.6 m. Cracks and deformation appeared in both houses due to insufficient burial depths.

Buried-pile foundations with ventilation were also used to build the electrical machinery room, oil pump house, and boiler room of an oil pump station on the Qinghai-Xizang Plateau in 1976. The burial depth of the piles was 5 m. No evident deformation has occurred since the completion of construction, demonstrating that this type of foundation is successful.

Another type, the so-called "deep mass foundation," had a dimension of 30.5 × 18 × 1.2 m and was buried to a depth of 6 m. It was used in the design of the coal storehouse of the Reshui coal mines in the Qilian Mountain permafrost region (Qinghai Design Inst. of Mines, 1978). The storehouse was completed and put into use in 1978 and the deformation has not yet exceeded permissible values.

Through these investigations and experience, poured, inserted, and buried pile foundations have recently come into general use in China in the engineering of railroad and highway bridges, as well as house construction, in permafrost areas. The use of deep foundations has thoroughly transformed the unstability of building in these areas, and it has greatly improved the confidence of engineering. It can be stated without a doubt that the 1980s will be years characterized by the general use of deep foundations in China's permafrost areas.

Since the 1970s our researchers have been investigating the problems of designing deep foundations for permafrost areas. The major problems include: the bearing characteristics of piles in permafrost, methods for computing the resistant frost-heaving stability of piles, ways to protect pile foundations from frost heave, and calculating the stability of the thaw disc beneath heated buildings. This research has extended the use of deep foundations. What follows is a brief introduction to some of our results.

BEARING CHARACTERISTICS OF PILE IN PERMAFROST

The bearing characteristics of a pile in permafrost depend upon the nature of the permafrost and the freezing properties of the boundary surface between it and the pile. The vertical bearing capacity of the pile involves the boundary adfreezing force and the reaction force at the pile end. The major bearing capacity depends on the former. The magnitude of the adfreezing force relates not only to the soil properties (water content and permafrost temperature), but also to the material used, the roughness of the pile surface, and the lateral pressure acting on the pile. Loading tests have been made on frequently used piles, such as reinforced concrete inserted piles, driven piles, and poured piles; the results are listed in Table 1.

TABLE 1 The critical adfreezing strength (T/m^2) of bored piles.

Average surrounding temperature (°C)	Type of pile: Inserted	Driven	Poured
-1.0	6.0	8.7	-
-0.5	4.1	8.2	12.2

It can be seen from Table 1 that the adfreezing strength of poured piles is the greatest because it has the roughest surface, that of driven piles is medium, since the lateral pressure that acts on it is the greatest, which brings it into tight contact with the surrounding frost, and that of inserted piles is the least of the three (Pile Foundation Res. Gp., 1978).

Other tests showed that while a pile is settling under external force, a reaction force appears at its bottom end, which increases with the external load. The magnitude of the reaction varies with the testing method, the load, and the pile type. For example, the reaction force under critical load appropriates 20% and 10% of the load for driven and poured reinforced concrete piles, respectively (Cheng et al., 1981).

The bearing capacity of a single vertical pile in frost is calculated by the following equation (NICARS, 1977):

$$P = \frac{1}{K} \sum \tau_i F_i + m_o A [R]$$

where

P - the permitted bearing force for a single pile, T

K - coefficient of safety

τ_i - the critical long-term and freezing strength between frost layer i and the lateral surface of the pile, T/m^2

F_i - the freezing area between frost layer i and the lateral surface of the pile, m^2

m_o - the decreasing coefficient of the bottom-end supporting capacity of the pile

A - the top-end supporting area of the pile

R - the permissible bearing capacity of the frost beneath the pile, T/m^2

The value of m_o has something to do with cleaning up the bottom of the hole. It is recommended, under ordinary circumstances, that 0.5-0.9 be adequate.

The above equation is basically similar to that being used by the frost engineering circle. Both equations suggest that bearing capacity is

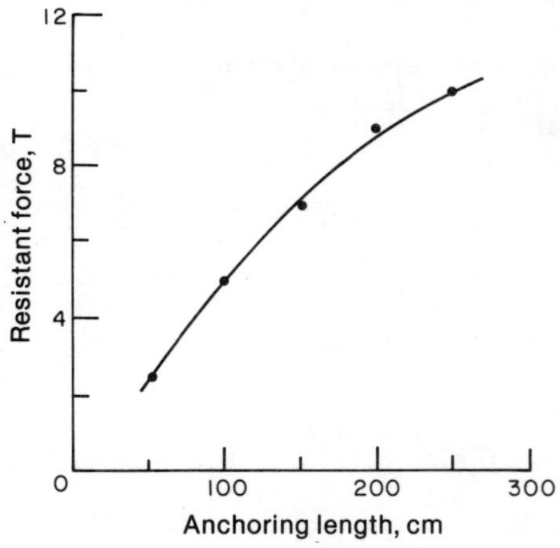

FIGURE 5 Resistant force of anchor-arm vs anchoring length.

directly proportional to the freezing area of the pile, i.e. to the length of the pile, provided the diameter remains unchanged.

In 1980, the Northwest Institute of the Chinese Academy of Railway Sciences made a series of tests on the resistance force of anchors in the permafrost region of the Qinghai-Xizang Plateau. Reinforced concrete anchors of 10 cm in diameter were adopted with burial depths of 2-4 m. Test results showed that the relationship between the resistant force and the length of the anchor tended to be nonlinear as the anchor reached a certain value (Figure 5) (Ding, 1981a). That is to say, there is no direct ratio between the resistant force of the anchor and its length.

The specific bearing capacity of the anchor is determined by gradual damage and the remnant adfreezing strength. The tests showed that under external load the distribution of shear stress along the freezing boundary surface of the anchor was uneven, attenuating exponentially with increasing depth. The destruction of adfreezing strength, therefore, does not take place simultaneously along the whole boundary, but occurs at a certain point in the boundary where shear strength has attained a critical state. The maximum shear stress then moves downward, bringing the freezing strength of the neighboring part to a critical state. This gradual destruction does not stop until the anchor loses its stability.

However, a remnant adfreezing strength will act on the freezing boundary of the anchor even if the adfreezing strength as a whole is destroyed. Continually increasing the length of the anchor can only serve to make further use of the remnant adfreezing strength, which is why the resistant force increases slowly, presenting a clearly nonlinear property.

Generally, the nonlinearity, as mentioned above, occurs in the relationship between bearing capacity and pile length if the length-to-diameter ratio of a single pile in permafrost reaches a certain value. Therefore, a revision factor of pile length should be taken into account in the formula for computing the bearing capacity of a single pile (Ding, 1981b).

Considering these bearing characteristics, we are sure that there must be an optimum length for a single pile in permafrost where the average adfreezing strength will be at the maximum. Moreover, it will be the most reasonable, both economically and technically.

The elastic foundation beam method is still used to calculate the horizontal load of a single pile in permafrost. After comparing results calculated on the basis of the assumed distribution pattern of ground coefficients with the test data, we found the distribution pattern achieved by the "value K method" was most similar to the test data; in other words, ground coefficients above the first elastic zero point are distributed in the form of a concave parabola and below it the distribution pattern with a constant is in relative conformance with the reality of piles in permafrost (Cheng et al., 1981).

REFREEZING GROUND SOIL SURROUNDING PILES

The time taken to refreeze the soil surrounding the pile is of great importance to the design and construction of the pile. It is controlled by such factors as the construction method, the type and dimension of the pile, the construction season, the temperature of the permafrost, and so forth. The refreezing times of ground soil surrounding driven, inserted, and poured reinforced concrete piles were measured in the permafrost areas of the Qinghai-Xizang Plateau. The results indicated that heat introduced by driven piles was the least, that is, refreezing occurred in the shortest period of time, from 5-11 days after the pile had been set. The ground temperature was disturbed the most by heat introduced by mixed materials along with the heat of liquefaction of concrete; as a result, refreezing took as long as 30-60 days. The time taken for refreezing the ground soil surrounding the inserted pile lay between the above two, and lasted from 6 to 15 days (NICARS, 1977).

In the permafrost region of the Great Xing'an Mountains, winter is usually chosen as the construction season to speed up the refreezing rate of the soil around poured piles. In this case, care must be taken to prevent the concrete from freezing and to guarantee the necessary strength of the concrete.

In the 1970s, along with the use of poured piles for highway bridges built in permafrost regions, the study of negative-temperature concrete was initiated. To reduce the heat effect of concrete, and increase its freeze-resistant durability and mildness and its hardening rate at low temperature, the researchers devised a way to add chemicals to the concrete. Different types and amounts of chemicals are added according to the construction season, the permafrost temperature, and other design specifications so the concrete would meet the needs mentioned. Trial house and bridge construction projects in the Great Xing'an Mountains and the Qinghai-Xizang Plateau have proved that the strength of concrete with chemical additives can attain or surpass the design grade regardless of the dimension of the poured pile (RCI, 1979; Heilongjiang Microtherm

TABLE 2 Strength of negative-temperature concrete at different stages.

No. of sample	Kind of sample	Design grade / Mixing ratio	Degree of slump (cm)	Compressive strength (kg/cm^2) R_7	R_{28}	R_{213}
10	without chemical additives	200 / 0.6:1.24:3.8	5	-- / 38.1	-- / 60.2	-- / 93
26	with chemical additives	200 / 0.6:1.24:3.8	10	--	141.6 / 155.6	207 / 162
80	with chemical additives	200 / 0.5:1.24:3.25	17	-- / 102.9	137.8 / 152.5	225.5 / 163

NOTE: Denominators refer to strength of 10x10x10 cm sample placed in permafrost. Numerators are strength of sample taken from test pile by drilling.

Res. Inst., 1979; He, 1981). In one of the tests, in the Great Xing'an Mountains, sodium nitrite and other chemicals were added to concrete and it was poured into two boreholes with diameters of 80 cm and 26 cm at burial depths of 2 m and 5.5 m respectively. Table 2 lists the strength of the concrete after it had been in the foundations at temperatures of -2°C to -5°C for 213 days. It can be seen that, under negative temperature, the concretes with additional chemicals all reached or exceeded the design strength.

RESEARCH ON THE FROST-HEAVE-RESISTANT STABILITY OF FOUNDATIONS

In the course of freezing the soil mass of the active layer that surrounds the upper part of a deep foundation, tangential frost-heaving force and the reaction forces will develop due to the resistance of foundation. These forces are equal in magnitude but opposite in direction. The reaction force is of greater importance in computing foundation stability.

Tests on the reaction force were made at a site in the Great Xing'an Mountains in 1978. Its distribution pattern around the foundation is shown in Figure 6 (Cui and Zhou, 1981; Heilongjiang Water Res. Inst., 1981). It can be seen that the reaction force around the foundation appears as a double hump, weaker near the foundation. This is because, during the course, part of the tangential frost-heaving force and the reaction force are relieved as the frost and the thawing soil move up and down the foundation circumference.

According to the theory that tangential frost-heaving force is equal to the reaction force, the tangential force measured in field tests by rigid frame counter-compression and load counter-compression must be equivalent to the reaction force shown in Figure 6. Here the work dissipated by friction can hardly be measured by the methods used.

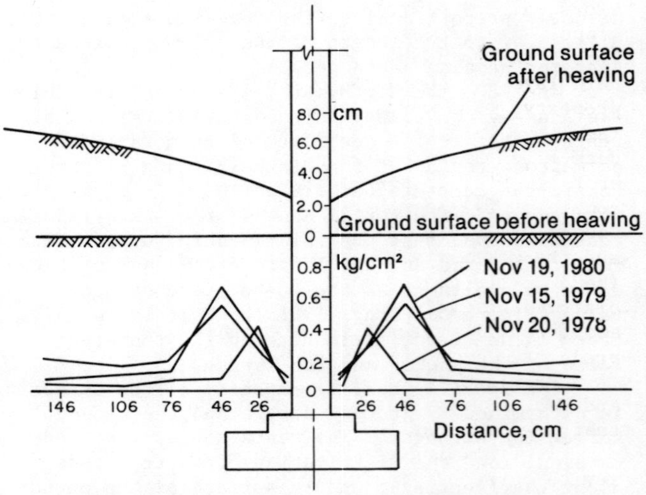

FIGURE 6 Distribution of heaving reaction.

To compute the frost-heaving stability of the foundation, the force system of stability must be added to the reaction force acting on the front edge of the foundation, while to compute the tensile strength of the foundation, the reaction force relieved by friction must be taken into account.

The distribution of tangential frost-heaving forces along the depth of the foundation is uneven, being distributed mostly within the upper part of the active layer at a depth two-thirds of its thickness. This can be taken as the base for calculating the tangential frost-heaving force when computing the frost-heaving stability of the foundation.

The following conversion factors are taken to determine the effect of foundation materials on tangential frost-heaving force: If the tangential frost-heaving pressure acting on the concrete foundation is 1, then that of steel is 1.66, of rock foundation with mortar is 1.29, and of wood is 1.06 (Ding, 1983c).

PROTECTING PILE FOUNDATIONS FROM FROST-HEAVE

The design of the burial depth of deep foundations in permafrost is often restricted by the requirements of frost-heaving stability. Frequently, piles must be buried more than 2.5 times deeper than the frost table to prevent pulling by frost-heaving. Even so, in some cases, deformation or destruction occurs because the foundation lacks the strength to resist the pulling.

There are two ways to reduce the tangential frost-heaving pressure acting on foundation surfaces. One way is to reduce the foundation material's susceptibility to water, to weaken ice cementing between the frost and the foundation; the other is to improve the soil body around the foundation to reduce frost heave.

A number of tests to reduce the tangential frost-heaving force on pile foundations have been conducted in the permafrost areas of the Qinghai-Xizang Plateau. In the tests, residuum and anion surface active agents were used to treat foundations comprehensively to meet the defined requirements. The process was carried out in this way: First, the foundation surface was daubed with a thin layer of residuum to change its hydraulic property, then certain parts of the soil surrounding piles were treated with active agents to reduce frost heave. This method reduced 95% or so of tangential frost-heaving pressure. After a couple of freeze-thaw cycles, no clear rise in tangential frost-heaving pressure was observed, and the stability has remained sound (Ding et al., 1982). This achievement was obtained in both laboratory and field model tests. Its effectiveness and durability remain to be verified in practical projects.

Problems still exist in design, construction, and use of deep foundations in permafrost regions that need further investigation and resolution:

1. Providing exact adfreezing strength is the key to correctly designing pile foundations: By tests, we see that there is a peak value and a remnant value of adfreezing strength. The average adfreezing strength used to design pile foundations relates not only to the constituents of soil, water content, ground temperature, and pile material, but also to the length and diameter of the pile. Of these, the first four factors influence peak and remnant adfreezing strength. Therefore, studying the strengths and making use of them along with the length and diameter of the pile to compute average freezing strength will give accurate parameters for computing adfreezing strength.

2. The rational distribution pattern of frost ground coefficients and the values of ground coefficients under different frost conditions.

3. The relationship between the bearing strength at the pile end and the adfreezing strength of the pile circumference: Although various methods of computing the bearing capacity of piles proposed in the past took into account the reaction force at the pile end, the action mechanism when both the pile end and the circumference are under load at the same time has not been fully illustrated.

4. The reasonable interval between piles, and tests and computions of pile groups.

5. The reasonable burial depth of buried-pile foundations: Buried-pile foundations are comparatively popular in China owing to their simple construction, the easy treatment to prevent frost-heave, and the low cost. It has been observed in earlier projects, however, that the annual variation of ground temperature for shallowly buried pile foundations is considerable, yielding frost creep as well as frost heave, which tends to cause deformation of the foundation. To prevent or reduce deformation, an allowable maximum temperature variation in the frost beneath such foundations seems to be inevitable.

6. Loading tests on pile foundations in permafrost: The step-by-step loading method currently used has some disadvantages, such as long testing time and considerable variations of boundary conditions. In addition, the criterion taken for deformation stability (0.3-0.5 mm/24 hr) lacks sufficient scientific support. Therefore, a method that not only reflects the varying characteristics of frost but also shortens testing time will be the goal of further study.

REFERENCES

Chen Xiaobai, Tong Boliang, Ding Jinkang, and Zhou Chanqing, 1979, Thirty years' study and engineering practice on frost in China: Journal of Glaciology and Cryopedology, v. 1, no. 2.

Cheng Zhohuai, Le Guoliang, Zhao Xisheng, and Wang Huaqing, 1981, Study on vertical and horizontal loading of pile foundation in permafrost area: Northwest Institute of China Academy of Railway Sciences.

Cui Chenghan, and Zhou Laijong, 1981, Study of reaction force against frost heave: Third Survey and Design Institute of China Railway Ministry.

Ding Jinkang, 1981a, Study on the long-term resistant force of anchor-arm in permafrost area: Northwest Institute of China Academy of Railway Sciences.

Ding Jinkang, 1981b, Study and test of several frost-mechanical problems.

Ding Jinkang, 1983, Outdoor experimental research on tangential frost-heaving force, _in_ Proceedings of the Second National Conference on Permafrost, Lanzhou (1981): Lanzhou, Gansu People's Publishing House, p. 251-256.

Ding Jinkang, Zhang Luxin, Ye Cheng, Yan Jingshan and Yu Changhua, 1982, Physical and chemical methods for decreasing the force of tangential frost-heave, _in_ Proceedings of the Symposium on Glaciology and Cryopedology held by the Geographical Society of China (Cryopedology), Lanzhou (1978): Beijing, Science Publishing House, p. 204-206.

FSDI (First Survey and Design Institute of China Railway Ministry), 1978, Report of study on house building in permafrost area of Qinghai--Xizang Plateau: Northwest Institute of China Academy of Railway Sciences.

He Changgen, 1981, Test and study on poured-in negative temperature concrete in permafrost: Third Survey and Design Institute of China Railway Ministry.

He Changgeng and Xiao Youming, 1981, Test and study of house foundations in permafrost areas: Third Survey and Design Institute of China Railway Ministry.

Heilongjiang Microtherm Research Institute, 1979, Test and study on poured-in pile foundation of negative temperature concrete in permafrost area.

Heilongjiang Water Research Institute, 1981, Study on reaction force against frost heave and self-anchoring of pile foundation in seasonal frost region.

Ma Zonglong, 1981, Study of explosion-expanding pile foundations in permafrost region of Qinghai-Xizang Plateau: Northwest Institute of China Academy of Railway Sciences.

NICARS (Northwest Institute of China Academy of Railway Sciences), 1977, Test and study of pile foundations in permafrost areas: First Survey and Design Institute of China Railway Ministry.

Qinghai Design Institute of Mines, 1978, Preliminary summary of construction and design in Reshui permafrost region.

Qiqiha'r Railway Administration Scientific and Technical Institute, 1975, A research report of the application of pile foundation and overhead ventilation foundation to trial buildings in Mangui permafrost region.

RCI (Railroad Construction Institute of China Academy of Railway Sciences), 1979, Technical specifications for designing and constructing bridges and culverts in permafrost regions of Qinghai-Xizang Plateau.

Research Group on Pile Foundations in Permafrost, 1978, Testing of pile foundations in permafrost areas, in Proceedings of the Third International Conference on Permafrost, Edmonton: Ottawa, National Research Council of Canada, v. 2, p. 179-185.

TSDI (Third Survey and Design Institute of China Railway Ministry), 1971, The application of explosion-expanding pile foundation to bridge and culvert engineering in severe cold and permafrost areas.

TSDI (Third Survey and Design Institute of China Railway Ministry), 1976a, Experience and lesson from erecting pipeline in permafrost area with explosion-expanded piles.

TSDI (Third Survey and Design Institute of China Railway Ministry), 1976b, A short summary on buildings of explosion-expanded pile foundations in De'rbur permafrost region.

Wang Jianfu, 1982, Specific features of ground temperature field of buildings furnished with heating in permafrost areas.

DESIGN AND PERFORMANCE OF ROAD AND RAILWAY EMBANKMENTS ON PERMAFROST

D.C. Esch

Alaska Department of Transportation and Public Facilities
Fairbanks, Alaska, USA

This review presents a summary of current practice, research efforts, and major research needs relating to road and railway embankments over permafrost, with emphasis on North American work. The stages of preliminary investigations, design, construction, and operation and maintenance of road and rail embankments are discussed in sequence. Embankments for roads and railroads necessarily encounter a wide variety of terrain types and permafrost conditions, and routings can seldom be selected to avoid all potential permafrost problem areas, due to origin and destination constraints. Because of the great distances covered by these facilities, subsurface information on any single permafrost unit is normally inadequate to predict accurately embankment performance. Embankments are therefore usually constructed without use of any intensive permafrost-related design analysis. Most engineering analyses are made only in the course of repair or reconstruction of the most severe post-construction problem areas. Road and railway embankment movement problems resulting from underlying permafrost are almost always of the chronic, long-term type, so the true economic consequences of inadequate design and construction practices are seldom if ever known.

CURRENT PRACTICE

Preliminary Investigations

Airphoto and geologic interpretations of subsurface conditions from surface features, with borings at intervals, are heavily relied upon in the route selection and subsurface investigation planning stages for roads and railways, but geophysical detection methods based on electromagnetic and radar soundings are currently under development. Theory and practice in geophysical detection of subsurface conditions are well developed in the mining and petroleum industries, but these technologies have not been readily adaptable to application by transportation engineers and geologists. There is considerable interest and hope, however, that these methods will provide the geologist with a new low-cost source of subsurface information on which to base his field boring and sampling program and even the preliminary route selection.

At the Fourth International Conference on Permafrost, 11 papers in the areas of remote sensing and geophysical sounding for the detection, mapping, and evaluation of permafrost terrain were presented. In view of the extreme variability of subsurface permafrost conditions, such as ice contents and active layer thicknesses, research is badly needed to speed the development of new methods of remotely detecting subsurface conditions that affect embankment performance. Information on the variability of soils and permafrost is needed on a continuous basis along a proposed road or rail route to predict the thermal and physical stability of alternative embankment designs.

Design of Embankments

Current design practice is not developed enough to predict distortions, ride roughness, travel speed, or maintenance costs related to alternative embankment designs. Engineers have generally adopted a negative view of such design features as insulation layers or passive heat-exchange systems, although the much higher initial costs will result in improved long-term performance. Unfortunately, researchers have not yet defined the overall life-cycle economics of such features. Insulation is not commonly used or even considered except under extreme arctic conditions, where shortages of gravel result in a direct savings in construction costs, such as the Alaska oil pipeline workpad (Wellman et al., 1976) of which about 112 km (70 mi) were designed and constructed as insulated embankments.

Simple and functional two- and three-dimensional thermal analysis methods, which would permit designers to analyze the long-term stability of alternative embankment designs, are not yet readily available, although there is considerable research activity in the area of thermal modeling. The "modified Berggren" calculation method is most commonly used for simple, one-dimensional, single-year thermal predictions.

Heat exchange of the embankment surface is handled by use of the simplified "n-factor" approach in design calculations. This factor, which is the ratio of surface to air temperature freezing or thawing indices, appears to be quite useful for analysis of paved surfaces where evaporative or latent heat exchange is a minor factor. N-factors for many different surfaces have been determined (Lunardini, 1978). Surface albedo, solar exposure, and wind effects are primary considerations in selection of an appropriate n-factor for a given site. The importance of traffic-generated wind effects and surface abrasion on the n-factor have been demonstrated (Berg and Esch, 1983).

Embankment slopes present a much more difficult surface temperature simulation problem than level paved surfaces. Ideally the effects of sun and slope angles, wind, snow cover, vegetation cover, evapotranspiration, and shading by adjacent trees and brush should all be taken into account in selection of a proper n-factor. A full energy balance approach should perhaps be attempted.

However, errors in the assumptions needed for an exact energy balance may become cumulative and lead to a large error in the end result (Goodrich, 1982).

Embankment shape considerations and their effect on embankment performance on warm permafrost have been studied by Esch (1978, 1983). Lateral berms, termed "thermal stabilizing berms," have been used extensively in Alaska in recent years, but their long-term benefits have not yet been demonstrated to justify the construction cost, except when used as waste-disposal areas. Berms result in a greater width of terrain disturbance and increased slope areas that have higher average surface temperatures than undisturbed forest or tundra. Therefore, the berms' benefit of increased insulation value over permafrost must be weighed against the increased heat intake of the entire embankment structure. For warm permafrost areas, some form of passive cooling system appears necessary, such as air or liquid convection pipes or two-phase "heat pipes," to increase wintertime removal of the heat that accumulates beneath embankment slopes and is trapped by the snow cover.

Embankment reinforcement, to reduce slope movements and to bridge smaller thermokarst pits, is currently a very active embankment design and research work area in Alaska. By installing multiple horizontal layers of selected "engineering fabrics" during construction, a more crack- and sag-free embankment may be possible. The ability of different fabrics to span voids of different widths under embankment and wheel loadings (Kinney, 1981) will be evaluated by field tests in 1984. To date, fabrics appear to show the most promise in eliminating the hazardous longitudinal embankment cracking that results from side-slope settlements and lateral spreading. The use of stiffer plastic mesh or steel reinforcement may prove even more beneficial as an embankment reinforcement. Design theory and calculation methods for predicting and analyzing the performance of fabric reinforced sections must still be developed and tested. Answers must also be developed to questions regarding the necessary extent of fabric lapping or sewing and the importance of preliminary fabric tensioning and fabric stiffness on performance.

Insulation of roadway embankments with foamed plastics to prevent thawing of underlying ice-rich permafrost has essentially remained in the experimental design stage since the first installations in 1969 (Esch, 1973). However, several airfield insulation projects have been successfully completed. Field studies have demonstrated that polyurethanes and sulfur foams lack the durability and moisture resistance for direct burial uses under wheel loadings and high water tables. Extruded expanded polystyrene foam insulations, in contrast, have generally proven very durable under these conditions. Economic considerations and concerns about differential surface frost formation on pavements over insulated areas have both acted to retard the use of insulation except for certain remote areas where gravel costs and other considerations have dictated the use of foam insulation. Recent competition in the production of low-cost, high-quality polystyrene foam insulation and developments in pavement surfaces that provide ice control, such as "Verglimit" and "Plus-Ride," should encourage increased use of embankment insulation.

Pavement design practice for roadways on permafrost is not significantly different from that for roadways in seasonal frost areas. The Alaska Department of Transportation has recently completed a major research study of pavement structure performance that evaluated pavement failures by flexural fatigue and rutting (McHattie et al., 1980). The presence of permafrost in the subgrade was not found to be a significant factor in performance. This is reasonable in view of the fact that vehicle wheel load stresses are carried almost entirely by the uppermost 1 m (3.3 ft) of the embankment, whereas the permafrost table is commonly at a depth of 2 to 4 m beneath the pavement. Pavement structures, defined as the load-carrying portion of an embankment, are therefore always included in the "seasonal frost" zone. The only special pavement design consideration applied for permafrost areas involves the use of a shortened design life period for pavements that are expected to be distorted severely by thaw settlements within a few years of new embankment construction. Thaw-settlement and frost-heaving problems are recognized and treated separately from pavement design considerations.

Embankment Construction

Embankments on permafrost are only rarely constructed in winter, except perhaps on the Arctic Slope. The major questions to be resolved relate to the compaction levels attainable with frozen soils and the detrimental effects of reduced soil-density levels and possibly trapped snow and ice layers on long-term performance. Coarse-grained soils are more amenable to compaction while in a frozen state and are less inclined to loss of strength upon thawing. Specifications that permit placement of certain frozen soil and rock types in winter could be developed.

Some performance aspects of winter versus summer construction of insulated embankments were investigated by constructing two adjacent test sections near Inuvik in 1972 (Johnston, 1983). Observations through 1978 showed little difference in performance due to the timing of construction. Construction timing specifications have been used successfully in Alaska for controlling construction of insulated road sections and for excavating frozen soils. Insulation placement is specified at the start of the thawing season, following surface preparation and snow-removal during the previous winter, to assure the lowest possible subsurface temperature. Benefits may result from construction timing specifications that require a 4-month period for prethawing and consolidation of ice-rich permafrost soils before embankment placement. These benefits have been analyzed at test sites (Esch, 1982) but this approach has not yet been used on a construction project.

An embankment was constructed on top of a mechanically placed layer of frozen peat in 1973 at a roadway site 80 km southeast of Fairbanks, Alaska. After an initial thaw stabilization period, the performance of this embankment has been excellent, and the thermal benefits of peat in the active layer beneath the roadway have been significant (McHattie, 1983). The great increase in the thermal conductivity of the peat upon freezing has resulted in lower average subsurface temperatures and preservation of the underlying permafrost. By comparison, the permafrost foundation of the adjacent normal roadway section has experienced annually increasing thawing, settlements, and degradation.

Standard construction practice in ice-rich permafrost requires preserving the surface vegetation mat to serve as an insulation layer, a filter medium, and a soil-reinforcing layer during and after construction. Machine-mulching or shredding of trees and brush with a wheeled machine, called a "hydro-axe," is commonly used in preference to hand-clearing. However, to avoid surface damage this machine work must be performed when the ground surface is frozen.

Maintenance of Embankments

Errors in embankment design and construction become the responsibility of the maintenance engineer. Sinkholes, sags in grades, longitudinal cracking, shoulder rotation and slumping, and culvert distortions and blockages are common manifestations of embankment instability caused by thawing permafrost. Sinks, sags, and shoulder slumps in paved road surfaces are routinely filled with asphalt patching material, with repairs required several times a year in the worst problem areas. For railroads, the rails must be shimmed periodically and eventually the ties lifted and ballast added to maintain grade where thaw-instability and settlement problems occur. By comparison, the frequent regradings performed on well travelled gravel road surfaces normally obliterate all sags and cracks. The extent of permafrost problems on gravel roads is not normally apparent.

Experimental installations of two-phase heat pipes to stabilize sinkholes in embankments have been made in Manitoba on the Hudson Bay Railroad (Hayley et al., 1983), and in Alaska on Farmers Loop Road near Fairbanks and at Bethel Airport. This new technology may prove very useful, as heat pipes can be installed by slant drilling to reach beneath areas that must remain open to traffic.

Maintaining drainage is often a major problem for embankments on permafrost. Culvert icing or aufeis problems may require almost daily efforts at steam thawing to maintain drainage. Methods in common use include steam discharge or hot water cycling through culvert thaw pipes, and temporary welder or permanent powerline connection to electric thaw wires located inside the culverts. Test installations of solar collectors and pumps have been made in Anchorage and Fairbanks, Alaska (Zarling and Miller, 1981) to thaw culverts automatically. This approach shows promise only for those areas where springtime ice buildup is the major problem. The construction of insulated subdrainage systems to intercept seepage water before it can reach the surface and freeze has done much to eliminate icing problems following new embankment construction (Livingston and Johnson, 1978).

RESEARCH EFFORTS

Permafrost area embankment monitoring in North America was started in 1953, with the instrumentation of five roadway sections near Glenallen, Alaska. Observations at these sites consisted of continuously recorded air and pavement temperatures and monthly observations of subsurface temperatures from thermistor strings. The stability of the early thermistors used cast some doubt on the long-term subsurface temperatures, and data reporting from these sites was minimal.

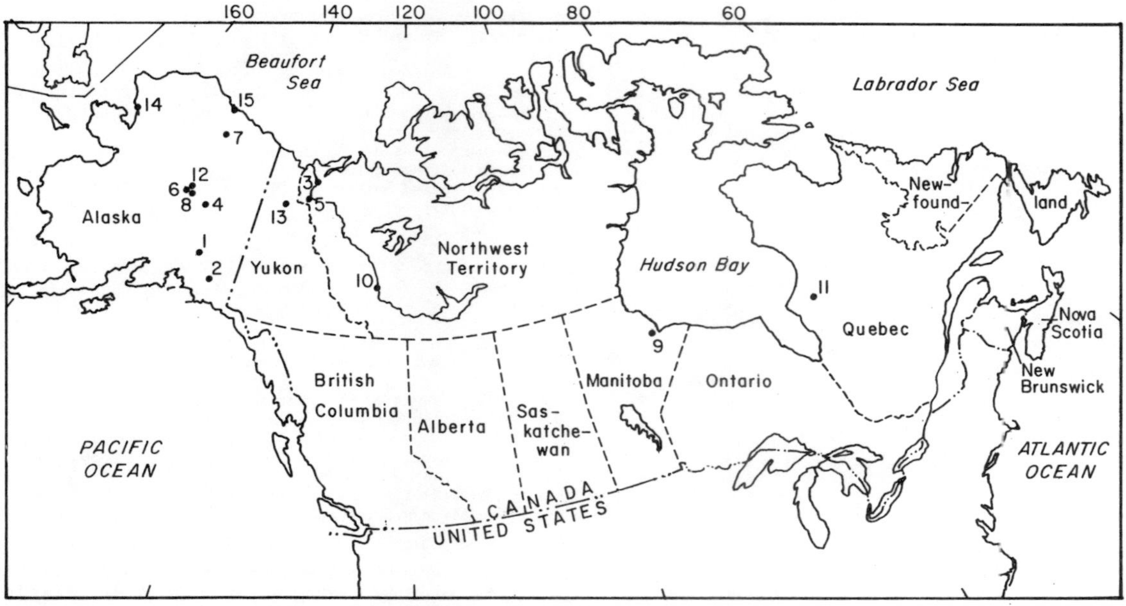

FIGURE 1 Experimental embankment monitoring sites on permafrost in North America. Numbers refer to sites listed in Table 1.

TABLE 1 Experimental embankments on permafrost in North America.

Figure 1 reference	Period monitored	Site description	Purpose	Reference
1	1953-60	Glenallen Area, Alaska Paved roadway embankments	Determine air and surface temperatures and thermal regime in normal road embankments.	Brewer et al., 1965
2	1969-83	Chitina, Alaska Insulated gravel road embankment, polystyrene foam	Compare 5 and 10 cm thick insulation layers with adjacent normal embankments by air and surface temperatures, thermal regime, and embankment settlements.	Esch, 1973
3	1972-78	Mackenzie Hwy., Inuvik, N.W.T. Polystyrene foam insulated gravel road embankment	Six road test sites with 3 insulation thicknesses (5, 9, and 11.5 cm) and winter and summer construction were compared by ground temperatures and settlements.	Johnston, 1983
4	1973-83	Richardson Highway, Alaska at Canyon Creek. Peat underlay placed beneath paved road in cut area	Evaluate benefits of placing 0.5 - 0.8 m of peat in active layer beneath road, by temperatures and settlements.	Esch, 1978 McHattie, 1983
5	1974-76	Dempster Hwy., N.W.T. 10 cm foamed sulfur insulation beneath gravel road	Evaluate performance of foamed sulfur as road insulation.	Raymont, 1978
6	1975-83	Parks Hwy., near Fairbanks Alder Creek insulated paved road cuts and Bonanza Creek experimental embankment	Compare benefits of insulation at 1.2 and 3.0 m depths, and analyze benefits of toe berms, toe insulation and air-cooling ducts on slope stability of 7-m high embankment.	Esch, 1978, 1983
7	1975-78	Yukon River to Prudhoe Bay, AK Pipeline Haul Road (Dalton Hwy.) Gravel surfaced roadway.	Evaluate the thermal regime beneath selected normal roadway embankment sections in discontinuous and continuous permafrost areas.	Brown and Berg, 1980
8	1975-83	Alaska Railroad, 50 km west of Fairbanks, insulated embankment	Monitor and evaluate the benefits of installing polystyrene foam insulation in a railway embankment with a history of permafrost thaw-settlement problems.	Trueblood, unpublished
9	1978-82	Hudson Bay Railway near Port Nelson, Manitoba	Evaluate the benefits of heat pipes installed to refreeze and stabilize permafrost and arrest thaw-settlement problems.	Hayley, 1983
10	1978-82	Mackenzie Hwy. near Wrigley, N.W.T., gravel surfaced, seasonal use, no snow removal	Evaluate the thermal performance of a thin (1.2 m) embankment on warm permafrost, including temperatures by data-logger, thermal conductivity by probes, and settlements by level surveys.	Goodrich, 1983
11	1981-83	Gravel test road in palsa field along Great Whale hydro project access road	Study probable performance and settlement vs predictions for road embankment areas overlying palsas, a permafrost feature surrounded by thawed muskeg. Insulated and geotextile reinforced sections were included.	Keyser and Laforte, 1983
12	1964-83	Goldstream, Moose, and Spinach Creek Bridge foundations in permafrost near Fairbanks	Examine the long-term thermal and physical stability of structures and embankments at stream crossings	Crory, 1975, 1978
13	1974-79	Eagle River Bridge on Dempster Highway, Yukon Territory	Analyze the thermal effects of bridge foundation piles and approach embankment fills insulated with 10 cm of polystyrene insulation.	Johnston, 1980
14	1969-73	Kotzebue Airfield Paved and insulated runway over permafrost.	Evaluate the thermal performance of an insulated runway and an uninsulated taxiway by use of a data logger to measure surface and subsurface temperatures.	Esch and Rhode, 1976
15	1980-83	Prudhoe Bay area, airport runway and roadway culvert monitoring	Compare temperatures around insulated and uninsulated road culverts and beneath Deadhorse Airfield pavement.	Brown, 1983 Eaton, unpublished

However, data did demonstrate that full annual subroadway refreezing was occurring beneath these paved roadway sections in spite of the warm (-1°C) permafrost temperatures. Monitoring of these sites was discontinued in 1960.

The first insulated road embankment on permafrost, located near Chitina, Alaska, was constructed and instrumented in 1969, and has been monitored continually since that time, providing the longest temperature data base for use in evaluating thermal models. Since that time, additional instrumented roadway and railway embankments have been constructed and monitored in Alaska and Canada. Details of all such installations cannot be given here, but Figure 1 and Table 1 show the locations and references for significant experimental data pertinent to roads and railways over permafrost.

Recently, three roadway sites in interior Alaska have been painted white to measure the effects of increased surface reflectance or albedo in reducing thaw-settlement problems. These sites provide a fair data base for predicting the performance benefits of insulated embankments and other experimental features such as berms, air-cooling ducts, and heat pipes, particularly in warm permafrost. Increased field monitoring efforts are needed to measure the thermal effects of culverts on embankment performance and the surface temperature effects of embankment slopes and slope vegetative covers, as well as the thermal stability of embankments overlying colder permafrost.

RESEARCH NEEDS

Because of the long thermal adjustment period that often results from new embankment construction over permafrost, thermal models are needed that will economically analyze alternatives over a period of 20 years or more. In fact, in view of the forecasts of a major climatic warming trend expected to occur in the Arctic as a result of the "greenhouse effect" due to increasing atmospheric carbon dioxide levels, even longer periods of thermal analysis are indicated. In areas of discontinuous permafrost, acceleration of thawing by preconstruction or construction operations may become the best design approach.

Research is needed particularly to quantify the thermal effects of alternative embankment side-slope surfaces and surface vegetation covers and to determine the role of snow-cover in embankment performance. Previous embankment research studies in discontinuous warm permafrost areas have demonstrated that progressive talik development beneath snow-covered embankment slopes is a major factor in embankment distress. Snow removal from roadway surfaces, by comparison, normally results in sufficient seasonal cooling to assure refreezing beneath paved roadways, even in warm permafrost areas. Although railway embankments have not been similarly instrumented and monitored, experience indicates that railways, because they lack a snow-free surface, should result in perpetual net warming and ongoing thaw-settlement problems in warm permafrost. Experience with continued maintenance on Alaska Railroad embankments constructed 60 years ago tends to verify this observation.

Embankment drainage design considerations relating to permafrost have not been studied in sufficient detail. Some roadway performance reports have criticized designers for channeling flows into culverts where cross-slope embankment routes intercept active-layer water drainage. No construction methods have been developed and tested that will assure water percolation through embankments without such channelization.

Drainage culverts may act as embankment air-cooling ducts or they may result in net warming, depending on seasonal air and water flows, snow cover, and so forth. Over a period of years, embankment settlements and frost heaving may distort or displace culverts and make them non-functional. These thermal aspects of culvert design have received essentially no research effort.

A more detailed listing of research needs related to embankments on permafrost, recently developed by the U.S. Committee on Permafrost, is titled "Permafrost Research: An Assessment of Future Needs" (1983). According to this report, the highest priority in permafrost research should be given to developing improved methods of detecting permafrost and ground ice and to mapping of critical permafrost parameters. The author shares this view, because lack of accurate knowledge of subsurface conditions prior to construction may be the major problem facing the embankment designer.

REFERENCES

Berg, R.L. and Esch, D.C., 1983, Effects of color and texture on the surface temperature of asphalt concrete pavements, _in_ Proceedings of the Fourth International Conference on Permafrost, Fairbanks: Washington, D.C., National Academy of Sciences, p. 57-61.

Brown, J., 1983, Interaction of gravel fills, surface drainage, and culverts with permafrost terrain: Hanover, N.H., U.S. Army Cold Regions Research and Engineering Laboratory, in preparation.

Brown, J. and Berg, R.L., eds., 1980, Environmental engineering and ecological baseline investigations along the Yukon River-Prudhoe Bay Haul Road: Hanover, N.H., U.S. Army Cold Regions Research and Engineering Laboratory, CRREL Report 80-19.

Crory, F.E., 1968, Bridge foundations in permafrost areas, Goldstream Creek, Fairbanks, Alaska: Hanover, N.H., U.S. Army Cold Regions Research and Engineering Laboratory, Technical Report 180.

Crory, F.E., 1975, Bridge foundations in permafrost areas: Hanover, N.H., U.S. Army Cold Regions Research and Engineering Laboratory, Technical Report 266.

Crory, F.E., 1983, Long-term foundation studies of three bridges in the Fairbanks area: Hanover, N.H., U.S. Army Cold Regions Research and Engineering Laboratory, in preparation.

Esch, D.C., 1973, Control of permafrost degradation by roadway subgrade insulation, _in_ Permafrost -- The North American Contribution to the Second International Conference, Yakutsk:

Washington, D.C., National Academy of Sciences.

Esch, D.C., 1978a, Performance of a roadway with a peat underlay over permafrost: Alaska Department of Transportation, Research Report 78-2.

Esch, D.C., 1978b, Road embankment design alternatives over permafrost, in Proceedings of the Conference on Applied Techniques for Cold Environments: New York, American Society of Civil Engineers, p. 159-170.

Esch, D.C., 1981. Thawing of permafrost by passive solar means, in Proceedings of the Fourth Canadian Permafrost Conference, Calgary: Ottawa, National Research Council of Canada, p. 560-569.

Esch, D.C., 1983, Evaluation of experimental design factors for roadway construction over permafrost, in Proceedings of the Fourth International Conference on Permafrost, Fairbanks: Washington, D.C., National Academy of Sciences, p. 283-288.

Esch, D.C. and Rhode, J.J., 1977, Kotzebue Airport, runway insulation over permafrost, in Proceedings of the Second International Symposium on Cold Regions Engineering: Fairbanks, Cold Regions Engineers Professional Association, p. 44-61.

Goodrich, L.E., 1982, An introductory review of numerical methods for ground thermal regions calculations: Ottawa, National Research Council of Canada, DBR Paper No. 1061.

Goodrich, L.E., 1983, Thermal performance of a section of the Mackenzie Highway, in Proceedings of the Fourth International Conference on Permafrost, Fairbanks: Washington, D.C., National Academy of Sciences, p. 353-358.

Hayley, D.W., Roggensack, W., Jubien, W.E., and Johnson, P.V., 1983, Stabilization of sinkholes on the Hudson Bay Railway, in Proceedings of the Fourth International Conference on Permafrost, Fairbanks: Washington, D.C., National Academy of Sciences, p. 468-473.

Johnston, G.H., 1980, Permafrost and the Eagle River Bridge, in Proceedings, Workshop on Permafrost Engineering: Ottawa, National Research Council of Canada, Technical Memorandum No. 130.

Johnston, G.H., 1983, Performance of an insulated roadway on permafrost, Inuvik, N.W.T., in Proceedings of the Fourth International Conference on Permafrost, Fairbanks: Washington, D.C., National Academy of Sciences, p. 548-553.

Keyser, J.H. and Laforte, M.A., 1983, Road construction in palsa fields: Paper given at poster session of the Fourth International Conference on Permafrost, Fairbanks.

Kinney, T., 1981. Use of geotextiles to bridge thermokarsts: Alaska Department of Transportation, Research Report 82-21.

Livingston, H. and Johnson, E., 1978, Insulated roadway subdrains in the Subarctic for the prevention of spring icings, in Proceedings of the Conference on Applied Technology for Cold Environments, v. 1: New York, American Society of Civil Engineers, p. 513-521.

Lunardini, V.J., 1978, Theory of N-factors and correlation of data, in Proceedings of the Third International Conference on Permafrost, Edmonton: Ottawa, National Research Council of Canada, p. 40-46.

McHattie, R.L. and Esch, D.C., 1983, Benefits of a peat underlay used in road construction on permafrost, in Proceedings of the Fourth International Conference on Permafrost, Fairbanks: Washington, D.C., National Academy of Sciences, p. 826-831.

McHattie, R., Connor, B., and Esch, D.C., 1980, Pavement structure evaluation of Alaskan highways.

Raymont, M.E.D., 1978, Foamed sulfur insulation for permafrost protection, in Proceedings of the Third International Conference on Permafrost, Edmonton: Ottawa, National Research Council of Canada, p. 864-869.

Trueblood, T., 1983, Railroad roadbed design in areas of discontinuous permafrost: Alaska Railroad Engineering Department, report in preparation.

Wellman, J.H., Clarke, E., and Condo, A., 1976, Design and construction of synthetically insulated gravel pads in the Alaskan Arctic, in Proceedings of the Second International Symposium on Cold Regions Engineering: Fairbanks, University of Alaska, p. 62-85.

Zarling, J. and F. Miller, 1982, Solar-assisted culvert thawing device: Alaska Department of Transportation, Research Reports AK-RD-82-10 and 83-36.

DESIGN AND PERFORMANCE OF WATER-RETAINING EMBANKMENTS IN PERMAFROST

F.H. Sayles

Cold Regions Research and Engineering Laboratory
Hanover, New Hampshire, USA
(Formerly with Office of Federal Inspector, Irvine, California, USA)

To date, the water-retaining structures constructed and maintained on permafrost in North America have been designed and built using a combination of soil mechanics principles for unfrozen soils and unproven permafrost theory. In the USSR, at least five sizeable hydroelectric and water supply embankment dams as well as several small water supply embankment dams have been constructed and maintained on permafrost. The larger dams are understood to have performed well, but the smaller dams have been a mix of successes and failures. (See Table 1 for examples of problems recorded in the literature.) Specific criteria are still lacking for design, operation, and post-construction monitoring of water-retaining embankments founded on permafrost. The purpose of this presentation is to review the current practice, point out how it is deficient, and note what major problems need attention.

CURRENT PRACTICE

General Considerations

In current practice, the designs of water-retaining embankments on permafrost can be divided into two general types, frozen and thawed. The frozen type of embankments and their foundations are maintained frozen during the life of the structure. The thawed type of embankments usually are designed assuming that the permafrost foundation will thaw during either the construction or the operation of the structure. In some locations where water is to be retained intermittently for short periods of time, thawed embankments have been designed assuming the permafrost is to remain frozen throughout the life of the embankment. In selecting the type of design for a particular site, many factors that are peculiar to cold regions must be considered, including:

- The anticipated type of service of the embankment; i.e. retain water continuously or only intermittently.
- The width, depth, temperature, and chemical composition of the body of water to be retained by the embankment.
- Regional and local climate conditions, especially temperature.
- The temperature of the existing permafrost.
- The extent in area and depth of the permafrost.
- The availability of the type of earth materials required for construction.
- The accessibility of the construction site for logistics involving man-made construction materials.
- The consequences to life and property in the event of embankment failure.
- The effects of the construction and operation of the embankment on the environment.
- The orientation of the downstream face (i.e. the dry face) of the embankment with respect to solar radiation.
- Frost action on the dry slopes and crest of the embankment.
- The economics of constructing a selected design in the cold region.

In addition to these rather general factors, each type of design has special requirements that must be taken into account in making the final selection of a particular design.

Unfrozen Embankment on Thawing Permafrost

The design for an unfrozen embankment founded on thawing permafrost is most suitable for sites where the foundation materials are thaw-stable; i.e. where the thawing strengths of the earth materials provide an adequate factor of safety against shear failure, and deformations resulting from thawing will not endanger the integrity of the embankment. This requirement usually restricts the use of the thawing foundation design to sites where permafrost soil is ice-poor or where reasonably sound bedrock can serve as the foundation. At sites where only a portion of the foundation contains ice-rich permafrost at shallow depths, this ice-rich portion is usually thawed before placing the embankment, or the frozen soil is excavated to a predetermined depth (Gluskin and Ziskovich, 1973). MacPherson et al. (1970) suggest a method of estimating the depth of excavation so that thaw consolidation can be limited to a predetermined amount during the operation of the embankment.

Where the permafrost is not removed and the foundation is expected to thaw during the life of the structure, the embankment design is similar in many respects to that of a water retaining embankment located in a temperate climate. However, special consideration is given to certain elements of the embankment. One such element is the impervious zone, which must be constructed of self-healing soils (Gupta et al., 1973) so that this zone can remain "impervious" even if cracking occurs during the settlement of the foundation. Soils that become stiff and brittle when compacted in the impervious zone are avoided. Other design provisions that are often included to accommodate the anticipated settlement are: the use of flatter embankment slopes; overbuilding the height of the embankment by an amount equal to the anticipated settlement; and periodically rebuilding portions of the embankments that settle below a tolerable limit (Johnston, 1969; MacPherson et al., 1970). Prethawing followed by preloading to consolidate the foundation before placement of the embankments has also been suggested as a means of reducing foundation settlements (Gluskin and Ziskovich, 1973). The concept of utilizing sand drains in a thawing foundation to increase the rate of consolidation and, hence, quickly improve the shear resistance and stability of embankments has been used successfully (Johnston, 1965, 1969; MacPherson et al., 1970). In addition, analytical methods have been developed for estimating the rates of thaw and settlements of dikes on permafrost during operations using simple heat conduction and heat balance equations for a one-dimensional transient condition that takes into account heat from water seepage (Brown and Johnston, 1970).

Grouting has been used successfully in cold regions to stabilize thawing foundations and to cut off seepage. At sites where the foundation material is expected to be weakened or where initial seepage is expected to present a problem when the ice melts from the voids, the foundation areas have been thawed and grouted before the embankments were constructed (Gluskin et al., 1974). As an alternative, the fissures or voids can be grouted in steps during the operational life of the embankment. In this case, the foundation is thawed by the heat from the impounded reservoir. As the thawing front progresses into the foundation, the fissures and voids are grouted periodically (Demidov, 1973). This method requires careful and continuous monitoring of the thawing front beneath the impervious zone of the embankment.

Essential elements of any thawed embankment are the filters and drainage systems for the safe control of seepage through and beneath the embankment. The design of these elements is similar to those for embankments in nonpermafrost areas. However, special provisions are required to avoid plugging the drainage system with ice, which could render it useless at a time when it may be needed most.

Thawed Embankment on Permafrost

The use of a thawed embankment on a nonthawing permafrost foundation is usually limited to regions of cold permafrost where the water is to be retained by the embankment for a relatively short period of time each year or less frequently. At these sites, artificial cooling of the foundation may be required during construction (Rice and Simoni, 1966; Kitze and Simoni, 1972) and maybe even during the operation period. As a further provision for keeping the foundation frozen, it is essential that a positive seepage cutoff be provided (Borisov and Shamshura, 1959; Trupak, 1970). Such a cutoff may take the form of sheet piling, a plastic membrane (Belikov et al., 1968), or other waterproof materials that are sealed to the frozen foundation and extend up to the embankment crest. The most economical and effective seepage cutoff often is a zone of frozen soil. This zone can be created from the surfaces of the embankment slopes during the winter season by natural freezing when water is not being retained. Effective temperature and water seepage monitoring systems are necessary in operating this type of water-retaining embankment in order to detect thawing that may occur that would initiate a seepage path through the embankment or in the foundation beneath it.

Frozen Embankment Design

A frozen embankment design is usually suitable for sites where the foundation becomes unstable upon thawing to a considerable depth and for regions where the permafrost is continuous. The most essential requirement of this type of design is that the embankment and foundation be completely impervious to seepage since heat from seeping water would eventually thaw the foundation. If left unchecked, the combination of thermal and mechanical erosion (piping) could breach the embankment.

Currently, frozen embankments operating with permanent reservoirs are located in regions where the mean annual temperature is -8°C or colder. Even at these temperatures, embankments 10 m or more in height require supplemental artificial freezing for at least part of each year to ensure that the embankment and foundation remain frozen. In addition to the climate at the site and the embankment height, the lateral dimensions of an impounded reservoir must be considered. With time, a deep reservoir develops a talik beneath it similar to that produced by a wide lake. The larger the talik, the greater the risk of lateral thawing under the embankment and of initiating seepage or stability difficulties.

It has been suggested that for a frozen embankment design almost any type of earth material can be used to construct the embankments, provided the pores of the material are filled with ice and mass is maintained frozen during the life of the structure (Johnston and MacPherson, 1981). It has also been suggested that ice be used as an impervious core (Tsytovich, 1973). Although these suggestions have merit, they also can cause difficulties. A large portion of the upstream section of an embankment that is required to retain a permanent reservoir and the foundation supporting it will thaw from the heat of the reservoir. Therefore, the upstream slope can become unstable if this portion of the embankment is not constructed of thaw-stable materials and if provisions are not made to accommodate the differential settlement that can occur between the core and the upstream shell.

Spillways and water outlet control structures founded in permafrost require special considerations with respect to their location and to the preservation of permafrost (Tsytovich et al., 1972; Gluskin and Ziskovich, 1973). When these facilities conduct water downstream near or through the embankment, the heat released into the structure by the flowing water can eventually melt the supporting permafrost, which may result in detrimental settlement of the structure and breaching of the seepage barrier. Table 1 reveals that most of the problems that have been associated with water-retaining embankments on permafrost have been related to seepage around and beneath outlet structures. To avoid these problems, spillways and other outlet structures usually are not located within or adjacent to ice-rich permafrost without special refrigeration or other measures to preserve the earth in a frozen state. Where possible, the outlet facilities are located at a remote site and preferably in competent bedrock. For small dams the use of siphons and pumps have been used to control reservoir levels (Gluskin and Ziskovich, 1973). A chute spillway elevated above the embankment surface (Bogoslovskii et al., 1966) is one possible solution, but the most common practice is to locate the spillway in one of the abutments on bedrock and to use refrigeration or grout to control the seepage. The importance of a positive seepage cutoff must be emphasized because of the possible formation and enlargement of a talik that can grow laterally from the outlet works into and beneath the adjacent embankment. As a result, the entire facility can be endangered if it is a frozen-type design.

Thermal Analysis

Essential to the design of water retaining embankments on permafrost is the determination of the thermal regime throughout the life of the structure (Bogoslovskii, 1958; Tsytovich et al., 1972). A number of methods have been developed and others are being developed for estimating the thermal regimes of these embankments and their foundations. Some of the methods currently in use include:

a. One-dimensional analyses that are applied at critical locations within the cross-sections of the embankment and foundation (Tsytovich et al., 1972). In areas where water does not change phase, simple heat-conduction equations are used in the analyses. The Stefan-Boltzmann equation or the Neumann solution to the Fourier equation is used to establish the position of the freezing and thawing front. This one-dimensional method can be used to analyze the transient heat flow condition.

b. Hydraulic analog computers to analyze two-dimensional simple heat flow conditions and to establish the location of the freezing or thawing front (Tsytovich et al., 1972).

c. Numerical techniques, such as the finite differences and finite elements methods, to analyze heat flow, including change of phase for two dimensions (Bogoslovskii, 1958).

d. Physical models constructed of soil in the laboratory to validate the analytical methods as well as to simulate the prototypes for heat flow.

e. The three-dimensional finite element method of analysis, which accounts for the heat transferred by the seepage water; this is still in the development stage. It is especially useful for analyzing the special conditions at dam abutments, areas adjacent to water outlet facilities, and short dams (Bogoslovskii, 1970).

Of these methods, the one-dimensional analysis is now used most often, although the two-dimensional finite element method, which accounts for the latent heat of fusion, is also frequently used.

Meaningful thermal analysis requires realistic information for defining the thermal properties of soil, the initial temperature distribution within the soil, the geometrical and thermal boundary conditions, and, where appropriate, information about the amount of heat carried by the seepage water. The analysis must also account for the change in the thermal properties of the soil due to a change in phase (i.e. frozen or unfrozen properties) and the change in density due to consolidation of the soil upon thawing. At a given site, the initial soil temperature distribution often is determined by in situ temperature measurements.

Where accurate meteorological data is available for several years, reasonable estimates of the initial temperature distribution of undisturbed sites have been made using heat balance equations that take into account, directly or indirectly, such heat sources and losses as solar radiation, air temperature, wind velocity, evapotranspiration, geothermal gradient, and flowing ground and surface water. The upper boundaries for thermal analyses are usually the interfaces of the atmosphere with the surfaces of the embankment and the natural ground. Some analyses consider the surface of the snow, if it exists, as the upper boundary. In such cases, the depth and thermal properties of the snow are required for the analysis. The lower thermal boundary often is considered to be isothermal, or a constant geothermal gradient is assumed to exist at this surface. In two-dimensional analyses, the lateral boundary surfaces are usually assumed to be adiabatic.

Although few details are published on the thermal analyses for artificially freezing water-retaining embankments on permafrost, the necessary information is available on methods of analysis developed for freezing shafts, tunnels, and walls to exclude water and soil from excavations during construction in nonpermafrost areas (Khakimov, 1963; Sanger and Sayles, 1978).

The efficiency with which heat passes between the atmosphere and earth materials is computed by using heat-transfer coefficients or factors. The values of the heat coefficients used at the upper boundaries depend upon the color and type of surface exposed to the atmosphere and sun; that is, whether the surface is light or dark and whether it is covered with gravel, stone, snow, vegetation, or other types of materials.

Methods for calculating heat transfer due to seepage and the convection of water are still quite crude, but numerical methods are being formulated that should eventually improve this computation (Brown and Johnston, 1970). The convection of air within a rockfill embankment can transfer large amounts of heat under certain conditions (Mukhetdinov, 1969; Melnikov and Olovin, 1983). Methods for analyzing convection in embankments have been developed and are still under development.

Stability

In evaluating the stability of a water-retaining embankment on permafrost, the functions of each zone within the embankment must be considered. Embankments designed to remain frozen throughout their life consist of an upstream thawed zone and a downstream frozen zone. The upstream zone serves as a thermal insulator (Bogoslovskii, 1958) to protect the downstream frozen zone from the heat of the water being retained as well as to protect against mechanical erosion from moving water or ice. The frozen downstream zone provides the impervious water barrier and resistance to the horizontal forces exerted on the embankment. The thawing front within the upstream zone and its foundation can be the seat of embankment failure when the shear resistance of the soil decreases due to development of excess pore water pressure. This excess pressure can occur when the thawing front advances faster than the meltwater can escape. The upstream shell of the embankment is especially susceptible to this type of shear or slide failure when the level of the water being retained by the embankment is lowered rapidly. In determining the stability of the upstream slope, the thawing front is included in the trial failure surfaces.

The stability of unfrozen embankments on permafrost is evaluated similarly to embankments in non-permafrost areas, except that special consideration is given to the low shear resistance that may exist in the foundation at the thawing interface of the permafrost. A portion or all of this interface is included as part of the trial failure surfaces that are investigated in the stability analyses. The shearing resistance at this interface is a major consideration in assessing the resistance to horizontal forces. To increase the shearing resistance at the thawing front, vertical sand drains terminating in an intercepting horizontal drainage system have been used successfully (Johnston, 1969; Gupta et al., 1973) to accelerate the dissipation of excess pore water pressure.

Maintaining the Frozen State of Water-retaining Embankments

Techniques for maintaining embankments in a frozen state include both natural and artificial cooling. Natural cooling is usually used for small embankments with heights of 10 m or less where the natural cold atmosphere freezes the top and the exposed slopes of the embankment. Techniques that have been used or proposed to encourage natural freezing include: removal of snow from the downstream face, by equipment or by constructing horizontal wind "vanes" that direct the wind parallel to the surface of the dam and blowing snow away from the downstream slope; placing a shelter over the downstream slope to keep snow and rain off the surface and to shade it from the sun; creating a thick snow cover on the downstream surface during the early spring and providing for winter ventilation at the embankment surface by constructing duct work beneath the snow and ice on this surface; protecting the downstream toe from the warming effect of the tailwater by using berms; and keeping the foundation frozen during construction (Borisov and Shamshura, 1959; Tsvid, 1961). Tsytovich (1973) has pointed out that the natural cooling of soils on the downstream slope of an embankment dam will reach a maximum depth of about 10 m in a severely cold climate if the surface is kept clear of snow. Therefore, auxiliary cooling is required if freezing is to be accomplished at greater depths.

Frozen earth dams with heights of up to about 25 m have been built in the USSR using artificial refrigeration. The refrigeration systems include: the circulation of natural chilled air, the circulation of artificially chilled liquid brine, and the installation of a series of automatic thermal devices such as thermal piles. The cooling elements in each of these systems consist essentially of vertical pipes installed along the axis of the embankment. Each pipe extends down through the embankment into the permafrost foundation. When air is used as a coolant, each vertical pipe has a smaller pipe located concentrically inside it. Air is circulated down the annulus between the two pipes to a point near the bottom of the outer pipe where the air enters the inner pipe, and is returned up to the atmosphere through an exit manifold (Biyanov and Makarov, 1978). The air is allowed to flow through the system only when its temperature is lower than some predetermined temperature, say about -15°C. Cooling by air can be used effectively at locations where the mean annual air temperature does not exceed -5°C (Trupak, 1970). In regions of high humidity, rust and ice can form inside the freezing pipe, resulting in reduced heat removal efficiency and even plugging the pipes completely (Sereda, 1959; Gluskin and Ziskovich, 1973; Biyanov and Makarov, 1978). When liquid brine systems are used, they consist of a refrigeration unit with a heat exchanger for cooling a heat transfer liquid or brine, which is circulated through the cooling elements in the embankment. Calcium chloride solution has been used as a brine, and in one instance the system had to be converted to an air-cooling system when the calcium chloride leaked into the frozen soil and melted it (Borisov and Shamshura, 1959; Tsvetkova, 1960). The cooling brine must not contain impurities or water that will deposit in the pipes and restrict the flow of brine or plug the pipes, so corrosion inhibitors are incorporated into the brine. In addition, the piping systems are made of materials that resist the attack of the circulating brine (Borisov and Shamshura, 1959). Thermal devices (e.g. thermal piles) that have been used in the USSR to freeze an impervious zone in a dam usually are of the single-phase gravity type that uses kerosene as a transfer liquid (Gapeev, 1967; Biyanov and Makarov, 1978), although the two-phase thermal device that uses ammonia as a transfer fluid is now used in the USSR. In North America, two-phase thermal piles are used to a limited extent to maintain river levees in the frozen state. Both the single-phase and the two-phase systems can perform satisfactorily.

Embankment dams on permafrost in the USSR that are higher than 25 m are usually designed as unfrozen rockfill dams on a thaw-stable bedrock foundation. The temperature of the foundation is monitored to follow the movement of the thawing front in the foundation, and in at least one instance cement grout is periodically injected into the thawed zone of the foundation as the thawing front advances in depth (Demidov, 1973). It has been observed that air circulating within the downstream rockfill section of an embankment dam by natural convection has resulted in ice deposits from moisture condensing on the cold rock surfaces near the downstream surface of the dam in its upper reaches, and that the foundation freezes rather than thaws near the downstream toe of the dam (Kamenskii, 1973). Preliminary studies (Melnikov and Olovin, 1983) indicate that cold winter air sinks through the downstream rockfill zone to the base of this zone, where it encounters the relatively warm, moist foundation. The air is warmed and picks up moisture at this point, then rises through the rockfill where the moisture is deposited on the colder rock surfaces located near the surface of the downstream slope. In one case, the heat removed from the foundation by this convection has frozen at least part of the talik (Melnikov and Olovin, 1983) that existed beneath the river bed before the dam was constructed.

Experience in Operation and Construction

Experience in the USSR and North America has demonstrated that water-retaining embankments can be constructed and successfully operated on permafrost if appropriate precautions are taken. In the USSR, at least five intermediate-size embankment dams with heights ranging from 20 to 125 m (Biyanov, 1973; Tsytovich et al., 1978; Johnson and Sayles, 1980; Johnston and MacPherson, 1981) are now in operation on permafrost. Several smaller embankments have been functioning for many years. The oldest known embankment dam on permafrost was built in 1792 at Petrovsk-Zabaykalskiy and continued in operation until 1929 without incident (Tsvetkova, 1960; Tsytovich, 1973). Many of the smaller dams performed satisfactorily; however, some have had problems, especially in controlling the seepage around the water outlet facilities (see Table 1).

Much experimentation to develop techniques for creating and maintaining frozen embankments on permafrost has been conducted in the USSR. One pioneer project is a 10-m-high frozen dam built in 1942 on the Dolgaia River at Noril'sk that was initially cooled by circulating a calcium chloride brine through vertical freezing pipes installed along the centerline of the dam. These pipes extended through the talik beneath the river bed. Brine leaks made it necessary to change the circulating fluid from brine to air. The dam performed satisfactorily; however, only after an ice sheet was created on the downstream slope to stabilize the thermal regime (Borisov and Shamshura, 1959; Tsvetkova, 1960).

Two water supply dams located in the Irelyakh River near Mirnyy, Yakutia, USSR, were built on permafrost using innovative methods for that area. The older one, a temporary dam built in 1957, is a 6-m-high earthfill dam with a timber crib filled with loam serving as an impervious core (Lyskanov, 1964). This core extends down to the fissured sedimentary bedrock. The spillway consists of a timber flume with wooden crib wing walls. Temperature observations over a 7-yr period of operation showed that the foundation thawed to a depth of approximately 50 m due to the flow of water through the spillway (Johnson and Sayles, 1980). The average temperature of the permafrost in this area was about -2°C. The second dam, completed in 1964, is a 20.7-m-high earthfill embankment located about 2 km upstream from the temporary dam (Biyanov, 1966; Semenov, 1967; Smirnov and Vasiliev, 1973). This main Irelyakh River dam is founded on about 8 m of Quaternary deposits consisting of frozen sandy gravel and silty clays having up to 60% ice content. Beneath this deposit lies fractured dolomitized limestone and marl with ice content up to 10%. The embankment cross section consists of a large silty clay central section with zones of sand beneath layers of gravel on both the upstream and downstream slopes. A silty clay cutoff extends through the Quaternary deposits to bedrock along the centerline of the dam.

Because thawing of the ice-rich foundation soils would cause unacceptable settlements, a line of vertical cooling pipes was installed at 1.5-m spacing along the axis of the embankment to establish a positive seepage cutoff. The cooling pipes were designed to extend down through the embankment and about 3 m into the foundation; however, during construction the permafrost thawed an additional 6 m and the cooling pipes were consequently extended to this depth.

After operating the cooling system the first winter, the dam foundation and core became almost entirely frozen along a line of merged ice-soil cylinders that froze around each freezing pipe. Only two unfrozen sections were detected and these became frozen during the second winter.

To reduce the velocity of seepage from the reservoir in the talik beneath the embankment during the initial operation of the cooling system, an additional row of cooling pipes was installed along the downstream toe of the dam across the old river bed. After the central ice-soil cylinders had merged, chilling at the downstream toe was no longer required. Horizontal cooling pipes were installed beneath the concrete spillway located on the left abutment. To protect the adjacent embankment from thawing as a result of heat from the water discharged over the spillway, a line of vertical cooling pipes was installed along the embankment side of the spillway discharge chute, which is paved with concrete slabs on a gravel filter. After about 10 years of operation with a permanent reservoir, temperature measurements in the spillway area showed that the permafrost was thawing beneath the spillway chute toward the embankment. As a result, not only was the thawing front advancing toward the dams, but the concrete slabs also became misaligned and moved from their original positions, thus inducing further degradation of the permafrost. To arrest this thawing, additional freezing pipes were installed and the spillway crest was raised to increase the pool capacity and thereby reduce the amount of water discharged during periods of high runoff (Anisimov and Sorokin, 1975).

The Irelyakh River dam has had a permanent pool behind it for over 18 years, and although degradation of the permafrost in the spillway area had to be arrested, the embankment has performed satisfactorily as a result of proper maintenance.

It is interesting to note that no especially high embankment dams (i.e., higher than about 50 m) have been built on permafrost with frozen soils as the water seepage barrier. The larger embankments are founded on incompressible bedrock in the USSR (Tsytovich, 1973) and are designed as thawed embankments. One such embankment that has impounded a reservoir since 1969 is the dam for the Vilyui Hydroelectric Station located on the Vilyui River at the town of Chernyshevskiy in Yakutia. The site of the dam and reservoir is underlain by Palaeozoic and Mesozoic sedimentary and volcanic rocks. Intrusive rocks are from the lower Triassic period. At the dam site, the rock to a depth of 30 m contains small fissures, some of which were open, containing only ice.

The average air temperature at the site is -8.2°C and permafrost temperatures vary from -2 to -6°C depending upon the orientation of the ground slope with respect to solar radiation. Permafrost thickness varies from 200 to 300 m, but a talik extends entirely through the permafrost beneath the river bed at the dam site. The cross section consists of a rockfill shell with an inclined clay zone protected by two-layer filters upstream and downstream. A concrete pad enclosing a grouting gallery forms the base of the impervious zone. Initial grouting into 4-m deep holes was performed to reinforce the blast-shattered rock and to reduce seepage beneath the concrete pad. Further grouting of the bedrock fissures is accomplished as the ice melts from these fissures. The requirement for further grouting is indicated by temperature sensors and pioezometric measurements (Demidov, 1973).

During the construction and operation of this dam, ice built up in the pores of the downstream rockfill (Melnikov and Olovin, 1983). Studies of the circulation of air within the rockfill indicated that the talik beneath the former river bed was freezing. Moisture is entering the rockfill pores from convection, from precipitation, and from occasional tailwater backup. The ice formed from the tailwater is solid and remains at the base of the fill, while that from precipitation is distributed in the rockfill depending upon its temperature distribution. As the voids in the backfill become filled with ice, air currents are damped and the freezing of the talik ceases (Melnikov and Olovin, 1983). The embankment dam has been operating successfully since its construction in 1969.

Other high water-retaining embankments constructed on bedrock permafrost in the USSR are those at Ust-Khantaysk and the Kolyma hydroelectric power station (Evdokimov et al., 1973; Gluskin et al., 1974; Tsytovich et al., 1978). Although there is some information describing the construction of these dams (Gluskin and Ziskovich, 1973; Tsytovich et al., 1974; Gluskin et al., 1974), the details of their performance are not available in western literature at this time. It is assumed that they are performing as designed.

Two specific problems that have been given special attention in the USSR literature are frost heaving and thermal cracking of embankments. Since frost action can occur to depths of a few meters at the crest of an embankment, non-frost-susceptible soils, anti-heaving salting, and heating have been suggested as measures to counteract this problem (Kronik, 1973; Tsytovich et al., 1978). Transverse thermal cracking of embankments has been observed during and after the winter season. Cracks in one of the irrigation dams on the Suola River near Yakutsk were investigated by Grechishchev and Sheshin (1973). They found the cracks occur within 1.5 m of the vertical walls of the wooden outlet structure. To protect against this type of cracking, a layer of gravel 2 m thick is being used at the crest of embankments on permafrost in the USSR. Grechishchev suggested embedding horizontal wooden rods near the crest parallel to the axis of the embankment, to act as reinforcement in the soil and prevent these cracks. A method of calculating the longitudinal thermal deformation of embankments has been developed (Grechishchev and Sheshin, 1973) and the calculated crack widths are in good agreement with those observed in the embankment under investigation.

In Canada, the literature does not record the construction of an embankment designed as a frozen structure on permafrost but it does reveal that several small dikes (Johnston and MacPherson, 1981) and a waste impoundment (Thornton, 1974) were designed and constructed as the thawed type of embankment on permafrost. In these designs, the amount of thaw consolidation that will occur in the foundation is estimated, and the embankment height is increased to accommodate the anticipated settlement. As an alternative to overbuilding the dikes, the design heights are maintained by periodically adding embankment material to the crest of the dikes as settlement progresses. In some instances, vertical sand drains (Johnston, 1969; MacPherson et al., 1970) are installed in the permafrost foundation beneath the embankment to reduce pore pressures and hence increase the shearing resistance of the soil while thawing occurs. The differential settlements associated with this type of design can lead to transverse cracking of the embankments. To accommodate the cracking, the embankments were constructed of soils that are self-healing in nature (Johnston and MacPherson, 1981). Essential to this type of design is a continuous observation program throughout the life of the structure.

Examples of thawed embankments founded on permafrost in northern Manitoba, Canada, are the dikes at Kelsey (MacDonald, 1966), Kettle (MacPherson et al., 1970) and the Long Spruce (Keil et al., 1973) hydroelectric generating stations. To date, these embankments appear to have been performing as envisioned by their designers.

In the United States, one 24-m-high water-retaining embankment, five small embankment dams less than 6 m high, and a few levees have been constructed on permafrost in Alaska and one in Thule, Greenland. The small water-supply dams at remote villages are of the frozen type of design and are without artificial cooling. Although some have been in operation for more than 30 years (Davis, 1966; Fulwider, 1973), no serious problems have been reported. The largest embankment dam founded on permafrost in Alaska is located near Livengood on Hess Creek (Rice and Simoni, 1966; Kitze and Simoni, 1972). This combination hydraulic fill and rolled earth fill structure was completed in 1946 to a maximum height of 24 m. The foundation consists of silt and ice to a depth of approximately 6 m overlying a 6.5-m thick deposit of a variable mixture of frozen clay, silt, sand, gravel, and fragments of chert rock. Beneath this deposit and overlying the fractured chert bedrock is a 6.5-m layer of frozen coarse sands and gravels containing considerable ice. After stripping the site, the hydraulic fill was placed directly on the frozen silt. Horizontal cooling pipes were installed at the surface of the foundation to provide artificial refrigeration to keep the foundation frozen during construction. Steel sheet piling embedded in permafrost and projected 2 to 3 m up into the hydraulic fill was used to improve the seepage cutoff. Since this dam only retained water during the short summer mining period each year, the reservoir was lowered every autumn to take advantage of winter cooling to preserve the permafrost foundation and refreeze the embankment. This dam performed satisfactorily from 1946 until 1958, when the mining operations were stopped for economic reasons. The major operating deficiency in the system was a water outlet tunnel located remotely from the dam. It completely collapsed some time after the mining operation ceased. Then in 1962, the overflow from the spring runoff washed out the timber spillway that was located on the right abutment of the dam. The embankment itself remains intact today.

Embankment Dams on Permafrost: Recorded Failures and Problems

Embankment dams on permafrost have been built and successfully operated in Canada, the USSR, and Alaska. A number of failures have been reported in the USSR and one in Alaska. Table 1 lists 15 embankment dams that encountered difficulties; some of the problems were corrected, but several ended in disaster. An examination of the table reveals that most of the difficulties arose because insufficient attention was given to establishing and maintaining a reliable frozen thermal regime and to controlling seepage. It should be noted that often the thawing and seepage in a frozen embankment or foundation are initiated adjacent to the spillway or outlet works. This, of course, indicates that inadequate cooling or cutoffs were established at these points. Other problems described in the table can be traced to the construction procedures and the control of earth placement. In some cases, adequate thermal protection was not provided for the foundation and adjacent areas during construction. In other cases, insufficient compaction was applied to the soil placed in the frozen state.

Table 1 includes only a portion of the problems that have been referenced in the literature. Although several papers on embankment dams on permafrost have been collected and translated, it is reasonable to assume that there are many either unreported or not available in our limited bibliography. However, the dams that are listed indicate the problems that must be addressed if safe and economical embankment dams are to be built on permafrost.

RESEARCH REQUIRED

The design, construction, and maintenance of safe, water retaining embankments in the Arctic and sub-Arctic is strongly dependent upon such considerations as structural stability, seepage control, the handling of materials, erosional control, and the environmental effect of the impoundment. The following is a list of areas that need further research:

a. The development of analytical methods for determining and predicting the thermal regime within and beneath water-retaining embankments on permafrost when ground water seepage occurs and where three-dimensional heat flow is important.
b. The development of effective methods for controlling the thermal regime and seepage, especially at locations contiguous to spillways and outlet works for embankments on permafrost.
c. The development of drainage systems that will not become plugged with ice.
d. The analysis of existing techniques and the development of new ones for constructing frozen and unfrozen embankment dams on permafrost during both the summer and winter construction seasons while maintaining a frozen foundation.
e. The development of more accurate analytical procedures for determining the stability of embankment and natural and cut slopes in permafrost during thawing.
f. The development of accurate methods that use electromagnetic, geophysical, and other types of tools for determining the extent and depth of permafrost, massive ice inclusions, and types of soils at proposed embankment sites.
g. The development of instrumentation for monitoring the construction and performance of embankments on permafrost, with special attention to early detection of seepage that occurs in a "frozen" embankment or its foundation.
h. The determination of the effects of impoundments on water chemistry and productivity in reservoirs and streams over permafrost.
i. The development of biological procedures for restoring reservoir landscapes in arctic and subarctic areas.
j. The determination of the tolerance of cold-dominated vegetation to periodic inundation.
k. The validation of existing methods for predicting the hydrological characteristics of catchment basins on permafrost.
l. The development of effective methods for controlling water and ice erosion of embankments, cuts, and natural slopes of reservoir walls in permafrost areas.

TABLE 1 Embankment dams on permafrost: Problems with existing dams

River name	Location	Embankment type	Height (m)	Length (m)	Problem/description	Reference
Thawing and Seepage at Spillway and Outlet Works						
Unknown	Northern USSR	Compact earth	21.4	230	Municipal water supply dam was completed in 1967. In 1970, a breach occurred through embankment at supply intake pipes due to thermal erosion and seepage.	Anisimov & Sorokin, 1975
Hess Creek	Livengood, AK	Hydro & compact earth fill	24	488	Completed in 1946 for water supply in mining operations. In 1962 breached between spillway due to thermal erosion and seepage.	Rice & Simoni, 1966
Myla River	Zarechnyy Region, USSR	Compact frozen sand		-	Constructed of uncompacted frozen sand during winter. Seepage through earth dam and joints in wooden spillway caused thawing and failure of dam in 1954.	Lyskanov, 1964
Vilyui River (Dam II)	USSR	Embankment w/crib cutoff, w/sand and silt	12	300	Constructed 1957-1960 on permafrost. During initial operating period, large seepage volume occurred and spill way was completely destroyed in first flood. After reconstruction in 1969, leakage was observed from the reservoir through caverns in foundation of the dam and at contact with the spillway. Causes of problems: (1) spillway too small for flood, (2) ice-retaining structures not located far enough upstream from dam, (3) poor spillway construction, (4) fissures in foundations not sealed, (5) poorly compacted runoff.	Biyanov, 1966
Vilyui River (Dam III)	USSR	Embankment with core	3	-	Clay-ice core constructed in winter with frozen clay and water. Seepage along spillway; contact between it and the embankment resulted in degradation of frozen core, which soon became nonfunctional.	Biyanov, 1965
Thermal Regime of Embankment and Foundation not Maintained						
Dolgaia River	Norilsk, USSR	Refrigerated earth	10	130	Constructed in 1942 with a "clay-concrete" (probably cement-stabilized clay) core with two rows of freezing pipes parallel to dam axis. The core was to reduce seepage during the freezing period. Calcium chloride brine, circulated through the freezing pipes, was not cooled sufficiently by atmosphere. Furthermore, the brine leaked into the dam, causing melting of embankment. Snow deposits impeded freezing from the downstream face. Disaster was averted by converting the cooling brine to air and running the circulating system only when air temperatures were colder than the ground. Only after an ice sheet was created on the downstream face by water applications in the winter was thermal regime stabilized throughout the year.	Tsvetkova, 1960, Borisov & Shamshura, 1959
Sredniy El'gen River	Kolyma River Basin, USSR	Earth	7.4	300	Built in 1944. Large deformations and cracks occurred along dam due to seepage and thawing. Seepage developed	Tsvetkova, 1960

TABLE 1 (cont'd) Embanmkment dams on permafrost: Problems with existing dams

River name	Location	Embankment type	Height (m)	Length (m)	Problem/description	Reference
					where timber piling was used as a cutoff. To prevent failure an upstream blanket was constructed.	
Myaundzha River	Kolyma Basin, USSR	Earth fill w/core	8	860	In 1952, initial attempts made to construct dam in the winter to obtain a frozen embankment, but thawing in the summer required artificial cooling. This was accomplished by installing "freeze" pipe in the embankment and circulating cold air down the pipes. The abutments were not protected by freeze pipes, thawing occurred at these locations, and the seepage ensued. (This article, in 1957, implied impending disaster.)	Tsvetkova, 1960
Amozar River	Near Mogocha on the Amur Railroad.	Crib-core Rock-earth fill	4	-	Failed due to seepage and thawing through body of the embankment. (Apparently built in 1910-1916.)	Tsvetkova, 1960
Kvadratnyy	Norilsk, USSR	Compact earth fill	6	-	Dam used for cooling water supply for electric power station. Destroyed within 1 yr after construction by thawing of foundations and abutment soils.	Borisov & Shamshura, 1959
Stake 89 (Picket Creek)	Norilsk, USSR	Compact earth fill	5.5	-	Destroyed within 2 yr after construction when seepage through the unfrozen soil thawed the frozen soil.	Tsvetkova, 1960
			Inadequate Construction Procedures			
Pravaya Magdagacha River	Northern USSR	Compact earth with concrete diaphragm	7.3	-	Failed after 2 yr of operation. Large deformation of dam resulted in cracks in the diaphragm all along the embankment dam and at the junction of the weir. Final failure occurred during heavy thunder storm when leakage appeared at the crest. Failure occurred over a 65-m length.	Tsvetkova, 1960; Saverenskii, 1950
Mykyrt River	City of Petrovsk-Zabaykalskiy, USSR	Earth	9.5	-	Built in 1792. In attempting to repair the wooden spillway of the 137-yr-old dam, proper measures were not taken to preserve the frozen embankment. The dam had to be completely rebuilt in 1945.	Tsvetkova, 1960
Bol'shay Never River	Skovorodino, USSR	Earth silt and gravel with clay core	9.6	530	Built in 1932. The clay-ice core became semi-liquid and the stability of the dam was threatened. In 1934 ballast was applied to the slopes, wooden piling was driven, soil behind the piling was replaced by more impervious material, and a wood gallery was constructed to catch the seepage. Deep thawing of foundation soil and bedrock in 1936 did not cause serious problems.	Tsvetkova, 1960
Vilyui River (Dam V)	USSR	Random earth fill with timber	16.8	332	Constructed on ice-saturated clayey silt and disintegrated rock overlying fissured clay-limestone. In the springs of 1965 and 1966 boils appeared downstream of the dam. Seepage was caused by thawing of ice in rock fissures during construction.	Biyanov, 1965

REFERENCES

Anisimov, V.A. and Sorokin, V.A., 1979, Repair work on a frozen dam (in Russian): Gidrotekhnicheskoe Stroitel'stvo, no. 5. (Trans.) Hydrotechnical Construction (1975), p. 24-25.

Belikov, M.P., Panasenko, G.A., and Antsiferov, A.S., 1968, Earth dam in permafrost region constructed with plastic waterproof barrier (in Russian): Gidrotekhnicheskoe Stroitel'stvo (Hydrotechnical Construction), no. 10, p. 878-880.

Biyanov, G.F., 1966, Discharge and blocking up the river during construction of the Vilyui Hydroelectric Power Station (in Russian): Gidrotekhnicheskoe Stroitel'stvo (Hydrotechnical Construction), no. 2, p. 1-5. (Trans.) National Research Council of Canada, Technical Translation 1353, 1969.

Biyanov, G.F., 1970, Experience in construction and operation of low-head dams on permafrost soils (in Russian): Gidrotekhnicheskoe Stroitel'stvo (Hydrotechnical Construction), no. 9.

Biyanov, G.F., 1973, Experience in building dams on permafrost in Yakutia (in Russian), in Proceedings of the Second International Conference on Permafrost, Yakutsk, v. 7, p. 125-132. (Trans.) Hanover, N.H., U.S. Army Cold Regions Research and Engineering Laboratory, Draft Translation 438.

Biyanov, G.F. and Makarov, V.I., 1978, Characteristics of construction of frozen dams in western Yakutiya, in Proceedings of the Third International Conference on Permafrost, Edmonton, v. 1: Ottawa, National Research Council of Canada, p. 779-784.

Bogoslovskii, P.A., 1958, Investigations on the temperature regime of earth dams under permafrost conditions (in Russian): Nauchnye Doklady Vysshei Shkoly, Stroitel'stvo, no. 1, p. 228-238. (Trans.) Hanover, N.H., U.S. Army Cold Regions Research and Engineering Laboratory, Draft Translation 22, 1966.

Bogoslovskii, P.A., 1970, Analytical calculation of a three-dimensional, stationary, steady-state temperature condition of a dam (in Russian): Sbornik trudov po gidrotekhnike i gidrostroitel'stva pri stroitel'stve plotin v suravykh klimaticheskikh usloviyakh. (Trans.) Hanover, N.H., U.S. Army Cold Regions Research and Engineering Laboratory, Draft Translation 601, 1977.

Bogoslovskii, P.A., Veselov, V.A., Ukhov, S.B., Stotsenko, A.V., and Tsvid, A.A., 1966, Dams in permafrost regions, in Proceedings of the First International Conference on Permafrost, Purdue University, 1963: Washington, D.C., National Academy of Sciences, p. 450-455.

Borisov, G.A. and Shamshura, G.Ia., 1959, Experience in the planning, construction and use of earth dams at Noril'sk (in Russian): Materialy po inzhenernomu merzlotovedeniiu, 7th Interdepartmental Permafrost Conference, USSR Academy of Sciences. (Trans.) Hanover, N.H., U.S. Army Cold Regions Research and Engineering Laboratory, Draft Translation 26, 1970, p. 110-119.

Brown, W.G. and Johnston, G.H., 1970, Dikes on permafrost: predicting thaw and settlement: Canadian Geotechnical Journal, v. 7, no. 4, p. 365-371.

Davis, R.M., 1966, Design, construction, and performance data of utility systems, Thule Air Base: Hanover, N.H., U.S. Army Cold Regions Research and Engineering Laboratory, Special Report 95.

Demidov, A.N., 1973, Dam in the Vilyuy hydraulic power system (in Russian): Trudy Gidroproekta, no. 34, p. 64-77. (Trans.) Hanover, N.H., U.S. Army Cold Regions Research and Engineering Laboratory, Draft Translation 526, 1976.

Evdokimov, P.D., Bushkanets, S.S., Zadvarny, G.M., and Vasiliev, A.F., 1973, Studies of foundations and earth rockfill dams under severe climatic conditions: Seminar No. 2, Experience of Hydro Project Construction under Severe Climatic Conditions, Leningrad, Joint Soviet-Canadian Working Group for Co-operation in the Electrical Power Industry, p. 67-86.

Fulwider, C.W., 1973, Thermal regime in an arctic earthfill dam, in Permafrost - The North American Contribution to the Second International Conference, Yakutsk: Washington, D.C., National Academy of Sciences, p. 622-628.

Gapeev, S.I., 1967, Consolidation of the frozen soils of foundations by cooling (in Russian): Gidrotekhnicheskoe Stroitel'stvo, no. 12. (Trans.) Hydrotechnical Construction, no. 12, p. 1082-1084.

Gluskin, Ia.E. and Ziskovich, V.Ye., 1973, Dam construction in the Northern USSR (in Russian), in Proceedings of the Second International Conference on Permafrost, Yakutsk, v. 7. (Trans.) Hanover, N.H., U.S. Army Cold Regions Research and Engineering Laboratory, Draft Translation 438.

Gluskin, Ia.E., Losev, E.D., Petrov, V.G., and Frumkin, N.V., 1974, Kolyma hydroelectric station (in Russian): Gidrotekhnicheskoe Stroitel'stvo, no. 8, p. 722-727.

Grechishchev, S.E. and Sheshin, Iu.B., 1973, Theoretical and experimental studies of thermal stresses and deformations of earth dams in winter (in Russian): All-Union Scientific Research Institute of Hydrogeology and Engineering Geology, USSR Ministry of Geology, Trudy, unpublished MS, no. 55 (S.E. Grechishchev, ed.), Moscow. (Trans.) Hanover, N.H., U.S. Army Cold Regions Research and Engineering Laboratory, Draft Translation 449.

Gupta, R.C., Marshall, R.G., and Badke, D., 1973, Instrumentation for dikes on permafrost, Kettle Generating Station: Canadian Geotechnical Journal, v. 10, no. 3, p. 410-427.

Johnson, T.C. and Sayles, F.H., 1980, Embankment dams on permafrost in the USSR: Hanover, N.H., U.S. Army Cold Regions Research and Engineering Laboratory, Special Report 80-41.

Johnston, G.H., 1965, Permafrost studies at the Kelsey Hydroelectric Generating Station, research and instrumentation: Division of Building Research, National Research Council of Canada, Technical Paper No. 178.

Johnston, G.H., 1969, Dykes on permafrost, Kelsey

Generating Station, Manitoba: Canadian Geotechnical Journal, v. 6, no. 2, p. 139-157.

Johnston, G.H., and MacPherson, J.G., 1981, Dams and reservoirs, ch. 9, in Permafrost Engineering Design and Construction (G.H. Johnston, ed.): New York, John Wiley & Sons, p. 393-413.

Kamenskii, R.M., 1973, Thermal regime of bearing ground and of the body of the Vilyui hydroelectric power plant (in Russian), in Proceedings of the Second International Conference on Permafrost, Yakutsk, v. 7, p. 228-235. (Trans.) Hanover, N.H., U.S. Army Cold Regions Research and Engineering Laboratory, Draft Translation 438, p. 300-308.

Keil, L.O., Nielsen, N.M., and Gupta, R.C., 1973, Thaw-consolidation of permafrost dyke foundations at Long Spruce Generating Station, Manitoba, in Proceedings of the 26th Canadian Geotechnical Conference: Toronto, p. 134-141, preprint.

Khakimov, Kh.R., 1963, Artificial freezing of soils - theory and practice (in Russian). (Trans.) A. Barouch and H. Heimann, NTIS: Springfield, Va. U.S. Department of Commerce, 1966.

Kitze, F.F., and Simoni, O.W., 1972, An earth fill dam on permafrost: Hess Creek Dam, Livengood, Alaska: Hanover, N.H., U.S. Army Cold Regions Research and Engineering Laboratory, Technical Report 196.

Kronik, Ia.A., 1973, Cryogenic phenomena in hydraulic structures built of earth (in Russian), in Proceedings of the Second International Conference on Permafrost, Yakutsk, v. 7, p. 316-320. (Trans.) Hanover, N.H., U.S. Army Cold Regions Research and Engineering Laboratory, Draft Translation 438, p. 240-243.

Lyskanov, G.A., 1964, Experimental construction of a frozen-type dam in Iakutiia (in Russian): Yakutsoe Knizhnoe Izd-vo. (Trans.) Hanover, N.H., U.S. Army Cold Regions Research and Engineering Laboratory, Translation 479, 1975.

MacDonald, D.H., 1966, Design of Kelsey dikes, in Proceedings of the First International Conference on Permafrost, Purdue University, 1963: Washington, D.C., National Academy of Sciences, p. 492-496.

MacPherson, J.G., Watson, G.H., and Kuropatnick, A., 1970, Dikes on permafrost foundations in northern Manitoba: Canadian Geotechnical Journal, v. 7, no. 4, p. 356-371.

Melnikov, P.I. and Olovin, B.A., 1983, Permafrost zone dynamics in the area of the Vilyuy River hydroelectric scheme, in Proceedings of the Fourth International Conference on Permafrost, Fairbanks: Washington, D.C., National Academy of Sciences, p. 838-842.

Mukhetdinov, N.A., 1969, Thermal regime of the downstream shoulder of rockfill dams (in Russian): Energiya, Leningrad, v. 90, p. 275-294. (Trans.) Hanover, N.H., U.S. Army Cold Regions Research and Engineering Laboratory, Draft Translation 586, 1977.

Rice, E.F. and Simoni, O.W., 1966, The Hess Creek Dam, in Proceedings of the First International Conference on Permafrost, Purdue University, 1963: Washington, D.C., National Academy of Sciences, p. 436-439.

Sanger, F.J. and Sayles, F.H., 1978, Thermal and rheological computations for artificially frozen ground construction: 1st International Symposium on Ground Freezing (H.L. Jessberger, ed.), Bochum, Germany, Ruhr University, v. 2, p. 95-117.

Savarenskii, F.P., 1950, Dams in permafrost regions (in Russian): Izbrannye sochenenia, p. 370-371. (Trans.) Arctic Construction and Frost Effects Laboratory (USAED), ACFEL Translation No. 29, 1960.

Semenov, N.G., 1967, Building dams in permafrost regions (in Russian): Gidrotekhnicheskoe Stroitel'stvo, no. 9, p. 14-15. (Trans.) Hanover, N.H., U.S. Army Cold Regions Research and Engineering Laboratory, Draft Translation 452, 1974.

Sereda, V.A., 1959, Experience in planning hydraulic structures with prolonged soil freezing (in Russian): Seventh Interdepartmental Permafrost Conference, USSR. Academy of Sciences, p. 120-128. (Trans.) Hanover, N.H., U.S. Army Cold Regions Research and Engineering Laboratory, Draft Translation 140, 1966.

Smirnov, E.A. and Vasiliev, S.F., 1973, Design and construction of earth and rock fill dams under severe climatic conditions, USSR experience: Seminar No. 2, Experience of Hydro Project Construction under Severe Climatic Conditions, Leningrad, Joint Soviet-Canadian Working Group for Co-operation in the Electrical Power

Thornton, D.E., 1974, Waste impounding embankments in permafrost regions: the oxidation pond embankment, Inuvik, N.W.T: Task Force on Northern Oil Development, Environmental Social Committee, no. 74-10, p. 159-193.

Trupak, N.G., 1970, Construction of earth dams on permafrost soils (in Russian): Gidrotekhnicheskoe Stroitel'stvo, v. 40, no. 9, p. 8-11. (Trans.) Hydrotechnical Construction, no. 9, p. 798-803.

Tsvetkova, S.G., 1960, Experience in dam construction in permafrost regions (in Russian): Materialy k osnovam ucheniia o merzlykh zonakh zemnoi kory, v. 6, p. 87-112. (Trans.) Hanover, N.H., U.S. Army Cold Regions Research and Engineering Laboratory, Draft Translation 161, 1966.

Tsvid, A.A., 1961, Freezing of an earth dam from the dry slope side (in Russian): Sbornik Nauchnykh Rabot, v. 1, p. 93-104. (Trans.) Hanover, N.H., U.S. Army Cold Regions Research and Engineering Laboratory, Draft Translation 430.

Tsytovich, N.A., 1973, The mechanics of frozen ground (in Russian): Moscow, Vysshaya Shkola Press. (Trans.) G.K. Swinzow and G.P. Tschebotarioff, eds.), New York, Scripta/McGraw-Hill, 1975.

Tsytovich, N.A., Ukhova, N.V., and Ukhov, S.B., 1972, Prediction of the temperature stability of dams built of local materials on permafrost (in Russian): Leningrad, Stroiizdat,

p. 143. (Trans.) Hanover, N.H., U.S. Army Cold Regions Research and Engineering Laboratory, Draft Translation 435.

Tsytovich, N.A., Kronik, Ya.A., and Biyanov, G.F., 1978, Design, construction, and performance of earth dams on permafrost in the Far North, in Proceedings of the Third International Conference on Permafrost, Edmonton: Ottawa, National Research Council of Canada, p. 137-149.

DESIGN AND CONSTRUCTION OF DEEP FOUNDATIONS IN PERMAFROST: NORTH AMERICAN PRACTICE

B. Ladanyi

Ecole Polytechnique
Montreal, Canada

In the last 20 years, the design and construction of deep foundations in permafrost in North America have seen quite remarkable developments, due to a combination of performance observations, field testing, and theoretical work. In particular, during this period, a great number of full-scale pile tests were performed in Alaska and northern Canada. These tests helped greatly to clarify the behavior of different types of deep foundations in permafrost under load and temperature change. Most of the work was done in the areas of pile installation methods, the behavior of piles and anchors under axial compression and tension loads, and the determination of adfreeze strength of various types of piles. As a result, we now have a relatively good idea of how a pile, installed by a given method, will behave under an axial load.

Nevertheless, there are still several problems in relation to deep foundations whose solution will require further experimental and theoretical work in the coming years. It is therefore useful to review the research areas that have been covered in the North American literature on deep foundations in permafrost over the last 20 years, to see where it may be necessary and justified to concentrate future research efforts.

For a faster survey of information, the review was subdivided into five subject groups, each covering one large research area. These are:

A. Pile installation methods
B. Pile behavior under load
C. Improvement of pile bearing capacity
D. Design methods for piles in permafrost
E. State-of-the-art reviews, literature surveys, and general design recommendations

Each subject group contains several subgroups, each dealing with a particular subject or an investigated effect on pile behavior.

The subgroups contain a list of all reviewed bibliographical references dealing with the subject matter. These references, listed at the end of the report, are subdivided into five groups:

1. Piles and deep footings (Ref. P1 to P48).
2. Deep anchors (Ref. AN1 to AN8).
3. Adfreeze bond (Ref. AD1 to AD11).
4. Frost heave effects on piles (Ref. FH1 to FH6).
5. Foundation materials (MAT1 to MAT3).

Under each subheading, there are only some brief statements on the state of knowledge on the subject, while the areas of needed research are listed at the end of the paper.

While a great effort has been made to include all published references, this does not mean that there are no inevitable omissions, such as unpublished internal reports and reports of a proprietary nature. It is hoped that this valuable information, especially on the performance of the vertical support members of the Alyeska pipeline, will become available in the near future.

LITERATURE REVIEW

A. Pile Installation Methods

A1. Freezeback of slurry around piles, theory and observations (P3, P32, P33)

The thermal problem has been solved theoretically, and the predictions appear to agree with field observations. The mechanical problem of freezeback pressures needs some further work, but it is known that

(a) The pressures dissipate quickly in warm permafrost;
(b) They have a positive effect on pile capacity in dense frictional frozen soil, but it is safe to neglect them in design;
(c) The pressures are important to consider when evaluating results of pile tests.

A2. Experience with driving piles in permafrost soils (P20, P22, P25, P26, P27, P38, P43, P44, P46)

(a) Steel piles, including H-shape, reinforced web, and sheet, can be driven in most frozen soils (except frozen gravels and cobbles) with impact, vibratory, or sonic hammers;
(b) When extremely hard driving is encountered, an impact hammer and cast-steel pile driving tips are necessary;
(c) Pile driving using thermally modified predrilled pilot holes is finding increasing use in practice;
(d) In plastic-frozen soils, driving in slightly smaller holes is possible and gives high pile capacities. The method becomes difficult in cold permafrost soils.

A3. Comparison of driven and slurried piles (P4, P14)

Steel H-piles when driven have 3 to 4 times higher capacity and much smaller settlement than when installed in slurried holes.

A4. Experience with different types of backfill slurries (P2, P46)

In order of increasing adfreeze strength of slurries is: clay, silt, sand, sand-vibrated, cement grout.

A5. Experience with cast-in-place piles in permafrost (P32, AN1 to AN5, MAT3)

(a) Cement-fondu-sand-water (1:1:1/2) grout used in Thompson and Gillam tests resulted in a corrugated, ice-free interface with the soil.
(b) A gypsum-based cement grout provides early strength at low temperatures, but forms a layer of saline unfrozen soil at the pile interface, reducing adfreeze bond.

B. Pile Behavior Under Load

B1. Compression tests on piles (P2, P4, P5, P10, P28, P40, P45, P46)

(a) The load-settlement curve and pile capacity depend on the rate of loading.
(b) There is disagreement on the yield or failure criterion for pile tests.
(c) Driven piles carry more and settle less than those installed in slurried boreholes.
(d) There is no standard method of testing.
(e) No design method gives the optimum L/d ratio for a pile.
(f) For long piles, only a small portion of the load is carried by the end bearing at service loads. The portion increases with time.
(g) Attenuating creep is usually observed at low stresses (settlements), and steady-state creep at high stresses (post-failure settlements).
(h) For piles in ice, steady-state creep can be attained at all loads.
(i) In slurried piles, end contact is not assured without driving the pile.

B2. Tension tests on piles (AD4, AN1 to AN5)

(a) Failure was observed in tension tests; there was no distinct failure in compression tests with end-bearing.
(b) Break of the bond occurs at a given displacement, probably depending on the roughness of the pile surface.
(c) When the bond is intact, there is a power-law relationship between creep rate and stress.

B3. Behavior of anchors and anchor systems (P13, P33, AN1 to AN7)

(a) The response to the load depends on the anchor type and the installation method.
(b) Deep plate anchors need more displacements for mobilizing the soil resistance than rod anchors, but may have higher ultimate safety factors because they rely on the weight of the soil cone lifted.

B4. Adfreeze bond (effects of soil type, pile material, temperature, rate of loading, shear displacement) (P2, P33, P34, P40, P41, P44)

(a) Tables and graphs of short- and long-term values are available for different types of soils and pile materials.
(b) Sometimes a layer of ice or unfrozen (saline) soil forms on the pile interface during refreezing.
(c) The adfreeze bond increases with ice content but decreases again after the soil is super saturated with ice.
(d) Bond creep is governed by a power law of stress relationship.
(e) The bond breaks at a given displacement, in a brittle manner, falling to a much lower residual value.

B5. Stress distribution along the pile and load transfer to the pile point (P2, P32, P40, P41, AD11)

(a) Stress distribution depends on installation method, pile type, load level, and time.
(b) For a given load, the bond stresses, initially decreasing with depth, become more uniform with time, and more load is transferred to the end.
(c) At a settlement of 5-10% of the pile diameter, 25-40% of the total load may be carried by the base. At service loads, the point-bearing load may be negligible.
(d) For driven piles, the load transfer to the pile point depends on the settlement rate under service loads.
(e) There is a limit of displacement for bond failure, depending on the soil and pile roughness.
(f) Dynamic loads lead to bond reduction and an increase in load transfer to the point.

B6. Theories of stress redistribution for compressible piles due to soil creep (P19, P32)

Under a constant load, lateral stress is redistributed along a compressible pile in the first 24 hours, with a tendency to make the distribution more uniform and to transfer more load to the base. Pile creep curves may not be representative within that period.

B7. Effect of seasonal soil temperature variation on pile settlement (P17)

Seasonal variation of the end-bearing load occurs due to freezing and thawing of the active layer and the temperature variation of permafrost within the embedded pile strength.

B8. Effect of slow load cycling on penetration rate (P30)

The penetration rate becomes practically independent of the loading history after the penetration resistance is fully mobilized (i.e., total settlement exceeds about 10% of pile diameter).

B9. Scale effects in pile loading tests (P18, P32, P42)

Results obtained with piles of different diameters are comparable at the same strain rate, but not at the same penetration rate.

B10. Frost-heave effects on piles (FH1 to FH6)

(a) Frost uplift stresses measured on vertical embedded columns vary mostly between 50 and 100 kPa, with short-term peaks up to 200 kPa. They are highest for steel columns, followed by concrete and wood.
(b) Frost uplift stresses are proportional to the heave rate.
(c) Peak values of uplift stresses occur during the early freezing period, when heaving rates are high, but maximum uplift forces often occur near the time of maximum frost penetration, late in the winter season.
(d) For steel columns, uplift stresses tend to decrease with increasing pile diameter.
(e) Frost-heave forces increase with pile inclination in the active layer.

B11. Lateral loading tests on piles (P8, P9, P16)

(a) By putting the results of tests in the form of a power law of time and load, a time-dependent modulus of lateral soil reaction and a p-y curve can be deduced from the tests.
(b) The pressuremeter test data have been used with good success to predict theoretically the behavior of laterally loaded piles in permafrost.
(c) Most published test results refer to stage-loaded tests with 24 h per stage. Behavior under long-term sustained loads and cyclic loads needs further study.

C. Improvement of Pile Bearing Capacity

C1. Thermopiles (P7, P12, P17, P21, P46)

To select the best system, a careful study of ground and temperature conditions should be made. With an appropriate system, excellent results can be obtained.

C2. New pile designs or installation methods for increased adfreeze bond and/or end-bearing (P12, P20, P21, P22, P27, P39, P40, P44, P45, AN8)

(a) Use of corrugated and helix-type piles considerably increases the bond strength (50-150%). The same was found for grooved and tapered concrete piles in slurried holes.
(b) Use of lugs around a pile considerably increases its bearing capacity.
(c) Both end-bearing and adfreeze bond are higher if piles are driven into partially slurry-filled oversized boreholes. Driving in slightly smaller holes, without slurry, also gives much higher bond strength, but it is limited to relatively warm plastic frozen soils. Thermally modified pre-drilled pilot holes facilitate driving without decreasing adfreeze strength.

C3. Deterioration of pile materials in permafrost (MAT1, MAT2)

(a) Steel piles: no corrosion was detected after 6-11 years of exposure to active layer conditions at three sites in Alaska.
(b) Timber piles: fungus attack was detected. Improved treatment procedures should be applied.

D. Design Methods for Piles in Permafrost

D1. Design methods based on laboratory-determined parameters (P13, P15, P16, P19, P23, P30, P31, P32, P33, P34, P36, P37, P41, P44)

(a) Although based on small strain assumptions, the theories for creep settlement of piles under axial and lateral loads seem to give reasonable predictions for short- and medium-term loads. More information and comparison with field data is needed for long-term behavior of piles under static and dynamic loads.
(b) End-bearing support is negligible for piles in all types of homogeneous frozen soils at service loads.
(c) It is suggested that pile design in ice--rich soils is governed mainly by settlement, while in ice-poor soils the design should satisfy both settlement and strength criteria.
(d) It is also suggested to neglect in design soils with temperatures above -1°C.

D2. Design methods based on field test results (P9, P16, P18, P42, P46)

(a) If properly generalized, the results of full-scale tests on axially and laterally loaded piles can be used directly for the design of other piles under similar conditions. Nevertheless, extrapolating such test results to the very long term remains uncertain until further verification.
(b) Well-controlled static cone penetration tests can be interpreted as small-diameter pile tests. Their results can be used for the design of driven piles, either directly through a strain rate scaling law, or indirectly after determining the frozen soil creep parameters.

E. State-of-the-art Reviews, Literature Surveys, and General Design Recommendations (P1, P3,

P6, P7, P11, P23, P24, P25, P31, P32, P33, P34, P35, P36, P37, P38, P40, P41, P43, P44, P48)

Several excellent state-of-the-art reviews of experience with piles in permafrost have been published in the last 20 years, and a complete coverage of current or recommended design procedures was described in special volumes (P31, P33, P37). Certain papers and volumes also contain recommended design parameters for piles in typical permafrost soils (P6, P23, P31, P32, P33, P34, P37, P41, P44).

RESEARCH REQUIRED

A. On Pile Installation Methods

(1) Driven piles

Further development and testing is needed of vibratory and sonic hammers, together with thermally treated, pre-drilled pilot holes.

(2) Drill-driven piles (in smaller boreholes)

In plastic-frozen soils, cylindrical piles driven in slightly smaller pre-drilled holes have shown increased adfreeze strengths. Dissipation of the mobilized lateral normal stress due to frozen soil relaxation should be studied. Driving tapered piles in tapered pre-drilled holes seems promising and should be tried in the field.

(3) Slurry-driven piles

Driving of piles in larger holes, partially filled with sand-slurry, was found to give high pile capacities. The method should be further investigated to find the optimum design and installation method for such piles and to evaluate their long-term performance.

(4) Slurried piles (in larger boreholes)

Freezeback pressure and its dissipation should be studied further. If a portion of this pressure can be preserved permanently, the bearing capacity of a pile in granular slurry could be considerably increased. The behavior of corrugated and tapered piles in slurry-filled boreholes should be studied.

(5) Cast-in-place piles

Piles should be made in such a manner that the concrete is compacted in the hole to produce a good end-bearing mobilization and a corrugated lateral surface for a better bond. A method similar to that used in making Franki piles seems promising.

B. On Pile Behavior Under Load

(1) There is a need to establish a standard testing method, especially with regard to the length of the loading stages, the loading sequence, and the criterion of yielding or failure. Possibilities for long-term extrapolation of test results should be explored.

(2) Most piles in permafrost have a very high L/d ratio. For that reason, even driven piles carry very little load in end-bearing. The portion of the load carried by the end-bearing is a function of the pile compressibility, the pile installation method, the aspect ratio L/d, and the total settlement of the pile. The design method should be able to give an optimum L/d ratio for which both the adfreeze bond and the end-bearing are properly utilized at an acceptable settlement or settlement rate of the pile.

(3) More information is needed about the long-term effects on the behavior of a pile under service loads of the seasonal temperature variation in the portion of permafrost in which the pile is embedded.

(4) Very little information is available on the long-term behavior of laterally loaded piles. Additional information is needed on the redistribution of lateral reaction with time and on the best method for its prediction. The behavior of such piles under cyclic loads is also needed.

(5) There is little information on pile groups in permafrost loaded vertically. Since the group action may be affected by the rate of penetration in frozen soils much more than in unfrozen soils, this effect should be investigated.

C. On Improvement of Pile Bearing Capacity

(1) Straight-shafted piles

Besides the soil type, ice content, and temperature, the bond also depends on the pile material and the roughness of the shaft. Adfreeze bond on a smooth shaft is regularly much smaller than the shear strength of frozen soil, and becomes still smaller if the shaft is covered with a thin layer of ice. In addition, test results show that after the bond fails a large portion of the bond strength is permanently lost, because the adhesion bond fails in a brittle manner. It is not clear whether the broken bond heals with time. To improve pile performance, which depends mainly on the bond, the use of special types of piles, as described below, is suggested.

(2) Corrugated piles

With a surface so rough that failure will be forced to occur within the frozen soil mass and not along the pile-soil interface.

(3) Tapered piles

By their displacements, tapered piles assure a good contact with the soil and a constant normal pressure on the shaft, furnishing a constant frictional portion of the total bond resistance if the pile is rough and surrounded by a frictional frozen soil. Because they are supported by the surrounding soil mass, such piles are expected to show an essentially plastic response without loss of strength.

(4) In general, the research should be oriented toward finding ways to increase the bond strength and make its response less brittle after failure.

(5) More R&D is needed on frost uplift reduction devices for piles.

D. General Design and Performance

(1) A number of methods for foundation design in permafrost have been proposed, but the systematic verification of their predictions against the performance of foundations under controlled conditions is still scarce. In addition, it is not yet clear how to determine relevant frozen soil parameters that properly represent long-term soil behavior without performing very long-term creep tests.

(2) Pile design under cyclic and dynamic loads needs further improvement.

(3) Extrapolation of full-scale loading test results to long-term performance should be investigated.

(4) Static cone and push-in pressuremeter devices are promising for getting the basic information needed for pile design in offshore permafrost. They can become practically useful only if they are used more frequently and their predictions are compared with actual pile performance.

ACKNOWLEDGMENT

A portion of the bibliographical data presented in this report was drawn from an earlier literature review (Ref. P35) made by the author for the National Research Council of Canada.

BIBLIOGRAPHY (1963-1983)

I. Piles and Deep Footings

P1. Johnston, G.H., 1966, Pile construction in permafrost, in Proceedings of the First International Conference on Permafrost, Purdue University, 1963: Washington, D.C., National Academy of Sciences, p. 477-481.

P2. Crory, F.E., 1966, Pile foundations in permafrost, in Proceedings of the First International Conference on Permafrost, Purdue University, 1963: Washington, D.C., National Academy of Sciences, p. 467-472.

P3. Crory, F.E., 1965, Pile foundations in discontinuous permafrost areas, in Proceedings of the Canadian Regional Permafrost Conference, Edmonton: Ottawa, National Research Council of Canada, Technical Memorandum No. 86, p. 58-76.

P4. Crory, F.E. and Matlock, C.S., 1967, Piles in permafrost for bridge foundations. ASCE Structural Engineering Conference: Seattle, Preprint 522.

P5. Crory, F.E., 1968, Bridge foundations in permafrost areas, Goldstream Creek, Fairbanks, Alaska: Hanover, N.H., U.S. Army Cold Regions Research and Engineering Laboratory, Technical Report 180, AD 667945.

P6. Sanger, F.J., 1969, Foundations of structures in cold regions: Hanover, N.H., U.S. Army Cold Regions Research and Engineering Laboratory, Monograph III-C4, AD 694371.

P7. Miller, J.M., 1971, Pile foundations in thermally fragile frozen soils, in Proceedings of the Symposium on Cold Regions Engineering: University of Alaska, Fairbanks, v. 1, p. 34-72.

P8. Watson, G.H. and Rowley, R.K., 1972, Inuvik pile load tests: Mackenzie Valley Pipe Line Research Ltd., Internal Report.

P9. Rowley, R.K., Watson, G.H., and Ladanyi, B., 1973, Vertical and lateral pile load tests in permafrost, in Permafrost -- the North American Contribution to the Second International Conference, Yakutsk: Washington, D.C., National Academy of Sciences, p. 712-721.

P10. Crory, F.E., 1973, Installation of driven test piles in permafrost at Bethel Air Force Station, Alaska: Hanover, N.H., U.S. Army Cold Regions Research and Engineering Laboratory, Technical Report 139, AD 774291.

P11. Linell, K.A. and Johnston, G.H., 1973, Engineering design and construction in permafrost regions: A review, in Permafrost -- the North American Contribution to the Second International Conference, Yakutsk: Washington, D.C., National Academy of Sciences, p. 553-575.

P12. Long, E.L., 1973, Designing friction piles for increased stability at lower installed cost in permafrost, in Permafrost -- the North American Contribution to the Second International Conference, Yakutsk: Washington, D.C., National Academy of Sciences, p. 693-699.

P13. Ladanyi, B. and Johnston, G.H., 1974, Behavior of circular footings and plate anchors in permafrost. Canadian Geotechnical Journal, v. 11, no. 4, p. 531-553.

P14. Crory, F.E., 1975, Bridge foundations in permafrost areas: Hanover, N.H., U.S. Army Cold Regions Research and Engineering Laboratory, Technical Report 266, ADA 013520.

P15. Ladanyi, B., 1975, Bearing capacity of strip footings in frozen soils. Canadian Geotechnical Journal, v. 12, no. 3, p. 393-407.

P16. Rowley, R.K., Watson, G.H., and Ladanyi, B., 1975, Prediction of pile performance in permafrost under lateral load. Canadian Geotechnical Journal, v. 12, no. 4, p. 510-523.

P17. Womick, O., Jr. and LeGoullon, R.B., 1975, Settling a problem of settling. Northern Engineer, v. 7, no. 1, p. 4-10.

P18. Ladanyi, B., 1976, Use of static penetration test in frozen soils. Canadian Geotechnical Journal, v. 13, no. 2, p. 95-110.

P19. Nixon, J.F. and McRoberts. E.C., 1976, A design approach for pile foundations in permafrost. Canadian Geotechnical Journal, v. 13, no. 1, p. 40-57.

P20. Rooney, J.W., Nottingham, D., and Davison, B.E., 1976, Driven H-pile foundations in frozen sands and gravels, in Proceedings of the Second International Symposium on Cold Regions Engineering, Fairbanks: University of Alaska, p. 169-188.

P21. Long, E.L., 1977, Thermo-Ring Pile and Thermal-Probe Foundation Systems. Anchorage, Arctic Foundations Inc.

P22. Bendz, J., 1977, Permafrost problems beaten in pile-driving research. Western Construction.

P23. Nixon, J.F., 1977, First Canadian Geotechnical Colloquium: Foundation design approaches in permafrost areas. Canadian Geotechnical Journal, v. 15, p. 96-112.

P24. Phukan, A., 1977, Pile foundations in frozen soils. Paper presented at the ASME Energy Technical Conference, Houston, Texas.

P25. Long, E.L., 1978, Permafrost foundation designs, in Proceedings of the Cold Regions Specialty Conference on Applied Techniques for Cold Environments, Anchorage, v. 2: New York, American Society of Civil Engineers, p. 973-987.

P26. Davison, B.E., Rooney, J.W., and Bruggers, D.E., 1978, Design variables influencing piles driven in permafrost, in Proceedings of the Cold Regions Specialty Conference on Applied Techniques for Cold Environments, Anchorage, v. 1: New York, American Society of Civil Engineers, p. 307-319.

P27. Myska, A.D. and How, G.T.S., 1978, Installation of pile foundations for a microwave tower system, Gillam-Churchill, Manitoba, in Proceedings of the Third International Conference on Permafrost, Edmonton, v. 1: Ottawa, National Research Council of Canada, p. 826-832.

P28. Black, W.T. and Thomas, H.P., 1979, Prototype pile tests in permafrost soils, in Pipelines in Adverse Environments: A State of the Art, Proceedings of the ASCE Pipeline Division Specialty Conference, New Orleans, v. 1: New York, American Society of Civil Engineers, p. 372-383.

P29. Tobiasson, W. and Johnson, P., 1978, The details behind a typical Alaskan pile foundation, in Proceedings of the Third International Conference on Permafrost, Edmonton, v. 1: Ottawa, National Research Council of Canada, p. 891-897.

P30. Ladanyi, B. and Paquin, J., 1978, Creep behavior of frozen sand under a deep circular load, in Proceedings of the Third International Conference on Permafrost, Edmonton, v. 1: Ottawa, National Research Council of Canada, p. 679-686.

P31. Phukan, A. and Andersland, O.B., 1978, Foundations for cold regions. Chapter 6 in Geotechnical Engineering for Cold Regions (O.B. Andersland and Anderson, eds.): New York, McGraw-Hill, p. 276-362.

P32. Weaver, J.S., 1979, Pile foundations in permafrost. Edmonton, University of Alberta, Ph.D. dissertation.

P33. Johnston, G.H. (ed.), 1980, Operation, Maintenance, and Performance of Structures in Permafrost Engineering Design and Construction. NRCC-ACGR: Toronto, John Wiley and Sons.

P34. Morgenstern, N.R., Roggensack, W.D., and Weaver, J.S., 1980, The behaviour of friction piles in ice and ice-rich soils. Canadian Geotechnical Journal, v. 17, no. 3, p. 405-415.

P35. Ladanyi, B., 1980, Field tests of foundations in permafrost: Literature review. Ottawa, National Research Council of Canada, Contract OSU79-00283.

P36. Phukan, A., 1980, Design of deep foundations in discontinuous permafrost. ASCE Convention and Exposition, Portland, Oregon, preprint 80-122.

P37. Linell, K.A., and Lobacz, E.F., 1980, Design and construction of foundations in areas of deep seasonal frost and permafrost: Hanover, N.H., U.S. Army Cold Regions Research and Engineering Laboratory, Special Report 80-34, ADA 090324.

P38. Thomson, S., 1980, A brief review of foundation construction in the Western Canadian Arctic. Quarterly Journal of Engineering Geology, London, v. 13, p. 67-76.

P39. Anon., 1980, Adaptation of overburden drill is ideal for piles in permafrost. Northern Development, March/April, p. 8.

P40. Thomas, H.P., and Luscher, U., 1980, Improvement of bearing capacity of pipe piles by corrugations. Collection of papers from the US-Soviet Joint Seminar on Building in Cold Climates and on Permafrost, Leningrad, USSR. Hanover, N.H., U.S. Army Cold Regions Research and Engineering Laboratory, Special Report 80-40: ADA 097516, p. 229-234.

P41. Weaver, J.S., and Morgenstern, N.R., 1981, Pile design in permafrost. Canadian Geotechnical Journal, v. 18, no. 3, p. 357-370.

P42. Ladanyi, B., 1982, Determination of geotechnical parameters of frozen soils by means of the cone penetration test, in Proceedings, ESOPT II: Amsterdam, v. 2, p. 223-234.

P43. Crory, F.E., 1982, Piling in frozen ground. Journal of Technical Councils, American Society of Civil Engineers, v. 108, no. TCl, p. 112-124.

P44. Nottingham, D., and Christopherson, A.B., 1983, Driven piles in permafrost: state of the art, in Proceedings of the Fourth International Conference on Permafrost, Fairbanks: Washington, D.C., National Academy of Sciences, p. 928-933.

P45. Manikian, V., 1983, Pile driving and load tests in permafrost for the Kuparuk Pipeline System, in Proceedings of the Fourth International Conference on Permafrost, Fairbanks: Washington, D.C., National Academy of Sciences, p. 804-810.

P46. Luscher, U., Black, W.T., and McPhail, J.F., 1983, Results of load tests on temperature-controlled piles in permafrost, in Proceedings of the Fourth International Conference on Permafrost, Fairbanks: Washington, D.C., National Academy of Sciences, p. 756-761.

P47. Dufour, S., Sego, D.C., and Morgenstern, N.R., 1983, Vibratory pile driving in frozen sand, in Proceedings of the Fourth International Conference on Permafrost, Fairbanks: Washington, D.C., National Academy of Sciences, p. 255-260.

P48. DiPasquale, L., Gerlek S., and Phukan, A., 1983, Design and construction of pile foundations in the Yukon-Kuskokwim Delta, Alaska, in Proceedings of the Fourth International Conference on Permafrost, Fairbanks: Washington, D.C., National Academy of Sciences, p. 232-237.

II. Deep Anchors

AN1. Barry, B.L, and Cormie, J.G., 1971, Constructional aspects of the 450 KVDC Nelson River transmission line. Manitoba Power Conference EHV-DC, Winnipeg, p. 459-492.

AN2. Reinart, I., 1971, Nelson River HVDC transmission line foundation design aspects. Manitoba Power Conference EHV-DC, Winnipeg, p. 422-440.

AN3. Walt, G.L., 1971, Testing of anchors in permafrost. Carleton University, Ottawa, M. Eng. thesis.

AN4. Johnston, G.H. and Ladanyi, B., 1972, Field tests of grouted rod anchors in permafrost. Canadian Geotechnical Journal, v. 9, no. 2, p. 176-194.

AN5. EBA Engineering Consultants Ltd., 1973, Installation and testing of rod anchors in permafrost at Norman Wells, N.W.T. Report to Canadian Arctic Gas Study Limited (unpublished).

AN6. Hironka, M.C., 1974, Anchoring in snow, ice, and permafrost: Port Hueneme, California, Naval Civil Engineering Laboratory, Technical Note No. 1344.

AN7. Johnston, G.H., and Ladanyi, B., 1974, Field tests of deep power-installed screw anchors in permafrost. Canadian Geotechnical Journal, v. 11, no. 3, p. 348-358.

AN8. Andersland, O.B., and Alwahhab, M.R.M., 1983, Lug behavior for model steel piles in frozen sand, in Proceedings of the Fourth International Conference on Permafrost, Fairbanks: Washington, D.C., National Academy of Sciences, p. 16-21.

III. Adfreeze Bond

AD1. Jellinek, H.H.G., 1959, Adhesive properties of ice. Journal of Colloid Science, v. 14, p. 268-280.

AD2. Stehle, N.S., 1970, Holding strength of piles in ice: Port Hueneme, California, Naval Civil Engineering Laboratory, Technical Report R-700.

AD3. Kaplar, C.W., 1971, Some strength properties of frozen soil and effect of loading rate. Hanover, N.H., U.S. Army Cold Regions Research and Engineering Laboratory, Special Report 159, AD 726913.

AD4. Paige, R.A., 1974, The McMurdo ice wharf, McMurdo Station, Antarctica. Bulletin of Associated Engineering Geologists, v. 11, no. 1, p. 1-14.

AD5. Laba, J.T., 1974, Adfreezing of sands to concrete. Transportation Research Record, No. 494, p. 31-39.

AD6. Rice, E.F., and Ferris, W., 1976, Some laboratory tests on creep of steel piling in frozen ground (Summary only), in Proceedings of the Second International Symposium on Cold Regions Engineering, Fairbanks: University of Alaska, p. 278.

AD7. Parameswaran, V.R., 1978, Laboratory studies of the adfreeze bond between small-scale model piles and frozen sand, in Proceedings of the Third International Conference on Permafrost, Edmonton, v. 1: Ottawa, National Research Council of Canada, p. 714-720.

AD8. Parameswaran, V.R., 1978, Adfreeze strength of frozen sand to model piles. Canadian Geotechnical Journal, v. 15, no. 4, p. 494-500.

AD9. Parameswaran, V.R., 1979, Creep of model piles in frozen soil. Canadian Geotechnical Journal, v. 16, no. 1, p. 69-77.

AD10. Frederking, R.M.W., 1979, Laboratory tests on downdrag loads developed by floating ice covers on vertical piles, in Proceedings, POAC 79, Trondheim, Norway, v. 2, p. 1097-1110.

AD11. Parameswaran, V.R., 1981, Displacement of piles under dynamic loads in frozen soils, in Proceedings of the Fourth Canadian Permafrost Conference, Calgary (Roger J.E. Brown Memorial Volume, H.M. French, ed.): Ottawa, National Research Council of Canada, p. 555-559.

IV. Frost Heave Effects on Piles

FH1. Crory, F.E. and Reed, R.E., 1965, Measurement of frost heaving forces on piles: Hanover, N.H., U.S. Army Cold Regions Research and Engineering Laboratory, Technical Report 145, AD 484955.

FH2. Penner, E. and Irwin, W.W., 1969, Adfreezing of Leda clay to anchored footing columns. Canadian Geotechnical Journal, v. 6, no. 3, p. 327-337.

FH3. Penner, E. and Gold, L.W., 1971, Transfer of heaving forces by adfreezing to columns and foundation walls in frost-susceptible soils. Canadian Geotechnical Journal, v. 8, no. 4, p. 514-526.

FH4. Penner, E., 1974, Uplift forces on foundations in frost heaving soils. Canadian Geotechnical Journal, v. 11, no. 3, p. 323-338.

FH5. Domaschuk, L., 1982, Frost heave forces on embedded structural units, in Proceedings of the Fourth Canadian Permafrost Conference, Calgary: Ottawa, National Research Council of Canada, p. 487-496.

FH6. Penner, E. and Goodrich, L.E., 1983, Adfreezing stresses on steel pipe piles, Thompson, Manitoba, in Proceedings of the Fourth International Conference on Permafrost, Fairbanks: Washington, D.C., National Academy of Sciences, p. 979-983.

V. Foundation Materials

MAT1. Romanoff, M., 1970, Corrosion evaluation of steel test piles exposed to permafrost soils, in Proceedings of the 25th Conference of the National Association of Corrosion Engineers, NBS Monograph 127: Washington, D.C., U.S. Department of Commerce, National Bureau of Standards.

MAT2. Sedziak, H.P., Shields, H.P., and Johnston, G.H., 1973, Conditions of timber foundation piles at Inuvik, N.W.T.: Department of the Environment, Canadian Forestry Service, Information Report OP-X-67, 27 p.

MAT3. Weaver, J.S. and Morgenstern, N.R., 1980, Considerations on the use of cast-in-place piles in permafrost. Canadian Geotechnical Journal, v. 17, no. 2, p. 320-325.

Frost Heave and Ice Segregation

PANEL MEMBERS

Edward Penner (Chairman), Division of Building Research, National Research Council of Canada, Ottawa, Canada

S.E. Grechishchev, All-Union Research Institute of Hydrogeology and Engineering Geology, Moscow, USSR

Chen Xiaobai, Institute of Glaciology and Cryopedology, Academia Sinica, Lanzhou, People's Republic of China

Robert D. Miller, Department of Agronomy, Cornell University, Ithaca, New York, USA

Peter J. Williams, Geotechnical Science Laboratories, Carleton University, Ottawa, Canada

Richard L. Berg, Cold Regions Research and Engineering Laboratory, Hanover, New Hampshire, USA

INTRODUCTION

E. Penner

The selection of panel members from the USA and Canada was based on their involvement in frost-action-related activities. S.E. Grechishchev of the USSR and Chen Xiaobai of the People's Republic of China were asked to describe briefly the frost-action research in progress in their countries. Unfortunately, the USSR representative was not able to attend the conference, but a summary of his paper was presented at the panel session by the panel chairman.

Miller presents a qualitative rationalization for a process of thermally induced relegation, a basic ingredient of his "rigid ice model" of frost heaving. While that model purports to explain well-known characteristics of the process, including the role of overburden pressure, its use for engineering purposes depends upon input functions that are not easily measured with the necessary accuracy at this time. So far, the model has been formulated only for air-free, solute-free incompressible soils. He mentions (unpublished) results of a similitude analysis of this model which imply that site-specific problems could be scale modeled using a geotechnical centrifuge.

Berg reviews the numerical models available in the literature in his presentation, with particular emphasis on the difficulties encountered. His review is particularly meaningful because of his considerable experience in modelling with applied engineering applications in mind.

Williams deals with moisture migration in frozen zones, a subject he has studied over many years. The presentation states both theoretical and experimental support of his approach to the subject. It also provides an overall evaluation of current knowledge on the subject.

The USSR and People's Republic of China panelists cover current research in their respective countries. Grechishchev presents his current concept of the thermorheology of cryogenic soils, drawing attention in particular to the inseparability of heat, moisture, stress and strain values, and the phase composition of pore moisture in frozen, freezing, and thawing soils. He states that thermophysical and mechanical problems cannot be resolved independently, and gives equations to describe the interrelationship. Chen Xiaobai covers current frost-action developments in the PRC, stressing engineering solutions to the frost heave problem. While different in detail, the approach is not unlike that followed in seasonally frozen and permafrost regions in North America. Both contributions will draw much attention, particularly from those concerned with solutions to frost action problems in soils.

The high annual cost of frost damage to engineering structures as a result of the freezing of foundation soils does not appear to have brought about a concerted, coordinated program of research in any of the four countries represented on the panel. It is clear from the literature that important studies are being carried out, but a national foci on the problem appears to be missing. At present the research on frost action in Canada and the USA is initiated primarily by interested individuals. While this certainly results in some excellent work, much more could be achieved if there were an organized approach. In both the applied aspects and numerical modelling studies, benefits from a systematic approach would accrue. The facilities and the effort expended would, however, have to be greatly increased to have a real influence. This may develop as the costs of annual damage increase.

THE PRINCIPLES OF THERMORHEOLOGY OF CRYOGENIC SOILS

S.E. Grechishchev*

All-Union Research Institute of Hydrogeology and Engineering Geology
Moscow, USSR

In this paper the term "cryogenic soils" is introduced as a shorter term for frozen, freezing, and thawing soils. Usually, thermorheology refers to a section in the mechanics of solid deformable media that represents the study of an interconnection in time between mechanical stresses, strains, and temperature. However, it is necessary to broaden the concept of thermorheology when it is applied to cryogenic soils, because any mechanical and thermal impact on these soils affects the phase composition of pore moisture in a complex way, namely, changes in the phase composition of pore water lead to variations in temperature and mechanical stresses or strains. Thus, the phrase thermorheology of cryogenic soils refers to the study of the interconnection and interaction in time between mechanical stresses and strains, temperature, and the phase composition of pore moisture in frozen, freezing, and thawing soils. This phrase was first proposed by Grechishchev (1969).

The thermorheology of cryogenic soils includes a consideration of the mechanical processes in soils under the impact of temperature changes, both within the range below 8°C (thermal strains, effects of an external overburden and unfrozen moisture content) and with the temperature passing through 0°C (heaving and moisture migration with freezing, thermal settlement and consolidation with thawing). With this approach, a combined investigation of the thermodynamic, thermophysical, physicochemical, and mechanical aspects of the cryogenic processes in soils appears necessary.

The study of physical regularities and the development of physico-mathematical theories of cryogenic processes has been intensive over the last 20 years. Physico-mathematical models based on Luikov's equations of heat-mass transfer have attained the widest development in geocryology. Being theoretically elaborated to match the conditions of phase transitions in porous media represented by soils, they were considered to be perspective for the investigation and the description of cryogenic processes. Theoretical and experimental works in this field have been performed by Martynov, Tyutyunov, Chistotinov, Feldman, Ershov, Zhestkova, Takagi, and many other Soviet and foreign specialists. The main results and the characteristic feature of this trend involve the establishment of a functional interrelationship of a thermal field and a moisture content field. Basically, it involves the study of the processes related to moisture migration under the influence of thermal gradients; namely, changes in soil temperature with moisture migration movement.

Although the given thermophysical trend of studying cryogenic processes has obvious merits, it also has a number of important limitations, which recently have become more and more evident as new experimental data are accumulated. First, the theory of heat-mass transfer in porous media, developed by Luikov and his disciples for positive temperatures, was less effective when applied to subfreezing soils due to an uncertain interaction at inner interphase boundaries where the phase transitions of pore moisture occur. Numerous attempts to avoid this difficulty by formally defining such an interaction have led to physically improbable results. The most obvious is shown by Feldman in the example of a boundary condition of the first type for describing moisture movement. Its use can lead to such a physically senseless concept as a negative moisture content. Second, the theory of heat-mass transfer does not include the mechanical phenomena that occur in soils under the effects of moisture migration and temperature changes. Finally, it is the mechanical process of soil deformation that causes various cryogenic structures to form in soils; these in turn determine the occurrence of cryogenic phenomena in macrovolumes of soils.

A mechanical trend has developed in the theory of cryogenic processes independent of the thermophysical one. Basically it consists of studying a stressed-strained state of soils; temperature and moisture content are investigated as the independent state parameters. This approach is fruitful for a certain range of engineering problems. The first fundamental results in this area were obtained by Tsytovich, one of the founders of geocryology, who investigated the mechanical characteristics of frozen soils as well as the process of their settlement when thawing and formulated the main principles of the mechanics of frozen soil. A significant contribution to this approach has been made by Vyalov, who described the principles of soil rheology at negative temperatures. Important studies have been performed by Voitkovskii, Zaretskii, Ter-Martirosyan, Orlov, Puzakov, Anderson, Andersland, Ladanyi, Morgenstern, Chamberlain, Sayles, and others.

While mentioning the important role and the value of investigations in mechanics of frozen soils for the development of the theory of cryogenic processes, it should be stated that a large part of these processes cannot be described within the framework of a mechanical trend alone. In the case of the mechanical approach, the stress-strain field depends on temperature and moisture content only in a parametric way, but temperature and moisture content themselves are stress-dependent.

A certain break between the thermophysical and the mechanical approaches to the construction of physico-mathematical models of cryogenic processes has taken place in developing the principles of the theory of these processes so far. General-

*Not present; summary presented by E. Penner.

ly, however, an interrelationship is evident between the stress-strain fields, temperature, and moisture content in soils at negative temperatures. Vyalov noted that mechanical stresses might impact the content and migration of unfrozen moisture in frozen soils. The most generalized form of the dependence of unfrozen moisture content on environmental effects (temperature, loading, etc.) is given by Tsytovich (1973) in his well-known principles of frozen soil mechanics: the equilibrium state of unfrozen moisture and moisture migration when the external parameters change. Important studies in this field have been performed recently by Ershov.

The study of thermal strains in frozen sandy and clayey soils, started by Fedosov, who was the first to discover that clayey soils,instead of expanding, shrank while freezing, and by Votyakov, who experimentally established a time lag of thermal strains after a temperature change, has played a significant role in developing the experimental principles of the thermorheology of cryogenic soils. The lag phenomenon of thermal strains occurring after the temperature change in soils has been also established experimentally. This group of phenomena has been called "time after-effect of thermal strains and stresses" (Votyakov and Grechishchev, 1969). Further study of thermal strains and stresses in frozen soils has been carried on by Votyakov, Grechishchev, Shusherina, Sheshin, Rachevskii, Simagina, and others.

The results of investigations on unfrozen moisture, texture, and state of pore ice obtained by Soviet scientists (Nersesova, Tsytovich, Andrianov, Bozhenova, Shumskii, Shvetsov, Saveliev, Dostovalov, Ananyan, Vtyurin, Pchelintsev, and others) as well as by foreign scientists (Jumikis, Yong, Anderson, Williams, Penner, Miller, and others) are of great importance to the development of the thermorheology of cryogenic soils.

The studies of moisture migration and frost heaving started by Taber, Beskow, Goldstein, and others are especially significant for developing the principles of thermorheology of cryogenic soils. Investigations in this field have become numerous and they are conventionally characterized by two approaches: the thermophysical approach including thermodynamics (Ananyan, Shvetsov, Chistotinov, Ivanov, Ershov, Merzlyakov, Winterkorn, Miller, Williams, Penner, Anderson, Hoekstra, and others) and the mechanical approach (Sumgin, Tsytovich, Fedosov, Pchelintsev, Puzakov, Goldstein, Orlov, and others).

Regardless of the number and the high level of the studies mentioned, the basic problems of frozen soil physics relating to the conditions of ice crystal growth and unfrozen moisture migration in soils at negative temperatures have not yet been solved. Current ideas of the motive forces of moisture migration are rather contradictory. Correlations between physical phenomena in frozen soils and the mechanical stresses and strains occurring in soils remain vague. This problem is complicated by the use of irreversible thermodynamics and thermodynamic potentials for describing moisture migration. A transition from these potentials to stresses leads to a physically complex conception, such as "negative pore pressure (suction)," which is bound to amount to thousands of atmospheres below the ideal vacuum.

Experiments to verify a new theory that would explain many thermorheological phenomena in cryogenic soils have started recently. It is based on the study of the physicochemical, thermodynamic, and mechanical aspects of interphase interaction at the boundary of phase transitions between frozen and unfrozen zones. The following expressions determining the correlations between mass flow, freezing rate, and temperature at the boundary of phase transitions were developed by the author from theoretical study of the kinetics of phase interaction in the vicinity of the phase transition boundary (Grechishchev, 1978, 1979):

$$C_1 (q_w - \varepsilon q_s) = L \Delta T_\xi / T_o - V_i \sigma_n^{sk} - (V_i - V_w) p \quad (1)$$

$$C_2 (q_w - \varepsilon q_s + V_\xi) = L \Delta T_\xi / T_o - (V_i - V_w) p \quad (2)$$

where q_w, q_s - effective rates of flow of pore moisture and mineral particles of soil skeleton; V_i, V_w - specific phase volumes of ice and water; V_ξ - advance rate of phase transition (freezing or thawing) boundary; σ_n^{sk} -- the normal stress in the soil skeleton, p_n - effective stress normal to the freezing boundary; ξ - pressure in the soil skeleton and pore water pressure; L - latent heat of water freezing; $T_o = 273°K$, $\Delta T_\xi = T_o - T_\xi$; T_ξ - temperature at phase change transition boundary, K; C_1, C_2 - empirical coefficients characterizing both the phase transition kinetics and the permeability of the phase transition zone, and decreasing with an increase in ΔT_ξ; ε - porosity.

The following physical backgrounds are accepted, when deriving expressions (1) and (2): (a) dispersed soil heteroporosity provides a difference over time of moisture freezing in pores of various sizes; (b) occurrences of fine and coarse pores hydraulically connected and the pores in which the walls (represented by the surfaces of mineral particles and ice crystals) interact through a water film are referred to as fine pores; (c) force interaction in fine pores occurs through the film, following the mechanism of disjoining pressure determined by Deryagin; (d) unfrozen zone strains are subjected to the theory of soil consolidation; (e) full stress components in dispersed soil can be represented by the sum of effective stresses in the mineral skeleton of soil and the pore pressure, i.e. Terzaghi's equation

$$\sigma_{ik} = \sigma_{ik}^{sk} + \delta_{ik} p \quad (3)$$

where i, k - direction of the axis, δ_{ik} - Kroneker delta.

Expression (1) characterizes the kinetics of phase transitions in fine pores and expression (2) that in coarse ones. As to the quantity, the proposed expressions show that moisture flow is formed ($q_w \geq 0$, when freezing), the movement of soil skeleton takes place ($q_w < 0$, when freezing), or both of them occur at the phase transition (i.e. ice and unfrozen moisture) boundary and under the impact of the latter (term with T). Opposed to this movement are an effective pressure in the soil skeleton (with coefficient $V_i \simeq 1.09$ $cm^3\ g^{-1}$) and the pore moisture pressure (with coefficient $V_i - V_w \simeq 0.09\ cm^3\ g^{-1}$), which are considered to be positive when compressing. Provided that condition (1) for the occurrence of migration flow q_w is not satisfied (e.g. when pressure gradient in unfrozen zone is insufficient for moisture flow formation with freezing, necessary for an increase in the ice band with the given ΔT_ξ), freezing occurs along the coarse pores at rate V_ξ, as determined by expression (2).

A full system of equations for the thermorheology of cryogenic soils can be constructed using expressions (1) and (2). For example, for the first approximation a phase transition zone can be regarded as a linear friction boundary between frozen and unfrozen zones, and expressions (1) and (2) can be taken for the boundary conditions at this boundary, remembering that according to Florin-Gersevanov's equation

$$q_w - \varepsilon q_s = -k_\xi \text{ grad } p_\xi \tag{4}$$

where k_ξ - effective coefficient of permeability (analogue of Darcy's ratio of an imagined boundary of freezing ξ).

Thus, a complete system of equations of thermorheology of cryogenic soils must consist of the thermoconductivity equations and equations of the mechanics of deformed media for unfrozen and frozen zones of soil. These equations have turned out to be fully combined with each other through environments (a) through (d) at the freezing (thawing) boundary. Hence, the distribution of heat, moisture, stress values, and strains in cryogenic soils happen to be closely connected. This accounts for the fact that thermophysical and mechanical problems cannot be solved separately in this case.

The proposed theory predicts that temperature and pressure in pore moisture at a freezing boundary depend completely upon the freezing conditions (i.e. the temperature of the outer boundaries and the external overburden), permeability, and compressibility of soils in an unfrozen zone. Therefore, they vary constantly in the process of freezing. When considering the problem theoretically, they cannot be given in advance and they are determined only as a result of the solution of the whole problem. This conclusion is born out by the experimental data, according to which the temperature of freezing soil is approximately proportional to the temperature of the cold environment in which a soil sample is frozen.

In conclusion, we can enumerate a number of known phenomena that can be explained with the help of the proposed theory in a natural way. They include: the discrepancy in phase equilibrium pressure when freezing and thawing; time-lag effect of thermal strains and stresses; hysteresis of unfrozen moisture and of volumetric strains related to it with cyclic freezing-thawing; discrepancy of thermal strains parallel and perpendicular to the isotherms; occurrence of optimum cooling rates (causing extreme stress values); frozen soil volumetric strains in time at constant temperature; discrepancy of heaving values, with soil freezing from the surface and from the bottom, and so forth. The theory also answers the fundamental question of why heaving does not occur in sands and unsaturated clays.

Further studies in the field of the thermo rheology of cryogenic soils must address mechanism of formation and growth of ice bands, which is one of the basic problems in the theory of cryolithogenesis.

REFERENCES

Grechishchev, S.E., 1978, Mezhfaznoe vsaimodeistvie v porovoi vlage i termoreologicheskie svoistva merzlykh gruntov (Interphase interaction in pore moisture and thermorheological properties of frozen soils): Dokl. AN SSSR, v. 242, no. 3 (in Russian).

Tsytovich, N.A., 1973, Mekhanika Merzlykh Gruntov (Mechanics of Frozen Ground): Moscow, Vysshaya Shkola Publishers (in Russian). (Trans.) New York, McGraw-Hill, 1975, 426 p.

Votyakov, I.N. and Grechishchev, S.E., 1969, O vremennom effekte posledeistviya temperaturykh deformatsii i napriazhenii v merzlykh gruntakh (Time effect of thermal deformations and stresses in frozen ground); Sb. Stroitel'stvo v Raionakh Vostochnoi Sibiri i Krainego Severa: Krasnoyarsk, v. 14, p. 41-60 (in Russian).

CURRENT DEVELOPMENTS IN CHINA ON FROST-HEAVE PROCESSES IN SOIL

Chen Xiaobai

Institute of Glaciology and Cryopedology, Academia Sinica
Lanzhou, People's Republic of China

INTRODUCTION

Up to now, because frost heave and ice segregation are still important problems in civil engineering construction in northeast and northwest China, the mechanism of frost heave and water migration and how to protect structures from frost damage have been studied in the last few years.

There has been good progress in the frost-susceptibility criterion of coarse-grained soils; the effect of penetration rate, surcharge stress, and depth to groundwater on frost heave; the basic characteristics of tangential, normal, and horizontal frost heave forces; the effect of allowable deformation of foundation and rheology of frozen soil on heaving force, the variation of normal heaving force under various foundation areas, and the application value of the three heaving forces. In the last two years, the energy theory for unsaturated soil has been used to establish a model of water migration and ice segregation in soil during freezing, and research experiments are being performed both in the laboratory and in the field.

Frost heave is one of the main problems for civil engineering and road construction in permafrost and seasonally frozen regions, especially for hydraulic structures, which usually suffer frost damage. In recent years, many investigators have studied the cryogenic process, frost heave, and the frost heave force, with special attention given to field observations. The author has prepared a general review of the problem in this paper.

ACTION OF FROST HEAVE

Because of the needs of engineering construction, the main factors affecting frost-heave processes, such as particle size, frost penetration rate, effective surcharge stress, and groundwater table, have been investigated. It is well known that, in saturated sand or gravel, water migration is mainly in response to a pressure gradient. However, for clayey soil, it results from the gradient of soil water potential. The current developments are described below.

Particle Size and Grading

According to the current frost susceptibility criteria for railway construction, a frozen coarse-grained soil in which the content of silt-clayey particles is less than 15% of the total weight does not consolidate during thawing. It is the author's experience, however, that because permafrost was formed over a geological period, the thawing settlement coefficient of a frozen gravel, whose content of silt-clayey particles is less than 2.5% of the total weight, is more than 12%. In an open system (Chen et al. 1982), if a saturated sandy gravel with uniform grading contains more than 6% silt-clayey particles, its heave ratio will exceed 1%; if the silt-clayey particle content is more than 8-10%, then its heave ratio will be more than 2%. If the sand or gravel has a non-uniform grading, corresponding with above heave ratio, the silt-clayey particle content is 3-4% and 6-7% of total, respectively, Wang (1983) indicated that if a coarse-grained soil contains 5% silt-clayey particles in open system, its average heave ratio is 1.8%.

In the current "Norm of Base and Foundation Design for Industrial and Civil Engineering Construction" (1975), fine sand is included in the non- or weakly frost-susceptible soil category. According to Wang's (1983) results, however, the above criterion is not complete. Compared with the content of 0.05-0.005 mm (X_2) and 0.1-0.25 mm (X_4), Wang Zhengqiu (1980) indicated that the size less than 0.005 mm (X_1) is the most important factor influencing the heave ratio η of fine sand. The relationship can be expressed by

$$\eta = 4.48 + 0.62X_1 + 0.07X_2 - 0.04X_4$$

in which the content of 0.05-0.1 mm particles (X_3) does not affect the heave ratio.

The results of field observations by the Low-Temperature Construction Institute, Heilongjiang Province, show that when a clayey soil contains more than 50% clay particles by weight, its permeability is very low and it might be classified as a weakly frost-susceptible soil.

Penetration Rate

The results of Chen et al. (1981) show that the penetration rate strongly affects the heave ratio. When the penetration rate is less than the first critical limit V_{f1}, the ratio of frost heave η is very large and ice lenses are strongly segregated. After the penetration rate exceeds the second critical limit V_{f2}, however, there is almost no water migration during freezing, and only in-situ pore water freezes. The relation might be expressed as follows:

when $V_f \geq V_{f2}$, then $\eta = \eta_o = \text{const.}$; i.e. no ice segregation;

when $V_{f2} \geq V_f \geq V_{f1}$,

then $\eta = \eta' = \eta_o + \Delta\eta \left[(1/\sqrt{V_f} - 1/\sqrt{V_{f2}})/(1/\sqrt{V_{f1}} - 1/\sqrt{V_{f2}})\right]^3$; i.e. a little ice segregation;

when $V_f \leq V_{f1}$,

then $\eta = \eta_o + \Delta\eta + C(1/\sqrt{V_f} - 1/\sqrt{V_{f1}})$; i.e. a lot of ice segregation.

Effective Surcharge Stress

As shown by the thermodynamic equation, the freezing point of soil decreases with an increase of overburden pressure or pore water potential. Chen et al. (1983) showed that the relation between freezing point drop ΔT and effective surcharge stress p for saturated loess is $\Delta T \simeq 0.07$ p. It was indicated that the increase of effective surcharge stress would prevent water from flowing toward the frost front, which might be shown (for saturated loess) by

$$\ln(\xi_\omega/\xi_{\omega o}) = -ap$$

where ξ_ω is specific volumetric suction water, when $p = o$, $\xi = \xi_{\omega o}$.

Because of the above relationship, under a surcharge stress the frost heave ratio η decreases exponentially (Chen et al., 1983), i.e.

$$\eta = \eta_o \exp(-bp) \ .$$

Observations for clayey loam obtained in the field by the Low-Temperature Construction Institute, Heilongjiang Province, are the same as the author's. Xu (1981a) has studied the characteristics of frost heave in overconsolidated, consolidated, and subconsolidated soil.

Chen et al. (1983) carried out a frost heave test of saturation loess under loading and obtained its shut-off pressure of about 4.6 kg/cm^2.

Groundwater Table

In the "Norm of Base and Foundation Design ..." (1975), the critical limit of the groundwater table is 1.5 m. However, in practice, the capillary height of various soils is quite different.

Wang Xiyao (1980, 1982) and Kong (1983) have provided some field results on frost heave under various groundwater table conditions in seasonal frost regions. For coarse sand, fine sand, light loam, and heavy loam, if the distance from the water table to the frost front is more than 40, 60, 120, and 160 cm, respectively, frost heave might not occur or it will be very weak. The distance is very close to the critical capillary height of these soils.

After analyzing the observed data, Zhu (1983) expressed the relation between the frost-heave ratio η and groundwater table H_ω for heavy silt loam by the following formula:

$$\eta = 64.4 - 29.6\ H_\omega\ (N = 20,\ r = 0.846)$$

Analysis of experimental and field data by Chen et al. (1982) has given results that are close to the above.

FROST-HEAVE FORCE

In recent years, many investigators have given more attention to doing field observations on tangential, normal, and horizontal frost-heaving forces.

Tangential Frost-heave Force

Chen and Wu (1978) presented a summary of the main results of this subject before the mid-'70s. Based on observations on the Qinghai-Xizang Plateau, Tong (1982a,b) provided its distribution with depth. After analyzing observed data on the Qinghai-Xizang Plateau, Ding Jinkang (1983) indicated that, under the same conditions, if the tangential frost-heave force for concrete foundation is taken as unity, the force for steel, stone, and wood foundations would be 1.66, 1.29, and 1.06, respectively. Assuming that a foundation is made of concrete and the distribution of the heaving force along its depth is uniform, Ding suggested that for coarse-grained soil $\tau = 0.5\ kg/cm^2$, and for clayey soil $\tau = 0.8\ kg/cm^2$. After summarizing observations from the Qilian Mountains, Chen (1976) divided the development process of the heaving force into slow increase, intensive increase, and relaxation; for coarse-grained soil the last step usually does not occur. Xiao and He (1981) and Cui and Zhou (1983) did a lot of field observation work from 1976 to 1978. They found that the maximum total tangential heave force was almost always measured at the permafrost table, and for clayey loam the maximum force was usually at a water content of about 50% by weight.

The Low-Temperature Construction Institute and the Fourth Engineering Construction and Assembly Company of Heilongjiang Province have investigated the frost-heave force in seasonally frozen regions since 1972. They found that the distribution of tangential heaving force with depth is similar to that of frost heave. The idea of an average tangential heaving force with depth is not suitable for designing shallow foundations buried within a seasonal frost layer.

During freezing, if a foundation is allowed to move and is subjected to heating, the following factors must be considered when determining tangential heaving force: foundation depth, allowable displacement, foundation material, and the effect of heating.

Normal Frost-heave Force

It is necessary to check a structure for heave stability, if its foundation depth is not greater than the active layer thickness, especially for a hydraulic structure with a mat foundation. Based on a lot of experimental results, Chen et al. (1977, 1980a) have provided a formula for describing the relation between normal heaving force σ_i and foundation area F_i($\times 10^2\ cm^2$):

$$\sigma_i = \sigma_o + a/F_i + b/F_i^2$$

Guan et al. (1981a) have completed a field study with a maximum foundation area of 400 x 400 cm. Using the above formula, they got a normal heaving force $\sigma_o = 1.75\ kg/cm^2$. After collecting the results at home and abroad, Tong (1982b) gave a relation between the normal force and foundation area F as follows:

$$\sigma = \beta F^{-\alpha}$$

where β and α are constants.

Guo and Han (1982) analyzed their field observation data collected from 1973 to 1978. Provided that the distribution of normal frost-heave force with depth is very similar to that of frost heave, and the heaving force σ is proportional to the free heave of ground surface Δh, i.e.

$$\sigma = k\Delta h$$

where k is the coefficient of the frost-heave force. k was found to vary with soil types and freezing conditions. In 1982, Guo and Han presented the variation of the force at different foundation depths.

The Low-Temperature Construction Institute (1981) analyzed their observed results in relation to foundation diameters of 10, 20, 40, and 80 cm, and provided a relationship between free heave Δh_i of ground surface and total normal heaving force P_i as follows:

$$P_i = I \cdot \Delta h_i$$

They also showed that the total heaving force N depends on diameter D_i expressed as

$$N = 0.283 D_i^{6.631}$$

After assuming that the deformation of the mat foundation is zero, Li and Dai (1981) provided a formula for calculating the normal force using elastic theory. Xu et al. (1981b) suggested that if a little deformation S of a foundation is allowable, then the heaving force σ decreases exponentially, i.e.

$$S = ae^{-b}\ \sigma$$

Zhou (1983) suggested that the force σ decreases with the increase of foundation area F_i and expressed it as follows:

$$\sigma_i = \sigma_o + a/F_i^b$$

Horizontal Frost-heave Force

In 1966, Chen et al. measured the horizontal heaving force against a shaft wall and the side surface of a foundation and found the maximum to be $2.8\ kg/cm^2$ in the Qilian Mountains. Shen Zhongyan et al. tested horizontal forces against a retaining wall in 1972. From 1977 to 1979, Ding Jinkang et al. built an L-shaped model retaining wall for field observation. They measured the displacement and horizontal heaving forces. It was found that the maximum force occurred at the middle of the wall or a little below for fine-grained soil, and at the base of the wall for coarse soil. They suggested a calculation scheme for permafrost regions with a value of 1-1.5 and $0.5\text{-}1\ kg/cm^2$ for fine-grained soil and coarse soil respectively.

Guan et al. (1981b) observed a developing process of horizontal forces against a 2.8-m-high retaining wall. Because a frost crack occurred at the top surface between the wall and soil, the heaving force was indicated to be at the base of the wall with a maximum of $1.25\ kg/cm^2$. Under shallow ground-water table, Yu Rici et al. built a test retaining wall filled with wet soil with over 40% moisture by weight. The maximums of 2.1, 1.63, and $1.12\ kg/cm^2$ occurred at the top, middle, and the base respectively.

Countermeasures for Frost-heave

In China it is common to prevent frost damage to a foundation by using available types of foundations, replacing clayey soil with gravel, paving insulator protectors, and so forth. After testing and applying these for five cycles, Ding et al. (1982) provided a comprehensive treatment using used oil and a surface active preparation with a positive ion. With the help of the treatment, the tangential frost-heave force around the foundation was reduced to about 5% of the original value.

Water Migration and Ice Segregation

In the early '60's, because of highway construction in cold regions, Rau Hongyen et al. (1965) suggested a method for calculating water accumulation in the subgrade during freezing. Chen et al. have done some experiments on water migration in the seasonal thaw-freezing layer in the permafrost regions of the Qilian Mountains since 1965, and have also given more attention to the continued growth of ice lenses at the permafrost table.

Wang Ping et al. (1982) measured water redistribution in the seasonally active layer and the permafrost in the Qilian Mountains and the Qinghai-Xizang Plateau by using a neutron moisture meter made by Lanzhou University. The results show that the water migration occurred not only in the active layer, but also in permafrost. They also recommended that this method could be used for determining the freeze-thaw boundary. The Northwest Institute of the Academy of Railway Sciences measured the dynamic equilibrium of water in soil during freezing on the Qinghai-Xizang Plateau with Lanzhou University (1976). Their results show that the water content, not only under the thawing boundary in the active layer but also in permafrost, usually increases in warm seasons. The effect may extend more than 1.4 m into the permafrost.

Studies have proceeded for the following reasons: 1) because as the active layer above permafrost freezes, water migrates in the frost front and changes into ice; 2) there is a non-equilibrium migration of unfrozen water in soil; 3) there is the phenomenon of material expulsion in ice lens formation; and 4) because of the deposition of ground on the surface, massive ice grows year by year. Cheng (1983) provided a view on the formation of massive ground ice produced by rhythmic ice segregation.

Zhou Youwu, Chen et al. (1964) found a new pingo that occurred in the spring in permafrost in the Da Xian Mountains. With the help of the theory of moisture movement in saturated sand under pressure gradient, they suggested a successful way to protect a railway from frost damage. Chen et al. (1979b, 1980a,b) have finished more tests on the characteristics of saturated gravel during freezing, and have presented the relation between pore water pressure, an increase or decrease in the gradient of water pressure, and the frost penetration rate. At the same time, they suggested a satisfactory method of using a gravel bed to protect structures from damage.

In recent years, many investigators have studied the distribution of ice lenses with depth in the field. Since 1980, using the theory of water migration in unsaturated soil, Chen and others have started to determine hydraulic conductivity, diffusivity, and soil water characteristic curves at 2° and 25°C, and measured the redistribution of water potential with depth beneath the frost front. They point out that there is an intensive changing zone of soil water potential or water content with a range of 20-30 cm. Since 1981, Zhu et al. (1983) have established an observation station for determining water migration during freezing in the Zhang Yi seasonally frozen regions. Using the theory of water migration in unsaturated soil, Zhang and Zhu (1983) provided some formulas for calculating the water migration during freezing from deep and shallow ground-water tables.

Using thermo-equilibrium, moisture-equilibrium, frost-heave and state-changing equations, and analogous test criteria, Ding Dewen (1983) presented formulas for determining the critical depth and total thickness of ice lenses. With the help of the thermodynamic method, Gao and Ding (1980) provided a means of evaluating the amount of frost heave in the base course. Using the same method, Gao and Guo (1983) have derived a formula for evaluating frost heave under loading.

This review is only a general outline of the current research into frost-heave processes in soil in China. We will continue to study the mechanisms of frost heave and ice segregation in detail in the search for viable countermeasures for frost damage in cold regions.

REFERENCES*

Chen Xiaobai, 1980, Observations of frost heave and frost-heaving force in the Jiangcan District, Qilian Shan: Journal of Glaciology and Cryopedology, v. 2, no. 2, p. 36-40 (in Chinese).

Chen Xiaobai, 1981a, Frost heave of seasonally active layer in Muli District, Qilian Mountains: Memoirs of Lanzhou Institute of Glaciology and Cryopedology, Academia Sinica, no. 2, Beijing, Science Press, p. 72-81.

Chen Xiaobai, 1981b, the Characteristics of thaw consolidation of permafrost, Muli District, Qilian Mountains: Memoirs of Lanzhou Institute of Glaciology and Cryopedology, Academia Sinica, no. 2, Science Press.

Chen Xiaobai and Wu Ziwang, 1978, Experimental research on principal mechanical properties of freezing and frozen soil in China, in Proceedings of the Second Conference on Permafrost, Lanzhou.

Chen Xiaobai, Tong Boliang, Ding Jinkang, and Zhou Changging, 1979a, Thirty years of research in cryopedology and engineering practices in China; Journal of Glaciology and Cryopedology, v. 1, no. 2.

Chen Xiaobai, Jiang Ping, Wang Yaqing, Yu Chongyun, and Zhang Xuezhen, 1979b, On antiheave measure by replacing clayey soil with gravel: Kexue Tongbao (Bulletin of Science), no. 20.

Chen Xiaobai, Jiang Ping, and Wang Yaqing, 1980a, A discussion of calculation method for easily determining normal frost-heaving force: Hydro-Electric Technique, no. 2.

Chen Xiaobai, Jiang Ping, and Wang Yaqing, 1980b, Pore water pressure of saturated gravel during freezing: Journal of Glaciology and Cryopedology, v. 2, no. 4, p. 33-37.

Chen Xiaobai, Jiang Ping, and Wang Yaqing, 1980c, Some characteristics of water-saturated gravel during freezing and its applications, in Proceedings of the Second International Symposium on Ground Freezing, Trondheim: Trondheim, Norwegian Institute of Technology, p. 692-701.

Chen Xiaobai, Wang Yaqing, Liu Jingsou, and Yu Chongyun, 1982, Preliminary study about the effect of loading on the frost heave, in Proceedings of the Symposium on Glaciology and Cryopedology held by Geographical Society of China (Cryopedology), Lanzhou (1978): Beijing, Science Publishing House, p. 126-127.

Chen Xiaobai, Wang Yaqing, and Jiang Ping, 1983, Influence of frost penetration rate and surcharge stress on frost heaving, in Proceedings of the Second National Conference on Permafrost, Lanzhou (1981): Lanzhou, Gansu People's Publishing House, p. 223-228.

Cheng Guodong, 1983, The forming process of thick-layered ground ice: Scientia Sinica, Series B, v. 25, no. 3, p. 777-788.

Cui Chenghan and Zhou Kaijiong, 1983, Experimental research of frost heaving reaction, in Proceedings of the Second National Conference on Permafrost, Lanzhou (1981): Lanzhou, Gansu People's Publishing House, p. 257-264.

Ding Dewen, 1983, Calculation of frost depth and moisture condition in an open system, in Proceedings of the Second National Conference on Permafrost, Lanzhou (1981): Lanzhou, Gansu People's Publishing House, p. 233-239.

Ding Jingkang, 1981, Outdoor experimental research on tangential frost-heaving forces, in Proceedings of the Second National Conference on Permafrost, Lanzhou (1981): Lanzhou, Gansu People's Publishing House, p. 251-256.

Ding Jingkang, Zhang L., Ye C., Yan J., and Yu C., 1982, Physical and chemical methods for decreasing the force of tangential frost-heave in Proceedings of the National Symposium on Glaciology and Cryopedology held by the Geo-

*This list does not contain all references cited in text.

graphical Society of China (Cryopedology), Lanzhou (1978): Beijing, Science Publishing House, p. 204-206.

First Design Institute of Railways, 1975, Engineering Geology in Railway: People's Railway Press.

Gao Min and Ding Dewen, 1980, Thermodynamic method used in the study of frost heave in natural soil, in Proceedings of the Second International Symposium on Ground Freezing, Trondheim: Trondheim, Norwegian Institute of Technology, p. 670-679.

Gao Min and Ding Dewen, 1983, Research on the amount of free frost heave of soils by thermodynamic method, Professional Papers on Permafrost Studies of the Qinghai-Xizang Plateau, Lanzhou: Lanzhou, Gansu People's Publishing House, p. 169-174.

Gao Min and Guo Xinming, 1983, Thermodynamic method in frost-heave calculation under load, in Proceedings of the Second National Conference on Permafrost, Lanzhou (1981): Lanzhou, Gansu People's Publishing House.

Guan Fengnian, Zhou Chanqing, Ben Yongwei, and Chen Xiaobai, 1981a, Experimental research of normal frost-heaving stress under mat foundation in situ: Journal of Glaciology and Cryopedology, v. 3, no. 3, p. 63-68 (in Chinese)

Guan Fengnian, Zhou Chanqing, and Chen Xiaobai, 1981b, Experimental research of horizontal frost-heaving forces on hydraulic retaining wall, in Proceedings of the Second National Conference on Permafrost, Lanzhou (unpublished).

Guo Minzhu and Han Huanguang, 1982, Experimental studies of vertical heaving strength of clay, in Proceedings of the National Symposium on Glaciology and Cryopedology held by the Geographical Society of China (Cryopedology), Lanzhou (1978): Beijing, Science Publishing House, p. 120-122.

Kong Qingquan, 1983, Classification of frost-heave soil of concrete channel bed and its anti heave measure, in Proceedings of the Second National Conference on Permafrost, Lanzhou (1981): Lanzhou, Gansu People's Publishing House, p. 462-470.

Lanzhou Institute of Glaciology, Cryopedology, and Deserts, Academia Sinica, ed., 1975, Cryopedology: Beijing, Science Publishing House.

Li Anguo and Dai Jingbo, 1981, Determination of maximum seasonal thawing depth of the permafrost in Northeast China: Journal of Glaciology and Cryopedology, v. 3, no. 4, p. 39-43.

Liu Hongxu, 1981, On calculation of normal frost-heaving force: Journal of Glaciology and Cryopedology, v. 3, no. 2, p. 13-17.

Norm of Base and Foundation Design for Industrial and Civil Engineering Construction, 1975: Chinese Construction Press.

Railway Ministry, 1976, Technical Standards for Engineering Construction in Railway: Traffic Press.

Third Design Institute of Railways, 1979, Handbooks of Engineering Design for Railway: People's Railway Press.

Tong Changjiang, 1982a, Frost heave of the soils in the seasonal thawing layer at the Fenhuo Shan permafrost region in the Qinghai-Xizang Plateau, in Proceedings of the Symposium on Glaciology and Cryopedology held by the Geographical Society of China (Cryopedology), Lanzhou (1978): Beijing, Science Publishing House, p. 123-125.

Tong Changjiang, 1982b, Frost-heaving force of foundation, in Proceedings of the National Symposium on Glaciology and Cryopedology held by the Geographical Society of China (Cryopedology), Lanzhou (1978): Beijing, Science Publishing House, p. 113-119.

Tong Changjiang and Yu Chongyun, 1982, On the relationship between normal frost heaving force and areas of bearing plates: Journal of Glaciology and Cryopedology, v. 4, no. 4, p. 49-54.

Wang Ping, 1982, The application of neutron moisture gauge to permafrost research, in Proceedings of the National Symposium on Glaciology and Cryopedology held by the Geographical Society of China (Cryopedology), Lanzhou (1978): Beijing, Science Publishing House, p. 109-112.

Wang Xiyao, 1980, Experimental research of frost heave under different levels of ground water in various soils: Journal of Glaciology and Cryopedology, v. 2, no. 3.

Wang Xiyao, 1982, Frost heave and its distribution in different layers influenced by shallow ground water: Journal of Glaciology and Cryopedology, v. 4, no. 2, p. 55-62.

Wang Zhengqiu, 1980, Influence of grain size constituent on frost heave in fine sand: Journal of Glaciology and Cryopedology, v. 2, no. 3, p. 24-28.

Wang Zhengqiu, 1981, Classification of frost susceptibility for fine sand, in Proceedings of the Second National Conference on Permafrost, Lanzhou (1981): Lanzhou, Gansu People's Publishing House, p. 218-222.

Wu Ziwang, Zhang Jiayi, Wang Yaqing, and Shen Zhongyan, 1981, Laboratory study of frost-heave of soils: Memoirs of Lanzhou Institute of Glaciology and Cryopedology, Academia Sinica, no. 2: Beijing, Science Publishing House, p. 82-96.

Xiao Youming and He Chanjgeng, 1981, Research on tangential frost heaving force of piles at the experimental houses in Jin-tao, Da Xinjganling, in Proceedings of the Second National Conference on Permafrost, Lanzhou (abstract): Lanzhou, Gansu People's Publishing House, p. 40.

Xu Shaoxin, 1981a, Freezing properties of soil under different consolidated states, in Proceedings of the Second National Conference on Permafrost (unpublished), Lanzhou.

Xu Shaoxin, 1981b, Relationship between allowable frost-heaving deformation and frost-heaving forces, in Proceedings of the Second National Conference on Permafrost (abstract), Lanzhou.

Zhang Shixiang and Zhu Qiang, 1983, An approach to calculation of frost heaving, in Proceedings of the Second National Conference on Permafrost, Lanzhou (1981): Lanzhou, Gansu People's Publishing House, p. 206-212.

Zhou Chanqing, 1980, A brief introduction to achievements in the study of bases and foundation on frozen ground in China: Journal of Glaciology and Cryopedology, v. 2, no. 1, p. 1-5.

Zhou Youcai, 1983, Analysis of the relationship between frost heave and normal frost-heaving force in clay loam and at Yangjagang, in Proceedings of the Second National Conference on Permafrost, Lanzhou (1981): Lanzhou, Gansu People's Publishing House, p. 245-250.

Zhu Qiang, 1983, Frost-heave characteristics of the concrete canals in Gansu Province and the measure of replacing subsoil, in Proceedings of the Second National Conference on Permafrost, Lanzhou (1981): Lanzhou, Gansu People's Publishing House, p. 455-461.

THERMALLY INDUCED REGELATION: A QUALITATIVE DISCUSSION

R.D. Miller

Department of Agronomy, Cornell University
Ithaca, New York, USA

A panel discussion before a general audience seems to be an appropriate occasion to offer a qualitative explanation of one distinguishing feature of one of the proposed models of frost heave in soils.

What has come to be called the rigid ice model of frost heave exploits the assumption that as water in soil freezes, ice forms as a continuous, inherently rigid phase that includes not only ice lenses but pore ice between the lenses and, specifically in the case of what has been called secondary frost heave, ice in pores on the warm side of the warmest ice lens. While the assumptions of continuity and inherent rigidity do not appear to be controversial, they have neither been accommodated nor exploited in models based on explanations that begin with ideas about ways in which a temperature gradient might induce movement of a liquid phase relative to soil grains.

Essential to the rigid ice model is the conclusion that a temperature gradient tends to induce movements of ice and soil grains with respect to one another. Specifically, it is concluded that when isolated grains or clusters of grains are embedded in ice that cannot itself move, imposition of a temperature gradient induces the grains to migrate up that gradient by a process that has been called thermally induced regelation. If that conclusion is correct, then Newton's Third Law implies that if ice largely fills interstices between a mass of grains that cannot move, the ice will move down the temperature gradient if such movement is feasible, thereby causing frost heave.

Since the time when Taber (1930) concluded that there must be an adsorbed film of mobile water that holds a growing ice lens at a distance from adjacent soil grains, few have questioned that idea. Acceptance is equivalent to accepting the proposition that for some reason, liquid water is more strongly attracted by a grain surface than is ice; there must be an "adsorption force," real or virtual, that is felt more strongly by liquid water than by ice. This is evidently a body force that acts across distances that are appreciably larger than the lengths of chemical bonds that would immobilize the molecules.

It may be helpful to call to mind another "grain," namely earth, which has a greater affinity for liquid water than for an equal volume of ice at the same elevation. Thus, as the ocean freezes, the ice formed is "held at a distance by an adsorbed film of mobile water." In the case of the earth grain, the attribute of liquid water that gives rise to preferential adsorption is the mass concentration (density) of the liquid phase, which is greater than that of ice, and the mechanism of attraction is gravitation. In the case of grain surfaces, various attributes of the surfaces, liquid water, and ice can be suggested as operative factors in various mechanisms; that matter will be set aside for another occasion.

Among the consequences of preferential adsorption of liquid water, two are important to a qualitative understanding of the process of thermally induced regelation. The first is that if the intensity of the force field felt by liquid water diminishes rapidly with distance from a grain surface, and if the amount of liquid is not enough to "fill" the potential force field, then, in the case of a symmetrical grain, mechanical equilibrium cannot be achieved until the geographic centers of volumes of grain and water, respectively, coincide. This may be accomplished by movement of either grain or water with respect to the other, or both.

This first conclusion implies that given any configuration of an interface between water and some alien substance (ice, air, or another grain), if the alien substance is less strongly attracted than liquid, then whenever the alien substance displaces liquid water from the force field, the grain and the alien substance will appear to repel each other. This will be true if movement would result in more complete filling of "adsorption space" close to the grain surface by liquid water. In the case of earth and its ocean, this "disjoining force" is called the buoyant force experienced by any alien body that displaces water from the "adsorbed film."

The second consequence of the preferential adsorption of liquid water relative to ice is the conclusion that when some of the water surrounding a grain is induced to freeze, the equilibrium thickness of the residual unfrozen liquid film decreases as the temperature is lowered below 0°C. This conclusion can be justified by considering the influence of temperature on the conditions for equilibrium between water and ice at an interface, including the role of the specific surface free energy (surface tension) of the interface and the influence of the force field on the free energies of water and ice.

If the conclusion with respect to film thickness and temperature is accepted, and if there is a macroscopic thermal gradient in ice containing an occluded grain, one would expect the film to be thicker on the warm side of the grain than on the cold side. While such asymmetry is compatible with conditions for phase equilibrium at the film/ice interface, it is incompatible with the symmetry that represents the condition for mechanical equilibrium of a grain surrounded by adsorbed water. The anomaly is resolved by continuous movement of the grain relative to the rigid ice, the process that has been called thermally induced regelation. Movement will be in a direction that tends to restore symmetry even though asymmetry tends to be continuously maintained by melting and freezing taking place concurrently on opposite sides of the grain.

The rate at which thermally induced regelation takes place is governed in part by the rate at which the heat that is required to melt ice on the warm side and to refreeze it on the cold side diffuses toward and away from those interfaces, locally extracting and returning sensible heat to the mainstream of the flux of sensible heat induced by the overall thermal gradient. The process therefore depends, in part, on the thermal conductivities of the substances through which heat is diffusing and on the geometries of the pathways of diffusion. These will not change much as temperature changes. The rate of migration also depends, however, on viscous resistance to the flow of water around the grain, and this will be highly dependent on film thickness. Film thickness is inversely related to temperature. Thus, one would expect that for a given thermal gradient, migration velocity would increase as a grain travels up a temperature gradient toward the 0°C isotherm.

A preliminary attempt to detect thermally induced migration of isolated grains produced scanty results (Hoekstra and Miller, 1967). An apparatus was especially designed for prolonged tracking of individual grains and was highly successful (Römkens and Miller, 1973).

As a moving grain passes through the macroscopic ice/water interface at the 0°C surface, asymmetrical filling of adsorption space by water increases dramatically; migration should end with a spurt in velocity that drives the grain completely beyond the 0°C surface, where the force field can be fully occupied by liquid water.

Mentally, the process described can be reoriented. Imagine, for example, that grains have been poured into a vessel partially filled with water. When the grains have settled to the bottom, leaving a shallow pond of supernatant water on top, freezing can be initiated by cooling the surface of the pond. As the freezing front descends, ice begins to replace water within the adsorption force fields of the uppermost grains and, as an alien body, that ice begins to feel a repulsive force. (Opposite voids between the grains, the interface begins to develop convexity, an important matter that can be treated in terms of the surface tension of the ice/water interface. Although surface tension effects have an important role in the rigid ice model, they will not be discussed here.) Continued extraction of heat will result in continued frost heave in the mode that has been called "primary frost heave" in which the ice/water interface does not intrude into pores beyond the first layer of grains. Primary heave can only be sustained if the water pressure in the pores at the base of the ice lens does not fall too far below the pressure induced in the ice at the base of the lens by the weight of overburden. Primary heave is reminiscent of the final stage of expulsion of isolated grains moving through stationary ice. In primary heave, however, it is the grains that remain stationary while the ice moves. This is a paraphrase of the mechanism of frost heave envisioned by most scientists in Western Europe and North America from the days of Taber (1930) through Everett (1961) and beyond.

In due course it was realized that if ice intruded into pores below the uppermost tier of grains, heave would not necessarily cease, as had been generally supposed. Instead of being an "anchor" to be dragged through stationary pores below an ice lens by pressure-induced regelation, ice in this "frozen fringe" would be driven down the thermal gradient by thermally induced regelation. Thrust developed in this process can lead to the initiation of a new ice lens within the frozen fringe. The lens represents ice driven by thermally induced regelation in warmer pores, ice which then carries with it grains that migrate too slowly in the colder ice to remain in contact with their warmer neighbors. This is what one would expect if all ice were a continuous rigid body, constrained to move at uniform velocity whatever the local temperature may be. That velocity will be the same as the velocity of ice in the ice lens itself, i.e., equal to the observed rate of frost heave. This is a powerful conclusion that made it possible to formulate a mathematical model of secondary frost heave in terms of otherwise well-known macroscopic equations but limited, for the time being, to air-free, colloid-free, non-colloidal soils (Miller, 1978).

Given these equations, strategy and tactics for generating computer simulations of heave over time for various boundary conditions have been developed (O'Neill and Miller, 1982) and are still being improved. Such exercises require four input functions characteristic of the soil in question but so far, acquisition of reliable data for a priori simulations seems to be quite difficult. Hence, it may be some time before it is known whether the results of computer simulations will compare well with the results of corresponding experimental measurements of frost heave for a given soil. Qualitatively, results appear to be very encouraging (O'Neill and Miller, 1984).

Gilpin (1980) paraphrased the rigid ice model (Miller, 1978) for a case in which pore ice with frozen fringe is evidently taken to be rigid since the model uses a concept of lens initiation (Miller, 1977) that would be valid only if that ice did indeed sustain "nonisotropic stresses." Gilpin asserts, however, that this ice is stationary whereas lens ice is obviously moving, a velocity discontinuity presumably rationalized by his stated assumption that the ice does not support nonisotropic stresses at scales of time and space relevant to the mechanism of frost heaving. Despite this inconsistency, his model generates simulations of heaving that qualitatively resemble those generated by the rigid ice model in which movement of pore ice is driven by thermally induced regulation.

Finally, in their book on centrifuge modeling in the construction industry, Pokrovsky and Fyodorov (1968) pointed out that, in principle, simultaneous processes involving viscous flow and thermal diffusion could be modeled in a geotechnical centrifuge. Although they specifically mentioned processes involving freezing and thawing, they did not mention frost heaving. A similitude analysis (Miller, 1984) of processes of frost heaving based on the rigid ice model as well as a "capillary sink" mechanism of freezing-induced redistribution of water in moist soils suggested that, of various schemes for scale modeling of such processes, centrifuge modeling is the most promising whether or not soils are air-free, solute-free, or colloid-free. In North America, the first attempt to test this conclusion was sponsored by Northwest Alaska Pipeline, Inc. but it has not been carried to completion. A new attempt is now under way at the California Institute of Technology.

REFERENCES

Everett, D.H., 1961, The thermodynamics of frost damage to porous solids: Transactions of the Faraday Society, v. 57, p. 1541-1551.

Gilpin, R.R., 1980, A model for the prediction of ice lensing and frost heave in soils: Water Resources Research, v. 16, no. 5, p. 918-930.

Hoekstra, P. and Miller, R.D., 1967, On the mobility of water molecules in the transition layer between ice and a solid surface: Journal of Colloid and Interface Science, v. 25, p. 166-173.

Miller, R.D., 1977, Lens initiation in secondary heaving, in Proceedings of the International Symposium on Frost Action in Soils, v. 2: Sweden, University of Lulea, p. 68-74.

Miller, R.D., 1978, Frost heaving in non-colloidal soils, in Proceedings of the Third International Conference on Permafrost, Edmonton, v. 1: Ottawa, National Research Council of Canada, p. 707-713.

Miller, R.D., 1984, Manuscript in preparation.

O'Neill, K. and Miller, R.D., 1982, Numerical solutions for a rigid-ice model of secondary frost heave: Hanover, N.H., U.S. Army Cold Regions Research and Engineering Laboratory, CRREL Report 82-13.

O'Neill, K. and Miller, R.D., 1984, Manuscript in preparation.

Pokrovsky, G.I. and Fyodorov, I.S., 1968, Centrifuge model testing in the construction industry: Moscow, Stroiidzat. (Trans.) Building Research Establishment, United Kingdom, 1975.

Römkens, M.J.M., and Miller, R.D., 1973, Vibration of mineral particles in ice with a temperature gradient: Journal of Colloid and Interface Science, v. 42, p. 103-111.

Taber, S., 1930, The mechanics of frost heaving: Journal of Geology, v. 38, p. 303-317.

MOISTURE MIGRATION IN FROZEN SOILS

P.J. Williams

Geotechnical Science Laboratories, Carleton University
Ottawa, Canada

A decade or two ago it was widely assumed that frost heave occurred because of the accumulation of water at the boundary between freezing and unfrozen soil. The idea that water might move within the frozen material appeared a contradiction in terms, although vapour movement with accumulation of ice crystals in colder parts of unsaturated soil seemed probable. Hoekstra (1966) reported redistribution of moisture in a frozen unsaturated soil subject to a temperature gradient in excess of that to be expected due to vapour movement. Cary and Mayland (1972) observed moisture and salt redistribution in frozen unsaturated soils over several weeks and explained this as being due to mobile unfrozen water films. Jame and Norum (1974) found little moisture migration in frozen sand and ascribed this to lack of unfrozen water in such material. Mageau and Morgenstern (1980) reported laboratory observations of substantial moisture movement in saturated silts between 0° and -2°C. The experiments of Radd and Oertle (1973) with saturated soils indicated that frost-heave pressure was related to the temperature of growing ice lenses, and Goodrich and Penner's (1980) experiments confirmed this.

These results imply that water passes through saturated frozen soil to freeze on a growing lens. They also support the applicability of a freezing temperature/pressure relationship based on the Clausius-Clapeyron equation. The relationship takes account of the difference in pressure between the ice and water in freezing soil. The distribution of stresses in frozen ground is complex and it is evident that the pressure of the ice in lenses (which equals the heaving pressure) is other than the water pressure (which constitutes the freezing "suction"). The relationship is discussed by many authors (notably Edlefson and Anderson, 1943); Koopmans and Miller (1966) and Williams (1964) used it to explain unfrozen water in frozen soil. Usually the equation is given in the form

$$\frac{dT}{dP_w} = \frac{TV_w}{\ell}$$

where

dT = freezing temperature, 0°C
dP_w = pressure (suction of water)
T = absolute temperature
V_w = specific volume of water
ℓ = latent heat of fusion of water

In this form, the equation requires the assumption that the pressure of the ice is atmospheric. It tells us that the lower the pressure of the water in the frozen soil (i.e. the greater the suction), the lower the temperature. More widely applicable is the form:

$$dT = \frac{(dP_w V_w - dP_i V_i)\ T}{\ell}$$

where dP_i is the pressure of the ice. These relationships demonstrate the gradient of potential that occurs along a gradient of temperature in frozen soil. Thus, water could be expected to flow to colder frozen soil at a rate depending on the permeability of the frozen material. The equations also facilitate analysis of migrations of moisture due to applied stresses. Vyalov (1959) demonstrated the disappearance of ice lenses from a zone of high stress with enlargement of lenses outside the zone.

Burt and Williams (1976) devised a permeameter for isothermal frozen soils, applying pressure to cause water movement in and out of samples at temperatures down to -0.5°C and in one case to -5°C. Between 0°C and about -0.5°C, the permeability coefficients fall, at first rapidly, from those of unfrozen soils, to about 10^{-10} or 10^{-12} m s^{-1}. Below -0.5°C the coefficients decrease only slowly. It is generally agreed that adjacent to the boundary of the frozen soil (the "frost line") there will be a "frozen fringe" (Penner and Walton, 1978) - and perhaps a -0.1 to -0.3°C zone of high unfrozen water content interspersed with ice, which includes growing ice lenses fed with water from adjacent unfrozen soil. This is known as secondary heaving (Miller, 1972), and the term is often used generally to refer to heave within frozen soil. Values of hydraulic conductivity for the frozen fringe have been derived in various ways by several authors (Horiguchi and Miller, 1980; Loch, 1980) and are fairly alike, although generally not precise enough for accurate prediction of moisture transfer.

An intriguing question is that of movement of the ice within frozen soils. Burt and Williams (1976) found that complete, transverse layers of ice did not block flow as was expected and therefore presumably participated in it. Miller (1970) has demonstrated the movement of ice as a regelation process with accretion of ice to the one side of an ice layer and melting on the other. The gradients of potential should be the same, but whether both ice and water are moving is important for the permeability coefficient. It is often assumed that the permeability coefficient is closely dependent upon the amount of unfrozen water, although the experimental evidence for this is uncertain.

Little is known from laboratory experiments

about water migration and secondary heaving at temperatures of -1° or -2°C. Ratkje et al. (1982) calculate permeabilities down to -30°C, these being about two orders of magnitude lower than for -0.6°C.

The ultimate magnitude of heave and of heaving pressures (greater at lower temperatures) and the phenomenon of continuing heave without further frost penetration (Goto and Takahashi, 1982) are obviously related to water migration and ice accumulation within the frozen soil.

Heaving caused by the accumulation of ice within frozen soil occurs slowly, but it can have great practical significance where temperature gradients persist for long periods. Pipelines and other structures causing long-term freezing may experience displacements, and very high heaving pressure may develop. Over geological time periods there may be much relocation of ice in permafrost (Harlan, 1974). Prediction of rates is difficult. The conventional grain size criteria for predicting sensitivity to frost heave appear to be of little use for secondary frost heave.

Theoretical analyses of secondary heave rates have been made, for example, by Harlan (1973) and Miller et al. (1975). A procedure involving laboratory freezing tests and a calculation procedure based on the concept of water migration through the frozen fringe has been proposed by Konrad and Morgenstern (1981).

Field studies of ice accumulation (and thus water migration) in already frozen ground suggest, in comparison with laboratory studies, remarkably large values (Parmuzina, 1978). Mackay (1983) reported strain rates in already frozen ground equivalent to 7% yr^{-1} and with measurable effects even at -3°C. Many authors, especially in the Soviet Union, have reported increases in ice content near the surface of already-frozen active layers (Mackay, 1983).

Because of the variation of permeability of frozen soils with temperature, most moisture migration in frozen soil must involve change of moisture content. This implies deformation of the frozen soil to accommodate changes in the amount of ice. The resistant strength, or creep, properties of the frozen ground therefore appear important in all cases. They may well be a controlling factor in rates of moisture migration because of the effects of confining pressure (arising from the soil resistance), as indicated by the melting point-pressure relationship when taking ice pressure and water pressures into account as discussed for the Clausius-Clapeyron equation.

REFERENCES

Burt, T.P. and Williams, P.J., 1976, Hydraulic conductivity in frozen soils: Earth Surface Processes, v. 1, no. 4, p. 349-360.

Cary, J.W. and Mayland, H.F., 1972, Salt and water movement in unsaturated frozen soil: Soil Science Society of America Proceedings, v. 36, no. 4, p. 549-555.

Edlefsen, N.E. and Anderson, A.B.C., 1943, Thermodynamics of soil moisture: Hilgardia, v. 15, no. 2, p. 31-298.

Goto, S. and Takahashi, Y., 1982, Frost heave characteristics of soil under extremely low frost penetration rate, in Proceedings of the Third International Symposium on Ground Freezing, Hanover, New Hampshire, p. 261-268.

Harlan, R.L., 1973, Analysis of coupled heat-fluid transport in partially frozen soil: Water Resources Research, v. 9, no. 5, p. 1314-1323.

Harlan, R.L., 1974, Dynamics of water movement in permafrost: A review, in Seminar on Permafrost Hydrology: Ottawa, Canadian National Committee for the International Hydrological Decade, p. 69-77.

Hoekstra, P., 1966, Moisture movement in soils under temperature gradients with the cold-side temperature below freezing: Water Resources Research, v. 2, no. 2, p. 241-250.

Horiguchi, K. and Miller, R.D., 1980, Experimental studies with frozen soils in an 'ice sandwich' permeameter: Cold Regions Science and Technology, v. 3, no. 2,3, p. 177-183.

Jame, Y., and Norum, D.I., 1976, Heat and mass transfer in freezing unsaturated soil in a closed system, in Proceedings of the Second Conference on Soil-Water Problems in Cold Regions, Edmonton, Alberta.

Konrad, J.-M., and Morgenstern, N.R., 1981, The segregation potential of a freezing soil: Canadian Geotechnical Journal, v. 18, no. 4, p. 482-491.

Koopmans, R.W.R., and Miller, R.D., 1966, Soil freezing and soil water characteristic curves: Soil Science Society of America Proceedings, v. 30, p. 680-685.

Loch, J.P.G., 1980, Frost action in soils, state of the art, in Proceedings of the Second International Symposium on Ground Freezing, Trondheim: Trondheim, Norwegian Institute of Technology, p. 581-596.

Mackay, J.R., 1983, Downward water movement into frozen ground, western arctic coast, Canada: Canadian Journal of Earth Sciences, v. 20, no. 1, p. 120-134.

Mageau, D.W., and Morgenstern, N.R., 1980, Observations on moisture migration in frozen soils: Canadian Geotechnical Journal, v. 17, no. 1, p. 54-60.

Miller, R.D., 1970, Ice sandwich: Functional semipermeable membrane: Science, v. 169, p. 584-585.

Miller, R.D., 1972, Freezing and heaving of saturated and unsaturated soils: Highway Research Record, no. 393, p. 1-11.

Miller, R.D., Loch, J.P.G., and Bresler, E., 1975, Transport of water and heat in a frozen permeameter. Soil Science Society of America Proceedings, v. 39, no. 6, p. 1029-1036.

Parmuzina, O.Y., 1978, Cryogenic texture and some characteristics of ice formation in the active layer (Trans. from Problemy Kriolitologii, v. 7): Polar Geography and Geology, v. 4, no. 3, 1980, p. 131-152.

Penner, E., and Walton, T., 1978, Effects of temperature and pressure on frost heaving, in International Symposium on Ground Freezing (H.L. Jessberger, ed.): Bochum, Ruhr University, Germany, p. 65-72.

Penner, E., and Goodrich, L.E., 1980, Location of segregated ice in frost-susceptible soil, in Proceedings of the Second International Symposium on Ground Freezing, Trondheim: Trondheim, Norwegian Institute of Technology, v. 1, p. 626-639.

Radd, F.J. and Oertle, D.H., 1973, Experimental pressure studies of frost heave mechanisms and the growth-fusion behavior of ice, in Proceedings of the Second International Conference on Permafrost, Yakutsk: Washington, D.C., National Academy of Sciences, p. 377-384.

Ratkje, S.K., Yamamoto, H., Takashi, T., Ohrai, T. and Okamoto, J, 1982, The hydraulic conductivity of soils during frost heave: Frost i Jord, no. 24, p. 22-26.

Vyalov, S.S., 1959, Rheological properties and bearing capacity of frozen ground (in Russian): Moscow, Izdatelstvo Akademia Nauka SSSR. (Trans.) Hanover, N.H., U.S. Army Cold Regions Research and Engineering Laboratory, Translation 74.

Williams, P.J., 1964, Unfrozen water content of frozen soils and soil moisture suction: Geotechnique, v. 14, no. 3.

STATUS OF NUMERICAL MODELS FOR HEAT AND MASS TRANSFER IN FROST-SUSCEPTIBLE SOILS

R.L. Berg

Cold Regions Research and Engineering Laboratory
Hanover, New Hampshire, USA

Taber (1930) stated, "Freezing and thawing have caused much damage to road pavements in cold climates, but the processes involved have not been clearly understood, and therefore some of the preventive measures adopted have proved to be of little or no value." Although considerable time and money have been expended to mitigate and repair damage due to freezing and thawing in soils, relatively little money has been used to develop a better understanding of the complex physical, chemical, and mechanical processes that occur during freezing and thawing.

In the last decade, most research efforts have concerned the development of mathematical models coupling simultaneous fluxes of heat and water in freezing soil-water systems. The development of these models has emphasized the necessity of improving our knowledge of the thermal, hydraulic, chemical, and mechanical processes involved in freezing and thawing of soil-water systems.

In most models, the soil-water system is considered to be a macroscopic continuum. This allows the modellers to assume Darcian-type moisture flux and to apply equations for conservation of mass and energy. The general equations describing heat and moisture flux and other simultaneous processes occurring in soil-water systems have been available for some time, for example, Bird et al. (1960) and Luikov (1966).

The high-speed digital computer became widely available in the late 1960's and early 1970's, and Harlan (1972, 1973) and Guymon and Luthin (1974) developed the earliest models of simultaneous heat and moisture flow in freezing soil-water systems. In most models, water pressures and temperatures are the dependent variables that are computed at several points in time and space. Both finite difference methods and finite element methods have been used to solve the resulting differential equations in space. The finite difference method has been used in time.

In most mathematical models, the soil-water system is divided into three regions:

- a frozen zone,
- a freezing zone, and
- an unfrozen zone.

The concepts and mathematics applied to the three regions vary within a particular model and from one model to another. The remainder of this paper discusses numerical methods used in current models; a recent paper by O'Neill (1983a) presents a similar review of frost-heave models.

INITIAL AND BOUNDARY CONDITIONS

All of the models for computing frost heave in soils are one-dimensional in space, although mention of two-dimensional models is made in a few papers, e.g. Guymon et al. (in prep.) and Crory et al. (1982). The models require a set of initial conditions for both temperatures and pore-water pressures (or water content) with depth. Only a few of the models have been applied to nonhomogeneous (layered) soil-water systems (Guymon et al., in prep.; Crory et al., 1982). The models generally assume zero moisture flux through the upper boundary. A time-dependent heat flux or time-dependent temperatures may be applied to the upper boundary in most models, but the temperature condition is usually applied. A time-dependent temperature condition has also normally been applied to the lower boundary, and time-dependent pore-water pressures or moisture fluxes are applied at the lower boundary as well.

THERMAL AND HYDRAULIC PROPERTIES

The thermal and hydraulic parameters are entered into the models in a variety of ways and are used in a variety of algorithms. For example, constant values for the thermal conductivity of frozen and unfrozen soil-water-ice mixtures are used in some models (Outcalt, 1976), others use the DeVries method (Berg et al., 1980; Guymon and Luthin, 1974), and others use different empirical relationships (Fukuda, 1982). Table 1 contains

TABLE 1 Algorithms used in coupled heat and mass transfer models.

Algorithm	Influencing parameters
Hydraulic conductivity	Moisture content Temperature Soil density
Moisture characteristic	Soil density
Thermal conductivity	Moisture content Soil density Temperature
Heat capacity	Moisture content Soil density Temperature
Latent heat of fusion	Soil density Moisture content Temperature
Frost heave	Moisture content Overburden Temperature Temperature gradients Pressure gradients

several of the algorithms used in the models. The table also lists the parameters that influence each algorithm. Generally each model uses a different equation, or equations, in the algorithms.

Frost-heave algorithms vary from very simple to extremely complex. O'Neill (1983a) provides a more detailed review. Results from the frost-heave algorithm "drive" the entire solution. The simplest frost-heave algorithms use pore-water pressure, predetermined or computed in one of several ways, to cause water movement, and frost heave occurs if the soil becomes more than approximately 90% saturated (Berg et al., 1980; Outcalt 1976; Taylor and Luthin, 1976, 1978). In the simple models, the mesh may or may not be deformed. In the complex models (e.g. Hopke, 1980, or O'Neill and Miller, 1982), an extremely fine mesh is needed to represent the phenomena adequately, and a mesh that moves with the freezing zone has been used (Lynch and O'Neill, 1981).

FROZEN ZONE

The exact boundaries of the frozen zone depend upon the type of soil, the stress conditions, the thermal regime, and the chemical composition of the soil water. None of the models includes movement of chemicals or changes in the chemical composition of the soil water with time or position.

In models where the nodal spacing is on the order of centimeters (e.g. Berg et al., 1980; Outcalt, 1976; Dudeck and Holden, 1979), the boundaries of the frozen zone are defined by temperatures. The temperatures may be 0°C or an estimated or measured freezing point depression of the soil water. The primary reason for this simplification is that the nodal grid is so coarse that the space between two nodes may actually contain an unfrozen zone, the entire freezing zone, and part of the frozen zone.

In models where the nodal spacing is on the order of a millimeter or less, the warm-side boundary of the frozen zone is assumed to be at the base (warm side) of the growing ice lens. Hopke (1980) and O'Neill and Miller (1982) use this method.

The developers of most models assume no moisture flux in the frozen zone, but others relate the hydraulic conductivity or diffusivity to the subfreezing temperature.

UNFROZEN ZONE

Of the three zones, the boundaries of the unfrozen zone are the most readily defined. Unfrozen soil and water are generally assumed to occur at all temperatures greater than the freezing point depression of the soil water. The freezing point depression is normally estimated to be from a few hundredths to a few tenths of a degree below the freezing point of bulk water.

Dempsey's (1978) model includes moisture flux in the vapor phase, but none of the others incorporates vapor flux. The hydraulic conductivity is generally related to the moisture content or pore-water pressure of the soil and is allowed to vary with time and position as pore-water pressures change.

The thermal conductivity may be considered constant in time and space, or it may vary depending on the moisture content of the soil.

None of the models allows consolidation of the unfrozen soil as the frozen mass reacts against it or as the soil dries due to moisture flux to the freezing zone.

The effects of hysteresis are neglected. Nearly all of the models have been used to simulate only a continuously freezing situation, and freezing a soil is assumed to be analogous to drying it. Therefore, the effects of hysteresis have not been important. The models that have expelled water during the early stages of freezing, i.e. Hopke (1980), O'Neill and Miller (1982), and Gilpin (1980), all assumed saturated conditions, and hysteresis is not important under conditions of saturated flow.

FREEZING ZONE

The freezing zone, or frozen fringe, is the region between the unfrozen zone and the frozen zone. O'Neill (1983b) defined it as "the zone over which the volumetric ice content increases from zero at the freezing front to 100% (neglecting trapped air) at the location of the warmest lens." This definition is adequate for most situations, but implies that a freezing zone does not exist unless an ice lens is present. The width, or thickness, of the freezing zone is dependent upon the unfrozen moisture content as a function of temperature for the soil and a variety of other factors. Large overburden pressures, or large loads, and small temperature gradients both tend to increase the width of the freezing zone.

The most complex models, Hopke (1980), Gilpin (1980), and O'Neill and Miller (1982), use the equations proposed by Miller (1978) to model processes in the freezing zone. Very close nodal spacing is required: O'Neill and Miller (1982) used 0.025 mm to represent the complicated relationships adequately.

In the less sophisticated models, where the nodal spacing is on the order of centimeters, processes in the freezing zone are included in the frost-heave algorithm or the latent heat of fusion algorithm. Generally, the hydraulic conductivity of the soils at subfreezing temperatures must be somewhat arbitrarily divided by an "impedance factor" ranging from 1 to 1000 to obtain reasonable agreement between computed and observed frost heave from laboratory or field tests. No systematic method of estimating the impedance factor has been developed.

CONCLUSION

The emphasis of this paper has been the discussion of numerical models that use coupled equations of simultaneous heat and moisture flow. At least two other methods have recently been developed to estimate frost heave in soils (Crory et al., 1982; Konrad and Morgenstern, 1980). In both of these methods a heat conduction model is used to predict the position of the 0°C isotherm over time. A factor, which is dependent on the stress conditions at the 0°C isotherm and is developed from laboratory tests on the soil, is used to compute the amount of frost heave over time. Both of these models eliminate the moisture flux equations from the numerical model and use an empirical parameter to estimate frost heave. These approaches have some advantages over the more complicated coupled heat and mass flow models.

Many tests must be conducted and additional theories must be developed, tested, and proved before a particular frost heave model is widely accepted. O'Neill (1983) states that "no model is completely successful in a strict test of frost heave prediction, or in the unequivocal verification of any physics which has been assumed," and Guymon et al. (1983) note, "It is reasonable to state that agreement on the complexity of frost heave models or the formulation of algorithms representing processes in the freezing zone is not widespread."

REFERENCES

Aitchison, G.D., ed., 1965, Moisture equilibria and moisture changes in soils beneath covered areas: A symposium in print: Sydney, Australia, Butterworth and Company Ltd.

Anderson, D.M., and Tice, A.R., 1972, Predicting unfrozen water contents in frozen soils from surface area measurements: Highway Research Board Bulletin 393, p. 12-18.

Anderson, D.M., Tice, A.R., and McKim, H.L, 1973, The unfrozen water and the apparent specific heat capacity of frozen soils, in Proceedings of the Second International Conference on Permafrost, Yakutsk: Washington, D.C., National Academy of Sciences, p. 289-294.

Bird, R.B., Stewart, W.E., and Lightfoot, E.N., 1960, Transport phenomena: New York, John Wiley and Sons.

Cary, J.W., and Mayland, H.F., 1972, Salt and water movement in unsaturated frozen soil: Soil Science Society of America Proceedings, v. 36, no. 4, p. 549-555.

Crory, F.E., Isaacs, R.M., Penner, E., Sanger, F.J., and Shook, J.F., 1982, Designing for frost heave conditions, preprint of a paper presented at the ASCE Spring Convention, Las Vegas, Nevada.

Dempsey, B.J., 1978, A mathematical model for predicting coupled heat and water movement in unsaturated soil: International Journal for Numerical and Analytical Methods in Geomechanics, v. 2, p. 19-34.

DeVries, D.A., 1952, The thermal conductivity of soil: Waneningen, Meded. Landbouwhogesch, v. 32, p. 1-72.

DeVries, D.A., 1966, Thermal properties of soils, in Physics of Plant Environment (W.E. Van Wijk, ed.): Amsterdam, North Holland Publishing Co., p. 210-235.

Dirksen, C., 1964, Water movement and frost heaving in unsaturated soil without an external source of water: Cornell University, Ph.D. dissertation (unpublished).

Dirksen, C., and Miller, R.D., 1966, Closed-system freezing of unsaturated soil: Soil Science Society of America Proceedings, v. 30, p. 168-173.

Dudek, S.J.-M., and Holden, J.T., 1979, Theoretical model for frost heave, in Numerical Methods in Thermal Problems (Lewis and Morgan, eds.): Swansea, United Kingdom, Pineridge Press, p. 216-229.

Edlefson, N.E., and Anderson, A.B.C., 1943, Thermodynamics of soil moisture: Hilgardia, v. 15, no. 2, p. 31-298.

Everett, D.H., 1961, The thermodynamics of frost damage to porous solids: Transactions of the Faraday Society, v. 57, p. 1541-1551.

Fuchs, M., Campbell, G.S., and Papendick, R.I., 1978, An analysis of sensible and latent heat flow in a partially frozen unsaturated soil: Soil Science Society of America Proceedings, v. 42, p. 379-385.

Fukuda, M., 1982, Experimental studies of coupled heat and moisture transfer in soils during freezing: Contributions from the Institute of Low-Temperature Science, Series A, no. 31, Hokkaido University, Sapporo, Japan, p. 35-91.

Gilpin, R.R., 1980, A model for the prediction of ice lensing and frost heave in soils: Water Resources Research, v. 16, no. 5, p. 918-930.

Gilpin, R.R., 1979, A model of the "liquid-like" layer between ice and a substrate with applications to wire regelation and particle migration: Journal of Colloid and Interface Science, v. 68, no. 2, p. 235-251.

Gray, W.G., and O'Neill, K., 1976, On the general equations for flow in porous media and their reduction to Darcy's Law: Water Resources Research, v. 12, p. 148-154.

Groenevelt, P.H., and Kay, B.D., 1974, On the interaction of water and heat transport in frozen and unfrozen soils: II. The liquid phase: Soil Science Society of America Proceedings, v. 38, no. 2, p. 400-404.

Groenevelt, P.H., and Kay, B.D., 1980, Pressure distribution and effective stress in frozen soils, in Proceedings of the Second International Symposium on Ground Freezing, Trondheim: Trondheim, Norwegian Institute of Technology, v. 1, p. 597-610.

Groenevelt, P.H., and Kay, B.D., 1977, Water and ice potentials in frozen soils: Water Resources Research, v. 13, no. 2, p. 445-449.

Guymon, G.L, Harr, M.E., Berg, R.L., and Hromadka II, T.V., 1981, A probabilistic-deterministic analysis of one-dimensional ice segregation in a freezing soil column: Cold Regions Science and Technology, v. 5, p. 127-140.

Guymon, G.L., 1976, Finite element analog of frost heave: Hanover, N.H., U.S. Army Cold Regions Research and Engineering Laboratory, Final report on CRREL Contract DACA 89-76-G-055 (unpublished).

Guymon, G.L., Berg, R.L., Johnson, T.L., and Hromadka II, T.V. (in prep.) Mathematical model of frost heave in pavements: Hanover, N.H., U.S. Army Cold Regions Research and Engineering Laboratory, CRREL Report.

Guymon, G.L., and Luthin, J.L., 1974, A coupled heat and moisture transport model for arctic soils: Water Resources Research, v. 10, no. 5, p. 995-1001.

Guymon, G.L., Berg, R.L., and Hromadka II, T.V., 1983, Field tests of a frost-heave model, in Proceedings of the Fourth International Conference on Permafrost, Fairbanks: Washington, D.C., National Academy of Sciences, p. 409-414.

Harlan, R.L., 1972, Ground conditioning and the groundwater response to winter conditions, in Proceedings of the International Symposia on the Role of Snow and Ice in Hydrology, Symposium on Properties and Processes, Banff, Alberta.

Harlan, R.L., 1973, Analysis of coupled heat-fluid transport in partially frozen soil: Water Resources Research, v. 9, no. 5, p. 1314-1323.

Hesstvedt, E., 1964, The interfacial energy, ice/water: Oslo, Norwegian Geotechnical Institute, Publication No. 56, p. 1-4.

Hoekstra, P., 1966, Moisture movement in soils under temperature gradients with the cold-side temperature below freezing: Water Resources Research, v. 2, p. 241-250.

Hoekstra, P., Chamberlain, E., and Frate, A., 1965, Frost-heaving pressures: Hanover, N.H., U.S. Army Cold Regions Research and Engineering Laboratory, Research Report 176.

Hopke, S., 1980, A model for frost heave including overburden: Cold Regions Science and Technology, v. 3, no. 2/3, p. 111-127.

Horiguchi, K., and Miller, R.D., 1980, Experimental studies with frozen soils in an 'ice sandwich' permeameter: Cold Regions Science and Technology, v. 3, no. 2/3, p. 177-183.

International Symposium on Frost Action in Soils, 1977, 16-18 February, Sweden, University of Lulea, v. 1 and 2.

Jame, Y., 1978, Heat and mass transfer in freezing unsaturated soil: Saskatoon, University of Saskatchewan, Ph.D. dissertation (unpublished).

Jame, Y.W., and Norum, D.I., 1976, Heat and mass transfer in freezing unsaturated soil in a closed system, in Proceedings of the Second Conference on Soil-Water Problems in Cold Regions, Edmonton, Alberta.

Jame, Y., and Norum, D.I., 1980, Heat and mass transfer in a freezing unsaturated porous medium: Water Resources Research, v. 16, p. 811-819.

Johansen, O., 1977, Frost penetration and ice accumulation in soils, in Proceedings of the International Symposium on Frost Action in Soils: Sweden, University of Lulea, v. 1, p. 102-111.

Kay, B.D., and Groenevelt, P.H., 1974, On the interaction of water and heat transport in frozen and unfrozen soils: I. Basic theory: the vapor phase: Soil Science Society of America Proceedings, v. 38, p. 395-400.

Kersten, M.S., 1949, Laboratory research for the determination of the thermal properties of soils: Hanover, N.H., U.S. Army Cold Regions Research and Engineering Laboratory, ACFEL Technical Report 23, AD 712516.

Keune, R., and Hoekstra, P., 1967, Calculating the amount of unfrozen water in frozen ground from moisture characteristic curves: Hanover, N.H., U.S. Army Cold Regions Research and Engineering Laboratory, Special Report 114.

Kinosita, S., 1975, Soil-water movement and heat flux in freezing ground, in Proceedings of the First Conference on Soil-Water Problems in Cold Regions, Calgary, Alberta, p. 33-41.

Konrad, J.-M., and Morgenstern, N.R., 1980, A mechanistic theory of ice lens formation in fine-grained soils: Canadian Geotechnical Journal, v. 17, no. 4, p. 473-486.

Konrad, J.-M., and Morgenstern, N.R., 1981, The segregation potential of a freezing soil: Canadian Geotechnical Journal, v. 18, p. 482-491.

Loch, J.P.G., 1978, Thermodynamic equilibrium between ice and water in porous media: Soil Science, v. 126, no. 2, p. 77-80.

Loch, J.P.G., 1980, Frost action in soils, state-of-the-art, in Proceedings of the Second International Symposium on Ground Freezing: Engineering Geology, v. 18, p. 213-224.

Loch, J.P.G., and Kay, B.D., 1978, Water redistribution in partially frozen saturated silt under several temperature gradients and overburden loads: Soil Science Society of America Proceedings, v. 42, no. 3, p. 400-406.

Low, P.F., Hoekstra, P., and Anderson, D.M., 1967, Some thermodynamic relationships for soils at or below the freezing point. Part 2: Effects of temperature and pressure on unfrozen soil water: Hanover, N.H., U.S. Army Cold Regions Research and Engineering Laboratory, Research Report 222.

Luikov, A.V., 1966, Heat and mass transfer in capillary-porous bodies (Trans. by W.B. Harrison): London, Pergamon Press.

Lynch, D.R., and O'Neill, K., 1981, Continuously deforming finite elements for the solution of parabolic problems, with and without phase change: International Journal for Numerical Methods in Engineering, v. 17, p. 81-96.

McGaw, R.W., and Tice, A.R., 1976, A simple procedure to calculate the volume of water remaining unfrozen in a freezing soil, in Proceedings of the Second Conference on Soil-Water Problems in Cold Regions, Edmonton, Alberta, p. 114-122.

McRoberts, E.C., and Nixon, J.F., 1975, Some geotechnical observations on the role of surcharge pressure in soil freezing, in Proceedings of the First Conference on Soil-Water Problems in Cold Regions, Calgary, Alberta, p. 42-57.

Miller, R.D., 1973, Soil freezing in relation to pore water pressure and temperature, in Proceedings of the Second International Conference on Permafrost, Yakutsk: Washington, D.C., National Academy of Sciences, p. 344-352.

Miller, R.D., 1978, Frost heaving in non-colloidal soils, in Proceedings of the Third International Conference on Permafrost, Edmonton:

Ottawa, National Research Council of Canada, v. 1, p. 708-713.

Miller, R.D., 1980, Freezing phenomena in soils, Chapter 11, in Applications of Soil Physics (D. Hillel, ed.): New York, Academic Press, v. 1, p. 254-299.

Miller, R.D., and Koslow, E.E., 1980, Computation of rate of heave versus load under quasi-steady state: Cold Regions Science and Technology, v. 3, no. 2/3, p. 243-251.

O'Neill, K., and Miller, R.D., 1982, Numerical solutions for a rigid-ice model of secondary frost heave: Hanover, N.H., U.S Army Cold Regions Research and Engineering Laboratory, CRREL Report 82-13.

O'Neill, K., 1983a, The physics of mathematical frost heave models: A review: Cold Regions Science and Technology, v. 6, no. 3, p. 275-291.

O'Neill, K., 1983b, Physical and numerical study of unsaturated heat and moisture flow: Significance of convection: CRREL Report, in review.

Outcalt, S.I., 1976, A numerical model of ice lensing in freezing soils, in Proceedings of the Second Conference on Soil-Water Problems in Cold Regions, Edmonton, Alberta.

Outcalt, S.I., 1980, A step function model of ice segregation, in Proceedings of the Second International Symposium on Ground Freezing, Trondheim: Trondheim, Norwegian Institute of Technology, p. 515-524.

Penner, E., 1957, Soil moisture tension and ice segregation: Highway Research Board Bulletin 168, p. 50-64.

Penner, E., 1959, The mechanism of frost heaving in soils: Highway Research Board Bulletin 225, p. 1-22.

Penner, E., 1979, Effects of temperature and pressure on frost heaving: Engineering Geology, v. 16, p. 729-39.

Penner, E., and Ueda, T., 1977, The dependence of frost heaving on load application -- preliminary results, in Proceedings of the International Symposium on Frost Action in Soils: Sweden, University of Lulea, v. 1, p. 92-101.

Philip, J.R., 1957, Numerical solutions of equations of the diffusion type with diffusivity concentration dependent. II: Australian Journal of Physics, v. 10, p. 29-42.

Philip, J.R., 1980, Thermal field during regelation: Cold Regions Science and Technology, v. 3, no. 2,3, p. 193-203.

Philip, J.R., and deVries, D.A., 1957, Moisture movement in porous materials under temperature gradients: Transactions of the American Geophysical Union, v. 38, p. 222-228.

Sheppard, M.I., Kay, B.D., and Loch, J.P.G., 1978, Development and testing of a computer model for heat and mass flow in freezing soil, in Proceedings of the Third International Conference on Permafrost, Edmonton, v. 1: Ottawa, National Research Council of Canada, p. 75-81.

Taber, S., 1930, Freezing and thawing of soils as factors in the destruction of road pavements: Public Roads, v. 11, no. 6, p. 113-132.

Takagi, S., 1978, Segregation freezing as the cause of suction force for ice lens formation: Hanover, N.H., U.S. Army Cold Regions Research and Engineering Laboratory, CRREL Report 78-6.

Takagi, S., 1980, The adsorption force theory of frost heaving. Cold Regions Science and Technology, v. 3, no. 1, p. 57-81.

Taylor, G.S., and Luthin, J.N., 1976, Numeric results of coupled heat-mass flow during freezing and thawing, in Proceedings of the Second Conference on Soil-Water Problems in Cold Regions, Edmonton, p. 155-172.

Taylor, G.S., and Luthin, J.N., 1978, A model for coupled heat and moisture transfer during soil freezing: Canadian Geotechnical Journal, v. 15, p. 548-555.

Tice, A.R., Burrous, C.M., and Anderson, D.M., 1978, Determination of unfrozen water in frozen soil by pulsed nuclear magnetic resonance, in Proceedings of the Third International Conference on Permafrost, Edmonton: v. 1: Ottawa, National Research Council of Canada, p. 149-155.

Vignes, M., and Dijkema, K.N., 1974, A model for the freezing of water in a dispersed medium: Journal of Colloidal and Interface Science, v. 49, no. 2, p. 165-172.

Vignes-Adler, M., 1977, On the origin of the water aspiration in a freezing medium: Journal of Colloidal and Interface Science, v. 60, p. 162-171.

Williams, P.J., 1967, Properties and Behavior of Freezing Soils: Oslo, Norwegian Geotechnical Institute.

Subsea Permafrost

PANEL MEMBERS

David M. Hopkins (Co-Chairman), U.S. Geological Survey, Menlo Park, California, USA

Paul V. Sellmann (Co-Chairman), Cold Regions Research and Engineering Laboratory, Hanover, N.H., U.S.A.

Steve M. Blasco, Geological Survey of Canada, Bedford Institute of Oceanography, Dartmouth, Canada

Felix E. Are, formerly of the Permafrost Institute, Siberian Branch, Academy of Sciences, Yakutsk, USSR

James A.M. Hunter, Geological Survey of Canada, Ottawa, Canada

Hans O. Jahns, Exxon Production Research Company, Houston, Texas, USA

Donald W. Hayley, EBA Engineering Consultants, Ltd., Edmonton, Canada

INTRODUCTION

D.M. Hopkins and P.V. Sellmann

Permafrost, in the sense of earth materials at temperatures below 0°C, is widespread beneath the seabed in Arctic continental shelf areas. Nevertheless, in some areas the presence of interstitial brines and temperatures just below 0°C causes many seabed materials to be unfrozen or only partially frozen. Thus, the extent of areas underlain by noticeably ice-bonded material is much more restricted than the extent of ice-free or partially bonded subsea permafrost.

The presence of subsea permafrost on the Siberian arctic shelf was established as early as 1953 (Are, 1976). In North America the first observations and modeling were based on studies made near Barrow, Alaska (Brewer, 1958; Lachenbruch et al., 1962; Lachenbruch, 1957) that suggested that ice-bearing permafrost would be found only near shore. This, combined with data from holes onshore but near the beach at Cape Thompson, indicated that permafrost would be thin to absent over much of the remainder of the Chukchi Sea, encouraging a belief that permafrost was generally lacking beneath the Alaskan shelves. The first evidence of ice-bonded material beneath the Canadian Beaufort Sea probably came from the offshore drilling program conducted by the Arctic Petroleum Operators Association (APOA) in the 1960's (Golden et al., 1970; Mackay, 1972). The existence of subsea permafrost beneath westernmost Beaufort Sea near Point Barrow was first established in boreholes drilled by Lewellen (1973, 1975).

Petroleum exploration in the Arctic soon precipitated requirements for a better understanding of subsea permafrost. During the years 1975-1982, an ambitious governmental program of geological, geophysical, geochemical, and geotechnical studies was mounted in advance of offshore leasing on the Alaskan segment of the Beaufort Sea shelf as part of the Outer Continental Shelf Environmental Assessment Program (OCSEAP) (U.S. National Oceanographic and Atmospheric Administration/U.S. Bureau of Land Management) of the Minerals Management Service, then the Conservation Division of the U.S. Geological Survey. Aspects of some of these studies are summarized by Barnes and Hopkins (1978), Barnes (1981), and Miller and Bruggers (1980). Since the lease sales, much more detailed, more geotechnically oriented borehole programs and modelling studies have been conducted by various petroleum companies and their consultants. In the different regulatory atmosphere of the Canadian segment of the Beaufort Sea, offshore permafrost studies have been conducted jointly by the Canadian government agencies and by the petroleum companies.

The contributions of our panelists reflect these regional differences. Sellmann and Hopkins, geologists, report ideas and concepts developed largely during the OCSEAP studies of the Alaskan shelf. Blasco, a soils engineer, and Hunter, a geophysicist, in their separate contributions describe results that are part of the outcome of government-industry cooperation on the Canadian segment of the Beaufort Sea shelf. Are's brief review of Soviet studies, of course, comes out of still another organizational context. Hayley is a geotechnical consultant well acquainted with both the Alaskan and Canadian segments of the Beaufort Sea shelf. Jahns, a petroleum engineer, shares with us some of the results of his company's extensive in-house field research on the Alaskan segment of the Beaufort Sea shelf.

An exciting aspect of permafrost studies in this environment is that current investigations are still producing surprising results. This suggests that we are still in the process of understanding the distribution, properties, and processes of permafrost in this geological setting and that opportunities remain for stimulating scientific and engineering studies. Specific recommendations for future investigations are in the individual panelists' sections.

Preparation of this panel report was supported by the Department of Interior's Mineral Management Service, through an interagency agreement with the National Oceanic and Atmospheric Administration.

REFERENCES

Are, F.E., 1976, The Subsea Cryolithozone of the Arctic Ocean, Regional and Thermophysical Studies of Perennially Frozen Rocks in Siberia: USSR Academy of Sciences, Siberian Branch, Yakutsk Book Publishing House, p. 3-26. (Trans.) Hanover, N.H., U.S. Army Cold Regions Research and Engineering Laboratory, Draft Translation 686, 1978.

Barnes, P.W. and Hopkins, D.M., eds., 1978, Geological sciences, in Outer Continental Shelf Environmental Assessment, Interim Synthesis Report; Beaufort/Chukchi: Boulder, Colo., U.S. National Oceanic and Atmospheric Administration and U.S. Bureau of Land Management, p. 101-133.

Barnes, P.W., ed., 1981, Physical characteristics of the Sale 71 area, in Outer Continental Shelf Environmental Assessment Program, Beaufort Sea (Sale 71) Synthesis Report, p. 81-110.

Brewer, M.C., 1958, Some results of geothermal investigations of permafrost in northern Alaska: Transactions of the American Geophysical Union, v. 39, no. 1, p. 19-26.

Golden, Brawner and Associates, 1970, Bottom sampling program, southern Beaufort Sea. Arctic Petroleum Operators Association, Report No. 3.

Lachenbruch, A.H., 1957, Thermal effects of the ocean on permafrost: Bulletin of the Geological Society of America, no. 68, p. 1515-1530.

Lachenbruch, A.H., Brewer, M.C., Greene, G.W., and Marshall, B.V., 1962, Temperature in permafrost, in Temperature -- Its Measurement and Control in Science and Industry: New York, Reinhold Publishing Co., v. 3, p. 791-803.

Lewellen, R.I., 1973, The occurrence and characteristics of nearshore permafrost, northern Alaska, in Permafrost -- the North American Contribution to the Second International Conference, Yakutsk: Washington, D.C., National Academy of Sciences, p. 131-136.

Lewellen, R.I., 1975, The occurrence and characteristics of subsea permafrost, northern Alaska: Arctic Institute of North America, Progress Report (AINA-ONR-454), p. 131-135.

Mackay, J.R., 1972, Offshore permafrost and ground ice, southern Beaufort Sea: Canadian Journal of Earth Sciences, v. 9, p. 1550-1561.

Miller, D.L., and Bruggers, D.E., 1980, Soil and permafrost conditions in the Alaskan Beaufort Sea, in Proceedings of the 12th Annual Offshore Technology Conference, Houston, p. 325-338.

SUBSEA PERMAFROST DISTRIBUTION ON THE ALASKAN SHELF

P.V. Sellmann
Cold Regions Research and Engineering Laboratory
Hanover, New Hampshire, USA

D.M. Hopkins
U.S. Geological Survey
Menlo Park, California, USA

The subsea permafrost environment is a unique geological setting that has no real analogue on land. Recent investigations in Soviet, Canadian, and American coastal waters have begun to document its widespread distribution and the great variation in the occurrence and properties of frozen ground beneath the seabed.

Permafrost on land is relatively stable, natural changes and modifications being associated either with slow climatic change or with very local phenomena. Offshore, the permafrost is more dynamic, particularly in the coastal zone where active coastal retreat is common. In the Beaufort Sea (Fig. 1), inundation can cause a 10°C increase in the mean annual temperature of the coastal permafrost (Lachenbruch et al., 1982). This warming can influence a block of coastal subsea permafrost up to 600 m thick that may extend up to 30 to 40 km from the shore. The size of this zone is controlled by transgression rate and shelf slope. Along the coast the ice-bonded permafrost near the seabed not only undergoes rapid warming but is also exposed to salt water, causing noticeable reduction in strength and possible settlement due to thaw of the shallow ground ice.

As deeper parts of the large coastal permafrost block warm, additional changes occur more slowly. These may include regional settlement, increased permeability, decomposition of hydrates, and redistribution of free gas. Even though these changes are slow, the rates and magnitudes are greater than is normally associated with perma-

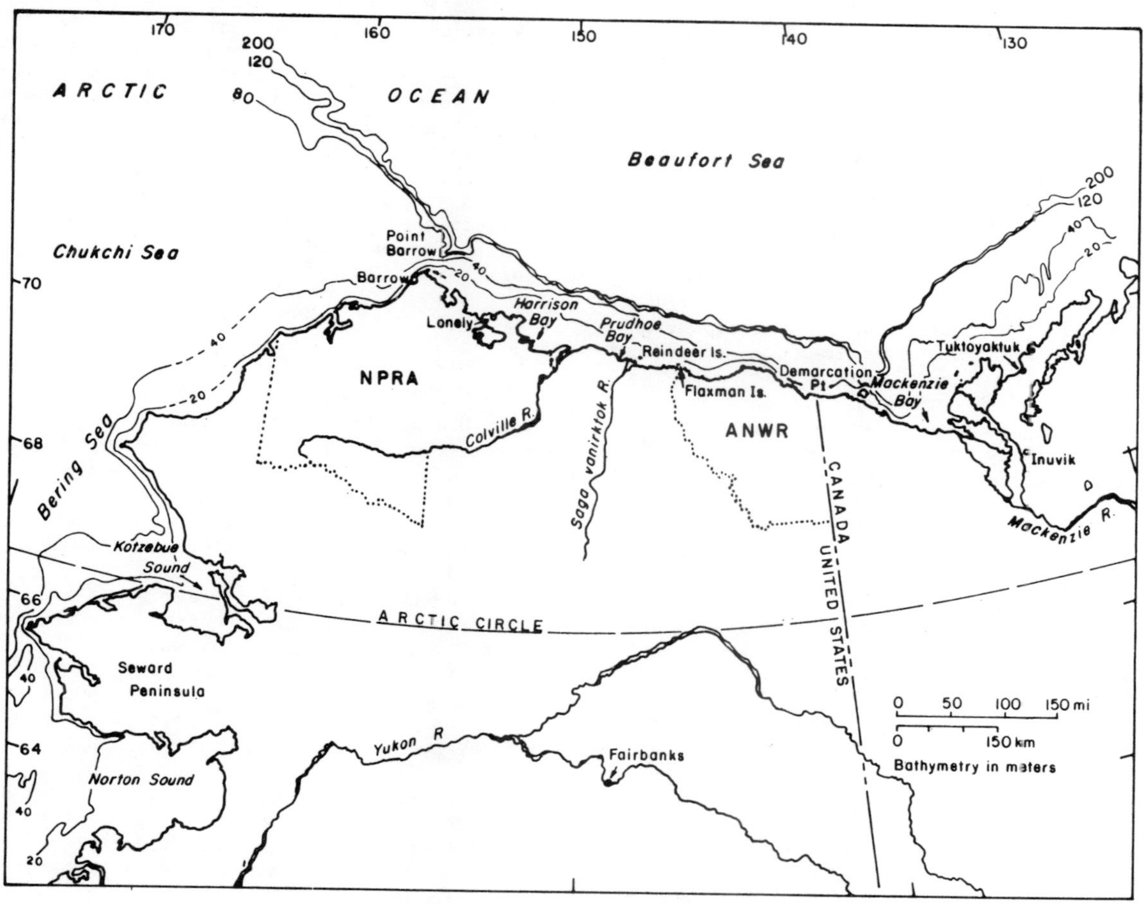

FIGURE 1 Index map of Alaska showing locations discussed in the text.

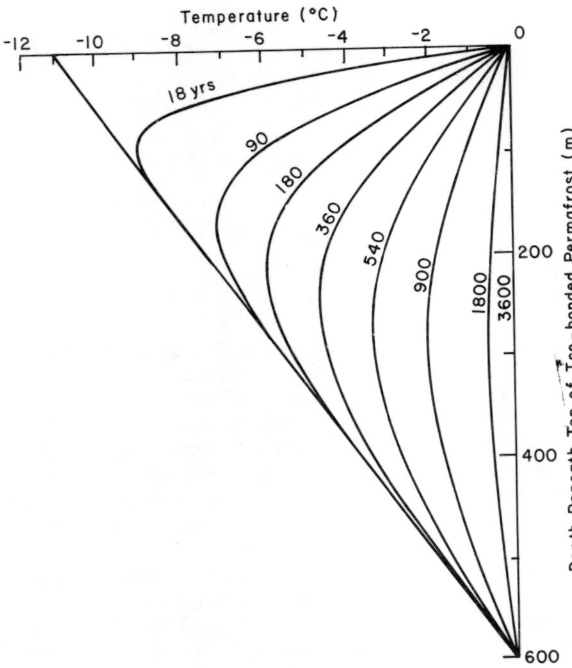

FIGURE 2 Marine temperature profiles for the Prudhoe Bay region from a tentative model proposed by Lachenbruch et al. (1982). The profiles noted by time in years indicate anticipated temperature modification caused by the sea covering an area for the noted period of time.

frost and nonpermafrost settings. An indication of the rate of warming in the Beaufort Sea near Prudhoe Bay, Alaska, can be obtained from the model shown in Figure 2. Offshore geological processes and events such as shoaling, barrier island migration, local scouring, and seasonal brine migration can influence subsea permafrost through changes in sea-floor temperature, water depth, and bottom-water chemistry. It is also now apparent that in certain situations, new ice-bonded permafrost can form in the marine environment, even though most offshore permafrost is relict and originated subaerially prior to submergence.

Another contrast between land and marine permafrost settings is the small size of the offshore zone in which noticeable seasonal freezing and thaw occur. Offshore an obvious "active layer" can only be found in very shallow water areas where significant seasonal freezing occurs when sea ice freezes to the bed. These shallow water areas are also subject to substantial warming during the summer. In contrast to land, the offshore "active layer" may be separated from deeper, well-bonded material by a layer of cold unbonded saline sediment.

Noticeable seasonal freezing can also occur at the seabed in deeper water where seabed sediments are freshened during the summer and frozen later in the year when cold, more saline water moves into an area (Sellmann and Chamberlain, 1979; Erik Reimnitz, pers. comm., 1983). However, in this zone, where water depth exceeds several meters and seabed temperatures often vary less than a degree Celsius, seasonal variations are usually very subtle. If "ice-bearing" material occurs at the seabed, it will usually be unbonded or only partially bonded. In this zone, deep, "well-ice-bonded" permafrost, usually relict, will be covered with a thick (10-200 m) layer of thawed saline sediments that can contain "ice-bearing" horizons. In some cases, the "ice-bearing" zones can be "ice-bonded."

The properties of subsea permafrost are more variable than those of permafrost found on land, since they are controlled not only by grain size, temperature, and ice volume, but also by the salinity of interstitial ice and pore water. The strength of subsea permafrost is greatly influenced by the increased salinity and higher temperatures, which weaken included ice and lower bonding forces between ice and soil because the bonding is affected by the small concentrations of unfrozen water that occur on soil particle surfaces even at temperatures several degrees Celsius below the freezing point of the material (Ogata et al., 1983). Fine-grained material containing fresh pore water has a higher unfrozen water content and correspondingly lower strength than coarse-grained material. Increased salinity in fine-grained material causes strength reduction similar to that caused by warming, the degree of warming being comparable to the amount the freezing temperature was depressed by salt. Low unfrozen water content in coarse-grained materials with fresh pore water causes the strength to increase rapidly with decreasing temperature. However, increased salinity greatly changes this pattern, extending strength reductions in coarse-grained materials over a much larger temperature range, more similar to what occurs in fine-grained material. In coarse-grained materials with large voids, strength reductions may be related not only to unfrozen water but to reduction in ice strength caused by increased salinity (Edwin Chamberlain, CRREL, Hanover, N.H., pers. comm.).

NOMENCLATURE

The year-round occurrence of sediments colder than 0°C at the seabed and the unbonded or partially bonded nature of parts of this permafrost section have created some nomenclature problems. The nomenclature issue is discussed here and certain solutions are proposed, more to call attention to the problems than to provide ultimate solutions.

The definition of permafrost is based upon temperature alone, and requires only that the material remain at temperatures below 0°C from year to year (Péwé, 1974). Permafrost, thus defined, occurs beneath a major part of the Beaufort Sea shelf, and in some coastal areas it extends from the seabed to depths as great as 600 m. However, because of moderate temperatures and salt in the pore water, ice-bonded sediments that can cause engineering problems are less abundant offshore. The nomenclature problem is evident in recent literature, where some authors quite properly apply the term "permafrost" to any material colder than 0°C, independent of properties, while others apply the term only to material with noticeable ice content and associated bonding of soil particles.

New qualifiers have been used to describe specific subsea permafrost properties. Hunter (this panel) speaks, for example, of subsea materials with seismic velocities above a specific threshold as "acoustic permafrost." "Ice-bonded" is a qualifier used by many to describe material containing ice that has sufficient bond with soil particles to cause a noticeable increase in strength properties, compared to otherwise similar unfrozen material. "Ice-bearing" has been used by some as a very general term to indicate that ice can be expected to occur or does occur in subsea material, while others have used it more specifically, indicating that ice quantity or bonding is not sufficient to influence strength properties.

Because of the complex nature of subsea permafrost, it appears that the nomenclature will be most useful if it is simple and self-explanatory. Since the use of the term "permafrost" may be very ambiguous and has led to confusion, it should be qualified or, if unqualified, used only to indicate that materials are below 0°C. More descriptive language such as "perennially frozen" would be most effective. We propose that "ice bearing" be used only in a very general sense, to indicate that ice exists in the material in some quantity with no inference drawn concerning ice volume or properties. "Ice-bonded," as discussed earlier, would then be a special case of "ice-bearing." More specific qualifications of bonding could include "well bonded" or "partially bonded" where adequate information is available. "Ice-bearing" material that has no apparent bonding might be referred to as "unbonded."

The bulk of the permafrost qualifiers used on land can also be used offshore, with the following representing some good examples: thaw-stable and thaw-unstable (Linell and Kaplar, 1966), thaw-sensitive and partially frozen (van Everdingen, 1976), and ice-rich (Brown and Kupsch, 1974).

DISTRIBUTION

Knowledge of the distribution of subsea permafrost off North American coasts is largely restricted to the Beaufort and Chukchi Sea shelves. Hopkins (1980) and Osterkamp and Harrison (1982) report no evidence for permafrost beneath the Bering Sea shelf and doubt that it will be found except along the most rapidly retreating coastal segments of Norton Sound and the northwesternmost Bering Sea. No ice-bonded permafrost was found during drilling for placer gold near Nome or at Daniels Creek on the south coast of the Seward Peninsula.

The eastern part of the Chukchi Sea is warmed by the Alaskan Current, which consists of relatively warm Pacific and Yukon River water that flows northward through the Bering Strait. Florence Weber (U.S. Geological Survey, written comm. to D.M. Hopkins, 1952) noted evidence of collapse over thawing ice wedges in shallow water off rapidly retreating parts of the west coast of the Baldwin Peninsula in Kotzebue Sound. Subzero temperatures were also encountered in Kotzebue Sound in nearshore boreholes drilled by Osterkamp and Harrison (1982).

Subzero temperatures have been observed in the few holes drilled in the coastal waters of the Chukchi Sea (Osterkamp and Harrison, 1982; Lachenbruch et al., 1962; Lachenbruch, 1957). A study of bottom temperatures and the limited borehole data from the Chukchi Sea basin indicates that ice-bearing subsea permafrost is probably thin or absent in most of the basin except along the coast and beneath the northernmost segment of the Chukchi shelf. Even in this northern segment, it is believed that ice-bearing permafrost becomes thin or absent approximately a kilometer offshore (Osterkamp and Harrison, 1982). No seismic evidence of permafrost has yet been recognized on the Chukchi Sea shelf (A.A. Grantz, U.S. Geological Survey, pers. comm.).

The situation is quite different beneath the Beaufort Sea. Point Barrow marks an important oceanographic boundary, with bottom waters being much colder to the east. Coastal processes are also much different to the east, since the coastline is less stable (Hopkins and Hartz, 1978; Hopkins, 1980). The erosion rate of coastal permafrost often exceeds several meters per year along the Beaufort coast where Lewellen (1977) reported an average rate of 4.7 m/yr for 68 observation points between Harrison Bay and Barrow. Short-term rates as great as 30 m/yr were observed by Leffingwell (1919).

Ice-bonded sediments have been observed in many of the holes drilled in the coastal waters of the Beaufort Sea. The distribution varies; in some areas, ice-bonded sediments occur at depths of 10-15 m below the seabed many kilometers from shore. In other offshore areas, occurrence of ice-bonded sediments has been confirmed at depths greater than 100 m below the seabed.

Aside from Lewellen's (1973, 1975) original borehole transects east of Point Barrow and Harrison and Osterkamp's (1981) data off Lonely where ice-bonded sediments were observed near the seabed, most ground-truth information on subsea permafrost off the Alaskan segment of the Beaufort Sea shelf comes from the area where petroleum exploration has been most active. This includes the 250-km stretch of Beaufort Sea coast from Cape Halkett through Harrison Bay and Prudhoe Bay to Flaxman Island and Brownlow Point at the northwest boundary of the Arctic National Wildlife Refuge. The information from widely spaced borehole and probe data in this region (Osterkamp and Harrison, 1976, 1982; Osterkamp and Payne, 1981; Miller and Bruggers, 1980; Blouin et al., 1979; Harrison and Osterkamp, 1981; Sellmann and Chamberlain, 1979) was extended considerably by analysis of seismic data (Neave and Sellmann, 1982, 1983, in press; Rogers and Morack, 1980; Morack and Rogers, 1982).

The seismic studies, borehole and probe data, and preliminary information from extensive industry studies suggest a patchy and irregular distribution of ice-bonded permafrost and the important roles of geologic history and seabed sediment type in determining permafrost properties and distribution. The above information suggests three common distribution patterns for ice-bonded permafrost in the Alaskan segment of the Beaufort Sea (Fig. 3). In the first, shallow, ice-bonded permafrost near the seabed extends many kilometers from shore. In the second, material with high seismic velocities,

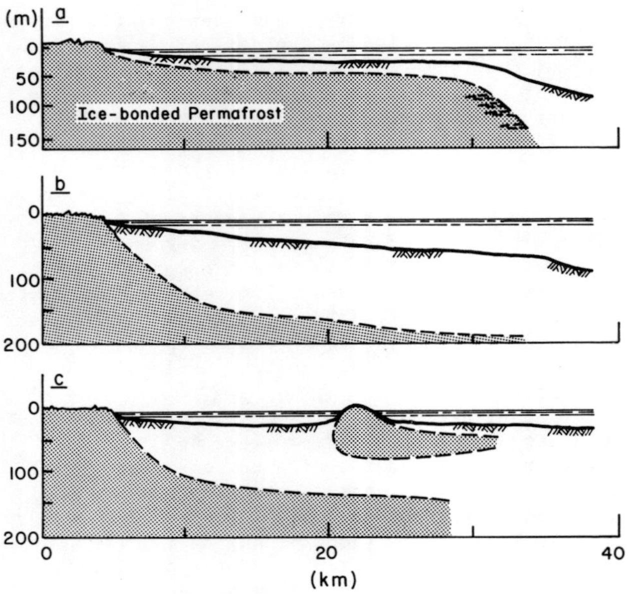

FIGURE 3 Three distribution patterns suggested for ice-bonded subsea permafrost based on borehole and seismic data: (a) shallow ice-bonded, (b) deep ice-bonded, and (c) layered ice-bonded materials (from Neave and Sellmann, in press).

as much as 200 m below the seabed offshore, can be traced into shallow, firmly bonded material near the shore. The third pattern, one that may turn out to be very common when we have more detailed subsurface information, comprises layered situations.

Shallow high-velocity materials are extremely widespread in at least two areas: Harrison Bay and an area off the mouth of the Sagavanirktok River near Prudhoe Bay (Fig. 4 and 5). Neave and Sellmann (1982) first inferred the wide distribution of firmly bonded permafrost in the western part of Harrison Bay on the basis of analysis of seismic data; this interpretation and an even greater distribution of shallow ice-bonded material was confirmed by drilling results reported by Walker et al. (1983), who in their oral presentation indicated that firmly ice-bonded material can be found at depths of 8 to 10 m below the seabed throughout Harrison Bay out to the 20-m isobath. Drilling has also confirmed that shallow high-velocity material recognized seismically off the Sagavanirktok River delta is firmly bonded permafrost (Miller and Bruggers, 1980).

Figures 6 and 7 show examples of the second pattern, in which the depth to strongly bonded permafrost increases rapidly with distance from shore. Both profiles are from the Prudhoe Bay area but are based on different techniques. Profile 6 is from analysis of seismic data presented by Neave and Sellmann (1983); the deeper part of this profile was confirmed by offshore borehole logs (Osterkamp and Payne, 1981). Profile 7, based on analysis of the transient electromagnetic data discussed by Ehrenbard et al. (1983) not only shows comparable depths to the top of bonded permafrost but also the great depth to the bottom of the ice-bearing material.

A Siberian example of a layered case is shown by Vigdorchik (1980, Fig. 3.13) and may be related to vertical differences in material properties. In the Canadian Beaufort Sea, ice-bonded coarse-grained material is often found surrounded by unfrozen fine-grained material.

Experience thus far indicates that two particular situations are likely to generate layered permafrost. One situation occurs off major river deltas, where Holocene sequences of stratified sand, silt, and clay have accumulated, where the overlying seawater may be freshened, and where fresh groundwater may enter the most permeable strata and freeze. The Canadian examples dis-

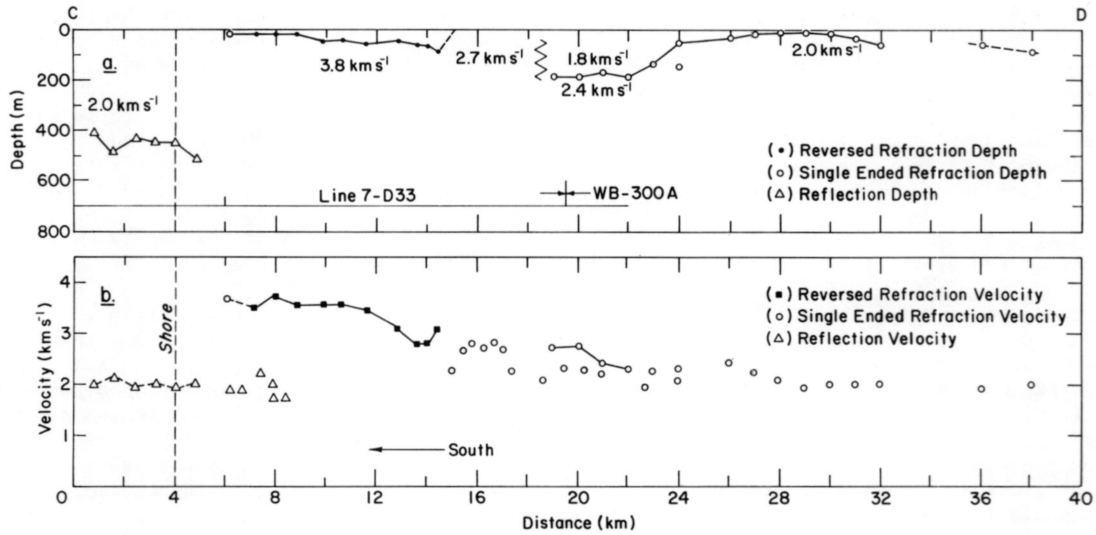

FIGURE 4 Profile showing the top of shallow high-velocity material in the western part of Harrison Bay interpreted to be ice-bonded permafrost (from Neave and Sellmann, 1982).

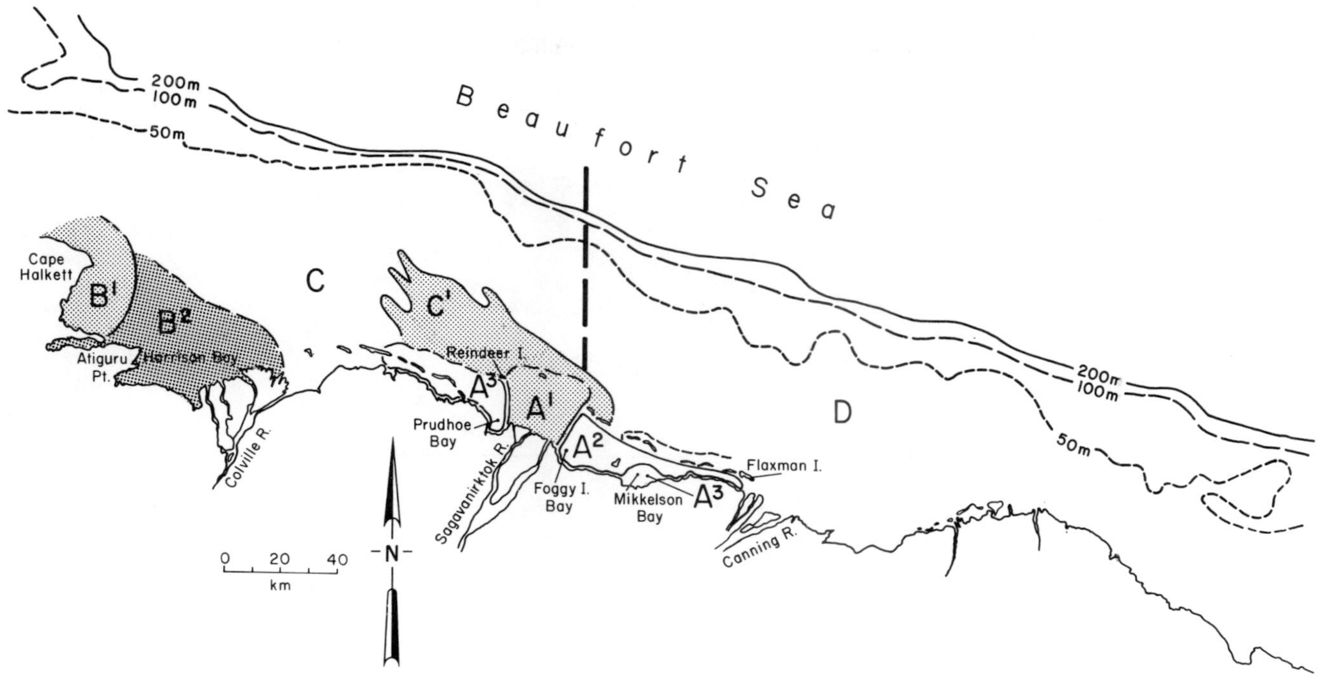

FIGURE 5 Shaded areas A^1, B^1, and C^1 show the distribution of shallow high-velocity zones near Harrison Bay and Prudhoe Bay, interpreted from seismic data to be ice-bonded permafrost (from Neave and Sellmann, 1983). Shallow ice-bonded material is probably also present in area B^2.

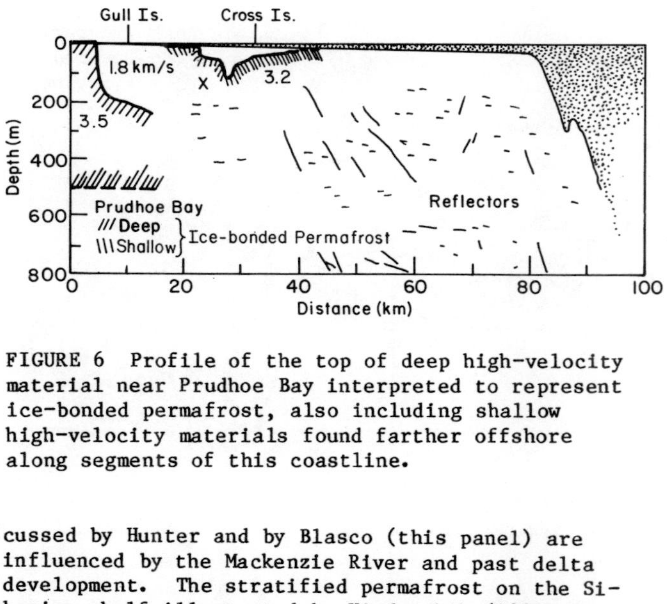

FIGURE 6 Profile of the top of deep high-velocity material near Prudhoe Bay interpreted to represent ice-bonded permafrost, also including shallow high-velocity materials found farther offshore along segments of this coastline.

cussed by Hunter and by Blasco (this panel) are influenced by the Mackenzie River and past delta development. The stratified permafrost on the Siberian shelf illustrated by Vigdorchik (1980, Fig. 3.13) may be another example of this case. An Alaskan example is found in Harrison Bay, which receives the sediment of the Colville River, the largest stream that enters the Alaskan segment of the Beaufort Sea. This layered case (shown in Fig. 8) was inferred from transient electromagnetic sounds near the Colville River delta and is interpreted to represent a shallow, irregular layer of ice-bonded material over a thawed zone, both overlying much deeper ice-bonded material.

The other situation consists of migrating barrier islands and shoals, which can leave a trail of newly formed permafrost to mark their passage. Profile C in Figure 3 is a diagrammatic representation of this mechanism. Reindeer Island (Fig. 9) may be an example of this case.

The important role played by material type in determining depth to ice-bonded material is demonstrated by borehole results in Prudhoe Bay and on the shallow shelf immediately to the north (Fig. 9). There, areas underlain by overconsolidated marine clay dating from the last interglacial period are ice-bonded at depths of less than 20 m, as seen north of Reindeer Island. Ice-bonded permafrost is much deeper south of the island, lying at depths greater than 100 m beneath a broad, shallow trough that extends out of Prudhoe Bay and then turns westward parallel to the present coast. This area is interpreted as a Pleistocene valley carved at times when sea level was low and the shelf was exposed. It is filled with alluvial gravel overlain by a few meters of soft Holocene marine silt and clay. Evidently seawater gained entry into the gravel during and since the drowning of the valley, destroying interstitial ice to depths of more than 100-200 m. Boreholes elsewhere have encountered older bodies of gravel, some frozen and some thawed, buried beneath ice-bonded, overconsolidated clay (Miller and Bruggers, 1980).

All of the Alaskan borehole data come from the inner shelf. The only data suggesting the presence of ice-bonded permafrost on the middle and outer shelf come from reflection seismic data discussed by Neave and Sellmann (in press). An extensive reflector 200 m below the seabed is

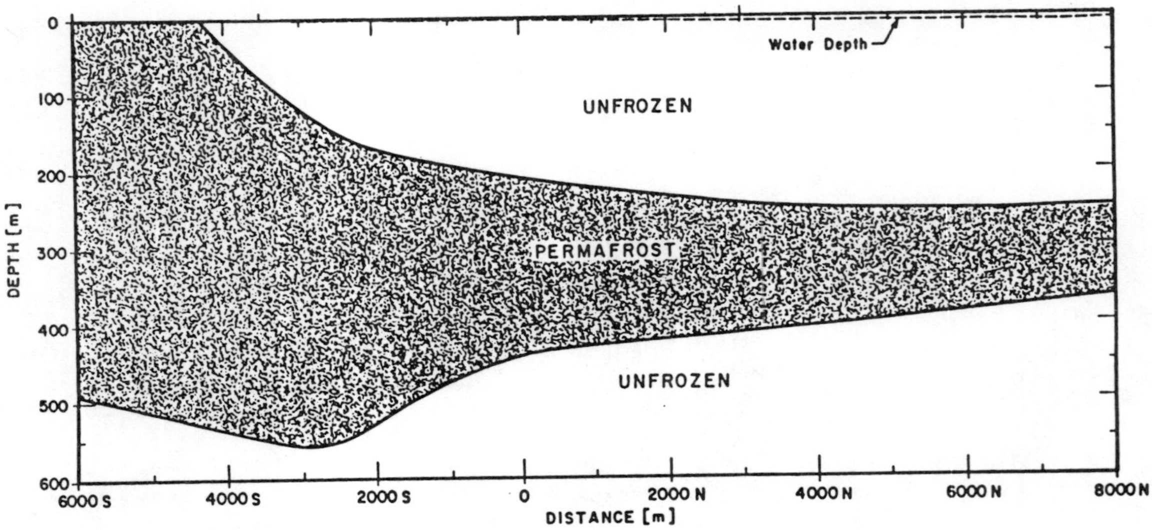

FIGURE 7 Cross-section of zone interpreted to be ice-bearing based on transient electromagnetic soundings (from Ehrenbard et al., 1983).

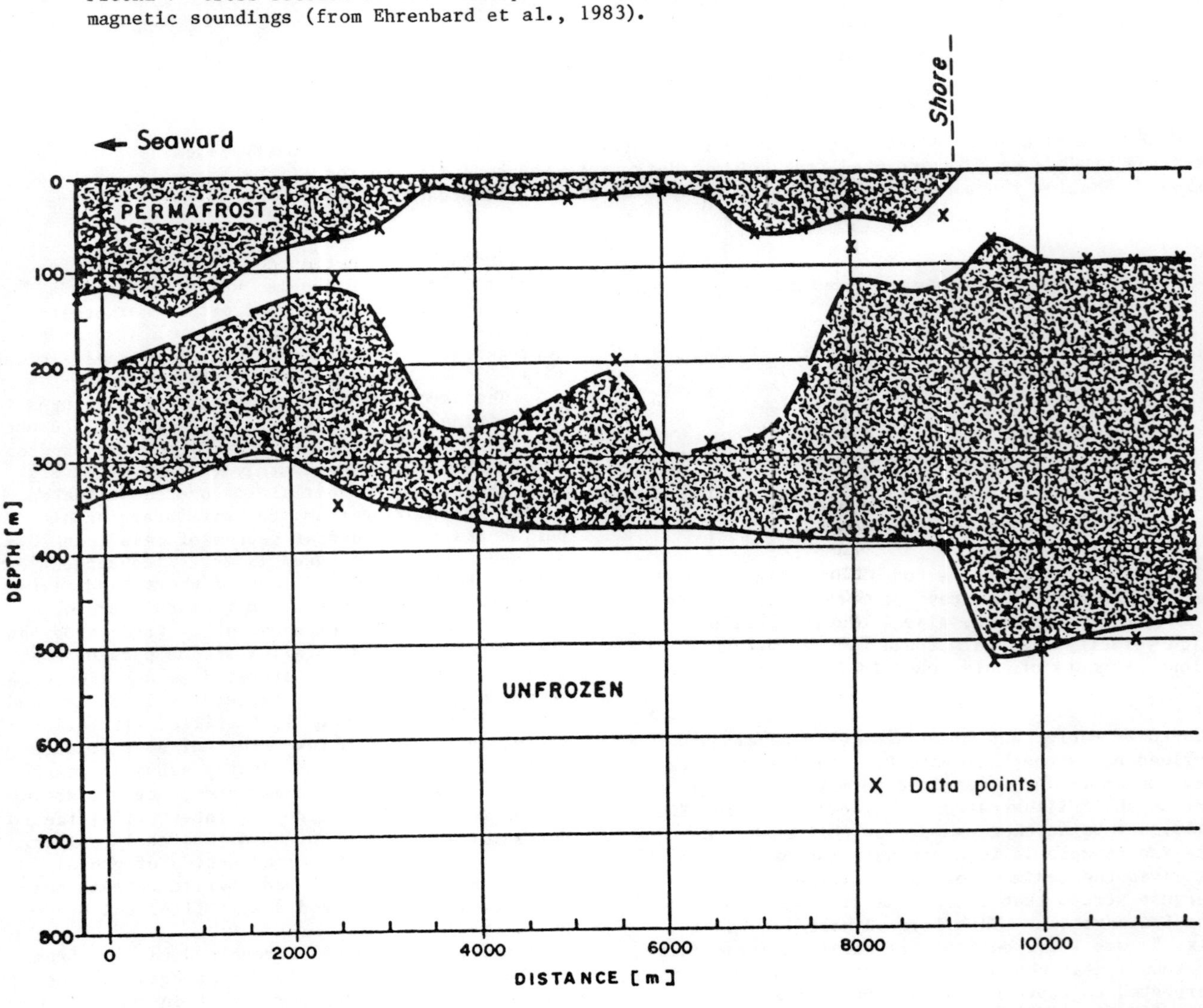

FIGURE 8 Layered ice-bearing material inferred from transient electromagnetic soundings off the Colville River delta (illustration part of oral presentation by Ehrenbard et al. [1983], provided by the courtesy of the authors).

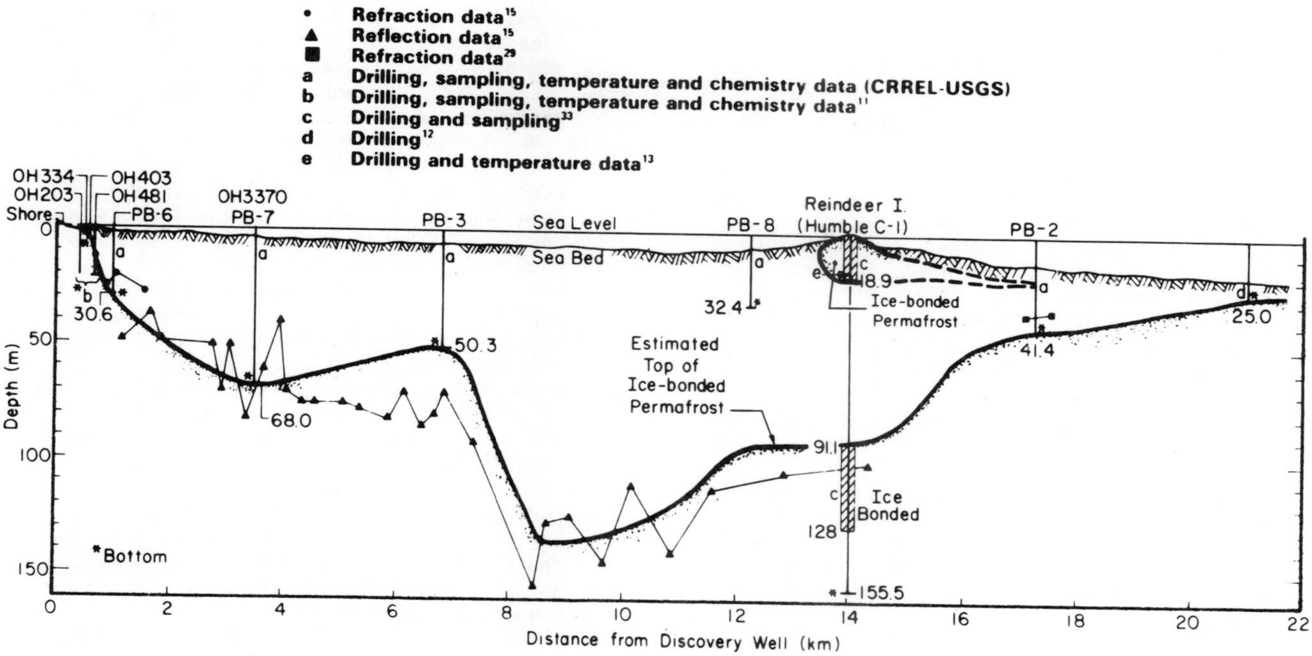

FIGURE 9 Variable position of top of ice-bonded permafrost controlled by variations in material type and geological history. Illustration was based on data from a number of sources listed in Sellmann and Chamberlain (1979).

thought to be traceable shoreward to a reflector representing bonded permafrost on the inner shelf. However, it must be acknowledged that the complex stratigraphic geometry of the inner shelf gives way seaward to an outer shelf sequence of smooth and widespread reflectors that Dinter (1982) interprets as sandy layers formed during successive Pleistocene marine transgressions; the 200-m reflector noted by Neave and Sellmann (in press) may be one of these.

The increasing literature, only some of which is mentioned, confirms the widespread distribution and the variable nature of ice-bonded subsea permafrost. Each new detailed study provides unexpected results. An example is the recent discovery of seasonal seabed freezing in areas well removed from the coast, and the extensive distribution of shallow ice-bonded materials even at great distances from shore.

REFERENCES

Blouin, S.E., Chamberlain, E.J., Sellmann, P.V., and Garfield, D.E., 1979, Determining subsea permafrost characteristics with a cone penetrometer - Prudhoe Bay, Alaska: Cold Regions Science and Technology, v. 1, no. 1.

Brown, R.J.E., and Kupsch, W.O., 1974, Permafrost terminology: Ottawa, National Research Council of Canada, Technical Memorandum No. 11.

Dinter, D.A., 1982, Holocene marine sediments on the middle and outer continental shelf of the Beaufort Sea north of Alaska: U.S. Geological Survey Miscellaneous Investigations Series, Map I-1182-B.

Ehrenbard, R.L., Hoekstra, P., and Rozenberg, G., 1983, Transient electromagnetic soundings for permafrost mapping, in Proceedings of the Fourth International Conference on Permafrost, Fairbanks: Washington, D.C., National Academy of Sciences, p. 272-277.

Harrison, W.D., and Osterkamp, T.E., 1981, Subsea permafrost: Probing, thermal regime and data analysis, in Environmental Assessment of the Alaskan Continental Shelf, Annual Reports, v. 7, p. 291-401.

Hopkins, D.M., and Hartz, R.W., 1978, Coastal morphology, coastal erosion, and barrier islands of the Beaufort Sea, Alaska, Open File Report 78-1063.

Hopkins, D.M., 1980, Likelihood of encountering permafrost in submerged areas of northern Bering Sea, in Smith, P., Hartz, R., and Hopkins, D., Offshore Permafrost Studies and Shoreline History as an Aid to Predicting Offshore Permafrost Conditions: Environmental Assessment of the Alaskan Continental Shelf, Annual Report, v. 4, p. 187-193.

Lachenbruch, A.H., 1957, Thermal effects of the ocean on permafrost: Bulletin of the Geological Society of America, no. 68, p. 1515-1530.

Lachenbruch, A.H., Brewer, M.C., Greene, G.W., and Marshall, B.V., 1962, Temperature in permafrost, in Temperature -- Its Measurement and Control in Science and Industry: New York, Reinhold Publishing Co., v. 3, p. 791-803.

Lachenbruch, A.H., Sass, J.H., Marshall, B.V., and Moses, T.H., Jr., 1982, Permafrost, heat flow, and the geothermal regimes of Prudhoe Bay, Alaska: Journal of Geophysical Research, v. 87, no. B11, p. 9301-9316.

Leffingwell, E. de K., 1919, The Canning River region, northern Alaska: U.S. Geological Survey Professional Paper 109.

Lewellen, R.I., 1973, The occurrence and characteristics of nearshore permafrost, Northern Alaska, in Permafrost -- The North American Contribution to the Second International Conference, Yakutsk: Washington, D.C., National Academy of Sciences, p. 131-136.

Lewellen, R.I., 1975, The occurrence and characteristics of subsea permafrost, northern Alaska: Progress Report (AINA-ONR-454), Arctic Institute of North America, p. 131-135.

Lewellen, R.I., 1977, A study of Beaufort Sea coastal erosion, northern Alaska: U.S. National Oceanic and Atmospheric Administration, Environmental Assessment of the Alaskan Continental Shelf, Annual Reports, v. 15, p. 491-527.

Linell, K.A. and Kaplar, C.W., 1966, Description and classification of frozen soils, in Permafrost, Proceedings of an International Conference, Purdue University, 1963: Washington, D.C., National Academy of Sciences, p. 481-487.

Miller, D.L. and Bruggers, D.E., 1980, Soil and permafrost conditions in the Alaskan Beaufort Sea, in Proceedings of the 12th Annual Offshore Technology Conference, Houston, p. 325--338.

Morack, J.L. and Rogers, J.C., 1982, Marine seismic refraction measurement of near-shore subsea permafrost, in Proceedings of the Fourth Canadian Permafrost Conference, Calgary: Ottawa, National Research Council of Canada, p. 249-255.

Neave, K.G. and Sellmann, P.V., 1982, Subsea permafrost in Harrison Bay, Alaska: An interpretation from seismic data: Hanover, N.H., U.S. Army Cold Regions Research and Engineering Laboratory, CRREL Report 82-24.

Neave, K.G. and Sellmann, P.V., 1983, Seismic velocities and subsea permafrost in the Beaufort Sea, Alaska, in Proceedings of the Fourth International Conference on Permafrost, Fairbanks: Washington, D.C., National Academy of Sciences, p. 894-898.

Neave, K.G. and Sellmann, P.V., in press, Determining distribution patterns of ice-bonded permafrost in the U.S. Beaufort Sea from seismic data, in The Beaufort Sea: Physical and Biological Environment (P.W. Barnes, D. Schell and E. Reimnitz, ed.): New York, Academic Press.

Ogata, N., Yasuda, M. and Kataoka, T., 1983, Effects of salt concentration on strength and creep behavior of artificially frozen soils: Cold Regions Science and Technology, v. 8, no. 2, p. 139-153.

Osterkamp, T.E. and Harrison, W.D., 1976, Subsea permafrost at Prudhoe Bay, Alaska: Drilling report and data analysis: Fairbanks, Geophysical Institute, University of Alaska, Report UAG R-247.

Osterkamp, T.E. and Payne, M.W., 1981, Estimates of permafrost thickness from well logs in northern Alaska: Cold Regions Science and Technology, v. 5, p. 13-27.

Osterkamp, T.E. and Harrison, W.D., 1982, Temperature measurements in subsea permafrost off the coast of Alaska, in Proceedings of the Fourth Canadian Permafrost Conference, Calgary: Ottawa, National Research Council of Canada, p. 238-248.

Péwé, T.L., 1974, Permafrost: New York, Encyclopedia Britannica.

Rogers, J.C. and Morack, J.L., 1980, Geophysical evidence of shallow near-shore permafrost, Prudoe Bay, Alaska: Journal of Geophysical Research, v. 85, no. B9, p. 4845-4853.

Sellmann, P.V. and Chamberlain, E.J., 1979, Permafrost beneath the Beaufort Sea near Prudhoe Bay, Alaska, in Proceedings of the 11th Annual Offshore Technology Conference, Houston, p. 1481-1493.

van Everdingen, R.O., 1976, Geocryological terminology: Canadian Journal of Earth Sciences, v. 13, no. 6, p. 862-867.

Vigdorchik, M.E., 1980, Arctic Pleistocene history and the development of submarine permafrost: Boulder, Colo., Westview Press.

Walker, D.B.L., Hayley, D.W., and Palmer, A.C., 1983, The influence of subsea permafrost on offshore pipeline design, in Proceedings of the Fourth International Conference on Permafrost, Fairbanks: Washington, D.C., National Academy of Sciences, p. 1338-1343.

A PERSPECTIVE ON THE DISTRIBUTION OF SUBSEA PERMAFROST ON THE CANADIAN BEAUFORT CONTINENTAL SHELF

S.M. Blasco

Geological Survey of Canada, Bedford Institute of Oceanography
Dartmouth, Canada

The distribution of permafrost in the upper 100 m below the seabed on part of the Canadian Beaufort Sea shelf has been defined, based on interpretation of several thousand line kilometers of high-resolution shallow seismic profiles, correlated with data from approximately 90 boreholes that encountered ice-bearing sediments (O'Connor, 1981) (Fig. 1).

Three types of subsea permafrost containing ice have been identified to date using analog reflection seismic techniques: hummocky permafrost (Fig. 2), with laterally and vertically discontinuous patches of ice-bearing sediments; stratigraphically controlled permafrost (Fig. 3), where the occurrence of ice is controlled by bedding, and ice is found only in some layers; and marginally ice-bonded permafrost (Fig. 4), where the distribution of ice within the sediments is so limited spatially and volumetrically that regional mapping of discrete horizons is difficult.

The patches of hummocky permafrost appear to be primarily associated with massive, coarser-grained sediment (fine to medium sand), while stratigraphically controlled permafrost is associated with discrete layers of sediment (sands, silts, clayey silts) that exhibit variations in ice content and bonding. Marginally ice-bonded permafrost is associated with the coarser fraction of finer-grained sedimentary sequences. In general, the distribution of shallow ice-bearing subsea permafrost is quite variable in lateral and vertical extent, being controlled by lithology, pore water salinity, and the thermal regime (Mackay, 1972; Hunter et al., 1976; MacAulay and Hunter, 1982).

The geological history and properties of the sediments in various depositional environments also have a major impact on distribution. Sedimentological, biostratigraphic, and seismic evidence (O'Connor, 1980; Hill et al., 1984) from the Canadian Beaufort Sea indicate that the upper 100 m of sediment consists of a thin veneer of Holocene Mackenzie River muds that grade downward into a transgressive sequence of interlayered sands, silts, and clays of varying thickness. These sediments lie unconformably upon an older complex of glaciofluvial deltaic sediments (Fig. 5). Radiocarbon dates suggest that this entire sequence is less than 25,000 years old, and that it records a rise in sea level from about 100 m below today's

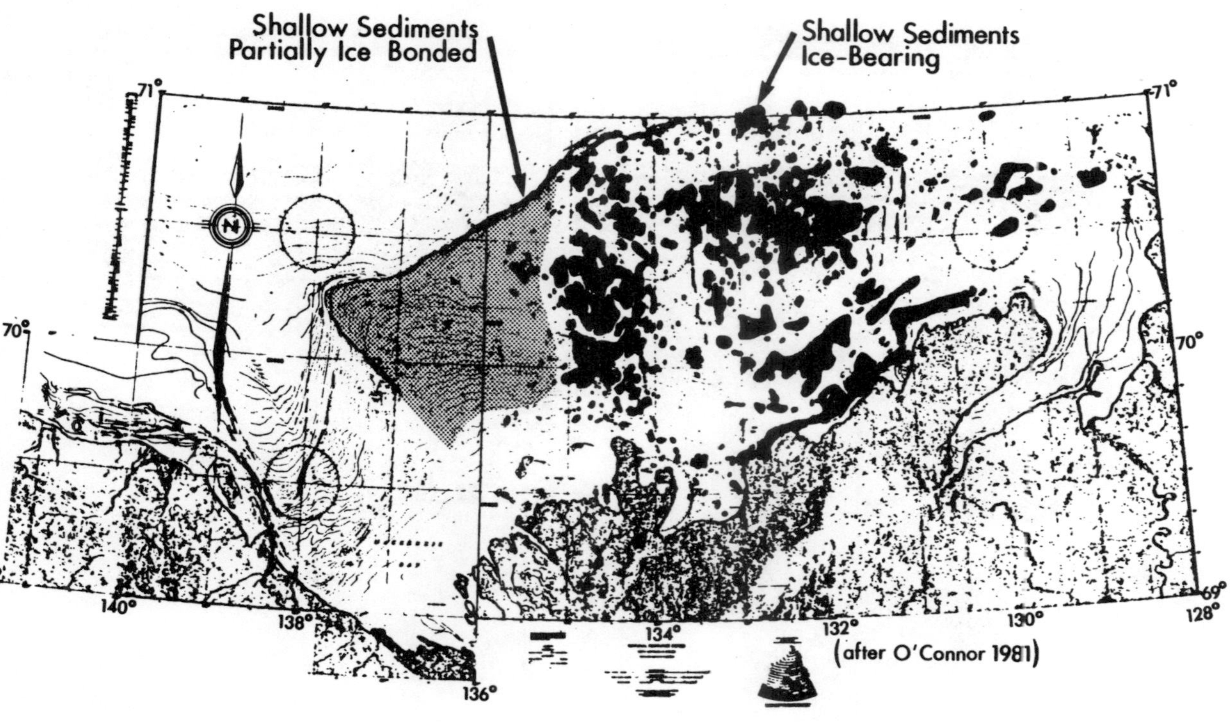

FIGURE 1 Distribution of shallow, acoustically defined permafrost (APF).

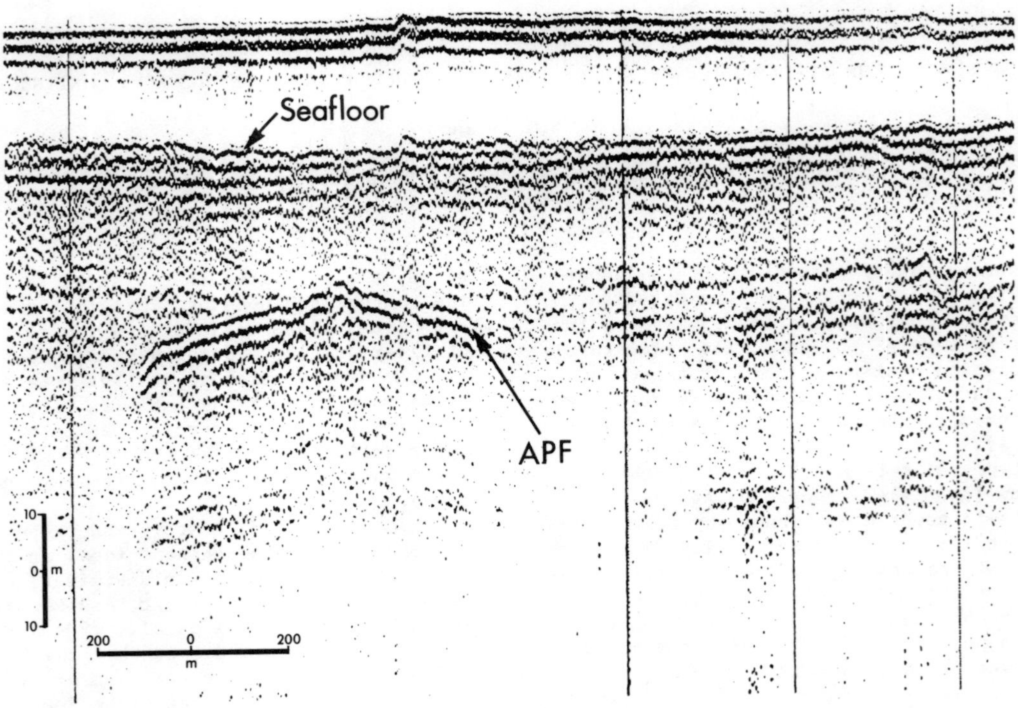

FIGURE 2 Hummocky acoustically defined permafrost (APF).

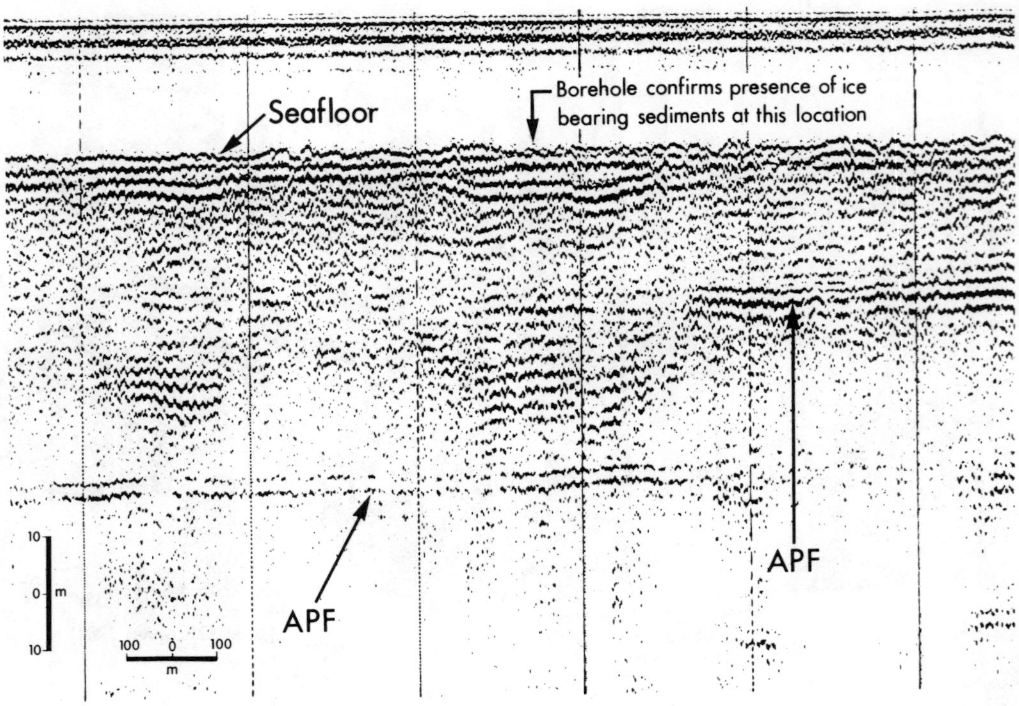

FIGURE 3 Stratigraphically controlled APF.

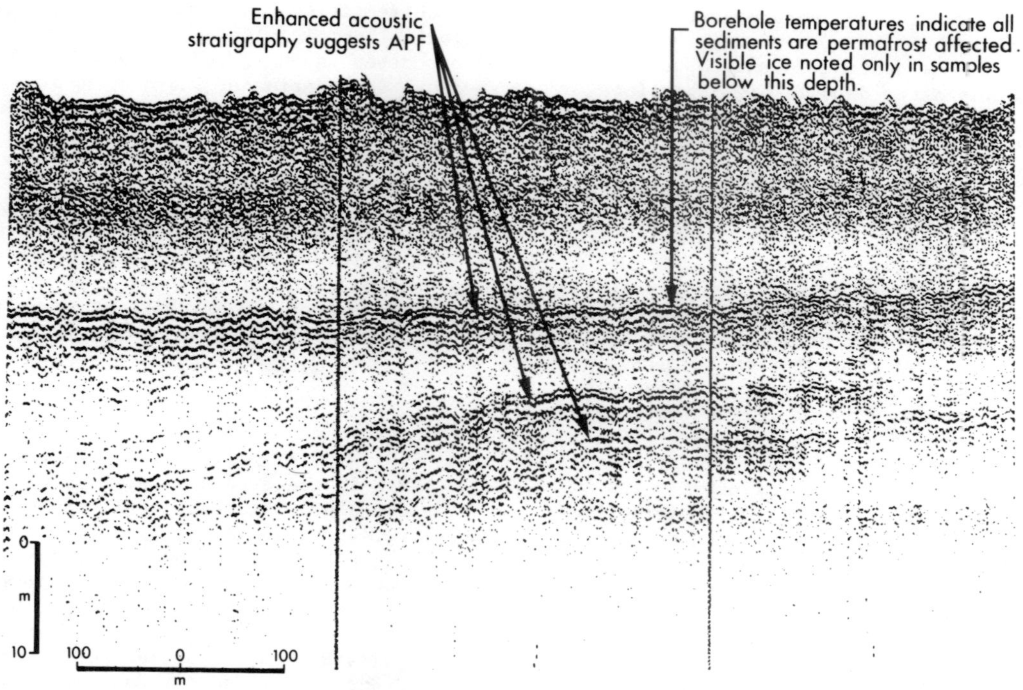

FIGURE 4 Typical seismic signature where shallow sediments are marginally or partially ice-bonded.

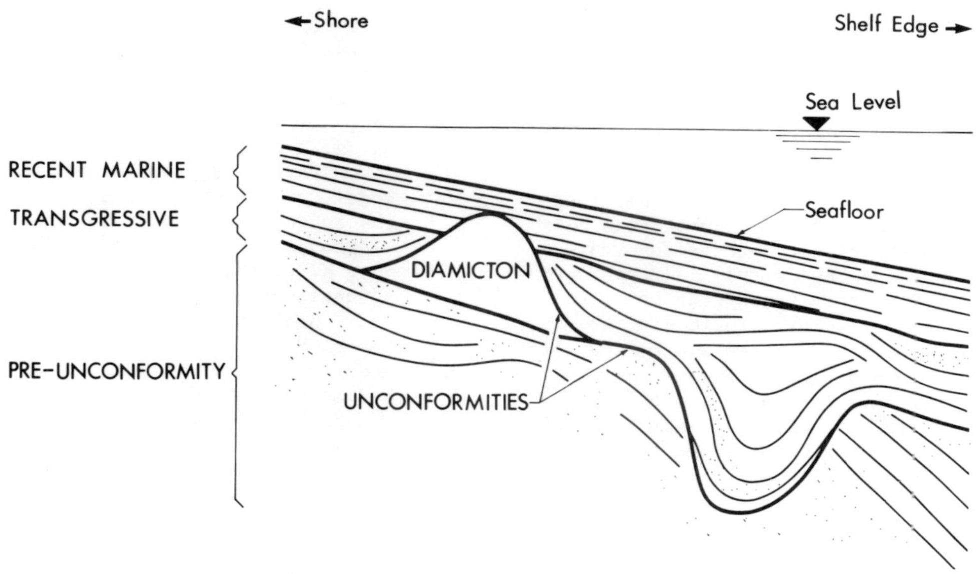

FIGURE 5 The general geological model.

level. The evidence also suggests major deltaic progradation associated with glacial retreat, and the formation of ice-bearing permafrost in these sediments shortly after deposition. At present, ice-bearing sediments are primarily found within the lower glaciofluvial delta complex, some 10 to 20 m below the unconformity, suggesting possible degradation during and after the inundation of these sediments.

Stratigraphy interpreted from the upper sections of conventional seismic reflection profiles suggests that within the study area the shelf has been consistently delta-dominated since the late Miocene (Willymsen and Cote, 1982). Sedimentary environments within a delta complex are many and varied, as deposition shifts from proximal to distal regimes. In addition, with time, the seaward advance of the delta plain over prodelta sequences deposits coarser-grained sediments over fine-grained material.

Permafrost in the delta complexes appears to have formed shortly after deposition. Therefore, the occurrence and properties of ice-bearing sediments should vary spatially and in degree of ice-bonding, depending on the properties of the depositional environment at the time of permafrost formation. Thus, subaerially exposed delta plains with coarser-grained sediments and fresh water are more likely to contain ice and to be well-bonded than are fine-grained prodelta sediments deposited contemporaneously under more saline shallow marine conditions. As depositional environments changed in the delta, permafrost distribution acquired its spatial variability. During periods of lower sea level, caused by glacial advance (either regional or otherwise), the shelf was exposed, permitting more extensive regional permafrost aggradation. Crustal downwarping due to the proximity of the ice front may also have left the shelf covered by shallow water at various times, limiting the possibility of extensive or prolonged subaerial exposure. However, there is no evidence to support a widespread presence of glacier ice over the shelf during the late Wisconsin period. If, as seems likely, these conditions persisted through the Pleistocene, the distribution of ice-bearing permafrost should be complex throughout the entire permafrost section.

This information on permafrost distribution and properties can be utilized to evaluate the impact of subsea permafrost on engineering activities. As a result of the variation in spatial distribution, volume of ground ice, and degree of ice-bonding within stratigraphic sequences, engineering problems become site-specific. From our observations, the permafrost section needs to be viewed as multilayered and containing ice-bearing layers with diverse properties. Thermal degradation of ice-bonded sediments may or may not lead to an unacceptable degree of thaw subsidence, depending in part on the history of the substrate. Sediments that are deposited slowly and that were exposed to repeated freeze-thaw cycles prior to deep burial may be thaw-stable, while similar sediments that were rapidly deposited and buried with contemporaneous growth of ground ice may not be. Perhaps through the integration of geological and geotechnical research, specific depositional environments that generate permafrost conditions that are potentially hazardous for engineering projects can be identified and evaluated. For example, in geological settings where transgressive and related processes (such as coastal erosion) lead to the regional degradation of ice-bearing permafrost to depths of 20 to 30 m below the seabed, thaw settlement may not have to be considered in the engineering design of some projects near the seabed, such as pipelines.

The incorporation of geological studies into geotechnical investigations will assist the engineer to differentiate subsea permafrost that will have no impact on offshore development and to recognize the extent and severity of permafrost that will influence development projects.

REFERENCES

Hill, P.R., Mudie, P.J., Moran, K.M., and Blasco, S.M., in press, A sea level curve for the Canadian Beaufort continental shelf.

Hunter, J.A., Judge, A.S., MacAulay, H.A., Good, R.L., Gagne, R.M., and Burns, R.A., 1976, The occurrence of permafrost and frozen sub-sea-bottom materials in the Southern Beaufort Sea: Geological Survey of Canada and Earth Physics Branch, Department of Energy, Mines and Resources, Canada, Beaufort Sea Project, Technical Report No. 22.

MacAuley, H.A., and Hunter, J.A., 1982, Detailed seismic refraction analysis of ice-bonded permafrost layering in the Canadian Beaufort Sea, in Proceedings of the Fourth Canadian Permafrost Conference, Calgary: Ottawa, National Research Council of Canada, p. 256-267.

Mackay, J.R., 1972, Offshore permafrost and ground ice, Southern Beaufort Sea: Canadian Journal of Earth Science, v. 9, p. 1550-1561.

O'Connor, M.J., 1980, Development of a proposed model to account for the surficial geology of the southern Beaufort Sea: Geological Survey of Canada, Open File Report 954.

O'Connor, M.J., 1981, Distribution of shallow acoustic permafrost in the southern Beaufort Sea: Geological Survey of Canada, Open File Report 953.

Willymsen, P.S., and Cote, R.P., 1982, Tertiary sedimentation in the southern Beaufort Sea, Canada: Canadian Society of Petroleum Geologists, Memoir 8, p. 43-53.

SOVIET STUDIES OF THE SUBSEA CRYOLITHOZONE

F.E. Are*

formerly of the Permafrost Institute, Siberian Branch, Academy of Sciences
Yakutsk, USSR

The emphasis of studies in the Soviet Union has been on the principles involved in the development of subsea permafrost. This information is being used to establish methods for predicting subsea permafrost distribution and for determining its significance in relation to resource development on the Arctic shelf. The geocryological studies made on land along the coast and on Soviet Arctic islands are based on borehole drilling, vertical electrical sounding, and geothermal and hydrological observations. Many boreholes have also been drilled in the near-shore shallows. Data on the subsea cryolithozone have been obtained in Sannikov Strait (New Siberian Islands), in Dmitri Laptev Strait, and in Van'kina Bay in the Laptev Sea. Information from the open ocean is obtained through geophysical studies. Considerable emphasis is being placed on paleogeographic considerations and mathematical modeling of subsea permafrost development.

During the Pleistocene and Holocene periods, part of the Soviet shelf was continually submerged while other parts alternately were submerged and emerged. During submergence, unfrozen sediment and rocks on the shelf became saturated with sea water and frozen materials degraded. During periods of emergence, the saline rocks and sediments froze under cold continental conditions. Freezing of the saline pore fluids produced weakly mineralized ice and brines. At temperatures below -8°C, mirabilite ($Na_2SO_4 \cdot 10\ H_2O$) precipitates out of brines. The brines can move downward along with the advance of the freezing front. Present-day ground water on the shelf is similar in composition to sea water, but it is slightly more mineralized.

During regressions, syngenetically frozen ice-rich deposits accumulated on newly exposed parts of the shelf. During transgression, these deposits were partly eroded in the littoral zone and, with inundation, underwent further erosion and thaw.

The mean annual temperature at the seabed depends on the water depth. In shallow water, where ice freezes to the bottom, it is usually negative; in water 2 to 7 m deep it may be negative or positive; and in 7 to 200 m of water it is negative on most parts of the shelf but higher than the freezing temperature of the bottom sediments.

The depth of thaw below the seabed at any particular point on the continental shelf depends in part upon the length of time the overlying water column was between 2 and 7 m deep. The time required for the sea level to rise through that depth range varied at different times during the post-glacial transgression and amounted to several centuries in some areas and one to three millenia in others. During that time, subsea permafrost degraded partially or completely; particularly intensive thawing occurred during the Holocene climatic optimum. Therefore, the top of frozen material on the shelf may be at a considerable depth below the seabed. However, in zones where ice can seasonally freeze to the bottom and in zones of recent thermoabrasion, frozen material can be close to the seabed.

In the Soviet Arctic, the temperature regime of materials on the shelf developed in different ways. The Barents Sea shelf and a major part of the Kara Sea shelf were never exposed, and consequently there is no frozen ground. Part of the Kara Sea shelf was exposed out to the present 30-m isobath for two short periods during the postglacial period; within this part of the shelf, except for the zone where ice can freeze to the bottom, the existence of relict perennially frozen layers is possible but unlikely. The most favorable conditions for preservation of frozen materials exist on the shelves of the Laptev and East Siberian Seas, where they may occur out to about the 140-m isobath. Permafrost is absent in the southeastern part of the Chukchi Sea, due to the positive water temperatures near the bottom. Along the Soviet Arctic coast, frozen seabed materials are widespread in zones where ice seasonally freezes to the bottom and within 10 to 30 km (and locally 50 km) of shorelines that are rapidly retreating due to thermoabrasion.

Research priorities in future investigations will be the study of the interaction between sea water and perennially frozen ground as controlled by geocryological, thermophysical, and physiochemical parameters. An approach to this study would be to investigate these parameters along a profile extending offshore from a rapidly retreating thermoabrasion coast.

*Not present.

GEOPHYSICAL TECHNIQUES FOR SUBSEA PERMAFROST INVESTIGATIONS

J.A.M. Hunter

Geological Survey of Canada
Ottawa, Canada

Much of our knowledge of offshore permafrost has been acquired from studies based on geophysical techniques. The detection of ice-bearing permafrost by geophysical means is of current interest and was the subject of both formal and informal discussions at the Fourth International Conference on Permafrost.

During the 1970s one of the major geophysical techniques used was seismic refraction. Several papers have been published by Canadian and US workers on this technique for mapping ice-bearing permafrost in the Beaufort Sea. Seismic velocities in unconsolidated sediments in excess of 2400 m/s were assumed, on the basis of laboratory studies in the USA, Canada, and the USSR, to represent ice-bonded sediment, and this velocity was used as a threshold for seismic mapping of permafrost. This material was termed "acoustic permafrost" (APF). Often, materials with velocities less than the threshold value were considered "unfrozen," "non-ice-bonded," "thawed material," and so forth. Most workers realized that ice-bearing material containing a lower percentage of ice than the total pore volume might exist and, if so, would have velocities less than 2400 m/s. However, emphasis was placed on the higher velocity ice-bonded APF, because geophysicists tend to be conservative in their interpretations, and because it was thought that high-velocity materials would be of more concern initially to geotechnical engineering. Much discussion of ice content and the velocity cutoff for ice-bonded material took place at the Third International Conference on Permafrost in Edmonton in 1978, but at that time we had little ground truth and lacked an adequate theoretical framework for guidance.

We are now able to extend our interpretations to lower seismic velocities, so that over much of the Beaufort Sea we can substitute "possibly ice-bearing" for the "unfrozen zones" noted in our early papers. We now have more ground truth; ice-bearing material has been recovered from strata having relatively low velocities. We have also been given a possible theoretical framework for the interpretation of velocities in terms of ice content (King, 1984).

The correlation between low velocities and material that may contain some ice can be illustrated with some recent Geological Survey of Canada data from the Canadian sector of the Beaufort Sea. Our early refraction seismic work demonstrated the presence of high-velocity ice-bonded permafrost at depth below the seabed. The geometry of the hydrophone array was such that the near-sea-bottom materials were often inadequately measured, and for interpretation purposes the upper "unfrozen" layer was usually assigned a velocity of 1600 m/s. During the last two seasons we have been measuring velocities of near-sea-bottom materials using a wide-angle 12-channel reflection eel array, specifically designed for engineering surveys, in conjunction with vertical incidence reflection surveys to obtain velocity measurements from wide-angle reflections to a depth of 60 m below the sea bottom.

Figure 1 is a compilation from 500 seismic records taken over a large part of the Canadian Beaufort Sea shelf. It is in the form of a two-dimensional histogram plotting depth below sea bottom vs average velocity from the sea bottom to the reflector. The contours are in frequency of occurrence with a contour interval of five observations. The data are presented in this manner so that a generalized velocity-depth function (the dashed line) can be selected for converting two-way vertical incidence travel time to depth.

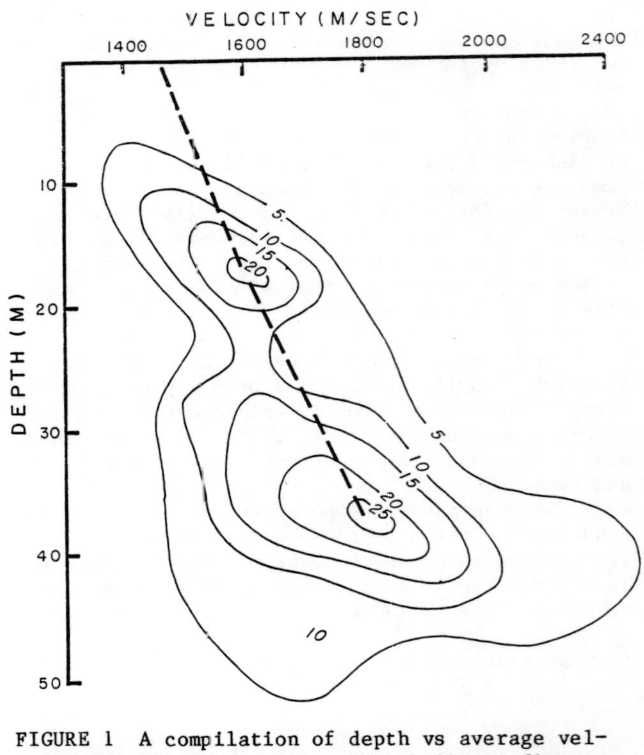

FIGURE 1 A compilation of depth vs average velocity between the seabed and a shallow reflector, based on 500 records from a large area in the Canadian Beaufort Sea. Contours are frequency of occurrence.

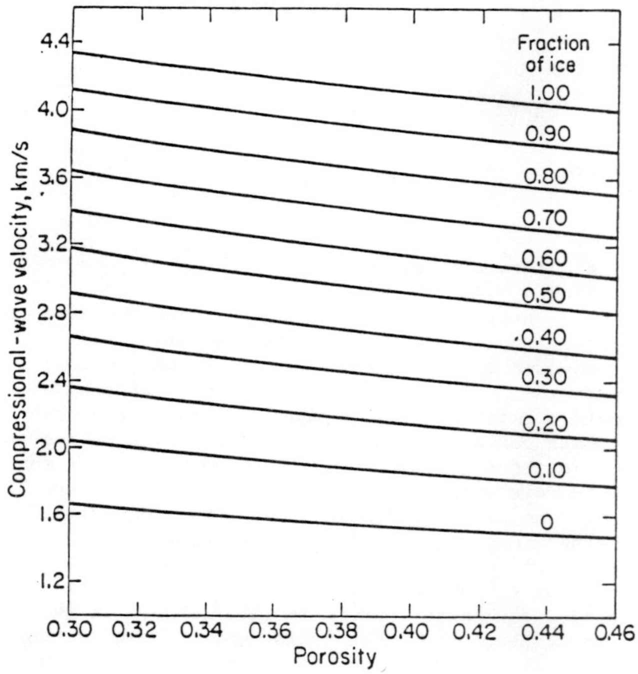

FIGURE 2 Theoretical curves relating compressional wave velocities, porosities, and fraction of ice in the pores of unconsolidated material (after King, 1984).

This histogram also indirectly demonstrates the occurrence of anomalously high velocities at shallow depths. The extremely low velocities at very shallow depths come from reflectors at the base of a thin layer of fine-grained material that blankets the shelf (unit A of S. Blasco, pers. comm.). There are a large number of reflectors at the depth of 17 m with an average velocity of 1600 m/s; this means that the interval velocity between the base of unit A and the reflectors at 17 m is somewhat higher than 1600 m/s.

A large number of reflectors can also be found centered at a depth of 37 m with an average velocity of 1800 m/s to that depth. The actual interval velocities between the upper reflectors and this reflector zone are often much higher: 2000-m/s interval velocities are not uncommon. This lower reflector grouping can be correlated with the top of the upper discontinuous ice-bonded layer recognized in our refraction surveying; it is also the reflector from the various forms of APF recognized in our earlier reflection surveying. Thus, surficial deposits with relatively high seismic velocities occur from shallow depths downward to the top of the APF.

This raises the question as to whether this relatively high-velocity material represents ice-bearing sediments. Figure 2 is a set of theoretical curves from King (1984). These curves relate compressional wave velocities, porosities, and the fraction of ice in the pores of unconsolidated materials. Notice that for the range of porosities given, the "non-ice-bearing" condition is less than 1650 m/s. Notice also that the threshold velocity of 2400 m/s used in early refraction interpretations to indicate ice-bonded material can be related to 21 to 33% ice-filling of pore spaces and that, according to King (1984), ice-bearing materials can exist between these velocities.

King notes that this model should be extensively tested, but employs experimental data to show a possible linear correlation between velocity and the remaining water-filled porosity. This is independent of original porosity in the unfrozen state, fraction of clay particles, or temperature below 0°C. King's work brings us a long way toward establishing a firm framework to interpret seismic velocities in terms of the index properties of perennially frozen materials.

The measurement of seismic velocities from marine or sea-bottom refraction or reflection, or by borehole seismic methods, should be an integral part of future site surveying in the Beaufort Sea and in other shelf areas where permafrost conditions prevail. An ideal site survey would combine multidisciplinary geophysical techniques with electromagnetic and electrical surveys and a complete suite of geophysical logs on all geotechnical boreholes as a complement to the seismic data.

It is obvious from presentations at this Fourth International Conference on Permafrost that great strides have been made in the application and development of geophysical techniques for use in detecting and describing marine permafrost for use in geotechnical site evaluations. The burden is now on the geotechnical engineer to make effective use of them.

REFERENCE

King, M.S., 1984, The influence of clay-sized particles on seismic velocity for Canadian Arctic permafrost: Canadian Journal of Earth Sciences, v. 21, p. 19-24.

SUBSEA PERMAFROST AND PETROLEUM DEVELOPMENT

H.O. Jahns

Exxon Production Research Company
Houston, Texas, USA

The following comments constitute a petroleum engineer's view of the significance of subsea permafrost in relation to two common heat sources introduced during offshore petroleum development: production wells and pipelines.

The question is often asked -- "Since we've solved production well problems on shore, what's different about offshore permafrost?" On land at Prudhoe Bay we are in fact producing through permafrost in a satisfactory manner using well casings designed to resist the stresses and strains that occur when the surrounding permafrost thaws. But there are some important differences between the onshore and offshore permafrost environments and how they are developed. Offshore, wells will have to be placed closer together, since space is at a premium on a man-made island. The resulting bundle of wellbores will constitute a greater heat source than the more widely spaced wells on an onshore drilling pad. Consequently, there will be more thaw offshore, and the thaw front will move farther out from the wells. A second difference is that offshore permafrost is warmer, on average, than onshore permafrost. In general, the entire column of offshore permafrost is relatively warm, and it can be near the freezing point of the overlying sea water. Therefore, with a given heat input there will be more thaw in the upper portion of the offshore permafrost zone than under otherwise comparable onshore conditions. A third difference is that offshore permafrost may have a higher salt content. Therefore, a smaller fraction of the pore water will be frozen at a given initial temperature. This means that less heat is required to melt the pore ice.

As stated, the thaw bulb around the wellbores will tend to be larger in the upper portion of the subsea permafrost section than for typical onshore wells. Therefore, surface subsidence is expected to extend farther from the wells into areas on an island where one would like to place production facilities. Another consideration is the volume change and thaw strain that will occur within the thaw bulb. Some of the same factors that cause the thaw bulb to be larger offshore can tend to reduce the thaw strain. In areas with high permafrost temperatures and salt contents, a smaller fraction of the pore water may be frozen, resulting in smaller volume change and less soil deformation when a given volume of permafrost thaws. However, if relict permafrost with fresh pore water is present, thaw strain may not be reduced.

In summary, the special characteristics of offshore permafrost can be detrimental in some respects and beneficial in others. Therefore, the properties of subsea permafrost need to be understood when predictions are being made of thaw radius and thaw subsidence for an offshore production island.

I will cover the next topic, offshore pipelines, in a little more detail. I briefly discussed this subject as a member of the panel on pipeline construction. The result of placing a warm pipeline in the thawed layer above ice-bonded subsea permafrost is shown in Figure 1. As indicated, it is assumed that the thawed zone has saline pore water. If the pore water were fresh, the zone would be frozen because the mean annual seabed temperatures are usually below 0°C in the Beaufort Sea. The rate of salt migration seems to control the natural degradation of offshore ice-bonded permafrost (Swift et al., 1983). It is assumed, therefore, that saline water replaces the original pore water in the sediments.

The natural process of salt migration and degradation of ice-bonded permafrost is altered when we introduce a warm offshore pipeline. As shown in Figure 1, permafrost degradation is accelerated below the pipeline. Permafrost degradation can now occur without involving salt migration. However, the thaw rate will still be influenced by the rate of salt migration within the thaw bulb. If salt migration is too slow to keep up with the advancing melt front below the pipeline, then the temperature at the melt front will be determined by the salinity of the original pore water. If the permafrost is essentially fresh, then the melting temperature will be near 0°C. If, on the other hand, salt migration is fast enough to keep up with the melt front below the pipeline, then we will have sea water in contact with the permafrost, and the effective melting

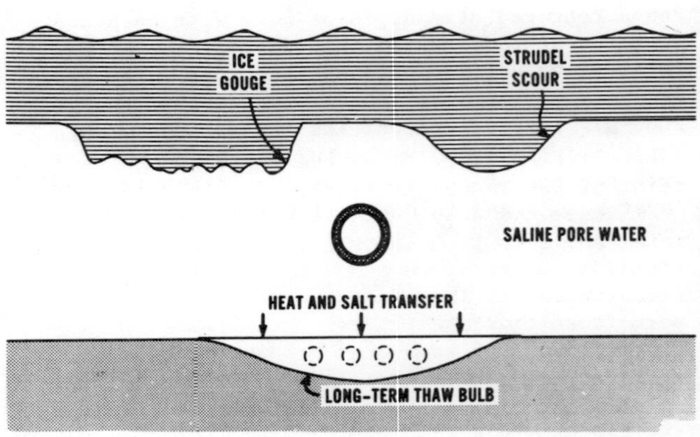

FIGURE 1 Thaw associated with placement of a warm pipeline in the thawed layer above ice-bonded subsea permafrost.

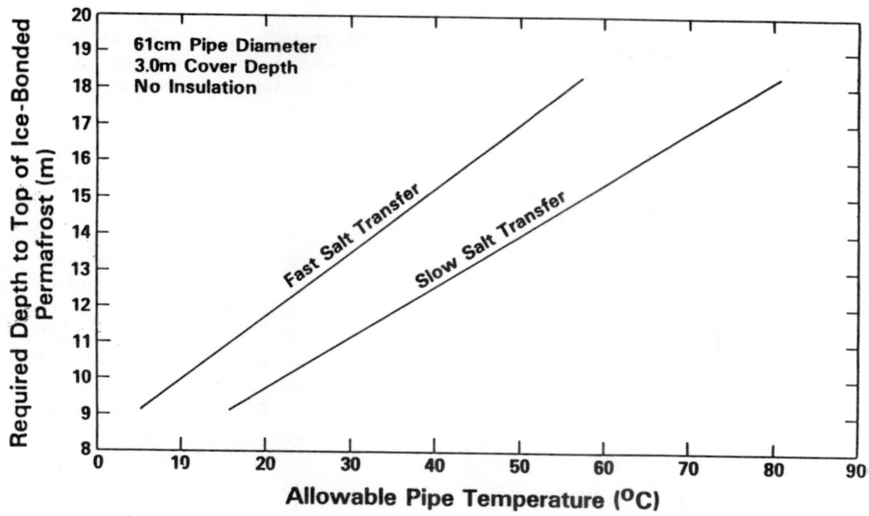

FIGURE 2 Influence of allowable pipe temperature on the required natural depth to top of ice-bonded permafrost for different salt transfer rates (from Heuer et al., 1983).

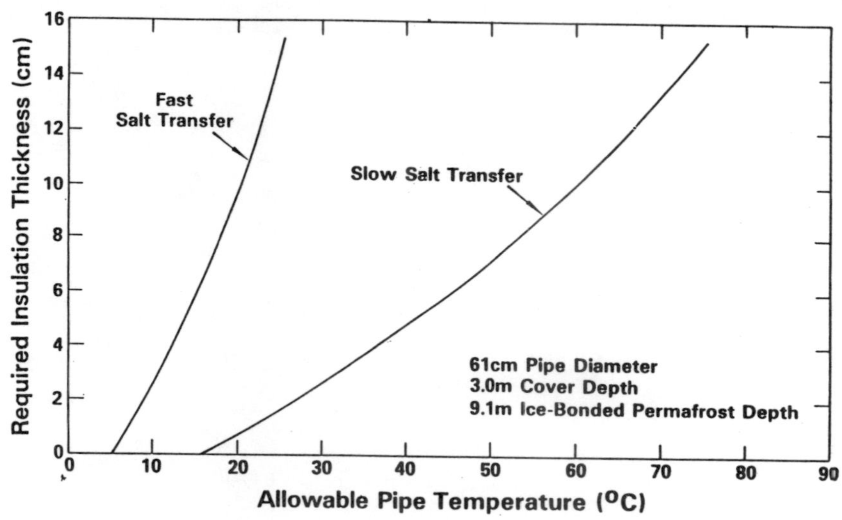

FIGURE 3 Example of the influence of allowable pipe temperature on required insulation thickness for different salt transfer rates, in an area of given depth to the top of ice-bonded permafrost (from Heuer et al., 1983).

temperature will be depressed to the freezing point of sea water.

These two cases will result in different rates and extents of permafrost thaw under the pipeline (Heuer et al., 1983). Figures 2 and 3 demonstrate the importance of salt migration in permafrost thaw predictions for offshore pipelines. Figure 2 shows the allowable pipe temperature, assuming certain soil properties and pipe deformation criteria, as a function of depth to the top of the permafrost before pipeline startup. The two curves were obtained by assuming "slow" salt transfer in one case and "fast" salt transfer in the other. In the calculations, this difference is expressed simply by the temperature specified at the melt front: 0°C and -1.9°C (corresponding to 35 ‰ salinity), respectively. Except for this difference, the calculations are carried out in exactly the same way for both cases, i.e. simulating only heat transfer by conduction. This simplification is permissible for the limiting case of either very rapid or very slow salt migration. Intermediate cases are more difficult to calculate, since both heat and mass transfer have to be simulated numerically.

For example, consider a case where the pipeline is at 50°C and salt transfer is slow relative to melting of permafrost below the pipe. The calculations indicate that the pipeline will not be over-stressed if the top of permafrost is initially 14 m or more below the sea floor. If, on the other hand, salt transfer is sufficiently fast to keep up with permafrost degradation below the pipeline, the required initial permafrost depth is shown to be about 18 m below the sea floor.

In situations where permafrost is shallower than the required depth indicated by this diagram, the pipeline may have to be insulated or the pipe temperature may have to be controlled in order to limit permafrost thaw.

Figure 3 shows the results of calculations for shallow permafrost (9 m below the sea floor) for a situation where an insulated line may have to be used. The thickness of insulation required, and the allowable pipeline temperature, depend again on the assumed rate of salt transfer. For example, with 10 cm of insulation, the allowable pipe temperature is shown to be 60°C in the case of slow salt transfer, but only 20°C for fast salt transfer.

It could be argued that the sediments must be of the type that only allows relatively slow salt transfer in those situations where the top of ice-bonded permafrost is found close to the sea floor. Therefore, the more optimistic right-hand curve on this diagram may be applicable for most situations. This point deserves further investigation.

In conclusion, it appears that mechanisms of salt transport in sea floor sediments can play a significant role in the thermal design of offshore pipelines. However, I must also emphasize that, as Heuer et al. (1983) and Walker et al. (1983) show, engineering design solutions are available to handle most, if not all, anticipated permafrost conditions in the Alaskan Beaufort Sea.

REFERENCES

Heuer, C.E., Caldwell, J.B., and Zamsky, B., 1983, Design of buried seafloor pipelines for permafrost thaw settlement, _in_ Proceedings of the Fourth International Conference on Permafrost, Fairbanks: Washington, D.C., National Academy of Sciences, p. 486-491.

Swift, D.W., Harrison, W.D., and Osterkamp, T.E., 1983, Heat and salt transport processes in thawing subsea permafrost at Prudhoe Bay, Alaska, _in_ Proceedings of the Fourth International Conference on Permafrost, Fairbanks: Washington, D.C., National Academy of Sciences, p. 1221-1226.

Walker, D.B.L., Hayley, D.W., and Palmer, A.C., 1983, The influence of subsea permafrost on offshore pipeline design, _in_ Proceedings of the Fourth International Conference on Permafrost, Fairbanks: Washington, D.C., National Academy of Sciences, p. 1338-1343.

GEOTECHNICAL AND ENGINEERING SIGNIFICANCE OF SUBSEA PERMAFROST

D.W. Hayley

EBA Engineering Consultants, Ltd.
Edmonton, Canada

My role on the panel is to buffer the state-of-the-art science being discussed with more pragmatic engineering considerations. I plan to do this by reviewing some of the implications of subsea permafrost in the exploration and production of offshore resources. This will be based on exposure to current standards of practice in both the Canadian and Alaskan sectors of the Beaufort Sea. I will discuss several concerns that have had little or no coverage at the Fourth International Conference on Permafrost, and I also wish to suggest how we can improve our approach to site investigation and data assimilation. Finally, instead of discussing how well we can deal with certain problems, I will highlight some deficiencies in our approach to offshore engineering when permafrost is present.

First, let us look at the SOHIO Arctic Mobile Structure (SAMS) that has been proposed for use in Harrison Bay, Alaska (Fig. 1). This self-contained structure would be floated into place and ballasted onto the seabed, and then large piles, or spuds, would be driven into the bed. The spuds will not carry a vertical load but are required to develop sliding resistance under an applied ice load. This innovative design evolved from a desire for a universal structure applicable for most seabed soils in Harrison Bay.

The soil profile shown in Figure 2 is from one of the more troublesome sites in Harrison Bay. Stiff, overconsolidated clay or silt is present at the seabed, and ice-bonded permafrost is present at a depth of about 10 m. The shear strength of the clay decreases with depth, and a distinct zone of soft clay with a shear strength of only about 25 kPa is present just above the ice-bonded perma-

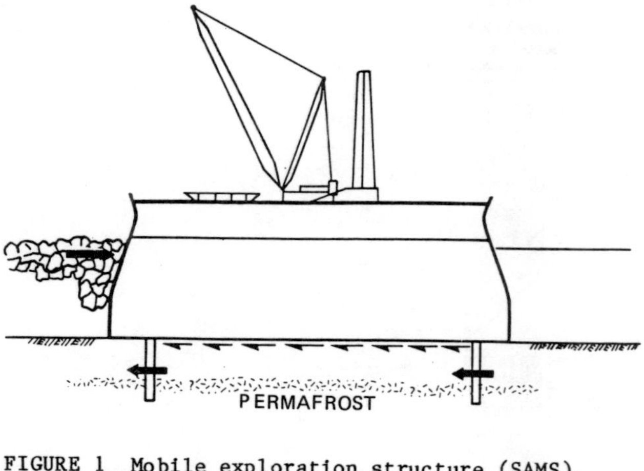

FIGURE 1 Mobile exploration structure (SAMS), from Gerwick et al. (1983).

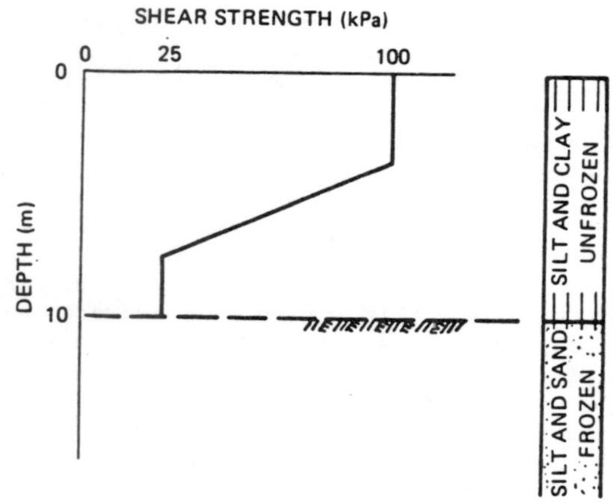

FIGURE 2 Shear strength of unfrozen soils, Site A, Harrison Bay, from Bea et al. (1983).

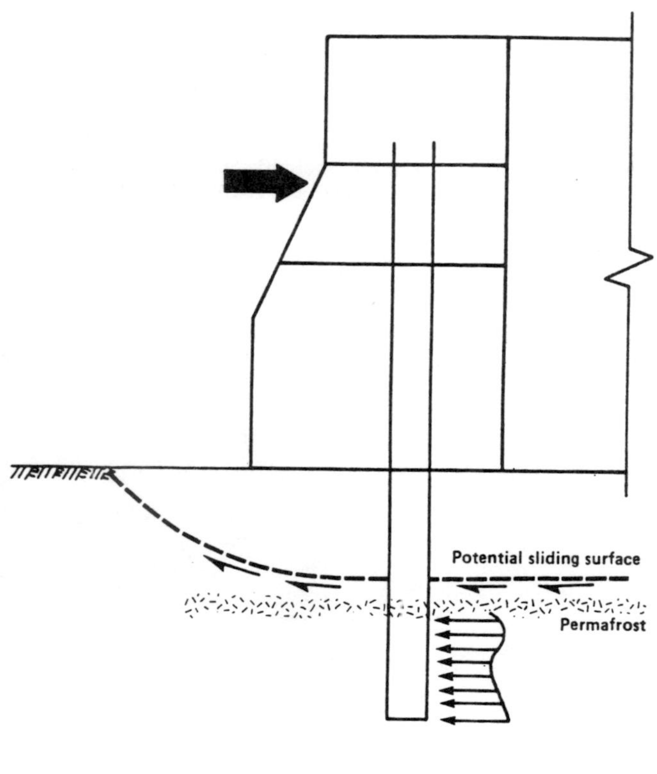

FIGURE 3 Stability evaluation.

frost. The significance of this soft zone in relation to the SAMS is shown in Figure 3. An overall stability evaluation for this structure suggests that the critical failure plane could extend down to the soft zone, with the obvious implication that much of the sliding resistance is picked up by the portion of the spuds embedded in the bonded permafrost. Fundamental soil parameters required for analysis of this structure include the shear strength of the ice-bonded permafrost as well as its modulus and creep characteristics.

The question of pile driveability or resistance to penetration also surfaces for a structure of this type and has not been adequately addressed. We have had some success with field methods for evaluating the bearing capacity of subsea permafrost by static cone penetration tests (CPT) using procedures similar to those described by Ladanyi (1976). But although the engineering properties of permafrost can be measured in the laboratory, acquiring, handling, and testing of frozen saline soil can be a problem. The properties of these materials are very sensitive to variations in temperature, movement, and drainage of saline pore water and deformation that may occur during sampling and handling. Therefore, there is a place for techniques that can be used to measure material properties in situ, such as the use of the pressuremeter discussed by Briaud et al. (1983).

I would like to address briefly the implications of thawing subsea permafrost as it affects production structures. Offshore production will require multiple wells in close proximity to each other (Fig. 4). In some cases, well deviations or slant hole drilling within the permafrost may be required to develop shallow reservoirs. Deviated wells will be subjected to complex casing loading; to my knowledge this problem has not been assessed. Soil movements at depth, furthermore, will be reflected by surface settlements. In some cases, it may be possible to locate foundations for production wells outside the zone of influence. However, this is probably not feasible for deep-water structures, and therefore foundation design will have to take surface subsidence into

TABLE 1 Observed permafrost features unique to offshore locations

1. Freshwater ice in nonbonded saline silt or clay (crystals and segregated lenses)
2. Alternating bonded sand and nonbonded silt or clay (identical temperatures)
3. Excess ice in sand above a contact with underlying nonbonded clay
4. Hypersaline conditions below bonded intervals in sands and clays (salt exclusion by freezing?)
5. Significant differences in degree of bonding with subtle textural change (silt/clay)

TABLE 2 Improvements needed in arctic offshore site investigation practice.

1. Look beyond immediate requirements when planning programs.
2. Improve wireline coring techniques in permafrost.
3. Log geologic details.
4. Adopt on-site testing of frozen soils
   - Time-domain reflectometry for unfrozen water content
   - Refractometer for salinity content measurements
5. Expand in situ testing methodology
   - Cone penetration testing in permafrost
   - Pressuremeter testing in permafrost

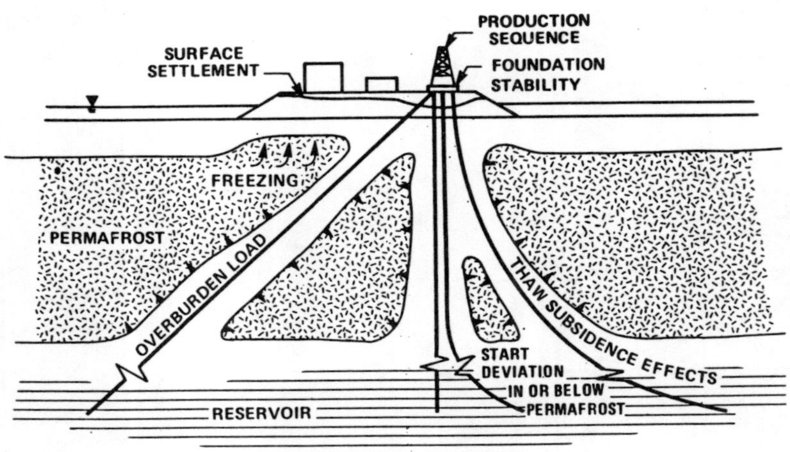

FIGURE 4 Some concerns arising from thaw around production wellbores.

consideration. The use of pad foundations with provision for re-leveling is not new for northern structures, although application of such drastic procedures in design of large production modules is not readily accepted. A prerequisite to rational design will be a proven capability to predict settlement magnitude and distribution, together with a knowledge of how the settlements might develop during the life of the structure.

A key element that has been missing in most analyses conducted to date has been a geotechnical knowledge of soil properties throughout the permafrost zone. Technology currently exists to obtain continuous cores to the bottom of permafrost and to characterize fully the engineering properties of this material through laboratory testing. However, surprisingly little has been reported in the literature on the properties of deep subsea permafrost.

Clearly, subsea permafrost presents a new set of engineering challenges. The requirements for careful determination of site conditions cannot be overstressed. Some of our observations resulting from the past decade of data collection in the Beaufort Sea are shown in Table 1. Routine site investigation practice is not adequate to deal with these conditions and to provide parameters of the type needed for design. Some of the improvements that we need to see in site investigation practice are listed in Table 2, in hope of stimulating applied research directed toward their understanding.

REFERENCES

Bea, R.G., Nour-Omid, S., and Coull, T.B., 1983, Design for dynamic loadings, Paper presented at Conference on Geotechnical Practice in Offshore Engineering, Austin, Texas.

Briaud, J.L., Smith, T.D., and Meyer, B.J., 1983, Using the pressuremeter curve to design laterally loaded piles: 15th Offshore Technology Conference, Paper OTC 4501, Houston, Texas.

Gerwick, B.C., Potter, R.E., Matlock, H., Mast, R., and Bea, R.G., 1983, Application of multiple spuds (dowels) to development of sliding resistance for gravity-base platforms: 15th Offshore Technology Conference, Paper OTC 4553, Houston, Texas.

Ladanyi, B., 1976. Use of the static penetration test in frozen soils: Canadian Geotechnical Journal, v. 13, no. 2, p. 95-110.

Pipelines in Northern Regions

PANEL MEMBERS

Oscar J. Ferrians, Jr. (Chairman)*, U.S. Geological Survey, Menlo Park, California, USA

Amos C. Mathews, Office of the Federal Inspector, Alaska Natural Gas Transportation System, Anchorage, Alaska, USA

Hans O. Jahns, Exxon Production Research Co., Houston, Texas, USA

Michael C. Metz, GeoTec Services Inc., Golden, Colorado, USA

Elden R. Johnson, Alyeska Pipeline Service Co., Anchorage, Alaska, USA

INTRODUCTION

O.J. Ferrians, Jr.

Because of the critical need to develop new sources of oil and gas, several major pipelines and related facilities have been constructed in the arctic and subarctic regions of the world during the last two decades, and undoubtedly more will be constructed in the future. These regions pose special engineering problems and are environmentally "sensitive." Consequently, they require careful consideration to ensure the integrity of pipelines and related facilities and to minimize adverse environmental impacts. One of the most important factors to be considered is permafrost.

Permafrost is a widespread natural phenomenon in northern regions, and it underlies approximately 20% of the land area of the world. The permafrost region of Alaska includes 85% of the state (Ferrians et al., 1969), the permafrost regions of Canada and the U.S.S.R. cover about 50% of each country, and the permafrost region of China covers about 20% of that country (see Frontispiece).

The most serious permafrost-related engineering problems generally are caused by the thawing of ice-rich permafrost, which results in a loss of bearing strength and a change in volume of ice-rich soil. Under extreme conditions that permit very rapid thawing, ice-rich, fine-grained soil can liquefy and lose essentially all of its strength. A more common occurrence is differential settlement of the ground surface.

Proposals to chill the natural gas in buried pipelines as a means of mitigating the problems caused by the thawing of permafrost pose a different problem--namely frost heaving, due to the freezing of pore water in soils surrounding the pipe and to the freezing of additional water attracted to the freezing front.

There are numerous topics that could have been treated in a discussion of pipelines in northern regions; however, for this report the following important and timely subjects have been selected:

- Pipelines in the northern U.S.S.R.
- Hot oil and chilled gas pipeline interaction with permafrost
- Pipeline thermal considerations
- Pipeline workpads in Alaska
- Performance of the trans-Alaska oil pipeline

*The panel chairman wishes to acknowledge the assistance of Fred Crory, Cold Regions Research and Engineering Laboratory, in organizing the panel. Unfortunately, Mr. Crory was unable to attend the conference.

PIPELINES IN THE NORTHERN U.S.S.R.

O.J. Ferrians, Jr.*

U.S. Geological Survey
Menlo Park, California, USA

The factual material in the following description of pipelines in the northern U.S.S.R. was obtained primarily from Ivantsov et al. (1983).

The northern regions of the U.S.S.R. contain huge oil and gas deposits, especially in western Siberia. The development of these resources requires the construction of pipelines, pump and compressor stations, roads, railroads, temporary and permanent settlements, and other facilities. Permafrost exists in the oil and gas fields, along the pipeline, and along other transportation corridors of western Siberia and in other northern regions of the U.S.S.R. Consequently, solutions to a variety of complex permafrost-related scientific and engineering problems are necessary to accomplish this development effort successfully with a minimum of adverse environmental impacts.

Western Siberia between the Ural Mountains and the Yenisey River is an extensive, poorly drained lowland. It is underlain for the most part by fine sand, silt, and clay. Sand deposits are restricted to the flood plains of medium-size and large rivers. Bedrock and gravel deposits are absent throughout most of the region. Since these materials are generally needed for roads, work

*This summary was prepared in the absence of O.M. Ivantsov, Ministry for Construction for Oil and Gas Industries, Moscow.

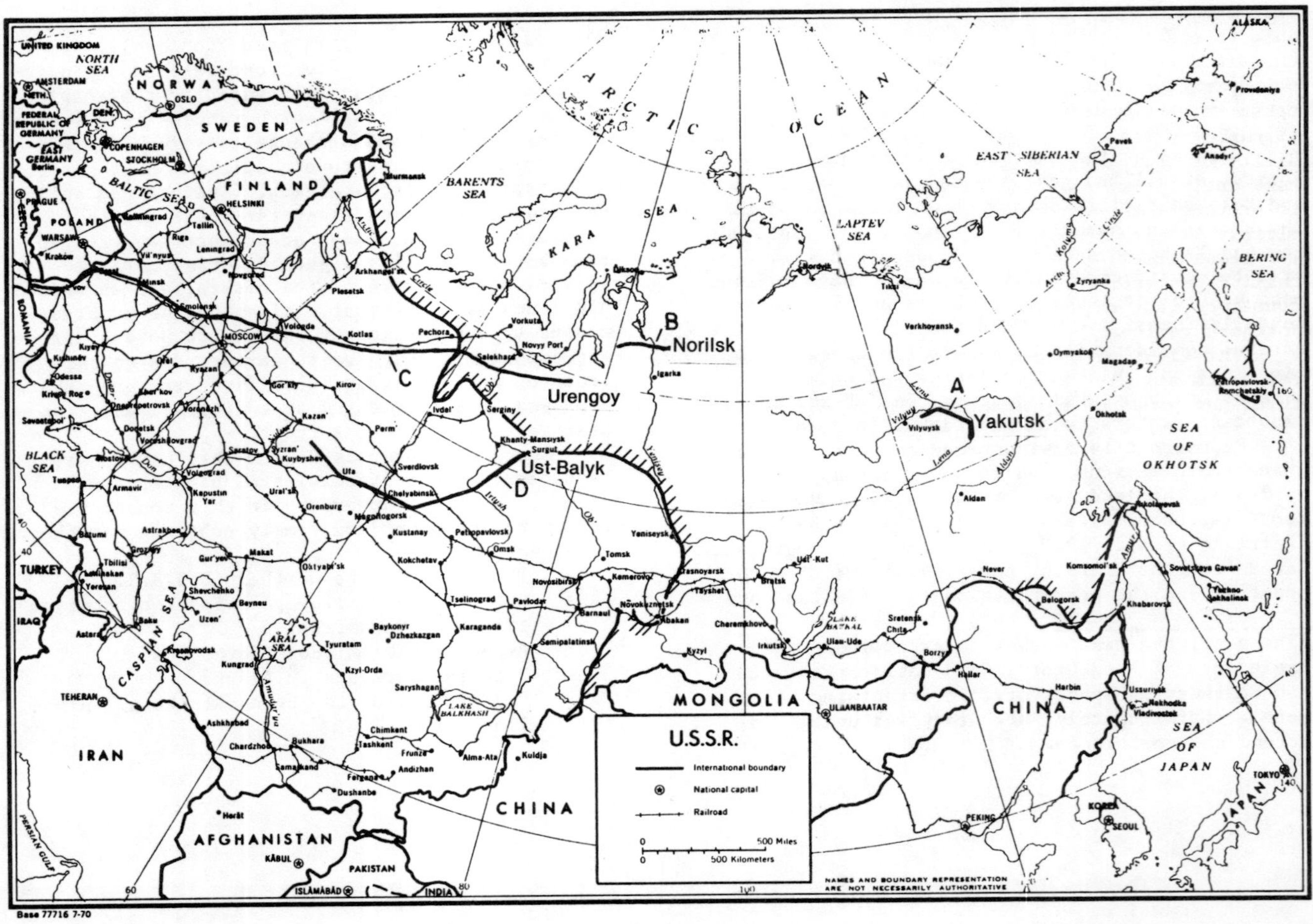

FIGURE 1 Index map showing the location of important pipelines in northern USSR: (A) the small-diameter gas pipeline serving Yakutsk, (B) the small-diameter gas pipeline serving Norilsk, (C) the large-diameter mainline gas pipelines originating in the vicinity of Urengoy, and (D) the large-diameter oil pipeline originating near Ust-Balyk. The southern boundary of the permafrost region is shown by the hachured line.

pads, bedding and padding of the pipe, riprap, and concrete aggregate, construction is difficult and expensive.

Perennially frozen ground, or permafrost, is widespread in the northernmost coastal lowlands, tundra, and forest tundra zones of western Siberia. In these areas, the thickness of the permafrost ranges from 500 m in the north to 50 m in the south. Permafrost is discontinuous south of the Arctic Circle, and it occurs only in peat bogs south of the 63rd parallel. The temperature of the permafrost in these more southerly areas generally ranges from 0°C to -1°C or -2°C and, consequently, it is very sensitive to disturbances. If the thermal regime at the ground surface is altered, thawing quickly takes place.

Pipeline construction in permafrost regions can cause thawing of permafrost, thaw settlement, and thermal erosion. These detrimental effects generally result from removal of the vegetation cover, placement of backfills or thin workpads, modification of surface and subsurface drainage, or changing the albedo due to the accumulation of dust. During construction, many of these man-induced changes can be severe; but they can be controlled by proper design, scheduling, and timely maintenance, including revegetation compatible with the new environment. It is necessary to be able to predict the potential environmental changes, since they can directly influence the successful operation of a pipeline system. The measures taken to protect the environment are equally applicable to safeguarding the integrity and reliability of the pipeline system. Therefore, the best engineering solutions and the best measures to protect the environment and the pipeline are closely interrelated and interdependent. Obviously, the thermal effects of warm oil or gas or of chilled gas must be considered in order to design, construct, and maintain pipelines properly in northern regions.

The first major pipeline constructed in the Far North was the 529-mm-diameter, 600-km-long natural gas pipeline serving Yakutsk (Fig. 1). This pipeline, which includes 250 km of aboveground line, has been operating successfully for more than 15 years. A system of gas pipelines serving the city of Norilsk was constructed entirely above ground except for river crossings. One line is more than 600 km long and 720 mm in diameter, another is 300 km long but has a smaller diameter. Construction of these pipelines started in 1968, and the first line was completed in 1969. These early gas pipelines in permafrost areas were built largely above ground to eliminate the possibility of adverse impacts on permafrost. As more experience is gained in pipeline construction in northern areas, underground placement has gained favor.

Large transcontinental gas pipeline systems were constructed in the 1970s. These pipelines have a diameter of 1420 mm and pressure of 75 kg/cm^2. The first giant pipeline, with a length of 3000 km, originated in the northern part of the Tyumen region near the Medvezhye gas field. During 1981-1985, a multiline system of five parallel pipelines will be constructed, each with a diameter of 1420 mm. Three of these lines, the Urengoy-Gryazovec-Moscow, Urengoy-Petrovsk, and Urengoy-Novopskov, have already been constructed. During this period, the Urengoy-Pomary-Uzhgorod gas pipeline for exporting gas also will be put into operation. The length of this pipeline is 4500 km. Oil pipelines from the northern regions also are under construction. One of the largest oil pipeline systems is the main pipeline Ust-Balyk-Kurgan-Ufa-Almetyevsk, having a diameter of 1220 mm and a length of more than 1800 km.

The construction of pipelines is most advantageously done in winter, when damage to the ground surface can be held to a minimum and the frozen ground ensures that heavy equipment can move about easily on the tundra and in swampy areas. Underwater crossings are built in the winter using the ice cover as a platform. To reduce the impact of gas pipelines on permafrost, a method of cooling the gas to a temperature of -2 to -3°C has been developed, and cooling facilities are currently being constructed at the Urengoy gas fields.

HOT-OIL AND CHILLED-GAS PIPELINE INTERACTION WITH PERMAFROST

A.C. Mathews

Office of the Federal Inspector, Alaska Natural Gas Transportation System
Anchorage, Alaska, USA

The complications arising from the interaction of pipelines and permafrost generally can be categorized as: (1) surface deterioration, (2) erosion, (3) drainage interruption, (4) thaw settlement, (5) frost heave, and (6) effects associated directly with the construction effort. In dealing with these complications, there is no clear-cut distinction between cause and effect--drainage interruption can cause aufeis, which can cause surface deterioration, which causes drainage interruption, etc. It is this circular, exacerbating chain of events that can cause such chaos if permafrost problems are not anticipated and designed for, or if mitigative measures are not taken immediately when a problem is identified. Thaw settlement occurs to a structure when heat is introduced causing ice-rich soils under its foundation to thaw. Frost heave is a complementary problem that occurs when subfreezing cold is introduced into non-frozen, frost-susceptible soil. Moisture, attracted to the cold, percolates through the soil and freezes at the cold front in quantities excess to the pore capacity of the soil. Over a period of time, a large frost bulb can build up, causing heaving in a structure above it. Finally, there are a number of complications arising directly from construction, such as excavation, the dormant period (when the pipe is completed and buried but not yet carrying product), the gravel work pads used in summer, snow/ice pads used in winter, and the massive requirements for gravel to ameliorate permafrost complications.

For reasons of both security and economy, pipelines generally are best kept below ground. Therefore, the trans-Alaska oil pipeline was buried wherever possible, in unfrozen soil and in thaw-stable permafrost. In areas of thaw-unstable soils, refrigerated piles were used to elevate the line and to preclude settlement of the piles due to permafrost thawing. At those few locations where it was necessary to bury the line in thaw-stable soil, the trench was refrigerated and/or the line was heavily insulated to provide stability. Refrigerated piles were used to elevate the line in high temperature permafrost to preclude thaw settlement of the piles due to permafrost thawing caused by the construction disturbance. Finally, all piles had to be designed against annual frost jacking. Designers of the proposed Alaska Highway natural gas pipeline plan to bury the line everywhere, with the exception of a few stream crossings. Because the gas is chilled, this should work well in permafrost areas, provided that thawing problems do not arise during the dormant period prior to startup. But in thawed frost-susceptible soils, the designers face a frost-heave problem whose magnitude is just now beginning to be fully understood. They plan to solve this problem by identifying the areas that contain these frost-susceptible soils and then employing suitable mitigative measures in construction. Underground stream crossings are apt to be a problem because the cold pipe could cause slow-moving groundwater to freeze and, in turn, affect surface drainage; faster flow rates seem to dissipate the effects of cold pipes. In any event, the pipe could be insulated to reduce the cold sink to a level that will not cause significant frost growth or, as a last resort, the pipe could be elevated over small streams where the effects are in question.

PIPELINE THERMAL CONSIDERATIONS

H.O. Jahns

Exxon Production Research Company
Houston, Texas, USA

The thermal aspects of pipelines in permafrost terrain are of paramount importance. In fact, it is difficult to talk about arctic pipelines without mentioning thermal interactions with the surrounding soil or, specifically, the consequences of a phase change in the soil's moisture content. In essence, this is what makes arctic pipelines different from conventional pipelines.

Figure 1 shows four schematic cross sections of a buried pipeline with different thermal regimes. A warm pipeline in permafrost is shown in Figure 1a, a warm pipeline in thawed soil in Figure 1b, a cold pipeline in permafrost in Figure 1c, and a cold pipeline in thawed soil in Figure 1d. It also illustrates the two basic mechanisms that give rise to engineering concerns: thaw settlement (Fig. 1a) and frost heave (Fig. 1d). The warm pipe in permafrost will cause a thaw bulb to grow around it. If the soil is ice-rich there would be subsidence, as indicated by the sag in the soil layers in Figure 1a. If the resulting settlement over-stresses the pipe, mitigative measures will be called for. Figure 1d, a mirror image of Figure 1a, shows a freeze bulb growing around the cold pipe in thawed soil. If the soil is frost-susceptible, it may heave, deforming the pipe. If the heaving is sufficient to overstress the pipe, mitigative measures would be needed. The two remaining cases (Figs. 1b and 1c) are trivial--not much happens with a cold pipe in frozen ground or with a warm pipe in thawed soil, except for some thinning of the active layer above the pipe. However, there are some complications at permafrost boundaries.

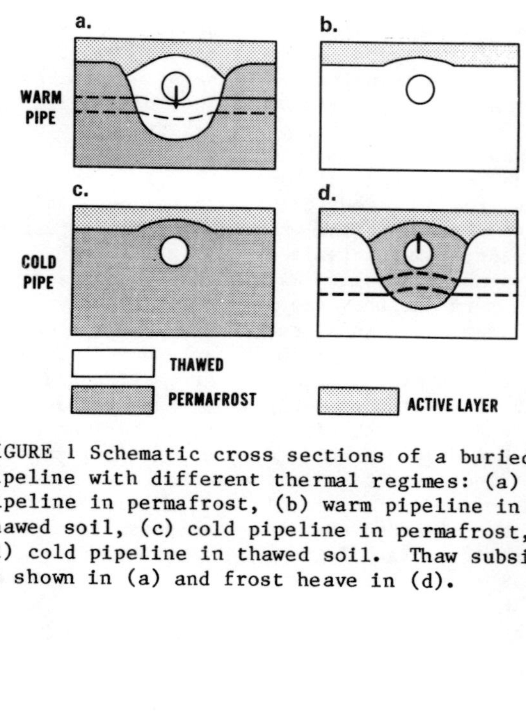

FIGURE 1 Schematic cross sections of a buried pipeline with different thermal regimes: (a) warm pipeline in permafrost, (b) warm pipeline in thawed soil, (c) cold pipeline in permafrost, and (d) cold pipeline in thawed soil. Thaw subsidence is shown in (a) and frost heave in (d).

For both the thaw settlement (Fig. 1a) and frost heave (Fig. 1d) cases, the allowable thaw or freeze bulb size depends on the mechanical interaction between the pipeline and the soil. Two design approaches are possible. First, the design could be based primarily on thermal considerations, and pipe deformation would be avoided by minimizing the size of the thaw or freeze bulb. Alternatively, pipe deformation could be allowed up to some operational limit. This second approach would have a less stringent thermal design, but more would have to be known about the mechanical behavior of pipes and soils.

The first method of thermal control that comes to mind is to insulate the pipeline. The effect can be dramatic. In fact, with good insulation, such as 10 to 15 cm of "jacketed" urethane, it is possible in many cases to prevent the formation of a thaw or freeze bulb, or to limit its size to only a few decimeters below the pipe, as shown in Figure 2. Design of the insulation system must consider the long-term effects of moisture absorption. If necessary, the natural soil below the pipe could be excavated half a meter or so and replaced with select material such as gravel that would undergo little thaw strain or frost heave. The term "overexcavation" is used for this construction mode.

There is a serious problem, however, with the construction mode shown in Figure 2a. This thaw bulb will grow slightly in size during the summer and will link up with the active layer above the pipeline. During the winter, the active layer freezes and the thaw bulb becomes isolated. As freezing continues, the pipe may be subjected to frost jacking due to pore water expansion, independent of whether or not the soil in the thaw bulb is frost-susceptible. This may be an unacceptable risk.

The buried insulated pipeline mode has another limitation: when the soil temperature is near the freezing point, insulation alone is not sufficient to stop the growth of a thaw or freeze bulb;

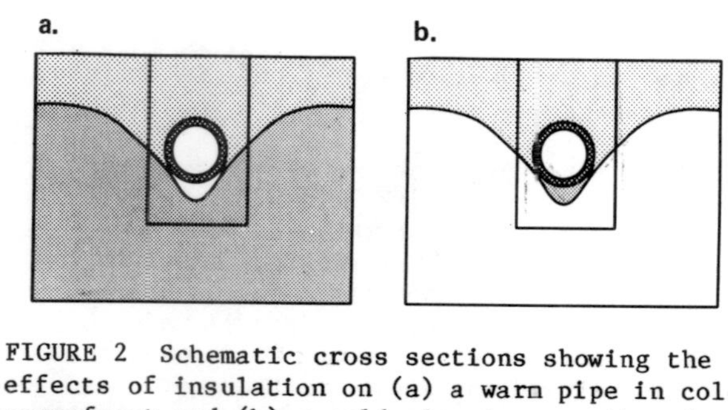

FIGURE 2 Schematic cross sections showing the effects of insulation on (a) a warm pipe in cold permafrost and (b) a cold pipe in warm thawed soil.

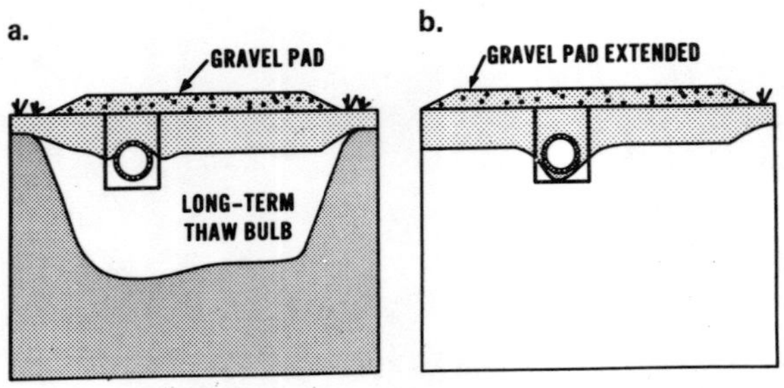

FIGURE 3 Schematic cross sections showing the effects of surface disturbance (gravel pad) on (a) a warm pipe buried in warm permafrost and (b) a cold pipe buried in cold thawed soil. Both pipes are insulated.

it will only slow it down. But during the 20- to 30-year lifetime of the pipeline, it will grow deeper than any practical amount of overexcavation. That is why Figure 2a is labeled "cold permafrost" and Figure 2b is labeled "warm thawed soil." As a rule of thumb, we may assume that these diagrams refer to permafrost at temperatures colder than -6°C and thawed soil warmer than +3°C. The intervening temperature regime from -6 to +3°C is the one that is most difficult to design for, for two reasons: (1) the fact that pipeline insulation is not sufficient to prevent long-term growth of a freeze bulb or a thaw bulb, as already discussed, and (2) the construction surface disturbance and its effect on the thermal regime in the soil.

The surface disturbance causes warming both in permafrost and in thawed soil. Therefore, the two cases shown in Figure 3 are no longer symmetrical. In warm permafrost (Fig. 3a), the gravel pad used for construction and pipeline maintenance causes long-term permafrost degradation, perhaps 6 to 9 m deep during the life of the pipeline. If the soil is ice-rich, this pipeline mode may not be acceptable. In the case of the trans-Alaska oil pipeline, the answer was to elevate.

The surface effect is very different for the cold pipeline in thawed soil (Fig. 3b). In fact, the surface disturbance is beneficial in that it tends to counteract the effect of the cold pipeline (Jahns et al., 1983). The disturbance can be enhanced by extending the gravel pad on both sides of the pipeline trench. Calculations indicate that this pipeline mode should be applicable in essentially all situations where the soil is already thawed and is covered with natural vegetation. In contrast to standard practice, this design does not call for revegetation. On the contrary, measures may have to be taken to limit revegetation.

The elevated pipeline mode shown in Figure 4 works equally well for both cold and warm pipelines. However, a distinction needs to be made between cold permafrost, which remains stable in spite of the surface disturbance, and warm permafrost, which degrades under the gravel pad. Designing an elevated pipeline for cold permafrost is straightforward, as shown in Figure 4a. Although the active layer depth increases, the pile supports are frozen in and remain in frozen soil

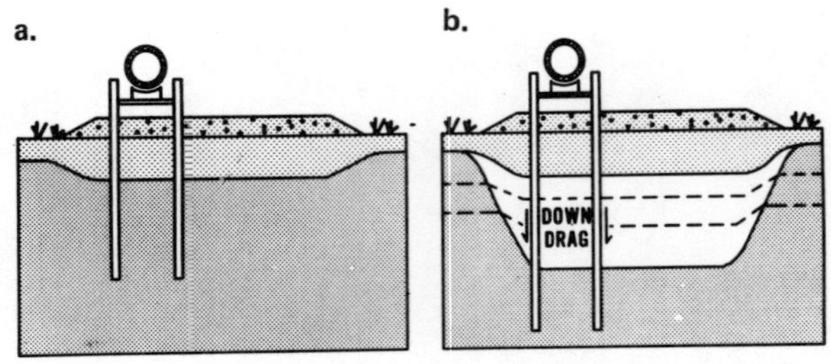

FIGURE 4 Schematic cross sections showing the thermal effects of an elevated pipeline and a gravel workpad on (a) cold permafrost and (b) warm permafrost.

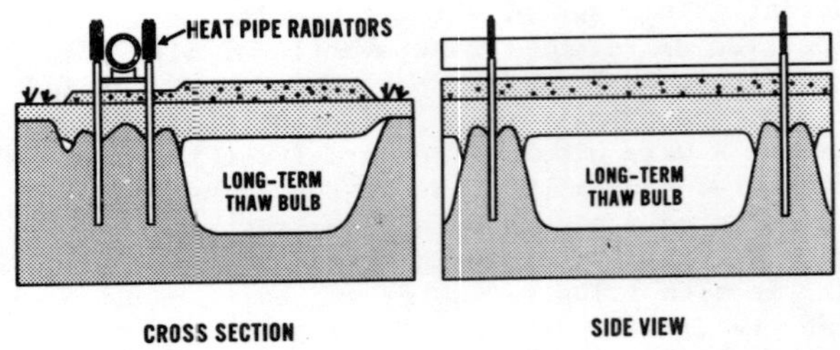

FIGURE 5 Schematic cross section and side view of an elevated pipeline with heat pipes (mitigative pipeline mode) on warm permafrost.

during the life of the pipeline. In warm permafrost, shown in Figure 4b, the situation is different: The permafrost degrades under the gravel pad and may subside and cause downdrag forces on the pipeline supports. The piles have to be designed to withstand these forces in addition to the weight of the pipeline. This may require very long piles and high installation cost. In some cases, the required pile lengths may become impractical. The thermal VSM design for the trans-Alaska oil pipeline was developed as a solution to this problem. A pair of heat pipes inside each pile removes heat from the surrounding soil during the winter to prevent thawing in the vicinity of the piles, out to a distance of about 3 m. Figure 5 shows the configuration as it may look 20-30 years after the installation: the pile supports are surrounded by pillars of frozen soil, while the soil under the gravel pad and between supports has thawed to depths of 6 to 9 m.

Under certain conditions, heat pipes may also be useful for a buried, cold gas pipeline. This may arise at transitions where the pipeline passes from frozen, thaw-unstable ground into thawed, frost-susceptible soil. Many such boundaries exist in nature and they are not always clearly defined. Therefore, insulated pipe may have to be extended some distance into frozen ground. Figure 6 shows two situations that may arise. Figure 6a shows the insulated cold pipeline where it has been extended into the frozen ground near the permafrost boundary. Since the cold pipe is insulated, it is not effective in preventing permafrost degradation under the gravel pad. The long-term result is similar to that for insulated, warm pipe shown in Figure 3a. Perhaps only a small frozen zone will remain below the pipe, not sufficient to

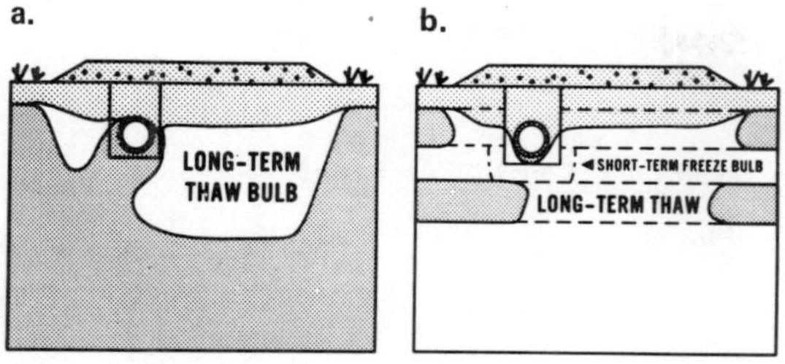

FIGURE 6 Schematic cross sections showing the thermal effects of an insulated, cold, buried pipeline at permafrost boundaries (a) in warm permafrost and (b) in mixed frozen/thawed soil.

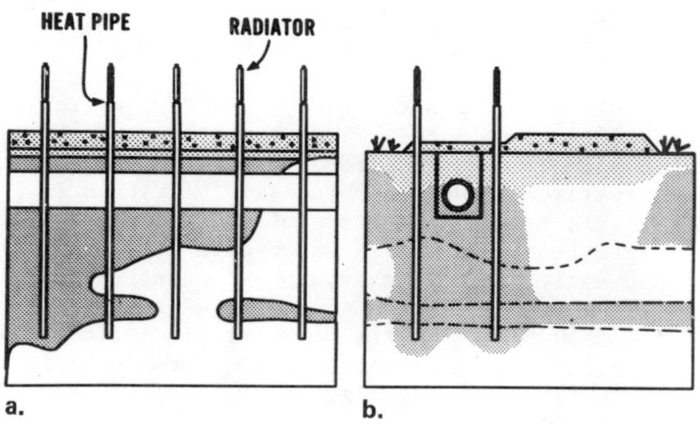

FIGURE 7 Schematic cross section and side view showing the thermal effects of an insulated, cold, buried pipeline with heat pipes (mitigative pipeline mode) at permafrost boundary at (a) time of installation and (b) long-term.

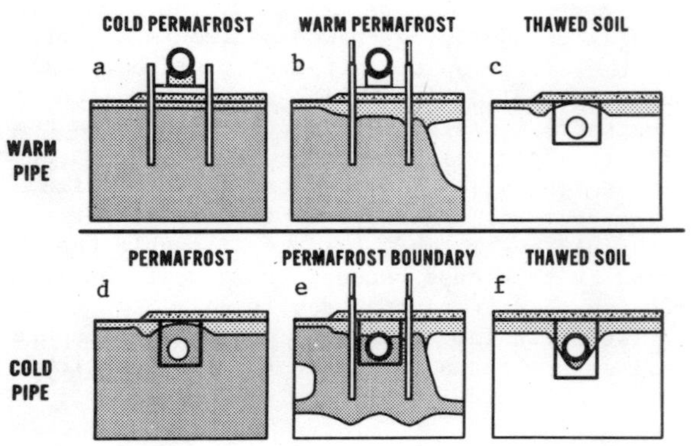

FIGURE 8 Schematic cross sections showing examples of mitigative pipeline modes for both warm and cold pipelines under various soil conditions.

give reliable support. Accurate predictions of thaw bulb geometry and pipe deformation are difficult to make in this case. Groundwater flow could lead to additional thaw. A more complex situation with alternating layers of frozen and thawed soil is shown in Figure 6b. In this case, the upper frozen layer would shield the soil below the pipeline from the effect of the surface disturbance during the early years of pipeline operation. A temporary freeze bulb could be formed during this time. Later on, once the near-surface permafrost layer has thawed, the newly formed freeze bulb and and relic permafrost below the pipeline would begin to degrade. Thus, the pipeline could undergo first heaving and then subsidence. To overcome these problems, the pipeline must be designed to be insensitive to the particular geometry of permafrost boundaries.

Figure 7 shows an example of a mitigative mode for permafrost boundaries based on the use of free-standing heat pipes placed in a row on either side of the buried cold pipeline. The function of the heat pipes is to freeze any thawed ground in the vicinity of the pipeline and to keep initially frozen ground frozen. The main direction of heat flow is horizontal, radially into the heat pipes, so the potential for frost heave is greatly reduced. This conclusion is supported by experience from the trans-Alaska oil pipeline where heaving of thermal VSM's (vertical support members) has not been a significant problem; however, some minor heave has occurred (Table 1). Depending on the timing of heat pipe installation and pipeline startup, it may be possible to eliminate pipeline insulation in this construction mode. Longitudinal spacing of the heat pipes, shown in the side view (Fig. 7a), would be much closer than in the design for the trans-Alaska oil pipeline, perhaps every 3 m or so, since a continuous freeze bulb is desired under the pipeline. The total number of heat pipes should not be excessive, however, since this special pipeline mode would be required only sporadically for short distances--perhaps 30 m--on either side of a permafrost boundary.

Transition problems also occur for warm pipelines. These problems were avoided on the trans-Alaskan oil pipeline by going from the buried to the elevated mode before entering frozen, thaw-unstable soil. The elevated mode was also used in areas of mixed frozen and thawed soil.

Figure 8 shows a summary of the pipeline construction modes that have been described: three for the warm pipe (Figs. 8a, 8b, and 8c), and three for the cold pipe (Figs. 8d, 8e, and 8f). The modes shown for the warm pipeline are the ones used for the trans-Alaska oil pipeline: simple elevation on cold permafrost, elevation with heat pipes in warm permafrost, and conventional burial in thawed soil. The three buried modes for a cold pipeline are conventional burial in permafrost, heat pipes at permafrost boundaries, and insulated pipe in thawed soils. An extended gravel pad, shown in Figure 8f, is used to take full advantage of the surface disturbance for preventing the growth of a freeze bulb. For the same reason, revegetation would be limited. These measures would, of course, not be required in areas where the thawed soil is not frost-susceptible, or where it can be shown that the expected amount of frost heave would not over-stress the pipe.

Some of the more important design considerations for warm offshore pipelines in areas of ice-bonded subsea permafrost are shown in Figure 9. Subsea permafrost is warm permafrost, not much colder than the overlying seawater, typically about -1 to -3°C. In that regard, it is similar to the warm permafrost in central Alaska, but the similarity ends there; there is no active layer and no organic layer, and therefore construction

TABLE 1 Performance of the various pile types used to elevate the Trans-Alaska Oil Pipeline.

Pile type	Soil conditions	No. installed	% settled (> 9 cm)	% jacked (> 9 cm)
Thermal	Warm permafrost	42,720	0.5	1.2
Thermal	Thawed ground	18,420	0.6	3.4
Adfreeze	Cold permafrost	13,473	0.3	0.0
Friction/end bearing	Thawed ground	1,992	2.1	0.3
Other	Mixed profiles	1,127	0.2	1.0
Total		77,732	0.6	1.9

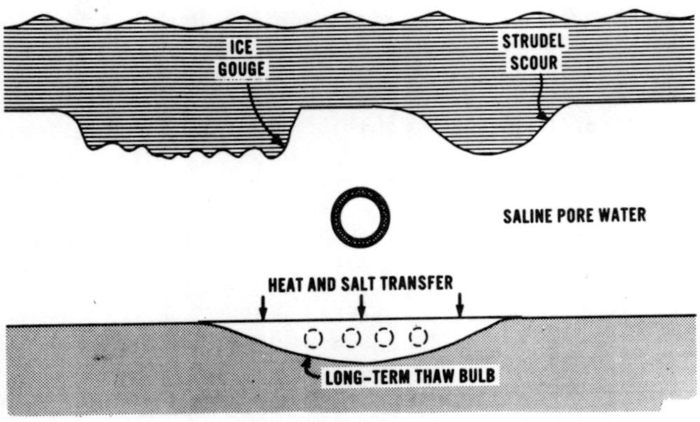

FIGURE 9 Schematic cross section showing critical elements of a buried pipeline underlain by offshore permafrost.

causes no thermal disturbance. A very drastic surface disturbance took place at the time of inundation by the sea, raising the surface temperature to near the freezing point and initiating a gradual process of downward salt migration as seawater replaced the original pore water in the sediment. Since the mean annual sea floor temperature is below 0°C, it is the interplay between the warming process and the salt migration that determines the rate at which the offshore permafrost degrades. Depending on the type of sediment and on the time since inundation, the top of the permafrost is found at greatly varying depths beneath the sea floor.

Other factors that are important for offshore pipelines are ice gouges and strudel scour depressions that could uncover the pipe and perhaps damage it. Thus, offshore pipelines in shallow arctic waters may require relatively deep burial up to 3 m or more. This brings the pipe closer to the permafrost table and tends to aggravate potential thaw-settlement problems.

Generally, offshore pipelines would be much smaller in diameter than large trunklines onshore, typically only 30 to 60 cm. Therefore, such lines would be more tolerant to thaw settlement, and a certain amount of permafrost thaw below the pipeline would be acceptable. On the other hand, predictions of permafrost thaw are complicated by the fact that the rate and effect of salt migration are difficult to model. Therefore, in some calculations (Heuer et al., 1983) two extremes have been evaluated: one in which salt migration is negligible in the time frame of the operating life of a pipeline, and one in which salt migration is sufficiently rapid to maintain a concentration similar to that of seawater at the advancing edge of the thaw bulb. In the computer simulation, the assumed melting temperature of the permafrost is near 0°C for slow salt migration and -1.9°C for fast salt migration.

Conventional burial of uninsulated pipe in a trench is probably acceptable when the top of the ice-bonded permafrost is more than 27 m below the seafloor. Calculations show that the warm pipe may still cause permafrost melting at this depth, particularly in sediments where salt migration is effective. However, any differential settlement at these depths would be masked by the substantial thickness of overlying thawed sediments. Therefore, the effect on the pipeline would be greatly reduced. Pipeline insulation may be required if the permafrost table is less than 27 m below the seafloor. This value depends, of course, on specific design values for thaw strain, pipe size, and temperature. Again, some thawing of permafrost would be allowed, but the allowable thaw is reduced in this case because there would be less dampening of differential settlement.

Two potential mitigative modes for offshore pipelines are shown in Figure 10. If ice-rich

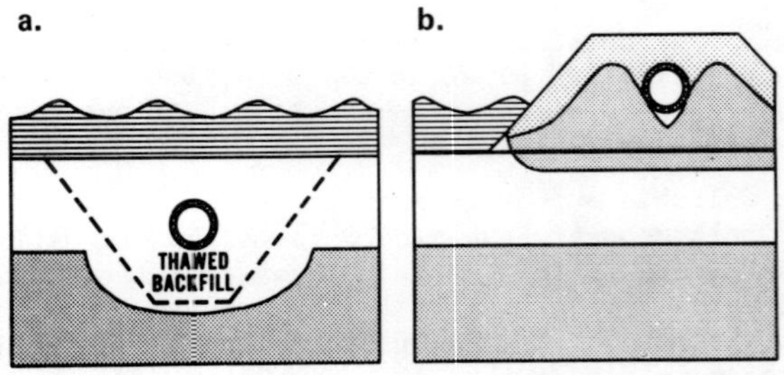

FIGURE 10 Schematic cross sections of two possible mitigative pipeline modes for offshore permafrost areas, (a) by overexcavating and backfilling with thawed material and (b) by using a causeway.

permafrost occurs near the seafloor, a certain portion of it could be removed from below the pipeline by overexcavation and backfilled with thawed soil before the pipe is laid into the trench. The required trench depth could be substantial, perhaps as much as 9 to 12 m in some areas where deep pipeline burial is required for protection against ice gouging (Fig. 10a). This kind of excavation would be costly, but it is feasible with modern dredging equipment. Laying the pipeline above water in a causeway is an alternative for shallow water. With the causeway surface exposed to the cold air, it will gradually develop a frozen core. Rather than thawing the subsea permafrost, this construction mode may actually lead to the formation of new permafrost below the original seafloor (Fig. 10b).

Feasible and practical pipeline modes have been identified for the three major types of pipelines that are needed for petroleum development in Alaska's permafrost areas: hot oil pipelines and cold gas pipelines onshore, and warm buried underwater pipelines offshore. Much work is required to develop the most cost-effective configuration and construction method for each particular pipeline. However, the technology is available now to design, construct, and operate onshore and offshore pipelines for developing Alaska's petroleum resources in a safe and environmentally acceptable manner.

PIPELINE WORKPADS IN ALASKA

M.C. Metz

GeoTec Services Inc.
Golden, Colorado, USA

The construction of oil and gas pipelines in remote northern regions generally requires the preparation of a working surface or workpad that will support construction traffic and provide access during the operation of the pipeline. Commonly, gravel fill is used for this purpose. A gravel workpad, however, usually causes significant thermal disruption and permafrost degradation, which makes the effects of the workpad a primary consideration in oil and gas pipeline design. In most instances the long-term thermal effects of the workpad are an integral part of the long-term design and performance expectations of a pipeline system.

Early coordination with construction and operations personnel is required to establish the short- and long-term performance criteria for any workpad. Normally there is no need to design a workpad capable of supporting heavy wheel-load traffic for the design life of a pipeline (25-30 yr). Significant cost savings can be realized if the workpad is designed to support heavy construction traffic only during the construction phase of a pipeline (2-3 yr) and, with periodic maintenance, to provide limited access for light surveillance and maintenance traffic during project operations.

The Trans-Alaska Pipeline System (TAPS), which connects the oil fields at Prudhoe Bay to the ice-free port of Valdez, had a construction workpad for its entire 800-mile length. Most of the TAPS workpad was designed and constructed as either an all-granular embankment or a polystyrene-insulated embankment. Two short segments of the pipeline were constructed utilizing snow workpads. The design intent of the TAPS workpad was to provide only the thickness of gravel, and in some cases polystyrene insulation and gravel, that would allow short-term passage of heavy construction equipment.

Generally an overlay workpad was utilized for the trans-Alaska hot oil pipeline, although there were some areas where cut and fill was used. For most of the permafrost areas the basic overlay embankment shown in Figure 1 was used. A structural embankment was constructed over both thawed ground and permafrost. In permafrost areas this embankment allows for the development of an increased active layer and allows for long-term deep thaw beneath the workpad. The structural embankment was for both the aboveground and belowground pipeline in warm permafrost areas and for the belowground pipeline in cold permafrost areas. For the aboveground pipeline segments of the trans-Alaska pipeline in cold permafrost areas, two types of thermal workpads were considered. An all-gravel thermal embankment design of sufficient thickness to prevent thaw beneath the embankment base was

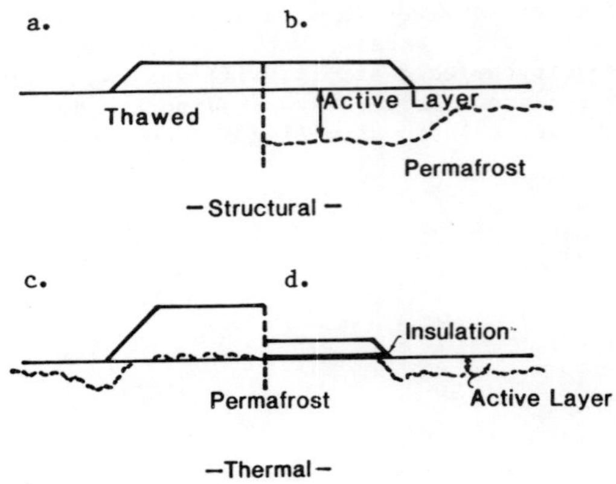

FIGURE 1 Examples of basic overlay embankments. Structural embankment (a) over thawed ground and (b) over permafrost, and thermal embankment (c) without insulation and (d) with insulation.

considered first. For TAPS, however, an alternative thermal workpad design, which utilized a synthetic insulation layer at the base of a thin gravel overlay, was developed as an effective alternative. The synthetic insulation material selected for this design was extruded polystyrene foam manufactured by the Dow Chemical Company. Consequently, gravel requirements were greatly reduced. The design of the synthetic insulation was based on maintaining a frozen subgrade for a construction period of approximately 2-3 years and on long-term thaw not exceeding one and a half times the original active layer depth. Performance monitoring, after 5 years, indicates that the insulated workpad is still maintaining frozen ground directly beneath the insulation.

The use of a polystyrene-insulated workpad on TAPS resulted in significant cost savings through a reduced requirement for gravel embankment materials and reduction in the embedment length of the vertical support members (VSMs). The reduction in gravel use also reduced terrain disturbance and environmental damage associated with the development of borrow sources. Except for areas of localized distress due to construction-related insulation placement problems and to post-construction removal of or damage to the insulation layer, the 80 miles of insulated workpad has performed as intended.

The thermal-insulated workpad was designed to prevent long-term thaw of permafrost and is an integral part of the VSM design on the North Slope portion of the pipeline. In contrast, the structural workpad was designed to support the short-term construction traffic loading but to allow

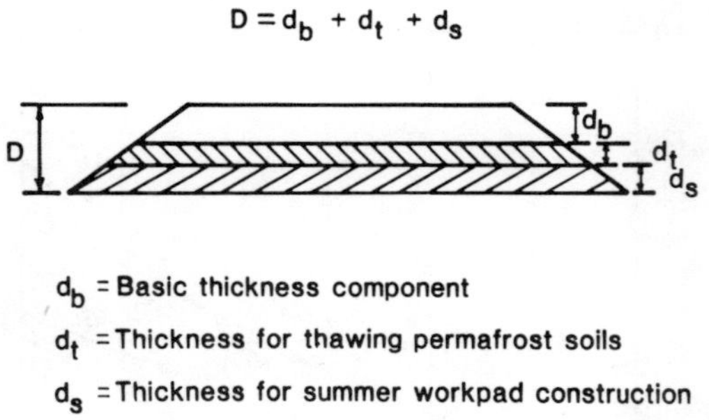

FIGURE 2 Components procedure for designing structural workpads.

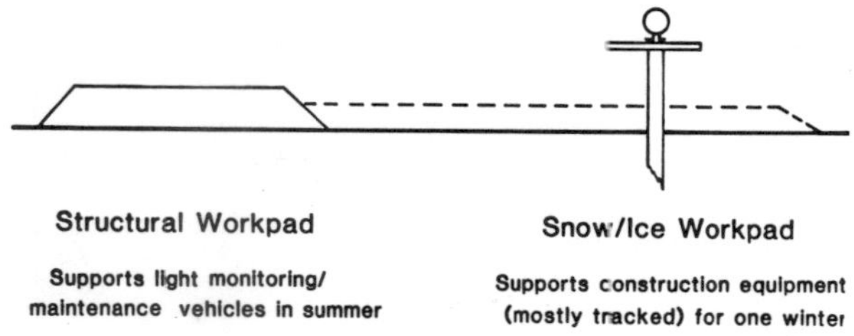

FIGURE 3 Example of a combination structural and snow/ice workpad used for construction of an elevated oil pipeline during the winter.

thawing of permafrost. Deformation of the surface and degradation of permafrost was anticipated and yearly maintenance programs were planned to maintain the workpad for use by light surveillance and pipeline maintenance vehicles.

The structural workpad design used on TAPS utilized the component procedure shown in Figure 2. This design procedure was used for workpads constructed in warm permafrost and for belowground pipeline areas in cold permafrost. The total thickness of the workpad D is equal to the basic thickness d_b plus an allowance d_t for thawing permafrost soils and an additional allowance d_s for summer construction of a workpad. The d_b component is the basic thickness required for structural support over a certain soil type for a design wheel load regardless of thermal state. d_t is an added thickness to account for thawing permafrost soils and is related to the predicted thaw-strain in the upper 5 ft of soil. d_s is the thickness added to account for summer construction over soils that are thawed in the natural active layer. Figure 2 is not meant to represent three different structural layers but to demonstrate the additive thickness of the three components. In thawed areas, only the d_b thickness was used. In permafrost areas, where the workpad was constructed in the winter, the combined thickness of d_b and d_t was specified. For summer construction in these same areas, the specified thickness was the total of all three components. The component procedure was verified by test sections constructed prior to pipeline construction.

The all-gravel structural workpad was designed to support the high bearing pressures associated with pipeline construction wheel-loads and to compensate for thaw strain in the underlying soils during construction. Loss of capacity to support heavy wheel-loads on the workpad due to thawing permafrost was expected to occur after construction, but was not considered to be a problem since operations traffic would consist of light-wheel-load surveillance and maintenance vehicles. As anticipated, deterioration of the permafrost in non-insulated gravel workpad areas has resulted in thaw settlement and a weakened subgrade. Fines have been pumped into the overlay material, resulting in a significant loss of trafficability, but yearly workpad maintenance allows passage of light-wheel-load vehicles.

Low-cost gravel resources are not present everywhere on the Alaskan North Slope. Gravel resources diminish in both quantity and quality west of Prudhoe Bay, and development activities are moving in that direction. The resulting higher cost of gravel necessitates the development of workpad configurations that reduce the demand for gravel resources.

Willingness to accept certain limitations on pipeline maintenance activities can result in additional savings. If there is no need to provide for year-round access during the operational phase, the designer can consider using a seasonal workpad constructed of snow, ice aggregate, or ice.

Variations of workpad design utilizing a narrow gravel workpad with an adjacent snow workpad as shown in Figure 3 have been successfully used for many years by the producing companies at Prudhoe Bay. This combination gravel-snow workpad provides for the needed width during construction as well as access during the life of the pipeline for surveillance and maintenance. This solution requires winter construction, but in Prudhoe Bay, for example, this does not represent a deviation from normal construction practice.

The gravel workpad associated with the combination workpads is usually approximately 3 to 3.5 ft thick and has a minimal working width. It is utilized mostly during the winter for pipeline construction. Heavy-wheel-load, rubber-tired vehicles are usually restricted to the gravel portion. During the late summer, thaw weakens the subgrade, but light-wheel-load monitoring and maintenance vehicles can still pass with no problems.

The snow or ice workpad constructed adjacent to the gravel embankment provides a winter working surface for installation of pipe supports and pipelines. This has worked successfully at Prudhoe Bay, although there is some risk to the construction schedule during late snow years. The embankment acts as a natural snow fence, however, and aids in trapping available snow by drifting.

In recent years, drawing on experience and research conducted by the Department of Transportation in Alaska, new workpad configurations have been developed. The utilization of geofabrics in workpad design is being tested in North Slope applications.

Two applications of geofabrics in workpad design are shown in Figure 4. The Atlantic Richfield Company has successfully employed geofabrics to encapsulate low-strength soils into embankments. Materials stripped from other operations that contain ice-rich silt and organic material are encapsulated in geofabrics within the embankment. This provides for effective use of other-

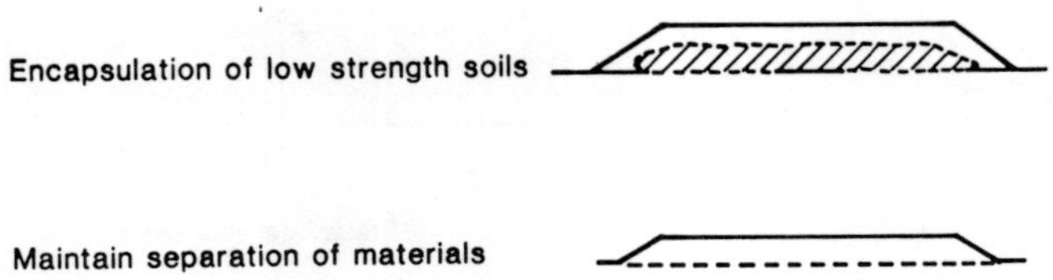

FIGURE 4 Two examples of the use of geofabrics in construction of workpads.

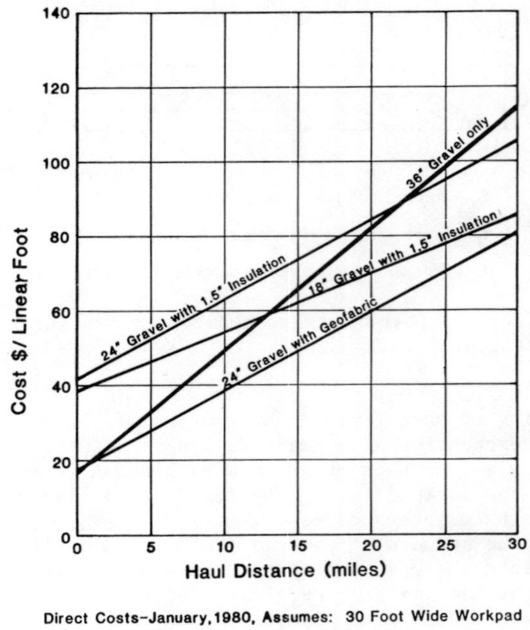

FIGURE 5 Chart comparing direct costs of constructing various types of workpads.

wise wasted materials while reducing the requirements for select gravel fill. The performance of these embankments, including a portion of at least one aircraft runway, has been good.

One option that has only been tested on a limited basis, however, is the use of geofabrics to maintain separation of the workpad embankment and underlying fine-grained permafrost soils. In the case of a structural workpad, the benefits that would be realized are the reduction of the d_s design component to zero and perhaps a substantial reduction of the d_t component. If only the d_s component is considered, which accounts for contamination of the lower portion of the embankment by underlying fines during summer construction, a savings of 1 to 2 ft of gravel can be realized by using geofabrics.

The capacity of geofabrics to maintain separation of materials is well known. The added benefit of reinforcement over thawing permafrost soils, which could reduce the d_t design component, is not as well known. However, if the reduction of a minimal thickness of only 1 ft is considered, then gravel usage and cost differential can be calculated. Comparative costs of workpad construction based on actual North Slope projects are given in Figure 5, with the estimated costs of a workpad uitlizing geofabric shown in comparison. Since the cost of a workpad is directly related to the haul distance from the source of gravel, the direct cost per linear foot of workpad is plotted versus gravel haul distance. It should be emphasized that these are direct costs and do not include indirect costs. As can be seen in Figure 5, we reach a break-even point between the 36-in. all-gravel pad and the 24-in. insulated gravel pad with a haul distance of 22 miles. This simply means that once the haul distance for gravel is greater than 22 miles, it becomes economically feasible to build an insulated workpad. At Prudhoe Bay where gravel sources are plentiful and haul distances short, the cost of insulation is probably not justified.

However, if by using geofabrics 1 ft of gravel overlay can be saved, the break-even haul distance is only 1 mile and, therefore, a geofabric-supported workpad constructed more than 1 mile from a gravel source costs less than a 36-in.-thick all-gravel workpad. Maintaining separation of frost-susceptible fine-grained soils from overlying sand and gravel with geofabrics makes problem settlement areas easier to maintain and upgrade. This additional long-term advantage makes this alternative well worth considering.

The continued study of the performance of existing workpads in Alaska and continued evaluation of their impacts are needed to provide design engineers with the information required to improve workpad design criteria. Development of new and smaller oil fields as well as other remote facilities in Alaska will require more cost-effective designs, and one of the most critical factors will be the availability of gravel resources. For example, gravel deposits are extremely scarce and haul distances are greater in the very large National Petroleum Reserve--Alaska and other areas on the North Slope than they are at Prudhoe Bay. The concept of an almost unlimited, cheap gravel resource must therefore by abandoned by the designer. The higher cost of gravel in areas of future development will require workpad designs that use less gravel.

PERFORMANCE OF THE TRANS-ALASKA OIL PIPELINE

E.R. Johnson

Alyeska Pipeline Service Company
Anchorage, Alaska, USA

The trans-Alaska oil pipeline is the first large-diameter pipeline transporting hot crude oil across permafrost terrain. Many technical questions faced designers at the time of pipeline design, but now, after 6 years of operating experience, many of the questions regarding pipeline performance can be answered (Johnson, 1981). The performance of the trans-Alaska oil pipeline has been exceptionally good, with only minimal operational and environmental problems. There have been some technical problems, to be sure, but they have been manageable and confirm the high quality of the design.

The fundamental factors differentiating the trans-Alaska oil pipeline from most other crude oil pipelines are those caused by an operating temperature of 48.6°C and the presence of permafrost. Approximately 70% of the pipeline route is underlain by permafrost. Three basic pipeline design modes were developed to cope with the various foundation conditions (Table 1). In the thawed ground and in thaw-stable permafrost, the conventional buried mode was used. In thaw-unstable permafrost, the aboveground mode was used. In thaw-unstable permafrost where there were access restrictions or where there were environmental constraints such as the need for extended animal crossings, a refrigerated buried mode was used.

In June 1979, two belowground pipeline failures occurred that resulted in leaks. These failures were caused by 1.2 m of thaw settlement in one location and 1.7 m of thaw settlement in the other. In both cases, settlement was limited to short 100- to 120-m-long zones underlain by previously undetected ice-rich permafrost.

Alyeska Pipeline Service Company routinely conducts an extensive monitoring and evaluation program designed to detect potential pipeline settlement problems. Visual surveillance detects evidence of subsurface settlement that may be related to settlement of the pipeline. Where problems are detected, monitoring rods are attached to the top of the pipe to allow survey monitoring at the top of the pipe for subsequent settlement. If the settlement is severe, the monitoring rods are installed less than 10 m apart. The curvature of the pipe can be calculated from the survey measurements and compared to known failure criteria.

In addition to the monitoring rod program, major geotechnical and thermal investigations have been carried out. Since startup, over 1,000 boreholes have been drilled to gather geotechnical information and more than 315 thermistor cables have been installed to gather information about the thermal regime of the permafrost. In addition, a pipeline deformation monitoring "pig" detects evidence of dents, buckles, and possible incipient pipeline cracking. This monitoring and surveillance program has been very successful in detecting several additional problem areas besides the two previously mentioned.

Several techniques have been used to repair the pipeline. In the Atigun Pass area, the settled pipeline was excavated, conventional elevated-pipeline hardware was installed to support the pipe, and the trench was backfilled. The piles were embedded in competent bedrock. Because of the high water table in the Dietrich River area south of Atigun Pass, a sling system was used to support the pipe and to minimize excavation below the pipe. The piles were driven to bedrock at 33.5 m. At another location a major through-cut was made, the pipeline was supported on elevated hardware, and the pipeline was left uncovered, so a buried pipeline was changed to a fully restrained elevated line.

TABLE 1 Lengths of the three basic construction modes and soil conditions along the route of the Trans-Alaska Oil Pipeline.

Mode	Conditions	Length mi	Length km
Conventional burial	Thawed ground or thaw stable permafrost	380	612
Aboveground	Thaw unstable permafrost	416	669
Refrigerated burial	Thaw unstable permafrost with access restrictions	4*	6
		800	1287

*1.8 mi (2.9 km) refrigerated piping at pumping stations 1, 2, 3, 5, and 6 in addition to mainline piping.

Another technique that is being used more consists of simply lifting the pipe up to even out the extreme downward curvature and backfilling with a weak sand/cement slurry underneath the pipe to give continuous support. Naturally, this remedial work is followed by careful monitoring. In most cases, the thaw bulb has already grown to a good portion of its total size, so any additional thawing that might occur would be deep seated. Consequently, the need for further correction of the pipeline in these areas should be limited. On the south side of Atigun Pass, an area of thawed ground was associated with flowing groundwater. The problem was solved by using mechanical refrigeration to refreeze the ground and heat pipes to maintain the frozen condition (Thomas et al., 1982).

The elevated portion of the pipeline consists of 77,732 pile supports. The piles are 46-cm-diameter steel pipes and are embedded to an average depth of 9.1 m. There are 17 different pile types and they can be roughly grouped into five categories (see Table 1 of paper by Jahns). The results of monitoring the piles supporting the aboveground pipeline show that 0.6% of the sample settled more than 9 cm, that 1.9% heaved more than 9 cm, and that the thermal piles installed in thawed ground had the greatest movement (Table 1 of Jahns paper).

The 9-cm threshold was chosen because it was thought that this much movement could be of concern; however, site-specific investigations have revealed that very few minor problems occur with movements of this magnitude.

There are 122,000 heat pipes installed along the pipeline to prevent excessive permafrost thaw or frost heaving. The heat pipes are of a two-phase, closed system containing anhydrous ammonia as a working fluid. Monitoring the heat pipes with infrared imagery has revealed that many of them are losing their effectiveness, probably because of blockage of the heat pipes by a non-condensable gas. Gas blockage could ultimately cause total failure of the heat pipe. Preliminary analysis indicates that the blockage is being caused by hydrogen gas presumably being formed by corrosion caused by water or other contaminants left inside the heat pipes during their manufacture. Figure 1 shows how the heat pipe functions and how the accumulation of hydrogen gas in the top portion of the pipe can limit the dissipation of heat into the atmosphere. A relatively simple and inexpensive repair technique has been developed, which uses a metallic halide ("getter" material) that absorbs hydrogen and, consequently, restores the heat pipes to a fully functional condition.

The buried refrigerated pipeline mode was utilized for 6 km of the main line and for approximately 2900 m of buried pipeline at the northern pump stations. This mode consists of a brine-loop mechanical refrigeration system and insulated pipeline. The pipeline is insulated with polyurethane half shells and then coated with a polyethylene or FRP waterproof barrier. In some cases the waterproof barrier has failed and water has infiltrated the insulation, causing long-term degradation of the insulation. The corrosion risk that develops because of the infiltrated water around the unprotected pipe is a more serious problem than thermal degradation. Repair of the refrigerated buried mode requires excavation of the pipe and replacement of the damaged insulation. A repair method has been developed that allows placement of an effective anti-corrosion system. Repair in and around pump stations and under buildings, is also quite difficult and expensive.

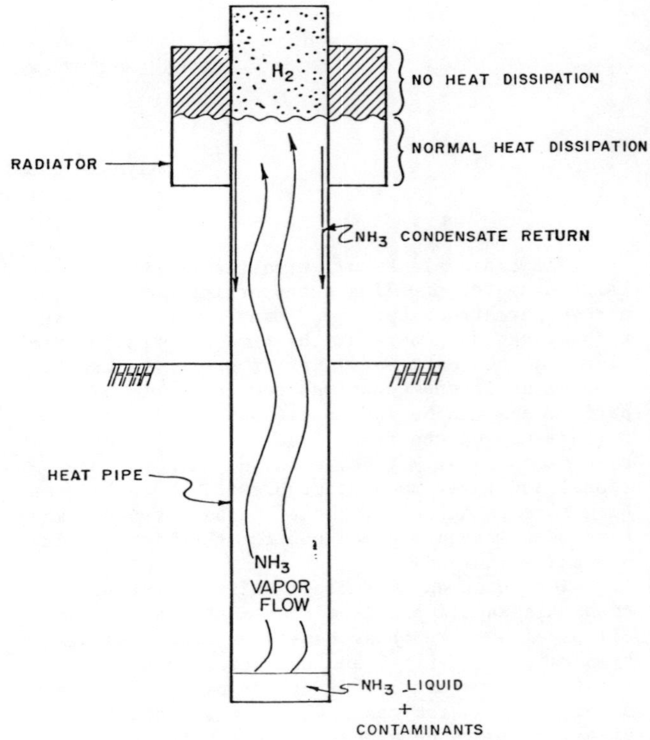

FIGURE 1 Schematic diagram of a heat pipe showing how noncondensable gas (hydrogen) restricts heat dissipation.

The problems encountered during the first 6 years of operation of the trans-Alaska oil pipeline have been the focus of this discussion. It should be stressed that these problems have been easily managed and involve only a small portion of the total pipeline system. The overall performance of the trans-Alaska oil pipeline has been exceptionally good, and no problems have been encountered that impair the system's capability to transport the required amounts of oil from the Alaskan North Slope to Valdez.

SELECTED PANEL SESSION REFERENCES

Clarke, E.S., Krzewinski, T.G., and Metz, M.C., 1981, Trans-Alaska Pipeline System synthetically insulated workpad--An evaluation of present conditions: Energy-Sources Technology Conference and Exhibition, Houston, Texas.

Ferrians, O.J., Jr., Kachadoorian, R., and Greene, G.W., 1969, Permafrost and related engineering problems in Alaska: Washington, D.C., U.S. Geological Survey Professional Paper 678.

Heuer, C.E., 1979, The application of heat pipes on the trans-Alaska pipeline: Hanover, N.H., U.S. Army Cold Regions Research and Engineering Laboratory, Special Report 79-26.

Heuer, C.E., Krzewinski, T.G., and Metz, M.C., 1982, Special thermal design to prevent thaw settlement and liquefaction, in Proceedings of the Fourth Canadian Permafrost Conference, Calgary: Ottawa, National Research Council of Canada, p. 507-522.

Heuer, C.E., Caldwell, J.B., and Zamsky, B., 1983, Design of buried sea-floor pipelines for permafrost thaw settlement, in Proceedings of the Fourth International Conference on Permafrost, Fairbanks: Washington, D.C., National Academy of Sciences, p. 486-491.

Hwang, C.T., 1976, Predictions and observations on the behavior of a warm gas pipeline on permafrost: Canadian Geotechnical Journal, v. 13, p. 452-480.

Ivantsov, O.M., Semyonov, L.P., Crory, F.E., and Ferrians, O.J., Jr., 1983, Problems of environmental protection during construction and operation of pipeline systems in northern regions: 11th World Petroleum Congress, London, preprint.

Jahns, H.O., Miller, T.W., Power, L.D., Rickey, W.P., Taylor, T.P., and Wheeler, J.A., 1973, Permafrost protection for pipelines, in Permafrost -- the North American Contribution to the Second International Conference, Yakutsk: Washington, D.C., National Academy of Sciences, p. 673-684.

Jahns, H.O., and C.E. Heuer, 1983, Frost heave mitigation and permafrost protection for a buried chilled-gas pipeline, in Proceedings of the Fourth International Conference on Permafrost, Fairbanks: Washington, D.C., National Academy of Sciences, p. 531-536.

Johnson, E.R., 1981, Buried oil pipeline design and operation in the Arctic - Lessons learned on the Alyeska Pipeline, in 37th Petroleum Mechanical Engineering Workshop and Conference, Dallas, Texas, preprint.

Lachenbruch, A.H., 1970, Some estimates of the thermal effects of a heated pipeline in permafrost: Washington, D.C., U.S. Geological Survey Circular 632.

Metz, M.C., Krzewinski, T.G., and Clarke, E.S., 1982, The Trans-Alaska Pipeline System workpad--An evaluation of present conditions: in Proceedings of the Fourth Canadian Permafrost Conference, Calgary, 1981 (Roger J.E. Brown memorial volume): Ottawa, National Research Council of Canada, p. 523-534.

Miller, T.W., 1979, The surface heat balance in simulations of permafrost behavior: Journal of Energy Resources Technology, v. 101, p. 240-250.

Myrick, J.E., Isaacs, R.M., Liu, C.Y., and Luce, R.G., 1982, The frost heave program of the Alaskan Natural Gas Transportation System, American Society of Mechanical Engineers paper 82-WA/HT-12.

Nixon, J.F. Stuchly, J., and Pick, A.R., 1984, Design of Normal Wells pipeline for frost heave and thaw settlement, in Proceedings of the Third International Offshore Mechanics and Arctic Engineering Symposium, v. 3: New York, American Society of Mechanical Engineers, p. 69-76.

Pearson, S.W., 1979, Thermal performance verification of thermal vertical support members for the trans-Alaska pipeline: Journal of Energy Resources Technology, v. 101, p. 225-231.

Thomas, H.P., Johnson, E.R., Stanley, J.M., Shuster, J.A., and Pearson, S.W., 1982, Pipeline stabilization project at Atigun Pass, in Proceedings of the Third International Symposium on Ground Freezing, Hanover, N.H., preprint.

Wheeler, J.A., 1978, Permafrost thermal design for the trans-Alaska pipeline, in Moving Boundary Problems (D.G. Wilson and A.D. Solomon, eds.): New York, Academic Press, p. 267-284.

Williams, P.J., 1979, Pipelines and permafrost, physical geography and development in the circumpolar North, in Topics in Applied Geography (D. Davidson and J. Dawson, eds.): London and New York, Longman.

Environmental Protection of Permafrost Terrain

PANEL MEMBERS

James E. Hemming (Chairman), Dames & Moore, Anchorage, Alaska, USA

Nikolai A. Grave, Permafrost Institute, Siberian Branch, Academy of Sciences, Yakutsk, USSR

Curtis V. McVee and Jules V. Tileston, U.S. Bureau of Land Management, Anchorage, Alaska, USA

Hugh M. French, Departments of Geography and Geology, University of Ottawa, Ottawa, Canada

Max C. Brewer, U.S. Geological Survey, Anchorage, Alaska, USA

Patrick J. Webber, Institute of Arctic and Alpine Research, University of Colorado, Boulder, Colorado, USA

INTRODUCTION AND A CASE STUDY

J.E. Hemming

During the past decade, accelerated development has occurred in the Arctic, resulting in a rapid evolution of environmental controls and techniques to minimize negative impacts during construction. Essentially all of the circumpolar nations have experienced oil and gas exploration and development, pipeline construction, and increased mining activity. The primary problems are related to the scarcity of gravel in some tundra areas, the sensitivity of permafrost soils to disturbance, and the disposal of drilling muds and other wastes in frozen soils. Recent research has emphasized better understanding of the biological and physical characteristics of tundra ecosystems and the development of low-impact techniques for transportation systems and facilities. As results are published and the significance of various development activities is better understood, regulations have been changed and new guidelines and standards have been developed.

The panel papers that follow emphasize various aspects of environmental protection of the permafrost areas in the United States, Canada, and the U.S.S.R. The panel acknowledges that environmental problems can be minimized or avoided by positive approaches to problem-solving early in the project planning process. This point was also discussed and described in a workshop held in 1980 in Fairbanks to review steps necessary for terrain protection during development (Brown and Hemming, 1980).

Several papers and reports present additional information on environmental conditions and problems in permafrost regions (Armstrong, 1981; Gerasimov, 1979; Grave, 1983; Hanley et al., 1981; National Research Council, 1980; Woodward-Clyde, 1980).

An example of the environment planning process is summarized below as an introduction to this panel report. The process includes inventories of both the natural resources and the engineering constraints, mapping of sensitive areas, a preliminary siting plan, and the detailed engineering design.

Natural Resources Inventory: Once a new project has been identified, but before the exact location of facilities has been confirmed and the detailed engineering design has been initiated, field resource inventories are conducted. Information is gathered on bird, mammal, and fish occurrence and sensitive habitat, endangered species, wetlands, and subsistence, recreation, and other land use patterns. These data are used to support applications for government permits and to provide a basis for siting the project facilities so that they will cause the least adverse impact.

Engineering Constraints Inventory: Simultaneously with natural resource inventories, data are collected by engineering geologists through a combination of aerial photo interpretation and field surveys in order to delineate engineering constraints. Information on geotechnical factors such as grades, organic soils, patterned ground, solifluction zones, and aufeis areas is used to delineate constraints.

Delineation of Sensitive Areas: When engineering constraints and natural resource inventories are complete, the results are integrated into detailed maps of the sensitive areas.

Preliminary Siting Plan: Preliminary siting of facilities involves avoiding sensitive areas. When sensitive areas cannot be avoided, special mitigation plans are prepared. The reduction of environmental impact and mitigation planning are

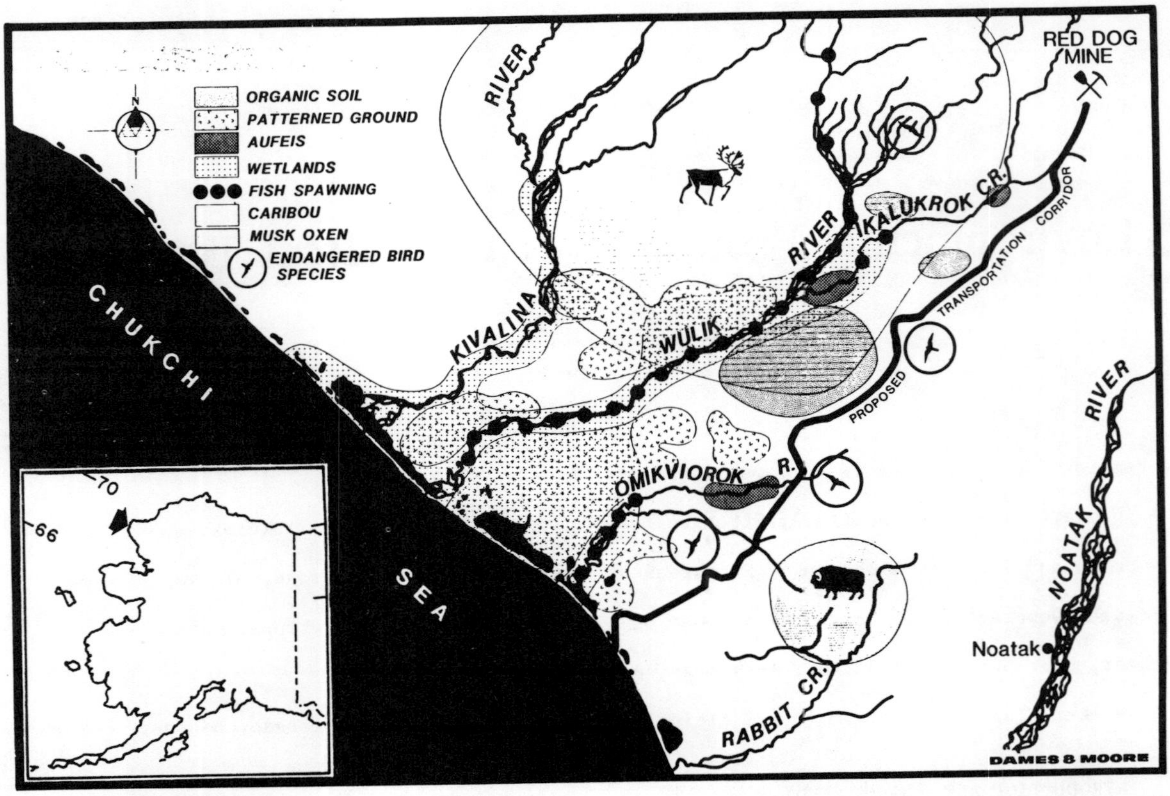

FIGURE 1 A map of environmental and engineering constraints that were avoided during selection of a transportation corridor alignment for the proposed Red Dog Mine Project, Alaska.

an integral part of the permit process in the United States.

Detailed Engineering Design: After preliminary project designs and mitigation plans have been reviewed and accepted by government agencies, a project can proceed to detailed engineering design.

The above process was applied to the proposed Red Dog Mine, a world-class lead-zinc deposit along the Alaskan coast of the Chukchi Sea (Dames & Moore, 1983). This project involved planning for 110 to 130 km of road plus a new port, airfield, water storage reservoir, mine tailings disposal, mill, and housing facilities. The entire area is underlain with permafrost with an active layer about 0.5 to 1.0 m thick. Wildlife and fisheries resources are abundant and subsistence activities by local Eskimos are well established.

Regulatory agencies were most concerned about the potential impacts of mine facilities on fish spawning and overwintering areas, endangered or rare wildlife, large-mammal overwintering areas, and wetland habitats. The mining company was most concerned about construction and maintenance costs related to the alignment of the transportation corridors and how the economic feasibility of the project would be influenced by the environmental and engineering constraints.

Natural resource and engineering constraints inventories involved review of the literature and extensive field surveys by engineers, geologists, and biologists. Fish spawning, rearing, and overwintering areas, as well as endangered species habitats, large-mammal overwintering areas, wetlands, patterned ground, areas of sheet flow, solifluction, and aufeis zones were identified and delineated on 1:63,360 scale topographic maps. Once the individual natural resource values and engineering constraints had been delineated, a composite map was prepared. It was then possible to select potential transportation corridors and to evaluate the comparative costs of construction.

The interdisciplinary process applied in this case resulted in the selection of a transportation corridor (Fig. 1) that would meet government permit requirements with the fewest potential environmental impacts and the lowest estimated construction cost. This approach is being adopted for other development projects within Alaska.

REFERENCES

Armstrong, T., 1981, Environmental control in remote areas: The case of the Soviet Northlands with an Alaskan comparison: Polar Record, v. 20, p. 329-335.

Brown, J. and Hemming, J.E., 1980, Workshop on environmental protection of permafrost terrain: Northern Engineer, v. 12, no. 2, p. 30-36.

Dames & Moore, 1983, Environmental baseline studies, Red Dog Project: Cominco Alaska, Inc.

Gerasimov, I.P., 1979, Protection of the environment in northern Siberia: Izv. of USSR Academy of Sciences, Ser. Geogr., no. 1, p. 45-52.

Grave, N.A., 1983, Stability of the surface in mechanical disturbance associated with northern development: Polar Geography and Geology, v. 7, no. 3, p. 175-186.

Hanley, P.T., Hemming, J.E., Morsell, J.W., Morehouse, T.A., Leask, L.E., and Harrison, G.S., 1981, Natural resource protection and petroleum development in Alaska: Washington, D.C., U.S. Fish and Wildlife Service, Biological Services Program, FWS/OBS-80/22, and FWS/OBS80/22.1.

National Research Council, 1980, Surface coal mining in Alaska: An investigation of the Surface Mining Control and Reclamation Act of 1977 in relation to Alaskan conditions: Washington, D.C., National Academy of Sciences.

Woodward-Clyde Consultants, 1980, Gravel removal studies in arctic and subarctic floodplains in Alaska: Washington, D.C., U.S. Fish and Wildlife Service, Biological Services Program, FWS/OBS-80/08.

DEVELOPMENT AND ENVIRONMENTAL PROTECTION IN THE PERMAFROST ZONE OF THE USSR: A REVIEW

N.A. Grave

Permafrost Institute, Siberian Branch
Academy of Sciences, Yakutsk, USSR

Certain results of research into this problem were presented at the Third International Conference on Permafrost in Canada and were published in the proceedings both in English (Brown and Grave 1979a, 1979b) and in Russian (Brown and Grave 1981).

Further progress in this direction has been achieved during subsequent years (1979-82) as this paper will demonstrate. This progress is reflected in particular by the passing of a number of government Bills and regulations. In accordance with Article 18 of the Constitution of the USSR, the Supreme Soviet of the USSR has enacted basic legislation on the use of waters, forests and mineral resources. In the Soviet Union we have begun formulating standards for environmental protection which will regulate the interaction of man and his environment; these standards are divided into sections on "Soil", "Landscape", and "Land". Within this last section the sequence of restoration and revegetation of land disrupted due to human activities is specified.

A series of special measures on environmental protection in the tundra zone and in the area of the Baikal-Amur railway was first passed in 1978 by a resolution of the Central Committee of the Communist Party and by the Council of Ministers of the USSR, entitled "On additional measures to reinforce environmental protection and to improve the use of natural resources." Responsibility for the state of the natural environment and implementation of the appropriate measures was entrusted to government bodies, and at the administrative level they are to be implemented by the local Soviets of People's Deputies.

At a time of accelerated development of the northern and northeastern regions of the USSR these measures have played a positive role in environmental protection. Environmental Protection departments have been created within the main ministries operating in the permafrost regions; their task is to implement the environmental protection measures with regard to industrial and transport projects and to monitor their effectiveness. The Soviet of Ministers of the Yakutsk ASSR has passed a resolution to regulate the movement of tracked vehicles in summer in the nine northern districts of the Republic. Voluntary societies for environmental protection have been established in the republics, krays and oblasts in the northern areas of the country and in some areas even Universities of Environmental Protection. They have been pressing for widespread popularization of the bases for rational land use in areas, settlements and towns experiencing development.

Widespread discussion of the results of scientific investigations into environmental protection in the North, which were undertaken in the 1970s at all-union, regional, and departmental meetings, conferences, and symposia (e.g. *Okhrana okrazhayushchey sredy* 1980, *Geokriologicheskiy prognoz,* 1980, 1982, *Inzhenerno-geokrilogicheskiye aspekty* 1981; etc.), has encouraged the appearance of a number of environmental protection recommendations and manuals. On the basis of the Permafrost Institute's investigations, VSYeGINGYeO and PNIIIS have produced an *Interim manual on the laying of pipelines in the North* (1980), which includes recommendations on the selection of construction sites, requirements for engineering preparations, the use of thermal insulators, recommendations on methods of biological restoration of disturbed areas, and the use of thermal piles on gas pipelines, and proposals as to the regulation of the construction and operation of installations. A map of the natural-area complexes (landscapes) of the north of Western Siberia is included in an appendix.

Specific needs in terms of environmental protection in the North were the cause for subsequent intensification of scientific research with a view to establishing more precise criteria as to the degree of impact caused by human activities on the environment, and the latter's response. It was also important to determine precisely the permafrost aspects of the environmental protection problem, embracing not only the land surface but also water, air, flora and fauna. This did not simplify the problem for permafrost scientists who are starting to focus special attention on the sensitivity of the surface in permafrost areas to human activities (Grave 1980), since the actual concept of the surface includes essentially all the main components of landscapes.

The main approach in solving the problems of surface stability involves prediction of changes in the permafrost conditions and the development of cryogenic processes as a consequence of human disturbance of the surface. The problem of geocryological, or more accurately geological, prediction in connection with the solution of problems of environmental protection in the North has become the main focus of permafrost studies in the past few years. Researchers are striving to obtain quantitative and not just qualitative predictions and, in connection with this, the quantitative characteristics of the original components of the landscape.

Theoretical investigations by Grechishchev (1982a,b), based on the results of 15 years of research in Western Siberia, allowed him to construct a functional thermodynamic model of the environmental/spatial complex (i.e. the landscape) in a permafrost area. It takes into account the relations between thermophysics (changes in the thermal and moisture conditions of the materials)

TABLE 1 Changes in northern taiga landscapes caused by disturbances of the surface during construction along the line of the Salekhard-Igarka Railway (adapted from Grechischev et al. 1983). Nature of disturbance and changes inthe major landscape components 22 years after building of the railroad.

Topography	Vegetation	Soils	Engineering-geocryological conditions			
			Permafrost distribution	Mean annual temp.	Active layer	Exogenic processes
Total removal of vegetation cover and levelling of surface	Replacement of sparse shrub/ sphagnum forest by sparse willow/ birch forest with shrubs, grass and *Politrichum* 1-3 m high	Removal of peat layer to depth of 0.4 m	Lowering of permafrost table by 3-3.5 m	Rise of 0.5°C	Increase in active layer depth by 2-3 times; in some places giving way to a seasonally frozen layer	Thermokarst at sites with segregation ice; seasonal frost heaving

and mechanics (development of cryogenic physical-geological processes). This model is seen by the author as a base for quantitative predictions of changes in permafrost conditions resulting from man-induced impacts on the environment.

The model includes such factors as hydrometeorological aspects, the composition and properties of the vegetation cover and soil, the local topography, and technogenic factors. The latter include: heat flux into the ground from engineering installations, disturbance of vegetation and soil cover during construction or operation of those installations; changes in the surface drainage due to artificial inundation or drainage; and changes in the surface topography (embankments or pits).

However, there are some problems associated with the construction of a computerized mathematical model in this case; these are associated with the fact that the interrelations between the temperature field, moisture content of the soils and thermokarst have not been determined and with the fact that information is also lacking as to some of the other interdependences between components of the landscape. Simplified mathematical models can be used in predictions where only one-way links between components are considered, e.g. in the case of the removal of the vegetation cover without any subsequent restoration.

Examples of qualitative, semi-quantitative (statistical-analog) and quantitative (physical-mathematical) predictions of changes in geocryological conditions in the main oil-and-gas fields of the north of Western Siberia, i.e. on the interfluve between the Pur and the Nadym and the southern part of the Taz Peninsula, are cited in the monograph by Grechishchev *et al.* (1983) as well as in a number of articles (Grechishchev and Mel'nikov 1982, Moskalenko 1980). The main method of geocryological forecasting of the sensitivity of the surface to man-induced impacts is that of the physical-geographical analog (Mel'nikov, P.I., et al. 1982, Moskalenko and Shur 1980). Man-induced disturbances of the landscape may be divided into permanent (e.g. repeated use of land for agriculture), periodic (removal of undergrowth along pipeline rights-of-way) and impulsive (on one occasion and for only a short time). The method of the physical-geographical analog consists of comparing the landscape under study with one already developed, and which experienced the same sequence of development and contains identical components. The method involves the following stages: assembly of base-line information by means of a geocryological survey; investigation of disturbed and undisturbed sections of the landscape; analysis of the landscape from aerial photos flown in different years; selection of comparable sites for long-term observation; and finally, analysis of the assembled data and compilation of predictive tables and maps.

In the previously cited monograph (Grechishchev et al. 1983) the methodology of studying man-induced changes in landscapes (environmental/spatial complexes) by way of the stages listed above is presented in detail. Predictions of changes in the temperature regime and of seasonal freezing and thawing of the ground both in the first few years after the start of development, and also 15-20 years thereafter, and predictions of cryogenic processes (thermokarst, thermally induced settlement, frost heave and frost cracking) are examined in detail.

Examples of predictions of changes in landscape conditions over the long term as a result of development in northern West Siberia have also been cited in articles (Moskalenko 1980, Grechishchev and Mel'nikov 1982, Grechishchev et al. 1983). Over an area of about 10,000 km² 60 types of geobotanical units were identified and studied using geocryological and landscape mapping and long-term field observations. Over part of the area the vegetation cover, the microrelief and the uppermost peat horizons were removed.

Table 1 presents examples of changes in landscape and in geocryological conditions as a result of disturbance of the surface along the line of the Salekhard-Igarka railroad in terms of the situation 22 years after construction. In practice the disturbances did not differ from those within the area under development being studied and may serve as a baseline for geocryological predictions.

The section of the abandoned railway is located on a marine plain, composed of loams covered with peat. Islands of permafrost coincide with peatlands, palsas and bogs. Several recommendations as to development of the area and as to protection of the environment have been formulated, with application to landscapes of a region with various degrees of sensitivity to human activities (Grechishchev et al. 1983). In accordance with these predictions development of even well drained sections—slopes, hillocks and frost-heave mounds—must lead to the development of thermokarst and thermal erosion; this will necessitate the fixing of the disturbed sections with vegetation or the avoidance of particularly ice-rich sediments.

The detailed landscape and geocryological characteristics of the West Siberian gas-bearing province have been presented in two monographs (Mel'nikov, Ye. E., et al. 1983a, b).

A detailed survey of geocryological and thermophysical site investigations of the upper horizons of the permafrost and of the active layer with a view to geocryological forecasting in connection with man-induced disturbances of the surface was published by Pavlov and Shur (1982). Long-term observations were carried out under various physical geographical conditions: on arctic tundra, forest tundra and taiga; in continental and maritime areas: and in areas of continuous permafrost and in areas near its southern boundary. These observations not only confirmed a number of earlier conclusions but also produced new results.

The results of regular, systematic, long-term observations on various sites, both natural and disturbed, in the area of the Urengoy gas field have also been published (Chernyad'yev 1980). Using Leibenzon's methods, fairly simple relationships for calculating the depth of thaw were derived, taking into account the ground surface temperature, the length of the freeze-thaw period, the insulating properties of the vegetation and soil covers, latent heat and the thermophysical characteristics of the sediments. Monograms were compiled which allow one to calculate the ground temperature depending on the thermal resistivity of the vegetation and soil covers and the temperature of the surface.

These observations confirmed the major role played by the snow cover in the thawing of the ground; snow accumulation on sites where peat and moss had been removed markedly increased the depth of thaw and raised the temperature of the ground. By contrast gravel pads dumped on a peat/moss cover cooled the ground. Beneath the pad the peat becomes moist and the snow blows off the pad in winter. This reduces the depth of thaw beneath the pad (Chernyad'yev 1980, Chernyad'yev and Abrashov 1980, Shur et al. 1982, Chernyad'yev 1982, Kondrat'yev et al. 1979).

Other observations also point to the important influence of the snow cover on the temperature of the surface layers of the ground. Thus Shvetsov (1982) reached the conclusion that the most unstable soils in terms of their thermodynamics are those in tundra areas with little or no snow cover. These sites cool drastically during the cold season but warm up drastically in summer. At latitudes of 66-72° and farther north the soil begins to thaw even with air temperatures well below zero in late spring. This is confirmed by field observations at Cape Kharasavey (Yamal Peninsula) over the period 1978-1980 (Sergeyev 1982). In the vicinity of the Kular mine in the north of Eastern Siberia (1967-1975) observations were carried out at experimental sites on solifluction and the destruction of the sides of the pits. Some recommendations were made as to measures to improve the technology of open-pit mining, especially with regard to thermal amelioration of ice-rich loams (Olovin 1979).

A classification of landscapes has been worked out for estimating engineering conditions in the permafrost areas along the Baikal-Amur Railway (Mikhaylov 1980). The first results of long-term observations in the eastern part of the BAM zone have been published (Zabolotnik and Shender 1982, Balobayev et al. 1979). The observations were made in 1978 in the Upper Zeya basin (elevation 300 m) in an area with widely developed mari, i.e. thick peat bogs underlain with ground ice. Removal of the moss cover led to an abrupt increase in the depth of summer thaw and to the development of thermokarst; there was also a tendency for the thin bodies of permafrost in this southern area to degrade. Removal of snow in winter not only halted this process but also provoked a marked lowering of the temperature of the permafrost, down to -4 and -5°C.

In the northern taiga zone of Western Siberia, i.e. near the southern limit of permafrost, experimental observations have revealed that removal or compaction of the snow cover on forested, well-drained divides and on swamps, led to the formation of *pereletoks*. Cutting of forests on high floodplains and levelling of sites on hummocky surfaces led to an increase in the thaw depth and to thermokarst. Frost cracking of the ground occurred only in very cold surfaces, at temperatures below -23° to -25°C on peatlands and sands, and below -15°C on loams (Grechishchev et al. 1980).

Investigations of landscape and permafrost in Central Yakutia associated with predictions as to the suitability of areas of the taiga for clearance for agricultural purposes, were carried out using cryolithological and thermophysical techniques. Despite the high ice content of the underlying materials, removal of the vegetation cover and ploughing of sites in meadowlands on the high floodplain of the Amga River did not produce intense thermokarst because of the poor insulating properties of a meadowland vegetation. By contrast removal of the forest on older, high terraces containing massive ice wedges led to an abrupt increase in the depth of summer thaw and to the development of thermokarst. Only a dry climate, as a rule, will stop this process in the early stages of development (Turbina 1982). Evidently dry summers are associated with generally minor disturbances of the relief along the buried gas pipeline running from Taas-Tumus to Yakutsk along the section in the vicinity of the Kenkeme and Khanchaly rivers (Turbina 1980).

Investigations of areas through which gas pipelines have been laid in the north of Western Siberia, in Yakutia and at Noril'sk have revealed that man-induced disturbances of the surface are limited to sections with ice-rich sediments, whereas along sections with ice-poor sands, which are widespread in these areas, serious surface disturbances did not occur, and for the greater part these sections are experiencing natural revegetation (Sukhodol'skiy 1980, Spiridonov and Semenov 1980).

A somewhat different approach to the compilation of predictive maps is also quite widespread. The object of the mapping is the ground materials rather than landscapes, although the physical geographical conditions affecting the properties of the sediments still play a crucial role in this method.

A method has been proposed for evaluating permafrost conditions for the early stages of planning above-ground structures and for assessing their impact on the landscape. The heat inertia of the frozen materials and the peculiarities of cryogenic processes both under natural conditions and under conditions of man-induced disturbance, form the bases for compiling evaluation maps. The cryogenic processes are divided into three categories: stable (weak activity), resiliently stable (showing seasonal activity) and unstable (progressively developing). The latter category involves thermokarst, thermal erosion, landslides, solifluction and frost heave. Here the ice content of the permafrost materials represents the main factor in evaluating the stability of the area (Garagulya and Parmuzin 1980, 1982, Demidyuk 1980, Garagulya et al. 1982, Baulin et al. 1982, Garagulya and Gordeyeva 1982).

A quantitative method for evaluating the complexity of engineering-geocryological conditions has been proposed; it examines these conditions as a system of inter-linked components. A master plan for evaluating these systems has been developed. The final stage of the operation involves special mapping of the engineering geology of the area.

This method has been used in one of the areas where pipelines are being built in the North of Western Siberia. It was possible to demonstrate the influence of various environmental factors on the cost of planning, construction and operation of a pipeline where special environmental protection methods were used. Areas with varying complexities of engineering-geological conditions were identified on a map and were categorized as simple, moderately complicated and very complicated (Bondariye and Pendin 1982).

Predictive investigations of permafrost are increasingly being used in areas of major development in the North (Kudryavtsev 1982). It has been revealed that to some extent the difficulties of development in the permafrost zone associated with the problems of environmental protection have been exaggerated. Both cartographic and field investigations have revealed that the areas where development would lead to intense development of cryogenic processes had been exaggerated (Pavlov and Shur 1982). The unstable areas, as a rule, occupy about 25-50% of the surface of the areas investigated in the most sensitive areas of the Arctic (Grave et al. 1982). This aspect testifies to the clear benefits of preliminary geocryological investigations in areas to be developed, in terms of reducing the costs of environmental protection measures.

It should be noted that very little attention is paid to the injurious impact of placer mining on the landscapes of the North. Both natural and artificial methods are used to thaw the placer deposits once the vegetation and the surficial soil layer have been removed; these include sluicing of the surface, intensification of percolation into the deposits, the introduction of chemical admixtures in the water added to thaw the deposits, the introduction of gases, etc. (Perl'shtein 1979, Balobayev et al. 1983). All of these operations disrupt the biological productivity of the ecosystem and aggravate erosional processes, slope denudation and stream pollution. Hence Domenik et al. (1979) have focussed particular attention on the need to plan for revegetation measures to prevent erosion during the planning of such mining operations. Treatment of the mining effluents and both engineering and biological reclamation must be seen as inalienable parts of mineral extraction operations. It is recommended in particular that dredge tailings be dispersed using auxiliary equipment. However, in a number of cases major changes in the surface near borrow pits due to slope processes and the development of lifeless "lunar landscapes" have not occurred. Thermokarst depressions have in some cases been colonized by *Arctogracis* and by *Equisetum arvense* while on slopes and *baydzherakhy* thick growths of arctic groundsel have appeared. Additional sprinkling of the surface leads to the appearance of grasses and after 6-8 years turf will appear in such areas in the Arctic (Olovin 1980).

The problem of revegetation of areas disturbed by development continues to absorb the attentions of northern researchers. The major investigations are being conducted in the tundra zone. Multi-disciplinary studies of the heat balance and of geobotany have been carried out in the arctic region of Western Siberia along the lower reaches of the Bol'shaya Kheta. They revealed that solar radiation and heat were adequate for the growth of sown grasses (*Poa*) and Gramineae; this proves the feasibility of revegetating disturbed landscapes in the tundra zone (Skryabin and Sergeyev 1980). Investigations of revegetation in the extreme north of the Yenisey region have demonstrated that the most active native anchoring plants on disturbed tundra are *Calamagrostis lapplandia* (reed bent-grass), *Festuca* etc. (Skryabin 1979). The results of multi-disciplinary investigations into soils, man-induced disturbances and the conditions for regrowth of various plants on the tundra of the Polar Urals have been published. Types of soils have been identified in terms of suitability for revegetation, and recommendations have been made as to improving the properties of the soil, the seeding of grasses and the planting of shrubs and trees (Vital' 1980, Liverovskaya 1980, Smironov et al. 1980).

Fires, the felling and rooting up of trees, ploughing, and many other types of development are among the causes of soil disturbance and degradation. In Magadan oblast' soil studies have been carried out in the tundra, the taiga and the alpine tundra. The soils of this region were classified both on the basis of biological productivity and on their degree of vulnerability to the types of disturbance listed above. The results of these studies are useful for planning developments and for achieving optimum land use (Sapozhnikov 1980).

Investigations are continuing on the impact of urban development on geocryological conditions. At Labytnangi, on the Arctic Circle on the lower Ob', construction began in 1949 with small, closely spaced buildings without ventilated foundations. Removal of the trees and grass cover, and the

TABLE 2 Landscape changes in the forest tundra zone caused by disruptions resulting from exploratory drilling for gas (adapted from Grechishchev et al. 1983). Characteristics of disturbances and changes in the major components of the landscape, 7-10 years after well was drilled.

Topography	Vegetation	Soils	Engineering-geocryological conditions			
			Permafrost distribution	Mean annual temp.	Active layer depth	Exogenic processes
Subsidence around well sites; gullying on slopes	Replacement of sparse forest with shrub and lichen communities with grass/moss communities	Removal of peat and organic horizon; reduction of soil moisture by 50%	No changes recorded	Rise of 1°C in depressions	Increase by 150%	Thermal erosion

accumulation of snow in yards led initially to the development of thermokarst, thermal erosion and the formation of swamps. Improvements in building technology, dispersal of the buildings, the use of ventilated foundations and gravel pads led to the elimination of the destructive processes cited above (Stremyakov et al. 1980). The reverse phenomenon, i.e. disappearance of the permafrost, was recorded in the recently-built town of Ust'-Ilimsk; this was associated with the southerly location of this new town and a better building design. The formation of *pereletoks* has been recorded in the case of temporary, poorly built buildings (Maksimova et al. 1980).

With the development of towns in the North, and also in association with agriculture, serious concerns are being provoked by the process of secondary salinization of soils and sediments by industrial and domestic waste water. Increased mineralization of suprapermafrost water and ground water (up to 300 g/ℓ) has led to the thawing of sediments to depths of several meters; this threatens the stability of buildings and menaces the natural vegetation cover as well as crops (Anisimova 1980, 1981).

Investigations are under way and recommendations have been made with regard to protecting the permafrost from thawing around drilling sites. As a result of thermokarst around drilling sites in the gas fields of Western Siberia, pits up to 10 m deep and 3-4 m in diameter have been recorded. Table 2 lists the changes in the landscape and in geocryological conditions around drill sites in the northern taiga zone, on well-drained forest tundra. This phenomenon is also known from the American North. In the USSR insulations have been developed which operate on the principle of natural cold accumulation to prevent the thermokarst phenomena just described. Drilling muds of a special composition and stricter regulations as to their use have also been recommended, in order to reduce pollution of the environment (Orlov et al. 1982, Akhmadeyev 1982). Disruption of the surface around gas flare-offs have been examined and calculations to determine the scale of the problem have been recommended (Polozkov et al. 1982).

In the case of the tundra zone preliminary preparation of areas to be developed has been recommended, prior to the erection of any structures; these precautions would obviate subsequent disruption of the surface due to cryogenic processes. Thus, for example, it is proposed that several years prior to construction one should remove the vegetation cover and drain the site in order to produce a safe construction site (Grigor'yev 1979, Sukhodrovskiy 1979). Alternatively it is proposed that one should place gravel pads about 1 m thick on the tundra surface and at the same time construct a surface drainage system; the pads should be mainly laid in winter (Konstantinov 1979).

For conditions prevailing in Central Yakutia experiments have led to optimal regimes for channel irrigation of meadows which will eliminate bog formation and the rise of the permafrost table, as well as salinization (Mel'nikov and Pavlov 1980). To combat the latter, both on agricultural land and especially in urban areas, it is proposed that cryopegs (brine solutions) be replaced with fresh water, which will freeze to produce an impermeable frozen layer (Mel'nikov, P.I., et al. 1982).

A schematic map has been compiled with a view to achieving a very general assessment of the degree of surface stability in areas under development in the permafrost regions of the USSR. It classifies areas as surfaces which are predominantly severely unstable, slightly stable and relatively stable, in terms of man-induced thermokarst, since this is the cryogenic process most hazardous to the environment (Grave 1982).

The map identifies areas most vulnerable to development, where one must pay greatest attention to the possible consequences; this mainly involves areas with ice-rich permafrost. Ground ice melts differently under different physical-geographical conditions. Thus under severe continental conditions, where in summer evaporation exceeds precipitation, or in the very cold conditions of arctic deserts where there is massive Pleistocene ground ice, melting of that ice when the vegetation and soil cover are destroyed, occurs only on a limited scale. Compilation of maps of ice content in the permafrost represents a major task for the immediate future. The maps of distribution of ground ice and of thaw-subsidence in permafrost, compiled for the uppermost 10 m layer for the north of Western Siberia (Trofimov et al. 1980) may serve as a model for such maps.

Future investigations must concentrate not only on the rather labor-intensive methods of physical geographical evaluation, which have been most widely used in Western Siberia, but also on perfecting mathematical models, taking into account both direct and inverse relationships between components of the landscape.

It is also essential to develop specific environmental protection measures for the permafrost areas.

LITERATURE

Anisimova, N. N., 1980, Cryogenic hydrogeochemical changes in alluvial deposits associated with development in Central Yakutiya, in *Okhrana okruzhayushchey sredy pri osvoyenii oblasti mnogoletnemerzlykh porod* [Environmental protection and development of the permafrost zone]: Moscow, Nauka, p 107-111.

Anisimova, N. N., 1981. *Kriogidrogeokhimicheskiye osobennosti merzloy zony* [Cryohydrogeochemical peculiarities of the permafrost zone]: Novosibirsk, Nauka, 152 p.

Akhmadeyev, R. G., 1982, Some questions of environmental protection associated with well drilling in areas of the Far North, in *Geokriologicheskiy prognoz v osvaivayemykh rayonakh Kraynego Severa* [Geocryological prediction in areas being developed in the Far North]: Abstracts of papers, Moscow, VSYeGINGYeO, p. 187.

Balobayev, V. T., Zabolotnik, S. I., Nekrasov, I. A., Shastkevich, Yu. G., and Shender, N. I., 1979, Dynamics of ground temperatures in northern Priamur'ye in association with development of the area, in *Tekhnogennye landshafty Severa i ikh rekul'tivatsiya* [Technogenic landscapes in the North and their revegetation]: Novosibirsk, Nauka, p 74-88.

Balobayev, V. I., Pavlov, A. V., and Perl'shtein, G. Z., 1983. *Teplofizicheskiye issledovaniya kriolitozony Sibiri* [Thermophysical investigations in permafrost in Siberia]: Novosibirsk, Nauka, 213 p.

Baulin, V. V., Dubikov, G. I., Uvarkin, Yu. T., and Minkin, M. A., 1982, Problems and methods of geocryological forecasting in prospecting, in *Geokriologicheskiy prognoz v osvaivayemykh rayonakh Kraynego Severa* [Geocryological prediction in areas being developed in the Far North]: Abstracts of papers, Moscow, VSYeGINGYeO, p. 14-19.

Bondariye, G. K., and Pendin, V. V., 1982, Methods of quantitative evaluation of engineering-geology conditions and of special engineering-geology mapping, *Inzhenernaya geologiya*, No. 4, p. 82-89.

Brown, J., and Grave, N. A., 1979a, Terrain disturbances and ground temperature regimes, in *Proceedings of the Third International Conference on Permafrost:* Ottawa, vol. 3.

Brown, J, and Grave, N. A., 1979b, Physical and thermal disturbance and protection of permafrost, *CRREL Special Report* 79-5.

Brown, J., and Grave, N. A., 1981. *Narusheniye poverkhnosti i eye zashchita pri osvoyenii Severa* (Disturbance and protection of the surface in connection with development in the North]: Novosibirsk, Nauka, 87 p.

Chernyad'yev, V. P., 1980, Prediction of geocryological conditions associated with the disruption of natural conditions, in *Geokriologicheskiy prognoz i sovershenstvovaniye inzhenernykh izyskaniy* [Geocryological prediction and perfecting of engineering exploration]: Moscow, Stroyizdat, p. 32-54.

Chernyad'yev, V. P., 1982, Investigations of heat exchange in freezing and thawing sediments at the Urengoy hospital, in *Termika pochv i gornykh porod v kholodnikh rayonakh* [Thermal aspects of soils and rock in cold regions]: Yakutsk, Permafrost Institute (IM), SO USSR Academy of Sciences (AN SSSR),p. 26-34.

Chernyad'yev, V. P., and Abrashov, B. A., 1980, The thermal regime of sediments on experimental sites, in *Geokriologicheskiy prognoz i sovershenstvovaniye inzhenernykh izyskaniy* [Geocryological prediction and the perfecting of engineering exploration]: Moscow, Stroyizdat, p. 5-31.

Demidyuk, L. M., 1980, The problem of rational utilization of engineering-geocryological resources, in *Okhrana okruzhayushchey sredy pri osvoyenii oblasti mnogoletnemerzlykh porod* [Environmental protection in association with development of the permafrost zone]: Moscow, Nauka, p. 42-50.

Domenik, O. P., Martsinyuk, B. F., and Payusova, Ye. A., 1979, Technology and ecological consequences of dredging operations, in *Tekhnogennye landshafty Severa i ikh rekul'tivatsiya* [Technogenic landscapes in the North and their revegetation]: Novosibirsk, Nauka, p. 113-117.

Garagulya, L. S., and Gordeyeva, G. I., 1982, Classification of changes in natural conditions for engineering-geology evaluations of terrain in connection with pipeline construction, *Inzhenernaya geologiya*, no. 2, p. 106-113.

Garagulya, L. S., Kondrat'yeva, K. A., and Maksimova, L.N. 1982, Predictive geocryological maps at the stages of planning and regional exploratory investigations, in *Geokriologicheskiy prognoz v osvaivayemykh rayonakh Kraynego Severa* (Geocryological prediction in developing areas of the Far North]: Abstracts of papers, Moscow, VSYeGINGYeO, p. 36-39.

Garagulya, L. S., and Parmuzin, S. Yu., 1980, Indices for evaluating sensitivity of land to changes in natural factors and to human activities in the permafrost zone, *Merzlotnye issledovaniya, Izd-vo MGU,* vol. 19, p. 53-59.

Garagulya, L. S., and Parmuzin, S. Yu., 1982, Methods of evaluating permafrost conditions at the early stages of planning above-ground structures, *Inzhenernaya geologiya*, No. 4, p. 98-107.

Geokriologicheskiy prognoz i sovershenstvovaniye inzhenernykh izyskaniy (Geocryological prediction and perfecting of engineering exploration], 1980: Moscow, Stroyizdat, 199 p.

Geokriologicheskiy prognoz v osvaivayemykh rayonakh Kraynego Severa [Geocryological prediction in developing areas of the Far North]: Abstracts of papers given at an Interdepartmental Conference, 9-11 February 1982: Moscow, VSYeGINGYeO, 204 p.

Grave, N. A., 1980, The status and direction of geocryological investigations into the problem of environmental protection and rational resource utilization in the permafrost zone, in *Ustoychivost' poverkhnosti k tekhnogennym vozdeystviam v oblasti vechnoy merzloty* [Sensitivity of the surface to human activities in the permafrost zone]: Yakutsk, IM SO AN SSSR, p. 6-12.

Grave, N. A., 1982, Sensitivity of the surface to mechanical disturbance associated with development in the North, *Izvestiya AN SSSR, seriya geograficheskaya*, no. 6, p. 54-62.

Grave, N. A., 1983, A geocryological aspect of the problem of environmental protection, in *Proceedings of the Fourth International Conference on Permafrost*, Washington, DC, National Academy of Sciences, p. 369-373.

Grave, N. A., Mel'nikov, P. I., and Moskalenko, N. G., 1982, Geocryological problems in environmental protection in association with the development of oil and gas fields in the permafrost zone, in *Geokriologicheskiy prognoz v osvaivayemykh rayonakh Kraynego Severa* (Geocryological prediction in developing areas of the Far North]: Abstracts of papers, Moscow, VSYeGINGYeO, p. 32-36.

Grechishchev, S. Ye., 1982a, Some problems in geocryological prediction, *Inzhenernaya geologiya*, No. 3, p. 3-13.

Grechishchev, S. Ye., 1982b, Theoretical questions in modelling cryogenic landscapes, in *Geokriologicheskiy prognoz v osvaivayemykh rayonakh Kraynego Severa* [Geocryological prediction in developing areas of the North]: Abstracts of papers, Moscow, VSYeGINGYeO, p. 46-47.

Grechishchev, S. Ye., Krishuk, L. N., Nevechera, V. P., Sheshin, Yu. V., and Shur, Yu. P., 1980, Thermokarst and other cryogenic processes developing under the impact of human activities in the northern tayga zone of Western Siberia, in *Okhrana okruzhayushchey sredy pri osvoyenii oblasti mnogoletnemerzlykh porod* [Environmental protection in association with development in the permafrost zone]: Moscow, Nauka, p. 12-19.

Grechishchev, Ye. S., and Mel'nikov, Ye. S., 1982, Problems in regional geocryological investigations and predictions, in *Geokriologicheskiy prognoz v osvaivayemykh rayonakh Kraynego Severa* [Geocryological prediction in developing areas of the Far North]: Abstracts of papers, Moscow, VSYeGINGYeO, p. 9-14.

Grechishchev, S. Ye., Moskalenko, N. G., Shur, Yu. P., Slavin-Borovskiy, V. B., Nevechera, V. P., Mel'nikov, Ye. S., Krizuk, P. N., Sheshin, Yu. B., Veysman, P. I., Malevskiy-Malevich, S. P., Lazareva, N. A., and Sagaydachnyy, Yu. A., 1983, *Geokriologicheskiy prognoz dlya Zapadno-Sibirskoy gazonosnoy provintsii* [Geocryological prediction for the West Siberian gas-bearing province]: Novosibirsk, Nauka, 180 p.

Grigor'yev, N. F., 1979, Preliminary destruction of the cover as one method of improving permafrost conditions, in *Tekhnogennye landshafty Severa i ikh rekul'tivatsiya* [Technogenic landscapes of the North and their revegetation]: Novosibirsk, Nauka, p. 45-50.

Inzhenerno-geokriologicheskiye aspekty ratsional'nogo ispol'zovaniya i okhrany geologicheskoy sredy [Engineering geology aspects of the rational utilization and protection of the geological environment], 1981: Moscow, Nauka, 240 p.

Kondrat'yev, V. G., Zontov, M. N., Chushkina, N. I., Zamolotchikova, S. A., and Poltev, N. F., 1979, Permafrost conditions in the Yenisey valley (Dudinka *rayon*) and changes in them associated with human activities, in *Takhnogennye landshafty Severa i ikh rekul'tivatsiya* [Technogenic landscapes in the North and their revegetation]: Novosibirsk, Nauka, p. 89-97.

Konstantinov, I. P., 1979, Gravel pads as a technogenic cover and as a building foundation, in *Tekhnogennye landshafty Severa i ikh rekul'tivatsiya* [Technogenic landscapes in the North and their revegetation]: Novosibirsk, Nauka, p. 62-73.

Kudryavtsev, V. A., 1982, Theoretical and methodological bases for geocryological prediction, in *Geokriologicheskiy prognoz v osvaivayemykh rayonakh Kraynego Severa* [Geocryological prediction in developing areas of the Far North]: Abstracts of papers, Moscow, VSYeGINGYeO, p. 3-6.

Liverovskaya, I. T., 1980, Changes in the properties of soils in the eastern foothills of the Polar Urals and adjacent plains, connected with pipeline construction, in *Okhrana okruzhayushchey sredy pri osvoyenii oblasti mnogoletnemerzlykh porod* [Environmental protection in association with development in the permafrost zone]: Moscow, Nauka, p. 116-121.

Liverovskiy, Yu. A., Popov, A. I., and Smironov, V. V., 1980, Revegetation of natural landscapes disturbed as a result of human activities, in *Okhrana okruzhayushchey sredy pri osvoyenii oblasti mnogoletnemerzlykh porod* [Environmental protection in association with development of the permafrost zone]: Moscow, Nauka, p. 111-115.

Maksimova, L. N., Boyarskiy, O. G., and Dubrovin, V. A., 1980, Study of the changes in permafrost conditions during the course of development of an area, in *Okhrana okruzhayushchey sredy pri osvoyenii oblasti mnogoletnemerzlykh porod* [Environmental protection in association with development of the permafrost zone]: Moscow, Nauka, p. 19-25.

Mel'nikov, P. I., Gubanov, B. A., and Konosavskiy, P. K., 1982, Substitution of cryopegs with fresh water in constructing a zero cycle: *Inzhenernaya geologiya*, no. 3, p. 117-119.

Mel'nikov, P. I., and Pavlov, A. V., 1980, Permafrost investigations in connection with the development of methods to protect the natural environment of the North, in *Merzlotnye issledovaniya v osvaivayemykh rayonakh SSSR* [Permafrost investigations in developing areas of the USSR]: Novosibirsk, Nauka, p. 3-8.

Mel'nikov, Ye. S., Anan'yev, T. V., Andrianov, V. N. et al., 1983a, *Geokriologicheskiye usloviya Zapadno-Sibirskoy gazonosnoy provintsii* (Geocryological conditions in the West Siberian gas-bearing province]: Novosibirsk, Nauka, p.

Mel'nikov, Ye. S., Veysman, L. I., Goral'chuk, M. I., et al. 1983b, *Landshafty kriolitozony Zapadno-Sibirskoy gazonosnoy provintsii* [Permafrost landscapes of the West Siberian gas-bearing province]: Novosibirsk, Nauka, p.

Mel'nikov, Ye. S., Moskalenko, N. G., and Tigunova, P. N., 1982, Landscape methods of geocryological prediction, in *Geokriologicheskiy prognoz v osvaivayemykh rayonakh Kraynego Severa* (Geocryological prediction in developing areas of the Far North]: Abstracts of papers, Moscow, VSYeGINGYeO, p. 49-50.

Mikhaylov, N. A., 1980, Classification of landscapes during evaluation of engineering geology conditions in the BAM zone, in *Geokriologicheskiy prognoz i sovershenstvovaniye inzhenernykh izyskaniy* [Geocryological prediction and perfecting engineering exploration]: Moscow, Stroyizdat, p. 149-166.

Moskalenko, N. G., 1980, Landscape identification of engineering geology conditions for predictions of changes due to the impact of human activities, in *Merzlotnye issledovaniya v osvaivayemykh rayonakh SSSR* [Permafrost investigations in developing areas of the USSR]: Novosibirsk, Nauka, p. 8-18.

Moskalenko, N. G., and Shur, Yu. P., 1980, Typical disturbances to environmental complexes in the North of Western Siberia under the influence of pipeline construction and predictions on their dynamics, in *Okhrana okruzhayushchey sredy pri osvoyenii oblasti mnogoletnemerzlykh porod* [Environmental protection in association with the development of the permafrost zone]: Moscow, Nauka, p. 71-78.

Okhrana okruzhayushchey sredy pri osvoyenii oblasti mnogoletnemerzlykh porod [Environmental protection in association with development in the permafrost zone] 1980: Moscow, Nauka, 110 p.

Olovin, B. A., 1979, The dynamics of re-arranging landscapes in connection with the exploitation of deposits in the tundra zone, in *Tekhnogennye landshafty Severa i ikh rekul'tivatsiya* [Technogenic landscapes in the North and their revegetation]: Novosibirsk, Nauka, p. 98-112.

Olovin, B. A., 1980, Some questions on environmental protection associated with placer mining in the North, in *Okhrana okruzhayushchey sredy pri osvoyenii oblasti mnogoletnemerzlykh porod* [Environmental protection in association with the development of the permafrost zone]: Moscow, Nauka, p. 50-55.

Orlov, A. V., Polozkov, A. V., Nikitin, V. N., and Gumenyuk, A. S., 1982, Measures to protect the permafrost around well-heads, in *Geokriologicheskiy prognoz v osvaivayemykh rayonakh Kraynego Severa* [Geocryological prediction in the developing areas of the Far North]: Abstracts of papers, Moscow, VSYeGINGYeO, p. 198.

Pavlov, A. V., and Shur, Yu. P., 1982, Thermophysical site investigations of the upper layers of the cryolithozone, in *Termika pochv i gornykh porod v kholodnikh rayonakh* [Thermal aspects of soils and rocks in cold regions]: Yakutsk, IM SO AN SSSR, p. 5-14.

Perl'shtein, G. Z., 1979, *Vodno-teplovaya melioratsiya merzlykh porod na Severo-Vostoke SSSR* (Warming and draining of permafrost materials in the Northeast of the USSR]: Novosibirsk, Nauka, 302 p.

Polozkov, A. V., Orlov, A. V., and Nikitin, V. N., 1982, Peculiarities of the effects of gas flare-offs and gas-bearing gushers in permafrost, in *Geokriologicheskiy prognoz v osvaivayemykh rayonakh Kraynego Severa* [Geocryological prediction in developing areas of the Far North]: Abstracts of papers, Moscow, VSYeGINGYeO, p. 194.

Sapozhnikov, A. P., 1980, Principles of assessing forest soils in the permafrost zone with a view to their preservation and rational utilization, in *Geografiya i genezis pochv Magadanskoy oblasti* [The geography and genesis of the soils of Magadan oblast']: Vladivostok, p. 161-168.

Sergeyev, B. P., 1982, Heat and moisture exchange between the soil and the atmosphere in the area of Cape Kharasavey (Yamal Peninsula), in *Termika pochv i gornykh porod v kholodnikh rayonakh* [Thermal aspects of soils and rocks in cold regions]: Yakutsk, IM SO AN SSSR, p. 15-26.

Shvetsov, P. F., 1982, Thermodynamic criteria for the stability of soils and sediment complexes in the Subarctic, in *Geokriologicheskiy prognoz v osvaivayemykh rayonakh Kraynego Severa* (Geocryological prediction in developing areas of the Far North]: Abstracts of papers, Moscow, VSYeGINGYeO, p. 6-8.

Shur, Yu. L., Pavlov, A. V., and Chernyad'yev, V. P., 1982, Some results of field investigations of the regime of permafrost conditions in developing oil-and-gas areas in the permafrost zone, in *Geokriologicheskiy prognoz v osvaivayemykh rayonakh Kraynego Severa* [Geocryological prediction in developing areas of the Far North]: Abstracts of papers, Moscow, VSYeGINGYeO, p. 24-28.

Skryabin, S. Z., 1979, Investigations into biological restoration of tundra disturbed by human activities in the Yenisey Far North, in *Tekhnogennye landshafty Severa i ikh rekul'tivatsiya* (Technogenic landscapes of the North and their revegetation]: Novosibirsk, Nauka, p. 51-61.

Skryabin, P. N., and Sergeyev, B. P., 1980, Water and heat regimes of soils associated with biological reclamation of landscapes disturbed by human activities, in *Merzlotnye issledovaniya v osvaivayemykh rayonakh SSSR* (Permafrost investigations in developing areas of the USSR]: Novosibirsk, Nauka, p. 23-31.

Smironov, V. V., Marakhtanov, V. P., Liverovskaya, I. T., Smirnova, T. I., and Vital', A. D., 1980, Environmental disturbance and the first stages of re-establishment of soils and vegetation covers in association with some forms of linear installations, in *Okhrana okruzhayushchey sredy pri osvoyenii oblasti mnogoletnemerzlykh porod* [Environmental protection associated with the development of the permafrost zone]: Moscow, Nauka, p. 121-127.

Spiridonov, V. V., and Semenov, P. P., 1980, Problems of environmental protection associated with pipeline construction, in *Okhrana okruzhayushchey sredy pri osvoyenii oblasti mnogoletnemerzlykh porod* [Environmental protection associated with the development of the permafrost zone]: Moscow, Nauka, p. 60-66.

Stremyakov, A. Ya., Chekhovskiy, A. P., and Pavlunin, V. B., 1980, The impact of urban structures on the heat regime of frozen sediments, in *Geokriologicheskiy prognoz i sovershenstvovaniye inzhenernykh izyskaniy* (Geocryological prediction and perfecting of engineering exploration]: Moscow, Stroyizdat, p. 54-67.

Sukhodol'skiy, S. Ye., 1980, Disturbance of the environment and its consequences in connection with building linear structures in the North of Western Siberia, in *Okhrana okruzhayushchey sredy pri osvoyenii oblasti mnogoletnemerzlykh porod*

[Environmental protection associated with the development of the permafrost zone]: Moscow, Nauka, p. 67-71.

Sukhodrovskiy, V. P., 1979, *Eksogennoye rel'yefobrazovaniye v kriolitozone* (Exogenic relief formation in the permafrost zone]: Moscow, Nauka, 279 p.

Trofimov, V. T., Bodu, Yu. B., and Dubikov, G. I., 1980, *Kriogennoye stroyeniye i l'distost' mnogoletnemerzlykh porod Zapadno-Sibirskoy plity* [Cryogenic structure and ice content in permafrost on the Western Siberian plate]: Moscow, Izdatel'stvo MGU, 245 p.

Turbina, M. I., 1980, Changes in natural conditions on a section of operating pipeline (Central Yakutiya), in *Merzlotnye issledovaniya v osvaivayemykh rayonakh SSSR* [Permafrost investigations in the developing areas of the USSR]: Novosibirsk, Nauka, p. 31-36.

Turbina, M. I., 1982, Evaluation of geocryological conditions in the valley of the middle Amga in connection with agricultural development of the area (based on the example of the Betyun' area), in *Termika pochv i gornykh porod v kholodnikh rayonakh* [Thermal aspects of soils and rocks in cold regions]: Yakutsk, IM SO AN SSSR, p. 135-141.

Vital', A. D., 1980, Root systems of turf-forming plants and their significance for re-establishing a vegetation cover destroyed by development of an area, in *Okhrana okruzhayushchey sredy pri osvoyenii oblasti mnogoletnemerzlykh porod* [Environmental protection associated with the development of the permafrost zone]: Moscow, Nauka, p. 128-133.

Vremennoye rukovodstvo po zashchite landshaftov pri prokladke gazoprovodov na Kraynem Severe [Interim manual on protection of the landscape during the laying of gas pipelines in the Far North], 1980: Yakutsk, IM SO AN SSSR, 44 p.

Zabolotnik, S. I. and Shender, N. I., 1982, Some results of an annual cycle of soil temperature measurements under natural conditions and where the vegetation cover has been destroyed on the Upper Zeya Plain, in *Termika pochv i gornykh porod v kholodnikh rayonakh* [Thermal aspects of soils and rocks in cold regions]: Yakutsk, IM SO AN SSSR, p. 50-57.

REGULATORY RESPONSIBILITIES IN PERMAFROST ENVIRONMENTS OF ALASKA FROM THE PERSPECTIVE OF A FEDERAL LAND MANAGER

C.V. McVee* and J.V. Tileston**

U.S. Bureau of Land Management
Anchorage, Alaska, USA

Until very recently all land and water resources in Alaska were owned by the federal government. There was very little local management and people could, without approval, search out resources and proceed to develop what was found with very little regulation. In 1959, the U.S. Congress began a division of land and responsibilities between the newly created State of Alaska and the U.S. government. This led to some changes in philosophy and made developers deal with two owners. This largely was the situation in 1968 when the Prudhoe Bay oil field was discovered and in 1972 when the trans-Alaska pipeline was proposed. In 1971, an old ownership situation was solved when Congress settled aboriginal rights claimed by Alaskan natives (Eskimos, Indians, and Aleuts). This settlement resulted in a decision to convert about 12% of the State of Alaska into private native ownership. Finally, in 1981, the remaining federal lands in Alaska were further subdivided among four major federal landowners. The net effect of these three federal laws was to create short-term uncertainties as to resource ownership and, therefore, uncertainty as to regulatory responsibilities.

Thus, a resource developer in Alaska must have specific goals and objectives that are translated directly into regulatory requirements for resource development in permafrost environments. This multiple-ownership regulatory approach requires unusual coordination and is often a cause for delay in project approvals. Special regulations that apply in permafrost environments appear onerous but reflect the fact that when surface disturbances are not done correctly there is substantial potential for long-term degradation of the permafrost. Federal agencies, on the other hand, are obligated to assure that the special requirements are necessary, fair, and uniformly applied.

NATIONAL ENVIRONMENTAL POLICY ACT

The National Environmental Policy Act (NEPA) is the basic national charter for protection of the environment. It requires federal decision-makers to use an interdisciplinary approach in the planning and approval processes and to seek comments from concerned citizens before decisions are made and before actions are taken. The NEPA process is intended to help federal officials make decisions that are based on understanding of environmental consequences and take actions that protect, restore, and enhance the environment. The environmental impact assessment is a systematic process that formally documents the decision-maker's rationale for the decision taken and the alternative courses considered. The basic stipulations, terms, and conditions used in the federal regulatory process are direct by-products of the NEPA analysis that focus upon how unwanted effects can be avoided entirely or at least reduced to an acceptable level.

LAND OWNERSHIP IN ALASKA: ITS PRESENT AND FUTURE

<u>1959 - Alaska Statehood Act</u>. This act provided for 103 million acres (42.7 million hectares) of unreserved federal land to be transferred to state ownership. The state has until 1994 to complete its selection of the acreage. The 1959 act also granted to the state title to all submerged lands under inland navigable waters. In all, ownership of about 110 million acres (44.5 million hectares), exclusive of offshore areas, will transfer to the state.

<u>1971 - Alaska Native Claims Settlement Act</u>. As a condition to settling aboriginal rights, Congress provided that 70 million acres (28.3 million hectares) of federal land and some state land be set aside from which Alaska natives could select 44 million acres (17.8 million hectares) for private ownership. The 1971 act established Alaska native village corporations to own lands around established native communities and Alaska native regional corporations to own other lands and subsurface resources under village corporation lands.

<u>1981 - Alaska National Interest Conservation Act</u>. This act placed final parameters on major federal land areas in Alaska as follows:

- 76 million acres (30.8 million hectares) in national wildlife refuges to be managed by the U.S. Fish and Wildlife Service.
- 52 million acres (21 million hectares) in national parks to be managed by the U.S. National Park Service.
- 68 million acres (17.5 million hectares) in public lands and special management areas to be managed by the U.S. Bureau of Land Management.
- 23 million acres (9.3 million hectares) in national forests and special management areas to be managed by the U.S. Forest Service.

This 1981 act also established two major provisions that require special attention by federal land managers in Alaska. The first requires the federal land manager to evaluate all proposals to use any federal lands for any significant restriction on subsistence uses. The second established

*Former State Director, BLM.
**Presented by J.V. Tileston.

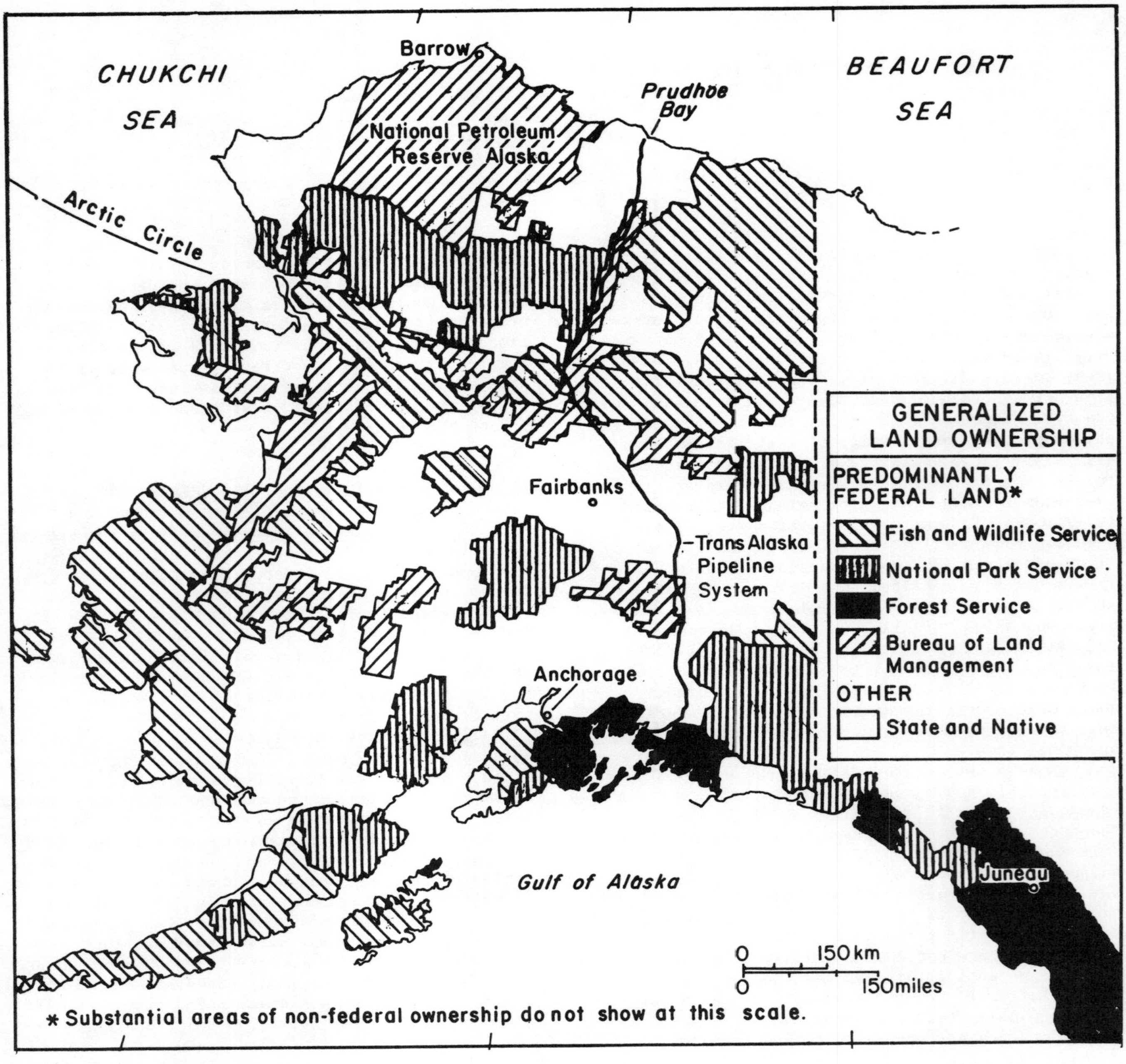

FIGURE 1 Distribution of federally managed and state and native lands in Alaska.

uniform federal transportation evaluation and approval procedures.

Including about 2.1 million acres of national defense lands, the federal government, principally the Department of the Interior, will retain direct proprietary regulatory responsibility for about 59% of Alaska.

As of June 1983, some 89 million acres (36 million hectares) were under exclusive jurisdiction of the state or Alaska native corporations. For federal lands selected but not conveyed to state or native entities, there is some form of mutual cooperation, coordination, or actual sharing of the regulatory responsibilities. Ownership is still fluid. Between June and October 1 of 1983, an additional 6 million acres (2.4 million hectares) will be transferred from federal to state or native ownership.

Land exchanges between federal, state, and native ownership also are becoming a prominent factor in transferring resource ownership to enhance both preservation and development opportunities. A major focus of recent exchange proposals has involved the Alaska Arctic Slope, where development of energy resources is the primary use.

A practical question of any major resource development project in a permafrost environment is, who owns the gravel? In Alaska, gravel is a subsurface resource; therefore, if on native village corporation lands, the native regional corporation owns the gravel. If the gravel source is in a non-navigable waterway, the adjacent landowner owns the gravel; if in a navigable waterway, the state owns the gravel.

LAND MANAGEMENT OBJECTIVES AND REGULATORY ACTIONS

Each native corporation has its own land management objectives, and the state also has land management objectives for land under its jurisdiction.

Each federal land manager is given specific ground rules by the U.S. Congress for resource protection and development in Alaska. These range from strict preservation to various forms of multiple-use and preservation. For example, the U.S. Bureau of Land Management manages the 23-million-acre (9.3-million-hectare) National Petroleum Reserve in Alaska for petroleum extraction. While just north of Fairbanks, Congress had directed the Bureau of Land Management to manage the highly mineralized 1-million-acre (0.4-million-hectare) White Mountains National Recreation Area for public recreation and the adjacent 1.2-million-acre (0.5-million-hectare) mineralized Steese National Conservation Area.

Each land manager, through its regulatory functions, is fulfilling its obligation to make wise decisions for the benefit of the owners. This is a balancing game where most options are neither good nor bad, but "better" or "worse." A federal agency's constituencies include 227 million stockholders or citizens of the U.S., the President of the U.S., the Secretary of the Interior, Congress (especially the Alaska delegation and members of the appropriations committees), the Governor, and all too frequently the federal courts.

The state and native landowners have similar constituencies to be considered in land-use authorization in permafrost environments.

REGULATORY ACTIVITIES TO PROTECT PERMAFROST ENVIRONMENTS

The degree of federal regulation is usually commensurate with the level of the expected activity and its duration, the sensitivity (physical and political) of the area to be disturbed, and the public and private interests at risk.

Federal regulation can, and does, take place during the design, operation, and maintenance stages of resource development at four distinct periods:

- Exploration - Locate the resource.
- Development - Prepare the resource to go to market.
- Production - Produce and transport the resource into the market.
- Shutdown - Economic life of the production phase is completed.

Certain regulations have universal application, while others are specific to the lands managed by a particular owner (federal, state, or native).

At the federal level in Alaska, there is a striking degree of similarity to the regulatory process in that Congress fixes the level of resource uses and determines broad priorities for which resource uses come first.

Each federal land manager prepares a land-use plan that outlines the criteria for resource uses and protection standards. Each land-use plan undergoes numerous local, state, and often national reviews before it is finally adopted.

There is an accelerating trend to place with the applicant primary responsibility for showing that there will be no undue or unnecessary harm to the existing socioeconomic values or to the natural environments as a result of the proposed resource development program. In Alaska, these include investigations on subsistence use, endangered species of plants and animals, and cultural resource avoidance as well as the more traditional permafrost engineering design questions.

In Alaska, federal regulatory actions expressed by stipulations, terms, or conditions in the land-use authorizations come in three principal categories:

1. Those designed to prevent or mitigate physical damage to permafrost environments. Examples include:
    - Avoiding sensitive permafrost areas where mass wasting, major drainage reconfigurations or untried engineering would be required.
    - Constructing temporary ice roads and ice airfields (natural or man-made) during the exploration phase in lieu of permanent facilities where thermal erosion is a likely by-product.
    - Requiring either all upland or aquatic gravel extraction sites according to thermal regimes and site restoration capabilities.
    - Requiring gravel or substitute insulation for work pads, airfields, roads, drilling areas, and housing areas, according to the permafrost regime and the commitment of available gravel resources for other uses.
    - Permitting surface movement of heavy equipment only when the ground is frozen to a minimum depth of 12 in.
    - Using air-cushion or low-ground-pressure vehicles as a means to avoid construction of either temporary or permanent transportation facilities.
    - Using air-transportable equipment where either temporary or permanent roads would be necessary.
    - Installing rip-rap or culverted drainages to mitigate or avoid thermal erosion.
    - Successfully testing new engineering methods before surface disturbance starts.
    - Requiring rehabilitation/restoration when the surface disturbance activities are suspended for any significant period or as part of the shutdown phase.
    - Restricting fuel storage and waste disposal sites and methods.
2. Those designed to maintain availability of habitats not physically disturbed. Examples include:
    - Establishing minimum flight altitudes for

aircraft crossing high-density waterfowl nesting areas during the nesting season.
- Avoiding activity near occupied bear sites.
- Avoiding caribou calving grounds during the calving season.
- Avoiding major caribou migration routes or requiring special engineering designs for mitigation.
- Avoiding peregrine falcon nest areas during the nesting period.
- Designing and locating snow-collecting structures in relation to winter habitat areas used by wildlife.
- Designing culverts for fish passage.
- Maintaining natural hydrologic regimes through fish spawning areas.
- Maintaining minimum flows in key overwinter fish habitats.

3. Those designed to prevent or mitigate sociocultural and aesthetic disturbance. These include:
- Locating facilities away from rural communities.
- Prohibiting families at construction or remote operatonal facilities.
- Avoiding local subsistence use areas (fish camps, hunting camps, traplines, caches, cabins, berry-picking areas, local firewood, or house log areas), especially during periods of use.
- Locating and avoiding cultural and historic places.
- Requiring employment of local workers.
- Instituting job training to provide skills needed by local residents to secure employment during the exploration, construction, or operational phases of the proposed project.
- Increasing employment opportunities of minority groups through minority contracts.
- Prohibiting hunting by personnel while located at remote facilities.
- Evaluating how the project will look when completed (color, form, fit to terrain).
- Evaluating whether temporary or permanent transportation facilities will be approved at the exploration phase.

TRANSPORTATION FACILITIES

Major resource developments in Alaska frequently require construction of new transportation facilities. While the mine site, timber harvest area, or oilfield may occupy a relatively small area, be in a single ownership, and impact only a few ecosystems, transportation systems involve much larger areas of impact, cross numerous ownerships, and frequently involve numerous ecosystems. The development and transportation decisions are usually separate.

The transportation system is frequently the focus of major controversy, as new transportation creates new development opportunities for other resources.

REGULATORY UNIFORMITY

All land managers, resource developers, and their respective interest groups desire uniform regulatory processes and fair stipulations, terms, and conditions.

The problem is, "Whose process?" and "Fair to whom?"

In Alaska, there are several cooperative ventures started or now going under the auspices of the Alaska Land Use Council, a federal/state/native entity created by Congress under the Alaska National Interest Land Conservation Act. These ventures include:
- Standards for wildfire suppression on state, federal, and native ownerships.
- A process for conducting subsistence evaluations.
- Uniform standards for oil and gas exploration, development, and production on state and federal lands.
- Management of federally reserved access easements across native ownerships.
- Establishment of natural control areas against which to measure the effectiveness of permit stipulations.

CONCLUSIONS

Conferences such as this pinpoint current state-of-the-art opportunities and outline potential solutions being tested. More importantly, they maintain established lines of communication and open new ones between land managers, researchers, and resource developers.

The regulatory process is an expression of the concerns landowners have in protecting resources entrusted to their care. In the case of federal agencies, these concerns are required by law. Requirements imposed in permafrost areas may appear particularly onerous, but they reflect the potential for long-term serious degradation of the permafrost environment. A principal factor in the decision process must be "When in doubt, be conservative."

Federal agencies, on the other hand, are also obligated to assure that requirements placed on industry are necessary, are fairly and uniformly administered, and are reasonable in relation to the values at risk.

Government agencies also have the responsibility to assure that proposed developments that cut across land ownership boundaries are evaluated under similar criteria. Terms, conditions, or stipulations should be reasonably compatible between adjacent landholders. Nothing strains the credibility of the regulatory process more than having radically different standards on opposite sides of an invisible ownership line.

TERRAIN AND ENVIRONMENTAL PROBLEMS ASSOCIATED WITH EXPLORATORY DRILLING, NORTHERN CANADA

H.M. French

Departments of Geography and Geology, University of Ottawa
Ottawa, Canada

Recent environmental concerns in northern Canada relate largely to the search for hydrocarbons in either the Northern Yukon, the Mackenzie Valley, the High Arctic islands, or offshore in the Beaufort Sea. Since the extraction of other nonrenewable mineral resources is still limited in extent (Indian Affairs and Northern Development, 1983), this paper restricts itself to land-based oil and gas exploration activity, and provides examples from the northern interior Yukon.

The last 15 years has witnessed a growing awareness of the sensitive ecological nature of much of northern Canada. Unfortunately, some of the recently discovered hydrocarbons occur in the more fragile tundra and marine environments where plant and animal life is relatively abundant and where ice-rich permafrost is widespread. A cornerstone of federal government policy in Canada is that northern development can only be sanctioned if all possible effort is made to minimize the environmental impact of these activities upon both the physical environment and the northern indigenous peoples (Chretien, 1972; Allmand, 1976).

One of the more important steps that have been taken to promote this general objective was the passing of the Territorial Land Use Act and Regulations in 1971. Other initiatives taken include terrain sensitivity and surficial geology mapping, undertaken primarily by the Geological Survey of Canada in areas of potential economic activity (e.g. Barnett et al., 1977), preparation of environmental impact statements by industry and their assessment by federal government agencies prior to the granting of permits (e.g. Slaney and Co., 1973; Beak Consultants, 1975), establishment of government-sponsored commissions of inquiry into the social and environmental effects of major development proposals (e.g. Berger, 1977; Lysyk et al., 1977), and support of research into arctic land use problems by the Federal Department of Indian Affairs and Northern Development (e.g. Kerfoot, 1972; Wein et al., 1974; Bliss, 1975; Smith, 1977; French, 1978a, 1981).

Other papers related to environmental problems of drilling in northern Canada include those by Bliss and Peterson (1973) and French (1978b; 1980). Similar studies dealing with northern Alaskan well sites are also available (e.g. Lawson et al., 1978; Lawson and Brown, 1979; Johnson, 1981).

GENERAL CONSIDERATIONS

Terrain problems in northern Canada frequently relate to the melt of ice-rich permafrost and subsequent ground instability. Other concerns relate to the geotechnical properties of certain materials, particularly those that are unconsolidated and susceptible to either natural or man-induced failure and mass movement.

With respect to oil and gas exploration, some of the early environmental concerns were related to seismic and other transportational activities. Recent studies in the Northwest Territories, however, suggest that the impact of modern transportational activities has been reduced to a minimum; improvements in industry operating practices (e.g. the use of vehicles equipped with low-pressure tires) and the strict application of land use regulations (e.g. restriction of cross-tundra vehicle movement to winter months) are important factors. Viewed in this light, the potential for the most serious terrain and environmental damage is now associated with the drilling operation itself, which often extends into the critical summer months, and the disposal of waste drilling fluids.

Several factors accentuate terrain disturbance problems adjacent to well sites. First, on many islands there is an absence of easily accessible gravel aggregate suitable for pad construction. This problem is particularly acute on the Sabine Peninsula of eastern Melville Island where the Drake and Hecla gas fields are located and where there is a near-continuous lichen-moss cover overlying soft, ice-rich shales of the Christopher Formation. Wherever possible, urethane matting is placed around the rig and beneath buildings to compensate for any gravel deficiency. Road construction is also a problem because of the lack of aggregate in many localities. Furthermore, although compacted snow is often used for winter roads elsewhere in northern Canada, the very low snowfall of the High Arctic limits this possibility, and strong winds leave many flat areas virtually snow-free during the entire winter.

A second factor indirectly promoting terrain disturbance is that an increasing number of wells are being drilled to greater depths as deeper geological structures are tested. Since the time needed to drill a hole increases with depth, activity at many arctic well sites often continues into the critical summer months when tundra is thawed. The movement of equipment and supplies around the site at this time of year can lead to considerable terrain disturbance, especially if there is a gravel shortage at the site.

Environmental problems of land-based wells often relate to the disposal of waste drilling fluids (French, 1980). These may be toxic in nature. As a consequence, the Territorial Land Use Regulations require that the fluids be buried in below-ground sumps or in reserve pits, so that they freeze in situ and become permanently contained within the permafrost. In itself, the construction of a sump is a major terrain disturbance. In addition, the influx of relatively warm drilling fluids can lead to significant changes in the thermal regime of the permafrost adjacent to and beneath the sump (French and Smith, 1980). Moreover, if a well is drilled deeper than antici pated for various technical or geologic reasons the sump may be too small to contain the fluids used. In some other instances, fluids and toxic materials have been either spilled on the tundra or allowed to enter water bodies. In recent years, the deeper drilling associated with many onshore wells and the larger volumes of drilling muds required have highlighted waste fluid disposal problems.

A HISTORICAL PERSPECTIVE

The progressive evolution of land use and environmental management associated with hydrocarbon drilling activity in Canada can be traced historically. Pre-Land Use Regulation operations (i.e. pre-1971) can be compared, in most instances unfavorably, to post-1971 operations. In addition, the most recent drilling operations, when compared to the early post-1971 operations, indicate both an increase in flexibility on the part of the regulatory agencies and an increased awareness of environmental and permafrost-related issues on the part of the companies involved, resulting in an overall decrease in environmental and permafrost concerns. Evidence of this desire to cooperate and work together was the recent informal agreement between Panarctic Oils Ltd. and Land Resources, DIAND, to drill a well without a sump in the High Arctic, on Ellef Ringnes Island, in winter 1981/82 as an experimental procedure to document the effectiveness of traditional waste-fluid containment procedures (i.e. sump use) vis-a-vis surface disposal upon the tundra (e.g. Panarctic, 1982a,b).

These broad trends can be illustrated with reference to a number of abandoned well sites in the northern Yukon. Each is representative of a different stage in the progression of land use and environmental management procedures. The northern Yukon is an area underlain by widespread discontinuous permafrost, and lies within the boreal forest-shrub tundra vegetation transition (Oswald and Senyk, 1977; Wiken et al., 1979). Since 1962, over 50 wells have been drilled in the northern Yukon, by far the majority on either Eagle Plain or Peel Plateau.

Pre-Land Use Operations

Typical of many early well-drilling operations in the northern Yukon was the Blackstone D-77, drilled in 1962 and 1963 to a depth of 4028 m. It was a two-season operation that commenced in March 1962 and finished in January 1963. In all probability, there was activity around the site during the summer of 1962, but the nature of this activity is difficult to ascertain from the well records. It would appear that the well was drilled from a pile-supported platform, remnants of which still exist at the site.

The well was located on gently sloping terrain occurring towards the bottom of a north-south trending tributary valley of the Blackstone River. Poor drainage conditions at this site together with silty ice-rich colluvial sediments produce an open forest woodland of black spruce (*Picea mariana*) and tamarack (*Larix* sp.) with a relatively dense shrub layer (dwarf birch, willow, ericaceous shrubs) and sedges (*Carex* spp.).

During site preparation, in the winter of 1961/62, this vegetation had been cleared and probably burned. Recent ecological and forest surveys of the northern Yukon now clarify the implications of such actions. For example, in the moister environments of the interior Yukon, a fire cycle occurs that is slightly different from that characteristic of well-drained sites. According to Zoltai and Pettapiece (1973), in moist environments, sedges and cottongrass tussocks survive the fire and thrive in the absence of competition and because of the increased moisture associated with the thickening of the active layer. This makes the re-establishment of black spruce and tamarack difficult. The result is the development of areas of fire-induced tundra composed of tussocks or tussock-dwarf birch. When re-examined in 1979, black spruce and tamarack had not succeeded in reclaiming the area. Instead, substantial thermokarst had developed within the area of the well site. Thermokarst mounds, 1-3 m in height, together with deep pools of standing water, 2-4 m deep in places, surrounded the old drilling platform. Thaw depths in the disturbed terrain commonly exceeded 90 cm while those in adjacent undisturbed terrain were less than 45 cm. Revegetation was in the form of sedges, mosses, dwarf birch, and cottongrass.

In the general vicinity, seismic lines reveal a similar history of physical disturbance and thermokarst. The removal of trees and surface organic material had led to the formation of water-filled trenches, 80-100 cm deep, colonized by aquatic plants (*Equisetum* spp., *Carex* spp., mosses).

In summary, the Blackstone D-77 well site illustrates how environmental issues were ignored before the introduction of the Territorial Land Use Act and Regulations. The permafrost-vegetation relationships were little understood and the physical (i.e. thermokarst) changes that resulted were long-lasting and permanent. In many ways, these early Yukon well-drilling operations were similar to some of the early operations on the Alaskan North Slope (e.g. Lawson et al., 1978).

Post-Land Use Operations

One of the first wells to be drilled on the upland surface of Eagle Plain following the introduction of the Territorial Land Use Regulations was the Chevron SOBC Wm N. Parkin D-61. Not only was it drilled completely in the winter of 1971/72, but the site chosen was relatively well drained and underlain by consolidated bedrock possessing low ground ice amounts. In contrast to the Blackstone D-77 well site, therefore, minimal terrain disturbance occurred at this site. This reflected (a) the one-season, winter drilling program and (b) the low terrain sensitivity of the site. The absence of a rigpad probably reflected the absence of easily accessible surficial aggregate and the reluctance of the land use regulatory bodies at that time to extract aggregate from local bedrock sources.

Recent Operations

Since the early 1970's, there has been a progressive refinement in both industry operating techniques and the application of the Territorial Land Use Regulations. This trend is best illustrated by the last well drilled on Eagle Plain. This was the Aquitaine Alder C-33 well, drilled during the late winter of 1977/78.

The well was located on a south facing, gently sloping (2-5°) interfluve, several kilometers south of the main Eagle Plain escarpment. Vegetation in the area consists of an open canopy of black spruce (*Picea mariana*) together with a shrub layer of birch (*Betula glandulosa*), alder (*Alnus* sp.), and some Labrador tea. A moss-lichen layer constitutes a surface organic mat more than 10 cm thick. At the well site the presence of white spruce (*Picea glauca*) and charred remnants of black spruce indicated a previous burn.

During June 1977, geotechnical and terrain investigations were undertaken by consultants at the proposed well site location. The depth of the frost table and active layer temperatures were measured at several sites. As was to be expected, the thaw depths bore a close relationship to the vegetation cover and type, and confirmed the important thermal role that vegetation plays in these northern boreal forest-shrub tundra environments. Shallow drilling revealed 10-24 cm of organic material overlying about 2.4 m of silty colluvium, which graded into highly weathered and fissile shale bedrock. The upper 2 m of colluvial sediments possessed excess ice amounts of between 10-25%, and natural water (ice) contents exceeded 20%. Ground ice amounts then progressively decreased with increasing depth. The deeper cores retrieved from underlying shale and sandstone bedrock indicated low ground ice content and high structural cohesiveness of the rock.

Following these observations, a land use permit was issued for the drilling of the well during the winter of 1977/78. A large gravel pad was constructed for the operation, to prevent thaw of the ice-rich surficial material. The aggregate was obtained by crushing sandstone from a borrow pit opened up along the crest of the main Eagle Plain escarpment to the north. Although a small sump was used during drilling, the majority of the waste fluids were transported to the borrow area, which served as a main sump. The operation was atypical, therefore, since the main rig sump was not at the site itself. During the summer of 1978, operations at the site terminated and site rehabilitation was accomplished the following winter.

As a result of these procedures, terrain disturbances adjacent to the rigpad were minimal. Moreover, no signs of thawing of the permafrost, as might be indicated by slumping of the pad, were visible in July 1979. It was concluded that the methods adopted to prevent thaw were sufficiently effective to offset any increase in thaw due to the removal of trees and the surface vegetation layer.

DISCUSSION

These three selected case histories illustrate the effectiveness and application of the Territorial Land Use Regulations in the boreal forest-shrub tundra environments of northern Yukon.

The changes effected by the introduction of the Territorial Land Use Regulations in minimizing disturbance are clearly apparent. The relative sophistication of the Aquitaine C-33 operation, where a large gravel pad was constructed and activity was limited to the winter months, can be compared to the earlier Blackstone D-77 well site where typical man-induced thermokarst occurred following land clearance and drilling activity in the summer in an ice-rich locality. Observations at the Blackstone site suggest that over the long term (i.e. 15-25 years) natural recolonization, similar to the changes following fire, is as successful as artificial seeding in site rehabilitation.

The containment of waste drilling liquids in below-ground sumps has been a standard operating condition for wells drilled under the Territorial Land Use Regulations in both the N.W.T. and the Yukon. Equally, it has been one of the conditions that has presented the most problems. The Aquitaine Alder C-33 illustrates how an innovative waste disposal program was undertaken and is representative of an increased awareness of the need for flexibility in the application of the Territorial Land Use Act and Regulations, especially as they pertain to the disposal of waste fluids. Alternate disposal methods are actively being investigated, including surface disposal, detoxification procedures, trucking to designated disposal sites, and dilution in surface water bodies. The surface disposal experiment on Ellef Ringnes Island in the High Arctic (Panarctic, 1982a, b) is the most recent approach. However, the variability of local terrain conditions in northern Canada and the inherent uncertainty involved in exploration drilling continues to prevent any one waste disposal procedure from being adopted in preference to others.

CONCLUSIONS

There appears to be no easy solution to the terrain and environmental problems of oil exploration in northern Canada. However, the continued application of the Territorial Land Use Act and Regulations, together with the positive attitudes increasingly adopted by the operators themselves, are leading to fewer problems that cannot either be resolved or minimized. In the boreal forest-shrub tundra environments of northern Yukon, terrain disturbances associated with well drilling in recent years have been minimized. Where there is a lack of ice-rich surficial deposits, the relative rapidity of natural revegetation is important in reducing terrain damage. In spite of this, terrain disturbance and sump-related problems do still occur in other localities in northern Canada, especially north of treeline in the ecologically more delicate tundra and polar, semi-desert environments. They emphasize the desirability of site-specific solutions to minimize impact and a thorough understanding of permafrost-vegetation relationships, in addition to the imposition of general land use operating conditions.

REFERENCES

Allmand, W., 1976, Guidelines for scientific activities in northern Canada, Advisory Committee for Northern Development, Department of Indian Affairs and Northern Development: Ottawa, Ministry of Supply and Services, Catalogue No. L 2-47/1976.

Barnett, D.M., Edlund, S.A., Dredge, L.A., 1977, Terrain characterization and evaluation: An example from Eastern Melville Island: Geological Survey of Canada, Paper 76-23.

Beak Consultants Ltd., 1975, Banks Island development, environmental considerations: 1974 research studies, Consultants report prepared for Panarctic Oils Ltd., Calgary, and Elf Oil Exploration and Production Ltd., Calgary.

Bliss, L.C., 1975, Plant and surface responses to environmental conditions in the western High Arctic, ALUR 74-75-73: Ottawa, Department of Indian Affairs and Northern Development.

Bliss, L.C. and Peterson, E.B., 1973, The ecological impact of northern petroleum development, in Arctic Oil and Gas: Problems and Possibilities; Proceedings of the 5th International Congress, Fondation Francaise d'Etudes Nordiques, Le Havre, p. 1-26.

Berger, T.W., 1977, Northern frontier, northern homeland: Report of the Mackenzie Valley Pipeline Inquiry: Ottawa, Ministry of Supply and Services.

Chretien, J., 1972, Northern Canada in the 70's, Report to the Standing Committee on Indian Affairs and Northern Development: Ottawa.

French, H.M., 1978a, Sump studies I: Terrain disturbances, Environmental Studies No. 6: Ottawa, Department of Indian Affairs and Northern Development.

French, H.M., 1978b, Terrain and environmental problems of Canadian Arctic oil and gas exploration: Musk-Ox, v. 21, p. 11-17.

French, H.M., 1980, Terrain, land use and waste drilling fluid disposal problems, Arctic Canada: Arctic, v. 33, p. 794-806.

French, H.M., 1981, Sump studies IV. Terrain disturbances adjacent to exploratory well sites, northern Yukon Territory, Environmental Studies, No. 19: Ottawa, Department of Indian Affairs and Northern Development.

French, H.M. and Smith, M.W., 1980, Geothermal disturbances in permafrost adjacent to Arctic oil and gas well sites, Environmental Studies No. 14: Ottawa, Department of Indian Affairs and Northern Development.

Indian Affairs and Northern Development, 1983, Canada's North: The Reference Manual, Catalog No. R71-31/1983E: Ottawa, Ministry of Supply and Services, p. 8.1-8.19.

Johnson, L.A., 1981, Revegetation and selected terrain disturbances along the trans-Alaska pipeline, 1975-1978: Hanover, N.H., U.S. Army Cold Regions Research and Engineering Laboratory, CRREL Report 81-12.

Kerfoot, D.E., 1972, Tundra disturbance studies in the Western Canadian Arctic, ALUR 71-72-11: Ottawa, Department of Indian Affairs and Northern Development.

Lawson, D.E. and Brown, J., 1979, Human-induced thermokarst at old drill sites in northern Alaska: Northern Engineer, v. 10, p. 16-23.

Lawson, D.E., Brown, J., Everett, K.R., Johnson, A.W., Komarkova, V., Murray, B.M., Murray, D.F., and Webber, P.J., 1978, Tundra disturbance and recovery following the 1949 exploratory drilling, Fish Creek, Northern Alaska: Hanover, N.H., U.S. Army Cold Regions Research and Engineering Laboratory, CRREL Report 78-28.

Lysyk, K.M., Bohmer, E.M., and Phelps, W.L., 1977, Alaska Highway Pipeline Enquiry: Ottawa, Ministry of Supply and Services.

Oswald, E.T. and Senyk, J.P., 1977, Ecoregions of the Yukon Territory: Canadian Forestry Service, Environment Canada Publication BC-X-164.

Panarctic Oils Limited, 1982a, Surface disposal experiment of waste drilling fluids, Hoodoo N-52 well, Ellef Ringnes Island, N.W.T., Initial Report, Phase One, Consultants Report: Calgary, Alberta.

Panarctic Oils Limited, 1982b, Surface disposal experiment of waste drilling fluids, Hoodoo N-52 well, Ellef Ringnes Island, N.W.T., Phase Two Report, Consultants Report: Calgary, Alberta.

Slaney, F.F. and Company Ltd., 1973, Environmental impact assessment study, Melville Island, N.W.T., Consultants Report: Calgary, Panarctic Oils Ltd.

Smith, M.W., 1977, Computer simulation of microclimatic and ground thermal regimes: Test results and program description, ALUR 75-76-72: Ottawa, Department of Indian Affairs and Northern Development.

Wein, R.W., Sylvester, T.W., and Weber, M.G., 1975, Vegetation recovery in Arctic tundra and forest tundra after fire, ALUR Report 74-75-62: Ottawa, Department of Indian Affairs and Northern Development.

Wiken, E.B., Welch, D.M., Ironside, G., and Taylor, D.G., 1979, Ecological land survey of the northern Yukon, Ecological Land Classification Series: Lands Directorate, Environment Canada, no. 7, p. 361-372.

Zoltai, S.C. and Pettapiece, W.W., 1973, Terrain, vegetation and permafrost relationships in the northern part of the Mackenzie Valley and Northern Yukon: Environmental-Social Program Northern Pipelines, Task Force on Northern Oil Development, Report 73-4.

PETROLEUM EXPLORATION AND PROTECTION OF THE ENVIRONMENT ON THE NATIONAL PETROLEUM RESERVE IN ALASKA

M.C. Brewer

U.S. Geological Survey
Anchorage, Alaska, USA

The National Petroleum Reserve in Alaska (NPRA) comprises almost 96,000 km^2 of the tundra-covered area of the North Slope. Its northern boundary is the Arctic Ocean, including almost equal portions of the coastline of both the Chukchi and the Beaufort Seas. A little more than one-half of its southern boundary follows the divide of the Brooks Range.

Except beneath a portion of Teshekpuk Lake, which is approximately 810 km^2 in area, the entire Reserve is underlain by permafrost ranging in thickness, to the best of our present knowledge, from about 200 to 400 m. The minimum temperature of the permafrost, below the depth (20 m) of measurable annual change (0.01°C), is -10.7°C. The thickness of the active layer, depending on the soil material, topography, location within the reserve, and vegetative cover, ranges from 0.3 to 1.5 m. The annual precipitation averages about 0.3 m and almost 80% of that occurs in the form of snow. Thus, the area has been termed a polar desert, in spite of the myriad of generally shallow lakes (1 to 3 m) that are found within the coastal plain area. Most of the Reserve, except near the mountain front, is almost devoid of gravel that could be used for construction, although approximately 13,000 km^2 of the area is covered by mostly stabilized sand dunes.

Exploration of the Reserve for oil and gas potential generally has been concentrated into three time frames. The first was a geological reconnaissance during the period 1923-1926 by personnel from the U.S. Geological Survey. This was followed by the Navy's exploration program, involving geologic mapping, geophysical prospecting, and the exploratory drilling of 36 generally shallow test wells and 44 core test wells from 1944 to 1953 (Reed, 1958). The most recent explorations occurred during the period 1974-1981, when first the Navy and then the U.S. Geological Survey accomplished 21,800 km of reflection seismic survey and drilled 28 exploratory and 10 development wells, the latter in the small, shallow natural gas fields near Barrow (Schindler, 1983a). Two of the exploratory wells were the deepest ever drilled in Alaska, approximately 6200 m deep.

The Navy's 1944-1953 exploration program began under wartime conditions and even used some military-type equipment, such as tracked landing vehicles for transporting men and equipment across the tundra. It was a pioneering effort in which one of the goals was to see what equipment would work in the rigorous arctic environment and what modifications might be required in exploratory techniques. The most important finding of the latter was the knowledge that overland travel could only be successfully accomplished during the winter. The environmental protection ethic was not emphasized in those days, nor was research far advanced concerning the impact of this type of operation on the arctic environment. In retrospect, there were two serious environmental results. The first involved overland transportation, especially during the summer periods when, in order to obtain traction and to keep from high-centering vehicles, the active layer was bladed down to permafrost. In some instances this resulted in erosion; in all cases it left ditches across the tundra that are expected to be visible longer than the old Roman roads in Britain. The second major environmental impact was that debris, especially empty fuel drums, was scattered across the tundra. The latter has been largely rectified, the debris gathered up and burned or buried, but at a very large financial cost (Schindler, 1983b).

One very positive result of that earlier exploration was that, in 1947, the Navy introduced a program of research in arctic environmental matters, including operating techniques in that environment. The research program continued at the Naval Arctic Research Laboratory at Barrow, so that by the time oil was discovered at Prudhoe Bay, in 1968, an impressive amount of knowledge had been accumulated (Reed and Ronhovde, 1971). This included information regarding casing collapse in wells; knowledge of the thicknesses of gravel required along the Arctic Coast to bring the frost line up into the gravel fill and thus prevent deterioration of the underlying permafrost in all-season roads and airstrips; building designs to fit the environment; parameters concerning fresh ice and sea ice; and a far greater appreciation of the physical and biological parameters to be found in the Arctic.

The oil development program at Prudhoe Bay generally has been cited as a model for accomplishing a task, where a decision has been made, in an environmentally acceptable manner. As a result, during the past few years whenever a program was projected for the Arctic, there has been a tendency for both engineers and many environmentalists to want to use methods and parameters shown to work at Prudhoe and to leave a neat and tidy appearance. The former has environmental advantages, but it also has many environmental disadvantages for an oil and gas exploration program.

Prudhoe Bay is a development complex with an established transporation system. The roads, pads, airstrips, and other facilities are permanent, at least for 40 to 50 years. However, in an oil and gas exploration program, one does not want permanence, at least until after a discovery is made. Impermanence not only provides greater environmental protection, but it is less expensive. In an exploration program, one wants to "tip-toe" into an area, run the seismic surveys, drill any exploratory wells deemed desirable, and then carefully move out, leaving no tracks or other evidence of the activity.

The Navy/U.S. Geological Survey exploration program during 1974-1981 tried to follow the approach of leaving as little evidence of their exploratory activities as possible. All seismic surveys were run during the winter-spring period when the snow cover was greatest, when overland transportation would be most efficient, and when the least numbers and species of wildlife were present. Exploratory drilling was confined to the winter season, except where wells deeper than about 3500 m were involved. Beyond that depth, the operation was faced with the trade-offs between the environmental and cost impacts of constructing all-season roads and airstrips and the environmental and cost impacts of operating over several winters as well as the risk of not completing a well when work had been suspended several times.

Lake ice was used for aircraft landing surfaces wherever possible. In two instances, airstrips 1600 m long to accommodate Hercules C-130 aircraft were constructed directly over a flat tundra expanse. In one of those instances the airstrip was constructed in the same location the following year. Even after the second season, it was extremely difficult to find any evidence of the 30-cm-thick airstrips constructed on the tundra. In the case of three deep wells, high-strength insulation was used in airstrips in order to reduce the thickness of gravel required to bring the frost line into the base of the fill.

Ice roads were constructed wherever heavy traffic, such as 20-m^3 dump trucks, might be making repeated trips. These were built to a minimum thickness of 15 cm before heavy traffic was allowed. The longest ice road constructed was 60 km. Water was applied daily to keep a smooth running surface, but by the time 85,000 m^3 of gravel had been hauled over the road, at speeds of up to 80 km per hour, the ice road had an average thickness of about 45 cm. Almost no damage, not even a "green" trail or trail discernible from a low-flying helicopter, was evident the following summer. Snow roads, per se, were not constructed because of the generally insufficient amount of snow available and because whenever snow is used it must be scraped up. This requires blading with tractor-type equipment and, no matter how careful the operator is, considerable tundra damage usually ensues.

While 1.5- to 1.7-m-thick gravel drilling pads will bring the frost line into the base of the pads along the coast of northern Alaska, a pad of that thickness will be quite visible in the long term, will be well-drained, and will be difficult to revegetate because of a lack of adequate soil moisture. A thick pad, however, is not necessary to support a winter-only drilling operation. A thin pad, 0.5 m thick, frozen during the winter months, satisfactorily supports a winter-only drilling operation; it requires no added material to be hauled to the site since sufficient material is available after digging a reserve (mud) pit; it retains sufficient moisture to enable grasses to become established, thus is easier to revegetate; and it rapidly recedes into the underlying landscape. These thin pads were used on the NPRA during the latter portion of the exploration program and proved to be more satisfactory than the thick pads from both the engineering and environmental standpoints. They eliminated the need to open material sites, and thus the need for hauling heavy loads of gravel and of rehabilitating material sites. In addition, they were economical to prepare.

The recent exploration program on the NPRA received few complaints from environmental groups, and since its completion it has occasionally been cited by such groups as the model way to accomplish an exploratory program in the Arctic. The program also was completed approximately 12% under its budgeted funding.

REFERENCES

Reed, J.C., 1958, Exploration of the Naval Petroleum Reserve No. 4 and adjacent areas, 1944-1953, Part 1, History of the exploration: U.S. Geological Survey Professional Paper No. 301.

Reed, J.C., and Ronhovde, A.G., 1971, Arctic Laboratory: Washington, D.C., Arctic Institute of North America.

Schindler, J.F., 1983a, The second exploration, 1975-1982, National Petroleum Reserve in Alaska (formerly Naval Petroleum Reserve No. 4): Anchorage, Alaska, Husky Oil NPR Operations, Inc.

Schindler, J.F., 1983b, Solid waste disposal - National Petroleum Reserve in Alaska, _in_ Proceedings of the Fourth International Conference on Permafrost, Fairbanks: Washington, D.C., National Academy of Sciences, p. 1111-1116.

TERRAIN SENSITIVITY AND RECOVERY IN ARCTIC REGIONS

P.J. Webber

Institute of Arctic and Alpine Research, University of Colorado
Boulder, Colorado, USA

The protection of vegetation, and its sensitivity to and recovery from various impacts, such as surface damage, oil spills, and dredge-and-fill activities, have been under investigation on the tundra of the Alaskan Arctic Slope for the past 15 years. Much of what I will discuss applies to similar permafrost terrain, both in Alaska and elsewhere.

There seem to be two generic problems or difficulties concerning the protection of tundra and the necessary development of its resources:

- The first problem is the complexity of land ownership, permitting, and regulatory matters.
- The second problem is the lack of knowledge of ecosystem dynamics and variability.

The first problem has been discussed by several of the previous speakers and I will confine my remarks to the second problem.

Lack of knowledge and understanding of the ecosystem is due in part to the vastness of the region under consideration, its large variability, and the complexity of its systems. Permafrost regions vary from polar and alpine deserts through tundra to forests. Their holistic and complex nature is well illustrated by the many effects triggered by seemingly simple events, such as compression of the tundra vegetation mat. This variability and complexity make generalizations difficult, and predictive capabilities are poor. At present, the problems are best analyzed and treated by examining each case of potential or actual impact in its specific site context. It is to be hoped that with increased knowledge, prediction based on established principles will be possible.

Our lack of understanding also rises from semantic confusions in the use and definitions of terms. There are at least three causes for the semantic problems associated with the protection of permafrost terrain. The first is careless use of terms and terminology. The second is the interdisciplinary nature of environmental protection, and the third is the language difficulties associated with this international concern.

The semantic confusion inhibits the acquisition of new knowledge. It serves to reduce communication among the various interest groups, such as developers and environmentalists, since they may use terms in different ways. A few examples of semantic problems will suffice to illustrate the prospects for confusion:

- The use of the term "fragile" for tundra vegetation can be misleading. While some tundra is fragile and easily disturbed, the majority of it, in my opinion, is reasonably robust. Of course, "robustness" is another term that would need defining!
- The concept of tundra fragility relates more to its slowness to recover rather than its ease of disruption or breakage. If we are to use the term "fragile," then it must be done with more precision.
- The use of the terms "resistance" and "resilience" of ecosystems to stress is another example of the confusion from semantic problems. These are two very different notions. "Resistance" variously means the robustness of the vegetation or system to disruption or the ease with which it returns to its former state after disruption, and "resilience" is a better term for recovery from disturbance.
- One final term that causes confusion is the term and idea of "cumulative impact." This idea is an easy one to envisage--just as is fragility--but a working concept of "cumulative impact" is not readily available: there is no theory of "cumulative impact." Since the notion is incorporated in the U.S. legislation to protect wetlands from dredge-and-fill activities, it requires clear, uniform understanding. This is an area in need of future research.

Until we exercise care with our use of terms and reduce semantic conflicts, our progress toward acceptable practices to regulate activities, minimize impacts, and enhance recovery is likely to be impeded or confused. I should like to make a plea for someone to make a thorough review of appropriate terms and terminology with recommendations for appropriate use in the protection of permafrost terrain. Until this is done, however, we must define our terms with each usage or refer to similar usage in the relevant literature.

A further concern in this area of definitions is that of landform, soil, and vegetation classification. There is a remarkable lack of rigor in doing this, which can only lead to misinformation or miscommunication.

Adequate protection and management of permafrost terrain requires good terrain description, adequate mapping, and inventory of the lands being affected. Adequate large-scale mapping (1:6000) will help in the planning, locating, permitting, and construction design of various structures. The North Slope Borough, the principal local government of the Alaska Arctic Slope, is encouraging the mapping and inventory of its lands. The North Slope Borough is currently mapping, at a variety of scales, the geology, soils, terrain, and vegetation of its lands. The data from this large project are being incorporated into an automated, computer-retrievable, geographic-based information system. This system will be of great value in the permitting process, for the researcher, the developer, and the landowner. Similar efforts have been undertaken by several organics and petroleum companies in the Prudhoe Bay region and the Arctic National Wildlife Refuge (Walker et al., 1980, 1982).

Our lack of knowledge can be remedied by more basic applied research and more discussion among specialists. This will, I believe, better enable us to protect cold-dominated ecosystems, but it requires time, funds, and people. Perhaps, of these three requirements, time is the most important. Time is also at a premium in the process of permitting specific activities. Thus, we see a conflict between the researcher who needs more time to document a potential impact and the developer and consumer who does not want to wait for these conclusions or to find ways to mitigate them. Clearly, the solution is a compromise, and both development and research need to be accomplished simultaneously. However, development that would potentially leave impacts should not proceed without adequate concurrent research or equivalent mitigative actions.

Research priorities to increase our understanding appear to fall into two related categories (see for example National Academy Press, 1983)

- The first priority is to establish sites for long-term studies of man-made and natural changes, to include changes in atmospheric carbon dioxide, permafrost temperature, and biological communities.
- The second research priority is to better document and monitor case histories of impacts and recovery from damage. This includes the development of better methods of measurement and prediction of impacts and recovery from damage.

Finally, the continuing development of permafrost lands can be done in a fashion compatible with preserving their biological wealth and heritage. Construction and maintenance costs to provide environmental protection may at times be considered excessive. Development and protection can be achieved through a better knowledge of the natural science of these systems and appropriate planning as evaluated in the introduction to the panel report. Such knowledge, once gained, should be exchanged freely.

REFERENCES

National Academy Press, 1983, Permafrost Research: An Assessment of Future Needs: Washington, D.C.

Walker, D.A., Everett, K.R., Webber, P.J., and Brown, J., 1980, Geobotanical atlas of the Prudhoe Bay region, Alaska: Hanover, N.H., U.S. Army Cold Regions Research and Engineering Laboratory, CRREL Report 80-14.

Walker, D.A., Acevedo, W., Everett, K.R., Gaydos, L., and Brown, J., 1982, Landsat-assisted environmental mapping in the Arctic National Wildlife Refuge, Alaska: Hanover, N.H., U.S. Army Cold Regions Research and Engineering Laboratory, CRREL Report 82-37.

Climate Change and Geothermal Regime

PANEL MEMBERS

Alan Judge (Panel Organizer) and Jean Pilon (Chairman), Department of Energy, Mines, and Resources, Ottawa, Canada

Cheng Guodong, Institute of Glaciology and Cryopedology, Academia Sinica, Lanzhou, PRC

Thomas E. Osterkamp, Geophysical Institute, Fairbanks, Alaska, USA

Michael Smith, Department of Geography, Carleton University, Ottawa, Canada

James Gray, Department of Geography, University of Montreal, Canada

INTRODUCTION

A. Judge and J. Pilon

The energy crisis in 1973 emphasized the need to find and develop arctic hydrocarbon reserves. A similar need for other mineral resources will develop in the next few decades. Exploitation of these resources and the associated infrastructures of buildings and transportation facilities will occur on and in permafrost, which underlies 24% of the land area of the globe. Much of the near-surface permafrost contains massive ice. Numerous reports and papers have documented the immediate impact of development through clearing vegetation, interrupting drainage, and operating hot oil and chilled gas pipelines. A 2-m-diameter pipeline buried in permafrost, for example, will thaw to 9 m in 5 years. Mean air temperature changes of several degrees can be expected to occur over the lifetime of such projects, with possible long-term regional effects on the terrain traversed by a pipeline or highway as important as the short-term localized impact of the development itself. Knowledge or prediction of such climate trends becomes a critical aspect of design for long-term performance.

The regional distribution of permafrost is a function of climate, although wide variations in ground thermal conditions do occur in small areas of uniform climate due to site-specific conditions. Where mean annual ground temperatures are close to 0°C, specific local factors can determine whether or not permafrost is present. Changes in the ground thermal regime, and hence in the distribution and thickness of permafrost, can result from changes in both climate and local conditions. The specific effects on permafrost of a macroscale climate change are not necessarily simple, depending on the complex interaction of climate, microclimate, surface, and ground thermal conditions. If, for example, a global warming, concentrated in the high latitudes, results from increasing CO_2 levels, the effect will certainly be profound. Several recent papers report on the possible consequences of increased temperature and changes in precipitation on permafrost conditions in Alaska (Goodwin et al., 1984; Osterkamp, 1984).

Assuming that a rise in mean annual air temperature is translated directly into a similar increase in mean annual ground temperature, permafrost at the southern limit may retreat northwards by as much as 100 km per degree warming. This has been observed in Manitoba where over the past 200 years the areal extent of permafrost has diminished from 60 to 15%. With the discontinuous permafrost a thicker active layer will develop, widespread thaw settlement in areas of massive ice will occur, south-facing slopes will become permafrost-free, and surface and groundwater patterns will change. Extensive thermokarst features will form from the thaw of massive ice leading to the formation of lakes, new drainage patterns, alas areas, etc. Slopes in general will become less stable, leading to extensive failures and erosion. Rivers less confined by frozen banks will change course and carry increased sediment loads. Coastal areas could well be inundated by the sea as the thaw settlement lowers elevations to below sea level, especially if a general rise in sea level occurs. The impact of differential thaw settlement of as much as a meter on construction such as roads, pipelines, and buildings is not difficult to visualize. In the continuous permafrost zone, the southern boundary of which will also move northward, ground temperature will also rise, leading to some thawing and related settlement problems. Most of the immediate impact of warming and thawing will occur close to the ground surface where human activities are most dramatically affected.

A decrease in air and ground temperature will lead to a reversal of the above features, including a southerly advance of the discontinuous and continuous boundaries of permafrost and corresponding vegetation zones. More importantly it will lead to the growth of massive ice bodies, a reactivation of the growth of ice wedges in more southerly areas, and a general thickening of permafrost and consequent changes in surface morphology and drainage patterns (see Osterkamp, 1984, and Brown and Andrews, 1982, for additional discussion).

Climatic fluctuations seem to be reflected in permafrost behavior, for example the northward retreat of the southern permafrost boundary, the initiation of extensive thermokarsting in the southern Mackenzie Valley in the past hundred years, and within the last decade, the reactivation of ice wedges on the arctic coast.

In general, direct observational measurement of climatic change in the Arctic is limited, so proxy methods such as palynology, oxygen isotope distribution, and geothermal analysis become important ways of extending the record of the past. Geothermal analysis does offer a direct, if resolution-limited, method of examining past changes in mean ground temperature.

Brown and Andrews (1982) have summarized the trends from many arctic series of proxy information as follows:

- A long-term cooling trend of the order of 0.7°C per thousand years.
- Superimposed on the long-term trend, a series of major oscillations with wavelengths of 500 to 600 yr.
- Evidence from the eastern Canadian Arctic that summer temperatures fell below the mean of the past 6000 years between 2500 and 1500 years ago and did not rise above the mean until the last 100 years.
- Temperatures in the 20th century that are, on average, as warm as they have been for several hundred years, if not thousands of years.

The last point may be an exaggeration, although certainly no warmer periods have occurred since the 11th century. Whether or not the general trends will continue or whether the impact of CO_2 in the atmosphere will reverse such trends remains a point of considerable debate.

Aside from the problems of interpreting the strongly attenuated events in the distant past, geothermal analysis of more recent climatic events has had notable success. In temperature logs throughout eastern North America, the Mackenzie Valley, and the arctic coastal plain, the evidence for a general warming trend starting a century ago followed by a recent downturn agrees well with meteorological data. In general, the evidence indicates that ground temperatures increased more than air temperatures. A detailed study in the uniform shield rocks of northern Ontario revealed perturbations caused by surface ground temperatures increasing by 1.5°C in the Little Climate Optimum from 1000 to 1200 A.D. and decreasing by 1°C during the Little Ice Age from 1500 to 1700 A.D.

One of the serious problems with using geothermal analysis as a proxy technique arises from the non-uniqueness of the solutions. Given a series of dates at which climate change occurred from other sources, the techniques will successfully resolve amplitudes of the change for up to about four events. Improved continuous or semicontinuous logging methods may eventually improve upon this. In summary, the present strengths and uses of this type of analysis of borehole temperatures are probably as follows:

- Given good meteorological information on climate change over the past 100 years, to determine how that change has been translated into ground temperature. The analysis will enable some prediction of how future trends will affect permafrost.
- In areas where meteorological data are non-existent or of short duration, to determine the long-term fluctuations of ground temperature over the past several centuries. Some answers will be derived as to whether areas such as the High Arctic follow global trends.
- Given proxy information on the dates and duration of climatic events, some ideas of amplitude can be derived from geothermal analysis.
- Promising areas of application to more remote climate events, including the derivation of ice-base conditions during Wisconsin glaciation, require further refinement.

REFERENCES

Brown, J. and Andrews, J.T., 1982, Influence of short-term climate fluctuations on permafrost terrain, in Carbon Dioxide Effects Research and Assessment Program, Environmental and Societal Consequences of Possible CO_2-Induced Climatic Changes, DOE/EV/10019-02, v. 2, pt. 3: Washington, D.C., U.S. Department of Energy.

Goodwin, C.W., Brown, J. and Outcalt, S.I., 1984, Potential responses of permafrost to climatic warming, in Proceedings of a Conference on the Potential Effects of Carbon Dioxide-Induced Climatic Changes in Alaska (J.H. McBeath, ed.): Fairbanks, University of Alaska, Misc. Pub. 83-1, p. 92-105.

Osterkamp, T.E., 1984, Potential impact of a warmer climate on permafrost in Alaska, in Proceedings of a Conference on the Potential Effects of Carbon Dioxide-Induced Climatic Changes in Alaska (J.H. McBeath, ed.): Fairbanks, University of Alaska, Misc. Pub. 83-1, p. 106-113.

STUDY OF CLIMATE CHANGE IN THE PERMAFROST REGIONS OF CHINA - A REVIEW

Cheng Guodong

Institute of Glaciology and Cryopedology, Academia Sinica
Lanzhou, People's Rebulic of China

LATE PLEISTOCENE STAGE

In China, there are no reports of double-layered permafrost or the typical ice wedge casts. The depth of the measured ground temperature in permafrost regions is less than 200 m. Therefore, most of the environmental reconstructions in geocryology depend on relict periglacial features, especially involutions and soil wedges.

Qinghai-Xizang Plateau

Cui (1980a) has suggested that the lower limits of permafrost on the Qinghai-Xizang Plateau in the Late Pleistocene are indicated by the widespread involutions on the second terrace (3900 m) of the Pingchu River at Xigaze in the south and the involutions on the second terrace (3600 m) at Naij Tal in the north (Fig. 1). The mean annual air temperature at Naij Tal is about -0.6°C at present, and the mean annual air temperature at the lower limit of the existing permafrost on the plateau is from -2 to -3°C. Accepting that the mountains have lifted 600-900 m since the end of the Late Pleistocene, it has been deduced that there was a probable temperature depression of 5-6°C on the plateau during the Late Pleistocene. But Pu et al. (1982) argued that the lower limit of permafrost during the Late Pleistocene reached the latitude of 30°N and an altitude of 4200 m in the south, along the Qinghai-Xizang Highway, but was lower than 3600 m in the north.

Polygons and soil wedges in corresponding profiles have been found in large quantities on the Qinghai-Xizang Plateau. Most of the soil wedges are located on the second terrace of rivers. Carbon-14 dating of the material at the bottom of a soil wedge located on the second terrace of Zuomaokong Qu, on the northern slope of Fenghuoshan, gives 23,500 ± 1200 years B.P. The mean

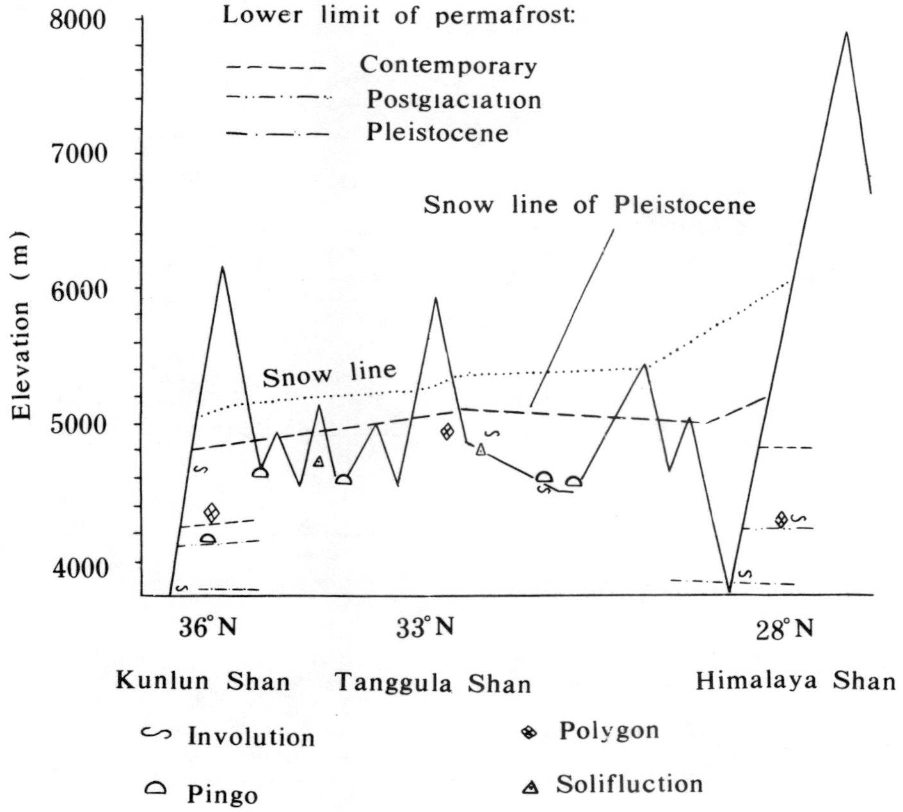

FIGURE 1 Lower limits of permafrost at various ages along Qinghai-Xizang Highway (after Cui, 1980).

annual air temperature there today is about -5°C. Assuming that an air temperature of -8°C is necessary for the formation of soil wedges in coarse-grained soils, and again accepting that the mountains have uplifted 600-900 m since the end of the Late Pleistocene, it is inferred that the temperature was 6-8°C lower when the soil wedge was formed than at present. This temperature approximates the temperature inferred from the lower limit of permafrost, based on the involutions, and also coincides with the conclusion inferred from the paleontologic evidence (carbon-14 dating) at 23,100 ± 850 years B.P. on the north slope of Qin Ling (Shi et al., 1979), that the temperature was 8°C lower during the last glaciation. Zhang (1979) argued that the wedge is an ice wedge, and that the temperature for the formation of the ice wedge cast ought to be -12°C.

Northeast China

Based on the southern limits of both the soil wedges and the involutions, Guo and Li (1981) have outlined the southern limit of permafrost during the last glaciation in northeast China (Fig. 1). This limit roughly coincides with the present day 7-8°C isotherm of mean annual air temperature. It is deduced that the temperature depression during the last glaciation in northeast China was also 7-8°C. Cui and Xie (1982) suggested that at that time the southern limit of permafrost in East China reached to 39°W-40°N, and the southern limit of continuous permafrost was located at 45°N, but Pu et al. (1982) stated that the southern permafrost boundary reached to 34°N. It is questionable to use involutions alone to indicate permafrost, until their origin is better known. Cui (1980b) has proposed that, to be an indicator of permafrost, an involution must show the following characteristics: (a) the disturbed layer consists mainly of sand or sand with gravel; (b) there is a clear boundary between a disturbed layer and an underlying horizontal layer; and (c) the "fold" is symmetrical and extends to a certain distance. There are also many problems of distinguishing ice wedge casts from soil wedges; the temperature required for the growth of soil wedges must be determined by further investigations.

TABLE 1 Comparison of climatic changes during postglaciation between Europe and East China.

	Europe		East China*		
Date	Stage	Period	Period	Temperature fluctuation (°C)	Years ago
--2000			Little Ice Age	-1	0-500
- A.D.		Sub-Atlantic	Pu Lan Dian little climate optimum	+1 - +2	-
--0				1400	2000-
			Zhou Han	-1	
- B.C.			Neoglaciation	3000	-
-- 2000		Subboreal			4000-
-	Post Glacial		Yang Shao	+2 - +3	-
-- 4000		Atlantic			6000-
-					
-- 6000		Boreal	Hypsithermal	8000	8000-
-					-
			Xie Hu	-5 - -6	
-- 8000		Preboreal		10,700	10,000-
-		Upper Dryas	Pre-Cold		-
		Allerod			12,000-
--10,000	Late Glacial	Lower Dryas			
-					-
		Bolling			14,000-
--12,000					
-		Oldest Dryas			-
--14,000	Full Glacial	Arctic			16,000-

*After Duan et al. (1981).

THE LAST 10,000 YEARS

Chu (1973), the outstanding Chinese meteorologist, has studied the climate changes in China during the last 5000 years based on meteorology and phenology records in Chinese historical literature and materials found in archaeological studies. On the basis of Chu's works, the climate change during the postglaciation in East China has been roughly outlined (Duan et al., 1981) as follows: the Xie Hu cold period (10,700-8000 B.P.), 5-6°C lower than today; the Yang Shao warm period (8000-3000 B.P.), 2-3°C higher than today; the Zhou Han cold period (3000-1400 B.P.), 1-2°C lower than today; the Pu Lan Dian warm period (1400-500 B.P.), 1-2°C higher than today; and the Little Ice Age (500 B.P.-present), 1-2°C lower than today. The climatic tendency over the last 10,000 years in East China roughly coincides with that of the rest of the world, but there are some differences in timing. Table 1 provides a comparative base for reconstructing the palaeoclimate in permafrost regions of China.

Qinghai-Xizang Plateau

From carbon-14 dating, we know that the peat buried in the middle section of Qinong at Yangbajain is 9180 ± 100 years old. The involutions in the overlying sand layer may have been formed in the Xie Hu cold period (Pu et al., 1982). In this period of dry climate and strong winds, dunes were widespread over the terraces of the Tongtian, Tuotuo, and Fenghuo Rivers, implying less than 200 mm of precipitation annually. A wind-eroded landscape developed in the East Kunlunshan, the east slope (3000 m) of Riyue Shan, and the upper part (4600 m) of the Gongba relict mountain in the Dingri basin. The stratified deposits on the west slope (4000 m) of Xiaonanchuan in Kunlunshan may have formed in this period (Cui, 1980a). The periglacial loess of this period is also well developed.

The carbon-14 age of most of the peat on the Qinghai-Xizang Plateau is between 8000 and 3000 B.P., implying a rise in temperature during the Yang Shao warm period, i.e. the Holocene Hypsithermal. The 4- to 6-m-thick solifluction deposits in Fenghuoshan and the solifluction terrace, up to 10 m, in Tanggulashan are also evidence of rising temperatures in this period. The periglacial loess, formed in the Xie Hu cold period, usually became part of the solifluction. The stabilizing of the dunes by vegetation reflected a warm, wet environment, with more than 400 mm of precipitation. Part of the dunes stabilized by vegetation was covered by active solifluction. According to Ding and Guo (1982), the average thawed depth in the Holocene Hypsithermal was 15 m, which means the permafrost had not disappeared entirely at that time, although it had largely degraded.

During the Zhou Han cold period, i.e. Neoglaciation, the temperature on the Qinghai-Xizang Plateau dropped again. A group of fossil pingos has been found at Xidatan (4300 m) in the north part of the plateau. The carbon-14 age of its equivalent layer is 7530 ± 300 B.P. It seems likely that these pingos formed in the Neoglaciation. If the altitude at which the pingos formed was 100 m higher than the permafrost lower limit (Cui, 1980a), and accepting that the mountains have uplifted 300-500 m since the end of the Hypsithermal, it is inferred that the temperature in the Neoglaciation was about 2°C lower than now. The involutions of this period are widely distributed on the plateau. In the north, they are encountered on the first terrace at Naij Tal (3600 m) with a carbon-14 age of 4190 ± 100 B.P.; in the south, they are found at Langkaze (4400 m) near Yamzho Yumco, Yali (4400 m), the Dingri basin (4300 m), and Yangbajain (4200 m), with a carbon-14 age of 3270 ± 70 and 6130 ± 90 B.P. All these indicate a cold environment at that time.

Northeast China

From pollen analysis, it is known that during the Hypsithermal broad-leaf forest, consisting mainly of oak, elm, alder, and birch grew from the Sanjiang plain to South Liaoning. The air temperature at that time on the Sanjiang plain was about 6-8°C, which is 2-4°C higher than today's. Taking 1°C per degree of latitude as the lapse rate of temperature, it is calculated that the temperature to the north of Mangui was about 0°C. Analysis of light and heavy minerals in samples taken from a drill hole at Amur shows a large content of scree material, which implies that the area to the north of Mangui had not been affected by the warm, wet climate. Based on this evidence, Guo and Li (1981) have concluded that there were relict permafrost islands to the north of Mangui during the Hypsithermal (Fig. 2). This has been confirmed by the calculations of Fu et al. (1983).

During the Neoglaciation, the temperature dropped again. Involutions of the black-grey fine sand and silt layer with humus (carbon-14 age of 3010 ± 80 B.P.) have been found in both the isolated taliks north of both the existing southern permafrost boundary and the seasonally frozen ground boundary south of it. In other words, during the Neoglaciation the permafrost exceeded today's southern permafrost boundary.

THE LAST 1000 YEARS

Zhang et al. (1981) have analyzed the rings of a 900-year-old cypress tree (<u>Sabina przewalskii kom.</u>) that is located at the upper tree limit (3760 m) at Tianjim (37°24'N, 99°56'E) in Qilianshan. Here variations of ring widths are mainly due to fluctuations in air temperature, so they reflect the temperature changes on the Qinghai-Xizang Plateau. The results of the analysis are shown in Figure 3.

Chu (1973) has shown that there were two obvious cold periods during the last 1000 years in East China. The first was in 1000-1200 A.D., and the second was in 1400-1900 A.D. with temperature ranges of 1-2°C in the mean annual temperature. The latter corresponded to the Little Ice Age in Europe (1541-1890 A.D.). It was also indicated by the results of the tree-ring analysis of the cypress tree at Tianjim in Qilianshan.

During the cold period of the last 500 years, the climate in China still fluctuated between cold and warm periods (Table 2). Based on the pheno-

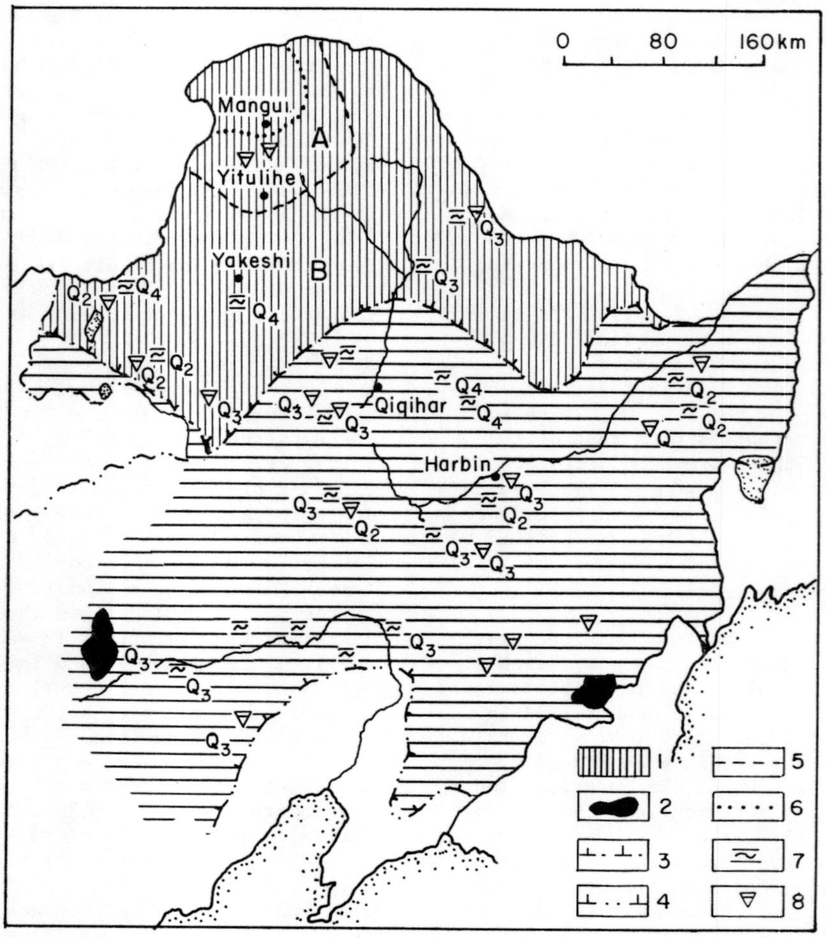

FIGURE 2 Boundaries of permafrost during the late Quaternary (modified from Guo and Li, 1981). 1) Zone of present-day permafrost; 2) Alpine permafrost; 3) Southern boundary of present-day permafrost; 4) Southern boundary of permafrost during late-Pleistocene; 5) Present-day boundary between "widespread continuous permafrost" and "island or valley bottom permafrost"; 6) Inferred southern boundary of permafrost during the Hypsithermal Interval (areas shown by horizontal cross hatch); 7) Involution layers; 8) Sand wedges. The Q symbols refer to different aged deposits.

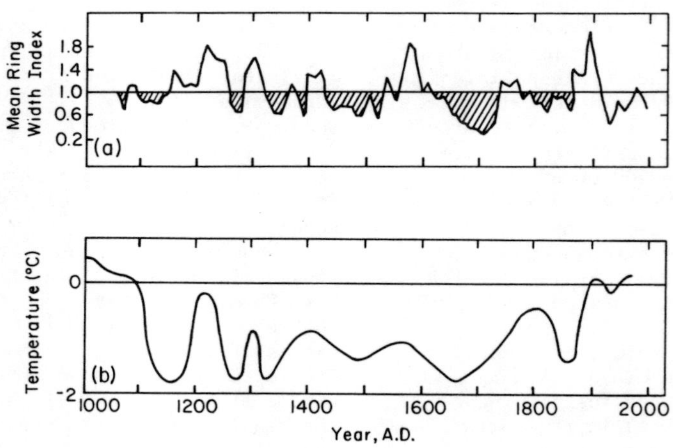

FIGURE 3 Climatic fluctuations during the last 1000 years. a) Running 10-yr means of tree ring width indices of Qilianshan cypress tree (Zhang et al., 1981). b) Temperature fluctuation during the last 1000 years in China (Chu, 1973).

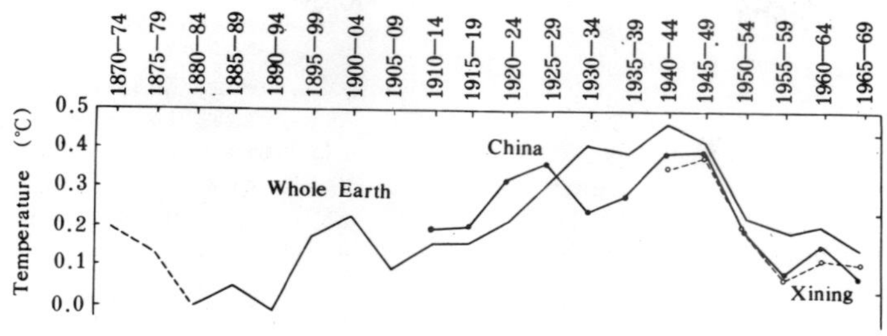

FIGURE 4 Comparison of average temperature changes since 1870 in the world and in China. Successive 5-yr means expressed as departures from the means for 1880-84, in the whole Earth (after Zhang Jiacheng).

TABLE 2 Comparison of cold and warm periods between East and West China.

Periods	East China (from phenology)	West China (from dendroclimatology)
	1470-1520 A.D.	1428-1537 A.D.
	1620-1720 (especially 1650-1700)	1622-1740
Cold	1840-1890	1797-1870
	After 1945 (especially 1967-present)	After 1924
	1550-1600 A.D.	1538-1621 A.D.
Warm	1770-1830	1741-1796
	1916-1945	1871-1923

logical records, the cold period of the last 500 years in East China can be divided into four smaller cycles, including four periods of cooling and three periods of warming with periodic lengths of 50-100 years and a temperature range of 0.5-1.0°C. All these oscillations were recorded in the ring width variations of the cypress tree at Tianjim, but the time phase of East China is slightly different from that of West China.

By correlating the tree-ring width indices and the air temperatures recorded at meteorological stations on the Plateau, it is deduced that during the last 500 years the temperature range was about 1.0-1.5°C, and the temperature in the coldest period was 0.8°C lower than today.

From the instrument records, the trends of climatic changes in Xining in West China coincide with those of both East China and the rest of the globe (Fig. 4). The warming in the Northern Hemisphere, which began in the 1880s, reached its maximum in the 1940s. But the ring-width index series shows that the warming trend in West China began in the 1870s, and reached its maximum in 1924. The time of transition from warming to cooling advanced by 10 to 15 years. The reason for this is not yet clear.

REFERENCES

Chu Kochen, 1973, A preliminary study on the climatic fluctuations during the last 5000 years in China: Scientia Sinica, v. 16, p. 226-56.

Cui Zhijiu, 1980a, Periglacial phenomena and environmental reconstructions on the Qinghai-Xizang Plateau, in Scientific Papers on Geology for International Exchange, No. 5: Beijing, Publishing House of Geology, p. 109-117.

Cui Zhijiu, 1980b, On the periglacial mark in permafrost area and the relation between glaciation and periglaciation: Journal of Glaciology and Cryopedology, v. 2, p. 1-6.

Cui Zhijiu and Xie Youyu, 1982, On Late Pleistocene periglacial environments in the northern part of China, in Quaternary Geology and Environment of China: China Ocean Press, p. 167-171.

Ding Dewen and Guo Dongxin, 1982, Primary research about the development history of permafrost in the Qinghai-Xizang Plateau, in Proceedings of the Symposium on Glaciology and Cryopedology held by Geographical Society of China (Cryopedology), Lanzhou (1978): Beijing, Science Publishing House, p. 78-82.

Duan Wanti, Pu Qingyu, and Wu Xihao, 1981, A preliminary study on climatic change during Quaternary in China, in Proceedings of the National Conference on Climatic Change (1978): Beijing, Science Publishing House, p. 7-17.

Fu Liandi, Ding Dewen, and Guo Dongxin, 1983, Numerical calculation about the evolution of permafrost in Northeast China, in Proceedings of the Second Second National Conference on Permafrost, Lanzhou (1981): Lanzhou, Gansu People's Publishing House, p. 165-169.

Guo Dongxin and Li Zuofu, 1981, Preliminary approach to the history and age of permafrost in Northeast China: Journal of Glaciology and Cryopedology, v. 3, p. 1-16.

Pu Qinguy, Wu Xihao, and Qian Fang, 1982, Historical changes of the perma-frost along the Qinghai-Xizang Highway, in Proceedings of the Symposium on Glaciology and Crypedology held by the Geographical Society of China (Cryopedology), Lanzhou (1978): Beijing, Science Publishing House, p. 74-77.

Shi Yafeng, Li Jijun, and Xie Zhichu, 1979, The uplift of the Qinghai-Xiang Plateau and its effect on China during ice age: Journal of Glaciology and Cryopedology, v. 1, p. 6-11.

Zhang Linyuan, 1979, The uplift of the Qinghai-Xizang Plateau and its effect on environmental evolution during Quaternary in China, in Contributions to Third National Conference on the Quaternary: Lanzhou, Lanzhou University Press, p. 24-33.

Zhang Xiangong, Zhao Qin, and Xu Ruizhen, 1981, Tree rings in Qilian Shan and the trend of climatic change in China, in Proceedings of the National Conference on Climatic Change (1978): Beijing, Science Publishing House, p. 26-35.

RESPONSE OF ALASKAN PERMAFROST TO CLIMATE

T.E. Osterkamp

Geophysical Institute
Fairbanks, Alaska, USA

Permafrost forms by freezing from the ground surface downward when mean annual surface temperatures (MAST) are less than 0°C. The MAST is controlled by climate, so in this sense, permafrost is a product of climate. Climate may not be the only controlling factor. For example, the geological setting, topography, geothermal heat flow, and vegetation may have a strong effect on local permafrost conditions. This paper is a review of past and present climate in Alaska and the response of permafrost to climate. Surface heat and mass transfer problems, the effect of air temperature on MAST through the intervening snow and vegetation cover and active layer, and the effects of precipitation, winds, and water bodies are discussed elsewhere in this panel and will not be considered here.

PAST AND PRESENT CLIMATE

An understanding of the present thermal regime of permafrost and its thickness requires consideration of climate trends for many tens of thousands of years. Figure 1 shows a composite of trends in global and Alaskan climate. The USC/GARP (1975) curve represents the generalized Northern Hemisphere air temperature trend based on mid-latitude sea surface temperature, pollen records, and worldwide sea-level records. The second curve was selected from Brigham (1984) where the effective diagenetic temperature (EDT) in the permafrost in the western Arctic was obtained from amino acid geochronology. The data of Brigham and Miller (1983) require an EDT ≈ -14°C for the past 125,000 yr, which is ≈ 4°C or more colder than the present MAST at Barrow and about 2°C colder than the MAST about a century ago (Lachenbruch et al., 1982). This suggests permafrost temperatures colder than -14°C between the present and past interglacial periods. The main features of the two climate models in Figure 1 are the two interglacial periods separated by a period of about 10^5 years that was characterized by colder, oscillatory temperatures. These models suggest that the glacial period 14,000-30,000 yr B.P. appears to have had mean annual air temperatures (MAAT) of about -18°C or colder, assuming temperatures of ≈ -9°C for the period from 8500-14,000 yr B.P. and ≈ -12°C for the last 8500 yr (Brigham, 1984).

These paleoclimate considerations suggest that the permafrost on Alaska's North Slope was thicker at the end of the last glacial period and that it may still be thawing from the base in response to the present, warmer MAST. This is a very simplified assessment, and a number of assumptions (e.g. the initial permafrost thickness at the starting point of the calculations) are critical for determining the present state of the permafrost. Obviously much more research on paleoclimates and the effect of climate on permafrost is required. However, it should be noted that, given the unknowns of paleoclimate models and permafrost/climate interactions, it may never be possible to do more than to arrive at general constraints on the paleoclimate models or a general assessment of the present state of the permafrost, particularly for time periods extending back to the last glacial period and beyond.

The above statements apply to the thick permafrost on Alaska's North Slope. The discontinuous permafrost south of the Brooks Range is much thinner and has much shorter response times to climate changes. Brown and Andrews (1982) have reviewed recent research on climate and permafrost in the North American Arctic, and Greenland and Alaskan weather records have been examined by Hamilton (1965) and Juday (1984). Hamilton (1965) suggests that the absence of pronounced regional trend differences indicates that a composite of the Alaskan data would be fairly representative for the state as a whole. Figure 2 shows his composite MAAT (mean annual air temperature) for Alaska based on 8-yr running means. Juday (1984) constructed the 5-yr running MAAT for Barrow and

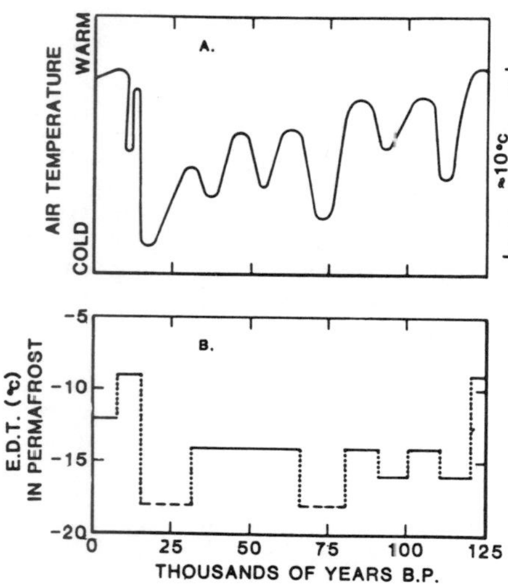

FIGURE 1 (A) Generalized Northern Hemisphere air-temperature trends based on mid-latitude sea-surface temperature, pollen records, and worldwide sea-level records. Modified from USC/GARP (1975). (B) A possible model for the effective diagenetic temperature in permafrost (EDT) on the western Arctic Alaska coastal plain based on amino acid geochronology. Modified from Brigham (1984).

the University Experiment Station near Fairbanks (Fig. 3). These graphs are open to interpretation; however, one interpretation is that there has been a general warming trend in Alaska since the late 1800s until the early 1940s amounting to ≈ 1-2°C. This warming trend was followed by a sharp cooling trend amounting to ≈ 1°C, which lasted for about three decades until the mid-1970s. Since the mid-1970s, there has been a sharp increase in MAAT amounting to ≈ 1-2°C. An examination of recent Alaskan weather records suggests that the climate warming since the mid-1970s has been statewide (Hoffman and Osterkamp, unpublished research).

While the interpretation of Alaskan weather records is far from complete, it seems clear that, during the past century, Alaskan climate has experienced a warming trend followed by a short cooler period and then a very recent sharp warming. Qualitatively, the Alaskan trend is similar to global climate trends. A key question is, what has been the response of Alaskan permafrost to these climate changes?

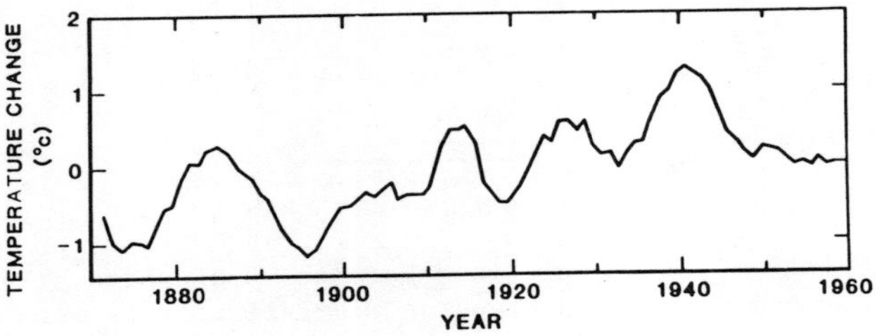

FIGURE 2 An Alaska composite of 8-yr running means for the mean annual air temperatures. Modified from Hamilton (1965).

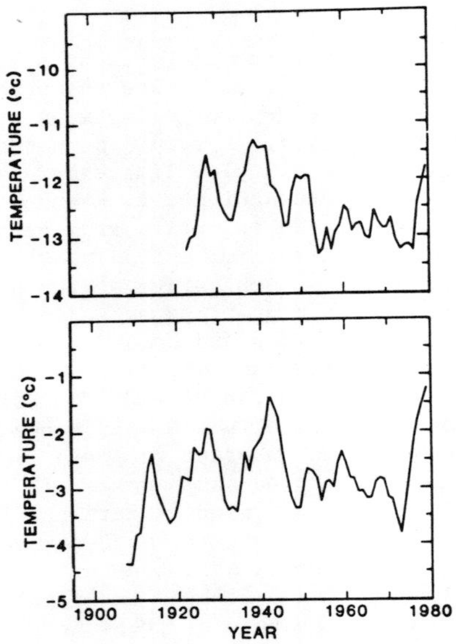

FIGURE 3 Five-year running means for the mean annual air temperatures at Barrow (upper graph) and the University Experiment Station, Fairbanks (lower graph). Modified from Juday (1982).

THERMAL REGIME OF ALASKAN PERMAFROST

There is little information available on the thermal regime of Alaskan permafrost. The most extensive and long-term data are from holes logged in the last three decades by the U.S. Geological Survey, mostly near Alaska's arctic coasts (Lachenbruch et al., 1982; Gold and Lachenbruch, 1973; Lachenbruch and Marshall, 1969; Lachenbruch et al., 1966; Lachenbruch et al., 1962; Brewer, 1958). Figure 4 shows four temperature profiles that illustrate the main features of the thermal regime of the permafrost in these coastal areas. These are a strong curvature towards warmer temperatures in the upper 100-160 m of the profiles, similar past MAST, similar and near-normal continental heat flows, differing gradients (and permafrost thicknesses) that appear to be related to large variations in the local thermal conductivity and small variations in MAST, gradient contrasts across the base of the ice-bearing permafrost for the porous formations near Prudhoe Bay, and what appear to be near-steady-state conditions (except for the top 100-160 m).

Osterkamp et al. (unpublished research) have recently established 15 drill holes in permafrost on a north-south transect of Alaska along the trans-Alaska oil pipeline for the purpose of investigating permafrost/climate questions. Figure 5 shows a preliminary temperature profile from a hole near Deadhorse Airport compared to a nearby hole of Lachenbruch et al. (1982). The straight line is an extrapolation of the deeper thermal gradient to the surface and represents average temperatures a century in the past. The present MAST ≈ -7°C or slightly colder, and the variation from past temperatures at the 50-m depth was ≈

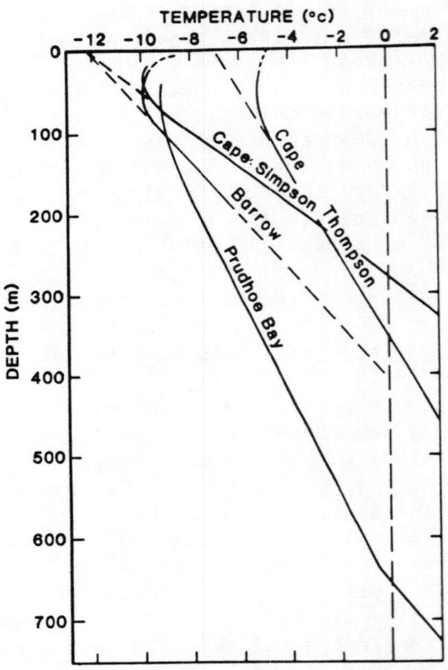

FIGURE 4 Generalized temperature profiles in permafrost along the arctic coasts of Alaska. Modified from Lachenbruch et al. (1982).

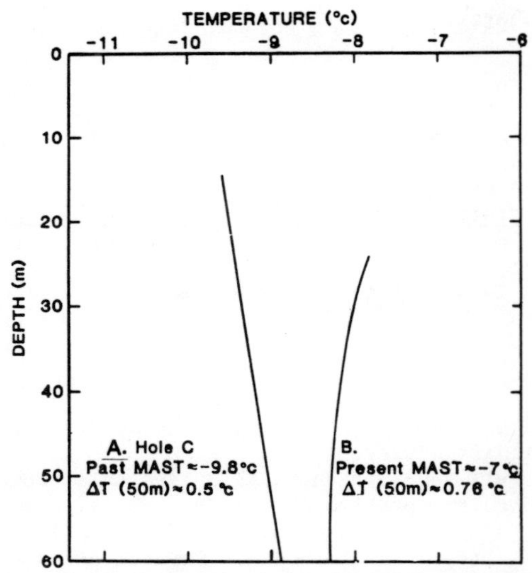

FIGURE 5 Recently measured temperature profile (B) in permafrost near Deadhorse. These data may be compared to the profile extrapolated from depth (A) of Lachenbruch et al. (1982). ΔT (50 m) represents the temperature difference between the profile extrapolated from depth and the measured profiles at the 50-m depth.

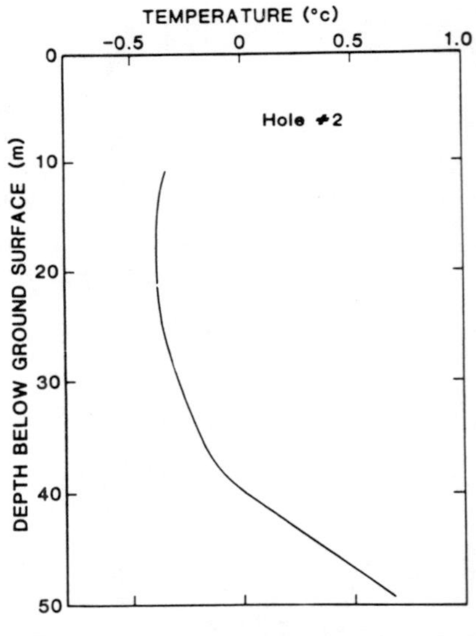

FIGURE 6 Recently measured temperature profile (June 1983) in permafrost on the University of Alaska, Fairbanks campus.

3/4°C, which may be compared to the recent values of ≈ -8°C and ≈ 1/2°C reported by Lachenbruch et al. (1982). These differences could represent local variations in the MAST and/or the effects of a very recent accelerated warming.

Figure 6 shows a preliminary temperature profile that was obtained in June 1983 as part of the permafrost/climate studies, noted above, from discontinuous permafrost at a site located on the University of Alaska, Fairbanks campus. This profile shows a strong curvature toward warmer temperatures in the upper 25 m. It is suggested that this curvature is a result of recent warmer MAST. Similar curvatures also occur in some of the temperature profiles from other drill holes in both continuous and discontinuous permafrost areas.

DISCUSSION

The Alaskan climate, as reflected in the MAAT, has had a very complex history over the past century with an initial oscillatory warming trend followed by a cooler period and then a very recent, sharp warming trend. Global climate changes appear to be similar to this pattern. Analysis of the weather data at Barrow, Fairbanks, and other sites in Alaska suggests that most, if not all, of the state was subject to this climate history (Hamilton, 1965; Juday, 1984; Hoffman and Osterkamp, unpublished research). Hamilton (1965) also noted that there was a minor latitudinal and longitudinal lag in some of the oscillations associated with the warming trend prior to 1940. Undoubtedly local variations from this general climate behavior must exist; however, their extent and magnitude are unknown. The response of the permafrost to these climate changes is contained in its thermal regime (temperature profiles and thickness). This response is buffered by the intervening snow and vegetation cover and the active layer so that MAAT cannot be directly converted to MAST. In addition, climate changes, as reflected by changes in MAAT, may be accompanied by changes in precipitation, wind, or snow cover, which can reinforce or negate the effect of changes in MAAT or MAST. Nevertheless, it is fruitful to investigate permafrost temperature profiles and thicknesses for evidence about past changes in the MAST.

A relatively simple permafrost/climate model that illustrates the response of the permafrost to sudden changes in the MAST and the time scales for the response when the new surface temperature does not exceed 0°C is shown in Figure 7. It is assumed that a steady-state temperature condition has prevailed until time $t = 0$ and that at $t > 0$ the surface temperature changed instantaneously from T_o to T_s, where $T_s < T_b$, the melting temperature at the base of the ice-bonded permafrost, which is at $x = X$ at time $t = 0$. A surface warming of 3°C is assumed. The response of the permafrost to the new boundary condition is described by the heat conduction equation

$$\frac{d^2T}{dx^2} - \frac{1}{\kappa}\frac{dT}{dt} = 0 \qquad (t > 0,\ 0 \leq x \leq X), \qquad (1)$$

where κ is the thermal diffusivity of the permafrost.

The boundary conditions require

$$T(0,t) = T_s \qquad (2)$$

$$T(X,t) = T_b \qquad (3)$$

with the initial conditions given by

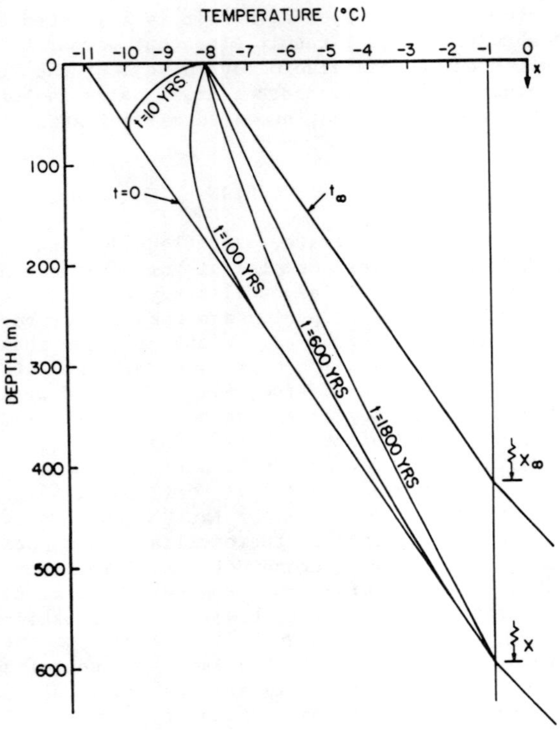

FIGURE 7 Thermal warming model for the continuous permafrost zone (e.g. Prudhoe Bay, $\kappa = 50\ m^2\text{-}a^{-1}$). The assumed initial equilibrium temperature profile ($t = 0$) and the final equilibrium temperature profile at $t = t_\infty$, with a permafrost depth of X_∞, are shown with the approximate temperature profiles for t = 10, 100, 600, and 1800 yr. Melting at the base of the permafrost amounts to ≈3.5 m during the first 1800 yr and is ignored in this graph.

$$T(x,0) = (T_o - T_s)\left(1 - \frac{x}{X}\right) + T_s + (T_b - T_s)\frac{x}{X} \,. \qquad (4)$$

Equation 1 can be solved by superposition with

$$T(x,t) = T_1(x) + T_2(x,t) \qquad (5)$$

where

$$T_1(x) = T_s + (T_b - T_s)\frac{x}{X} \,. \qquad (6)$$

The remaining time-dependent part of the solution can be obtained by separation of variables and calculation of the Fourier coefficients consistent with the boundary conditions so that

$$T_2(x,t) = \frac{2(T_o - T_s)}{\pi} \sum_{n=1}^{\infty} \frac{1}{n} \cdot \exp\left(\frac{-n^2\pi^2}{4\tau}\, t\right) \cdot \sin\frac{n\pi x}{X} \qquad (7)$$

TABLE 1 Values of τ, in yr, for fine-grained soils with $\kappa = 38\ m^2\text{-}a^{-1}$ and coarse-grained soils with $\kappa = 50\ m^2\text{-}a^{-1}$ where a, annum, is the time in yr.

Thickness X (m)	25	50	100	400	600
Fine-grained soils	4.1	16.4	66	1053	-
Coarse-grained soils	3.1	12.5	50	800	1800

where n is an integer and the time constant

$$\tau = \frac{X^2}{4\kappa} \qquad (8)$$

depends only on the permafrost thickness and its thermal diffusivity.

A full description of this type of approach, applied to the problem of thawing subsea permafrost, is given by Lachenbruch and Marshall (1977). T(x,t) is graphed for several elapsed times in Figure 7. Table 1 shows values for τ calculated using eq. 8 for several permafrost thicknesses. The time constant τ is associated with the time required for permafrost with fixed thickness X and fixed basal temperature T_b to respond to an instantaneous surface temperature change from T_o to T_s.

The heat balance at the permafrost base is

$$-L\phi \frac{dX}{dt} = q - K_f \left.\frac{dT}{dx}\right|_{x=X} \qquad (9)$$

where L is the volumetric latent heat of the permafrost, ϕ is the porosity, q is the geothermal heat flux, K_f is the thermal conductivity of the permafrost, and a constant heat flow, q, to the base has been assumed.

Substitution of eqs. 5, 6, and 7 in eq. 9 gives the rate of melting of the permafrost base:

$$-\frac{dX}{dt} = \frac{1}{L\phi}\left[q - \frac{K_f(T_b - T_s)}{X} - \frac{2K_f(T_o - T_s)}{X} \cdot \sum_{n=1}^{\infty} (-1)^n \cdot \exp\left(\frac{-n^2\pi^2}{4\tau}\, t\right)\right] . \qquad (10)$$

Figure 8 is a graph of eq. 10 for conditions similar to those at Prudhoe Bay with $q = 1.72\times10^6\ J(m^2a)^{-1}$, $L\phi = 1.2\times10^8\ J\text{-}m^{-3}$, $K_f = 10.6\times10^7\ J(ma°C)^{-1}$, X = 600 m, $T_b = -1°C$, $T_s = -8°C$, $T_o = -11°C$, and $\tau \approx 1800$ yr. This graph shows that melting at the permafrost base (x = X) does not begin until the surface thermal disturbance has propagated to X, which requires a time period on the order of 0.1τ or more. The melting rate increases rapidly for $0.1\tau < t < \tau$. For $t > \tau$, the summation of the right-hand side of eq. 10 approaches zero and, for times much longer than τ, $-dX/dt \approx 0.43\ cm \cdot a^{-1}$. For $t < \tau$ the total melting amounts to about 3-4 m, which justifies using this approach. The new equilibrium thickness,

$$X_\infty \approx \frac{K_f}{q}(T_b - T_s) = 431\ m \qquad (11)$$

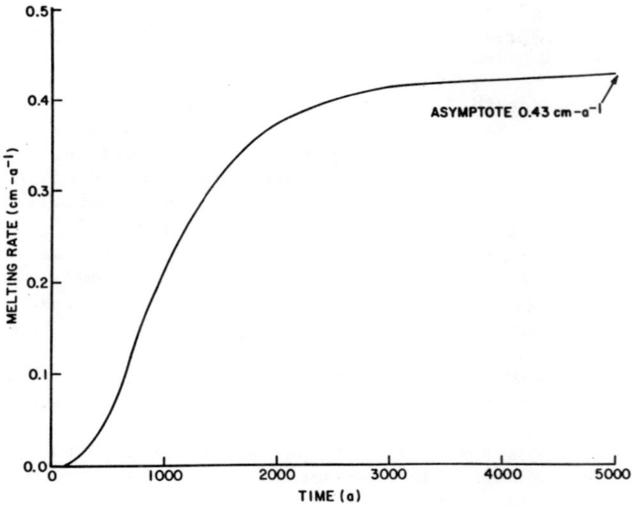

FIGURE 8 Melting rate $-dX/dt$ at the ice-bonded permafrost base for the model and thermal parameters shown in Figure 7.

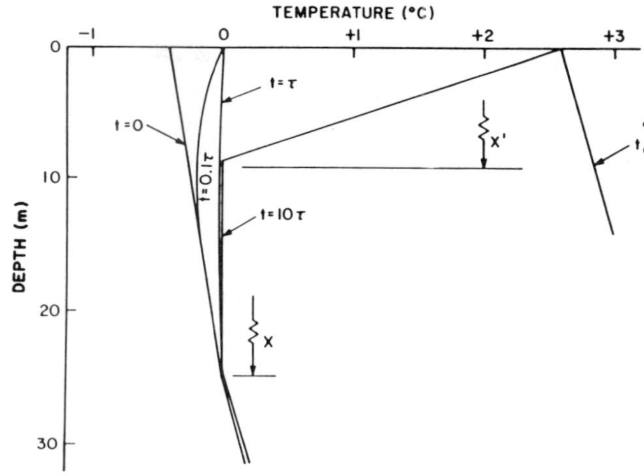

FIGURE 9 Thermal warming model for the discontinuous permafrost zone (e.g. Fairbanks, $\kappa = 50$ m^2-a^{-1}, X = 25 m). It is assumed that the warming occurs in two stages where the ground surface warms from -0.4° to 0°C in the first stage and from 0 to 2.6°C in the second, where the first stage lasts for $\tau \approx$ 3-4 yr. The temperature profiles are shown for time t = 0, 0.1τ, τ, 10τ, and t_∞.

would be reached in about 30,000-40,000 yr, assuming no further changes in surface temperature during this period.

This thermal analysis shows that the major changes expected in the 600-m-thick permafrost at Prudhoe Bay, as a result of a ground surface temperature increase, are a general warming of the permafrost at all depths and melting at the base of the permafrost until it thins to a new equilibrium thickness. The warming of the permafrost would begin immediately, the temperature profile would be nearly linear after a time $t \approx \tau$, and the time required to reach a new equilibrium thickness would be several tens of thousands of years.

This analysis is not very realistic in that it assumes a constant heat flow to the permafrost base and a sudden change in surface temperature. Lachenbruch et al. (1982) used a more realistic model that accounts for the possible time variation in surface temperature. If the temperature change at the surface of the permafrost can be represented by a power law of the form

$$T(0,t) = At^{n/2} \quad n = 0,1,2 \ldots \tag{12}$$

where A is a constant and t is the time since the start of the warming, then the temperature disturbance at depth x and time t is

$$T(x,t) = T(0,t) \cdot 2^n \, \Gamma(n/2 + 1) \cdot i^n \operatorname{erfc}(x/2\sqrt{\kappa t}) \tag{13}$$

where $\Gamma(\beta)$ is the gamma function of argument β and $i^n \operatorname{erfc}(\beta)$ is the repeated integral of the error function of β. Application of this type of approach to permafrost temperature profiles along Alaska's arctic coasts (Lachenbruch et al., 1982; Lachenbruch and Marshall, 1983) suggests that the curvature found in the upper portion of the profiles represents a surface warming of 1.5-3°C during the past century. Lachenbruch and Marshall (1969) note that this surface warming was greater at Barrow than at Prudhoe Bay and also that a slight cooling occurred at Barrow in the decade prior to 1962.

A model for permafrost warming at sites in the discontinuous zone, in which the new surface temperature exceeds the melting temperature of the permafrost, is shown in Figure 9. The warming is considered to occur in two stages. In the first stage, the surface temperature warms instantaneously from T_o to T_s = 0°C, with the permafrost base at T_b = 0°C, and remains there for a time $t \approx \tau$ followed by a second warming from T_s = 0°C to T_s > 0°C, which occurs for the time $t > \tau$. A total surface warming of 3°C is assumed. During the first stage, the temperature profiles in the permafrost can be obtained by solving the heat conduction equation as before. The solution

$$T(x,t) = \frac{2T_o}{\pi} \sum_{n=1}^{\infty} \frac{1}{n} \cdot \exp\left(\frac{-n^2\pi^2}{4\tau} t\right) \sin \frac{n\pi x}{X} \tag{14}$$

is graphed in Figure 9 for several times during the first warming stage for conditions characteristic of coarse-grained soils in interior Alaska. Except for surface thawing, the value of τ, and other time and length scales, the results for thin and thick permafrost are qualitatively similar. Table 1 shows that, for thin permafrost in interior Alaska (X = 25 m), the time scale required to bring the permafrost to temperatures very near its melting point is $t = \tau \approx$ 3-4 yr.

The rate at which the permafrost is thinned from below can be obtained from the heat balance at the permafrost base (eq. 9). Substitution of eq. 14 into eq. 9 yields

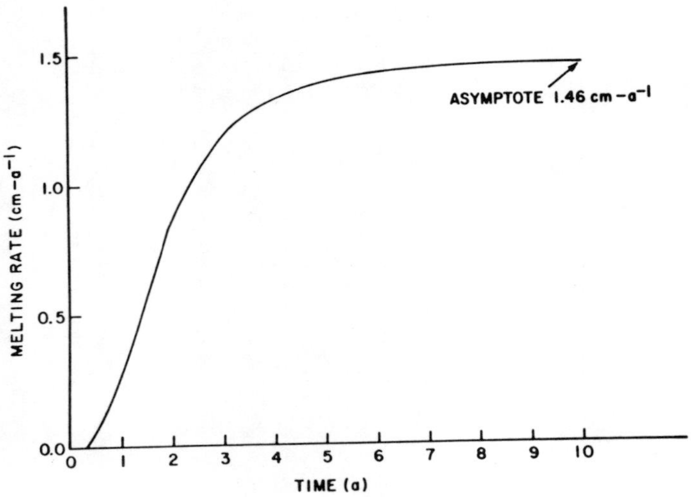

FIGURE 10 Melting rate -dX/dt at the ice-bonded permafrost base for the model and thermal parameters shown in Figure 9.

$$-\frac{dX}{dt} = \frac{1}{L\phi}\left[q - \frac{2K_f T_o}{X}\sum_{n=1}^{\infty}(-1)^n \exp\left(\frac{-n^2\pi^2}{4\tau}t\right)\right] \quad (15)$$

Figure 10 is a graph of eq. 15 for coarse-grained soils and conditions appropriate to some interior Alaska sites (i.e. $L\phi = 1.2x10^8$ J m^{-3}, K_f = $10.6x10^7$ J $(ma°C)^{-1}$, X = 25 m, T_b = 0°C, and T_o = -0.4°C). It is assumed that the heat flow q is the same as at Prudhoe Bay. Apparently only 4-5 cm of the permafrost may be expected to melt at its base during the time period $0 < t < \tau$. For times much greater than τ,

$$-\frac{dX}{dt} \approx \frac{q}{L\phi} = 1.46 \text{ cm a}^{-1} \quad (16)$$

During the second stage, $t > \tau$, it is assumed that the surface temperature warms instantaneously to $T_s > 0°C$. The depth of thaw during this stage can be calculated using the Stefan formula

$$X' = \sqrt{\frac{2K_t}{L\phi}(T_s - T_b)\,t} \quad (17)$$

which reduces to

$$X' \approx 1.64\sqrt{t} \quad (18)$$

for $K_t = 6.2x10^7$ J $(ma°C)^{-1}$, T_s = +2.6°C, and T_o = 0°C, and the conditions assumed above, where X' is in m and t is in yr. Equation 17 describes thawing from the ground surface downwards when the MAST exceeds 0°C. The model assumes constant surface temperatures during both stages and linear temperature profiles near the surface ($0 \leq x \leq X$) during the second stage; however, in reality, variable surface temperatures and the presence of an active layer would make near-surface temperature profiles highly nonlinear.

The thickness of the remaining permafrost at any time can be obtained by combining eqs. 15 and 18. Apparently 185 yr or about two centuries would be required to thaw completely 25 m of coarse-grained permafrost under the assumed conditions. The final temperature profile would be that shown for t_∞ in Figure 9, which would be parallel to the subpermafrost temperature profile at t = 0. This assumes no further modifications of surface temperatures during that period.

The above analyses show that the permafrost response to climate variations includes changes in the temperature profile, melting at the base, and, if the MAST > 0°C, melting at the table. For time periods $t \leq \tau$, information on the changes in the MAST is contained in the temperature profile. The thickest permafrost in Alaska, 629 m at Mikkelsen Bay (Osterkamp and Payne, 1981), would have a $\tau \approx$ 2000 yr assuming thermal parameters similar to Prudhoe Bay. For time periods $t > \tau$, climate information is available in the thermal gradients in the permafrost and in its thickness. These relatively simple models suggest that climate information for time periods of several tens of thousands of years may be available in the thermal gradients and thicknesses of the permafrost in the Prudhoe Bay area. Harrison and Bowling (1983) have begun to analyze the data of Lachenbruch et al. (1982) in an effort to extract paleoclimate information.

Preliminary analysis of temperature profiles from interior Alaska such as that shown in Figure 6 suggests that this warm, thin, discontinuous permafrost is responding to warmer MAST. Curvature in the top portions of the profiles appears to be a result of a very recent warming (within the last decade). This permafrost (Fig. 6) is extremely warm, within 0.4°C of thawing, with an approximate past MAST near -1°C. If the surface warming shown in Figure 6 is characteristic of other sites in interior Alaska then, given the range of permafrost conditions, somewhat warmer and thinner permafrost is presently thawing at both the permafrost table and the base.

The quantitative nature of these tentative conclusions is difficult to ascertain, partly because of the non-uniqueness of the solutions and the accuracy of the data, but primarily due to the sparsity of data. For example, with the MAST variation of eq. 12, Gold and Lachenbruch (1973) show that each of the cases n = 0,1,2,3 reproduces the temperature observations with a standard error of 0.01°C. Therefore, while all the models indicate a warming of the MAST during the last century, it is not possible to select between them. While the accuracy of the available temperature data is very good, the determination of thermal conductivity and thermal gradients leads to relatively large uncertainities in the heat flow data. For example, Lachenbruch et al. (1982) found that the heat flow at Prudhoe Bay is 55 ±8 mW m^{-2}. The corresponding uncertainty in the equilibrium thickness of the ice-bonded permafrost is ≈ 80 m.

The precise relationships between changes in climate and in the thermal regime of Alaskan permafrost have not been established. However, the foregoing suggests that the permafrost is responding to changes in the MAAT during the last century and may still be responding to earlier climates. In addition, changes in the MAST over the past century, as determined by analyses of the permafrost temperature profiles, are of the same magnitude as the MAAT changes determined from Alaska weather records.

SUMMARY

The implications of climate change for permafrost have already been noted by Brown and Andrews (1982), Goodwin et al. (1984), and Osterkamp (1984). Washburn (1980) has summarized most of the recent information on the use of fossil permafrost features as temperature indicators. If the warming trend of the past century and particularly of the past decade continues, then substantial changes noted by the above authors may occur in permafrost areas.

The data obtained from amino acid geochronology studies show that the effective diagenetic temperature of Alaskan permafrost was ≈ -14°C for the past 125,000 years. Present and past interglacial periods have been warmer, suggesting that the intervening glacial period was colder, with a mean annual air temperature of ≈ -18°C or colder. Alaskan weather records suggest an oscillating warming trend in air temperatures from the late 1800s until the early 1940s of ≈ 1-2°C, followed by three decades of cooler temperatures (≈ 1°C) until the mid-1970s, when a sharp warming of ≈ 1-2°C occurred. This accelerated warming trend is still under way.

Temperature profiles from thick permafrost along Alaska's arctic coasts all show a strong curvature in the upper 100-160 m and near steady-state conditions otherwise. This curvature suggests a mean annual surface temperature warming of 1.5-3°C during the last century. A cooling of the MAST occurred in the decade prior to 1962. Recent temperature profiles obtained in areas of discontinuous permafrost in interior Alaska suggest that there has been a warming in the MAST since the mid-1970s. The discontinuous permafrost may be melting or freezing at the base, depending on its thickness and thermal parameters. The main features of the climate trends of the past century appear to be reflected in the thermal regime of the permafrost.

First-order estimates of the time scales and magnitudes of past changes in the mean annual surface temperature of permafrost areas can be made with relatively simple models. These models show that the permafrost response consists of curvature in the temperature profile progressing downward with time, followed by slow melting at the permafrost base and establishment of a new thermal gradient. If the MAST exceeds 0°C, then melting at the permafrost table would also occur. The results of applying these models to Alaskan permafrost show that the time scales for curvature in the permafrost temperature profiles range from a few years up to about 2000 yr. Melting at the permafrost base proceeds very slowly, and the time scales required to establish an equilibrium thickness of permafrost in response to surface temperature changes range from a century or less to many tens of thousands of years. When the MAST exceeds 0°C, melting at the permafrost table also proceeds very slowly, and a time scale of about a century is required to melt 10-20 m. These considerations show that a record of past climate is contained in the temperature profiles and thickness of the permafrost.

An understanding of the exact nature (magnitude, timing, regional trends) of past climate changes and their effect on mean annual surface temperatures, permafrost features, and the thermal regime and thickness of the permafrost requires much more detailed data, analyses, models, and interpretation. This understanding of permafrost-climate problems is needed to assess the influence of the present climate warming trend on permafrost and to anticipate changes in the permafrost that will have an impact on the environment and on human activities.

ACKNOWLEDGMENTS

I wish to thank J. Gosink, K. Kawasaki, and G. Weller for reviewing a draft of this manuscript. This research was supported by the Alaska Council on Science and Technology, Grant No. W-81-03.

REFERENCES

Brown, J., and Andrews, J.T., 1982, Influence of short-term climatic fluctuations on permafrost terrain, in Carbon Dioxide Effects Research and Assessment Program, Environmental and Societal Consequences of Possible CO_2-Induced Climate Changes, DOE/EV/10019-02, v. 2, pt. 3: Washington, D.C., U.S. Department of Energy.

Brewer, M.C., 1958, Some results of geothermal investigations of permafrost in northern Alaska: Transactions of the American Geophysical Union, v. 39, no. 1, p. 19-26.

Brigham, J.K., 1984, Marine stratigraphy and amino acid geochronology of the Gubik formation, western Arctic Coastal Plain, Alaska: University of Colorado, Boulder, Colorado, Ph.D. dissertation.

Brigham, J.K., and Miller, G.H., 1983, Paleotemperature estimates of the Alaskan Arctic coastal plain during the last 125,000 years. Paper presented at the poster session of the Fourth International Conference on Permafrost, Fairbanks, Alaska.

Gold, L.W., and Lachenbruch, A.H., 1973, Thermal conditions in permafrost - A review of North American literature, in Proceedings of the Second International Conference on Permafrost, Yakutsk: Washington, D.C., National Academy of Sciences, p. 3-23.

Goodwin, C.W., Brown, J., and Outcalt, S.I., 1984, Potential responses of permafrost to climatic warming, in Proceedings of a Conference on the Potential Effects of Carbon Dioxide-Induced Climatic Changes in Alaska (J.H. McBeath, ed.): Fairbanks, University of Alaska, Misc. Pub. 83-1, p. 92-105.

Hamilton, T.D., 1965, Alaskan temperature fluctuations and trends: Analysis of recorded data: Arctic, v. 18, no. 2, p. 105-117.

Harrison, W.D., and Bowling, S.A., 1983, Paleoclimate implications for Arctic Alaska of the permafrost at Prudhoe Bay. Paper presented at the poster session of the Fourth International Conference on Permafrost, Fairbanks, Alaska.

Juday, G.P., 1984, Temperature trends in the Alaska climate record: Problems, update, and prospects, in Proceedings of a Conference on the Potential Effects of Carbon Dioxide-Induced Climatic Changes in Alaska (J.H. McBeath, ed.): Fairbanks, University of Alaska, Misc. Pub. 83-1, p. 76-91.

Lachenbruch, A.H., Brewer, M.C., Greene, G.W., and Marshall, B.V., 1962, Temperatures in permafrost, in Temperature - Its Measurement and Control in Science and Industry, v. 3, pt 1: New York, Reinhold Publishing Co., p. 791-803.

Lachenbruch, A.H., Greene, G.W., and Marshall, B.V., 1966, Permafrost and the geothermal regimes, Environment of the Cape Thompson Region, Alaska, Chapter 10: Washington, D.C., U.S. Atomic Energy Commission, Divison of Technical Information, p. 149-165.

Lachenbruch, A.H., and Marshall, B.V., 1969, Heat flow in the Arctic: Arctic, v. 22, p. 300-311.

Lachenbruch, A.H., and Marshall, B.V., 1977, Subsea temperatures and a simple tentative model for offshore permafrost in Prudhoe Bay, Alaska, U.S. Geological Survey, Open File Report 77-395: Menlo Park, California.

Lachenbruch, A.H., Sass, J.H., Marshall, B.V., and Moses, T.H., Jr., 1982, Permafrost, heat flow and geothermal regime at Prudhoe Bay, Alaska: Journal of Geophysical Research, v. 87, no. B11, p. 9301-9316.

Lachenbruch, A.H. and Marshall, B.V., 1983, Permafrost and the geothermal regime, U.S.G.S. Circular 911: Alexandria, Va., U.S. Geological Survey.

Osterkamp, T.E., 1984. Potential impact of a warmer climate on permafrost in Alaska, in Proceedings of a Conference on the Potential Effects of Carbon Dioxide-Induced Climatic Change in Alaska (J.H. McBeath, ed.): Fairbanks, University of Alaska, Misc. Pub. 83-1, p. 106-113.

Osterkamp, T.E., and Payne, M.W., 1981, Estimates of permafrost thickness from well logs in northern Alaska: Cold Regions Science and Technology, v. 5, p. 13-27.

USC/GARP, 1975, Understanding Climatic Change, A Program for Action: Washington, D.C., National Academy of Sciences.

Washburn, A.L., 1980, Permafrost features as evidence of climatic change: Earth-Science Reviews, v. 15, p. 327-402.

CLIMATE CHANGE AND OTHER EFFECTS ON PERMAFROST

M. Smith

Department of Geography, Carleton University
Ottawa, Canada

This is not a review of climate and permafrost relationships. Other contributors have addressed this, but it might be useful to restate a number of basic ideas about climate and permafrost given the considerable interest in the implications of climate change for permafrost conditions. After a brief summary of some basic ideas, three broad questions are posed.

Obviously, the occurrence of permafrost depends on climate, since climate, in some way or another, determines the temperature at the surface of the earth. Many discussions of climate and climate change seem to focus almost exclusively on the factor of temperature, but it must be realized that climate is far more complex than this, and that changes in precipitation and other factors of climate are also very important. In addition, more than just climatic conditions prescribe the surface temperature regime, under which permafrost may or may not be present. However, when climate does change as shown in Figure 1 -- the temperature change over roughly the last century or so -- then ground temperatures can also change (Gold and Lachenbruch, 1973; Osterkamp, this volume). However, ground thermal conditions can change for a variety of other reasons, some of which are nonclimatic, and some of which have implications for, or feedbacks with, climatic processes.

For instance, fire can produce disruptions of surface cover and change ground thermal conditions. Exposure of ice-rich sediments by river erosion can initiate spontaneous and continuing thaw of the materials. Snow banks can form either through changes in climate, including precipitation and wind, or by geomorphic or vegetational changes. A poster paper by Bowling (this conference) illustrates the effects of snow cover on the ground thermal regime in the absence of other kinds of climate change. Vegetation differences can produce substantial variations in ground thermal conditions between sites. Figure 2 shows data from two adjacent sites, where organic material is present at one and not at the other. As can be seen, quite different ground thermal regimes develop under the same climatic conditions. Lastly, there is a range of human disturbances that bring about changes in the ground thermal conditions. Jahns and Heuer (1983) have suggested how this could be exploited in mitigative frost heave, by utilizing the disruption in surface conditions to counteract freezing beneath a pipeline. Thermokarst can be initiated by surface disturbance during construction and such features can then have feedbacks to natural climatic and microclimatic processes that may amplify or diminish the influence of such disturbances.

There is of course considerable debate over the possible course of the earth's climate over

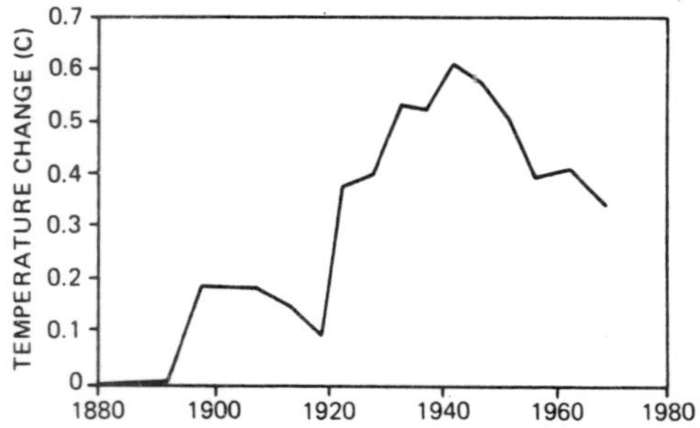

FIGURE 1 Mean hemispheric (0-80°N) changes in air temperature since 1880 (from Schneider, 1976).

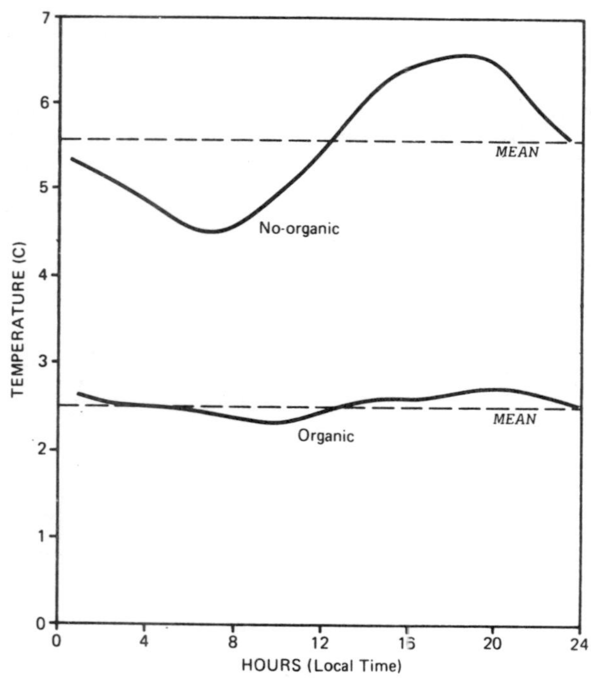

FIGURE 2 Mean diurnal 10-cm ground temperature regimes at adjacent sites in the Mackenzie Delta (from Smith, 1975).

the next 50 to 75 years. Conventional wisdom does point to at least a doubling of atmospheric carbon dioxide by the middle of the next century, but the implications of this for climate are less certain and there is no unanimity on just what will happen. We do know from the past that changes in arctic climate, especially temperature, are three to five times greater than for the northern hemisphere as a whole. In coastal regions of the Arctic, climate changes are complicated by simultaneous changes in the ice regime, but the general implications of CO_2-induced climate change would be a warming in the Arctic with a likely increase in precipitation (Harvey, 1982). If we accept this as a hypothesis valid for the next 100 years, then the following questions might be posed.

First, how can we assess the effects of climate change on permafrost conditions, given that local conditions exert such strong influences on ground thermal regime, as discussed by Smith and Riseborough (1983). Unfortunately there is a dearth of published information on ground temperature regimes at neighboring sites over long periods of time. There appears to be no report in the literature of any monitoring network of shallow permafrost temperatures that has been continued on a regular basis for any length of time. Figure 3 is based on a few years' data from Brown (1978); it illustrates that, within a small area, there is considerable variation in the thermal regime from site to site and from year to year. For example, the warmest year was 1976 at some sites, but 1975 at others. Thus we cannot expect individual locations, where permafrost is present, to respond in the same way to the same climate-forcing function; there are intervening processes that have to be understood before we can predict the influence of climate and climate variation on permafrost conditions. How can we best address this particular problem of the complexity introduced by the variation in local conditions? Goodwin et al. (1984) have addressed this issue, using a numerical energy balance model.

Secondly, is the likely rate and magnitude of permafrost change due to climate likely to be less, equally, or more important than changes due to other natural processes, such as the spontaneous thawing of ice-rich sediment exposed by riverbank erosion or fire? A corollary to this is, how will climate change affect the rate of such natural processes? We have initiated some work in the central Yukon to monitor the year-by-year retreat of retrogressive thaw flow slides to see if there is any relationship between the rate of retreat and climatic conditions. Such studies could be repeated in various places. Lastly, are changes in the permafrost regime resulting from possible climate change likely to be important considerations in northern geotechnical engineering? Given

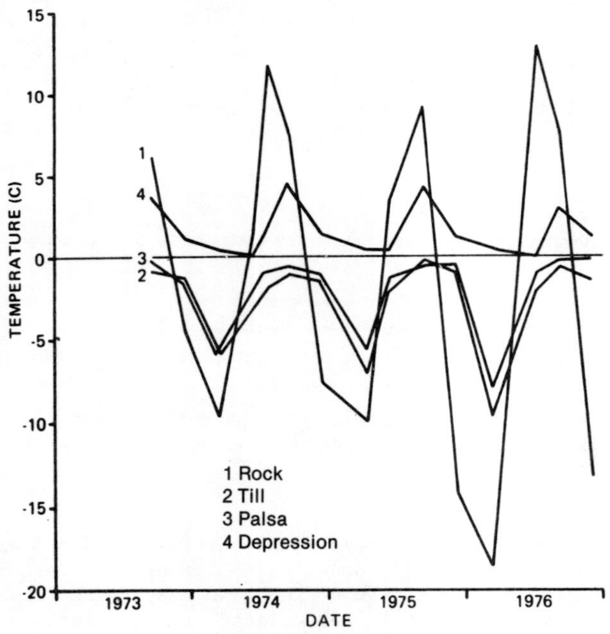

FIGURE 3 Ground temperatures at 1.06-m depth for four sites at Churchill, Manitoba (from Smith and Riseborough, 1983).

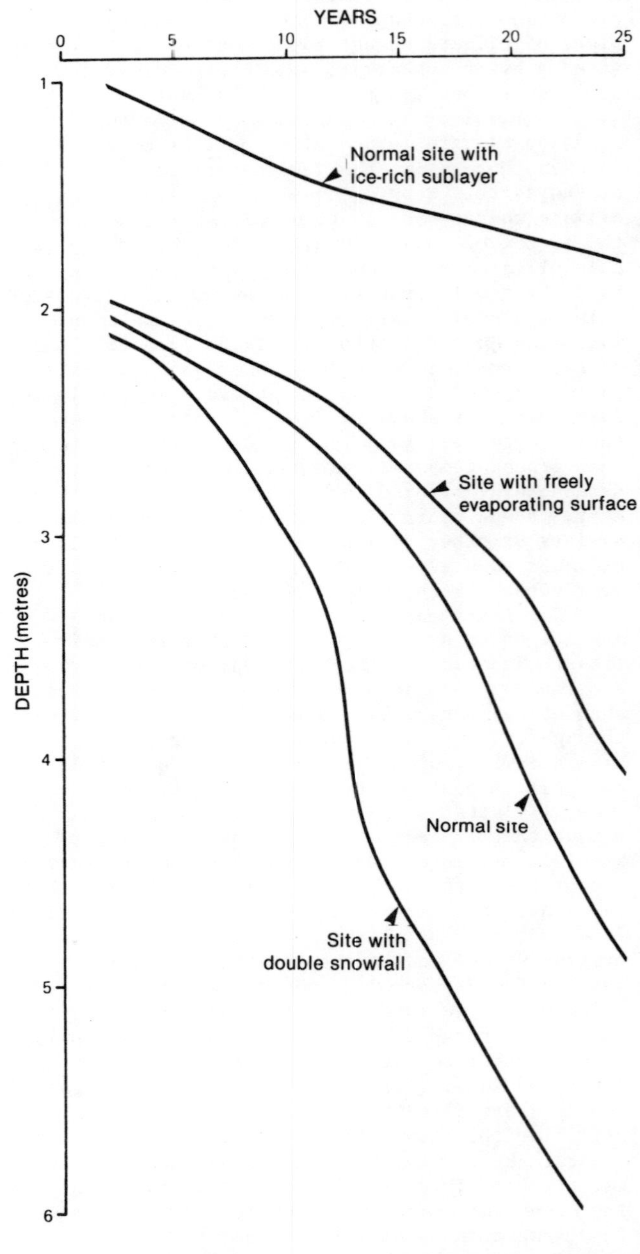

FIGURE 4 Predicted annual maximum active layer development for different sites under a uniform climatic warming trend (from Smith and Riseborough, 1983).

that structures in such environments are designed to be in some equilibrium with the natural thermal conditions, what are the implications if the natural background conditions themselves change? Figure 4 shows some attempts to predict the rate of permafrost recession under assumed climatic warming over the next 25 years at a number of sites deemed to be representative of discontinuous permafrost near Whitehorse, Yukon. As can be seen for various kinds of site conditions, the degradation of permafrost due to natural warming may be between 2 and 5 m over the period, surely not an insignificant amount for thaw-sensitive structures.

REFERENCES

Brown, R.J.E., 1978, Influence of climate and terrain on ground temperatures in the continuous permafrost zone of northern Manitoba and Keewatin District, Canada, in Proceedings of the Third International Conference on Permafrost, Edmonton: Ottawa, National Research Council of Canada, v. 1, p. 15-21.

Gold, L.W. and Lachenbruch, A.H., 1973, Thermal conditions in permafrost - A review of North American literature, in Proceedings of the Second International Conference on Permafrost, Yakutsk: Washington, D.C., National Academy of Sciences, p. 3-23.

Goodwin, C.W., Brown, J., and Outcalt, S.I., 1984, Potential responses of permafrost to climatic warming, in Proceedings of a Conference on the Potential Effects of Carbon Dioxide-Induced Climatic Changes in Alaska (J.H. McBeath, ed.): Fairbanks, University of Alaska, Misc. Pub. 83-1, p. 92-105.

Harvey, R.C., 1982, The climate of arctic Canada in a $2\times CO_2$ world. Canadian Climate Centre Report No. 82-5.

Jahns, H.O., and Heuer, C.E., 1983, Frost heave mitigation and permafrost protection for a buried chilled-gas pipeline, in Proceedings of the Fourth International Conference on Permafrost, Fairbanks: Washington, D.C., National Academy of Sciences, p. 531-536.

Schneider, S.H., 1976, Is there really a food-climate crisis? in Atmospheric Quality and Climatic Change, University of North Carolina at Chapel Hill, Studies in Geography No. 9, p. 106-143.

Smith, M.W., 1975, Microclimatic influences on ground temperatures and permafrost distribution, Mackenzie Delta, N.W.T.: Canadian Journal of Earth Sciences, v. 12, p. 1421-1438.

Smith, M.W. and Riseborough, D.W., 1983, Permafrost sensitivity to climate change, in Proceedings of the Fourth International Conference on Permafrost, Fairbanks: Washington, D.C., National Academy of Sciences, p. 1178-1183.

RECONSTRUCTIONS AND FUTURE PREDICTIONS OF THE EFFECTS OF CLIMATE CHANGE

J.T. Gray

Department of Geography, University of Montreal
Montreal, Canada

There are three main purposes for studying the relationships between climate change and the geothermal regime in cold environments.

In the first place, we may wish to interpret, for a given region, how documented climate change in the past ought to have affected ground temperatures and hence permafrost distribution and thickness, ground ice content, and active layer depths. Variations in these properties may then be inferred to have played a role in the physical, chemical, and biological processes active at and immediately below the ground surface. Thus, good documentation of past climate change has the potential of explaining fluctuations in these processes when interpreted by Quaternary specialists in such fields as geomorphology, pedology, hydrology, geochemistry, and plant ecology.

In the second place, we may wish to reverse the procedure and try to use fluctuations in the various physical, chemical, and biological processes as indicators of changes in the geothermal regime and hence as imprints of climate oscillations for a given region. The aim would then be to try to establish the age, trend, and amplitude of these oscillations.

In the third place, we may wish to predict the effects of future climate change on the thermal regime, and more specifically on the permafrost and active layer conditions. At one level we may wish to assess the effects of such future change in terms of the possible thermal disturbance to the near surface environment of man-induced micro-climatic modifications, resulting from engineering projects, on various scales. The planned reservoir construction in the Great Whale River basin of Northern Quebec, although presently shelved, is an interesting example of such a project on a very large scale. The Trans-Alaska Pipeline and the pipeline planned in the Mackenzie Valley in northwestern Canada are examples of such modification on a somewhat smaller scale. At a second level, we may wish to assess the long-term effects on a regional or even a continental scale of lower atmosphere warming and cooling trends. In this regard reference may be made to a possible warming trend related to the very significant increase in the CO_2 content of the atmosphere in recent years (Watts, 1980).

Whichever of these three goals interests us as researchers, it is clear that a key element is the quality and continuity of the climate data itself. In this regard, I would like to briefly raise a number of points about the different types of climate data available for reconstructions of goethermal regimes for the Holocene interval in periglacial regions, and then I would like to make a few remarks concerning the kind of data we will need to predict future climate change.

The climate data can be conveniently subdivided into three categories. In the first place, data are available for a sparse network of arctic meteorological stations for a period rarely exceeding the last 50 years. The most useful data for studies of the ground thermal regime that are often available are on air temperatures, precipitation, snowfall, snow cover, and global radiation. However, the very useful parameter of surface temperature for a variety of cover types is almost never available. It must be derived using corrective indices that vary according to radiation and the snow cover characteristics of the various surfaces. Lunardini's "n" factor correction (Lunardini, 1978) is one such corrective index.

The second type of data available to us is recorded in temperature logs that have been obtained from deep drill holes in arctic regions (Lachenbruch and Marshall, 1969; Taylor and Judge, 1979; Judge et al., 1981; Osterkamp and Payne, 1981; Poitevin and Gray, 1982; Pilon, 1982). The fluctuations in the ground temperature profiles beyond the zone of zero annual amplitude reflect changes in the mean annual surface temperatures at the site, but interpretation of the amplitude and periodicity of these temperature fluctuations is not straightforward. A number of corrections have to be applied to the temperature profiles to take account of thermal conductivity contrasts as well as of terrain variations in the vicinity of the drill hole, and assumptions have to be made concerning the geothermal flux from the interior of the crust (Taylor and Judge, 1979; Pilon, 1982).

The third possibility is the use of proxy climate data obtained from a series of indices at or below the ground surface. These indices are based on geomorphologic, pedologic, geochemical, and vegetational information. They can take the climatic record much farther back than the meteorological data, generally using radiocarbon dating for chronological control. However, use of such proxy data results in problems of interpretation, which necessarily imply a certain degree of subjectivity and a fairly high margin of error in derivation of the timing, direction, and amplitude of climate fluctuations. In this regard, let us examine one of the best established indices of climate change - pollen analysis.

Pollen spectra from successively deposited layers of organic material in peat bogs and lake beds have frequently been used to reconstruct former vegetation assemblages at given sites (Faegri and Iversen, 1964; Short and Nichols, 1977; Richard, 1977, 1978). Because of the relationship between vegetation associations and climate zones, it becomes possible to draw some inferences about former climates from the pollen stratigraphy and, through dating of the organic material, to have some chronological control on climate transitions. A number of problems must be taken into account, however. In the first place, the pollen spectra in arctic and subarctic regions do not only reflect the pollen rain derived from the local vegetation community, but also significant and perhaps temporally variable long-distance transport from forest regions to the south (Nichols et al., 1978; Elliott-Fisk et al., 1982). Secondly, it is presumed that there is a synchronism between vegetation change as reflected in the pollen stratigraphy and climate change acting as the causal factor. Richard (1977), in discussing climatic interpretation of pollen diagrams, suggests that vegetation succession towards a state of equilibrium with the climatic milieu is relatively rapid, well within the temporal precision of the technique for most basins of sedimentation. Finally, although palynologically determined changes in vegetation cover may give an index of the direction of climate change in terms of a deterioration or amelioration of growing conditions, it is very difficult to translate such change into meaningful terms for geothermal studies. Andrews et al. (1980) and Andrews and Nichols (1981) have attempted to use pollen transfer functions, using knowledge of modern pollen rain at control sites near meteorological stations across a range of vegetation zones in the Eastern Canadian Arctic, to accurately reconstruct temperature conditions in the growing season during the late Holocene interval. However, the geothermal regime is also related to winter temperature conditions and to snow cover, and these parameters cannot as yet be modelled by the pollen transfer functions.

Other proxy data from vegetation, such as $^{18}O/^{16}O$ ratios in buried tree trunks or in peat, tree-ring studies in living vegetation (Cropper and Fritts, 1981) and in buried macro fossils (Gagnon and Payette, 1981), geomorphological criteria such as the distribution of active and inactive periglacial landforms (Harris, 1982), all present similarly serious obstacles to the straightforward interpretation of climate change as it influences the geothermal regime. In addition to the problems of interpretation, there are often serious problems resulting from difficulties in establishing a time scale for the observed changes, either due to a total lack of in situ organic material for ^{14}C dating or due to the large errors inherent in ^{14}C data obtained from some organic materials--such as soils.

As a result of the various problems associated with proxy data, it may be appropriate to ask the following question when considering the use of paleoclimatic data for given sites in arctic regions for geothermal studies; Is it preferable to use a fragmentary record of locally derived proxy data with moderate to poor chronological control, or is it better to extrapolate to the given site from a distant region, relatively continuous records of climate change, perhaps using some kind of transfer function based on present-day climate records? In this regard, we might mention the very complete climate records based on the $^{18}O/^{16}O$ cores available from Camp Century on the Greenland ice cap (Dansgaard et al., 1970) and from the Devon Island ice cap in the Eastern Canadian Arctic (Koerner and Fisher, 1981). The latter path might appear at first glance to be a reasonable one, but must be followed cautiously. Recent climate records for eastern and western Arctic regions of North America (Fig. 1) do not show perfect synchronism in terms of warming and cooling trends. I would venture to suggest that while the regionally available proxy data are undoubtedly difficult to relate to mean annual ground surface

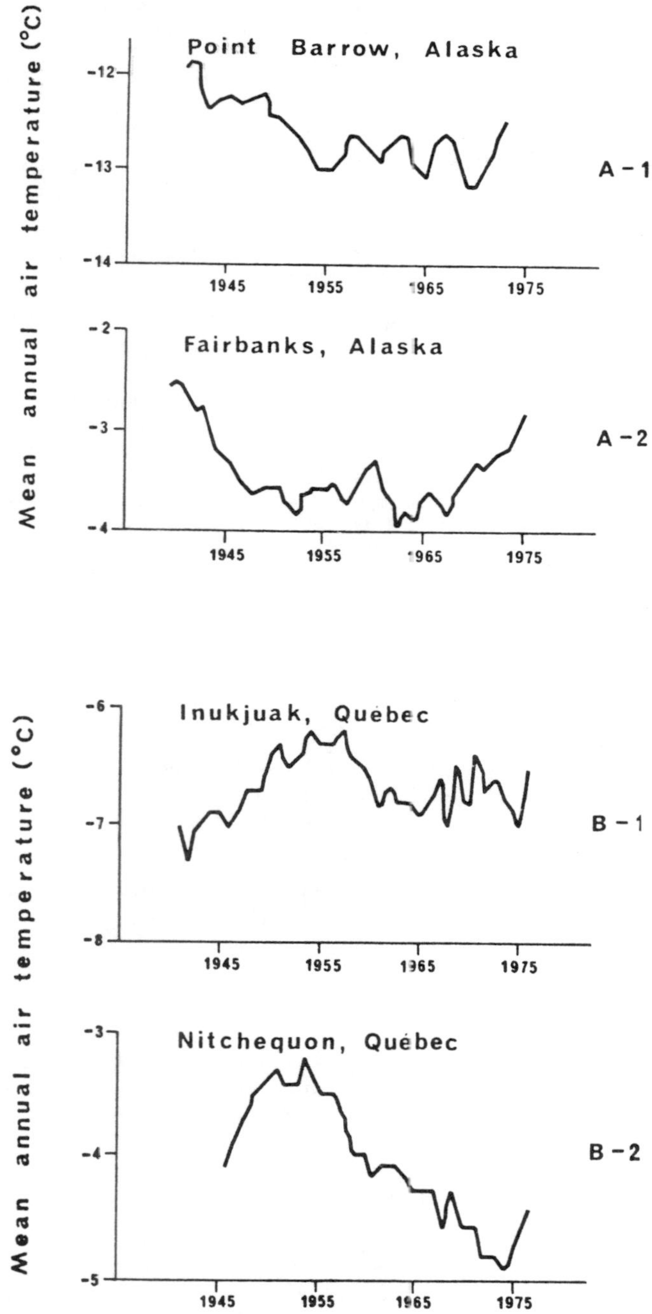

FIGURE 1 Eight-year running mean for annual air temperatures from 1940 to 1980 for selected stations in arctic North America. A-1 and A-2 show data for stations in the tundra and boreal forest zones of Alaska. B-1 and B-2 show data plotted for corresponding vegetation zones in northern Quebec.

temperatures, they do offer potentially valuable tools for the reconstruction of climate change if they can be combined with a few deep hole temperature logs. Taylor and Judge (1979) have shown that knowledge of the general direction and age of climate trends can, through application to the geothermal data, be used to reconstruct the amplitude of such trends.

One major difficulty will always remain in the use of the proxy data for paleoclimate reconstructions. The record, as regards detection of change in the various indices, the data of such change, and the derivation of paleotemperature curves, will always become progressively more imprecise as one goes backward through time.

With regard to prediction of future effects of climate change on permafrost and active layer conditions, it might be appropriate to suggest the establishment of a series of monitoring stations associated either with arctic weather stations or with long-term construction projects. Brown's research in Keewatin (Brown, 1978) and the data obtained at the McGill Subarctic Research Laboratory at Schefferville in Northern Quebec (Adams and Barr, 1973; Nicholson, 1978, 1979) represent very isolated examples where multi-year acquisition of ground surface temperatures through the various seasons has been attempted. The Schefferville data have been obtained for a 25-yr period, but even there the record is rather broken. What is needed for the North American Arctic is a series of recording stations from the eastern and western arctic regions for both the continuous and discontinuous permafrost zones. As well as standard meteorological data, each installation should be equipped to measure ground surface temperature, snow pack conditions, and ground temperature profiles (at least to a depth of about 5 m) for a variety of surface cover types in the vicinity of the installation. In this regard, long-term monitoring activities planned in association with the future Mackenzie Valley pipeline in the vicinity of Norman Wells in the Northwest Territories (Pilon, pers. commun.) should provide good data required for future prediction of the effects of climate change, at least for the western arctic regions.

REFERENCES

Adams, W.P. and Barr, W., 1973, Annotated bibliography of the McGill Subarctic Research Papers (1954-1970): Revue de Géographie de Montréal, v. 27, no. 4, p. 391-412.

Andrews, J.T., Mode, W.N., and Davis, P.T., 1980, Holocene climate based on pollen transfer functions, eastern Canadian Arctic: Arctic and Alpine Research, v. 12, no. 1, p. 41-64.

Andrews, J.T. and Nichols, H., 1981, Modern pollen deposition and Holocene paleo-temperature reconstructions, central northern Canada: Arctic and Alpine Research, v. 13, p. 387-408.

Brown, R.J.E., 1978, Influence of climate and terrain on ground temperatures in the continuous permafrost zone of Northern Manitoba and Keewatin District, Canada, in Proceedings of the Third International Conference on Permafrost, v. 1: Ottawa, National Research Council of Canada, p. 15-21.

Cropper, J.P. and Fritts, H.C., 1981, Tree-ring width chronologies from the North American Arctic: Arctic and Alpine Research, v. 13, no. 3, p. 245-260.

Dansgaard, W., Johnsen, S.J., Clausen, H.B., and Langway, C.C., Jr., 1970, Ice cores and paleoclimatology, in Radiocarbon Variations and Absolute Chronology (I.U. Olsson, ed.), Stockholm: Almquist and Wiksell, p. 337-351.

Elliott-Fisk, D.L., Andrews, J.T., Short, S.K, and Mode, W.N., 1982, Isopoll maps and an analysis of the distribution of the modern pollen rain, eastern and central northern Canada: Géographie Physique et Quaternaire, v. 36, no. 1-2, p. 91-108.

Faegri, K. and Iversen, J., 1964, Textbook of pollen analysis: Copenhagen, Monksgaard.

Gagnon, R. and Payette, S., 1981, Fluctuations Holocènes de la limite des forêts de mélèze (Rivière aux Feuilles, Nouveau-Québec: Géographie Physique et Quaternaire, v. 35, no. 1, p. 57-72.

Harris, S.A., 1982, Identification of permafrost zones using selected permafrost landforms, in Proceedings of the Fourth Canadian Permafrost Conference (Roger J.E. Brown Memorial Volume) (H.M. French, ed.): Ottawa, National Research Council of Canada, p. 49-58.

Judge, A.S., Taylor, A.E., Burgess, M.M., and Allen, V.S., 1981, Canadian Geothermal Data Collection - Northern Wells, 1978-1980, Geothermal Series No. 12: Ottawa, Geothermal Service of Canada, Earth Physics Branch, E.M.R.

Koerner, R.M. and Fisher, D.A., 1981. Studying climatic change from Canadian high arctic ice cores: Syllogeus (C.R. Harrington, ed.), no. 33, Museum of Natural Science, p. 195-218.

Lachenbruch, A.H. and Marshall, B.V., 1969, Heat flow in the Arctic: Arctic, v. 22, p. 300-311.

Lunardini, V.J., 1978, Theory of n-factors and correlation of data, in Proceedings of the Third International Conference on Permafrost, v. 1: Ottawa, National Research Council of Canada, p. 40-46.

Nichols, H., Kelley, P.M., and Andrews, J.T., 1978, Holocene paleowind evidence from palynology in Baffin Island: Nature, v. 273, p. 140-142.

Nicholson, F.H., 1978, Permafrost distribution and characteristics near Schefferville, Québec: Recent studies, in Proceedings of the Third International Conference on Permafrost, v. 1: Ottawa, National Research Council of Canada, p. 427-433.

Nicholson, F.H., 1979, Permafrost spatial and temporal variations near Schefferville, Nouveau-Québec: Géographie Physique et Quaternaire, v. 33, no. 3-4, p. 265-278.

Osterkamp, T.E. and Payne, M.W., 1981, Estimates of permafrost thickness from well logs in northern Alaska: Cold Regions Science and Technology, v. 5, p. 13-27.

Pilon, J.A., 1982, Etude de la couche active et du pergélisol dans la région de Baie aux Feuilles, Ungava. Unpublished Ph.D. dissertation, Dépt. de Géographie, Université de Montréal.

Poitevin, J. and Gray, J.T., 1982, Distribution du pergélisol dans le bassin de la Grande Rivière de la Baleine, Québec: Naturaliste Canadien, v. 109, p. 445-455.

Richard, P., 1977, Végetation tardiglaciaire du Québec méridional et implications paléoclimatiques: Géographie Physique et Quaternaire, v. 31, no. 1-2, p. 161-176.

Richard, P., 1981, Paleoclimatic significance of the late Pleistocene and Holocene pollen record in south central Quebec, in Quaternary Paleoclimate (W.C. Mahaney, ed.), Norwich, Geo. Book, p. 335-360.

Short, S.K. and Nichols, H., 1977, Holocene pollen diagrams from subarctic Labrador-Ungava; Vegetational history and climatic change. Arctic and Alpine Research, v. 9, p. 265-290.

Taylor, A. and Judge, A., 1979, Permafrost studies in Northern Quebec: Géographie Physique et Quaternaire, v. 33, no. 3-4, p. 245-251.

Watts, R.G., 1980, Climate models and CO_2 induced climatic changes: Climate Change, v. 2, p. 387-408.

Soviet Contributions

This section of the volume contains additional Soviet contributions to the conference. These include invited (p. 163) and contributed (p. 195) papers and translations of abstracts from a special conference volume of papers published in the Soviet Union (p. 315). (Problems of Geocryology, P.I. Melnikov, Editor-in-Chief, Nauka, Moscow, 1983, 280 pp.) The translated abstracts were provided by the National Research Council of Canada. The invited papers were presented in a special session of the conference. The present version of the invited papers was translated by William Barr, University of Saskatchewan. As was the case with the first proceedings volume, the camera-ready copy for the Soviet contributed papers was prepared under the direction of the U.S. Organizing Committee, following revision by English-speaking readers. However, no attempt was made to standardize the transliteration, terminology or quality of the translations. The Soviet papers were accepted for publication on the basis of Soviet review prior to submission to the U.S. Organizing Committee.

Academician P.I. Melnikov (right) and Nikolai Grave, senior representatives from the USSR, examining the polar bears carved in ice for the opening reception of the Fourth International Conference on Permafrost, University of Alaska Fairbanks Campus.

Invited Soviet Papers

MAJOR TRENDS IN THE DEVELOPMENT OF SOVIET PERMAFROST RESEARCH

P. I. Mel'nikov

Permafrost Institute, Siberian Branch
Academy of Sciences, Yakutsk, USSR

The author presents an overview of trends and achievements in Soviet permafrost research in recent years. He stresses both the progress in "pure science" areas and in practical applications of permafrost research. Within the first category would fall a major study (including mapping) of the geothermal heat flux in the permafrost zones; complex studies of energy exchange between soil, vegetation and atmosphere; and predictive studies of the possible impact on the permafrost of the widely publicized global warming trend due to the accumulation of CO_2 in the atmosphere. A wide range of studies with a more practical flavor has also been undertaken. These include a study of *naledi* (icings) in the mountains of Soviet Central Asia with a view to possible water supply; mapping of permafrost and prediction of the impact of development in the BAM zone; recommendations as to environmental protection in the gas fields of northern West Siberia; and studies of the potential use of underground storage chambers in permafrost, to name only a few.

Geocryology, the science of frozen earth materials, studies the composition, structure, conditions of formation and history of development of frozen, freezing and thawing earth materials, and also the processes and phenomena occurring in this zone of the earth's crust. Perennially frozen materials occupy one quarter of the area of the continents, including half of the territories of the USSR and Canada, three quarters of the area of Alaska, considerable parts of China and Mongolia, the northern parts of the Scandinavian countries, and the whole of Greenland and Antarctica.

As a discipline geocryology emerged in the 1920s, i.e. during the Soviet period. Geocryological research in the USSR began to develop intensively in the thirties in connection with the economic development of the northern areas where a range of natural resources is concentrated, including minerals which are rare, very valuable, unique or in very short supply in the country as a whole.

The development of the productive capacity of Siberia was critically complicated by the presence of permafrost. It was necessary to develop basic techniques and to work out special methods for constructing housing and industrial buildings, pipelines, roads for various purposes and airfields, for the rational extraction of minerals, to develop research into groundwater and to find new, effective techniques for exploiting it, and to perfect techniques for improving agricultural lands in terms of both water balance and temperature.

At the present time the Permafrost Institute of the Siberian Section of the Academy of Sciences of the USSR plays a leading role throughout the country in this area of science. Moscow State University, the Institutes of the Ministry of Geology of the USSR, Gosstroy, and of other ministries and organizations have made a significant contribution to the development of this science.

Co-ordination of the geocryological research carried out by academic and departmental research institutions and planning organizations in the USSR is effected by the Scientific Council on Terrestrial Cryology of the Academy of Sciences of the USSR.

In our country geocryological research is planned on a multi-disciplinary basis; it involves the integrated combination of projects of a general, thermophysical and engineering nature. Basic research into either theoretical or regional geocryology embraces several problems.

We study the thermophysical principles of the formation and development of the cryolithozone. Investigations are under way into the relationship between the earth's thermal fields and geological structures, into the evolution of the thermal state of the cryolithozone under natural conditions and under the influence of human activities, and into the laws governing the processes of heat exchange in the atmosphere/vegetation/soil system.

As a result of the research which has been carried out we have identified the conditions controlling surface temperatures and the freezing of the ground under the influence of the climate and of heat exchange with the atmosphere under various landscape conditions and in various environmental zones. The thermal regime of the materials has been defined for the main geostructures of Northern Asia: the West Siberian plate, the Siberian platform and the Verhkoyansk-Chukotka fold mountain area.

For all the permafrost zones in these regions we have compiled the first map of the magnitude of the internal geothermal heat flux. We have identified areas with very low geothermal heat flux (the center of the Siberian platform) and with very high geothermal heat flux (the Verkhoyansk mega-anticlinorium), and also the peculiarities of the thermal regimes of depressed structures within which are deposited the deposits of oil and gas on the Siberian platform. It has been established that disequilibrium permafrost is widely developed; it is anomalously thick and has an anomalous temperature regime, which does not correspond to the present climate and retains features of the cold periods of the past.

Theories of geothermal fields are successfully being worked out. We have created analytical models of the geothermal fields for homogeneous and non-homogeneous environments in areas with a complex relief. We have also developed methods for interpreting geothermal data and the structure from a limited amount of data.

For the first time, either in the USSR or in the world, we have carried out complex regime investigations of the process of energy exchange in the soil/vegetation/atmosphere system in various permafrost and climatic zones. They have made it possible to describe quantitatively the input-output processes of heat and moisture exchange between the atmosphere and the permafrost, and to establish the climatic conditions for the formation of seasonally and perennially frozen ground. We have achieved a number of results, new in principle, as to the physics of heat exchange in the snow pack and the vegetation cover. A new branch of the earth sciences has emerged, namely "thermophysics of landscapes," the basis of which rests on the energy balance techniques which allow one to characterize quantitatively the heat and moisture exchange between the individual components of the landscape.

Important investigations have been carried out into the thermophysics of cryogenic phenomena. Theories have been developed as to the thermal abrasion of coasts which will allow us to solve a number of scientific and practical problems as to the proper exploitation of the natural resources of coastal areas. Basic principles are being formulated for predicting thermal abrasion processes, and on this basis, for the first time in the world, we are working out a universal methodology for predicting the modification of shorelines of reservoirs in the permafrost zone. This research will allow us to solve urgent scientific and practical problems associated with the exploitation of the natural resources of the near-shore zone of arctic seas and artificial water bodies.

As a result of permafrost/geological investigations we have taken cryolithology in a new direction, focussing on the distribution in the permafrost of ground ice, depending on the genesis, accumulation conditions and freezing of the sediments.

The method of permafrost-facies analysis has been introduced into the practical side of geocryological research. Cryogenic textures are classified according to geological-genetic principles and a geological model of how they formed as terrestrial and submarine freezing proceeded has been proposed.

On the basis of permafrost facies analysis of paleontologically dated deposits in type sections in Yakutiya it has been shown that the age of the permafrost in the Aldan basin considerably exceeds 300,000 years and that within the boundaries of its present distribution permafrost has never totally disappeared. There are data to support the existence of stable permafrost in the Kolyma lowland in the early Pleistocene, i.e. approximately 1.5 to 2 million years ago.

The data cited on the age of permafrost represent an important discovery by Soviet scientists. We have to expand the investigations in this direction and give them special emphasis. We have to link the evolution of the permafrost to climatic changes, and for this we must work in collusion with the climatologists.

At the present time a great deal of attention is being paid to global climatic change. A warming of the global climate has been predicted due to the accumulation of CO_2 in the atmosphere. It is hypothesized that by 2100 the mean temperature of the planet will have risen by 2-3° as compared to present. Such a significant warming of the planet must provoke concern among us geocryologists since this undoubtedly would affect the thermophysical conditions of the permafrost materials. We have to predict what changes might occur in the permafrost in the various regions of the country when they occur. Will there simply be an increase in the temperature of the frozen materials or will they start to degrade? The geocryologists in every country must become involved on such predictions.

Major investigations are under way to study the laws governing the formation and peculiarities in the utilization of groundwater in the permafrost zones. The Permafrost Institute has participated in the compilation of Volume 20 in the monograph series *Gidrogeologii SSSR* [Hydrogeology of the USSR] which includes three maps: one on geocryology, one on engineering geology and one on hydrogeology.

A multi-disciplinary study of groundwater in the permafrost zones and the analysis and generalization of a vast amount of hydrogeological data accumulated by various organizations have allowed us to achieve the mapping of the permafrost hydrology of Eastern Siberia. For the first time in Soviet experience we have compiled for this vast region, with an area in excess of 7 million km³, a map of permafrost hydrology regions at a scale of 1:2,500,000 which contains a vast amount of information.

During the past few years at the Institute we have been developing in a rational fashion geophysical methods for studying the dynamics of physico-chemical processes. By using them one can study the regime of changes in the physico-chemical composition of the rocks in the upper layers of the permafrost during the annual cycle. We have established that intense physico-chemical processes are going on not only in the active layer but also at greater depth, down to the zone of zero annual amplitude.

The Institute's geochemists have achieved some interesting results. They have discovered the effect of selective adsorption of organic compounds on the surface of oxides and of ice. It has been established that materials possessing strictly defined potentials for ionization are adsorbed from water solutions on the surface of oxides and ice. A connection has been demonstrated between the electron structures of anions and their capability for becoming integrated into the crystal lattice of ice during the freezing of electrolytic solutions. It has also been established that the rate of migration of salts in frozen sands where no temperature gradient exists is determined by the size of the charge on the surfaces of the mineral skeleton and of the ice.

Geocryological conditions in the alpine regions of Kazakhstan and Central Asia are being systematically studied. It has been established that the area of distribution of permafrost in these regions amounts to 170,000 km², which is approximately six times greater than the total area of glacier cover in the mountains of Central Asia. Permafrost thickness in unconsolidated materials does not exceed 200 m but in bedrock it reaches 360 m in places. The volume of ice included in these frozen materials amounts to 687 km³. In the northern Tien Shan the lowest areas of permafrost appear at heights of 2500-2700 m above sea level and, in special locations, sometimes even as low as 1800-2000 m. In the extreme south of the region, in the southern Pamirs, the lower boundary of permafrost occurrence rises to a height of 3600-3800 m. Studies are under way of the permafrost conditions associated with the formation of glacial mud flows which are a constant threat to many towns in Central Asia and Kazakhstan. A direct connection has been demonstrated between the glacial mud flow activity during the last decade and the retreat of the glaciers which has exposed extremely ice-rich till masses which, on thawing, are potential sources of glacial mud flows. The Institute has recommended methods of stabilizing the unstable moraines with taliks by artificially freezing them.

Methods for mapping rock glaciers have been developed; data have been gathered on their structure and their rate of movement is being determined. The role of rock glaciers in the transport of rock detritus is being assessed. And finally the need to take the movement of rock glaciers into account during construction in alpine areas has been demonstrated.

A major object of permafrost investigations is *naledi* (icings); considerable volumes of water accumulate in icings in the Tien Shan and the Pamir during the cold season. The volume of ice in the largest *naledi* in the alpine areas of the Tien Shan and Pamirs in some years reaches many hundreds of thousands and even millions of cubic meters. Artificially increasing the volume of these *naledi* could be of great practical interest since induced accelerated melting of these same icings in summer could increase the discharge of the mountain rivers during the growing season when there is a particularly acute need for irrigation water in the foothills area.

Investigations connected with the protection of the environment are being intensified. The first map of the Yakutsk ASSR has been compiled to show degrees of sensitivity of the surface to human activities. Our Institute participates in the research into and construction of almost all major projects in Siberia, the Far East and the Far North and contributes to the development of recommendations concerning rational use of the environment and to the creation of progressive techniques of construction on permafrost.

At the present time we have established the permafrost conditions obtaining at all the diamond pipes which are being exploited, at all the tin deposits in the Northeast of the USSR, at the coal deposits of Southern Yakutiya and at all bedrock occurrences of gold and non-ferrous metals.

Permafrost investigations extending over a number of years in the area of the Baykal-Amur Mainline have been completed. We have compiled and published geocryological maps at a scale of 1:5,000,000 and a monograph describing this region; the latter reflects current ideas on the main permafrost characteristics of the northern Amur oblast and of Khabarovsk kray, and of the Baykal and southern Yakutiya districts. It has been proven that over a very wide area permafrost is surviving thanks to the specific nature of the surface cover, rather than due to climate. Hence when these areas are developed and the surficial insulating cover is destroyed degradation of the permafrost, with all its negative consequences, is inevitable. The Permafrost Institute has published predictions as to changes in permafrost conditions associated with the development of the eastern part of the BAM zone. On the basis of these data a permafrost-seismic map was compiled by the Institute of the Earth's Crust of the Siberian Branch of the Academy of Sciences of the USSR; the data were also widely used by planning organizations in compiling technical and working plans for the route of the BAM.

On the basis of fundamental investigations in the area of engineering geocryology, major applied tasks have been solved with regard to controlling the temperature regime of foundation materials, the improvement of agricultural lands and the restoration of disturbed landscapes. The effectiveness of using a styrofoam insulating cover to reduce the depth of thaw in frozen materials has been determined. Completion of these investigations has allowed us to begin developing an extremely promising method for controlling the depth of thaw and the ground temperature. This will open the way for more effectively increasing the bearing capacity of structural foundations and will considerably increase the operating safety of foundations.

On the basis of long-term multidisciplinary investigations in the North of Western Siberia we have developed recommendations as to the protection of landscapes from disturbance and we have proposed methods for selecting building sites and methods for restoring areas disturbed during the laying of northern gas pipelines. We have compiled an "Interim manual for the protection of landscapes during the laying of gas pipelines in the Far North." This is essentially the first attempt at a multidisciplinary approach to environmental protection in association with land use in the Far North.

We have carried out a large number of applied investigations directed at solving the Food Program. A series of projects have been completed, aimed at optimizing methods of both channel and sprinkler irrigation of hay lands and areas of fodder production in Central Yakutiya. Methods of irrigation regimes have been worked out and recommendations have been made as to increasing soil fertility.

We have been pursuing research into surface deformations provoked by the thawing of ground ice as a result of the breaking-in of new arable land. In Central Yakutiya 20-30% of newly broken arable fields become uncultivable within 2-3 years due to the formation of the thermokarst trenches up to 1.5 to 2 m deep along the line of ice wedges. Recommendations have been made as to the selection of areas suitable for cultivation.

Much attention has been paid to the problems associated with using groundwater as a source of water supply. New principles for estimating reserves of groundwater and new methods for establishing water-supply installations have been proposed. Methods of protecting operating water wells from freezing have been put into practice, to great effect.

Along with *Yakutniproalmag* (Yakutsk Diamond Prospecting Research Institute) the Permafrost Institute has developed a new, progressive type of foundation for permafrost areas: reinforced concrete piles equipped with built-in refrigeration devices using a naturally circulating coolant. They are being successfully used in building projects and have led to a considerable reduction in labor costs, in the cost of building materials, and the overall weight of structures and have increased the operating safety of the structures.

Drilling-and-filling piles are ready to be introduced. Our studies have shown that their bearing capacity in permafrost materials is twice that of ordinary prefabricated reinforced concrete piles. Experimental studies into increasing the efficiency of refrigeration devices and the freezing of foundation materials by means of an insulating cover on the surface have also been successful. It is recommended that one use seasonally operating refrigeration devices in building cheap, relatively easily constructed earth dams up to 20 m high, with a frozen, impermeable core, for impounding reservoirs for drinking water and for the purposes of irrigated agriculture. The effectiveness and reliability of such structures has been demonstrated by those built for the diamond-mining industry.

Long-term thermophysical investigations by the Institute on the dam at the Vilyuy hydropower development have established basic laws governing heat and mass transfer in the coarse rock material of the flanks of the dam and in the fissured bedrock of the dam's foundation; the intensity of water percolation was also determined. The valuable data accumulated have been used in planning hydroelectric schemes in other permafrost regions.

In view of the accelerating pace of development in the northern regions the problem of using permafrost materials as an environment for engineering structures is of particular interest. This has arisen from demands for rational use of resources and from economic expediency. One means of reducing the negative impacts of human activities on the natural environment is the effective use of underground chambers. The Permafrost Institute has developed a new technology for creating underground reservoirs in frozen sediments. This technology is safer, simpler to use and considerably more effective as compared to existing methods. The underground storage chambers in frozen sediments are designed for storing oil products, water, and for other purposes. In the agricultural areas of Yakutiya they have been effectively used for refrigerating milk in summer and for farm water supply.

I should like to emphasize the need for geocryological studies in the following directions: 1) continuation and intensification of studies into geocryological processes in the permafrost areas; 2) expansion of research into the complex impact of natural factors on the freezing and thawing of the ground and the creation of generalized models of cryological processes; 3) development of research into the theory of geocryological prediction, since it is very important to know of changes in permafrost conditions provoked by human activities and to develop reliable methods of geocryological forecasting in order to exclude the possibility of undesirable consequences; 4) expansion and intensification of the scientific fundamentals in terms of methods to protect the environment in permafrost areas, to which end we must establish reliable criteria for assessing the sensitivity of the land surface to anthropogenic activities and make recommendations as to the rational use of natural resources; 5) focussing of attention on geocryological research on the arctic continental shelf in connection with the intensification of prospecting work in the search for minerals and with their impending exploitation; 6) continuation of research dealing with the construction of oil and gas pipelines in permafrost areas and the development of models of the optimal conditions for the laying and operation of pipelines with a view to ensuring safe operation and the maximum economy in terms of construction; 7) improvement of foundation construction to achieve the most effective foundations in terms of both performance and cost.

Soviet geocryologists are directing their efforts to the further elaboration of basic problems in permafrost research and to the solution of practical problems which are of primary significance in the development of the northern areas of our country. Soviet geocryology has achieved major results in fundamental and applied research which have permitted us to assume a leading position in the world in this area of scientific knowledge.

DRILLING AND OPERATION OF GAS WELLS IN THE PRESENCE OF NATURAL GAS HYDRATES

A. I. Gritsenko and Yu. F. Makogon

All-Union Scientific Research Institute of Natural Gas
Moscow, USSR

Natural gas hydrates have been found to occur widely where reservoirs of hydrocarbons occur in permafrost areas. Exploitation of hydrocarbons lying beneath the gas hydrate formations is likely to encounter serious problems. Specifically the heat transmitted from the hot drilling mud can result in the decomposition of the gas hydrates adjacent to the drill hole; this results in the release of large volumes of gas which can lead to a violent "gas kick" and the ejection of the drilling mud from the hole. The author demonstrates how this problem may be overcome by increasing the density of the drilling mud, reduction of its temperature, or by a combination of these techniques.

Reserves of natural gas in a solid hydrate state located on land exceed 1×10^{14} m^3, while those beneath the ocean amount to about 1.5×10^{16} m^3 (Makogon, 1964, 1981; Trofimik et al., 1981; Judge, 1981). Exploitation of the reserves located on land requires the solution of some radically new technical problems. These problems arise from the unique properties of gas hydrates which saturate the pore spaces of gas bearing strata located in association with permafrost in the sedimentary cover of the earth's crust.

The most important of the properties of gas hydrates in terms of the peculiarities of exploiting the deposits of hydrocarbons in areas where the gas hydrates occur include: an abrupt change in the specific volume of water and gas during phase changes; the extremely low permeability of gas hydrates and of porous rocks saturated with hydrates; the low thermal conductivity of the hydrates; the significant lowering of the strength of rocks containing hydrates as they decompose; a sharp increase in pore pressure during their composition within a closed system; and an increase in capillary pressure as the hydrates accumulate in porous rocks. All these properties of gas hydrates dictate to a large extent the technology of drilling and operating wells in strata where gas hydrates are concentrated.

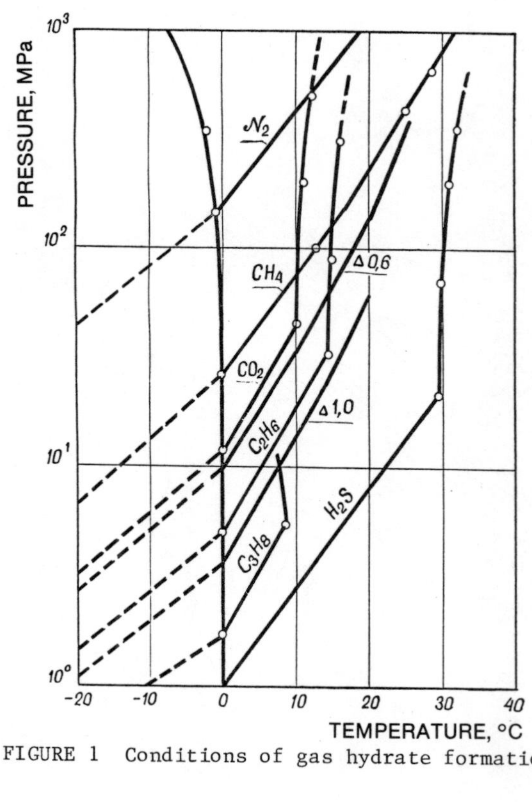

FIGURE 1 Conditions of gas hydrate formation.

Brief description of the hydrate formation zone and of gas hydrate deposits

Theoretical calculations, geophysical prospecting, and drilling have revealed the widespread distribution of gas hydrate formations and deposits in frozen sedimentary rocks in the earth's crust. The depth of occurrence of the hydrate formation zone is determined by hydrostatic pressure, rock temperature, composition of the gas and mineralization of the water in the reservoir. As regards conditions on land the hydrate formation zone extends from depths of a few tens of meters to 1-2 km. Figure 1 shows the equilibrium curve of hydrate formation for some common gases; Figure 2 presents a graph of the distribution of the hydrate formation zone for methane, CO_2, H_2S and natural gas with a specific gravity of 0.6 depending on the geothermal gradient. Figure 3 is a map of the distribution of the hydrate formation zone within the USSR (Barkan and Voronov, 1982).

Beneath the ocean the hydrate formation zone may reach thicknesses of several hundred meters, depending on latitude, ocean depth, bottom water temperature, the geothermal gradient in the sedimentary cover, hydrostatic pressure, composition of the gas and mineralization of the water, and in a number of areas exceeds 2 km. Figure 4 presents an approximate profile of the hydrate formation zone for conditions beneath the Beaufort Sea, compiled

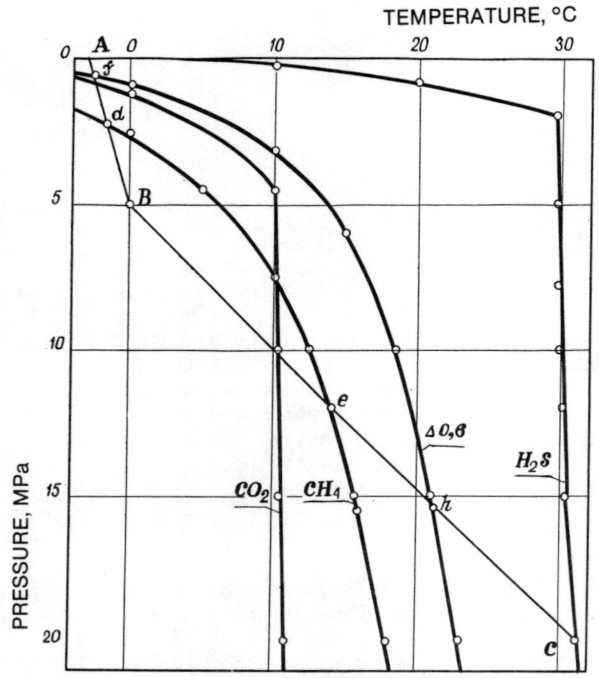

FIGURE 2 Depth intervals of the hydrate formation zones for some gases in the sedimentary cover of the earth's crust.

from data provided by Dome Petroleum (Weaver and Stewart, 1981).

Gas hydrate deposits, whose distribution may be local in nature where secondary gas hydrate deposits have formed from free gas deposits during cyclical changes in rock temperatures, are found in the hydrate formation zone under favorable conditions. Such deposits are usually found beneath permafrost areas of the continents or beneath young seas where conditions do not permit the temperature of the rocks to rise above that of hydrate decomposition, which would result in free gas formation.

Initial formation of gas hydrates beneath the ocean usually coincides with the upper horizons of the sedimentary cover. The upper boundary of such hydrate deposits may lie at a depth of only a few tens of centimeters and the lower boundary at depths of several hundred meters. In some situations gas hydrates accumulate in pagoda-shaped masses (Emery, 1974; Makogon, 1982).

Gas hydrate deposits contain vast reserves of gas in a solid hydrate form. The problems of exploiting these gas reserves are not discussed in the present paper; this is a separate, major topic.

At considerable depths beneath the gas hydrate there may be reserves of hydrocarbons at high temperatures, namely oil, gas and gas condensates. Development of these deep, high temperature hydrocarbons necessitates the drilling and operation of wells which extend through the hydrate zone. Well drilling and operation in the zone of hydrate-saturated rocks involves a number of specific problems which originate with the properties of the hydrates during the phase transition to water and gas.

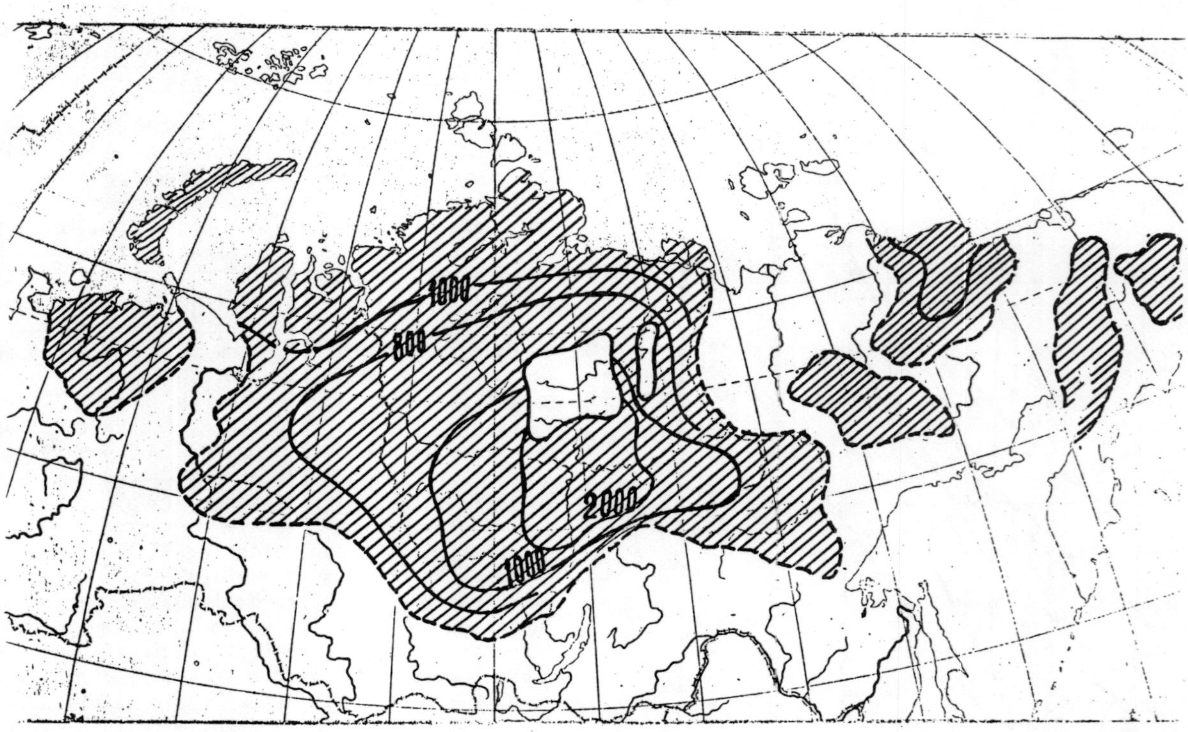

FIGURE 3 Schematic map of possible hydrate formation ($H \geq 400$ m) in the USSR.

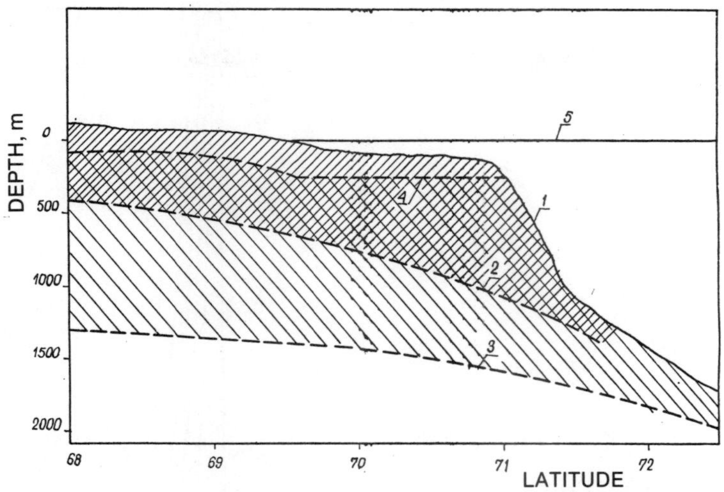

FIGURE 4 Approximate profile of methane hydrate formation along longitude 134-132°W.

The hydrates themselves act as a cementing agent in granular rocks which they saturate. With decomposition of the hydrates in some cases the rocks are converted into high pressure unstable zones.

When the phase change from water to hydrate or vice-versa occurs the specific volume varies by 26-32%; by contrast, of course the water-ice (or vice-versa) phase change involves a change in specific volume of only 9%.

The specific volume associated with the gas-hydrate (or vice versa) phase change varies by a hundred times. In this situation, if the process of hydrate decomposition occurs in a confined space the pressure may increase by several orders of magnitude. For example when methane hydrate forms at 0°C, and then decomposes due to heating to a temperature of 100°C within a closed cell, the methane pressure increases from 2.6 MPa to 2500 MPa.

Well Drilling Problems

Strata saturated with hydrates are characterized by extremely low permeability which prevents loss of fluids from the drilling muds by seepage into the rock and the formation of a protective coating of clay on the walls of the drill hole. The absence of a protective film on the walls of the drill hole where one is drilling with high temperature muds leads to intense heating of the adjacent rocks containing the hydrates and to decomposition of the hydrates. The latter phenomenon is accompanied by a marked weakening of the intergranular bonds and the resultant disintegration of the rocks adjacent to the drillhole. Associated with this extensive cavity development occurs; rock slippages may lead to bits jamming. Discharge of gases under high pressure during decomposition of the hydrates may lead to intense gas kicks and to the ejection of the mud from the hole.

Where mud temperatures, which in turn are affected by the hydrostatic pressure in the mud column, are higher than the temperature of hydrate decomposition, intense decomposition will occur, accompanied by the release of large volumes of gas.

Figure 5 presents curves which show the relationship between the volume of various gases contained in a unit volume of hydrates under equilibrium conditions. For comparison it also shows curves of the specific content of gas in a free state under identical temperature and pressure conditions. Comparison of the curves presented reveals that at hydrostatic pressures within the gas hydrates of up to 13.0 MPa (i.e. at depths of about 1300 m) the volumes of methane and CO_2 released due to gas hydrate decomposition significantly exceed the volumes of gas which could be held in a free state under similar conditions. For heavy gases which form structure II hydrates the comparable depth is about 550 m. For example in rocks at a depth of 750 m and a temperature of ≧10°C, 1 m³ of pore space filled with methane hydrate will yield 157 nm³ of gas, while 1 m³ of pore space filled with free methane will yield 85 nm³.

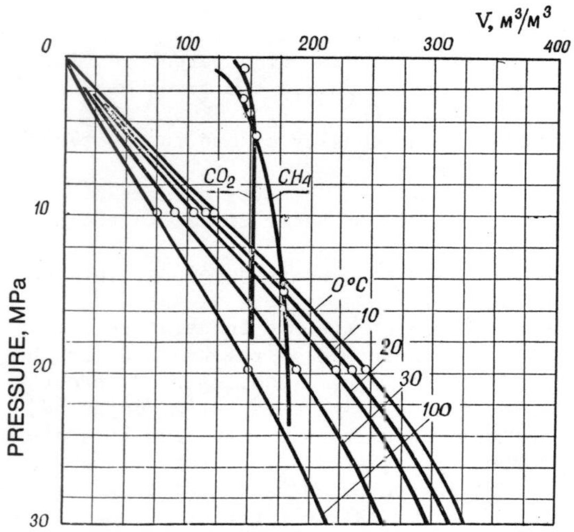

FIGURE 5 Gas content in a unit volume vs. pressure and temperature in a free and hydrate state for CH_4 and CO_2.

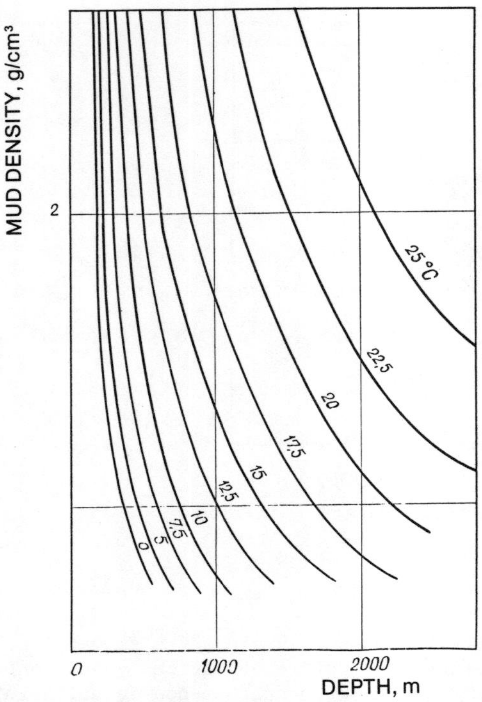

FIGURE 6 Mud density required to prevent methane hydrate decomposition vs. temperature and depth of gas-hydrate reserve.

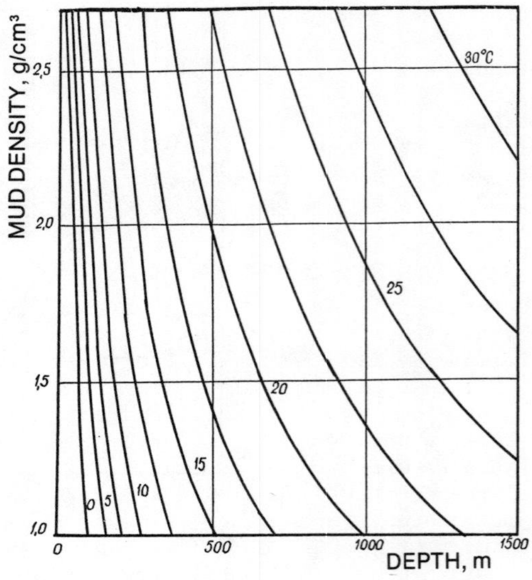

FIGURE 7 Minimum density of drilling mud to prevent decomposition of natural gas (spec. gravity 0.6) hydrate as a function of mud temperature and depth of hydrate deposit.

The depth at which the volumes of gas in the free and hydrate states contained in a unit volume of rock are equal is determined by the equation:

$$H = \frac{22.4\rho_H P_o T_z}{T_o \gamma (Mg + 18n)} \cdot M$$

where: ρ_H = hydrate density in g/cm³
P_o = atmospheric pressure in MPa
T = temperature at depths H and K
z = coefficient of supercompressibility of the gas in the reservoir conditions
T_o = 273.16K
γ = mean density of mud in reservoirs in g/cm³
Mg = molecular density of hydrate gas
n = ratio of water to gas in the hydrates.

The intensity of hydrate decomposition and blowouts of the drilling muds are dictated mainly by the amount by which the mud temperature exceeds the equilibrium temperature of the hydrates in the rocks. Released gases are discharged into the drillhole, the drilling muds become degassed, lowering their density and resulting in a violent mud blowout from the hole. In this situation the pressure of the released gas is dictated by the temperature at which the hydrates decompose rather than by hydrostatic pressure.

However a mud blowout is accompanied by an abrupt intensification in the process of hydrate decomposition which leads to significant cooling of the rock in the area where the decomposition occurs and to a dampening of the decomposition process (the process of hydrate decomposition is accompanied by heat absorption to an amount exceeding 400kJ per kg of the hydrate).

Intense cavity formation, jamming of drill bits, degassing and blowouts of the mud as a result of hydrate decomposition may be avoided by increasing the density of the drilling mud, lowering its temperature below the equilibrium temperature by a combination of these measures.

The appropriate density of the drilling mud P_γ which will eliminate the problem of hydrate decomposition in strata which are being drilled may be determined by the equation:

$$P_\gamma = \frac{10^2 P_{pt_\gamma}}{H} \text{ g/cm}^3$$

where P_{pt_γ} is the equilibrium pressure corresponding to mud temperature t_γ
H is the depth of the gas hydrate deposit in meters
P_p may be determined from the equation:

$\log P_p = 1.415 + 0.0417(t_\gamma + 0.01\, t_\gamma^2)$ in the case of methane

$\log P_p = 1 + 0.497\ (t + 0.005 t_\gamma^2)$ in the case of natural gas with a specific gravity of 0.6.

Figures 6 and 7 show the curves of drilling mud densities necessary to ensure hydrostatic pressures which will exceed the decomposition pressures of hydrates of methane and of natural gas with a specific gravity of 0.65, depending on the depth and on the mud temperature.

By applying Figures 6 and 7 to the examples of drilling into methane hydrate under the following conditions:

depth of gas hydrate deposit (N) = 500 and 1200 m;
temperature of gas hydrate = 2° and 18°C;
equilibrium temperature t_p = 6.5° and 19°C;
hydrostatic pressure in the hydrate P_h = 5 and 12 MPa;
and mud temperature t_γ = 0°, 5°, 10°, 15° and 20°C;

we can determine the density of the drilling mud necessary to prevent hydrate decomposition.

In the case of the upper hydrate deposit where t_γ = 0°, 5°, 10°, 15° and 20°C equilibrium pressures will be 2.6, 4.5, 7.5, 13.0 and 23 MPa respectively. The mud density necessary to prevent decomposition of the hydrates will be 0.52, 0.9, 1.5, 2.6 and 4.6 g/cm³ respectively. As can be seen from the values just quoted if the mud temperature is maintained between 0° and 5°C this will totally eliminate hydrate decomposition in the reservoir and will ensure safe drilling conditions. If the mud temperature is maintained between 10° and 15°C it is necessary to increase its density to 1.5 to 2.6 g/cm³; and at a mud temperature of 20°C it is effectively impossible to prevent hydrate decomposition by increasing the mud density (a mud density of at least 4.6 g/cm³ would be necessary). One can ensure normal, trouble-free drilling in this latter situation by using a combination of techniques, namely by lowering the mud temperature to 12.5°C and by increasing its density to 2 g/cm³.

In addition I would recommend a closed system of drilling whereby high wellhead pressures are maintained. For example when P_γ = 2 g/cm³ and t_γ = 20°C it is necessary to maintain a wellhead pressure of 36 MPa, although this will greatly complicate the drilling operation.

In the case of a gas hydrate deposit lying at a depth of 1200 m and with mud temperatures of 0°, 5°, 10°, 15° and 20°C the equilibrium pressure will be similar to those of the upper part of the deposit (with identical gas composition) i.e. 2.6, 4.5, 7.5, 13.0 and 23 MPa. The mud pressure necessary to prevent hydrate decomposition will be 0.22, 0.38, 0.63, 1.1 and 1.92 g/cm³ respectively.

Where one is estimating permissible temperatures and the requisite densities of drilling mud for drilling in gas hydrates one has to know the composition of the hydrate-forming gas and the equilibrium values for pressure and temperature. In each specific case, on the basis of technological limitations and economic considerations, one can then recommend certain values for the density and temperature of the drilling mud.

Problems Arising During Well Operation

During the exploitation of high-temperature oil or gas deposits located beneath a gas hydrate deposit the rocks surrounding drill holes heat up. If the temperature in the hydrate deposit rises above the equilibrium temperature, decomposition of the hydrate occurs, whereby the hydrate changes phase to free gas and the ice into water. If the hydrate deposits are confined top and bottom between impermeable strata and the rocks are highly saturated with hydrates, when the rocks near the drillhole heat up, the decomposition will occur in a closed cell. It is well known that hydrate decomposition is accompanied by an abrupt increase in pressure (Makogon, 1964).

Under certain conditions the gas pressure provoked by hydrate decomposition in a closed cavern around a drillhole may considerably exceed the rock pressure and may result in the collapse of the drillhole with all the resulting consequences.

Increases in pressure and temperature follow the equilibrium curve to the point where the hydrates are either totally decomposed or where the stress produced in the surrounding rock and in the sides of the drillhole reaches a critical value and failure occurs.

Figure 8 presents curves of the relationship between pressure and gas temperatures associated with decomposition of methane hydrate in a confined space, with various levels of dynamic collapse of that space in which the hydrate decomposition is occurring.

Prevention of the collapse of drillholes being operated through hydrate-saturated rocks may be achieved by reinforcing the well, by using casing with a high thermal resistivity through the hydrate deposit which would eliminate the heating of the adjacent rocks above the equilibrium temperature of the hydrate, by using either active or passive insulation techniques. In some cases the necessary effect may be achieved by lowering the temperature of the mud at the well bottom.

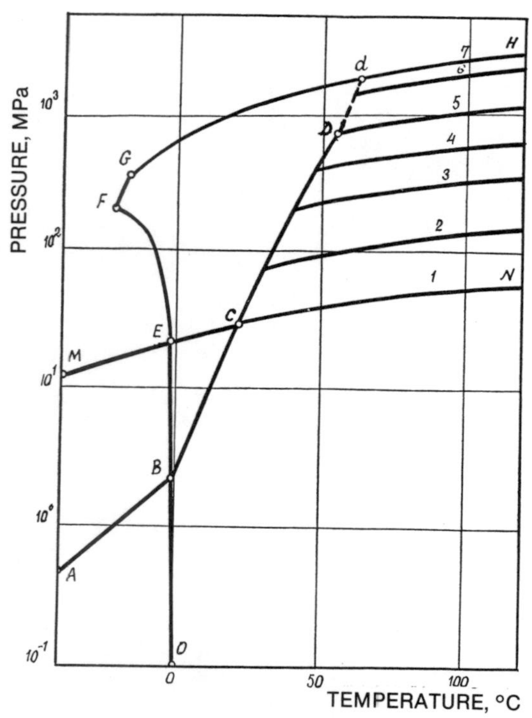

FIGURE 8 Pressure vs. gas temperature during methane hydrate decomposition in a limited volume.

REFERENCES

Barkan, E.S. and Voronov, A.N., 1982, Zones of possible hydrate formation on U.S.S.R. territory: Sovetskaia geologiia, no. 7, p. 37-41.

Emery, K.O., 1974, Pagoda structure in marine sediments, in Marine Science, v. 3: N.Y. Plenum Press.

Judge, A., 1981, Natural gas hydrates in Canada, in Proceedings of the 4th Canadian Permafrost Conference, Calgary: Ottawa, Ont., National Research Council of Canada, 1982, p. 320-328.

Makogon, Yu.F., 1964, Hydrates of natural gases: Moscow, Nedra; (English transl.) L.C. no. M880.M3213: Tulsa, Okla., Penn Well Publishing Co., 237 p.

Makogon, Yu.F., 1981, Perspectives for the development of gas hydrate deposits, in Proceedings of the 4th Canadian Permafrost Conference, Calgary: Ottawa, Ont., National Research Council of Canada, 1982, p. 299-304.

Makogon, Yu.F., 1982, Drilling of wells opening gas hydrate layers: Gazovaia promyshlennost', no. 4, p. 30-32.

Trofimuk, A.A., Makogon, Yu.F. and Tolkachev, M.V., 1981, Gas hydrate deposits - a new source of energy: Geologiia nefti i gaza, no. 10, p. 15-21.

Trofimuk, A.A., Makogon, Yu.F. and Tolkachev, M.V., 1983, Role of gas hydrates in processes of hydrocarbon accumulation and formation of their deposits: Geologiia i geofizika, no. 5, p. 3-15.

Weaver, J.S. and Stewart, J.M., 1981, In situ hydrates under the Beaufort Sea Shelf, in Proceedings of the 4th Canadian Permafrost Conference, Calgary: Ottawa, Ont., National Research Council of Canada, 1982, p. 312-319.

THE IMPACT OF INSTALLATIONS FOR THE EXTRACTION AND TRANSPORT OF GAS ON PERMAFROST CONDITIONS IN WESTERN SIBERIA AND COMPREHENSIVE ENGINEERING GEOLOGY FORECASTING OF RESULTANT CHANGES

Yu. F. Zakharov, L. D. Kosukhin, and Ye. M. Nanivskiy

Ministry of the Gas Industry
Tyumen', USSR

The bulk of the installations associated with the extraction and transport of natural gas represent major, constant sources of heat emission. The gas fields now being developed in northern West Siberia are in areas of ice-rich sediments such that settlement on thawing would greatly exceed acceptable limits. Hence the need for prediction of the possible interaction between a variety of structures built using various techniques and the underlying frozen sediments is paramount. The article outlines the steps in developing such a forecasting model with a view to selecting the optimum design for gas field structures and to developing an effective program of monitoring of their performance.

The development of the gas fields of Western Siberia has been accompanied by a wide range of intensive mechanical, hydrostatic, thermal, chemical, electrical and other forms of operations on frozen, thawing and freezing materials. The types of operations listed lead to severe and irreversible man-induced changes in the geobotanical environment, some of which may be anticipated in planning decisions, but many of which have turned out to be unforeseeable. Evaluation of changes in the geological environment, and especially in permafrost conditions, represents the basis for selecting design options in terms of engineering preparation of a site, minimizing temperature oscillations, and the choice of many aspects of design and technological aspects of buildings, installations and communications, and even in terms of plans for exploitation of the gas deposits.

An essential element in the evaluation of changes in the permafrost environment is long-term quantitative prediction of such changes. Interactions between the producing installations associated with gas extraction and distribution and the geological environment occur at many levels; moreover they may occur only once, or may be ongoing. As a result it is extremely difficult to isolate the results of each impact, its component parts, and the geological processes which it provokes. It is equally difficult to make specific forecasts, either general or local, of these impacts and processes, taken separately. But without such specific forecasts, general or local, it is impossible to achieve complex predictions of the consequences of the aggregate of all the interactions between the gas industry and the geological environment during the design stage for gas field operations.

The overwhelming majority of structures associated with gas extraction and transport represent major and constant sources of heat emission, and hence critically alter the thermal balance of the geological environment, leading to significant thawing of the permafrost materials and to loss of bearing strength.

Throughout the gas fields now being developed in Western Siberia the ice content of the permafrost materials is such that settlement on thawing would considerably exceed acceptable limits in the case of the overwhelming majority of the gas production installations. This has led to widespread use of preliminary thawing of the frozen sediments, and ubiquitous use of pile foundations and gravel pads during construction on the gas fields. Thus construction and operation of the installations associated with gas extraction and transport simultaneously lead to fundamental changes in the hydrogeological, geodynamic and geochemical conditions, the parameters of which depend entirely on the geological and geocryological peculiarities of the gas-bearing region.

Present designs of installations for extraction and transport of gas on frozen, thawing and freezing sediments which take into account only their thermal and mechanical interactions are inadequate. It is possible to increase the reliability of gas producing and gas transport systems by identifying, predicting and taking into account at the planning stages changes in all elements of the geological conditions, without exception. The omission or underestimation of any change, even in an apparently secondary element, may lead to the prediction being less reliable and to its being ignored during the design process. Hence the design for the construction and operation of gas fields in the permafrost zone must rest on the basis of comprehensive engineering and geological forecasts. Most importantly the gas field design must be seen as a single dynamic system, which we may refer to as a geotechnical system. Such a system possesses a whole range of specific and latent properties which effectively combine to reflect the entire aggregate of interactions (on various timescales and at various levels) between the geological environment and the production facilities of the gas industry during the entire designed lifetime of the gas field. For the purpose of comprehensive engineering-geological predictions the functioning of this geotechnical system must be examined as an aggregate of various ongoing changes in the geological environment along with, in particular detail, all the elements of the original engineering-geological conditions existing at the time when the design for the operation of the gas field was first formulated. During the process of this examination it is essential to obtain and analyze various types of information and, on the basis of an acceptable scheme for extraction, and of acceptable recoverable reserves of gas and gas condensates, to carry out the necessary predictive calculations of the following parameters: 1) the drop in reservoir pressures and dehydration of the strata being exploited; 2) extraction of mineral particles from the producing strata along with the gas, condensates and water; 3) compaction of the producing strata; 4) warping of the overlying strata; 5) surface subsidence; 6) raising of the groundwater table due to surface subsidence and development of the field; 7) accumulation of surface drainage in depressions due to surface subsidence; 8) changes in the temperature, nature and properties of the surface sediments due to waterlogging and changes in other engineering-geological processes; 9) changes in any other elements in the engineering-geology conditions due to the planned construction and to disruptive human activities in terms of the specific planned installations for gas extraction and transport.

The complexity of such forecasts is further increased by the need to take into consideration scientific and technical advances in the gas industry which are not yet common property and have not been included in operational designs but have already proved their effectiveness in industrial experiments, and hence can and must be brought into operational use during the designed lifetime of the gas field. In addition to those parameters derived by the process of traditional engineering exploration, the attainment of such long-term, comprehensive quantitative engineering geology forecasts demands the identification and evaluation of a whole range of parameters concerning geological conditions: 1) the coefficient of linear temperature-related deformation of mineral and organic soils and peat; 2) maximum sustained strength of the frozen materials; 3) coefficients of thermal conductivity and apparent electrical resistivity of frozen and thawed materials; 4) specific electrical conductivity and dielectrical penetrability of frozen and thawed materials; 5) thixotropic and rheological properties of thawed materials; 6) the phase composition of H_2O in the materials at various temperatures; 7) composition of organic materials, soluble salts and exchangeable cations in thawed materials; 8) content of ferrous oxides, aluminum oxides, oxides and protoxides of iron, amorphous acids and carbonates in thawing sediments; 9) capillary vacuum, capillary moisture capacity, water loss and the coefficient of filtration anisotropy in thawing and freezing materials; 10) flow lines in bog water, and other parameters.

To provide comprehensive engineering geology predictions using both traditional sources and these additional sources of information it is essential to carry out comprehensive long-term geological investigations at many levels. The focus, content, time frame and scale of these investigations can be determined by using a flow chart of objectives, closely related to a geodynamic model of the functioning of a predictive geotechnical system. The methodological basis for predicting changes in conditions affecting the engineering geology is the comparative geological approach based on analysis of formations, facies and paleogeographical and paleo-geomorphological conditions and on other methods of studying the geological environment, and based on constant and widespread use of spatial and time analogs. In studying the elements of the engineering-geology conditions and the technology of the gas industry a qualitative and quantitative evaluation is made of the parameters of both aspects with a view to determining the limits of possible acceptable and optimal changes in both. Special attention is paid to studying the mobility and stability of links between elements and to the identification of the roles of individual factors affecting those elements. The study proceeds by identifying the character and form of the functional relationship between the obligatory structures of geotechnical functions. At the same time studies also focus on the capacity for resisting the altered engineering geology conditions and restoration of the equilibrium, upset as a result of human activities. One of the tasks of planning investigations is to establish the symptoms for the changes occurring in the engineering geology conditions and any portents of critical parameters in order to provide a monitoring service to engineering geologists which will give accurate and reliable signals and diagnostic signs of the onset of any developing defects or failures in installations associated with gas extraction and distribution. All information on engineering geology conditions and on changes in those conditions, correlated with potential planning decisions *re* development and construction of a gas field, is examined from the point of view of systems and factor analysis during the process of comprehensive engineering geology predictions.

The algorithm of such predictions may be divided into seven steps, each of which may in turn be divided into a number of stages. Comprehensive engineering geology predictions begin with the following: 1) a scientific hypothesis as to possible changes in engineering geology conditions and the consequences associated with the implementation of specific activities during the development of gas resources in a specific area; 2) the creation of a working hypothesis, a plan for the operation of the designed geotechnical system, and a sequence of geological processes in association with implementation of potential design decisions.

The second step involves the process of studying the structure and clarifying the limits of the geotechnical system being forecast and the testing of the working hypothesis as to how it functions. This stage ends with the translation of the hypotheses into a conceptual, graphic information model of the system and the execution of a qualitative comprehensive forecast.

On the basis of such prediction, a program of investigations, experiments and of practical and analog modelling is implemented; a geodynamic cause-and-effect model of the functioning of the geotechnical system with all the anticipated variants is built and a corresponding computer program is created. The geodynamic model represents a hierarchical system of models: imitative and optimizational, balanced and dynamic, graphic and matrix-type, reflecting the structure of the system being modelled and its interrelations with other systems, and describing the interaction of installations for the extraction and distribution of gas with frozen, thawing and freezing sediments and all the other components of the geological environment. The analytical expression of such a system of models is synthesized by a number of determined, stochastic, euristic and dialogic algorithms and is aimed at achieving a predictive evaluation of the development and operation of the gas field.

The geocryological components of such a model include the possible, anticipated and inevitable quantitative changes in temperature, ice content of the sediments, depth of seasonal freezing and thawing, and the cryogenic processes (thermokarst, frost-heave, frost cracking, thermal erosion, solifluction, *naled* (icing) formation, etc.) which develop in association with well drilling or with the erection and operation of buildings, installations and communications, in association with repair and maintenance work, or in association with various forms of economic or non-productive human activities in the cryolithozone of Western Siberia. The incompleteness of our knowledge as to the nature of complex thermophysical processes and of heat and mass exchange in sediments in association with their interaction with heat-emitting installations does not allow us to formulate adequately the dynamics or the results of cryogenic processes in mathematical expressions. Analytical relationships used at present describe these processes only very sketchily and require a large number of assumptions and considerable generalization of the information being predicted. Hence it is impossible to make geocryological predictions without field studies of the dynamics of heat and moisture exchange and the changes they provoke in the sediments, in order to obtain many numerical parameters in predicting the interaction of specific planned installations for the extraction and distribution of gas with frozen, thawing and freezing sediments. Owing to the exceptionally high mobility of the elements of the geocryological component in the model, evaluation of the magnitude of the changes occurring when one builds and operates gas field installations is a labor-consuming task which can be handled only by means of a prolonged program of regular observations at various levels and by large-scale field experiments. These investigations are carried out in both artificial and natural sediments and also in materials lying beneath covers, installations, gravel pads, soil or vegetation covers and under peat accumulations. A study of the dynamics of seasonal freeze-thaw in sediments and of the peculiarities of the details of their temperature regime is carried out under natural conditions and where the conditions have been disturbed by man, and also during the course of the associated cryogenic processes as they occur, namely frost heave, solifluction, thermokarst, frost cracking, *naled* (icing) formation, thermal erosion, etc. In the process evaluations are made of the impact on the nature and magnitude of such processes of soil, vegetation and artificial coverings and of engineering operations associated with planned installations for the extraction and distribution of gas. At the same time detailed observations must be made of the heat balance of the atmosphere/soil system in different landscapes, of the formation, distribution and melting of the snow cover, and of a wide spectrum of climatic parameters in order to identify the extremes and annual mean values.

The operations listed above comprise the third and fourth steps in comprehensive forecasting. At the same time one has to undertake an analysis of the current state and trends in the development of the gas industry and of the associated branches of construction and of the economy. In the case of long-range forecasting it is essential to take into consideration social/political factors which are not conducive to easy prediction but which have an impact both on the entire process of operating the geotechnical system and on the development of its individual elements. In the first instance it is necessary to predict specific policies of state limitations on utilization of the environment, on allocation of resources, and on rigorous application of the norms for environmental protection. Such predictions of the development of the technological environment represent the fifth step in comprehensive engineering-geology forecasting.

The penultimate step involves morphological analysis of the data, utilization of techniques of "matrix interaction" (which the authors of the method, Gordon and Helmer (1969) consider to be "a method of analysis which allows systematic investigation of the results of the potential interaction of elements in an aggregate of events"), organization of the data in a form suitable for computer and graphic handling, analysis of the comprehensive qualitative forecast and a synthesis of specific and local forecasts into a unified forecast, and finally the predictive calculations themselves. The seventh and concluding step in a comprehensive engineering-geology forecast encompasses forecasting expertise, the development of recommendations as to its application, and also the injection of the results into planning any engineering-geology monitoring of the geotechnical system for which the forecast is being prepared.

For comprehensive engineering geology purposes three types of methods of scientific-technical prediction may be applied: that of technical indicators and expert evaluation; extrapolation and analogy; and modelling, whether analytical analog, or natural. Analytical methods are the most widely used, both in constructing the geodynamic model and in the predictions themselves. They are based on calculations selected from geomechanics, geodynamics, hydrodynamics, thermodynamics, whether applied in that sequence or concurrently, which occur in the course of man-induced changes in engineering geology conditions. The procedures of comprehensive engineering geology prediction consist of multi-step integration of the selected schemes of calculation which make up the aggregate of discrete and differential equations in a single forecast, based on such cybernetic methods of prediction as the construction of predictive flow diagrams, morphological analysis and "matrices of interconnections."

The results of comprehensive engineering geology forecasting may be stated in four ways: graphically, in tabular form, verbally, or cartographically. The most effective method of presenting a forecast is as a scenario. I have suggested presenting a forecast in the form of such a scenario where the results of the forecast are expressed along with the technology for implementing the accepted design solutions and with the development plans for the technical components of the system. On being compared with the characteristics of development of the technical components, the results of the forecast are translated into a dynamic means of determining the most effective possibilities for eliminating the negative consequences of intense human activities in the area of environmental modification in the permafrost zone. Using this scenario one can easily formulate alternative variants of the forecast, constrained by the non-simultaneity of the intended design solutions at the early stages of planning and the ambiguity of the forecast data.

On the basis of the results of both specific and general geocryological forecasts and of comprehensive engineering-geology forecasts one may specify the technical design for exploiting gas fields, compile technical designs for the operation and building of gas pipelines, and also formulate a program for engineering-geology monitoring of the operation of gas extraction and distribution systems.

One may ensure the reliability of installations for the extraction and distribution of gas on frozen, thawing and freezing sediments by means of: 1) changes in the content and technology of designs, whereby not only the technical designs of individual installations and communications must be developed, but also integrated designs for entire geotechnical systems (gas well installations; main gas pipelines, etc.) on the basis of comprehensive engineering-geology forecasts; 2) purposeful and effective operation of such systems; and 3) engineering-geology monitoring of the building and operation of such systems.

ENGINEERING GEOCRYOLOGY IN THE USSR: A REVIEW

S. S. Vyalov*

Research Institute of Foundations and Underground Structures
Moscow, USSR

In this review of engineering strategies in permafrost areas the author stresses that in view of the rapidly expanding scale of developments in the North, the engineer commonly is unable to select the most favorable site, in terms of permafrost conditions and hence must be prepared to deal with *any* permafrost conditions. From this premise the article proceeds to a study of building techniques involving either the preservation of permafrost or its elimination. In each case a variety of techniques and examples of their adoption are presented, along with a discussion of methods of calculation of the forces and bearing capacities involved. Examples of the more innovative techniques discussed are the use of artificially refrigerated piles in the case of techniques aimed at preserving the permafrost, and the use of localized prior thawing, specifically around the supporting piles, in an area of "warm" permafrost with a high ice content.

Engineering geocryology, as an applied branch of the science of geocryology, arose in response to practical needs, especially in connection with the building of the Transsiberian Railway in the late nineteenth-early twentieth centuries. In its modern form engineering geocryology embraces the study of the interactions between various types of engineering structures and permafrost, whether in the role of foundation, environment or building material, and also the development of methods for controlling these interactions. The theoretical fundamentals of engineering geocryology are the physics, thermophysics and mechanics of frozen (or more accurately frozen, freezing and thawing) sediments. The physics of frozen sediments examines the physical and chemical relationships between the sediments as a multi-component system and studies the mechanics of the freeze-thaw processes. Thermophysics embraces the conditions of heat and mass exchange in the system comprising environment/structure/ground and describes the process mathematically. Mechanics examines the laws governing the behavior of frozen/freezing/thawing sediments under load and establishes the relationships, expressed mathematically, between stress, strain, temperature and time. Inclusion of the time factor is dictated by the fact that the processes of the impact of heat and load on the sediments change with time, while the frozen ground itself possesses clearly expressed rheological properties (creep and long-term strength); accordingly its stress-strain condition changes with time. Hence the modern mechanics of frozen ground is based on the theory of rheology.

Each of the above-mentioned aspects of engineering geocryology has become quite well developed and has attained a number of major achievements. But the development of each aspect has proceeded autonomously without proper consideration of the interdependences which exist, although the connections between the physical, thermophysical and mechanical processes operating within the sediments are incontestable. In fact temperature is one of the main factors determining both the mechanical properties of frozen ground and its behavior under load. But the temperature regime of frozen ground experiences constant change both as a result of changes in air temperature and as a result of the thermal impact of installations. The properties of the ground will accordingly change with time and these changes can be identified only by examining the thermal and mechanical processes together.

It is evident that an integrated examination of this type can best be effected on the basis of the thermodynamics of irreversible processes inasmuch as it is precisely the thermodynamic equations that embrace expenditures of both thermal and mechanical energy. At the same time equations of thermodynamics derive from phenomenological relationships without examining the physical essentials. Hence, with a view to achieving greater generalization it is convenient to produce baseline laws governing the deformation and destruction of frozen materials on the basis of examining physical concepts pertaining to the mechanics of the process. Such an attempt has been made by the author (Vyalov 1978a) and by others. On the basis of an experimental study of the microprocesses occurring in the materials during deformation and protracted destruction, peculiarities and laws governing changes in the structure were identified. The changes included the displacement and reorientation of particles and the development of structural faults, while on the basis of kinetic theory they were considered to be thermally activated processes. By using Boltzmann's equation, which relates the displacement of particles to the activating force, a strain equation was derived which established the relationship between the strain rate and its magnitude and the stress, temperature and time. With certain simplifications this equation may be converted into the well-known phenomenological

*Not present; summary presented by N.A. Grave.

exponential equation of deformation, although with physical parameters.

On the basis of laws governing the accumulation of structural defects revealed by micro-structural investigations, an equation of long-term failure was derived and was subsequently generalized on the basis of thermodynamics. It was assumed that ground failure begins when the influx of entropy reaches some critical value dependent on reaching some degree of structural damage to the ground (intensity of accumulation of faults). As a result a physically based equation of long-term strength was derived which relates the magnitude of the destructive load to the time to failure and to the temperature.

The resultant equations of strain and long-term failure, based on physics, allow one to make calculations as to strain and strength in frozen ground at any given moment; this includes predictions for the entire operating life of the installation. The above-mentioned equations allow one to take into account the variability in time of both thermal and force components.

In discussing the simultaneous examination of both thermal and force components one may note the following. At the present time the thermophysical and mechanical problems are being solved separately, and in the solution of the mechanical problems parameters determined by the solution of the thermophysical problems on the same site are being substituted in solving the mechanical problems. Thus the mechanical state of materials is being determined on the basis of some field temperature (e.g. the annual maximum temperature). However ground temperatures vary with time, both as a result of variations in air temperature and as a result of the thermal impact of the installation. Mechanical processes in the materials (creep, long-term strength changes) are also proceeding with time. Hence it is essential to be able to describe the synchronous operation of these processes. Unfortunately such a combined equation has not yet been achieved. But once one has solved the theoretical thermophysical equation (and this has been achieved both in the USSR, by G. V. Porkhayev et al. and abroad) it is possible to regard this solution as the starting point for solving the mechanical problem. However to do this it is necessary to derive the strain and long-term strength equations with parameters which vary with time and which determine the dependence of the mechanical process on temperature. As mentioned earlier the author has achieved these solutions; they can also take into consideration a load which varies with time.

It is extremely important to take the above-mentioned circumstances into account in making practical calculations, as will be demonstrated further.

Control of the Interactions between Structures and Permafrost

The marked differences in the properties and behavior under load of sediments in the frozen and thawed states have dictated that one identify two separate principles for using these sediments in terms of their interaction with structures, i.e. whether the sediments are maintained in the frozen state (Principle 1), or whether they are allowed to thaw (Principle 2). This distinction appeared spontaneously during the early period of construction on permafrost and was incorporated in the *Building Standards and Rules for Design of Foundations on Permafrost* (1976) and the *Manual* for these standards (1978), both of which apply throughout the USSR. In the following discussion I shall examine the use of permafrost materials as a foundation for buildings and installations in accordance with the above-mentioned documents.

In the case of construction based on Principle 1, the main problem is to maintain the sediments in a frozen state throughout the operating life of the structure. The simplest way involves building a naturally ventilated foundation. This method has been used by Russian engineers since early in the century and was described by Stotsenko (1912). The theoretical basis of the method was presented in the 30's by N. A. Tsitovich. The first major concrete building to use this method was a thermal power station built in Yakutsk in the 30's; it is still operating quite successfully.

Subsequently the method of construction whereby the permafrost is preserved has been widely adopted, especially in the northern areas of permafrost occurrence. The buildings in cities such as Yakutsk, Mirnyy, Noril'sk, etc. have all been built using this method.

The method of construction whereby the permafrost is first thawed also emerged early this century, with reference to construction in the southern areas of permafrost occurrence, in cities such as Chita, Petrovsk-Zabaykal'skiy, etc. The main problem connected with the use of this method was to prevent major damage to buildings as a result of ground settlement during thawing. Hence the buildings were reinforced with rigid girders, straps, etc. and attempts were made to select building sites where thaw settlement of the ground would be minimal.

But in recent years, in the light of the large scale of the operations, conditions affecting construction in the North have become substantially more complicated. In the North we have begun to build multi-storey (9-12 storey) buildings, to erect industrial installations with large spans and hence with heavy foundation loadings. Hence the need for strength and durability in these structures has increased proportionally, yet on the other hand the need for economic effectiveness has also increased. But the main thing is that nowadays we cannot select the most favorable site in terms of the permafrost conditions. We have to build on sites dictated by the needs of technology or design; this is especially important in terms of building transport installations (railroads, highways, pipelines, electricity transmission lines, etc.). Hence one has to be able to deal with any permafrost conditions, including the least favorable conditions. In particular, in terms of construction in areas near the southern limits of permafrost one invariably encounters plastic-frozen "warm" permafrost sediments (very close to 0°C), which at the same time possess very high ice contents. In many cases, in addition, the permafrost here is discontinuous in terms of both vertical and horizontal distribution. It has also become necessary to build on sites with massive ground ice bodies, whereas previously such sites were generally avoided, and also on mineralized frozen ground, whose bearing strength is sharply reduced, but whose susceptibility to deformation is increased, and on frozen peat deposits. Finally, we have had to build on permafrost in areas with high seismicity.

All of this has called for a new approach to construction on permafrost. While earlier we attempted to adapt our construction techniques to the behavior of the permafrost this method is now found to be ineffective. In fact we are now everywhere encountering situations where due to permafrost conditions ("warm" permafrost with an unstable regime) or due to technological or construction peculiarities of the structure (large spans, heavy loadings on the floor, or a "wet" technological process) it is difficult to maintain the sediments in a frozen state by the simplest methods (i.e. the normal ventilated foundation), and where it is impossible to allow the sediments to thaw while the structure is in use, since this would lead to serious settlement, which would be inadmissible even if the structure were reinforced. Hence instead of adapting the construction techniques to the behavior of the permafrost beneath the structure by one means or another, we have to provide the frozen materials with specific geotechnical properties which will ensure the conditions necessary for the operation of the installation. In other words it is essential to be able to control the interaction of the permafrost and the structure erected upon it.

Where one uses the technique of preserving the sediments in the frozen state, control of those properties is achieved by regulating the thermal regime of the foundation materials and especially by effecting supplemental refrigeration where these sediments are "warm" plastic-frozen materials. The cooling is usually achieved by means of the external winter air, i.e. by utilizing the severity of the northern climate. The principle on which this method is based involved the difference in mean annual temperature between the air and the frozen ground, the air temperature being a few degrees lower. For example, cooling of the ground from -0.5°C to -2.5°C will triple its strength.

Where materials in the thawing condition are utilized, control of the properties involves regulating the thaw regime. Once the structure is in operation unregulated thawing is permitted only in rare cases where the permafrost materials possess only a low ice content and hence the settlement on thawing will be only minor. But in the majority of cases one encounters ice-rich materials which result in drastic subsidence on thawing. If one uses the thawing technique in this situation one can reduce the magnitude and variability of the subsidence by means of prior thawing of the sub-foundation sediments, in which case the bulk of the settlement will have occurred prior to erection of the structure.

One should note that in order to control the forces involved in frost heave, which affects foundations during seasonal freezing of the ground, physical-chemical methods have been used quite successfully; in particular, treatment of the surfaces of the foundations or of the sediments themselves with special greases and additives.

Design formula derived on the basis of the ideas presented above were presented by the author in a review article at the Third International Conference on Permafrost (Vyalov 1979). Here I shall discuss examples of construction where these techniques have been applied.

Construction with Preservation of the Permafrost

The traditional method of preserving the permafrost by means of a naturally ventilated subfloor (which represents the simplest technique of controlling the temperature regime) is quite efficient when the sediments are solidly frozen (i.e. very cold) and where the building is not very wide. But right now in Yakutsk, where almost all the modern buildings have been built using this method, construction has begun on the headquarters of the building industry, a structure whose main floor area occupies 160 × 200 m. It turned out to be impossible to provide natural ventilation to the subfloor of such a structure and hence they had recourse to building a system of air extraction shafts inside the building, along its central longitudinal axis. In other cases (garages, etc.) the supply of cold air to the sub-foundation sediments is achieved by using ventilation ducts placed in a sand-and-gravel pad beneath the floor, either extending below the entire building or just adjacent to the foundations. This method has been used effectively (both in the USSR and abroad) for quite a long time. More promising in the long term is a new method of refrigerating the ground using hollow ventilated foundations. They are constructed in the form of hollow, corrugated pallets and are laid on sand-and-gravel pads beneath the floor of the building, whereby the cold external air circulates through the hollow ducts of this foundation (Figure 1). Hence these foundations combine the functions

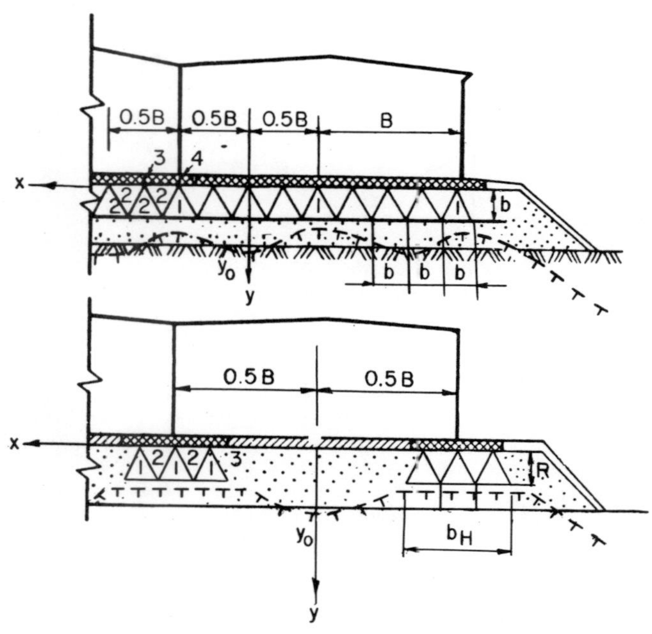

FIGURE 1 Corrugated ventilated foundations combining the functions of a load bearing unit and a refrigeration device.

of both load-bearing design and of a cooling device.

One should note that while in the case of solidly frozen "cold" permafrost it is sufficient just to maintain its natural temperature, in the case of plastic-frozen "warm" permafrost it is necessary to lower its temperature in order to convert the sediments from the plastic-frozen to the solidly frozen condition. Accordingly, when making thermal-technological calculations of the system embracing a building and its foundations the required temperature of the foundation is predetermined and on the basis of that temperature the construction and thermal regime for the operation of the subfloor is dictated. The method of calculation developed by G. V. Porkhayev is specified in the *Standards* and in the associated *Manual*. A similar approach is also used for cooling frozen sediments using ventilated ducts, hollow foundations, etc.

It has been demonstrated in practice that by using various cooling devices, including ventilated subfloors, it is possible to lower the mean annual temperature of sub-foundation permafrost by several degrees. But it will take several years (1-3) after erection of the structure to achieve this lowering of the temperature. Hence in calculating the bearing strength of a foundation this lowering of temperature cannot be taken into account and enters the calculations only as a safety margin. In order that the bearing strength of the foundation has been adequately increased by the time the full load of the structure has been placed upon it, it is necessary to lower the temperature of the ground prior to or during construction. Various methods are used for this. They include clearance of snow from the building site prior to the start of construction, the blowing of cold air into holes drilled for subsequent emplacement of piles, and the refrigeration of those piles with CO_2. But the use of thermal piles is particularly effective.

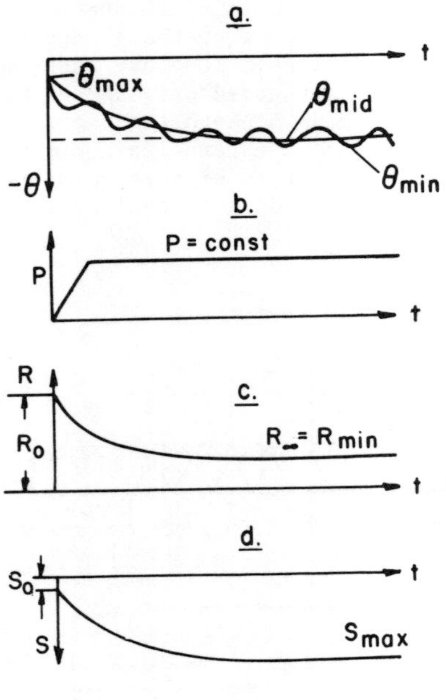

FIGURE 2 Processes, varying with time, which should be taken into account in calculating the bearing capacity and strain on frozen sediments forming the foundation of a structure:
a) changes in temperature of frozen sediments ϕ with time;
b) variations in load P with time;
c) reduction, with time, of the long-term strength of sediments R where θ = constant and P is also constant;
d) development over time of creep strain S where θ is constant and P is also constant.

In the Soviet Union use has been made of thermal piles, both with a liquid cooling agent—namely kerosene (this type of pile was first proposed by the Leningrad engineer, S. I. Gapeyev) and with a gas/liquid cooling agent—namely freon. In some cases the thermal piles are built into the structure of the foundation. This method, i.e. installation of reinforced concrete piles with liquid refrigerating devices (thermal siphons) built into them, was used in the construction of nine-story buildings in one of the subdivisions of Mirnyy at the suggestion of the Permafrost Institute of the Academy of Sciences of the USSR. A paper presented by Mel'nikov (1979) at the Third International Permafrost Conference was devoted to this topic.

In other cases the thermal piles are installed independently of the foundation and operate simply as cooling devices. This type of thermal pile (using freon as a cooling agent) has been widely used in Vorkuta in particular. The practical problems of using thermal piles in the North, their design and methods of relevatn calculations are considered in Vyalov (1983) and in Buchko et al. (1978).

Taking into Account Variations in Temperature and Loading

In examining the problem of regulating the temperature regime of foundations in permafrost it is necessary to proceed from the following statements.

In general the temperature regime of the refrigerated sediments is dictated by the initial cooling prior to construction, cooling during the course of operation of the building by means of a ventilated crawlspace (or ducts) and additional cooling by means of special devices such as thermal piles. All of this cooling proceeds against the background of the natural regime of the ground temperature, which is variable over time and is dictated by air temperature fluctuations. Thus the resultant mean annual temperature of the frozen ground for any point in the foundation can be expressed as a curve which will reflect the lowering of the mean temperature and subsequent attainment of a minimum constant value; this curve will also reflect seasonal fluctuations dictated by variations in the natural air temperature (Figure 2). Both the strength characteristics and mechanical properties of the sub-foundation materials will also vary accordingly with time.

Under the influence of an external load the processes of creep deformation and reduction of the strength of the materials will develop simultaneously with time. At the same time the external load itself may change with time, both during the construction period and while the structure is in operation (Figure 2). The challenge is to design a method of calculating the bearing strength and the settlement of the foundation which will take all these processes into account. The method described was worked out by the author and presented in his work *Termosvai v strotel'stve na Severe* [Thermal piles in construction in the North] (1983) on the basis of equations of long-term strength and of creep deformation of frozen sdeiments with temperature variations and load variations over time, which he had derived earlier. These equations were presented at the Third International Permafrost Conference (Vyalov 1978a, 1978b).

Figure 3 presents the results of determinations of the bearing capacity of piles in permafrost under various temperature regimes. If one regards the maximum mean monthly temperature of the sediments $\theta = 0.3°C$ as being a calculated one, the value for the bearing capacity P will be least and we will assume it to be unity (Diagram 1). If one takes into consideration the sinusoidal change in the mean monthly temperature along with a simultaneous lowering of the temperature due to the provision of a ventilated crawl-space and installation of thermal piles, then the value of P will increase by 1.51 times (Diagram 2). One should note that if in order to simplify the calculation one substitutes maximum values, changing in step-wise fashion (Diagram 3) or linearly (Diagram 4) for the sinusoidal oscillations in temperature, the value of P will increase 1.18 or 1.48 times. If at the same time the load is taken to be increasing throughout the construction period rather than constant, the value of P will increase 1.78 times (Diagram 5). And finally, if the sediments are pre-cooled from -0.3° to -1.2°C and one also takes into consideration subsequent cooling then the value of P increases by a factor of 4.2.

Thus the proposed method of taking into consideration the temperature variations in the sediments allows one to determine the bearing capacity of the foundations under a wide range of temperature regimes in the subfoundation materials and to increase substantially (by a factor of from 1.5 to 4) the designed bearing capacity of the foundation.

One should note the great significance of taking into account the variability of the load on the foundations. Even if one considers only the increase in load during the construction period (as occurs in reality) then the designed value of the bearing capacity increases by a factor of 1.64 (see diagrams 4 and 5).

But an even greater effect is achieved when one takes into account the short-term impact of climatic loads, namely snow, wind, or the forces of frost heave. This is particularly important for light structures, e.g. pipelines, electricity transmission lines, and also for buildings constructed of light materials, in that for such installations these loads are critical. At the present time the short-term impacts of loads are not taken into consideration and the strength characteristics of the sediments built into the design are calculated on the basis of the maximum mean monthly temperature in summer. In fact the maximum climatic loading as a rule does not coincide in time with the maximum ground temperature, which in fact coincides with the minimum bearing capacity of the materials. Thus, for instance, the maximum forces associated with frost heave develop in winter, when the ground temperature is certainly not at its

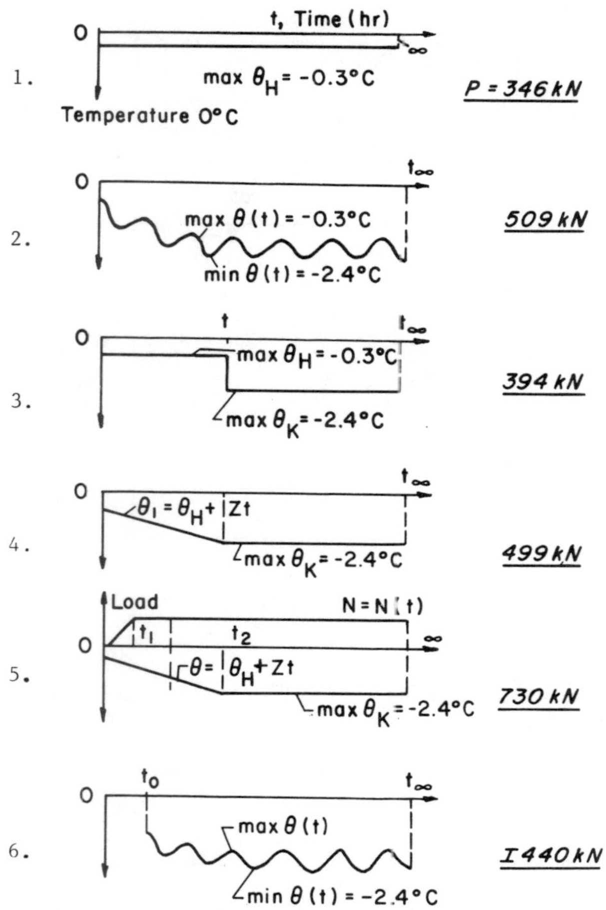

FIGURE 3 Bearing capacity of pile P with various temperature regimes in permafrost.

maximum, but rather is close to its minimum value. An analogous situation may be observed when one considers snow and wind-induced loads. Thus Figure 4 presents graphs (in relative units) of the change over time in the wind load (N) acting on an electricity transmission pylon in the Vorkuta area (curve 2) and changes over time in the bearing capacity ϕ of its foundations, which involve thermal piles (curve 1). The maximum value of N is reached in March, whereas the minimum value of the bearing capacity (ϕ) is reached in July-August. The curve of variation in the ratio N:ϕ over time, which dictates the necessary designed area of the thermal piles supporting the transmission pylon (F) (curve 3) reveals that the lowest value for this ratio occurs in August. This ratio is half what it would have been had one taken a maximum value for N and a minimum value for ϕ.

Foundations

Without dwelling in detail on the type of foundations used where the method of preserving the permafrost has been adopted in construction, in that this question will be discussed at a panel session at this conference, I wish to focus only

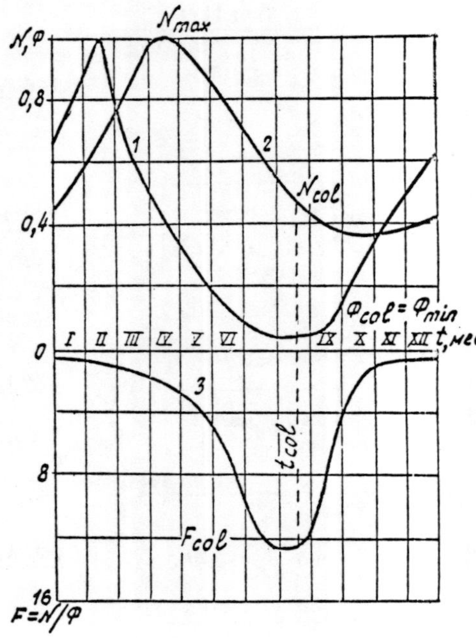

FIGURE 4 Changes over time in wind loading N on an electricity transmission pylon (2) and in the bearing capacity ϕ of a thermal pile (1) and the designed value for the area of a thermal pile F (3).

on the following aspects. The major type of foundation used in construction according to the first principle involves the use of piles. Perfection of the technique has been achieved by the composition of the slurry grouted into the bore hole; the slurry composition now selected (a sand-lime mixture) is designed to achieve optimal adfreezing strength and to ensure conditions of equal shear strength between the slurry and the surface of the pile and at the contact with the surrounding frozen sediments. In order to increase the bearing capacity of the pile we have begun using (e.g. at Noril'sk) gravel pads are placed beneath the butt end of the pile and tamped down as the pile is installed. In the case of soft materials (saline or plastically frozen) concrete expansions of the foot of the pile are being used (e.g. at Yakutsk). Methods of testing pile performance in permafrost have been considerably improved by means of test loadings; standards for such tests have been developed.

Methods of calculating and installing foundations in saline, peaty or ice-rich sediments and even in ground ice have been developed in fair detail. These topics are discussed in the following works: Vyalov et al. (1976, 1980), LenZNIEPP (1977), Gaydayenko (1978), and Roman (1981).

One should note that where ferro-concrete pile foundations have been used, cases of frost-damage to the concrete of the pile within the active layer have been found. This can be explained by the extreme temperature and moisture conditions affecting the upper part of the emplaced pile during seasonal freeze-thaw in the active layer. Measures have been developed to prevent this phenomenon; they mainly involve prevention of the penetration of soil moisture into the concrete of the pile, and increasing its frost-resistance. These topics are discussed by Poluektov (1983).

Construction where the Permafrost has Been Allowed to Thaw

Where one uses the method whereby the frozen sediments beneath a foundation have been encouraged to thaw, this introduces the need to solve the problems: 1) of determining the shape of the thaw zone beneath the structure and its evolution over time; 2) of determining the strain and strength characteristics of the thawed sediments; 3) of developing methods of reducing the magnitude and variability of settlement in the thawed foundation materials; and 4) of developing methods of calculating the combination of forces involved in the interaction between the structure , its foundation and the supporting thawed sediments.

The methods for making the thermal-technical calculations involved in thawing foundations are presented in the *Building Standards and Rules* and in the *Manual* to them, as well as in the books by G. V. Porkhayev and V. V. Dikuchayev. Values for the thermophysical properties (e.g. thermal conductivity, heat capacity) for various types of sediments in both frozen and thawed conditions and supplementary monograms for calculating such values are also presented in these works. To obtain more accurate results a computer was used and programs for such calculations have been developed.

A coefficient of thawing, A, which determines settlement due to thaw and is independent of load So, and a coefficient of compaction a = Sp/p, describing the settlement due to compaction beneath the load and the weight of the sediment itself, are used as the calculated deformation characteristics of the thawing sediments, i.e. S = So + Sp = h(A+ap) where h is the thaw depth at a given point. Laboratory determinations of the coefficients A and a were made by means of testing sediment samples as they thawed under pressure, but without any possibility of lateral expansion. But these experiments were of a supplementary nature. The major experiments were field tests using a heated press.

The methodologies of both laboratory and field tests complied with the appropriate standards. Strength characteristics (adhesion, angle of internal friction) were determined for the usual shear tests although the physical characteristics (density, moisture content) which the sediments assumed after thawing were also taken into account.

Reduction in the amount of settlement is achieved by controlling the thawing process. The most effective method involves prior thawing. As a result of this thawing the bulk of the settlement, i.e. settlement due to thaw (S = Ah) and part of the settlement due to compaction under the sediments' own weight ($Sg = a\gamma h^2$), occurs prior to erection of the building. Subsequent settlement, occurring during operation, is reduced to a minimum. The prior thawing may involve only part of the total depth of the potential depth of thaw, the amount being determined by calculation (Figure 5). The thawing is usually achieved by using either hot water or electrical heating. A number of buildings has been erected using this method in Vorkuta, Magadan oblast, and elsewhere, but sometimes solar heat alone is sufficient.

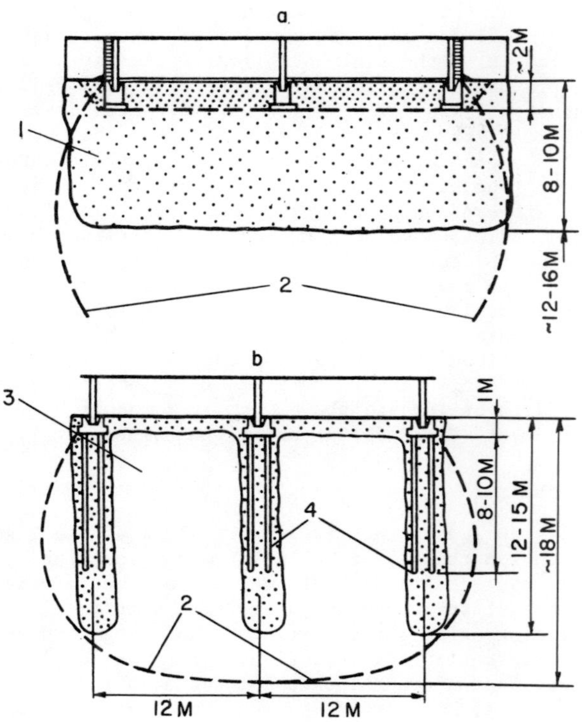

FIGURE 5 Prior thawing of sub-foundation permafrost sediments:
a) thawing beneath the entire building using a strip foundation;
b) local thawing on sites where piles are to be emplaced.
1. previously thawed sediments.
2. designed zone of thaw throughout operational life of the building.
3. permafrost.
4. local zones of thawing.

An interesting example is provided by the design of the foundations of buildings in Severobaykal'sk where the permafrost conditions in the construction site were unfavorable. The permafrost here is unstable, with a temperature close to 0°C. Although the sediment consists of gravel, it is distinguished by a high content of ice inclusions and on thawing it subsides drastically. But once thawed and compacted the gravel can serve as a good foundation. These sediments are 10-14 m thick, underlain by good, ice-free gravel. In such a situation it was difficult to maintain the frozen condition of the foundation materials since this would require specialized, expensive methods, and it was impossible to allow thawing during the course of operation of the buildings in that this would lead to excessive settlement. Moreover the situation was complicated by the high seismicity of the area. It was decided to excavate a deep pit, leave it open for a summer, thereby encouraging the thawing of part of the underlying sediments by solar heat, on the assumption that subsequent settlement of the entire sedimentary stratum would not exceed acceptable values. The foundation used took the form of a rigid, boxlike system which would not only support the above-ground structure but would also tolerate any remaining settlement and any seismic activity. All of this was calculated using a computer program which modelled the system encompassing the building, its foundation and the sediments beneath.

In the example cited the volume of sediments to be thawed was minor, and hence it could be achieved by natural processes. In the majority of cases prior thawing requires a great deal of expensive and protracted work. Hence a method of local prior thawing has been developed whereby the thawing is limited only to where the foundations are to be emplaced. In this case sill or pile foundations are used (see Figure 5).

The design of the foundations in one of the new towns of Western Siberia may serve as an example. The sediments beneath the building site consist of ice-rich silty sands. Just as in the previous example the permafrost here is unstable, with a temperature close to 0°. Construction using the technique of preserving the permafrost was technically and economically inexpedient. But it would be impermissible to let the sediments thaw out during the operation of the structures, since this would lead to major settlement. The solution was prior thawing, but in view of the great thickness of the permafrost, thawing the sediments beneath the site of every building in town would be extremely labor-intensive. Hence the technique of local thawing was adopted, the foundations used being single piles or groupings of piles. When the piles were being emplaced the thawed sediments beneath them were compacted, thus increasing the bearing capacity of the sub-foundation sediments. Thus piles are being widely used in construction not only where the permafrost is preserved, but also where it is allowed to thaw.

A description of this and other methods of construction on thawed sediments is presented in Mel'nikov and Vyalov (1981).

Combined Operation of the System Encompassing a Building, its Foundation and the Underlying Thawed Sediments

The erection of buildings on permafrost sediments which have been allowed to thaw, even when this thawing has occurred prior to construction, is always associated with the possible incidence of increased settlement. Hence along with measures to reduce settlement it is also essential to take measures to ensure that the design of buildings can absorb such settlement. One method sometimes used for buildings of frame construction involves a design which provides great flexibility, e.g. by means of an arrangement of hinged joints, pliable connections, etc. In such cases uneven settlement will not provoke excessive strains.

Another, more common method, involves the opposite approach, namely increased structural rigidity so that the structure can absorb the excess stresses from uneven settlement. This is achieved by reinforcement of all structural components of the building, the introduction of additional reinforcing members, and the breaking up of the structure's stress components into individual, rigid blocks. The foundations are also usually built with extra rigidity, in the form of transverse girders, a boxlike construction system, rigid slabs, etc. But where one uses prior thawing it is also possible to use foundations, including pile foundations, which stand separately from each other.

Foundations involving girders, slabs, etc. must be built in conjunction with the design of the above-ground structure. Calculation of the details of the total system embracing the building, its foundations and the sediments beneath, involves both determining the degree of settlement and its progress through time, and determining the forces invoked by that settlement in the foundation itself and in the above-ground structure, whereby the foundations are seen as a component part of the structural system of the building. The method of computer calculation of these aspects has been developed for civil buildings by the Leningrad institute LenZNIIAP and has been presented in articles by L. E. Neymark, V. V. Dokuchayev, etc.

When one examines the interaction between thawed sub-foundation sediments and the foundation of a building the following initial assumptions are made. Taking into consideration the curvilinear outline of the zone of thaw beneath the building the settlement in the sediments beneath the strip foundation will be variable along its length. In keeping with this the reactive forces transferred by the sediments to the underside of the foundation will also vary. In the marginal parts of the foundation strip, where the depths of thaw and the amount of sediment settlement are least, a concentration of pressure will occur and the reactive forces will reach maximum values $P = P\phi$. In the central part of the foundation strip where thawing

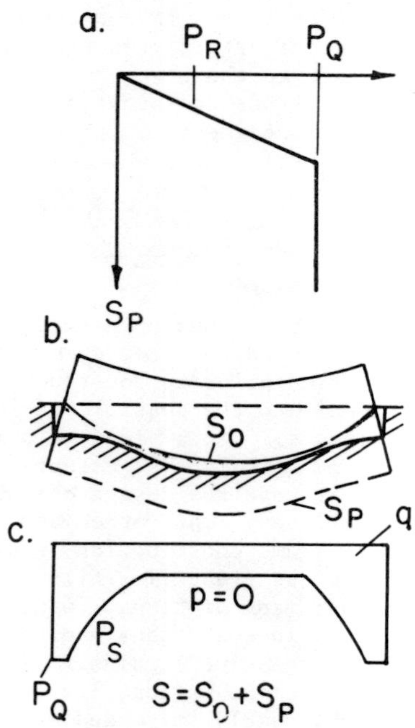

FIGURE 6 Diagram of foundation design.
a) elastic-plastic model of sub-foundation sediments showing relationship of settlement compaction Sp to load;
b) diagram of interaction between the foundation strip and the sub-foundation sediments;
c) diagram of reactive pressures in the sediments.

and settlement are at a maximum the reactive forces are minimal and may even equal zero if the thaw settlement in the sediments (So) is greater than the flexure of the foundation strip and the sediments "pull away" from it (Figure 6). The curved outline of the surface is determined by the change in the amount of settlement (So) along the length of the foundation beams (where prior thawing has been effected this value is close to zero). The amount of reactive pressure can be determined by solving the equation for the flexure of beams on an elastic foundation. In this connection the design model used for thawed sediments is the elastic-plastic model, the elastic strains in which are described by Winckler's theory of local strains. The bed coefficient in this model is determined as the ratio of load to the amount of compaction settlement ($C = P/Sp$) at any given point.

Conclusions

In concluding this review article I should like to mention that the development of engineering permafrost studies has allowed us to erect a wide variety of structures on permafrost materials, including multi-storey buildings and large industrial installations, under any permafrost conditions. But of course this demands a thorough preliminary engineering-geology reconnaissance and careful design decisions which observe all the requirements of building standards and regulations and which are based on technological and economic comparisons of possible variants; and naturally construction must be carried out so that all the requirements of this type of industrial work are met. One should also note that despite the achievements I have listed there are still a number of questions which require further investigation. We hope that the solutions of these problems will be achieved through the joint efforts of academic geocryologists and construction engineers in all the countries where permafrost occurs. The organization of international conferences on permafrost such as this will undoubtedly facilitate such joint efforts.

REFERENCES

Buchko, N. A. et al., 1978, Isskustvennoye zamorazhivaniye gruntov [Artificial freezing of sediments].

Gaydayenko, Ye. I., 1978, Ispol'zovaniye l'dov v kachestve osnovaniy sooruzheniy [Use of ice as a structural foundation].

LenZNIIEP, 1977, Fundamenty na sil'nol'distykh i zasolennykh vechnomerzlykh gruntakh [Foundations in ice-rich and saline permafrost materials].

Mel'nikov, P. I., and Vyalov, S. S., eds, 1981, Ottaivayushchiye grunty kak osnovaniya sooruzheniy [Thawing sediments as foundations for structures].

Poluektov, V. Ye., 1983, Ustroystvo fundamentov v vechnomerzlykh gruntakh [Construction of foundations in permafrost materials].

Roman, L. T., 1981, Fiziko-mekhanicheskiye svoystva merzlykh torfyanykh gruntov [Physico-mechanical properties of frozen peat].

Stotsenko, V., 1912, Chasti zdaniy [Parts of buildings].

Vyalov, S. S., 1978a, Reologicheskiye osnovy mekhaniki gruntov [The rheological basis of the mechanics of sediments]: Moscow, "Vysshaya shkola," 448 p.

Vyalov, S. S., 1978b, Long-term settlement of foundations in permafrost, in Proceedings of the Third International Conference on Permafrost, v. 1: Ottawa, National Research Council of Canada, p. 899-903.

Vyalov, S. S., 1979, The interaction between permafrost and buildings, in Proceedings of the Third International Conference on Permafrost, v. 2: Ottawa, National Research Council of Canada, p. 117-129.

Vyalov, S. S., ed., 1983, Termosvai v stroitel'stve na Severe [Thermal piles in construction in the North].

Vyalov, S. S. et al., 1976, Podsemnye l'dy i sil'no l'distye grunty kak osnovaniya sooruzheniy [Ground ice and ice-rich materials as foundations for structures].

Vyalov, S. S. et al., 1980, Stroitel'stvo promyslovykh sooruzheniy na merzlom torfe [Construction of industrial installations on frozen peat].

THE CONSTRUCTION OF DEEP PILE FOUNDATIONS IN PERMAFROST IN THE USSR - A SUMMARY

A.V. Sadovskiy

Research Institute of Foundations and Underground Structures
Moscow, USSR

The results of scientific research and experience related to the construction of pile foundations in the U.S.S.R. have been put forth in the U.S.S.R. Building Codes and Regulations under the headings "Building Pads and Foundations in Permafrost" (1977) and "Design Guidelines for Construction Pads and Foundations in Permafrost Soils" (1980).

There are a number of different types of piles and techniques for driving them. Some of these different types of piles are: reinforced concrete, metal, wood, combined, casing, and column piles.

The basic principles for pile selection are based on calculations of load-bearing capacity and pile deformation (i.e. settling in compliant frozen and icy soils).

Calculations of frozen piles with respect to load-bearing capacity are based on long-term shear resistance along the lateral surface of the piles (i.e. adfreeze resistance) and long-term resistance to pressure under the lower end of the pile. The recommended design values for these strength characteristics are given in the construction codes. Deformation calculations are derived from consideration of nonlinear settling creep. The problem is solved using computer-generated finite elements and is subsequently reduced to the simplest formulas with auxiliary nomograms. In addition, it is also solved for a general formula that includes equations for settling creep and long-term integrity. The use of a generalized formula makes it possible to determine, simultaneously, both deformation and integrity over time.

Horizontal load is calculated by looking at the lateral fixation of the pile in permafrost under varying conditions, such as depth of seasonal thaw and the temperature of the frozen soils.

Calculations for end-bearing piles are based on the pressure resistance of the underlying rock and the integrity of pile materials with regard to longitudinal bend. Additional negative friction loads are also taken into consideration.

The physical nature of adfreeze strength was studied with emphasis on the formation of an ice film at the point of contact of the pile with the frozen soil. The frozen soil's shear resistance was taken as the sum of the adfreeze strength at point of contact and the friction forces occurring due to radial soil compression. An equation for long-term frozen soil shear resistance was derived. A method of pile calculation in weak soils (i.e. those having high saline or ice content) was developed for a condition during which the resistance strength of the ground solution is equal along the lateral surface of the pile and along the perimeter of the borehole.

Temperature-dependent values for maximum shear resistance along the surface of the piles (reinforced concrete, metal, wood) were calculated for various types of frozen soils.

Increased load-bearing capacity of piles and cost-effectiveness are the result of:

- the use of composite wood/reinforced concrete and wood/metal piles;
- the use of pebble or gravel fill in a well with piles that have expanded concrete bases;
- selection of the optimal soil solution that is poured into the well;
- artificial cooling of high-temperature permafrost soils, including the use of thermo piles;
- the use of drilling piles in which concrete comes into contact with the frozen soil;
- the improvement of calculation methods, beginning with variable temperature and loads.

Existing methods for calculation of variable temperatures and loads are based on consideration of the worst possible conditions (highest frozen soil temperature) and the worst combination of loads. Actually, the soil temperature is variable over time due to fluctuations of the air temperature, underground cooling systems, and thermopiles. Pile loads are also variable over time. This is related to the pile's construction period and to short-term forces, specifically climate (snow, wind, heave). Maximum loads, however, do not coincide with the minimum load-bearing capacity. Snow, wind loads, and heave forces, for example, attain their maximum values in the fall and winter, while the smallest load-bearing capacity of the frozen earth coincides with the summer months. Consideration of this factor is particularly important for structures such as pipelines, ramps, transmission lines, and buildings made of light materials.

Calculation methods that allow us to account simultaneously for variability of loads over time, soil temperature, long-term integrity, and settling have already been developed.

Calculation methods for the construction of piles under special frozen conditions such as heave-ice soils, underground ice, and salty frozen soils have been developed. When building on soils where thawing may occur, as a rule, end load piles are used, which look like large deep supports or drilling piles. Suspended piles are also used in conjunction with preliminary local thawing of the frozen soil and compression of submerged piles.

During laboratory trials of pile models a relationship based on kinetic strength theory was determined between specific adfreeze strength and model dimensions.

Field trials of piles are conducted at large-scale sites. These tests have been standardized. A new quick (approximate) test method in which the assigned pile load is attained using a dynamometer has also been developed. The initial dynamometer stress is assigned, which then becomes weaker as compression of the buried pile increases. Stress is determined based on stabilization of pile settling.

Future research will involve the following:

- Improvement of pile-driving techniques in permafrost and coarse-grained soils;
- Improvement of pile construction;
- Improved usage of piles in frozen soils;
- Increased longevity of concrete piles;
- Improvement of calculations, tables, and nomograms;
- Standardization of international testing procedures;
- Collection of data related to adfreeze strength and load-bearing capacity in a wide range of conditions, which can then be compared in order to establish recommended characteristics.

CRYOGENIC PROCESSES ASSOCIATED WITH DEVELOPMENTS IN THE PERMAFROST ZONE

N. A. Grave

Permafrost Institute, Siberian Branch
Academy of Sciences, Yakutsk, USSR

The article first reviews the type of cryogenic processes (thermokarst, thermal erosion, thermal abrasion, frost heave, frost cracking and solifluction) likely to be induced by disturbance of the surface in permafrost areas, focussing on the probable variations in intensity under different conditions of climate, vegetation and ice content. The impact of the abandoned Salekhard-Igarka railway on the environment, and the degree of recovery achieved after 22 years are presented in a case study from near the southern limit of continuous permafrost, and are compared with the results of a study of the Fish Creek drilling sites (Alaska), 28 years after abandonment. Finally the author presents maps of surface sensitivity to disruption by human activities (with four classes of sensitivity) for both the USSR and North America. Using these he compares the areas with varying sensitivities in the two cases, and concludes that while some 7% of the total permafrost area in the USSR is extremely sensitive to disruption, the equivalent percentage in North America is less than 2%.

Human activities in the permafrost regions inevitably lead to the disturbance of the vegetation, soil and snow cover and even of the underlying sediments and bedrock. Cryogenic geomorphological processes such as thermokarst, thermal erosion, thermal abrasion, solifluction, frost heave and frost-cracking of the ground are thereby activated, dampened, or reactivated at sites where they had not occurred previously.

The primary prerequisite for the development of cryogenic geomorphological processes is the presence of H_2O which may be converted from the solid to the liquid phase or vice-versa under the influence of changing external conditions. The physics of the processes of ice formation, water migration, and of the phase changes in frozen, freezing and thawing materials has still not been adequately studied, although some theories as to these processes do exist (Kudryavtsev 1978). At the same time generally accepted explanations of the cryogenic processes which produce specific landforms have been established.

Cryogenic geomorphological processes usually operate together rather than separately, although one main process will be dominant in shaping the relief. The lead role will be assumed by different processes under different environmental conditions. The major factors determining the leading process are the ice content and grain-size of the material and the local topography. The water content of the soil and the depth to the permafrost table play a major role in frost heave. Climatic conditions exert a major impact on the intensity of the geomorphological processes.

These processes operate according to the same laws whether under natural conditions or in areas disturbed by human activities, although the rates at which they operate may vary substantially. Removal of vegetation and soil cover in areas with ice-rich sediments will provoke active development of thermokarst, thermal erosion, thermal abrasion and cryogenic slope processes, and this active stage may last from 2-3 to 10 years. Thereafter the processes will gradually die away and as material accumulates from the slopes and becomes vegetated the surface will stabilize. If no further disturbance of the surface occurs, over the subsequent 10-20 years the cryogenic processes will cease or will continue to operate at a rate comparable to that in an undisturbed area. Frost heave or frost cracking will be activated where snow cover is removed (Sukhodrovskiy 1979; Grechishchev et al. 1983).

Processes resulting in the disturbance of the surface and of the topography, namely thermokarst, thermal erosion, thermal abrasion and, in certain cases, solifluction (in the widest sense of the term) are the most hazardous in terms of human activities in permafrost areas. Thermokarst develops in areas with low relief. The immediate cause is an increase in the depth of the active layer as a consequence of the removal of the vegetation and soil cover, or of the waterlogging of the surface as a result of the ponding of water where surface flow has been intercepted.

The probability of thermokarst increases with increasing ice content in the soil. But the appearance of thermokarst also depends on the nature of the vegetation which was removed since this will dictate differences in the depth of the active layer as between the natural and the disturbed conditions. Thus, for example, this difference is greater in areas with a moss tundra than in those with non-sorted circles; or in areas with a larch *tayga* with a moss ground cover as compared to meadow areas. As a result the probability of thermokarst will be greater where larch forest or a moss tundra cover is removed than in areas with covers which are less effective insulators, although the ice content of the soil is the same in both cases (Grechishchev et al. 1983; Turbina 1982).

The probability of thermokarst and the intensity of its development will vary under different environmental conditions even in areas with extremely ice-rich sediments. The outer coast and interior lowlands of Northern Siberia and Alaska may serve as examples. A considerable proportion of their surfaces is underlain with old clay-silt loams which include very large Pleistocene and Holocene ice wedges and massive ground ice bodies. Extremely intense thermokarst develops with the removal of the vegetation and soil cover in the tundra and forest-tundra where the summers are relatively warm and the surface is sufficiently moist. In polar deserts, which embrace the arctic islands of both America and Eurasia and the extreme northern areas of the Siberian mainland and are characterized by a brief, cool summer and a discontinuous lichen cover, removal of the latter usually does not provoke deep thawing of the ground ice masses. But in Central Yakutiya when, in connection with agricultural activities, the taiga vegetation is removed from the surface of ancient terraces which are also underlain with clay-silt loams with massive ice inclusions, thermokarst develops only weakly and dies away at a very early stage. This is associated with the aridity of the relatively long, hot summers, during which evaporation exceeds precipitation (Grave 1983). Evidently an identical pattern may be observed in Central Alaska.

In areas where ground ice masses have developed, either as wedges or sheets, the development of thermokarst usually reflects the shape of these masses, especially the most widespread form, namely ice wedges associated with tundra polygons. In the initial stages this leads to a lowering of the surface and the production of a hummocky microrelief, i.e. *baydzharakhy* (cemetery mounds) separated by water-filled trenches. The greatest amount of thawing, resulting in small ponds, is observed at the intersection of ice wedges. More acute stages of thermokarst development are represented by thaw lakes. On the tundra such lakes commonly migrate; young ice wedges develop in the exposed areas of their beds and then, without any human interference, soon begin to melt. As a result a micro-relief of frost-heave polygons will emerge (Tomirdiaro 1980).

Thermokarst assumes different forms in areas where segregation ice has formed thin ice streaks in the frozen sediments. Here the depressions which develop after the removal of the vegetation and snow cover take the form of small, flat-bottomed pits. Thermokarst pits and ponds also form where peat and earth hummocks are sliced off during the levelling of a construction site.

The necessary conditions for the development of thermal erosion are a surface gradient and the concentration of drainage. In ice-rich sediments thermal erosion is closely related to thermokarst and thermal abrasion. The slopes of thermal erosional forms are further complicated by solifluction.

Thermal-erosional gullies of man-induced origin have been investigated in detail on the tundra in the northern parts of Krasnoyarsk *kray* (Sukhodrovskiy 1979). The direct cause of the appearance and growth of thermal-erosional gullies in a construction site was the increased surface runoff due to the melting of snowdrifts near buildings, the discharge of industrial and domestic waste water onto the surface, and the seepage of water from water lines.

Where these flows were eliminated or reduced, the gullies partially filled with deposits which had crept down from the side slopes and their growth was checked; but when the flows were renewed or intensified they began growing again. Nonetheless 10-20 years after the active phase of development, the gullies stabilize and become vegetated. Typically these forms reach dimensions of 1 km in length, several tens of meters in width, and up to 20-30 m deep. They commonly have a dendritic pattern in plan view.

Thermal abrasion due to human activities most often reveals itself in the destruction of the shores of reservoirs developed in ice-rich sediments in permafrost. Cases are known where intense destruction of sections of the sea coast has occurred due to human activities; here thermal abrasion was generally accompanied by thermal erosion, thermokarst and solifluction. Thermal abrasion induced by human activities has been investigated along the shores of the Vilyuy Reservoir, produced by the damming of the Vilyuy River (Sukhodrovskiy 1979). The shores of the reservoir are developed in trap rocks, broken by vertical tectonic fissures into blocks and underlain by plastic rocks. Here thermal abrasion assumes a peculiar character. The plastic rocks beneath the blocks of trap rock began to absorb water once the level of the water had risen and began to be compressed when loaded. The water entering the cracks between the blocks would freeze in winter, wedging the sides of the fissures apart even more. In summer, as a result of thermal abrasion, the ice in the cracks melts, the blocks slide into the water, are broken up, and are carried away into the reservoir in the form of gravel. Deep clefts in the shoreline result.

Where the vegetation and soil are disturbed by human activities on the sea coast or on the shores of lakes or rivers in the Arctic, and where those shores are developed on ice-rich materials, the resultant thermokarst significantly exacerbates thermal abrasion processes. Shores such as these break up into blocks which collapse into the water and the shoreline retreats relatively rapidly, threatening any structures built on it. Concurrently with this retreat ground ice masses forming the coastal cliffs gradually become covered with slump deposits which ultimately put a halt to the action of the waves by forming thermally eroded terraces and an extensive beach. Further destruction of the coast occurs due to thermokarst and thermal erosion.

In the case of slope processes, removal of the vegetation and soil from gentle slopes developed on ice-rich sediments leads to an increase in the depth of the active layer and an increase in its moisture content. The slow processes of solifluction accelerate into mud flows. Observations have shown that this process may attain velocities of 1 m per minute. Flat-bottomed troughs and areas where the surface has been lowered by thermally induced mass movement, up to 50 m in width form on slopes as a result of these processes. The processes do not persist actively for long. Thus on the slopes in cuttings on the Tayshet-Lena railway developed in ice-rich sediments this process ceased after 5-6 years (Sukhodrovskiy 1979).

Frost heave occurs during the freezing of the

active layer or during the formation of a seasonally frozen layer in areas where no permafrost exists. Human activities in areas with heave-prone materials, the associated destruction of the vegetation and soil covers and the compaction or removal of the snow cover, in general lead to differential heaving of affected areas. Pipelines, whether buried or above ground, may experience differential heave where geocryological conditions are particularly variable; to prevent this thorough preliminary predictive investigations are essential.

Frost-cracking of the ground, sometimes called cryogenic cracking, is widespread in areas where developments are being undertaken, as investigations in Western Siberia have shown (Grechishchev et al. 1983). Where the snow cover is removed cryogenic cracks may occur in areas where they had not developed beneath a snow cover. The depth of cryogenic cracks increases with decreasing ground surface temperatures and decreases as the temperature at the depth of zero annual amplitude increases. The depth of the cracks also increases with a decrease in the moisture content of the frozen ground. Thus, for example, in northern Western Siberia the depth of cracks in sandy-loamy soils with a moisture content of 20-25% reaches 9 m but where the moisture content is 25-30% it reaches only 6.5 m. The dimensions of the polygonal crack-bound blocks decrease as the ground surface temperature decreases, and vary between 10 and 20 m in northern Western Siberia. Calculations have shown (Grechishchev et al. 1983) that the highest stresses in pipelines as a result of cryogenic cracking in areas where snow has been removed can be expected in clayey materials.

Predictions of changes in geocryological conditions and of the development of geomorphological processes as a result of man-induced disturbances of the vegetation, soil cover and sediments may be arrived at either by calculation (Fel'dman 1977) or by the recently developed method of geocryological engineering analogy (Grechishchev et al. 1983). This latter method is of particular interest since it is based on investigations of developed areas continued over many years. By selecting landscapes undisturbed by human activities, but otherwise identical to the developed areas, one has an opportunity to follow and predict the probable changes which will result from various kinds of development. As an example the work cited above presents the results of investigations along a section of the route of the now abandoned Salekhard-Igarka railroad. This section is located in the sporadic permafrost zone, the permafrost islands coinciding with hummocky peatlands and bogs. Large-scale air photos of this section of the line, flown prior to the start of construction (1949), and 15 years later after work on the line had been abandoned (1966) were studied, and data from engineering site investigations were analyzed. Control studies were made in the summers of 1973 and 1974.

During construction of the railroad the forest was cut down, both winter roads and permanent corduroy roads were built, the vegetation cover was removed, and the soils were stripped off to depths of 0.2 to 0.3 m. Trenches and ditches were dug, temporary buildings and installations were erected, drainage was blocked and lakes lowered, while individual areas were either drained or became boggy.

In areas without permafrost where the forest was removed and where snow was removed in the building of winter or corduroy roads, the depth of seasonal freezing initially increased and frost heave was observed; this led to the destruction of the corduroy roads. After a few years these sections began to become overgrown and after 22 years a sparse birch forest with a ground cover of grasses and mosses had developed on the surface, while bogs had developed in cuttings. The seasonal freezing depth had decreased to values comparable to those under natural, undisturbed conditions.

Elimination of the vegetation and destruction of the microrelief during the laying of temporary roads across flat, boggy areas and peat plateau areas where permafrost was present led to an increase in the depth of summer thaw, sometimes to depths of 4-5 m, and in these areas thermokarst ponds appeared. Later, however, the topography became covered with vegetation and 22 years after the disturbance the depth of summer thaw had decreased again and thermokarst had almost totally ceased. These observations allow one to predict the results of man-induced disturbances of the surface in areas where industrial developments are taking place.

Comparable investigations have been pursued along the routes of operating gas pipelines and on drilling sites. Similar investigations of the consequences of disrupting the vegetation and soil covers while drilling oil wells in the area of Fish Creek in Northern Alaska have been carried out by American specialists, 28 years after disturbances occurred (Lawson et al. 1978). The wells were located on the tundra in the continuous permafrost zone in an area with widely developed ice wedge polygons. The disturbance to the surface here was due to the same causes as in the case of the route of the railroad mentioned earlier, with the addition of the spillage of diesel fuel. The surface disruption provoked thermokarst and thermal erosion, which are still persisting at the present time.

These examples cited from the continuous permafrost zone with ice-rich sediments (Alaska) and from a zone of sporadic permafrost near its southern limit (Salekhard-Igarka) demonstrate that the "healing" of man-induced thermokarst and restoration of the vegetation occurs considerably more slowly in the Far North than in the more southerly parts of the permafrost zone.

On the basis of both experimental data and calculations, predictive maps showing the distribution of the parameters of frost cracks which will result from removal of the snow cover have been compiled for the northern part of Western Siberia (Grechishchev et al. 1983). The distribution zones delineated illustrated both the depths of the cracks and the dimensions of the polygons, both of these parameters being related to the distribution of average January air temperatures in the area. Thus the dimensions of the polygons increase from 10 to 20 m from northeast to southwest, while the depth of the cracks increases in the opposite direction from 3 to 9 m.

Cryogenic geomorphic processes initiated by human activities within the permafrost region are conditioned by the sensitivity of the surface to technogenic operations. Analysis of the data in the literature on ice contents, thermal settling, grain sizes of the frozen materials and physical geographical conditions in the permafrost zone

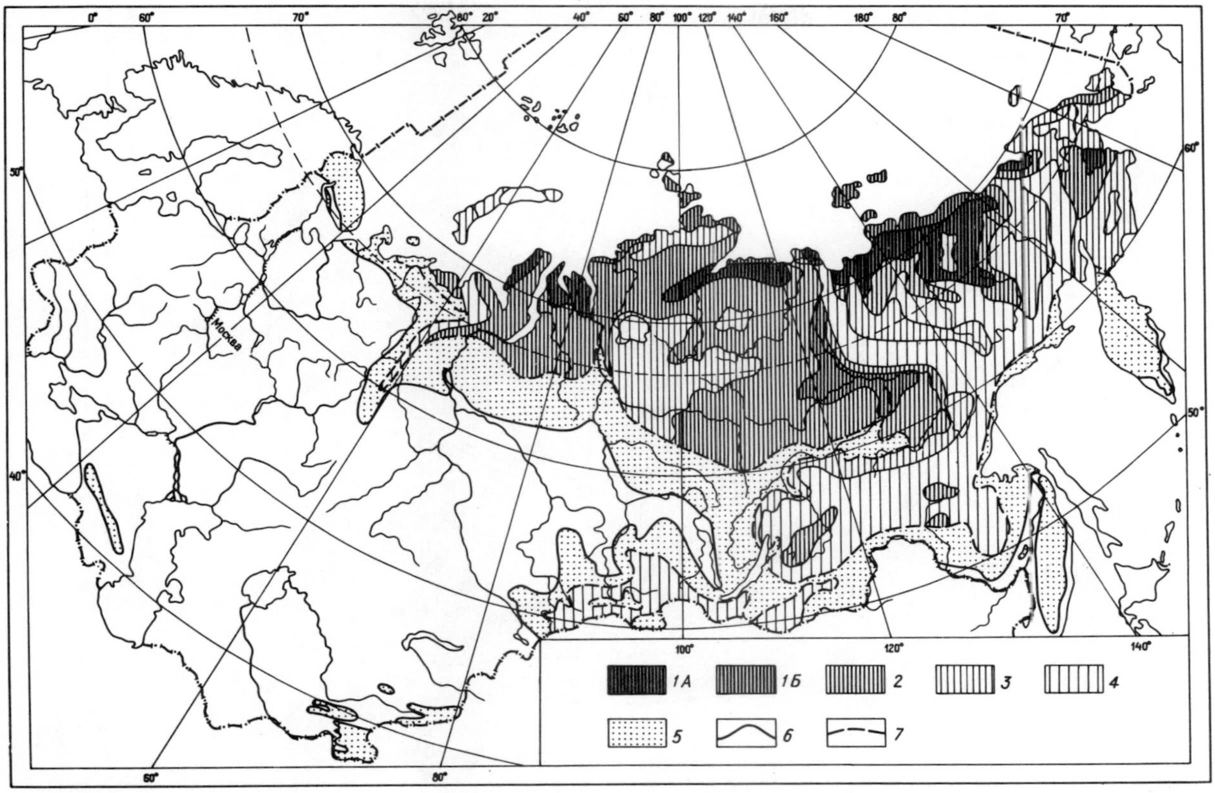

Legend:

1-4: Regions with different degrees of surface stability predominating:
- 1A - very unstable
- 1B - very unstable
- 2 - unstable
- 3 - slightly stable
- 4 - relatively stable

5: Regions with permafrost islands and sporadic permafrost

6: Southern boundary of the permafrost region

7: Southern boundary of the region of continuous permafrost and massive permafrost islands.

FIGURE 1 Map of the permafrost zone in the USSR illustrating sensitivity of the surface to removal of the vegetation and soil covers during development

within the USSR has permitted me to delimit four classes of surface sensitivity to removal of the vegetation and soil covers during industrial developments, and to illustrate their distribution in terms of a schematic map (Grave 1983). A comparable map has also been developed for North America. The two maps are presented as Figures 1 and 2.

To a certain degree these maps also illustrate the dominant cryogenic geomorphological processes operating on the surface, which would be activated or reactivated by man-induced disturbances.

In regions 1A and 1B with highly sensitive surfaces, the dominant process would be thermokarst, with the addition of thermal erosion and thermal abrasion in the north. In region 1A this would involve deep thermokarst associated with the melting of massive ice wedges, whereas in region 1B the thermokarst would be limited only to the early stages of development.

In zone 2, also with an unstable surface, the dominant process would again be thermokarst, but operating on less massive ice wedges and on segregation ice to produce primarily shallow, flat depressions in the surface. The role of solifluction and creep would be more significant

In region 3, with only a mildly unstable surface, solifluction and mass movement phenomena, including rock glaciers, assume a more important role. In region 4, with relatively stable surfaces, a major role is played by slope processes, including catastrophic processes such as rock slides, mudflows, etc.

The geographical distribution of the predominant surface types, in terms of their sensitivity to the removal of vegetation and soil covers, is shown in a generalized, schematic fashion on the regionalized maps of the permafrost zones of the USSR and North America presented as Figures 1 and 2.

It is interesting to calculate the areas occupied by the surfaces with various degrees of susceptibility to thermokarst; this has been done by S. P. Gribanova using a planimeter for the zones of continuous and massive-insular permafrost in the USSR and North America (Table 1).

The accuracy of the figures presented is relative, and corresponds to the accuracy of the chosen zone boundaries delimited on the maps. At the same time the interrelationships of the areas of the delimited zones are fairly obvious.

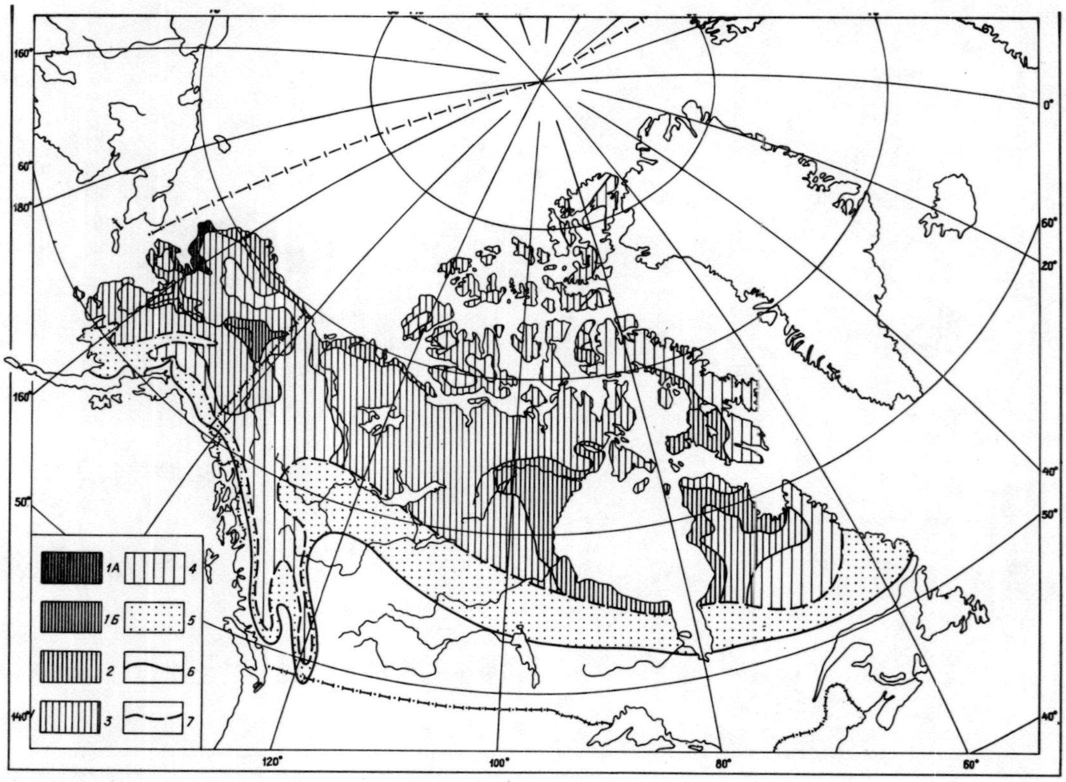

Legend

1-4: Regions with different degrees of surface stability predominating:
- 1A - very unstable
- 1B - very unstable
- 2 - unstable
- 3 - slightly stable
- 4 - relatively stable

5: Regions with permafrost islands and sporadic permafrost

6: Southern boundary of the permafrost region

7: Southern boundary of the region of continuous permafrost and massive permafrost islands.

FIGURE 2 Map of the permafrost zone of North America illustrating sensitivity of the surface to the removal of the vegetation and oil covers during development

TABLE 1 Approximate areas of varying degrees of surface susceptibility to man-induced thermokarst (thousands of km²)

Degree of surface susceptibility	Area	
	USSR	North America
1A Extremely susceptible	559	40
1B Extremely susceptible	161	55
2 Susceptible	2,800	750
3 Slightly susceptible	1,600	2,600
4 Relatively stable	1,900	1,410
Total area of continuous and massive-insular permafrost	7,020	4,864

In the USSR widespread extremely sensitive surfaces occupy an area in the order of 720,000 km², representing about 7% of the entire permafrost zone (including the discontinuous and sporadic zones), or about 10% of the continuous and massive-insular permafrost zones. In North America extremely sensitive surfaces occupy approximately one seventh of this amount (about 100,000 km²), representing less than 2% of the area with continuous and massive-insular permafrost. It is obvious that only some parts of the total area of the permafrost zone is susceptible, if developed, to catastrophic settlement and disruption of the surface and to irreversible changes in the landscape in a direction which is undesirable from the human viewpoint.

Further investigations should be directed to pinpointing the consequences of human activities in the permafrost zone and to perfecting the mapping of areas which are particularly hazardous in this respect, both with regard to the environment and to society.

LITERATURE

Fel'dman, G. M., 1977, *Prognoz temperaturnogo rezhima gruntov i razvitiya kriogennykh protsessov* [Prediction of soil temperature regimes and of the development of cryogenic processes]: Novosibirsk, Nauka, 191 pp.

Grave, N., 1983, A geocryological aspect of the problem of environmental protection, in: *Proceedings of the Fourth International Conference on Permafrost:* Washington, D.C., National Academy of Sciences, p. 369-373.

Grechishchev, S. Ye., N. G. Moskalenko, Yu. P. Shur, et al. 1983, *Geokriologicheskiy prognoz dlya Zapadno-Sibirskoy gazonosnoy provintsii* [Geocryological prediction for the West Siberian gas-bearing province]: Novosibirsk, Nauka, 180 pp.

Kudryavtsev, V. A., ed., 1978, *Obshchee merzlotovedeniye (geokriologiya)* [General permafrost studies (geocryology], second edition, revised and enlarged: Moscow, Izdatel'stvo MGU, 464 pp.

Lawson, D. E., J. Brown, K. N. Everett, A. W. Johnson, V. Komarkova, B. M. Murray, D. F. Murray and P. J. Webber, 1978, Tundra distrubance and recovery following the 1949 exploratory drilling, Fish Creek, Northern Alaska: *CRREL Report* 78-28, Hanover, NH, 91 pp.

Sukhodrovskiy, V. P., 1979, *Eksogennoye rel'yefobrazovaniye v kriolitozone* [Exogenic relief formation in the cryolithozone]: Moscow, Nauka, 277 pp.

Tomirdiaro, S. V., 1980, *Lessovo-ledovaya formatsiya Vostochnoy Sibiri v pozdnem pleystotsene i golotsene* [The icy-loess formation in Eastern Siberia in the late-Pleistocene and Holocene]: Moscow, Nauka, 184 pp.

Turbina, M. I., 1982, Evaluation of geocryological conditions in the valley of the middle Amga, in connection with the agricultural development of the area (based on the example of the Betyun' area, in: *Termika pochv i gornykh porod v kholodnykh rayonakh* [Thermal aspects of soils and bedrock in cold regions]: Yakutsk, Permafrost Institute, SO AN SSSR, p. 135-141.

Contributed Soviet Papers

THE PHYSICOCHEMICAL NATURE OF THE FORMATION OF UNFROZEN WATER IN FROZEN SOILS

Yu. P. Akimov, E. D. Yershov, and V. G. Cheveryov

Geology Department, Moscow State University
Moscow, USSR

The paper examines how the liquid phase of H_2O is influenced by the basic physicochemical factors in permafrost: average size of particles and pore-spaces, polydispersity and heteroporosity, specific surface, surface energy, structural bonds, composition of the solutions in pore-spaces, temperature, etc. The trend of the curve showing the relationship between the unfrozen water content and temperature is described theoretically; physics-based constants are an integral part of the formula used in calculating this relationship. The surface energy of the mineral skeleton of the soil is found to be dominant in dictating the presence of H_2O in the liquid phase; the contribution made by the quasi-liquid film on the ice surface was proved to be non-critical. As one proceeds from macro to microporous soils and from coarse to fine ones, the percentage of capillary water increases and that of film water decreases. It was shown that the relaxation time of unfrozen water does not exceed one hour. The correlation between typical soil moisture contents and the amount of unfrozen water was explained on a physicochemical basis; the thermodynamic characteristics of the unfrozen water are also discussed. The relationships between the physicochemical and petrographic parameters of frozen soils were demonstrated and on this basis the rules governing the formation of liquid H_2O in soils of differing genesis, age and granulometric, chemical and mineralogical compositions are demonstrated and explained.

The present-day physics, chemistry, and mechanics of frozen soils are impossible without studying the interaction between soil skeleton particles and water. Many aspects of this interaction are the same as or similar to the interaction of water with mineral particles at positive temperatures.

It was a great achievement for soil physical chemistry when changes in the properties and structure of bound water were proved to be influenced by mineral particles. The reverse influence of water on structural parameters of the particle surface, for example, on parameter "b" of their crystal lattice (Osipov 1979) also has an important effect. Minerals, having great surface energy, make water molecules nearest to their particles reorient in such a way that the structure of the nearest water layers most fully corresponds to the location of active surface centres of particles, i.e. to the surface structure of the mineral. Naturally, lower water temperatures result in further transmission of this type of structuring to water at greater distances from the surface (since an increase in temperature promotes diffusion of the oriented water structure due to thermal motion), increased differentiation between the surface structure and that of associates and more surface energy at the disposal of the mineral particles (Saveljev 1971). This is well-illustrated by the quantitative increase of loosely and tightly bound water that occurs as the temperature drops (Yershov et al. 1974).

Lowering the temperature below 0°C leads to a new water phase—ice. It is acknowledged now that ice in frozen soil is not an inert body. Having considerable surface energy and water-absorbing capacity, ice itself causes the formation, close to its surface, of a thin layer of intermediate-phase water at both the ice-air and ice-liquid water interfaces. Likewise, the structuring activity of unfrozen water also has an impact on ice since its 0-0 atomic gaps get larger (Kurzayev et al. 1977). Thus, it turns out that unfrozen water has multiple layers; the water close to the surface of soil particles is bound mostly by them and water close to the ice surface is bound mostly by ice, while its interstitial portion is under the orienting effect of both surfaces. This explains the nonmonotonic relationship between unfrozen water density and aqueous film thickness (Anderson and Morgenstern

1973) and the minimal heat of the ice water phase transition at the initial freezing temperature of tightly bound water (about -7°C), where enthalpy variation at ice-water transition varies from tabular values by a factor of 6 (Anisimov and Tankaev 1981).

Absence of an orienting ice surface causes monotonic change in the orientation of water molecules as they move away from the particle surface and a monotonic decrease of water-ice transition heat to zero when the moisture in unfrozen water is close to hygroscopic and the temperature is about -70°C.

A series of interesting publications, dedicated to the liquid phase content in frozen soils, has appeared in the last five years. The use of methods such as the NMR-spin echo permits us to deduce that unfrozen water may have at least two categories, which differ sharply in their molecular mobility (Ananyan et al. 1977). The mobile liquid phase content is not permanent and decreases as the temperature drops according to the relationship between the unfrozen water content (W_u) and temperature (t). The amount of water in the liquid phase showing very low mobility varies little with temperature and is close to maximum hygroscopic moisture.

One of the new research trends in this field relates to the relaxation time of unfrozen water (Danielyan and Yanitsky 1979, Yershov et al. 1979). The non-equilibrium of water transitions is caused by three major factors: delay in water transitions (discounting the effect of soil minerals), delay in transformation of soil texture and structure, and delay in mineral skeleton-water interaction. It has been shown that after the temperature has been set, it takes unfrozen water an hour to reach an equilibrium value, with the relaxation time depending on the mineralogical composition, texture, and structure of the soil.

Using soils of different genesis, composition, texture, structure, and properties a series of comprehensive studies has revealed the physicochemical content of the liquid phase in frozen soils. For the first time in world practice the structure of soil pore spaces has been studied using mercury porosimetry; the unfrozen water content has also been determined by the contact method and others.

The contact method for determining the liquid phase content is simple, reliable, applicable to a wide range of temperatures and accurate to approximately ±0.3%. This method does not require special equipment—the only thing needed is a refrigerating cabinet with a temperature control device to maintain isotropic distribution and to record temperatures with an accuracy of ±0.1°C. For investigation purposes, 3×4×0.5 cm cooled plates, made of dried-up soil, are covered on the two largest faces with plates of ice or the same soil, which has been water-saturated and frozen. Using polyethylene film they are made waterproof, pressed on top and put into the refrigerating cabinet at the required temperature. When the dry plates come into contact with the ice, liquid exchange and saturation of the dry soil with water take place due to intensive air, film, and capillary water transfer. The thermodynamic equilibrium of three liquid phases and the discontinuation of water exchange occur some time later; depending on mineraological composition, the texture of the soil, and the temperature, this time interval may last from several to 24 hours. The total moisture content in the initially dry plates corresponds to the equilibrium content of unfrozen water at a given temperature (Yershov et al. 1979).

The importance of structural and adsorptive characteristics in the formation of the liquid phase content in frozen soils depends on a synthesized interdependent relationship between pore space structure and the specific active surface of soils.

The results of experiments carried out on frozen soils having different moisture and size of ice crystals allows us to conclude that the quasi-liquid film on the ice-air interface in the total unfrozen water content is nonessential.

It has also been proved experimentally that the surface energy of the mineral skeleton can be reduced by using water repellants, resulting in a sharp drop in the soil's liquid phase content (Yershov et al. 1979).

This indicates that the unfrozen water content is almost entirely dependent on the liquid phase of water in small-diameter nonfreezing capillaries and on film water close to the surface of mineral particles.

As macroporous soils are gradually replaced by microporous ones, the specific surface becomes less important, while the pore space structure acquires greater significance. Thus, the pore space structure is the most important factor in formation of unfrozen water and ice in microporous soils, especially at high temperatures.

To estimate the content of water not freezing in small-diameter capillaries we can consider the fact that the distribution of pole volumes by radii can be well-described by lognormal law (Yershov et al. 1979):

$$V = V(0 < r \leqq r_1) = V_o \frac{1}{\sigma_r \sqrt{2\pi}} \int_{-\infty}^{\lg r_1} \exp\{- \frac{(\lg_r - m_r)^2}{2\sigma_r^2} d \lg r \qquad (1)$$

Here, V_o is total pore volume; V is the volume of pores having a radius smaller than r_1; σ_r, m_r are root-mean-square deviation and mode, respectively.

We can then use the relationship between the absolute drop of freezing temperature (ΔT) of water in capillaries and their radius (r):

$$\lg r = \lg \frac{2\omega_{wa} T_o}{\rho L} \cdot \frac{\omega_{sa} - \omega_{sw}}{\omega_{wa}} - \lg \Delta T, \qquad (2)$$

where: ω_{wa}, ω_{sa}, ω_{sw} are specific interfacial energies (water-air, soil-air, and soil-water respectively); T_o is a normal freezing temperature; ρ, L are density and specific heat of water transitions, respectively.

On substituting $\lg\Delta T$ for y, lg r for x and $\lg \frac{2\omega_{wa} T_o}{\rho L} \cdot \frac{\omega_{sa} - \omega_{sw}}{\omega_{wa}}$ for b, equations (1) and (2) may be combined to give the volume of water not frozen at ΔT_1:

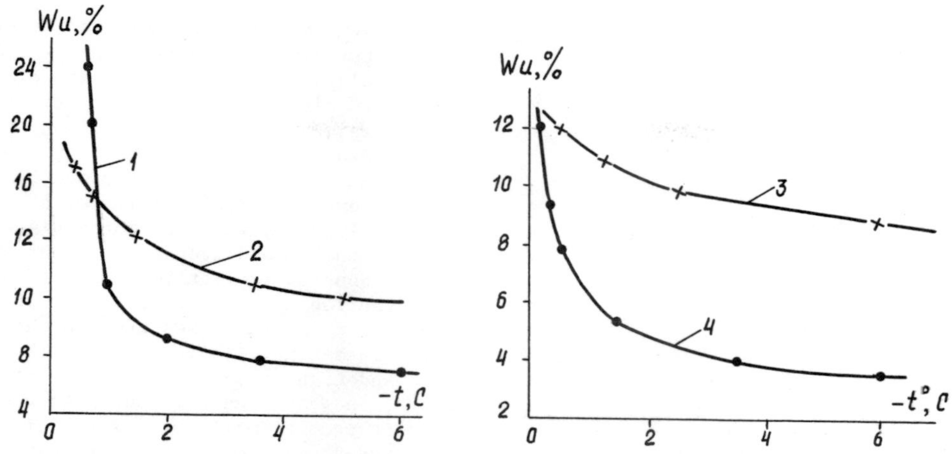

FIGURE 1 Influence of heteroporosity (a) and average pore radius (b) on unfrozen water content-temperature relationship in frozen soils: 1 - kaolin (σ_r = 0.18, $\bar{r} = 10^2$ nm; s = 15 m^2/g); 2 - Cambrian illite clay (σ_r = 0.41, $\bar{r} = 10^2$ nm; s = 16 m^2/g); 3 - light loam ($\bar{r} = 0.4 \cdot 10^3$ nm; σ_r = 0.58); 4 - heavy sandy loam ($\bar{r} = 2 \cdot 10^3$ nm, σ_r = 0.58).

$$V = V_o \frac{1}{\sigma_y\sqrt{2\pi}} \int_{b-y_1}^{\infty} \exp \{-\frac{(y-m_y)^2}{2\sigma_y^2}\} dy , \qquad (3)$$

with $m_y = b - m_r$. Assuming that unfrozen water density is a tabular value, W_u can be set equal to the sum of all its components (capillary and film water):

$$W_u = \frac{\rho V_o}{\sigma_y\sqrt{2\pi}} \int_{b-y_1}^{\infty} \exp \{-\frac{(y-m_y)^2}{2\sigma_y{}^2}\} dy + (\frac{2T_o\kappa^2}{\rho L\Delta T_1})^{1/3}$$

$$(\omega_{ia}-\omega_{iw}-\omega_{wa})S_i + (\frac{2T_0\kappa^2}{\rho L\Delta T_1})^{1/3} (\omega_{sa}-\omega_{sw}-\omega_{wa})S, \qquad (4)$$

where s_i and s are specific ice-air and mineral-air interfaces; ω_{ia}, ω_{iw} are specific ice-air and ice-water interfacial energies; κ is the minimum size of a drop at which one can begin to detect liquid properties.

The second term of the right member determining the share of the quasi-liquid film on the ice surface is of minor importance in the formation of unfrozen water. The first term determining the W_u-pore space relationship of the structure may be replaced by an approximate linear form: W_u vs lgΔT. Actually, the W_u-temperature relationships in these coordinates are linear, which indicates the particular importance of pore space structure in unfrozen water. The specific surface size, therefore, does not exert direct influence on the liquid phase content. The reason is that, even at low moisture contents, ultracapillary pores, which determine the specific surface size, turn out to be full of water. This immediately reduces (by 1-3 orders) the specific surface size, i.e. the size of the surface that is capable of further water binding.

These experiments have shown that when soils are mixed, the proportional summation of pore space structures and specific active surfaces leads to additivity of the unfrozen water content (Yershov et al. 1979), i.e. W_u is equal to the mean-weighed content of unfrozen water produced by calculating the W_u^i values for the i-th soil composing the mixture:

$$W_u = \sum_{i=1}^{n} W_u^i a_i$$

where a_i is a share of the i-th soil in the mixture.

The material presented above helps in understanding the physicochemical peculiarities of W_u-temperature relationship. The range of intensive phase transitions is observed to be broader in soils characterized by more uniform pore space, small average radius of pores and larger specific active surface (Figure 1). In addition, the unfrozen water content at low temperatures is larger in nonuniformly porous soils and smaller at high temperatures, when compared to monoporous ones (Figure 1). The nature of pore volume distribution by radii probably accounts for the S-shaped relationship between the unfrozen water content and temperature obtained in the experiments of various authors. These conclusions made it possible to determine the physicochemical nature of unfrozen water formation in soils of different mineralogical composition, texture and structure.

To make a pure assessment of the influence of soil dispersity on the W_u value, two cases should be considered. In the first case polydispersity is the same (estimated by the root-mean-square deviation (σ_d) of the particle size) while the average size of particles ($\bar{d}$) differs. This allows for a more precise analysis of the effect of particle size. Representative soils are heavy silty loam ($\bar{d}$ = 10 mcm), light silty loam ($\bar{d}$ = 30 mcm), and coarse-grained light sandy loam ($\bar{d}$ = 100 mcm). The σ_d value is equal to 0.65 in all soils. The reduction of particle size can first be detected by

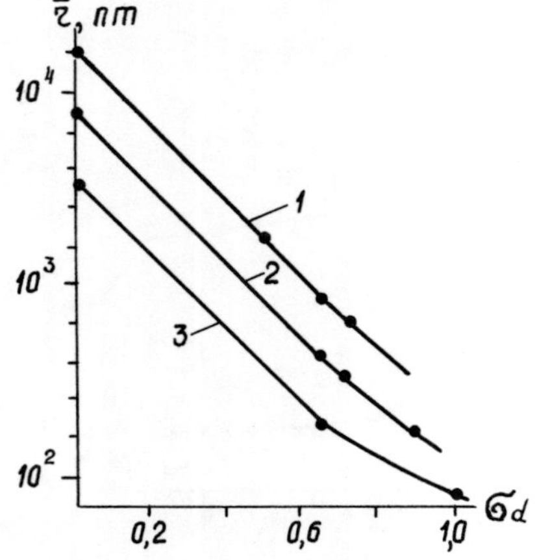

FIGURE 2 Relationship between average pore radius ($\bar{r}$) and root-mean-square deviation (σ_d) of particle-size distribution at different average sizes of particles ($\bar{d}$): $1 - \bar{d} = 1 \cdot 10^4$ nm; $2 - \bar{d} = 2 \cdot 10^4$ nm.

a reduction of the average pore radius (approximately 10 times), as seen in Figure 2, and secondly, by an increase in specific surface. The sharpest increase in the unfrozen water content occurs when the particle size changes from 100 to 30 mcm, and in heavy loam the liquid water phase content is 3-4 times greater (depending on temperature) than in sandy loam. It should be noted that due to availability of montmorillonite in light loam, its specific surface is larger than that of a heavy one.

In the second case, the influence of polydispersity is determined by selecting soils similar in $\bar{d}$ value and different in polydispersity (σ_d).

Increased polydispersity makes the pore space structure much thinner (Figure 2) while the specific surface expands at a slower rate. Therefore, in heavy sandy loam, fairly uniform in grade, the unfrozen water content is 1.5 times smaller than in light loam. The greater importance of unfrozen pellicular water and the smaller, compared to loam, importance of the capillary component of unfrozen water results in a more even $W_u(t)$ relationship in sandy loam, since the content of unfrozen film water varies slowly at $t < -4°C$ (Figure 1).

With a decrease in average particle size and an increase in polydispersity (these being the main features of sand-clay transition) the unfrozen water content grows larger and the range of intensive phase transitions wider due to proportional reduction of the average pore size (Figure 2) and the specific active surface of soils. During sand-clay transition, the capillary water increases, while unfrozen pellicular water decreases. In this case, the film thickness decreases by approximately one order, which is caused by a decrease in the specific surface energy (Rusanov 1967), and by a more rapid, compared to large particles, lowering of the force field stress as the mineral soil particle moves away from the surface (Andrianov 1946). Nevertheless, W_u increases as the transition to soils having small average size particles takes place, due to reduction in the size of the pore spaces and a rapid increase in the specific active surface of soils.

It is known that the unfrozen water content at a temperature lower than -1°C is maximum in montmorillonite clay, minimum in kaolinite clay, and medium in illite clay, although the specific water-absorbing capacity of montmorillonite clays is lower than that of kaolinite clays (Osipov 1979). The given soils are characterized by small specific surface energy and plasticity of the montmorillonite crystal lattice, which promotes high particle dispersity and an additional, in contrast to minerals with a nonplastic lattice, mode of pore volume distribution by radii in the area of $r < 10$ nm; this, in turn, increases the content of unfrozen water in montmorillonite clays. The small particle size and high polydispersity of montmorillonite clays, in which the high unfrozen water content results from a small average radius (about 30 nm) of intra-aggregate pores, facilitate this effect. Despite its large particle size, high polydispersity of illite clay resulting from its mineralogical composition, as well as the smaller average size of pores and very nonuniform pore space, make its unfrozen water content much larger than in kaolinite clay, even though their specific surfaces are practically the same (Figure 1).

A typical feature of the capillary mechanism of unfrozen water formation is its behavior at temperatures close to 0°C. In this case the W_u rate of change attains hundreds of percent per degree. Moreover, in soils with different mineralogical composition high initial rates of change occur in the water-phase content at various temperatures. For kaolinite clay it comprises approximately -0.6°C, for montmorillonite clay—approximately -0.3°C. This attests to the importance of the second (the third for bentonite) mode of pore volume distribution by radii, which depends on the size of interaggregate pores. The average size of bentonite microaggregates is twice that of kaolin despite very disperse grading. Accordingly, the size of interaggregate pores in bentonite is greater. This explains the rapid increase in the unfrozen water content in kaolin which at a temperature higher than -0.5°C becomes considerably larger than in bentonite (Figure 1).

Interaction of salts with the mineral skeleton surface and water governs the relationship between the unfrozen water content and soil salinity. Soils with a relatively small specific surface (both sands and clayey soils) are characterized by great dependence of the water phase content on salinity because particles of the diffusion layer of ions are but slightly suppressed by the diffusion layer close to the surface. At the same time suppression of the diffusion layer of ions is quite noticeable in soils having well-developed specific surface. This results in three ranges of the W_u-salinity relationship (Yershov et al. 1979). The lower the cryoscopic constant of this salt solution, the broader the range of weak salinity impact. When the diffusion layer of soil particles is completely suppressed by the diffusion layer of ions, an increase in W_u from salinity is in line with the theoretical relationship. The influence of pore solution concentration on the water phase content in frozen soils can be presented as a tangent variation of the slope angle of the straight line showing the relationship between the unfrozen water content and temperature and plotted according to a

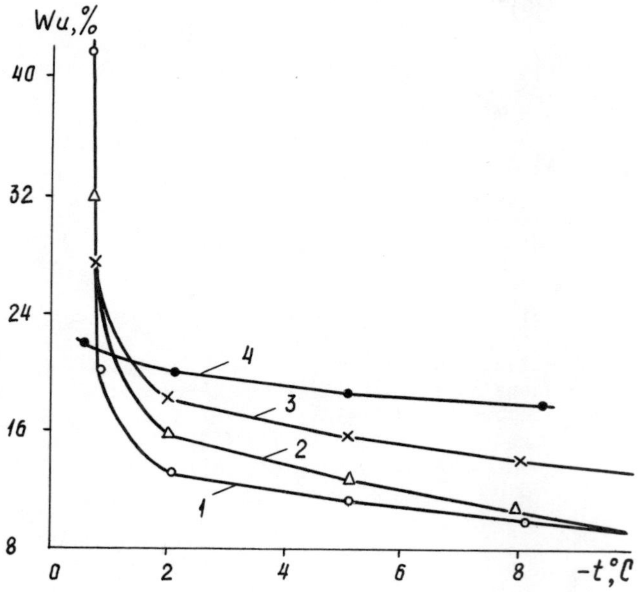

FIGURE 3 Relationship between unfrozen water content and temperature in siliceous soils: 1 - diatomite; 2 - diatom clay; 3 - tripolite; 4 - opoka.

logarithmic dependence. This influence can be estimated using cryoscopic constant of the solution, which is dependant upon the ionogenic capacity of salts and the interaction of ions with each other and with water molecules.

Thus, there exist complex inter-relationships between the unfrozen water content, structural and adsorptive parameters of soils, their grading and chemical and mineralogical compositions. These relationships are dictated by the conditions of formation and lithogenesis of dispersed deposits. Research into these relations should help us to explain the behavior of the unfrozen water content in soils of different geogenetic types (Yershov et al. 1979).

The explanation of the nature of unfrozen water formation is applicable not only to loose soils but also to tightly bound ones—sandstones, argillites, aleurolites, and siliceous soils (diatom clays, diatomites, tripolites, and opokas). Siliceous soils clearly show how their age influences their unfrozen water content. In the course of gradual transition from diatomites to tripolites, resulting from diagenesis and catagenesis, the volume of $\bar{r} = 4.10^2$ nm pores decreases and, accordingly, the volume of those with a radius smaller than 10 nm increases. The next stage of lithogenesis results in disappearance of $\bar{r} = 4 \cdot 10^2$ nm pores typical for diatomites and the formation of distinct bimodal distribution of pores by radii that changes for unimodal in opokas, where the pore radius is less than 10 nm. Thus, the volume of pores with this radius increases from zero in diatomites to 20% in tripolites and to 80% or greater in opokas, while the total pore volume decreases in all these soils. Accordingly, the unfrozen water content decreases at high temperatures ($t > -1.5°C$) and increases at low ones. In opokas, however, the change in W_u values is less noticeable (Figure 3).

REFERENCES

Ananyan, A. A., Golovanova, G. F., and Volkova, E. V., 1977, Issledovanija sistemy kaolin-voda metodom spinovogo eha. Transactions "Svyazannaja voda v dispersnyh sistemah," issue 4: Moscow, Moscow University Publications, p. 172-177.

Anderson, D., and Morgenstern, N., 1973, Physics, Chemistry and Mechanics of Frozen Ground. Proceedings of the Second International Permafrost Conference: Yakutsk: Washington, D.C., National Academy of Sciences.

Andrianov, P. I., 1946, Svyazannaja voda pochv i gruntov: Moscow-Leningrad, USSR Academy of Science Publications.

Anisimov, M. A., and Tankaev, R. U., 1981, Plavlenije l'da vblizi ghidrofil'noi poverhnosti. Journal of Experimental and Theoretical Physics, v. 81, issue 1(7), p. 217-225.

Danielyan, Yu. S., and Yanitsky, P. A., 1979, Neravnovesnye effekty v protsessah promerzanija vlazhnyh gruntov. Transactions of the Siberian Research Institute of Petroleum Industry, issue 47, p. 171-182.

Kurzayev, A.B., Qulividze, V.I., and Ananyan, A. A., 1977, Spetsifika fazovogo perehoda vody na poverhnosti biologhicheskih i neorganicheskih dispersnyh tel pri nizkih temperaturah. Transactions "Svyazannaja voda v dispersnyh sistemah," issue 4: Moscow, Moscow University Publications, p. 156-166.

Osipov, V. I., 1979, Priroda prochnostnyh i deformatsionnyh svoistv glinistyh porod: Moscow, Moscow University Publications.

Rusanov, A. I., 1967, Fazovyje ravnovesija i poverhnostnyje yavlenija: Leningrad, "Himija" Publications.

Saveljev, B. A., 1971, Fizika, himija i strojenije prirodnyh l'dov i myorzlyh gornyh porod: Moscow, Moscow University Publications.

Yershov, E. D., Akimov, Yu. P., Cheveryov, V. G., and Kuchukov, E. Z., 1979, Fazovyi sostav vlaghi v myorzlyh porodah: Moscow, Moscow University Publications.

Yershov, E. D., Shevchenko, L. V., and Yershova, G. G., 1974, Vlijanie temperatury na energhiju svyazi gruntovoi slaghi. Transactions "Merzlotnyje issledovanija," issue XIV: Moscow, Moscow University Publications, p. 151-160.

EFFECTS OF VARIATIONS IN THE LATENT HEAT OF ICE FUSION IN SOILS ON INTERACTION OF BOREHOLES WITH PERMAFROST

M. A. Anisimov, R. U. Tankayev, and A. D. Ulantsev

Research Institute of Petrochemical and Gas Industry
Moscow, USSR

Experimentally derived values of varying latent heats of fusion of ice ($\Delta\tilde{H}_m$) in samples simulating permafrost over a temperature range of -30°C to 0°C are cited. The value $\Delta\tilde{H}_m$ was shown to be heavily temperature-dependent and at -5°C it is more than six times less than the heat of fusion of homogeneous ice. Calculations of the thermal interaction of boreholes with permafrost have demonstrated that for clayey materials adjustments in the calculated radii of warming due to changes in $\Delta\tilde{H}_m$ may reach 150-300%.

Thermal calculation is a foundation for predicting the interaction of boreholes with permafrost. In computer calculations of thermal interaction of boreholes with permafrost one should take into account that at temperatures below zero, heat conduction of frozen rocks is complicated by water migration to the frozen front and that heat capacity is complicated by heat of ice fusion being absorbed in the temperature range between 70°C and 0°C. Simple estimations show that for the actual frozen soils and rocks in temperature range between -10°C and 0°C, 80-99% of heat capacity is determined by the heat of ice fusion occurring in the rock pores. The heat of ice fusion in the frozen rocks is one of the most important physical properties that governs, in the final analysis, the choice of methods for protecting the structures against harmful effects of permafrost.

Litvan (1966) was the first to demonstrate experimentally the possibility of change of ice fusion heat in permafrost rocks. Plooster and Gitlin (1971), Brun et al. (1973), Berezin et al. (1977) and Mrevlishvili (1979) investigated the integral heat of ice fusion in disperse systems of different origins. Their papers show that as the moisture content in the samples decreases, the integral molal heat of fusion is reduced. When the samples have a water content corresponding to 1-2 effective monolayers the heat becomes equal to zero, However, as far as we know, the data obtained by these authors were not used for thermal calculations since it was impossible to relate the integral heat of ice fusion to "mean" fusion temperature.

Figure 1 schematically illustrates a heat capacity anomaly stemming from the fusion of ice mass unity in permafrost rocks. The integral molal heat of fusion is equal to the shaded area bounded above by the relation curve of molal heat capacity of water in the sample and temperature and below by the consistent part of this relationship. According to more precise calorimetric (Mrevlishvili 1979, Anisimov and Tankayev 1981) and dilatometric (Litvan 1978) data the temperature at which fusion starts is about -70°C for different samples. Generally, a low-temperature branch of the heat capacity anomaly curve is not a smooth function of

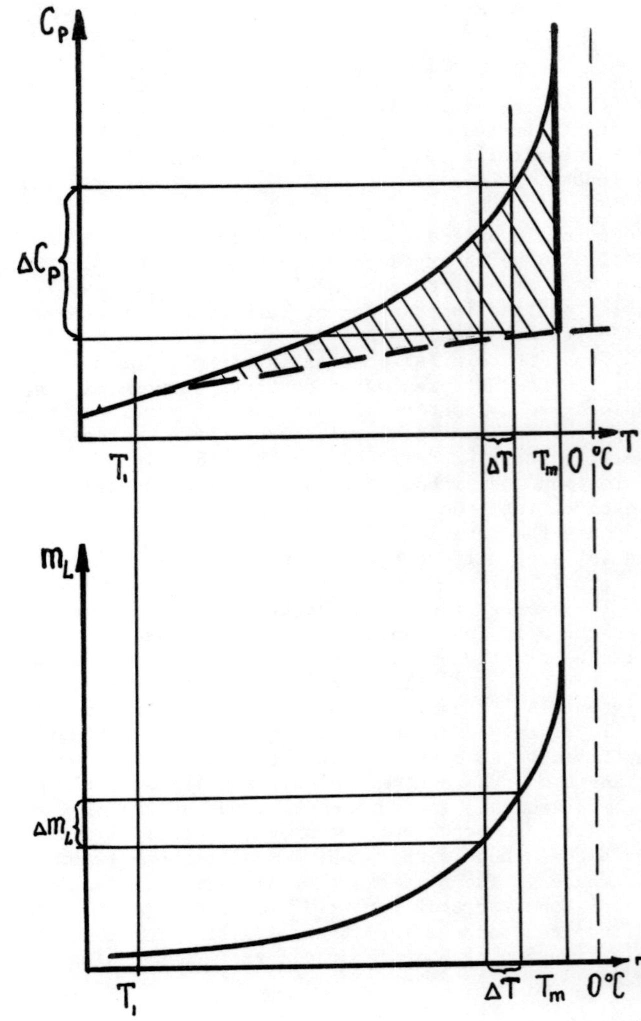

FIGURE 1 Schematic relationships between heat capacity (Cp) and liquid water content (M_L) and temperature (T) in a sample of frozen rock.

temperature and may be complicated by the peaks, which are due to either characteristic features of the crystalline structure of the mineral rock skeleton (Anderson and Tice 1971) or eutectics of salts dissolved in water (Mrevlishvili 1979). Evidently, integral molal heat of fusion is too crude a characteristic for such a complicated process to take place within a range of some tens of degrees.

To describe ice fusion in disperse systems the authors have introduced the concept of differential molal heat of fusion (Anisimov and Tankayev 1981) and developed a method of experimental determination. Thermodynamic definition of this value is expressed as follows:

$$\Delta\tilde{H}_m = \frac{d\Delta H_m}{dm_L} = \left(\frac{\partial\Delta H_m}{\partial P}\right)_T \cdot \frac{dP}{dm_L} + \left(\frac{\partial\Delta H_m}{\partial T}\right)_p \cdot \frac{dT}{dm_L},$$

at constant pressure:

$$(\Delta\tilde{H}_m)_p = \left(\frac{\partial\Delta H_m}{\partial T}\right)_p \cdot \frac{dT}{dm_L} \approx \frac{\Delta Cp \cdot \Delta T}{\Delta m_L}, \quad (1)$$

where $(\Delta\tilde{H}_m)_p$ is the isobaric molal differential heat of ice fusion in rocks; ΔH_m is variation of the soil sample enthalpy due to fusion; T and P are temperature and pressure; ΔCp and Δm_L are respectively the excess isobaric heat capacity of the sample due to fusion and the variation of moisture content in the sample at ΔT.

Figure 1 schematically illustrates the method of determination of initial data for the calculation $(\Delta\tilde{H}_m)_p$ according to formula (1) from experimentally derived relationships between Cp(T) and $m_L(T)$. When determining $\Delta\tilde{H}_m$ the following conditions should be observed:

1. Both initial relationships should correspond to the equilibrium state of a sample.
2. The condition $\Delta T << (T_m - T_1)$ should be observed.
3. Independent methods should be used to obtain both relationships for one and the same sample.
4. When calculating ΔCp, the data on the sample's phase relationship should be taken into account, i.e. $m_L(T)$.

In our experiments the heat capacity of samples was obtained by adiabatic calorimetry. The error of heat capacity measurements was less than 0.2%. The dependence of the sample's (m_L) moisture content upon the temperature was obtained by the wide line NMR method suggested by Kvlividze and Kurzayev (1979), who kindly provided the authors with a research sample and a table of their own results.

Just like Litvan (1966, 1978) and Plooster and Gitlin (1971) in our investigation we used a quartz glass-water system as a model sample. To make the sample aerosil, tiny spheres of quartz glass, diameter 200Å, specific surface 140 m^2/g, was mixed with distilled water or, in the case of sample 3, with a weak NaCl solution, and then degassed.

While calculating $(\Delta\tilde{H}_m)_p$, the value ΔT averaged to about 0.2°, which corresponds to about 1/300 of the entire range of fusion. The total error of the values $(\Delta\tilde{H}_m)_p$ varied from ±15% at -2°C to ±50% at -30°C. The relationship derived for $(\Delta\tilde{H}_m)_p$ and T is given in Figure 2. The dotted line indicates the limits of possible variations of $(\Delta\tilde{H}_m)_p$ caused by measurement errors $(\Delta\tilde{H}_m)_p$.

The relationship $(\Delta\tilde{H}_m)_p(T)$ has been obtained for the first time and, consequently, it is impossible to compare it directly with the data obtained by other authors. Nevertheless, indirect confirmation of the data obtained is possible.

The values of differential molal heat of fusion can be calculated based on differential molal heat of water vapor adsorbtion on quartz glass according to the formula:

$$\Delta\tilde{H}_m = T(\tilde{S}_v - \tilde{S}_s) - \Delta\tilde{H}_a, \quad (2)$$

where $\tilde{S}_v$ and $\tilde{S}_s$ are the values of differential molal entropy of water vapor and ice respectively at temperature T. Calculation of $\Delta\tilde{H}_m$ from formula (2) has certain drawbacks: it has to be assumed that differential adsorbtion molal heat $(\Delta\tilde{H}_a)$ is not dependent upon the temperature. There were no data taken at different temperatures because of experimental difficulties. The values $\Delta\tilde{H}_a$ are obtained for the samples of low moisture content only, which corresponds to low freezing points and results in large errors. Nevertheless, Figure 2 shows clearly that the values $\Delta\tilde{H}_m$ calculated in this way agree satisfactorily with our data. The results of the ΔH_a determination are taken from the paper of Yegorova et al. (1963).

Along with the relationship $(\Delta\tilde{H}_m)_p(T)$ a summary relationship between the integral molal heat of ice fusion (Q_m) in disperse systems of different origin and the temperature of ice fusion completion (T_m) is given (see Figure 2). Our relationship $(\Delta\tilde{H}_m)_p(T)$ allows the relationship $Q_m(T_m)$ to be completely explained which confirms the reliability of our data. However, comparison of data on the integral heat of ice fusion allows for an interesting conclusion.

As seen from Figure 2, in different disperse systems such as crystalline Al_2O_3, porous quartz glass, aerosil and biological object (DNA), the values of integral heat of ice fusion are almost the same at similar temperatures T_m. However, the scatter of each author's experimental data still corresponds to their deviation from the common relationship, which may be constructed from the data obtained by all the authors. This permits a well-founded hypothesis on the universality of relationship $Q_m(T_m)$ and, consequently, the relationship $(\Delta\tilde{H}_m)_p(T)$, at least in the temperature range from -15°C to 0°C. In this temperature range pre-melting is likely to be controlled by ice surface properties and is not dependent on underlying properties. This conclusion is of great practical significance, because it allows one to use a common relationship for any geological genetic type of permafrost rock and soil.

Thus, differential fusion heat allows one to describe in detail the heat capacity anomaly within a wide range of temperatures resulting from ice fusion in permafrost rocks. It is necessary to estimate the extent to which the variations of fusion heat affect the results of calculations for the thermal interaction of structures with permafrost. To achieve this, we must first assume that the derived relationship is universal and that only its specific surface needs to be known in order to describe the ice fusion in frozen rocks. There is doubt that this assumption is applicable where

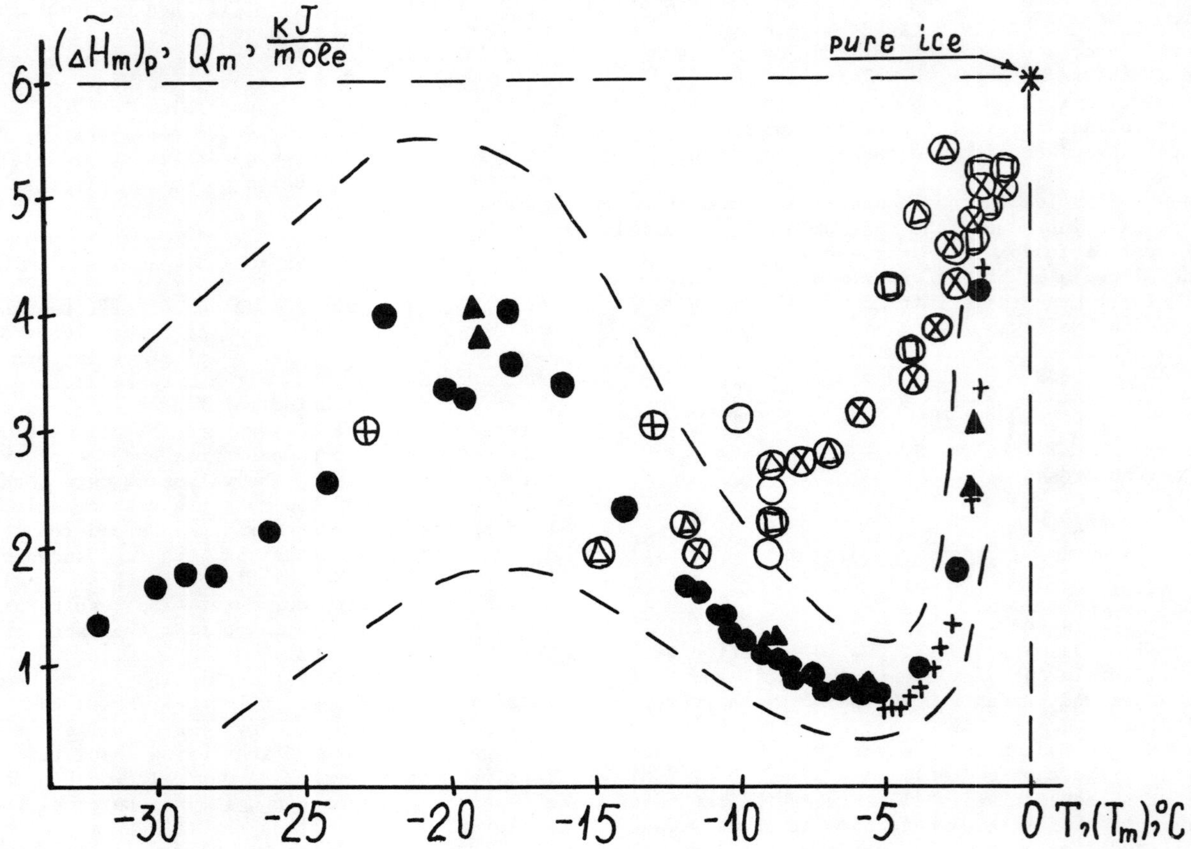

FIGURE 2 Relationship between differential heat of ice fusion $(\Delta H_m)_p$ and temperature T: +, •, ▲ represent samples 1, 2, 3 containing 31.6; 4.8; 6.2 grams of water per gram of aerosil respectively (our data). The values for ΔH_m have been calculated from formula (2) - ⊕. Relationship between integral molal heat of ice fusion (Q_m) and temperature of ice fusion completion (T_m): o - porous glass (Litvan 1966); ⊡ - aerosil (Plooster and Gitlin 1971, ⊿ - in Al_2O_3 (Brun et al. 1973); ⊗ - DNA (Mrevlishvili 1979).

there is only a comparatively narrow temperature range but the initial temperature of permafrost is rarely -10°C. Besides, mineralization of formation water has not yet been taken into account. This is a matter for future study.

Computers have been used in the calculation of thermal interaction of the borehole with permafrost. The following initial data were taken into account in the two-dimensional solution for heat conductivity: the borehole is surrounded by a homogeneous stratum of frozen rocks. Density of the mineral skeleton is 2.5 g/cm³. Complete water and ice content in the rock varies between 30% (by vol.) at the surface to 20% at a depth of 300 m. The initial temperature of permafrost varies between -4°C at a depth of 25 m to 0°C at a depth of 300 m. Temperature on the hole wall shows linear variation from 20°C at the well head up to 30°C at a depth of 300 m. In addition, various seasonal temperature variations at the surface and the heat flow from the formation have been examined with rock specific surface varied.

The computations have shown that a correction to the size of the thawing zones in case of rocks whose specific surface is less than 5 m²/g is not essential for engineering calculations. Such rocks include sand and sandstone. For the rocks with the specific surface 10 m²/g, however, the thawing zone radii, where changes of heat of ice fusion in the rock have not been considered, are underestimated by 1.5 times. With an increase in specific surface the difference in calculation results increased and, in the case of rocks with high clay content and a specific surface of 30 m²/g, the thawing zone radii differed by 3 times. The calculation program has allowed computation of both the temperature profile in the vicinity of the borehole and the position of the ice fusion front. In the case of particularly clayey rocks with low moisture content the ice fusion front and 0°C temperature profile did not coincide.

Figure 3 illustrates the calculation results of the position of the 0°C isotherm 1,245 days after start-up of the borehole "operation" for a rock specific surface of 20 m²/g.

It is common knowledge that the specific surface of clays may attain several hundred square meters per gram of soil. Calculations of thermal interactions of structures with permafrost soils and rocks made without considering the actual physical

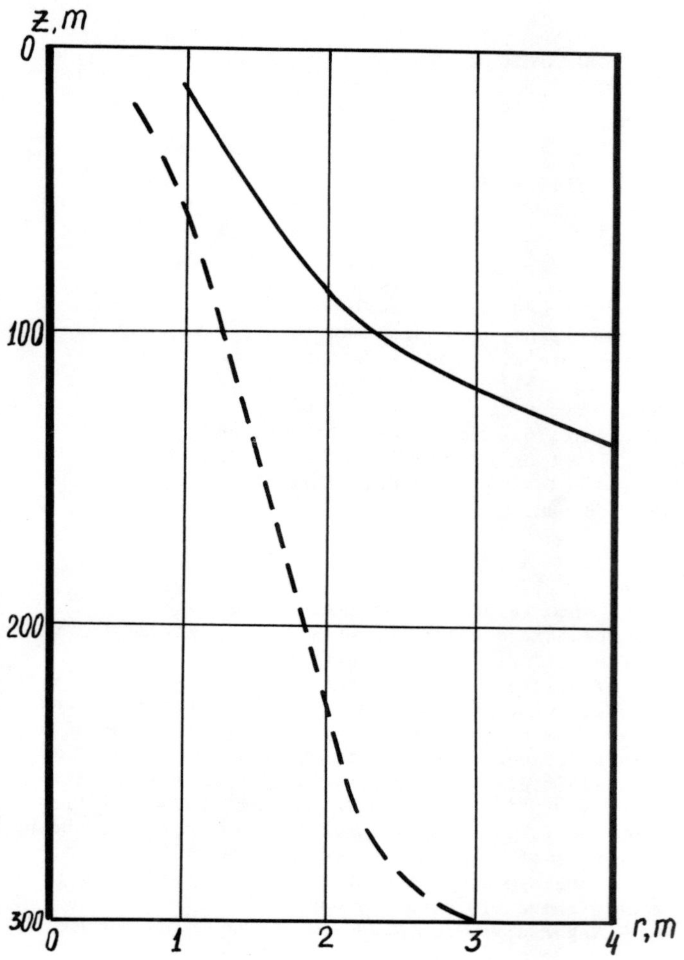

FIGURE 3 Position of the computed 0°C isotherm 1,245 days after the beginning of permafrost rock warm-up. Solid and dotted lines respectively show the position of the isotherm with allowance for $(\Delta\tilde{H}_m)_p(T)$ and without such an allowance. Z is the borehole depth (meters); r is the distance from the borehole centerline (meters).

pattern of ice fusion in frozen rocks are unrealistic. However, since layers of pure montmorillonitic clays rarely occur, such a case is atypical. atypical.

The conclusion may be drawn that a detailed study of ice-water phase transitions in frozen rocks made by modern techniques of experimental physics would provide a considerable increase in the precision and reliability of thermal calculations in permafrost.

REFERENCES

Anderson, D. M., and Tice, A. R., 1971, Low temperature phases of interfacial water in clay-water systems, Soil Science Society Amer. Proc., v. 35, p 47-54.

Anisimov, M. A., and Tankayev, R. U., 1981, Ice fusion near a hydrophilic surface: Zh. Eksp. Teor. Fis. v. 7, p 217-225 (Sov. Phys. JETP, v. 45, p. 110).

Berezin, G. I., Kozlov, A. A., Kooznetsova, L. V., 1972, Calorimetric study of capillary condensate phase transition. In Surface forces in thin films and disperse systems. Moskva, "Nauka" Publishers, p 202-209.

Brun, M. M., Lallemand, A., Loret, G., Quinson, I. F., Richard, M., Eyrand, L., Eyrand, C., 1973, Chengement d'etat liquidsolid dans les milteux poreux: Journal d'Chimie Physique, v. 70, no. 6, p 973-978.

Kvlividze, V. I., Kurzayev, A. B., 1979, Properties of thin water layers according to NMR method. In Surface forces in thin films: "Nauka" Publishers, p 211-215.

Litvan, G. G., 1966, Phase transition of adsorbates, I Specific heat and dimensional changer of the porous glass-water system: Canadian Journal of Chemistry, v. 44, p 2617-2622.

Litvan, G. G., 1978, Adsorption systems at temperatures below the freezing point of the adsorptive: Advances in Colloid and Interface Science, v. 9, no. 4, p 253-302.

Mrevlishvili, G. M., 1979, Low temperature calorimetry in biological macromolecules: Uspekhi Fisicheskich Nauk (Sov.), v. 128, no. 2, p. 273-312.

Plooster, M. N., and Gitlin, S. N., 1971, Phase transitions in water adsorbed on silica surfaces: The Journal of Phys. Chem., v. 75, no. 21, p. 3322-3326.

Yegorova, T. S., Kvlividze, V. I., Kiseliev, V. F., Iyevskaya, N. M., Sokolov, N. D., 1963, On nature of silica surface and water bond. In Modern views on bound water in rocks: Moscow, The USSR Academy of Sci. Pub., p 35-54.

MASS TRANSFER IN THE SNOW COVER OF CENTRAL YAKUTIA

A. L. Are

Permafrost Institute, Siberian Branch
Academy of Sciences, Yakutsk, USSR

On the basis of ten years of experimental research it has been established that the intensity of mass transfer in the snow pack is temperature-dependent and reaches its maximum at snow temperature above -20°C. The mean rate of vapor diffusion in snow layers near the base of the pack varies between $(2.0\text{-}2.5) \times 10^{-3}$ g/cm^2 per day. The "norms" were obtained for snow sublimation for all types of landscape in Central Yakutia, e.g. 12 mm on meadows; 5.5 mm in pine forest; 6.2 mm in larch forest; and 8.5 mm on lake ice. The value for mean daily sublimation of snow is determined on the basis of the air humidity deficit and the wind velocity. Ten-day sublimation totals were determined on the basis of the absorbed radiation.

During the last decade, mass transfer in the snow cover of Central Yakutia has been explored by the Permafrost Institute of the Siberian Branch of the USSR Academy of Sciences in studies of the snow cover effect in the permafrost thermal regime. The methods and first results of these investigations are reported in Pavlov (1965) and Are (1972, 1978a, 1978b). The snow cover is regarded as an independent natural component in conjunction with the entire variety of wintertime natural conditions. A study is made of the aqueous vapor diffusion within the snow cover thickness, of its surface evaporation and snow recrystallization processes, resulting from mass transfer under naturally occurring conditions.

There are presently very few in-situ measurements of the aqueous vapor diffusion rate in the snow cover: in the European part of the USSR these have been made in the Caucasus (Kuvaeva 1961, 1972) and in localities near Moscow (Pavlov 1965), whereas for continental regions which are characterized by longer snow cover, only data obtained in Transbaikalia (areas south of Lake Baikal) and Western Siberia are available (Kolomyts 1966, 1971).

Experimental measurements of snow sublimation are considerably more numerous. These have been carried out in various physiographical regions. Empirical formulae have been derived for sublimation assessment. But as far as Siberia and the North-East of the USSR are concerned, the data are still very meager. However, in such drought-ridden areas as Central Yakutia the determination of snow sublimation losses is of vital importance for calculating water resources.

Years of stationary observations have revealed that in Yakutia the shallow snow cover is characterized by a high degree of recrystallization, during which in a very short time the greater part of the snow depth (up to 70-90%) assumes a deep-hoarfrost structure (Are 1972). Deep-hoarfrost crystal formation begins in the early days of the formation of a stable snow cover, whose density does not exceed 0.09-0.13 g/cm³. It assumes a continuous course and within 3 or 4 weeks the original snow structure disappears completely in the snow layers at the base of the pack and a deep-hoarfrost horizon with a well-defined "fibrous" structure develops.

The investigations carried out by Kuvaeva (1972) in the Caucasus have shown that the formation and development of deep-hoarfrost crystals occur at a rate of aqueous vapor diffusion in excess of $1.3\text{-}1.5\times10^{-3}$ g/(24 hr·cm^2) and a mean air temperature below -10°C. Such conditions are characteristic of the formation period of the snow cover in the regions of Yakutia, where only one event of a 10-day-average air temperature was observed to be well above -10°C in the past decade.

Vapor diffusion measurements have indicated that the mean diffusion rate in snow layers at the base of the pack varies from (2.0-2.5) 10^{-3} g/(24 hr·cm²) to (3.0-4.0) 10^{-3} g/(24 hr·cm²), depending on soil moisture and weather conditions during the cold season. The mean diffusion rate, as measured in Western Siberia (Kolomyts 1977), is of the same order of magnitude, but varies over a wider range from (1.5-2.5) 10^{-3} to (3.5-5.0) 10^{-3} g/24 hr cm²). Near Mt. Elbrus the mass transfer rate is significantly lower (Kuvaeva 1972), and its maximum, equal to 3.0 10^{-3} g/(24 hr·cm^2) was observed only during one of five winter seasons during the period of lowest air temperatures.

It is known that when the snow's temperature is comparatively high, a thin quasi-liquid layer is formed on the surface of its crystals, whose molecules have considerable mobility (Ushakova et al. 1968). By increasing the snow's temperature the thickness of this layer increases and the molecular mobility intensifies; under such conditions the rate of sublimation-induced growth of crystals also increases. The maximum rate of sublimation recrystallization is probably in a temperature range of from -5 to -15°C. The temperature-dependence of the snow cover recrystallization rate was confirmed through experimental observations in Khibiny Mountains (Savelyev et al. 1967). A relationship between the vapor diffusion rate and the snow cover temperature was found also in a study of mass transfer in the snow cover of Western Siberia, where at snow temperatures below -10°C the

TABLE 1 Intensity of Water Vapor Diffusion i · 10^{-3} g/(24 hr · cm²) and Thermal Properties of the Snow Cover in the Winter of 1976-1977

Period	h = 5 cm						h = 10 cm		
	t	$\frac{\Delta t}{\Delta z}$	i	t	$\frac{\Delta t}{\Delta z}$	i	t	$\frac{\Delta t}{\Delta z}$	i
In a larch forest (near a lake)									
23-29 Dec	-31.2	1.4	2.0	-35.1	0.8	3.4	-33.6	1.0	2.6
22 Dec-2 Jan	-33.4	-	1.9	-33.2	0.6	1.6	-33.6	0.7	1.3
2 Jan-10 Feb	-28.0	0.9	1.9	-31.3	0.7	1.7	-30.2	0.7	1.6
10-20 Feb	-26.2	0.5	2.5	-26.8	0.4	2.0	-27.2	0.3	1.9
20 Feb-1 Mar	-26.8	0.5	2.0	-27.1	0.5	2.3	-30.2	0.5	0.8
1-10 Mar	-27.4	0.4	2.7	-26.0	0.4	2.0	-27.0	0.4	2.0
On ice of a lake									
23-29 Dec	-18.7	2.0	14.1	-19.4	2.2	19.0	-27.8	1.8	8.3
29 Dec-2 Jan	-25.2	2.2	7.4	-26.2	0.9	6.9	-29.6	1.4	4.2
2 Jan-10 Feb	-22.6	1.6	7.2	-18.9	1.5	7.3	-22.5	1.2	5.3
10-20 Feb	-14.6	0.9	18.3	-16.8	1.2	18.2	-18.6	1.4	12.6
20 Feb-1 Mar	-15.7	0.8	18.3	-14.9	1.1	21.4	-17.6	0.9	13.0
1-10 Mar	-17.3	1.4	12.9	-14.3	1.4	5.6	-20.3	0.9	8.0

Note: t is the mean temperature of snow layer (°C) in which diffusion is being measured; $\Delta t/\Delta z$ is the temperature gradient of this layer (°C/cm).

increase in the mass transfer rate was detected only up to gradients of 0.55-0.60 °C/cm. At higher differences in temperature the diffusion rate is practically independent of the temperature gradient (Kolomyts 1977).

The most favorable conditions of mass transfer are created within the warmest near-soil snow layer, where the vapor concentration gradient is greatest. The mass transfer rate in the snow cover decreases both with height and in the process of subsequent snow cooling during the winter insofar as the recrystallization rate is determined not only by the temperature gradient, but also by the temperature itself.

The highest rate of mass transfer in Yakutia was measured in the snow cover on lacustrine ice (Table 1). The obtained data show that in the snow cover at a level of 5 cm above forest soil the average winter diffusion rate comprises 2.2 10^{-3} g/(24 hr·cm^2). At the same level in the snow cover on lacustrine ice it comprises 13.0 10^{-3} g/(24 hr·cm²), i.e. six times the former. In addition, the winter-average temperature gradients in the same snow layer are 1.4°C/cm on the lake and 0.6°C/cm in the forest, i.e. they differ by a factor of 2 only. Such a high rate of mass transfer is due to the warming effect of a lake not frozen to the bottom. The temperature in the lower snow layer here averages about 8-10°C higher than in the same layer in the forest and on individual days this difference reaches 12-15°C. The relationship between the water vapor flow rate and the snow cover temperature is shown in the diagram (Figure 1), according to which a significant increase in the rate takes place at snow temperatures above -20°C. Figure 2 shows vapor diffusion rate versus temperature and temperature gradients. The largest rate of diffusion flow at the same temperature gradients, as follows from the diagram, corresponds to the range of lower snow layer temperatures from -6 to -15°C.

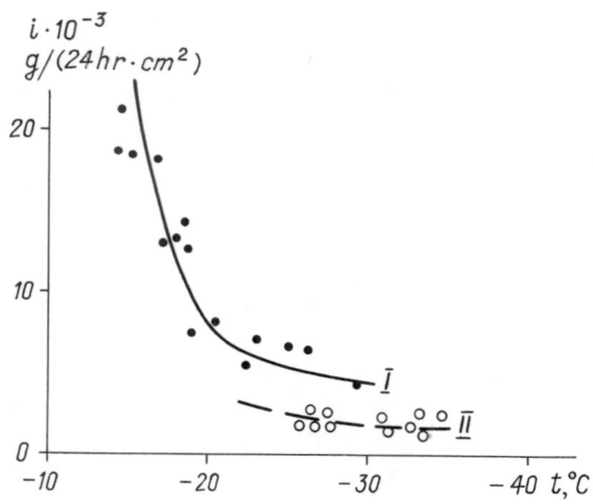

FIGURE 1 Vapor diffusion intensity as a function of snow temperature based on depth.
I - on lake ice
II - on the shore of a lake

During the winter of 1976-1977 (from 4 October to 14 March) 0.430 g/cm² of material was carried away from the soil surface of a dry glade while that removed from a snow layer 10 cm high above soil-top amounted to 0.209 g/cm², i.e. 2 times less.

Years of observations of snow sublimation have revealed a high annual variability in the total amount, associated with the annual variability of the weather. A norm of the snow sublimation rate has been obtained for all the basic types of landscape of Central Yakutia. For wide-open, forest-free regions it is about 12 mm or 16% of the amount of solid precipitation during the cold season. In some years the evaporated amount of moisture can exceed one-quarter of the entire winter moisture

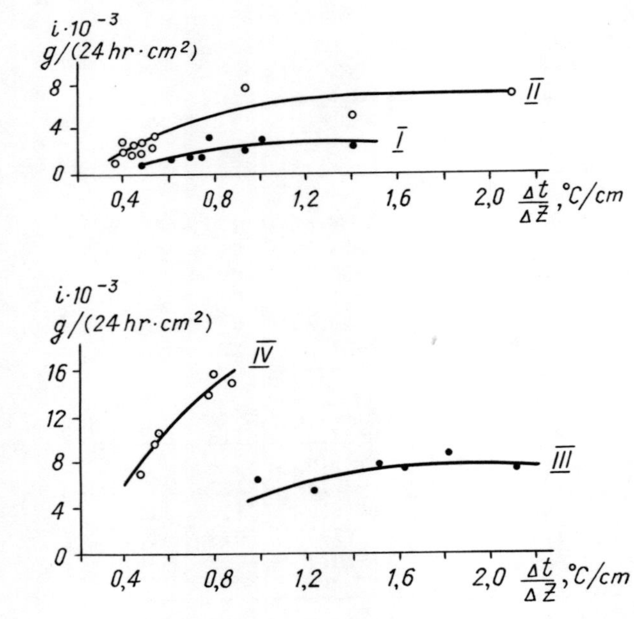

FIGURE 2 Relationship of the water vapor diffusion intensity i with temperature and temperature gradient $\Delta t/\Delta z$ in a snow layer 0-10 cm thick. The range of negative temperatures is: I - 30-35°C, II - 24-29°C, III - 18-23°C, IV - 6-15°C.

storage, i.e. it can substantially diminish, in the overall budget, the contribution from solid precipitation to melt flow and soil imbibition. Sublimation under the canopy of a pine forest is less than in open areas by a factor of 2.5.

In a larch taiga the sublimation in closed glades averages 9.7 mm while under a forest cover whose canopy is less dense than that in a pine forest due to the shedding of the needles, it is less by a factor of 1.5. The snow sublimation on the lacustrine ice surface is 36% less than that in a glade that stems from air thermal inversion in the deep basin of the lake.

The seasonal variation of snow sublimation, which has been studied using daily measurements, demonstrates the following features. In the coldest season from the beginning of winter till mid-March the sublimation intensity is low which is associated with the low temperature of the air and, subsequently, the snow cover's surface. From the second half of March the air temperature rises appreciably and snow sublimation intensifies. For the last 10-day period of March it constitutes 60% of the monthly total, or 1.9 mm. In April the sublimation rate continues to increase and its 10-day-averaged sums are 2.7, 3.7, and 5.3 mm. Figure 3 shows a plot of sublimation intensity versus ambient temperature. The rather high relative humidity and a weak wind, which characterize Yakutia winters, result in the lowest sublimation (curve 1). As the ambient temperature

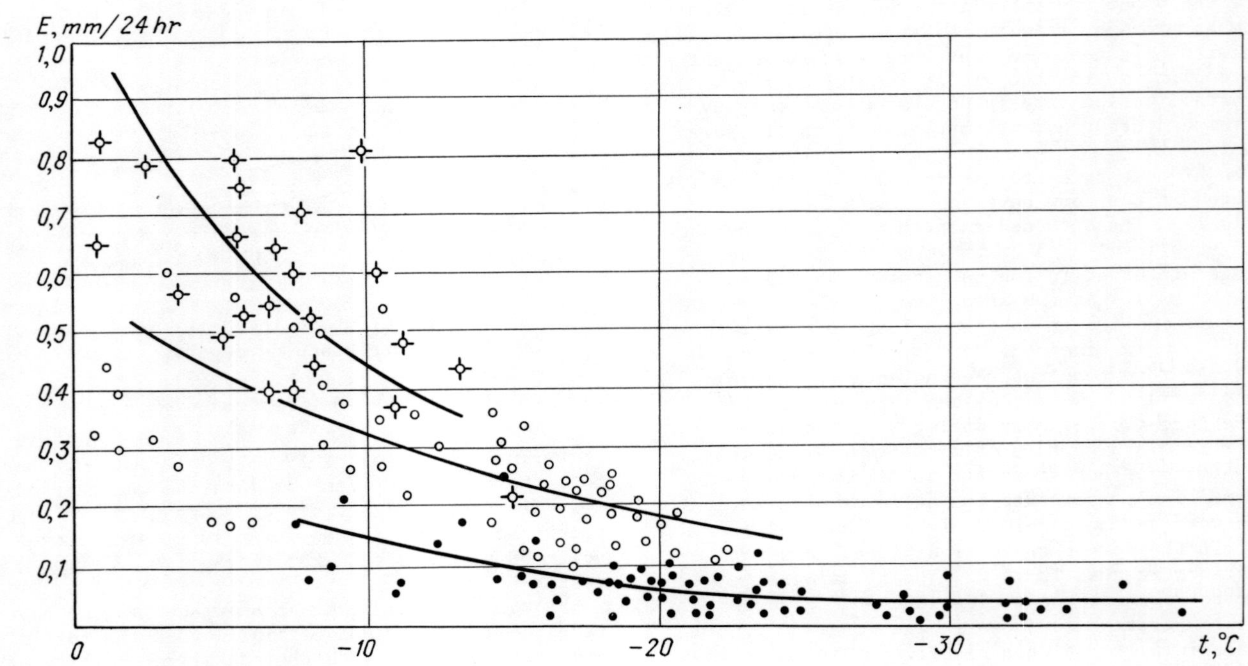

FIGURE 3 Relationship of snow sublimation with ambient temperature, daily mean relative humidity of air W(%) and daily mean wind speed V (m/s).

1 - W = 60 - 75%	V = 1 - 4 m/s
2 - W = 50 - 59%	V = 5 - 6 m/s
3 - W < 50%	V > 6 m/s

increases there is an increase in snow's surface temperature and the humidity gradient of the air layer closest to the surface layer rises which leads to increasing sublimation. Within the range of air temperatures from -20 to 0°C the sublimation rate is greater, the lower the relative humidity and the stronger the wind (curves 2 and 3). The figure shows that the best conditions are created at air temperature above -10°C.

The data from 190 daily measurements of evaporation from the snow surface have been used to obtain the empirical relationship

$$E = (193 + 18V) \cdot 10^{-5} \cdot d, \qquad (1)$$

where E is the sublimation rate from the snow surface, mm/24 hr;
V is the daily mean wind speed at 17 m altitude, m/s;
d is the daily mean deficit of air humidity, Pa.

The correlation coefficient for the relationship obtained is 0.75±0.05. The error of determination of sublimation value by the formula (1) as compared to the measured value is ±0.01 mm/24 hr.

According to analysis of the existing regional formulae, made by Kuzmin (1974), errors in calculation of the value for daily snow sublimation are on the average 2-3 times greater than those of its in-situ measurements. However, as the time interval increases, the errors decrease. Therefore formula (1), like other regional formulae, may be recommended for calculation of the mean sublimation rate for extended periods, a ten-day period, a month, and more.

During the typically radiant spring weather in Central Yakutia, when because of fair weather with some clouds and the high transparency of the dry air the solar radiation intensity is very great, the absorbed solar radiation plays a decisive role in snow sublimation and melting. In analytic terms this relationship can be roughly expressed as:

$$E = 72 \cdot 10^{-5} Q_n - 0.927, \qquad (2)$$

where E is the 10-day total of evaporation, mm;
Q_n is the 10-day total of the absorbed radiation, cal.

The correlation coefficient of this relationship is 0.74 ±0.11. The calculation error for the 10-day total sublimation, as compared to the measured sum, is ±0.16 mm.

Research results indicate that despite extremely severe climatological conditions the intensity of mass transfer is high in the snow cover of Central Yakutia. The temperature of the snow cover itself makes a crucial contribution to this process. A long-term "normal" of sublimation from the snow cover surface has been obtained for the major types of landscape in Central Yakutia: wide-open areas, coniferous and larch forest, lacustrine ice. An empirical relationship between the daily mean evaporation of snow and the humidity deficit of air and wind in addition to the relationship between a connection of 10-day sublimation totals and absorbed radiation have been established. These, then, can be used to calculate the water balance.

REFERENCES

Are, A. L., 1972, Evaporation and evolution of the snow cover in localities near Yakutsk. In Experimental investigations of mass transfer processes in frozen rocks: Moscow, Nauka, p. 160-167.

Are, A. L., 1978a, Snow cover of Central Yakutia, properties of its radiation and hydrothermal regime. In Heat exchange in permafrost landscapes: Yakutsk, p. 30-42.

Are, A. L., 1978b, Experimental investigations of the hydrothermal regime properties of the snow cover of Central Yakutia. In Geothermalphysical investigations in Siberia: Novosibirsk, Nauka, p. 41-47.

Kolomyts, E. G., 1966, Snow cover of mountain-taiga landscapes of the North of Zabaikalia: Moscow-Leningrad, Nauka, 183 p.

Kolomyts, E. G., 1971, Structure and regime of the snow cover of the West-Siberian taiga: Leningrad, Nauka, 174 p.

Kolomyts, E. G., 1977, Methods of crystallomorphological analysis of snow structure: Moscow, Nauka, 198 p.

Kuvaeva, G. M., 1961, Water vapor migration and structural changes of snow. Proceedings of the First Transcaucasian Scientific Conference on the Study of Snow Cover, Snow-Slips and Glaciers of the Caucasus. Proceedings of the Tbilisi Hydromet. Research Inst., v. 9: Leningrad, Gidrometeoizdat.

Kuvaeva, G. M., 1972, The conditions and time required for deep hoarfrost to evolve in a snow cover. In Proceedings of the High-Altitude Geophysical Institute, v. 18, "Snow and snow-slips": Leningrad, p. 61-70.

Kuzmin, P. P., 1974, On the calculated and experimental methods of determining the evaporation from the snow cover surface. Proceedings of the Fifth Transcaucasian Scientific Conference on the Study of Snow Cover, Snow-Slips and Glaciers of the Caucasus: Leningrad, Gidrometeoizdat, p. 25-43 (Proceedings of the Transcaucasian Hydromet. Research Inst., v. 58).

Pavlov, A. V., 1965, Heat exchange of freezing and thawing earth materials with the atmosphere: Moscow, Nauka, 254 p.

Savelyev, B. V., Laptev, M. N., Lapteva, N. I., 1967, Structure, composition, physical-and mechanical properties of snow in Khibiny Mountains and their variation in the process of metamorphism. In Snow and avalanches of Khibiny Mountains, p. 201-239.

Ushakova, L. A., Kvlividze, V. I., Sklyankin, A. A., 1968, Advances in Science. Hydrology of dry land, glaciology: Moscow, VINITI.

POSSIBLE APPLICATIONS OF FOAM CEMENTS IN PROTECTING THE ENVIRONMENT IN CONNECTION WITH WELL DRILLING

V. Bakshoutov

Research Institute of Petrochemical and Gas Industry
Moscow, USSR

The article presents the results of investigations into the application of crystallo-chemical intensification of the oil-well foam cements of the "Aerotam" type (aerated plugging material) with densities of 800 to 1600 kg/m^3 and thermal conductivities of 0.35 to 0.55 W/m^2 °K at temperatures from -10°C to +160°C. It demonstrates that cement slurries of the "Aerotam" type may find wide industrial application for protecting the environment in well construction and as "passive" thermal insulation for cementing the casings of oil and gas wells in permafrost areas, and also as a super-light plugging material for cementing casings under conditions of abnormally low well pressure and intense grout absorptions. Thawing of the permafrost during drilling and operation of wells in oil and gas condensate fields leads to irreversible ecological damage. Various methods of "active" and "passive" thermal protection are used; to prevent such thawing, in either case practice has shown that the best results have been achieved by the application of special plugging materials with high thermal insulating properties, namely foam cements, for well cementing.

In 1974 for the first time we obtained a plugging foam cement slurry with density of 800-1300 kg/m^3 and thermal conductivity factor of 0.35 W/m^2 °K, which normally hardened at temperatures as low as -10°C with the formation of the cement stone, which after two days of consolidation had a bending strength of from 0.3 to 1.2 MPa (Bakshoutov 1976, 1977).

Further work made it possible to modify foam cement slurries of the "Aerotam" type for low and high temperatures. Composition, properties, and the results of pilot studies under different conditions are cited in the works (Bakshoutov 1977-1981).

Plugging foam cement slurries consist of the following components:

1. Mineral viscous—commercial products or special oil-well cements and mixtures for specific applications.
2. Liquid for mixing the cement-calcium-alkalinity solution (K_2CO_3 + KOH) for low temperatures, salinated water and drilling mud for high temperatures and salt aggression.
3. Liquid emulsifier (foamer)—SAA (surface acting agents) or surfactants of the alpha-olefin sulphonate or "Afrox-200" (USA) type, and also sulphanol, liquid soaps, thermoresistant SAA (for example, polymetilensulphometilen-phenolate natrium, alkylethoxylated sulphate) and others, that are selected based on the type of mixing liquid and the thermal conditions to provide the foaming of the slurry and to obtain a stable foam with high foam quality and compressive strength.
4. Hard emulsifier (the foam stabilizer)—synthetic (aerosil, butoxy-aerosil) and natural (mould, flask, tripoli) preparations of silica with high dispersion ability, and various mud powders. In addition, it stabilizes the bubbles of air in the cement slurry.
5. The pores colmatater—the polymer silicorganic liquid of the HSW-type (used in a calcium-alkaline medium), silicate soluble or kasein (for water and mineralized liquids for cement mixing). It colmatates the permeable capillary pores in the structure (texture) of the cement stone.
6. Air (Bakshoutov 1976-1981) or nitrogen (Montman 1982)—to decrease the density of the slurry and of the cement stone.

To obtain a highly stable foam cement slurry and to optimize its composition a Box-Wilson experiment is conducted, which, with a total number of tests $n = 16$, makes it possible to describe the mathematical model of the process and to find its optimum. As the optimization parameter the Sp—stability of foam cement slurry—was used. For the low-temperature "Aerotam" modification the adequate regression equation looks like

$$y = 89.9x_0 + 2.88x_1 + 1.13x_2 + 4.63x_3 + 0.38x_4 + 2.13x_1x_2 - 0.38x_1x_3 - 1.63x_1x_4;$$

$$x_1x_2 = x_3x_4; \quad x_1x_3 = x_2x_4; \quad x_1x_4 = x_2x_3,$$

where y = parameter of Sp in %;
x_0, x_1, x_2, x_3, x_4 = concentration of cement, of calcium-alkalinity solution for cement mixing, of liquid and of hard emulsifiers, and of the polymeric additive, in mass %.

Planning which was done shows that to obtain a foam cement slurry with Sp = 100% its composition must include:

5% water solution of calcium-alkalinity solution (K_2CO_3 + KOH); 0.5% of liquid, 0.05% of hard emulsifiers, and up to 0.9% of the polymeric additive.

The results of research on the relationship between the strength of the foam cement stone and its density, processed with the assistance of a "Vang-2200" micromodule computer according to the "Basic" programme, showed that the curves are described using an equation of the second degree:

$$\bar{\sigma} = a_0 + a_1\rho + a_2\rho^2 ,$$

where $\bar{\sigma}$ is the compressive strength, in MPa and ρ is the density of the cement stone, in kg/m^3, with a reliability equal to 0.93-0.99; the coefficient of correlation (correlation factor) is equal to 0.96-0.99.

Research on the water absorption kinetics of the foam cement stone has shown that the samples with the polymeric additive of the optimum composition have a water absorption factor about three times less than the same samples without the colmatating additive, which helps the stone keep its good thermoinsulating properties in the well both during drilling and while the well is in operation.

The most important characteristic of the foam cement is its structural porosity, i.e. the distribution of pores in the volume of the cement stone according to their sizes, mainly influences its strength, permeability, thermoinsulating properties, etc.

The study of the structural porosity of the foam cement stone by measurement of mercury porosity at low and high pressure shows that the increase of hardening pressure decreases the total porosity from 79 to 55%.

Analysis of the research results allows us to distinguish three groups of pores: impermeable micropores with a radius of 50-500 Å, medium radius of 500-10,000 Å, and the permeable micropores with a radius of more than 10,000 Å. As the hardening pressure increases, the volume of impermeable micropores grows, and the volume of the medium-radius pores and micropores drops (Table 1).

TABLE 1 Porosity

P (MPa)	ε (%)	$r_1 >$ 10^4Å	10^4Å$>r_2$ $>10^3$Å	10^3Å$>r_3$ $>10^2$Å	10^2Å$>r_4$ >50Å
3.0	55	30.2	21.0	35.4	13.4
2.0	58	33.6	26.0	28.4	12.4
1.0	64	44.2	14.9	29.5	11.4
0.5	68	45.7	23.2	22.1	9.0
0.1	69	56.9	16.0	21.4	5.7

This provides for the formation of a denser cement stone structure with increased strength and decreased permeability.

The structure of the foam cement stone was also comprehensively studied by raster and translucent electron microscopy to obtain the photos of the pores with high depth and image definition.

Air pores of the foam cement stone arranged in the volume of hardening system are separated by viscous partitions of varying thicknesses, from 1.5 μm and less, up to the thickness of the particles of hard emulsifier and the surfactant adsorption layer.

If the liquid film of the foamer covers the surface of an air bubble immediately after the foam cement is mixed, then during further hydration of mineral particles its thickness may gradually decrease until it disappears, when the air pores will freely communicate, forming easily permeable micropores and channels 10 to 18 μm in size. Aldrich and Mitchell (1976) define their range of use by the magnitude of foam cement porosity with the inclusion of surfactants: Highly porous with relative porosity more than 52% are well permeable for liquid and gas can be used for plugging the bottomhole zone of sand-drifting (sand-flowing wells; cements with low porosity where the capacity for bubble communication is low may be used for cementing casings.

As the foam cement hardens, its permeable micropores become covered with the products of new hydrate formations, and the stone's permeability drops. To provide minimum permeability in the early stages of hardening it is necessary to add the modified additive to the foam cement slurry that colmatates the opened pores in their partitions. In the "Aerotam" low temperature slurry, a silicate-organic liquid is used that rapidly polymerizes in the alkaline medium of the cement mixing liquid.

The formation of the permeable macropore proceeds in such a way that after 12 hours of hardening it is 6 μm in size and may be filled with free capillary water. Over time the pore becomes covered with products ot silicate-organic liquid polymerization with silica gel gelation and topochemical sedimenting.

To determine the nature of the interconnection between the size of pores in the foam cement stone and its main properties (Table 2) (density ρ, porosity ε, permeability K, the bond force with metal $\bar{\sigma}_c$, and heat conductivity factor λ) the $\bar{R}$ parameter is used, which is a relatively medium pore radius.

TABLE 2 The foam cement properties

$\bar{R}$ (MPa)	$\bar{\sigma}$ (MPa)	$\bar{\sigma}_c \cdot 10$ (MPa)	$K \cdot 10^{-15}$ (m^2)	ρ (kg/m^3)	λ ($BT/m^2 {}^\circ K$)
3.0	3.66	2.4	1.5	1380	0.59
2.0	3.35	2.2	3.4	1290	0.53
1.0	2.30	0.8	5.6	1113	0.45
0.5	1.58	0.6	5.5	980	0.38
0.1	2.70	0.5	4.9	960	0.36

The results of the computer data processing show that changing of θ_n-stone properties ρ, $\bar{\sigma}$, $\bar{\sigma}_c$, K from $\bar{R}$ are described by means of square ($y = Q_0 + Q_{1x} + Q_2x^2$), and ε and λ - linear ($y = Q_0+Q_1x$) functions; the correlation factors are correspondingly equal to 0.98-0.99 and 0.95-0.99 (Table 3).

TABLE 3 The regression equations

θ_n	K_n^1: Q_0	Q_1	Q_2	KK
ρ	-10.75	6.18	-0.79	0.99
ε	- 0.46	0.25	-	0.99
K	23.3	-15.6	2.6	0.98
$\bar{\sigma}$	19.9	34.6	-7.6	0.99
$\bar{\sigma}_c$	-40.9	23.2	-3.1	0.99
λ	1.41	-0.25	-	0.95

The regression equations that describe the interconnection between the given parameters are cited in Table 3. By controlling the size of pores in the foam cement stone it is possible to change its main technological properties in order to solve serious tasks.

The good technological properties of foam cement stone are mainly due to the phase composition of the minerals that form it.

The phase composition of "Aerotam" foam cement stone differs from usual types in the following:

1. Due to the potassium alkaline mixing liquid a great number of OH^- ions are introduced into the hardening system which form: stable lowbasic hydro silicates of calcium of tobormarite group of the CSH (B) type, composed of $[CaO_x \cdot (SiO_2)_y \cdot (H_2O)_z]$; potassic hydrocarbonates of the $[K_4 \cdot H_2(CO_3)_3 \cdot 1.5 H_2O]$ type; potassic-calcium hydrocarbonates $[K_2 \cdot Ca (CO_3)_3] \cdot 2 \div 5H_2O$; potassic-calcium hydrocarbosilicates $[K_2O \cdot CaO \cdot CaCO_3 \cdot nSiO_2] \cdot aq$; potassic hydroaluminosilicates $[x(K_2O \cdot Al_2O_3) \cdot ySiO_2 \cdot zH_2O]$; and calcium hydrocarbonates $[CaCO_3 6H_2O \cdot 2KOH]$, which provide the stone with high strength both during the early and latter periods of hardening.

2. The presence of highly polymerized silicon-oxygen radicals [Si-O] leads to Si-O-Si connections in the alkaline medium in the presence of organosilicon compounds and synthetic silica agents, which are the "catalysts" for the polymerization of the silicon-oxygen radicals of the $[Si_6O_{17}]^{10-} \cdot aq$, $[Si_{12}O_{31}]^{14-} \cdot aq$ and $[Si_6O_{15}]^{6-} \cdot aq$ types.

Highly polymerized silicon-oxygen radicals are the basic structural units of low-basic hydrosilicates of calcium of the tobermarite type $Ca_{10}[Si_{12}O_{31}] \cdot (OH)_6 \cdot 8H_2O$ (riversideite, plombierite), ksonotlite $Ca_6 \cdot [Si_6O_{17}] \cdot (OH)_2$ (gyllenbrandite, phoshahyte) and gyrolite $Ca_4 \cdot [Si_6O_{15}] \cdot (OH)_2 \cdot 4H_2O$ (ocenite, necoite, truscotite)—the main inducers of cement stone strength.

"Aerotam" plugging foam cement slurry and the stone derived from it are characterized by the following technological properties: protection of the mixing liquid and of the cement slurry from freezing; the technologically necessary period of cement setting and time of cement thickening; a normal schedule for cross-linking and the development of durability, high strength, adhesiveness with permafrost rocks and low permeability of the foam cement stone; protection of permafrost rocks from thaw for up to 6 months during constant drilling and up to 3 months during well operation as compared to 1-2 months with the common plugging cements; decreased, specifically optimum, heat release during cement hydration, conservation of heat of the producing fluid and hydration prevention; improvement of the degree of mud expulsion; an increase of cold-resistant factor; absence of corroding effects on the casing metal; the possibility of employment of standard cementing equipment; pumps and compressors of the usual rig equipment.

The "Aerotam" foam cements can be used under conditions where other plugging materials don't provide quality well cementing (considerable strata absorption, underlifting of the cement slurry to the wellhead, plus and minus temperatures, salt aggression, zones of abnormally low pressure formations, the necessity of heat conservation during well fluid and thermal flow and the injection of hot agents, etc.).

The properties listed above indicate that foam-cement slurries are a promising substance for well construction and drilling from the standpoint of environmental protection.

The low-temperature modifications of "Aerotam" foam cements were successfully applied in 1977-1983 for cementing surface casings in oil and gas wells in the permafrost rocks in the fields of the North of Tyumen region, and Yakutia. High-temperature "Aerotam" modifications were successfully employed for cementing intermediate and production casing strings in high-temperature oil and gas wells in Turkmenia fields under conditions of high mineralization and abnormally low formation pressures (Bakshoutov 1976-1982).

In all cases of the "Aerotam" application no ecological damage to environment, characteristic for these regions, caused by human interference was observed.

REFERENCES

Aldrich, C. A., and Mitchell, B. I., 1976, Strength, permeabilities and porosity of oilwell foam-cement: Journal of Engineering for Industry, v. 98B, no. 3, p. 1103-1106.

Bakshoutov, V. S., and Danyushevsky, V. S., eds., 1976, Possibility of application of surfactant-base foam-cement for well cementing, in Proceedings of the Seventh International Congress on Surfactants, Moscow, Neftianoye Khoziastvo, no. 7, p. 48-50.

Bakshoutov, V. S., 1977, On the possibility of crystallochemical intensification of the foam cement slurries hardening processes: Moscow, Neftianoye Khoziastvo, no. 11, p. 16-19.

Bakshoutov, V. S. et al., 1980, Author's certificate of the USSR, N726306: Biulleten Izobreteniy SSSR, no. 13.

Bakshoutov, V. S. et al., 1982, Author's Certificate of the USSR, N927973: Moscow, Biulleten Izobreteniy SSSR, no. 18.

Bakshoutov, V. S. et al., 1978, The experience of the conductor casing cementing with "Aerotam" foam cement: Moscow, Neftianoye Khoziastvo, no. 6, p. 53-55.

Bakshoutov, V. S. et al., 1979, The application of standard equipment for conductor casing cementing with the plugging foam cement: Moscow, Neftianoye Khoziastvo, no. 1, p. 58-59.

Bakshoutov, V. S. et al., 1980, The application of lightened foam cement slurry for the high-temperature well cementing: Moscow, Neftianoye Khoziastvo, no. 7, p. 70-74.

Bakshoutov, V. S. et al., 1981, Design and research of the plugging foam cements for various conditions of application: Moscow, Neftyanoye Khoziastvo, no. 6, p. 22-27.

Montman, R. ed., 1982. Low density foam cements solve many oil field problems: World Oil, v. 194, no. 7, p. 171-186.

INVESTIGATION OF AREAS OF ICING (NALED) FORMATION AND SUBSURFACE WATER DISCHARGE UNDER PERMAFROST CONDITIONS USING SURFACE GEOPHYSICAL TECHNIQUES

S. A. Boikov,[1] V. E. Afanasenko,[1] and V. M. Timofeyev[2]

[1]Geology Department, Moscow State University
[2]All Union Research Institute for Hydrology and Geology, Northern Regions
Moscow, USSR

The article identifies the major tasks faced by the permafrost hydrologist in areas of icing (naled) formation and of subsurface water discharge, using surface geophysical techniques. A comparative evaluation of the possibilities and potential usefulness of various geophysical methods, both traditional and modern, is attempted. The peculiarities of geophysical procedures under varying conditions, both winter and summer, are discussed and methods of integrating various geophysical methods are considered.

The areas of icing (naled) formation and the discharge of different types of subsurface water constitute the principle object of inquiry in permafrost-hydrogeological investigations. These studies provide extensive data on the present subsurface drainage, hydrochemical characteristics of subsurface water formation, its interaction with perennial frozen rocks (PFR) and ultimately facilitate estimation of subsurface water resources.

During long-term, comprehensive, frozen ground hydrogeological and geological engineering studies within the development zone of the central Baikal-Amur Railway, methods of surface geophysics were widely used to study the areas of naled formation and subsurface water discharge since they were considered to be among the most informative, productive and cheap means of comprehensive analysis.

During preliminary analysis most icings in the test territory were mapped using space and aerial photo-surveying data, and classified by genetic types. The icing types were classified in greater detail during subsequent comprehensive field studies. The icings were differentiated according to the sources of the subsurface water, the time of formation, the size of icing glades, their shape, and the amount of ice accumulation. The icings of water in the seasonal thawing layer (STL), of ground water in blind subchannel and sublake taliks, of the pressure water of subfrozen ground drainage, and finally, ice crust of mixed recharge on account of the ground water of blind subchannel and sublake taliks and pressure water of the subfrozen ground drainage were distinguished.

Considerable accumulations of big and huge icing fields were recorded within the Usmunskaya and Chulmanskaya Mesozoic depressions, the Aldan shield and near the North Stanovaya overthrust zone. In the Baikal-Charskaya hydrogeological region, huge icing fields are concentrated in the Udokan cryohydrogeological, deep and superdeep freezing massif, as well as throughout the Imangrskaya and Tas-Yurekhskaya regional fracture zones. In the Daurskaya hydrogeological region, particularly in the Olekminsky cryohydrogeological deep freezing massif, medium and small icings are grouped near the mineral water sources of deep subfrozen ground drainage. A great number of large and medium-size icings occur in the Dzhugdzhuro-Stanovaya hydrogeological region, particularly in the so-called Unakhinsky hydrogeological massif of spotted PFR distribution.

The icing types and groups were classified according to the results of aerovisual observations and landings on the most typical naled formations. The activities of the landing parties were aimed at identifying the regularities of naled formation; typification; estimation of thickness, areas and amount of ice accumulation in the ice crusts; locating subsurface water sources and hydrochemical sampling using shallow test drilling and extensive geophysical exploration.

Geophysical investigations in areas of naled formation and subsurface water discharge are aimed at solving such permafrost-hydrogeological survey problems as identification of talik boundaries, in which the areas of subsurface water discharge are confined; detection of concentrated subsurface water discharge areas within them, of dislocations with a break in continuity or high jointing zones, to which the water-show areas are usually confined. These are used to study the structure and thickness of taliks and perennial frozen rocks framing these taliks, etc. Icing studies also include identification of ice crust thickness, the structure and condition of the bedrock by investigation of their interrelationship with taliks, etc. To solve the problems listed above we used various traditional and modern methods of geophysical research. They included electrometry, electro-prospecting, frequency techniques, seismic survey, magnetometry and radiometry.

It would be useful to begin geophysical exploration of the test areas, using the least labour-intensive, most productive methods capable of providing operational data about the planned distribution of the water discharge taliks and concentrated subsurface water discharge areas, to provide specific hydrochemical samples of the subsurface water sources and information on the geological-permafrost and hydrogeological conditions, which can later be studied using more labour-intensive and comprehensive geophysical techniques. Traditional geophysical methods such as longitudinal electroprofiling (most frequently two-differential symmetrical EP) and resistivimetry are the most widely used techniques in the initial stage of investigation (Dostovalov et al. 1979).

Longitudinal profiling in the areas of naled formation is conducted along available stream channels, since experience shows that the development of through taliks is most probable. The electroprofiling is conducted along the water line, while for shallow streams and the channels it follows the bends, crossing the headbands of the icings or subsurface water sources discovered earlier. Thus, in the areas of subchannel talik icings the electroprofiles are established either along the primary channel or along several channels, exiting beyond the outline of the naled formation area. In the areas of flange and slope icing of subfrozen ground drainage, longitudinal profiling is usually done from the primary river bed up the stream running along the icing area and exiting beyond its boundaries. In studies of those icing areas, potentially related to the drainage of water from mountain lakes located in upper valleys, the electroprofiles are established beginning with the shallow lake waters or the water line in the direction of suspected drainage through the icing area, etc.

According to the results of longitudinal profiling, the areas of permafrost development are sufficiently distinguished at ice crust base, as well as the through and blind taliks formed under the streams (Figure 2b). Preliminary analysis of the results of longitudinal profiling is followed by the establishment of individual non-extended transverse profiles through the discovered taliks to delineate the taliks in the plan. Resistivimetry data from longitudinal profiling, which provides information about the planned location of subsurface water sources, are used for the efficient distribution of these profiles.

Resistivimetry, one of the most productive and informative geophysical techniques, makes it possible to distinguish among numerous occurrences of water in the test areas the ones that are related to subsurface water discharge. It allows more specific hydrochemical sampling, significant reduction of hydrosamples and preliminary estimation of the rate of subsurface water mineralization (Afanasenko et al. 1974).

During the permafrost-hydrogeological investigations carried out in icing and subsurface water discharge areas, resistivimetry is carried out not only as point measurements of individual sources and other water manifestations (as an auxiliary method of hydrochemical sampling) but also as a variation of stream electroprofiling. In this case resistivimetry, like longitudinal profiling, is done continuously along the stream channels, exiting beyond the icing area. Well differentiated curves of resitivimetric profiling (Figures 1 and 2), which distinctly fix the sites of concentrated subsurface water discharge against a total background of depleted solutions, can be obtained. The rate of mineralization of the surface, subsurface and mixed stream waters, together with the relative output of subsurface water sources and a number of other evaluation data, can be obtained using simple calculations derived from these curves. Resistivimetric profiling of a series of profiles having a dense river network makes it possible to trace the flooded zones of individual fractures and tectonic disturbances. These problems are more successfully solved by magnetometry (Figure 2).

Resistivimetric observations are carried out using special portable meters based on combined instruments or electrothermometers and water scoops with built in 4-electrode units (for water resistance measurement) and thermometers or thermoresistors (for temperature measurement). The results of temperature measurements are used for qualitative estimation of surface and subsurface water mineralization rate and for their more efficient identification. Benthonic resistivimeters are used to detect the sources of subaqueous discharge under deep river and lake basins.

In order to conduct traditional electroprofiling for permafrost-hydrogeological studies in the icing areas, natural field (NF) electroprofiling, which allows more accurate interpretation of other EP types and supplementary information on the zones of subsurface water charging, movement and discharging, has been tested. On the whole, the information value of this method is not very high and it serves only an auxiliary function in the overall electroprospecting operations.

In addition to traditional profiling methods used in geophysical operations, more modern methods of electroprospecting, such as DEMP (dipolar electromagnetic profiling) and HFEP (high-frequency electroprofiling), have been tested. A substantial advantage of these methods, as compared with conventional EP, is their higher mobility combined with lower labor requirements (a team of 2-3 persons); this is important for the large-scale and time-limited operations carried out in the zone of Baikal-Amur Railway. The noncontact measurement method, which ensures the application under specific conditions in areas of large-block deposits (stone streams, stone taluses and polygons, etc.), on surfaces covered with ice bodies and particularly in operations carried out in winter, is equally important.

The experience of DEMP application in different geological-frozen conditions is on the whole indicative of its lower information content and resolution as compared with the EP method using direct current. Using DEMP data, the position of different geological and permafrost boundaries in the plan can be identified, but the qualitative interpretation of these data is impeded. Other substantial disadvantages of the method are the high screening effect of the thawed rocks, which makes it impossible to use it for studying deep taliks, and the strong dependence of the measured values on the earth's surface relief, which results in false anomalies, unrelated to the geological-permafrost characteristics of the section. For these reasons the DEMP method has not been used independently, but as an auxiliary method combined with other methods of electroprospecting.

The HFEP method, elaborated in VSEGINGEO and tested earlier in Western Siberia (Timofeyev 1977),

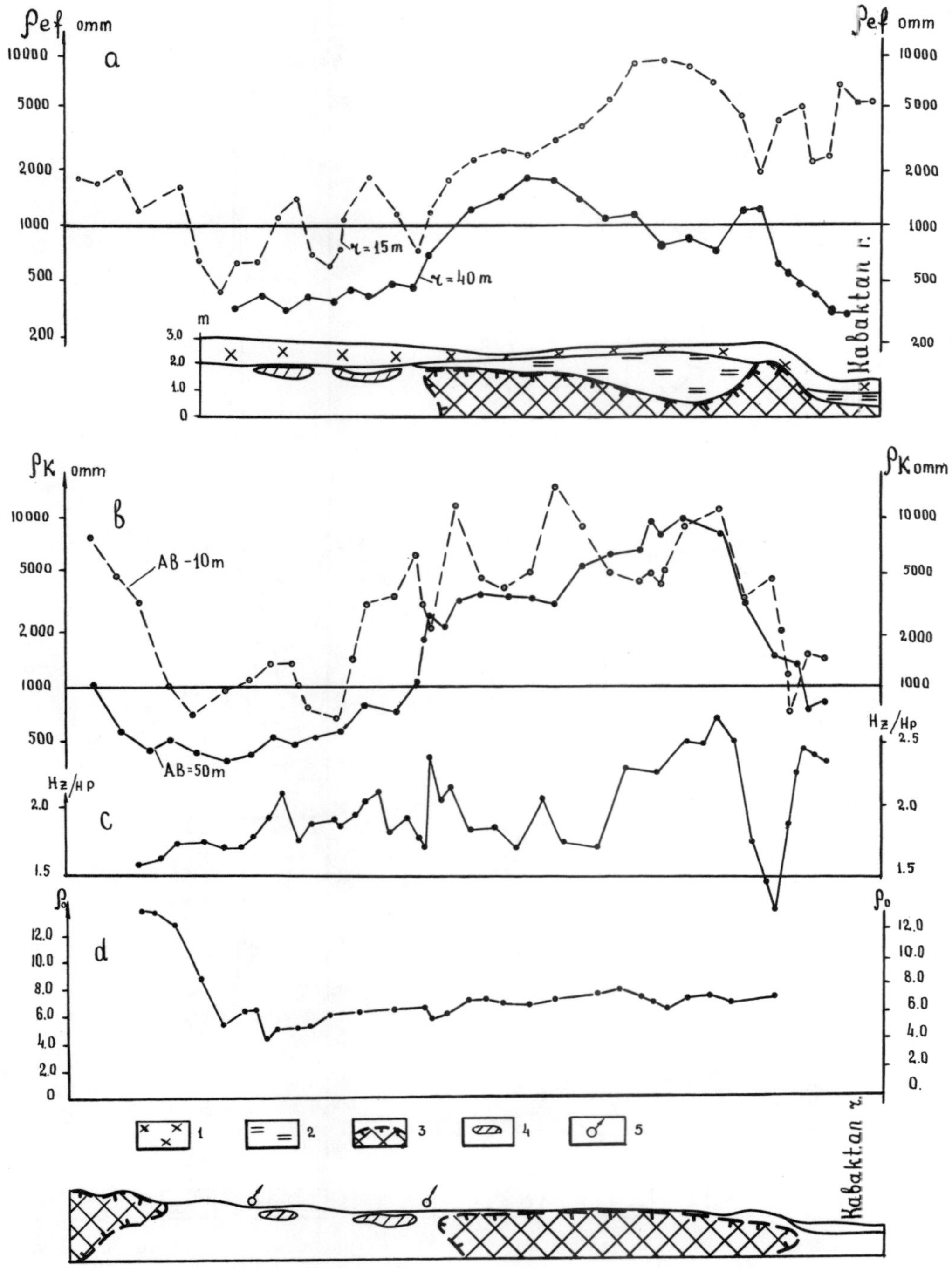

FIGURE 1 A comparison of the results of different geophysical operations performed in winter (a) and summer (b,c,d) periods in the ice crust formation area in the valley of the Kabaktan river: a - ρ_{ef} graphs of HFEP electroprofiling, b - graphs of ρ_k of symmetrical EP electroprofiling, c - graphs of H_z/H_p of dipole profiling DEMP, d - graph of resistivimetry, specific electric water resistances in terms of conventional units; 1 - snow, 2 - ice, 3 - frozen rocks, 4 - seasonal-frozen rocks, 5 - subsurface water sources.

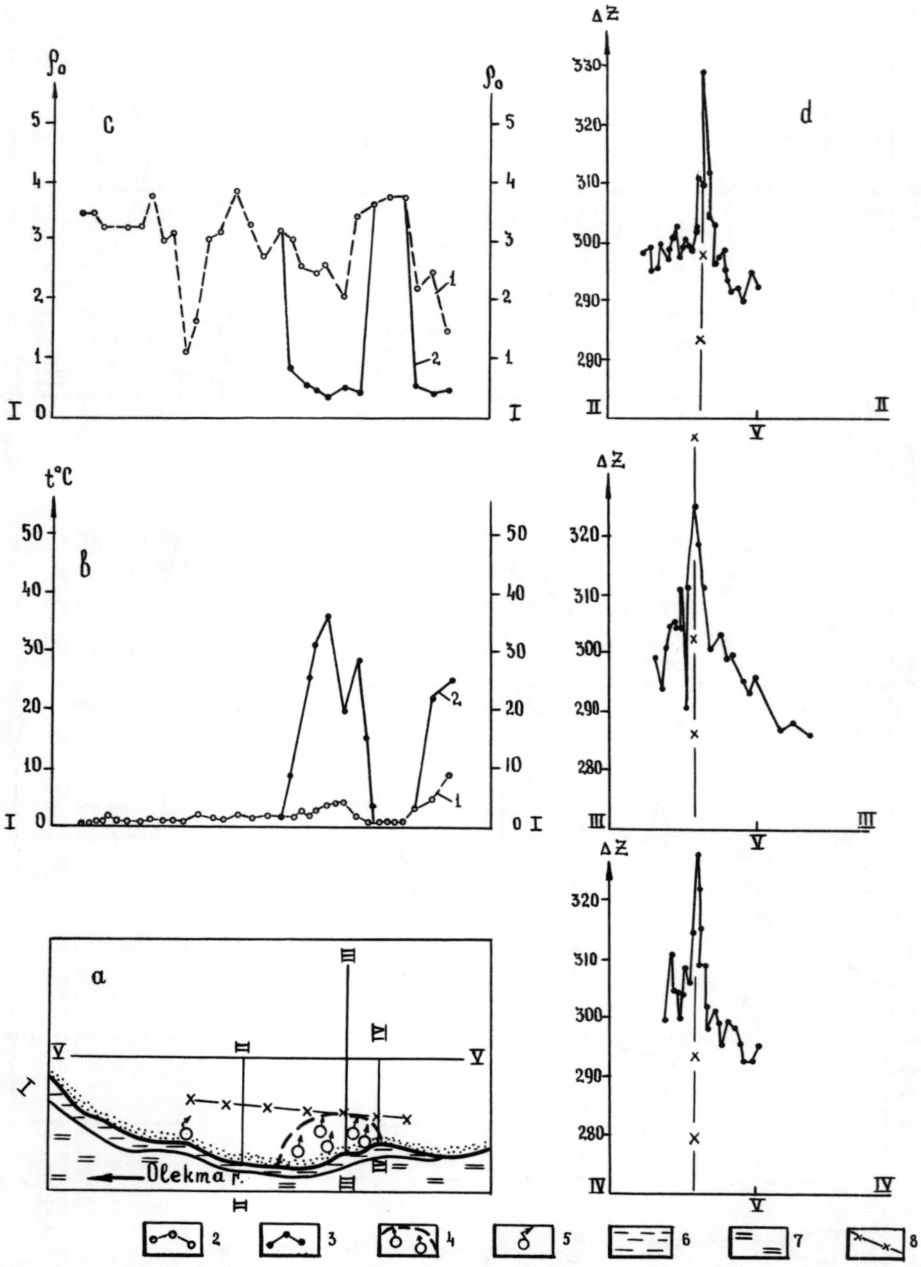

FIGURE 2 Graphs of resistivimetry, thermometry, and magnetometry of winter geophysical operations in the area of thermal subsurface water discharge in the valley of the Olekma river: a - schematic profile layout; b - water temperature in the riverbed (1) and spring (2); c - specific electric resistances of water in the riverbed (1) and springs (2) in terms of conventional units; d - graphs of magnetometry in terms of conventional units; 3 - subaerial subsurface water discharge area, 4 - subaqueous subsurface water discharge area, 5 - a non-freezing air hole in ice, 6 - river ice, 8 - the line of fracture strike, controlling subsurface water discharge.

has proved far more efficient. It approximated the EP method, when used under various permafrost-geological conditions of the Baikal-Amur Railway zone, and greatly surpassed it under certain conditions, particularly in winter and icing operations. Direct ρ_{ef} measurements provide good correlation of acquired data with the data from conventional electrometric investigations (EP and VEP). The variation in dimensions of the feeding and measuring dipoles and the spacing between them allows alteration of the test depth. A disadvantage of the method is the dependence of ρ_{ef} on the closeness of the capacitive antenna to the ground surface which is difficult to achieve during summer operations in well-developed brush, hummocky topography, shallow hummocky-sedge bogs. Nevertheless, the HFEP method can be considered as promising for such investigations, similarly to CEP (continuous electroprofiling), which has been elaborated on recently in PNIIIS.

Geophysical methods, their sequence and relationship to various field operations, vary according to specific research objectives, the complexity of the permafrost and hydrogeological conditions in areas of naled formation and subsurface water discharge, time and conditions of the operations being conducted.

In studies carried out in the areas of subsurface water discharge during the summer, the geophysical methods used include, as a rule, direct current, electroprofiling, resistivimetry, VEP and magnetometry. Several specific techniques, including longitudinal profiling along riverbeds and stream-channels with simultaneous resistivimetric and thermometric observations, provide accurate solutions to a number of problems, including detection of the areas of subaqueous discharge of subsurface water, identification of different water manifestations, etc. Additional information about feeding, transit and discharge of subsurface water can be obtained from electroprofiling, using the natural electrical field technique.

Alternating current (DEMP, NFEP, CEP) electroprofiling methods, which are more productive and efficient as compared to EP, are often used as auxiliary methods during summer regional studies.

The vertical structure of taliks and frozen rocks in the summer is basically estimated from VEP, VEP IP data. In cases of distinct horizontal heterogeneity of the section and steeply dipping boundary surfaces between the frozen and thawed rocks, which is typical for areas of icing and subsurface water discharge, special techniques are needed for VEP operations and the interpretation of the results obtained (Dostovalov et al. 1979).

During research on subsurface water discharge areas in the winter or early spring, the operational conditions change substantially, as well as the structure and parameters of the test sections and the problems being solved by geophysical methods. The limited application of electrometry, particularly EP and VEP, stems from the difficulty of achieving galvanic contact with ice or frozen rock and substantial measurement errors in winter; in the spring, when contact is realizable, the screening effect of the ice body and seasonal-frozen rocks interferes. All of this stipulates a need for a wider use of noncontact electroprospecting methods.

HFEP is a very informative method, providing results which are relatively simple to interpret. The practical application of this method on the ice bodies of Southern Yakutiya, the Amur region etc. is indicative of its possible wide use in mapping geological and permafrost boundaries under the snow or ice or during seasonal freezing. The HFEP method can also be used for tracing large water discharge channels by naled thickness and for estimating the thickness of naled bodies (Figure 1a), particularly as a form of geometrical sounding. More precise estimation of ice body thicknesses and the structure of the underlying frozen rocks is done by longitudinal seismic profiling using modern portable seismic stations.

Geophysical methods, mainly electrometry, are especially important in discovering and outlining the disintegration zones of cryogenic rock which are attracted usually to the upper or lower PFR boundaries (Afanasenko et al. 1980). Cliff massifs of these rocks, subjected to perennial freezing, gave rise to high pressure in the saturating subsurface waters, and their pressure discharge is accompanied by icing.

In the northern setting of the Verkhne-Zeiskaya depression the regional weathering crusts of discontinuous freezing often constitute a unique cap on the subsurface waters of the Unakhinsky hydrogeological massif (Alekseyeva 1978), which is characterized by considerable accumulation of ice body fields. The builders encountered difficulties in laying track for the Baikal-Amur Railway in such areas. Many road excavations made in these rocks in winter constitute foci of intensive ice crust formation.

The survey of subsurface water discharge areas and particularly of those with icing was followed by specific hydrosampling for tritium, that allowed evaluation of the water exchange in different hydrogeological structures and of the charging level (Afanasenko 1981).

The experience gained by using geophysical methods in permafrost-hydrogeological investigations in areas in which icing subsurface water discharge occurs indicates, on the whole, their high effectiveness in providing reliable information; this is contingent upon the use of a comprehensive approach in solving multi-faceted problems and strict follow-up of the acquired results at all stages of the operations.

REFERENCES

Afanasenko, V. E., and Boikov, S. A., 1974, Characteristics of chemical composition development and specific electric resistances of subsurface and surface waters of the northern part of Chulmanskoye plateau, in Merzlotniye issledovaniya, no. 14, Publ. Moscow State University, p. 122-131.

Afanasenko, V. E., Boikov, S. A., and Romanovsky, V. E., 1980, New data on the frozen ground-hydrogeological structure of Dzhagdy-Sokhtakhan block, in Geokriologicheskiye usloviya zony Baikalo-Amurskoi magistrali: Yakutsk, p. 141-150.

Afanasenko, V. E., 1981, Alteration of tritium concentration in subsurface water as an indicator of water exchange conditions in various hydrogeological structures of Baikal Amur Railway development zone, Thesis of the paper at All-Union symposium, in Isotopes in hydrosphere: Moscow, p. 56-57.

Alekseyeva, Z. I., 1979, Weathering crust manifestations in the eastern area of Baikal Amur Railway (Dzheltulaksky region), in Kory vyvetrivaniya, no. 16: Moscow, Nauka, p. 206-212.

Dostovalov, B. N., Boikov, S. A., and Pugach, V. B., 1979, Electrometric studies of subchannel and sublake taliks in the Extreme North for the solution of engineering and geological problems, Papers of the III International Conference on Geocryology: Novosibirsk, Nauka, p. 247-257.

Timofeyev, V. M., 1977, A new method of accelerated electroprofiling for engineering and geocryological survey, Papers of a scientific seminar: Methods of engineering and geological investigations and mapping of the permafrost region, no. 2: Yakutsk, p 19-20.

PRINCIPLES OF TERRAIN CLASSIFICATION FOR PIPELINE CONSTRUCTION

L. M. Demidjuk, G. S. Gorskaya, and V. V. Spiridonov

The All-Union Research Institute for Transmission Pipeline Construction (VNIIST), Moscow, USSR

As a basis for designing, constructing, and operating a safe system of long-distance pipelines it is essential to compile baseline assessment maps depicting engineering-construction conditions. These must take into account, on the one hand, the complete complex of environmental factors pertinent to the project and, on the other, technological and engineering peculiarities. Construction regionalization involves terrain classification on the basis of the most important environmental factors based on a classification in terms of construction and engineering geology developed by the authors.

The main characteristics of modern pipeline construction are as follows: the enormous length of pipelines (4,000 km and greater), the installation of several pipelines in one corridor, the passage through vast territories having varying complex natural conditions, and finally the high technical parameters of the pipelines themselves, that is, large diameters (1220-1420 mm) and high pressure (7.5-10 MPa), which define the category and responsibility of the installation.

All these determine the basis for engineering, constructing, and operating long-distance pipelines.

The methodological basis for engineering and geological investigations, especially in permafrost regions, should be terrain classification based on comprehensive environmental factors which affect both the construction process and selection of a mode of construction which takes into consideration future pipeline design-environmental interaction.

Such classification involves dividing the territory into spacial regions and results in a map or a series of maps of engineering and construction regionalization.

These maps should include the entire complex of environmental factors specific to the given construction mode as well as its design, technical features, etc.

As a theoretical basis for the classification by compressive engineering and geological analysis the relationship "geological environment-engineering project," we suggest a systematic approach to the principles of regionalization.

A systematic approach suggests the development of an industrial classification of environmental conditions with regard to predictions of change as well as technical standards for the given types of installation that contain technological requirements, available design solutions and the production technology rates. The choice of the territory estimating elements deals with its regional peculiarities and specific features of its development.

In view of the above stated problems, VNIIST concludes the work on the USSR terrain classification with regard to pipeline construction tasks. The authors have developed and incorporated into the design a construction and geological classification system that also constitutes a classification of engineering and geological conditions according to the complexity of the construction project and a standardized classification of pipeline construction modes and schedules. On the basis of this classification system, we have developed sketch maps of engineering and construction regionalization that include climatic, engineering-geological, and social-economic regionalization.

This system of construction, engineering, and geological classification consists of two parts: engineering-geological classification and classification of construction elements.

The engineering-geological aspect of the classification includes the scale of the main environmental factors (surface and frozen ground) which define the construction process and the environmental impact of pipelines. In addition the classification includes permafrost and anthropogenic processes and phenomena.

The surface factors pertaining to pipeline construction are as follows:

- character and types of vegetation, which influence movement of machines and equipment, in addition to determining the heat-moisture exchange in soils, their thermal state, the extent of seasonal freezing and thawing;
- hydrological factors: watercourses (rivers, channels, springs), water basins (lakes, water storage basins), swamps. Size and depth of water basins and swamps, the content and thermal conditions of sedimentary peat affecting construction decisions, methods of laying pipe, feasibility of moving equipment and the complications in the pipeline construction process;
- relief morphometrical characteristics (horizontal and vertical disintegration, surface inclination) affecting construction decisions, pipeline construction mode technology and equipment, as well as types and magnitude of engineering and geological processes.

The surface factors of natural environments also comprise special climatic features with zonal distribution. Air temperature, atmospheric precipitation, snow distribution and thawing, wind regime—all these factors define the construction work schedules and season. In addition, climatic conditions are taken into account in calculating the pipe and soil temperature behavior and the stability of the above-ground pipelines under wind loading. The frozen ground factors governing the thermal and mechanical interaction of pipelines with the ground in permafrost territories are as follows:

- type of permafrost region: continuous and discontinuous permafrost, massif island permafrost and permafrost islands, persistent snowbank and ice-lenses, relic permafrost;
- mean annual ground temperature at a depth of its annual fluctuation;
- composition, cryogenic structure and physical and mechanical properties of frozen and thawing ground (lithological features, moisture, ice content, heaving properties, thaw settling, salinity and corrosive activity of ground);
- thickness, structure and properties of ground in seasonally thawed and seasonally frozen layers;
- permafrost and physical-geological processes and phenomena (thermokarst, thermal erosion, frost fracturing, heaving, icing, solifluction, ground ices, surface waterlogging, new formation of frozen ground, seismicity, and gravitation processes, etc.).

Design and construction of long-distance pipelines require detailed prediction of changes of all engineering-geological factors, because when natural environmental conditions are disturbed, changes in the permafrost temperature regime occur. As a result, the composition and properties of the permafrost are altered and the intensity of engineering-geological and permafrost processes increases.

The above-mentioned factors are assessed from the standpoint of pipeline installation with regard to difficulties of construction and reclamation activities, as well as ground characteristics affecting the pipelines (ground content, ground thermal state, flooding, soil subsidence, development and intensity of permafrost processes and phenomena).

A detailed engineering and geological classification is presented in Table 1. This classification may be used at the working drawing stage and, given the necessary data, at the technical project stage. For small-scale investigations it may be expanded according to the initial data.

This classification assesses engineering and geological conditions according to five categories (or degrees) of complexity. The following surface conditions are included:

- vegetation, availability of forests, vegetation composition, thickness and height of tree trunks, cover, crown density;
- lakes and swamp cover (type of peat deposits, terrain passability, extent (%) and depth of lakes);
- watercourses (their nature, width of high and low waters);
- relief (morphometric properties - surface inclination, vertical and horizontal disintegration, slope length).

Frozen ground conditions are classified according to the following factors: soil thermal condition by thawing-frozen soil ratio and mean annual soil temperature); ice content of frozen grounds (by relative thaw subsidence); frost ground heaving (by relative heaving); soil salinity (by total content of easily dissolved salts in % dry soils; soil corrosion activity (by specific electrical resistance of soils); soil flooding (by the depth of the first water-bearing horizon).

Permafrost processes and phenomena include: frost fracturing and underground ice (by intensity of area development); icing (by presence and possible development of icing of different origins-river or ground); solifluction (by intensity of development depending on the steepness of slopes); sereis crumblings (by intensity of development depending on slope steepness); landslides, rock streams (by intensity of development depending on slope steepness); seismicity (by magnitude).

Anthropogenic factors include: railways, highways, farming lands, populated areas, reservations, animal migration routes, etc.

Classification of structural elements includes factors, defining types of pipeline laying, pipeline design, construction technology, and operation.

Depending on where the sections are located relative to ground surface they are differentiated as follows:

- underground, when the upper centerline of the pipeline is below the ground surface;
- on-ground, when the lower centerline of the pipeline lays on consolidated natural or artificial foundation, and the upper above the ground surface (either a berm or out in the open).
- above-ground—when the pipeline is laid upon separate piles or pile trestles at a height of 250 mm above the ground surface.

Depending on the operating temperatures that characterize the pipe-environment interaction (ground, water, air), the pipeline sections are differentiated as follows (Table 2):

- negative—with constant operating temperature below zero;
- cold—with a plus or minus temperature, but with a mean annual temperature equal to or lower than -1°C;
- warm—with a plus or minus temperature, but with a mean annual temperature higher than -1°C;
- positive—with a temperature higher than zero.

TABLE 1 Pipeline classifications

Temperature regime		Construction season		
		Year round	Summer	Winter
Negative	N	NAY	NS	NW
Cold	C	CAY	CS	CW
Warm	W	WAY	WS	WW
Positive	P	PAY	PS	PW

Possible technical decisions for every engineering-geological type of pipeline section (Table 1) are selected on the basis of technical requirements and construction conditions (Table 2). On the basis of technical decisions and standards, tentative economic indices can be derived as criteria for comparison of various routes.

By using letter designations for various classi-

TABLE 2 Classification by construction, engineering, and geological factors

Engineering and geological factors	Classification of engineering and geological conditions				
	Complexity category				
	I	II	III	IV	V
1	2	3	4	5	6
Vegetation	Area without vegetation (beaches, low-lying flood plains, drifting sand) or grass-covered (CC<20%) V_1	Area covered with moss lichen, dwarf shrubs; uneven or flat peat-land, shrubs with CC<50% V_2	Bushes with CC 50% and greater; forest cover 10% and less, or small forest up to 5 m high with trunk thickness 10 cm and less V_3	Sparse forests; mixed, conifers, birch, with 10 to 50% cover or small forest 5-10 m high with trunk thickness 11-20 cm (CD<50%) V_4	Shaded forest with mixed undergrowth and conifers with 50% cover and greater, trunk thickness >20 cm height > 10 m CD ≧ 50% V_5
Extent of swamps and lakes	Peats passable for special construction equipment with specific gravity 0.02-0.03 MPa or for ordinary machines by means of mats (or pads), rip-rapping roads reducing specific gravity to 0.02 MPa L_1	Peats passable for construction only by means of mats, rip-rapping of roads, reducing the specific gravity to 0.01 MPa L_2	Swampy areas with open wales surface; extent of lakes up to 40%; equipment moveable by flotation L_3	Small and large lakes up to 1.5 m deep covering more than 40% area L_4	Small and large lakes 1.5 m and more deep, covering more than 40% of area L_5
Watercourses	Channels up to 10 m width, small springs, permanent watercourses with high water < 50 m and low water up to 10 m W_1	Channels of 10 m and more width, small and medium rivers with high water < 150 m and low water 10-25 m wide W_2	Large rivers with high water width 250-500 m and low water width 50-100 m W_3	Large rivers with high water width 250-500 m and low water width 50-100 m W_4	Large rivers with high water width more than 500 m and low water width 100 m or more W_5
Relief	Flat, with surface inclination $\alpha < 2°$, length (L) > 300 m R_1	Flat and hilly $2° < \alpha < 5°$ L > 300 m R_2	Hilly-steep slope $5^o < \alpha < 7^o$ L > 300 m R_3	Steep slope and mountainous $7° < \alpha < 20°$ L > 300 m R_4	Mountainous $\alpha > 20°$ L > 300 m R_5
Ground content	Rock G_1	Rock debris G_2	Sand, sandy soils G_3	Sandy loam, clay G_4	Peat G_5

TABLE 2 (continued)

1	2	3	4	5	6
Ground thermal state	Unfrozen (Thawing) ground $T > 0$ M_1	Thawing ground with 20% discontinuous permafrost $0 > T > -1^o$ M_2	Thawing ground with more than 20% permafrost "islands" $0 > T > -1^o$ M_3	Continuous frozen ground $-1° > T > -3°$ M_4	Continuous frozen ground $T < -3°$ M_5
Ice content (relative ice content)	Poor ice content $i < 0.1$ i_1	Low ice content $0.1 < i < 0.2$ i_2	Moderate ice content $0.2 < i < 0.3$ i_3	Ice-rich $0.3 < i < 0.5$ i_4	Extremely ice-rich ground $i > 0.5$ i_5
Relative thaw subsidence of soils	With slight subsidence $S < 0.1$ S_1	With small subsidence $0.1 < S < 0.2$ S_2	With moderate subsidence $0.2 < S < 0.3$ S_3	With high subsidence $0.3 < S < 0.4$ S_4	With extreme subsidence $S < 0.4$ S_5
Thermokarst, thermal erosion	None E_1	Poor indications <10% E_2	Moderately affected 10-30% E_3	Greatly affected 30-50% E_4	Extremely affected > 50% E_5
Frost fracturing; ground ice	None F_1	Poor indications <10% F_2	Moderately affected 10-30% F_3	Greatly affected 30-50% F_4	Extremely affected >50% F_5
Icing (seasonal and perennial)	None I_1	Seasonal river icing I_2	Perennial river icing I_3	Seasonal, ground surface I_4	Perennial, ground surface I_5
Frost mounds, hillocky peatland	None M_1	Poor indications <10% M_2	Moderately affected 10-30% M_3	Greatly affected 30-50% M_4	Extremely affected >50% M_5
Relative frost heave susceptibility of soils	Slightly frost-susceptible $H < 0.01$ H_1	Low frost-susceptible $0.01 < H < 0.1$ H_2	Moderately frost-susceptible $0.1 < H < 0.2$ H_3	Highly frost susceptible $0.2 < H < 0.3$ H_4	Extremely frost susceptible $H > 0.3$ H_5
Soil salinity (in % of dry soil weight)	Slight salinity $Z < 0.1$ Z_1	Low salinity $0.1 < Z < 0.5$ Z_2	Moderate salinity $0.5 < Z < 2$ Z_3	High salinity $2 < Z < 5$ Z_4	Extreme salinity $Z > 5$ Z_5

TABLE 2 (continued)

1	2	3	4	5	6
Soil corrosion activity (in cm)	Slight $C > 100$ cm C_1	Low $100 > C > 50$ C_2	Moderate $50 > C > 20$ C_3	High $20 > C > 10$ C_4	Extreme $C < 10$ C_5
Soil flooding in m; minus (-) indicates below ground; plus (+) above ground surface	Slight $D > -2.5$ O_1	Low $-2.5 > D > -1.2$ O_2	Moderate $-1.2 > D > -0.7$ O_3	Heavy $-0.7 > D > +0.5$ O_4	Extremely heavy $D < +0.5$ O_5
Slope processes	None L_1	Solifluction (at slope steepness 2-7°) L_2	Slides, rock streams (at slope steepness 7-12°) L_3	Screes, avalanches (at slope steepness 12-18°) L_4	Earth flows, avalanches (at slope steepness > 18°) L_5
Seismicity	No seismic activity s_1	Earthquake force 6 and less s_2	Earthquake force 6-7 s_3	Earthquake force 7-8 s_4	Earthquake force >8 s_5
Anthropogenic factors	None A_1	Railroads, highways A_2	Farm land A_3	Infrequently used territories A_4	Populated areas, reservations, migration routes, etc. A_5

fication factors and structural elements, conditions of varying complexity for the various types of engineering and geological regions (or provinces, districts and regions depending on the degree of detail and the stage of design) are encoded.

The construction, engineering and geological classification, developed by VNIIST on the basis of the environmental factor analysis mentioned above and pipeline construction practice, may be used as a guideline for the organization of permafrost geological studies on pipeline construction, scientific analysis and systematization of preliminary permafrost data; it was also designed to assist in making well-grounded decisions on the construction and technological scheme of product transport with regard to possible environmental changes during construction and operation of an installation. In addition, the classification and engineering assessment and construction maps have been used by design organizations to select optimum routes for oil and gas pipelines, to make precautions, to mitigate the environmental impact, to assist reclamation within the right of way, and to provide reasonable estimation of the construction term.

Using the classification system for preliminary investigations permits standardization of the mapping procedure for engineering and geological regionalization, that up to now has depended on the qualifications and creativity of the compiler.

Engineering and construction classification of territory on the basis of construction, engineering and geological factors has already been used to design a number of large-diameter oil- and gas-pipeline projects.

REFERENCES

Beliy, L. D., 1964, Theoretical principles of engineering and geological regionalization: Moscow, Nauka.

Demidjuk, L. M., Gorskaya, G. S., and Spiridonov, V. V., 1982, The classification of territories based on construction conditions: A basis for rational use of geology in pipeline construction, Papers of the Interagency Conference: Geocryological prediction in developing regions of the Far North: Moscow, p. 66-67.

Spiridonov, V. V., Demidjuk, L. M., and Gorskaya, G. S., 1981, The principles of civil engineering regionalization in linear construction: The conference theses, Geocryological problems in the Transbaikal, Chita, p. 136-138.

Sulakshina, G. A., 1979, Engineering and geological classification of sites as a basis for predicting man induced geological change: Engineering Geology, v. 3, Moscow, p. 49-54.

ON THE PHYSICOCHEMICAL PROPERTIES OF THE SURFACE OF DISPERSED ICE (SNOW)

N. F. Fedoseev, V. I. Fedoseeva, E. A. Nechaev

Permafrost Institute, Siberian Branch
Academy of Sciences, Yakutsk, USSR

The interaction between a dispersed ice (snow) surface and toluene and hexanol solutions in a large number of organic compounds was studied. It was found that the ice surface adsorbed only substances with an ionization potential of 9.6 eV. Formation of a liquid-like layer on the ice surface occurred in 48, 15 and 2.5 hours at temperatures of -10, -5, and -2°C respectively. The solubility of carbon acids in the liquid-like layer was shown to be lower than in the water.

All large-scale processes occurring in nature and associated with the existence of water in a solid dispersed state are in some way or other connected with small-scale processes occurring on the surface of dispersed ice, whether it is substance migration over thin films in permafrost or sublimation phenomena during the metamorphosis of snow, etc. Therefore, the study of physico-chemical properties of a dispersed ice surface is of interest to a large group of specialists. Until now, strong evidence has been acquired (Kvlividze et al. 1974, 1978) for the existence on the dispersed ice surface of a so-called liquid-like film, representing an intermediate structure between ice and water.

The properties of this film have been the subject of a considerable number of papers (Kvlividze et al. 1974, 1978). There are, however, practically no data on the interaction of the ice surface with substances introduced into the system. We have studied the interaction of dispersed ice (snow) with toluene and hexane solutions of many organic compounds. Samples of fresh snow were taken on the leeward side far outside the limits of a town. The snow cover held for a considerable time and all the experiments were conducted at a temperature of -4°C under conditions when a liquid-like film is present on the surface of dispersed ice.

It turned out that of the numerous (more than 50) organic compounds sampled, representing different classes of molecules, only 1,2-naphthahynone and O-brombenzoic acid are absorbed by the ice surface. It should be noticed that the ionization potential of these compounds is 9.6 eV.

In this respect, adsorption regularities on ice do not differ from those on oxides. As has been shown in Nechaev et al. (1978, 1979), out of aqueous solutions on the oxide surface, organic substances are adsorbed which have a strictly defined ionization potential, referred to as the oxide resonance potential ($I_{res.}$). When the ionization potential of a substance deviates from the I_{res} value by more than 0.1-0.2 eV, adsorption is not detectable.

On the basis of the data obtained it seems possible to conclude that for ice I_{res} = 9.6 eV. This inference is also supported by the fact that on the ice surface anions having an ionization potential in between 9 and 10 eV are specifically adsorbed.

By processing data on O-brombenzoic acid adsorption (Figure 1) using a transformed Langmuir equation it becomes possible to determine the specific surface of samples of non-metamorphized (S_{sp} = 110 m²/kg) and metamorphized (S_{sp} = 25 m²/kg) snow.

In addition to the afore-mentioned substances, those highly soluble in water and having a high water-toluene distribution factor (inferior alcohols and acids) are markedly absorbed by snow, exposed at the temperature of -10, -5, -2°C. The nature of the interaction, however, is quite different; the sorption ability is independent of the ionization potential and the amount of sorption increases linearly with increasing solution strength.

The above-mentioned facts can be explained by assuming that the sorption of substances highly soluble in water is due to the transition of their molecules from the layer of the organic solvent to the liquid-like film.

We studied the absorption by snow of formic (I), chloracetic (II) and acetic (III) acids from solutions in toluene (I,II) and hexane (III). The concentration variation of substance after introducing a weighed snow sample was determined by the neutralization method. At high concentrations melting of the snow sample occurred to produce a bulk phase of aqueous solution.

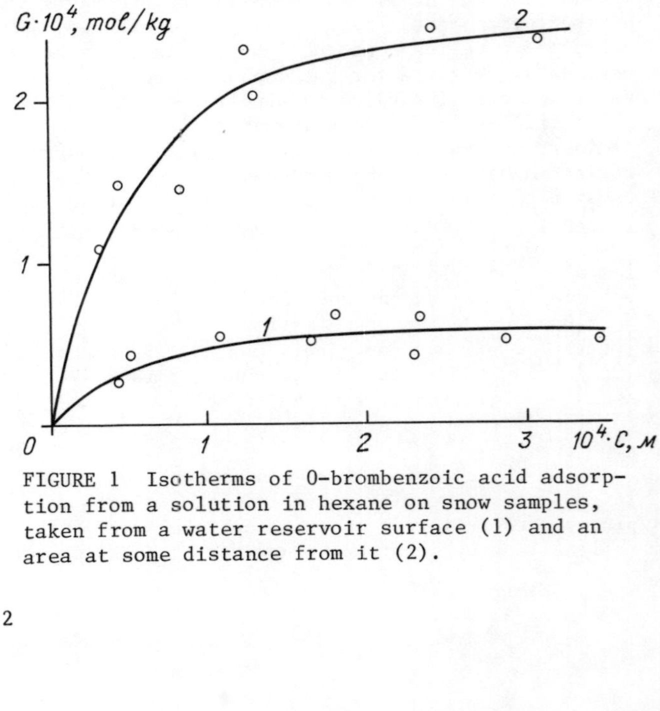

FIGURE 1 Isotherms of O-brombenzoic acid adsorption from a solution in hexane on snow samples, taken from a water reservoir surface (1) and an area at some distance from it (2).

According to current understanding, two liquid layers may be present on the surface of ice crystals introduced into a non-aqueous acid solution. On the one hand, as ice comes in contact with an acid solution, it may melt to produce a layer of acid aqueous solution of volume V, in which the acid concentration will be determined by temperature and can be inferred from a phase diagram. On the other hand, it can be assumed that part of the molecules will pass into a liquid-like layer, thus producing a solution whose properties may differ from the bulk phase. This is evidently due to the fact that the water structure within this layer is a transition stage between the water structure in the bulk and the ice structure. We will denote the volume of this layer by V_x.

The bulk phase V can exist if its acid concentration is at equilibrium with ice and toluene. The equilibrium concentration with respect to ice (C_e^i) is derived from the phase diagram, while with respect to toluene (C_e^t) it is determined by the water-toluene (K_o) distribution factor.

Let us assume that the initial acid concentration in toluene is so small that a concentration at equilibrium with ice, i.e. $C_e^t < C_e^i$, cannot arise in an aqueous solution. In this case the bulk phase does not occur. If, however, $C_e^t > C_e^i$, then ice dissolution will continue until it reaches equilibrium concentrations with respect to ice and toluene ($C_e^i = C_e^t$).

To find the concentration of toluene solution (C_{sc}) below which the bulk phase does not arise, i.e. $C_e^t < C_e^i$, we determined the distribution factors of formic and chloracetic acids between water and toluene at temperatures over 0°C and extrapolated to a low-temperature region. The respective values of the distribution factors K_o and C_{sc} are listed in Table 1. Apparently, a study of the properties of the liquid-like film would be worthwhile at concentrations smaller than C_{sc}.

TABLE 1

t,°C	C_e^i,M	K_o	$10^3 C_{sc}$,M	C_e^i,M	K_o	$10^2 C_{sc}$,M
	formic acid (I)			chloracetic acid (II)		
-2	0.87	688	1.26	0.77	54.5	1.41
-5	2.50	734	3.41	2.50	56.2	4.45
-10	4.90	830	5.90	5.53	59.4	9.31

The investigations indicate that the sorption properties of snow, stored at -35 to -50°C, depend on its exposure time at a given temperature. Measurements taken as soon as the snow attained a temperature equilibrium with ambient air provided isotherms, which exhibited a well defined inflection, the so-called point B (Greg and Singh 1970), and which permitted determination of the specific surface of dispersal ice on the assumption that the molecules form a dense adsorption monolayer. In the case of measurements carried out on snow that had been exposed for a definite time at a given temperature in a moisture-saturated air atmosphere, the sorption isotherm changed shape. The amount of absorbed substance, to a first approximation, was linearly dependent on concentration.

Experimental data have demonstrated that at a decreasing temperature, more time is required for snow sorption properties to change. It has been established that this necessitates an exposure of samples in a moisture-saturated air atmosphere for 48, 14, and 2.5 hours at -10, -5, and -2°C, respectively. At -20°C, a change in sorption properties occurred only after a three-week exposure.

These facts can be explained, if one accepts that considerable time, dependent on temperature, is required for a liquid-like layer to develop on the ice surface.

According to the available data (Nechaev et al. 1978, 1979) on adsorption of organic compounds from aqueous solutions on oxides, one may assume that the appearance of a liquid-like layer, and therefore of a layer of water molecules adsorbed on the ice surface, causes the acid molecules to be displaced from the surface. This is due to the fact that the H_2O molecules produce a hydrogen bond with the surface atoms; on the ice surface only those substances capable of producing a bond stronger than a hydrogen bond can be adsorbed. As has been shown (Nechaev et al. 1978, 1979), such conditions apply to the adsorption of substances having an ionization potential that corresponds to a resonance potential of a solid surface.

Therefore, it can be concluded that isotherms 1,1' in Figures 2-4 characterize adsorption of acid molecules from a toluene solution on the ice surface, while isotherms 1", 1"', and 2" characterize the acid molecule distribution between the toluene volume and liquid-like film. In this case at small concentrations the amount of absorbed substance must be a linear function of equilibrium concentration in toluene which is observed experimentally.

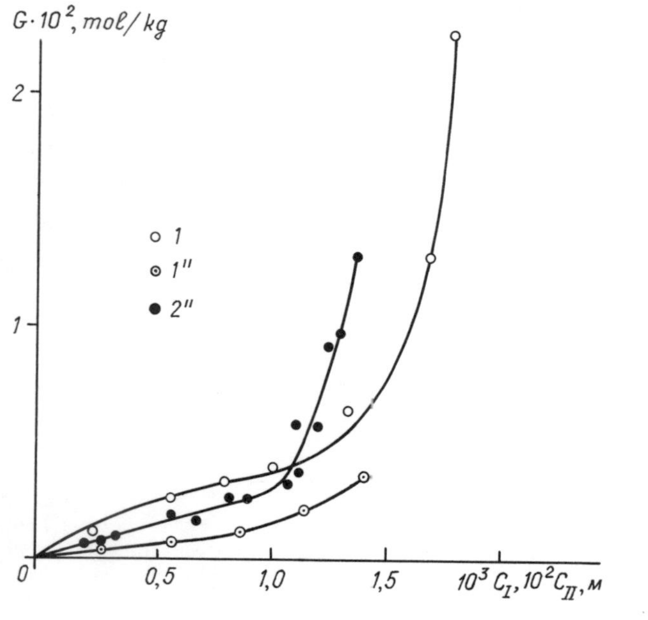

Figure 2 Isotherms of sorption II (1,1") and I (2") on snow at -2°C without preliminary exposure (1) and with a 2.5-hour exposure (1", 2").

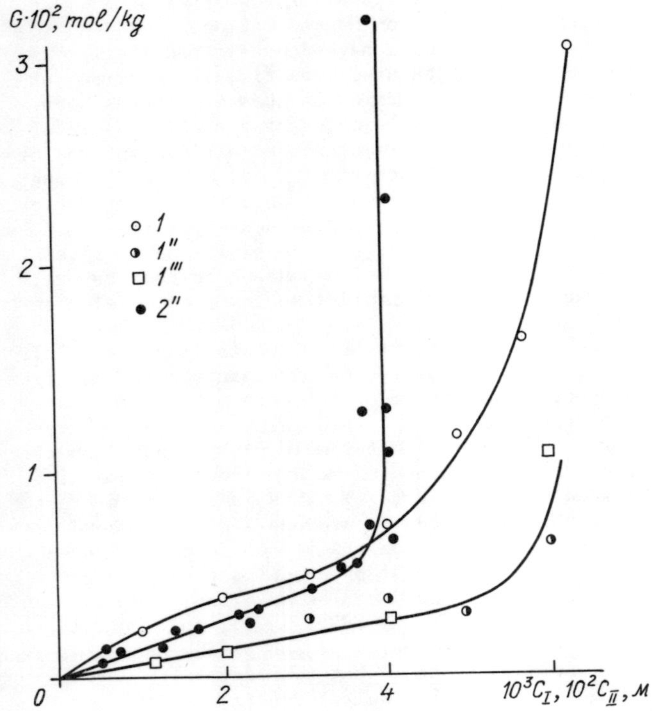

Figure 3 Isotherms of sorption II (1,1", 1''') and I (2") on snow at -5°C without preliminary exposure (1) and with 15-(1", 2") and 25-hour (1''') exposures.

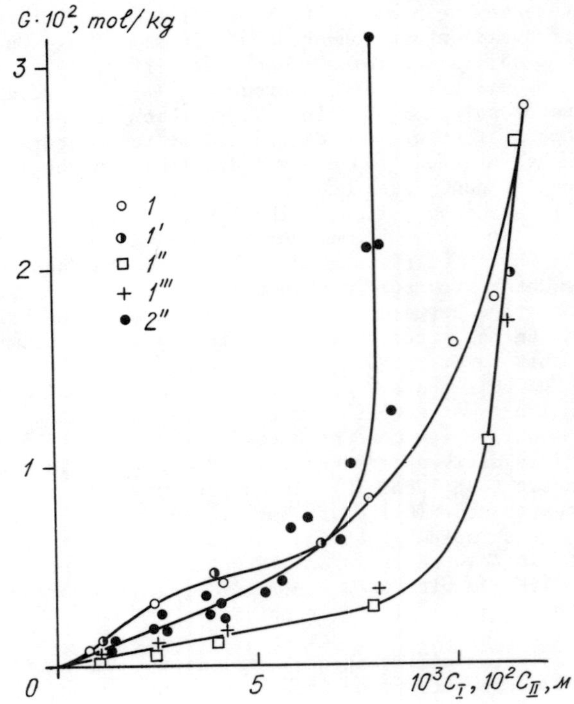

Figure 4 Isotherms of sorption II (1,1''') and I (2") on snow at -10°C without preliminary exposure (1) and with 25-(1'), 48-(1") and 120-hour (1''', 2") exposures.

Deviations from the linear dependence are observed at concentrations $C \geqq C_{sc}$, which indicates the accuracy of the distribution factor, K_o, values extrapolated to subzero temperatures.

Figures 2-4 (curve 1) and Figure 5 show that the amount of adsorbed molecules of formic, chloracetic and acetic acids at monolayer filling is the same ($0.4\ 10^{-2}$ mole/kg) and is independent of temperature. These facts make it possible to conclude that the same orientation on the surface of molecules of these acids is obtained, with the carboxylic group towards the surface. Assuming the area occupied by one molecule to be 0.205 nm² (Adam 1947), we have obtained a specific surface S_{sp} = 490 m²/kg.

This quantity exceeds the values, inferred from adsorption of O-brombenzoic acid (S_{sp} = 110 m²/kg). This is presumably due to the fact that snow contains narrow pores, accessible to small molecules of inferior acids and inaccessible to the large molecules of O-brombenzoic acid.

The specific surface value of 490 m²/kg can be used to estimate the thickness of the film. Assuming that the water-toluene distribution factor and the liquid-like film-toluene factor are the same, it is possible to calculate a specific volume of the film, $V_x = G/K_o\ C_e^t$. By dividing the value of V_x by S_{sp} we can derive the thicknesses d, listed in Table 2. The differences in thickness values

TABLE 2

t,°C	d,nm	K_f	d,nm	K_f	d_f,nm*
	formic acid (I)		chloracetic acid (II)		
-2	8.3	519	5.1	25.0	11.0
-5	3.8	557	2.4	27.0	5.0
-10	2.0	-	1.2	-	

*Kvlividze et al. (1974)

of the liquid-like film, inferred from sorption I and II with the use of the distribution factor for the water-toluene, suggest that the properties of bulk water and water within the film are differing.

On the basis of the data acquired we can conclude that the solubility of formic and chloracetic acids differs in the liquid-like film and bulk phase of water. We have calculated the film thicknesses at -2 and -5°C using data reported by Kvlividze et al. (1974) from the amount of the movable phase at these temperatures for snow with a specific area of 1000 m²/kg. (the value of d_f in Table 2). These were used to calculate the pertinent distribution factors $K_f = G/d_f\ S_{sp}\ C_e^t$, also listed in Table 2.

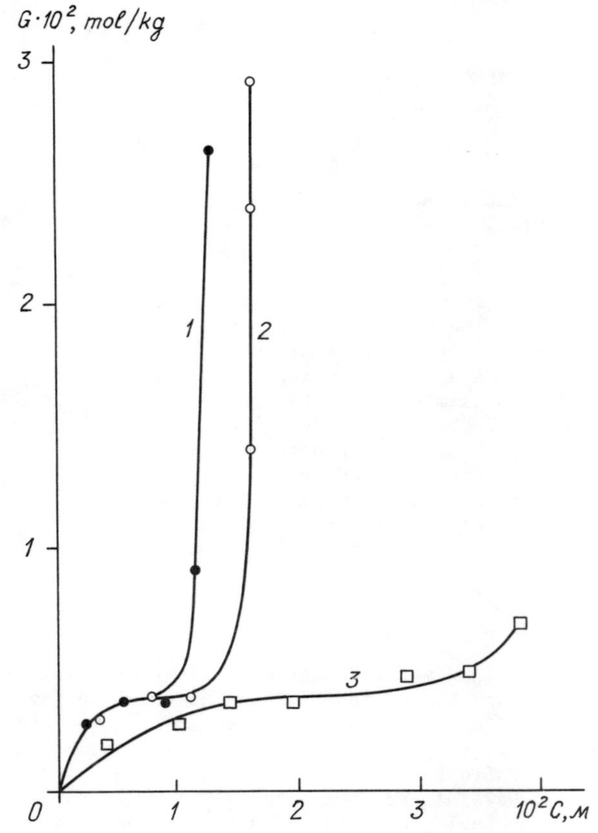

Figure 5 Isotherms of sorption I at $-15^{\circ}C$ (1), $-20^{\circ}C$ (2) and III at $-25^{\circ}C$ (3) without preliminary exposure of snow.

The comparison between the extrapolated and calculated values of the distribution factor K_0 and K indicates that the acid solubility in the liquid-like film is lower than in water. There is a considerable decrease in solubility of chloracetic acid, for which the distribution factor is an order of magnitude lower.

In conclusion, it should be noted that these results must be taken into account when conducting various tests of permafrost, snow and compact ice at temperatures close to that of ice melting, especially for samples subjected to preliminary deep cooling. If delayed formation of a liquid-like film is not taken into consideration, then one may obtain values which are far from equilibrium.

The study of the dissolving power of the film will be extended to other organic compounds, in particular, to inferior alcohols, after relevant techniques have been developed. Research is presently under way on the film's dissolving power with respect to inorganic compounds.

REFERENCES

Adam, N. K., 1947, Physics and chemistry of surface: Moscow-Leningrad, Gostekhizdat.

Greg, S. and K. Singh, 1970, Adsorption, specific surface porosity: Moscow, Mir.

Kvlividze, V. I., V. F. Kiselev, A. V. Krasnushkin et al., 1978, Proceedings of the Third International Conference on Permafrost, Edmonton, Alberta, Canada, v. 1, p. 113.

Kvlividze, V. I., V. F. Kiselev, A. B. Kurzaev and Z. A. Ushakova, 1974, The mobile water phase on ice surfaces, Surface Science, v. 44, No. 1, p. 60.

Nechaev, E. A. and N. F. Fedoseev, 1978, Adsorption of the organic matters from water solutions on the oxides of 4-valent metals. Zhurnal phisicheskoj khimii, v. 52, No. 5, pp. 1250-1254.

Nechaev, E. A. and V. P. Romanov, 1981, Adsorption of anions on ice and freezing potentials, Zhurnal phisicheskij khimii, v. 55, No. 7, p 1818-1821.

Nechaev, E. A., R. Ja. Shaidullin and V. A. Volgina, 1979, Adsorption of the organic matters from water solutions on the oxides and hydrooxides of 2-valent metals. In Sorption and chromatography: Moscow, Nauka, p 75-78.

THE THERMAL REGIME OF PERMAFROST
SOILS IN THE USSR

M. K. Gavrilova

Permafrost Institute, Siberian Branch
Academy of Sciences, Yakutsk, USSR

The author has compiled maps of heat flux through the ground surface and its relationship to the radiation balance during the thaw season. The correlation between the radiation balance and the heat flux is slightly higher in permafrost areas. Thus, in the arctic regions the ratio is 20% but in areas of seasonal freezing of the ground it is less than 10%. A ratio of 12.5% corresponds approximately to the southern limit of the continuous permafrost, a ratio of 10% to that of the discontinuous zone and one of 7.5% to that of sporadic permafrost. The total heat flux through the ground surface is greatest in the continental northeastern region of the USSR reaching 200 MJ/m^2. Northwards, westwards, and southwards it decreases to 100 MJ/m^2. This does not mean, however, that the greater heat flux results in a greater depth of thaw. Here a considerable amount of heat is being expended on warming the ground from extremely low winter temperatures up to 0° and on the ice-water phase change.

Ground soil thermal flux is one of the major components of the thermal balance of the earth's surface and represents the upper boundary condition of the thermal status of rocks. It serves as a link between climate and perennial freezing of the upper lithospheric layers. However, experimental investigations on thermal flux in ground soils are scarce. The existing methods for computing these flows (Manual etc. 1977, Averkiev 1960, and others) do not always yield reliable results for permafrost and deep seasonal freezing of rocks. This is associated with the difficulty of estimating the heat of water ice phase transitions as well as with the complexity of an accurate determination of initial parameters involved in calculation formulas.

Direct observations of thermal flux in ground soils in the huge territory of the USSR are chiefly conducted only by permafrost researchers working at the former V. A. Obruchev Permafrost Institute, the USSR Academy of Sciences (Golubev, Pavlov) and the Institute of Permafrost, Siberian Branch, USSR Academy of Sciences (Balobaev, Gavrilova, Gavrilyev, Molochushkin, Pavlov, Olovin, Skryabin, Tishin, and others). Thermal flux measurements in ground are carried out at the Institute of Permafrost, Siberian Branch of the USSR Academy of Sciences, by means of homemade heat flow meters, usually of acrylic plastic (Ivanov 1961, Pavlov and Olovin 1974) or by small-thermal flow meters designed at the Leningrad Technological Institute for Refrigeration Industry (LTIKP). The thermal-balance research group, headed by this author, utilizes the latter in their research. According to our findings (as compared to calculated values for a detailed determination of humidity, temperature, and thermal-physical properties), the LTIKP rubber thermal flow meters (30 cm in diameter and about 1 cm thick) do produce errors for individual times, but are quite acceptable as far as the 24-hour totals are concerned (reciprocal compensation of displaced maxima and minima). This was also confirmed by comparative measurements with these thermal flow meters under conditions existing in areas near Moscow and Vorkuta (Golubev and Pavlov 1961). Pavlov and Golubev have also proved the desirability of using LTIKP thermal flow meters in permafrost.

We have already completed a temporal and geographical generalized description of this important index of the thermal regime of the climate (Gavrilova 1981).

In the annual derivation, the radiation balance (R) in the permafrost region is the main source of heat to the earth's surface (500-1500 MJ/m^2). This heat is used up mainly in evaporation and warming the atmosphere. In regions with a higher moisture content, a greater amount of heat is used up by evaporation. Therefore, in the country's European territory (ETC) evaporation consumes nearly three-quarters of the radiation balance heat; in the Arctic part, about two-thirds; in droughty Central Yakutia, about half; and in dry areas of Mongolia, less than one-quarter. The ratios of heat expenditure on turbulent heat exchange with the atmosphere are reverse: the greatest loss occurs in the dry areas of Mongolia (in excess of three-quarters of the radiation balance); for Central Yakutia it is slightly less than half; in the Arctic, about one-third; in ETC, less than one-quarter (Pavlov 1979, Molochushkin 1969, Gavrilova 1973, 1974, and others).

Part of the radiation balance's annual heat is used to thaw snow, the greatest amount being expended in the Arctic regions of the country (about 10%). Although the snow there is not the deepest occurring in Euroasia (40-50 cm), by the end of the winter it is very dense (0.30-0.35 g/cm^3); on the other hand, the yearly radiation balance totals for the Arctic are the smallest (within 500-600 MJ/m^2). In the northern part of ETC and West Siberia, snow thawing consumes about 7% of the radiation balance heat (by the end of winter the snow depth is 60-80 cm), in Western ETC, about 5% (50-60 cm), and in continental Central Yakutia with a comparatively shallow snow cover (40 cm by the end of winter), about 2%. In semi-arid regions of Southern Mongolia, where the snow cover lies between 1 and 4 cm, the heat expended on snow thawing with respect to the yearly total of radiation balance (more than 1700 MJ/cm²) is negligible.

The ground thermal flux total through the earth's surface (B) during the year is, for the most part, close to 0. For individual years it can be within 20-60 MJ/m², which comprises 0 to ±5% of the radiation balance's yearly value.

Other correlations between the thermal balance components of the earth's surface occur during warm and cold seasons. From the standpoint of geocryology, the total accumulation and release of heat during the freezing and thawing seasons are of greatest interest (the criteria of the seasons are the dates of stable transition of temperature on the earth's surface through 0°C). Naturally, seasonal ground freezing increases in duration when moving from south to north and from west to east, while seasonal thawing, on the contrary, decreases. Thus, within the USSR's Asian territory, the duration of the freezing season is as follows: Arctic coast—250 days, the north of Western and Eastern Siberia—230 days, Central Siberia and Central Yakutia—200-215, the near-Baikal area—185 days, etc.

The season of surface ground soil freezing covers not only the wintertime, but also part of autumn and a substantial portion of spring. During the autumn and spring, a large re-organization occurs in the thermal regime of the atmosphere, the earth's surface and in the upper layer of rock. A considerable expenditure of the thermal balance during spring freezing is due to heat consumption by snow thaw. In western ETC it consumes more than one-quarter of the heat, in the Arctic, about one-fifth, and in Central Yakutia, about one-sixth. Only in Southern Mongolia, where the snow cover is shallow and disappears early, this fraction is a mere 3%. In the spring there is also an increase in the heat used up by evaporation. This consumption, for the freezing season together with autumn evaporation is, in Western ETC between 35 and 40%, and in Siberia from 20 to 30%. The lowest values apply for the Arctic and dry areas in Mongolia (5-10%).

Despite increased solar radiation in the spring, the total radiation balance for the freezing season is mostly negative or close to zero (it is positive only at some locations). Thermal flux for radiation from the surface is: in Western ETC about 40%, in Northwestern Siberia and in Central Yakutia 60-70%, and on the Arctic coast 75%.

The thermal loss by the earth's surface due to snow thaw, evaporation, and radiation is compensated for by the heat input from the atmosphere: 65-70% in Western ETC, 45-50% in the Arctic, and 20-35% in Central Yakutia.

Ground thermal flux provides a substantial replenishment of the earth's surface heat over the freezing season. In this case it makes the largest contribution in regions of perennial and deep seasonal freezing. Thus, in Western ETC (not deep freezing) the ground thermal flux constitutes 30-35%, in Northwestern Siberia 40-45%, in Yakutia 55-75%, and in Southern Mongolia (deep freezing) 75-80%.

Characteristically, ground thermal flux through the earth's surface changes its sign in autumn while still at positive values for the radiation balance and long before the beginning of freezing. This suggests that a specific amount of heat and time are required for the thermal field to reverse. Thus, the heat flow into soil traverses zero before the beginning of freezing: in Western ETC and in Southern Mongolia 2 months, in the south of Central Siberia, one month, in Central Yakutia 2 weeks, and in the Arctic one week before. In the spring the ground soil thermal flux reverses its sign, but mainly at positive radiation balance and with snow still present, i.e. prior to the onset of thawing. This has the following delays: in Western ETC 10 days, in the northern half of Siberia 1 month, and in Southern Siberia and in Mongolia, with their frequent recurrent snowfalls, 1.5-2 months.

The structure of the thermal balance of the earth's surface for the thaw season is well defined. The heat used to thaw the rest of the snow, since this season extends into part of the spring, is an additional factor. The overall radiation balance is positive and varies in the territory under consideration from 1200 to 1400 MJ/m² in Western ETC and in Central Yakutia from 750 to 800 MJ/m² in the Arctic.

In Western ETC the main thermal flux is used for evaporation (two-thirds to three-quarters of the total radiation balance) and then for warming the atmosphere (about one-quarter to one-third). Four percent of the heat is used to warm the soil and about 1-2% to thaw snow. In Northern ETC, in Western Siberia, and in Central Yakutia the ratios of thermal flux used for evaporation and turbulent heat exchange with the atmosphere are approximately equal (40-45%). Heat flow used to warm the soil comprises 10-12%, while that used to thaw snow averages 1%.

Analysis of the available actual data shows that the totals for soil heat flow for the thaw season (as well as for the freezing season) and their relationship to the radiation balance turn out to be fairly stable in separate regions and vary as one moves from north to south, thereby reflecting the properties of the permafrost regime. This allows for, on the one hand, representation of their spatial distribution (which we actually did for the thaw period using the most numerous and reliable data available) and, on the other hand, indicates the possibility of subsequent permafrost-climatic zoning.

For example, the distribution of the ratio of soil thermal flux to radiation balance of the earth's surface (Figure 1). The greatest share (over 20%) occurs in the coldest and moist (icing) regions of the Arctic islands and sea coast. To the south the ratios diminish, increasing only in the mountains (up to 15%). Typically, in many regions (except for dry Central Yakutia) the 14.5% isoline ratio passes close to the boundary of continuous permafrost, the 10% isoline runs along permafrost islands in the west and discontinuous permafrost in the east, and the 7.5% isoline, along sporadic permafrost in the Far East.

Since there are more observations on the radiation balance than on thermal flow into the soil, by knowing the characteristic B/R ratios for individual regions, one is also able to evaluate

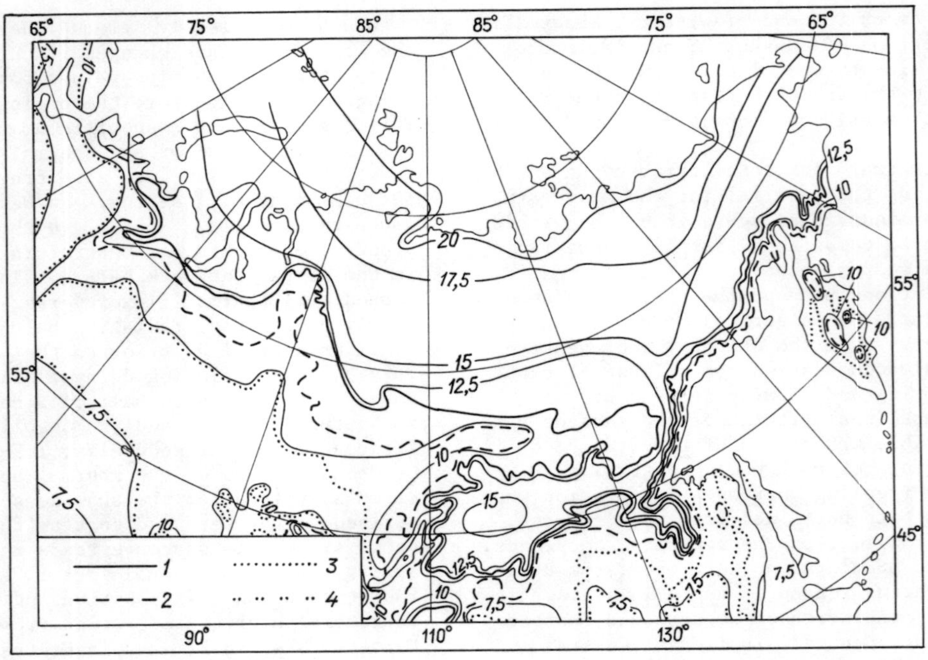

FIGURE 1 The ratio (%) of the sum of heat flow for soil to the sum of radiation balance for the thaw season. Boundaries of permafrost zones: 1, continuous; 2, discontinuous, 3, islands; 4, sporadic.

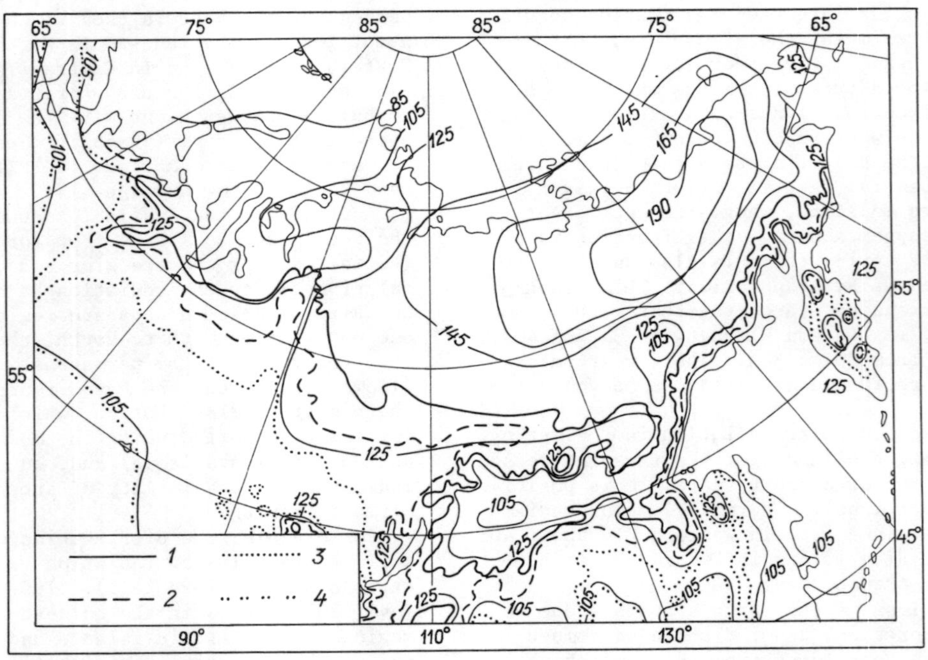

FIGURE 2 The sum of heat flow for soil (MJ/m^2) for the thaw season. Boundaries of permafrost zones: 1, continuous; 2, discontinuous; 3, islands; 4, sporadic.

the thermal flux totals for the season. Of course, total ground thermal flux depends both on the season's duration and on the temperature differences in the upper layer of soil. In addition, temperature gradients are greater in regions of greatest freezing. Figure 2 shows that in the permafrost region the smallest amount of heat flow to soil (on the order of 65 MJ/m^2) occurs on the most remote near-Atlantic islands of the Arctic, where the thaw period is short (60-65 days). To the south, as the duration of the season increases, the sum total of flow increases to 105-115 MJ/m^2 in northern ETC. But further in ETC (not shown on this map) this flow again decreases, but now in conjunction with a decrease in the temperature gradient, to values on the order of 40-45 MJ/m^2 in

TABLE 1 Monthly Thermal Balance Constituents in the Above-Ground Cover and Within Soil in MJ/m² Neryukteitsy 1965-1967

Low	Jan	Feb	Mar	Apr	May	June	July	Aug	Sept	Oct	Nov	Dec	Freeze season	Thaw season	Yearly
B	-22.79	-16.09	-11.73	-7.84	36.03	58.32	38.76	16.84	3.52	-27.02	-30.88	-34.19	-150.55	153.48	2.93
B_{sn}	0.04	0.42	0.84	0.75						-0.04	-0.04	-0.17	1.34		1.34
B_{gr}	-22.83	-16.51	-12.57	-8.59	36.03	58.32	38.76	16.84	3.52	-26.98	-30.38	-34.02	-151.89	153.48	1.59
B_+	0	0	0	0	4.02	10.81	4.74	-6.83	-10.73	-2.01	0	0	-2.01	2.01	0
B_{ph}	0	0	0	0	19.15	25.47	16.80	10.35	0.55	-72.32	0	0	-72.32	72.32	0
B_-	-22.83	-16.51	-12.57	8.59	12.86	22.04	17.22	13.32	13.70	47.35	-30.38	-34.02	-77.56	79.15	1.59

Notes: B - thermal flow through Earth's surface (surface of vegetation, snow); B_{sn} - heat, accumulated by snow cover; B_{gr} - thermal flow through soil surface, i.e. the one directly penetrated into or escaped from soil; B_+ - heat, accumulated in unfrozen layer (in the positive temperature layer zone); B_{ph} - heat of phase transitions, consumed for soil thawing or released in the freezing process; B_- heat, penetrated into underlying frozen layers (negative temperature zone). The area of Neryukteitsy is situated in the right bank of the Lena River valley. Diverse-grass, thin-leg carex steppe; sandy loam, sand. It is comparatively moistened in the first half of the summer, the mean moisture for the season in a 0-20 cm layer is 35%, below—23%. Maximum thawing is 180 cm. The mean height of snow varies between 5 cm in October and 31 cm in March.

the vicinity of Leningrad and Moscow, i.e. in regions where the ground is not deeply frozen.

As one moves from the European part to the East or, putting it another way, as the climate becomes increasingly more continental, the totals for ground thermal flux increase. Their greatest values (on the order of 165-190 MJ/m²) occur in the northern half of the USSR North-East, i.e. the region having the coldest winter or the greatest temperature differences in the soil during warmup.

Since the annual total thermal flux to ground soils through the Earth's surface can, for practical purposes, be assumed to be zero, the map in Figure 2 should be extended also to the freezing season, with the opposite sign only. As a result, the largest loss of heat by ground soils (165-190 MJ/m²) occurs in the region of continuous permafrost in North-Eastern Yakutia with its severe climate, after which it gradually decreases towards the areas of deep and shallow seasonal ground freezing. This agrees quite well with the actual situation.

The absolute values of thermal flux through the surface determine neither the absolute temperature indices in the upper ground layer, nor the depth of seasonal thawing. Considerable flux during the warm period of the year in northern regions only indicates that more heat was expended to restructure the thermal field during winter freezing. A substantial portion of this heat in permafrost regions is used for ice-water phase transitions, while a smaller share goes to warm up the thawed layer. Thus, according to our investigations (Gavrilova 1966, 1967a, 1967b, and others) in Central Yakutia with its continuous permafrost and annual temperature fluctuations, the warming-up of permafrost consumes nearly one-third of the thermal energy, while in the North-Transbaikal's regions which only experience deep freezing, only 2%. In both cases, more than half the heat is used for thawing (2 and 5 m). The remaining heat is used to change the temperature in the thawed layer: less than 10% in Central Yakutia and over 40% in North Transbaikal.

A general picture of the structure of thermal balance of soil in the region of permafrost and deep seasonal freezing of soil has been presented. However, in the same region there may occur individual departures from sector to sector which are associated with local physiogeography, the character of the moisture, and other local conditions. There is also an annual variation of all constituents of the thermal flux in soils, i.e. intraseasonal variations (Gavrilova 1981). Table 1 presents a detailed study of ground thermal balances using observations from the Ulakhan-Taryn research station (Gavrilova 1972).

It can be noticed that, with the disappearance of snow cover in Central Yakutia in late April the thermal flows in all of the layer become positive. At that time, a recombination of the entire structure of the soil's thermal balance takes place. The heat coming through the Earth's surface is largely used to thaw from above the soil frozen during the winter, i.e. ice water phase transitions. A part of the heat goes to warm up the thawed layer, another part the frozen layer. Since the thickness of the frozen soil up to the depth of attenuation of annual temperature amplitudes (10 m) is much greater than the thawing layer (less than 2 m), then B_-, in general, is larger for the season than B_+, although the level of the thawing boundary (flow of B_{ph}) somewhat impedes the downward penetration of the heat.

In the second half of the summer the upper soil layers begin to cool. Thus, in an unfrozen layer the flow at a depth of 0-10 cm changes its sign by the end of July, though the total ground mass continues to absorb heat. In September the heat loss increases in the upper 2 m, ground thawing stops, while the underlying frozen layers are still being warmed up.

Intense freezing from above soils thawed during the summer begins in the beginning of October. The soil (about 2 m thick), unfrozen at the beginning of the month, becomes completely frozen by the end of October, i.e. it requires one month, while thawing takes place over 5 months. Also, large amounts of heat are released as water changes into ice.

The winter thermal regime begins on completion of the freezing process of the ground which thawed during the summer. Large losses of surface heat, at a negative radiation balance, extend to even deeper layers of soil, and the entire comparatively thick layer of permafrost in the temperature fluctuation zone returns its heat over the following 6 months (November-April).

REFERENCES

Averkov, M. S., 1960, Annual variation of heat exchange in soils near Moscow, Vestnik Moskovskogo universiteta, Ser. V, Geography, No. 2, p. 15-27.

Gavrilova, M. K., 1966, Thermal regime of the Earth's surface and field and forest soils during thaw. In Proceedings of the VIII All-Union Interdepartmental Meeting on Geocryology, v.4, Yakutsk: Yakutknigoizdat, p. 194-206.

Gavrilova, M. K., 1967a, Thermal balance of a larch forest on the Lena-Amgin inter-river region. In Hydro-climatic investigations in Siberian forests. Moscow: Nauka, p. 28-52.

Gavrilova, M. K., 1967b, Microclimatic and thermal regimes of the Earth's surface and soils in the Chara hollow. In Geocryological conditions of Transbaikalie and Near-Bailal areas. Moscow: Nauka, p. 148-161.

Gavrilova, M. K., 1972, Temperature of soils under different physiogeographical conditions in the Ulakhan-Taryn region of Central Yakutia. In Permafrost and soil, v. 1, Yakutsk: YF SO AN SSSR, p. 35-51.

Gravilova, M. K., 1973, Climate of Central Yakutia, 2nd Edition: Yakutsk, Yakutknigoizdat, 120 p.

Gavrilova, M. K., 1974, Climatic factors in the formation and development of permafrost rock. In Geocryological conditions of the Mongolian People's Republic: Moscow, Nauka, p. 12-29.

Gavrilova, M. K., 1981, Present-day climate and permafrost on mainlands: Novosibirsk, Nauka, 112 p. + Suppl.

Golubev, A. V. and A. V. Pavlov, 1961, Investigation of thermal flux in soil for some types of active surface. In Materials for the principles of theory of permafrost of the Earth's crust, v. 7: Moscow, Izd. AN SSSR, p. 66-118.

Ivanov, N. S., 1961, Methods of measuring thermal flux in rocks. In Thermal and mass exchange in permafrost and frozen rock: Moscow, Izd, AN SSSR, p. 91-104.

Manual on thermal-balance observations, 1977: Leningrad, Gidrometeoizdat, 150 p.

Molochushkin, E. N., 1969, Results of a year of observations of soil temperature on the Muostakh Island. In Questions of Yakutiya's Geography, v. 5: Yakutsk, Yakutknigoizdat, p. 127-137.

Pavlov, A. V., 1979, Thermal physics of terrain: Novosibirsk, Nauka, 284 p.

Pavlov, A. V. and B. A. Olovin, 1974, Artificial thaw of frozen rock by solar radiation in developing placers: Novosibirsk, Nauka, 180 p.

LAWS GOVERNING THE COMPACTION OF THAWING SOILS TAKING DYNAMIC PROCESSES INTO EFFECT

V.D. Karlov and V.K. Inosemtsev

Institute of Civil Engineering
Leningrad, USSR

Comprehensive compression investigations of the deformation properties of thawing sandy soils were carried out under static (up to 0.4 MPa) and dynamic (up to 4000 mm/s^2) loading. The experiments were made with the artificially prepared samples of frozen sand whose porosity ratio varied between 0.544 and 1.046. The samples were 113 mm in diameter and 45 mm high. Maximum compaction of the thawing soil occurred in the early stages of vibration with vibration accelerating from 0 to 1000 mm/s^2 and from 1000 to 2000 mm/s^2. The impact of the static load is significant. Extremely intense compaction occurs under the static loads of up to 0.05 MPa. There is a linear relationship between the static load and the critical vibration acceleration. The amount of settling of thawing sandy soils under dynamic loads is greatly affected by the grain-size composition and other soil conditions (e.g. compactness, moisture content). It is suggested that settlement of thawing soils under dynamic loading may be estimated on the basis of the soil compaction coefficient when subjected to vibration, the thickness of the soil layer involved, and the values of the vibration acceleration both actual and critical. Taking into account both static and dynamic loading the amount of settlement in a thawing soil layer was calculated to be the sum of the settlements due to thawing and compaction under static loading, and of compaction settlement due to vibratory action. Settlement of foundations may be determined by summation of settlements in individual soil layers down to foundation depth within each of which the values of dynamic actions and the compressibility characteristics may be considered to be constant.

Thawing soils may undergo great deformations due to their particular postcryogenic texture which is extremely susceptible to sign-variable dynamic stresses. The laws governing thawing sandy soil compaction under dynamic loading have not yet been sufficiently studied and the available results of post investigations are poorly presented in both the domestic and the foreign literature.

To develop a reliable method for calculating the settlements of thawing foundation soils it is necessary to establish the principle laws governing their compressibility allowing for dynamic impact.

Comprehensive investigations on the deformation properties of thawing soils have been carried out in the Leningrad Civil Engineering Institute. Experimental studies of the compaction of thawing sandy soils were carried out using a specially designed installation. The main parts of the installation are: a low heat-conducting consolidometer, a specially designed hollow settlement, a liquid thermostat, a device for static and dynamic loading, a control console, a set of measuring instruments, and other equipment. The basic diagram of the installation is presented in the authors' work (Inosemtsev and Karlov 1980). This installation permits samples of thawing and thawed soils, both natural and with structural damage, to be tested under wide range of static and dynamic loading.

When the foundations are arranged on thawing soils, the dynamic effect may appear at different stages of thawing and stabilization of ground deformation. Therefore, when developing a procedure for estimating the compressibility of thawing soils, it is very important to ascertain the right moment to apply the dynamic load.

For this purpose three series of experiments were conducted. In each series of experiments frozen soil samples having identical physico-mechanical characteristics and under equally intense loads were tested. In the first series of experiments dynamic impact was produced directly during the process of thawing a soil sample loaded with a 0.2-MPa static pressure. In the second series of experiments dynamic loads were applied after thawing the sample and stabilizing the settlement from the subsequent static loading, which was also equal to 0.2 MPa. In the third series of experiments, the dynamic effects were produced immediately arter thawing the sample under a pressure of 0.2 MPa with partial stabilization of the settlement of the thawing soil.

Statistical analyses of the results of these investigations were conducted based on the criteria put forth by Kohran and Fisher (Pustilnik 1968), which established that the difference between the average values for the total relative settlements in the three series of experiments, with a 5% level of significance, should be considered to be insignificant. This means that the value of the final total deformation of thawing sandy soils is especially affected by the moment (of time) when the dynamic load is applied.

This conclusion makes it possible to carry out dynamic tests of frozen sandy soils after thawing and stabilization of the settlements due to static loading. Such a procedure has some important advantages as compared with the procedure in which dynamic impact is produced during thawing of the soil sample. In the latter case only the total value of sample settlement is determined, with no possibility of dividing it into components such as: thawing settlement, compaction settlement under static loading, and settlement due to the dynamic action.

If dynamic effects are produced after thawing and stabilization of sample settlement, then such a test procedure provides all the compressibility characteristics needed to calculate thawing foundations in the presence of thawing of a single soil sample. Therefore to study the laws governing deformability occurring during thawing of sandy soils, the procedure for testing frozen soils with the thawing of a single sample and its subsequent static and dynamic loading was adopted. Such a test procedure is much more economical and provides more certain results than testing several samples thawing under various intensities of dynamic action.

The laboratory investigation of the compressibility of the thawing sandy soils has been carried out in a wide range of static and dynamic loading depending upon the composition and the condition of sandy soils. The experiments used artificially prepared sandy soil samples having various degrees of compaction (the void ratio of the soil varied from 0.544 to 1.046), soil moisture (the pore ice volume varied from 0.14 to 1.0), and grain size (coarse-grained, medium-grained, fine-grained sands).

The test results were analyzed using the proximate statistical processing (Goldstein 1973). To establish probable interconnection between compressibility of the thawing soils and the factors under investigation, computer-assisted regression and correlation analysis were run.

The results of the investigation of the relationship between relative settlement of thawing, coarse-grained, water-saturated sand and the initial soil compactness at the moment in which varying degrees of dynamic action are applied are shown in Figure 1. This relationship is non-linear. From Figure 1

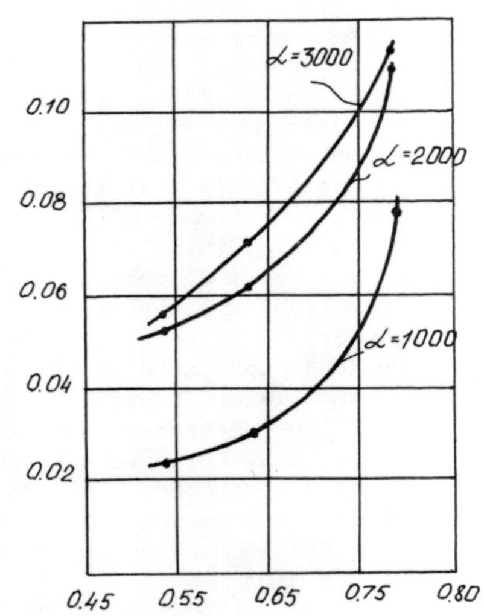

FIGURE 1 Relationship between the relative settlement of the thawed sandy soil and the compactness at various values of vibration acceleration.

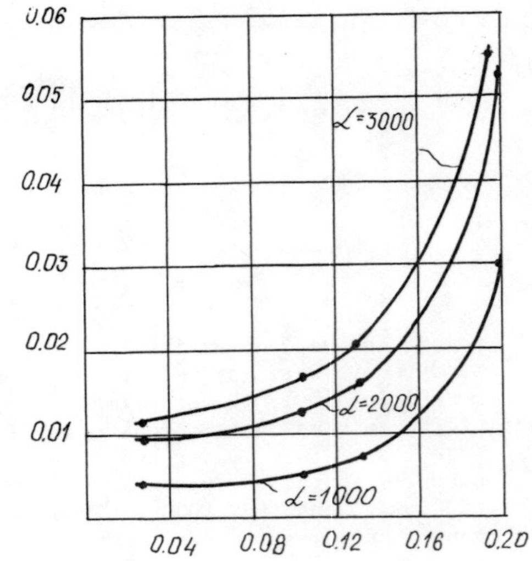

FIGURE 2. Relationship between relative settlement and moisture at various values of vibration acceleration.

one can see that the main part of the settlement of thawing sandy soils occurs in the early stages of dynamic impact. A further increase in the intensity of the dynamic impact causes less settlement. Thus, an increase in vibration acceleration from 2000 mm/s² to 3000 mm/s² causes an increase in settlement several times less than the increment of the vibration acceleration value from 1000 mm/s² to 2000 mm/s² (Figure 1). This is completely understandable: the thawed soil being compacted in the early stage of dynamic impact changes into a more compact state, which is characterized by less and smaller pores at the site of the thawed ice crystals; consequently, the conditions for particle displacement during compaction at the next stage of dynamic loading become worse.

Figure 2 shows the curves for the relationship between the relative settlement of the thawed sandy soil samples (with the same initial compactness in the frozen state) and the moisture content by weight at different values of vibration acceleration. An increase in the initial moisture content by weight increases the compressibility of the thawed soils under dynamic impact and maximum increase in the compressibility is observed with a variation of the moisture content from 0.13 to 0.20, which corresponds to the degree of saturation G = from 0.7 to 1.0. This specified influence of moisture content on settlement of the thawed sandy soils is explained by the action of the capillary forces of cohesion which prevent the soil from compacting during dynamic loading in slightly moist and in moist sands. And the presence of gravitational water, which serves as a lubricant, reduces friction between the sand particles, thereby increasing, during vibrations, the compressibility of the saturated sandy soils whose water is in a free state.

The results acquired from experiments made with sands of different grain size showed that with an increase of sand grain-size the compressibility of the thawing soils under dynamic action decreases. Under dynamic action with a vibration intensity of

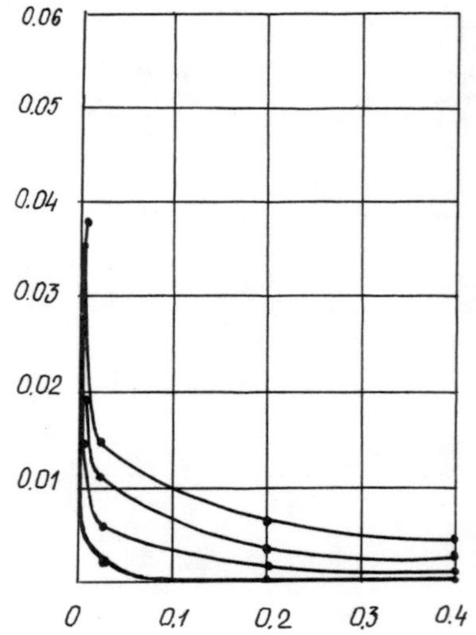

FIGURE 3 Relationship between the relative settlement and the value of static loading at various values of vibration acceleration.

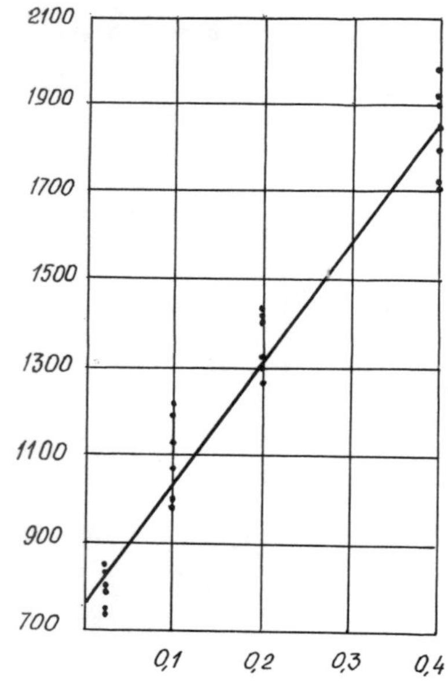

FIGURE 5 Relationship between the critical vibration acceleration of the thawed soil and the static loading.

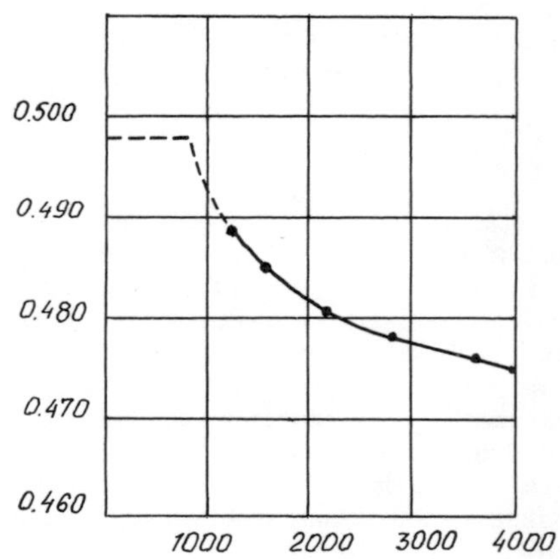

FIGURE 4 Impact-compression relationship of the thawed soil.

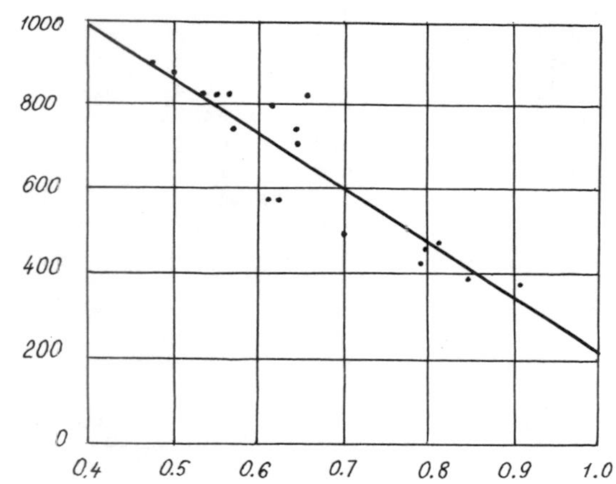

FIGURE 6 Relationship between the critical vibration acceleration and the void ratio of the thawed soil.

$\alpha = 0.6 - 0.7$ g, the medium-grained thawing sandy soils demonstrated a relative compaction settlement of 1.27 times and the fine-grained soils 1.62 times greater than the coarse-grained sand.

In addition to the factors examined above (density, moisture, grain size distribution) the values of static and dynamic loading highly affect the compressibility of the thawing sand.

Figure 3 shows the relationship between relative settlement of the thawed sandy soil being compacted by the dynamic action (δ_α) and of various values of static load (P). This relation is curvilinear. The relation curve $\delta_\alpha = f(P)$ drops sharply when the value for static loading is small; with a further increase in the load, the compressibility slightly decreases. The character of the effect of static loads on the compressibility of thawing sandy soil, when the dynamic load value is invariable, can be explained by the fact that as the static pressure increases, the internal friction forces between sand particles increase, thereby affecting the value of non-cohesive ground settlement (Ivanov 1967).

As a result of compression tests in the presence of dynamic action, it has been established that change in the porosity coefficient of thawing soil is a function of the intensity of the dynamic effect (vibration acceleration). Figure 4 represents the impact-compression relationship based on the results of experiments with coarse-grained, saturated sands thawed under pressure at 0.025 MPa. One can see that the soil compaction begins after vibration acceleration attains its limit value, which is called the critical vibration value by Savinov (1979).

The influence of static pressure and initial soil compactness on the value for the critical vibration acceleration α_{cr} of thawed sandy soil is shiwn in Figures 5 and 6. The regression equation of the relationship $\alpha_{cr} = f(P)$ for thawed coarse-grained saturated sandy soil is:

$$\alpha_{cr} = 775 + 2713P \qquad (1)$$

where α_{cr} is the critical vibration acceleration, mm/s², and P is the static pressure, MPa.

The correlation factor of the relationship $\alpha_{cr} = f(P)$ equaled 0.975, while the verification of the hypothesis as to the existence of a link between α_{cr} and P showed that the correlation relationship exists at the level of significance $\beta = 0.001$.

The correlation equation for critical vibration acceleration as a function of the void ratio $\ell_{o\alpha}$ of the thawed coarse-grained saturated sandy soil at the moment of applying the load $\alpha_{cr} = f(\ell_{o\alpha})$ is as follows:

$$\alpha_{cr} = 1462 - 1230\, \ell_{o\alpha} \qquad (2)$$

The correlation coefficient proved to be 0.876. By comparing the correlation coefficients it becomes clear that the relationship between critical vibration acceleration values and the void ratio of thawed sandy soil at the moment of dynamic loading appears to be less close than that between critical vibration acceleration and static load.

The relationship between change of the void ratio and the vibration acceleration α (Figure 4) of thawed, coarse-grained, saturated, sandy soil was investigated using pair correlation analysis.

The degree of closeness between parameters ℓ and α was evaluated based on the correlation coefficient. Its application is possible only when the values (ℓ, α) under consideration are random and distributed normally. Verification of the hypothesis related to the laws of void ratio distribution ℓ and vibration acceleration α was performed using Pearson's coordination criterion (x^2 - criterion). The calculations showed that at a 5% level of significance, the accepted hypothesis regarding normal distribution of ℓ in the general population is not rejected, but the hypothesis regarding the normal distribution of α is negated.

In this connection another hypothesis was adopted: "the distribution of α is logarithmically normal." The comparison of an empirical criterion with its critical value shows that the adopted hypothesis for the normal distribution of lg α is not rejected. Consequently, the correlation relationship of the void ratio and the vibration acceleration value may be considered in the ordinate system $y = \ell$, $x = \lg\alpha$, and the correlation coefficient may be used to characterize the degree of closeness in the relationship.

The graph of the relation between void ratio and vibration acceleration is presented in Figure 7 in semi-logarithmic coordinates. The correlation coefficient $\ell - \lg\alpha$ equaled 0.914.

Estimation of the certainty of the $\ell - \lg\alpha$ relationship showed that a correlation exists at the significance level of $\beta = 0.001$.

Statistical verification of the hypothesis related to the rectilinear form of the relationship $\ell - \lg\alpha$ was performed using the Fisher criterion.

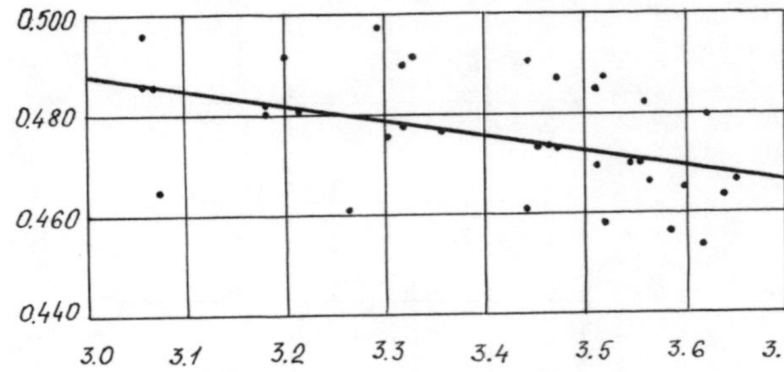

FIGURE 7 Relationship between the void ratio of the thawed sandy soil and the vibration acceleration.

Comparison of criterion F_ξ with the critical one F showed that the hypothesis adopted is negated at the significance level $\beta = 0.05$.

The regression equation for the relationship $\ell - f(\alpha)$ usually may be written as follows:

$$\ell = a - b\lg\alpha \qquad (3)$$

where a is a free term of regression and b is the regression coefficient.

Using the boundary conditions (with $\alpha = \alpha_{cr} \rightarrow \ell \rightarrow \ell_{o\alpha}$) eq 3 can be rewritten:

$$\ell = \ell_{o\alpha} - b\lg \frac{\alpha}{\alpha_{cr}} \qquad (4)$$

where $\ell_{o\alpha}$, α_{cr} are void ratio at the moment dynamic loading and the critical vibration acceleration of a sample of thawed, sandy soil, respectively.

The laws governing the compaction of thawing, sandy soil established on the basis of comprehensive investigations suggest an engineering procedure for estimating foundation settling values at static and dynamic loading.

The total settlement value of a thawing base at combined static and dynamic loading can be determined according to the formula:

$$S = S_{th} + S_p + S_\alpha \qquad (5)$$

where S_{th} is the settlement caused by thawing of the frozen soil and S_α is the compaction settlement of the thawed soil at dynamic action.

The settlement values S_{th} and S_p can be determined by a number of well-known methods (Build. Stand. II-18-76, 1974, Tsytovich 1973).

The settlement value of the thawed base under the dynamic actions can be determined by layer summation according to the formula (Inosemtsev and Karlov 1981):

$$S_\alpha = \sum_{i=1}^{n} k_{vi}\, h_i \ln \frac{\alpha_i}{\alpha_{cri}} \qquad (6)$$

where h_i is the thickness of the layer, i is the soil layer; n is the number of layers into which the zone of dynamic compaction is divided; α_{cr} is the critical vibration acceleration estimated from laboratory testing of the samples of the i - soil layer; α_i is the average vibration acceleration in the i - soil layer, equal to half of the sum of the vibration acceleration on the upper and lower boundaries of the layer; k_{vi} is the coefficient of relative compressibility of the i - soil layer, which is determined by the formula:

$$k_{vi} = \frac{k_i}{1+\ell_{oi} - (1+\ell_{oi})(A_i + a_i P)} \qquad (7)$$

where k_i is the compaction coefficient of the i - soil layer under the dynamic actions; ℓ_{oi} is the initial void ratio of the frozen soil; A_i is the coefficient of thawing; a_i is the coefficient of relative compressibility of the thawing soil and P is the static load.

Calculation of settlement S_α can also be made by another procedure using not the values of the vibration acceleration in the solution, but the areas of their depth curves. For this purpose, in eq. 6 the sign of the sum changes for the integral, in which case, the number of layers into which the base should be divided greatly decreases. The soil layers will differ only in the deformation properties. Then, in the case of uniformly distributed loading having an intensity of P, the settlement of the thawed homogeneous base is calculated by the formula:

$$S_\alpha = k_v \ (C - B) \qquad (8)$$

The values of parameters C and B are obtained from expressions:

$$C = Z(\ln\alpha_o - \frac{\delta_z}{2}) \qquad (9)$$

$$B = \int_o^z \ln \{ \frac{\alpha_{cro} + m \{\frac{P-\gamma h}{\pi} [2\mathrm{arctg}\frac{b}{2Z} + \sin(2\mathrm{arctg}\frac{b}{2Z}) + \gamma(h+z)]\}}{1+\frac{m}{g}\{\frac{P-\gamma h}{\pi}[2\mathrm{arctg}\frac{b}{2Z} + \sin(2\mathrm{arctg}\frac{b}{2Z}) + \gamma(h+z)]\}}\} dZ \qquad (10)$$

where z is the ordinate of the point for which the settlement is calculated; α_o is the vibration acceleration at the base of the foundation; δ is the attenuation factor of soil vibration; α_{cro} is the critical vibration acceleration of the unladen soil sample; γ is the specific weight of the thawed soil allowing for suspension caused by water; h is the depth of the foundation base; b is the width of the foundation base; m is the empirical coefficient derived from the dependency curve of the critical vibration acceleration and the static load g is the gravity acceleration.

CONCLUSIONS

On the basis of these investigations one can draw the following conclusions:

1. The total deformation of frozen, sandy soil samples caused by thawing, static and dynamic loading essentially does not depend upon the moment (time) that dynamic impact is applied.
2. The settlement value of the thawing sandy soils at impact loading is greatly influenced by the composition and the condition of the sands under investigation. As moisture content decreases and compactness and grain size increases, the compressibility of thawing sandy soils is reduced.
3. A nonlinear relationship was established between the compressibility of thawing, sandy soil and vibration acceleration values.
4. The settlement of thawing, sandy soils is greatly affected by static loading an increase in which causes a decrease in settlement due to dynamic impact.
5. A procedure for calculating the settlement of thawing bases of the foundations which takes into consideration dynamic actions was developed. It was recommended to calculate settlement of thawing bases using layer summation by eqs 6 and 8.

REFERENCES

Build. Standards II-18-76, 1974, Footings and Foundations on Permafrost, Moscow.

Goldstein, M. N., 1973, Mechanical properties of soils, Moscow.

Inosemtsev, V. K., and V. D. Karlov, 1980, On the procedure of laboratory estimating the thawing soil compressibility with consideration of the dynamic effect: Soil Mechanics. Footings and Foundations, Mezhvuz. Proc. LISI, Leningrad.

Inosemtsev, V. K., and V. D. Karlov, 1981, On the estimation of the thawing soil settlements with consideration of the dynamic effect. Dynamics of bases, foundations and underground constructions. Proc. of the All-Union Conf., Moscow.

Ivanov, P. L., 1967, Compaction of loose soils by impact, Leningrad.

Pustilnik, E. I., 1968, Static procedures of analyzing and processing the observations, Moscow.

Savinov, O. A., 1979, Present-day foundation constructions under machines and their design. Leningrad-Moscow.

Tsytovich, N. A., 1973, Mechanics of frozen soils, Moscow.

A THEORY OF DESICCATION OF UNCONSOLIDATED ROCKS IN AREAS WITH NEGATIVE TEMPERATURES

I. A. Komarov

Geology Department, Moscow State University
Moscow, USSR

The article presents the results of theoretical and experimental investigations of the process of desiccation in unconsolidated rocks in areas with negative temperatures. On the basis of a thermodynamic assessment of the phase equilibrium of water in the rocks it was demonstrated that the desiccation process occurs through sublimation and desorption of ice as well as through the combined effects of these processes, and occurs in two forms: voluminal and surficial. The instability of the surficial form is attributed to the presence of ultrapores. A capillary model of the process is proposed, based on an arbitrary subdivision of moisture present in the rock into three categories, which provide a qualitative description of the kinetics of desiccation in rocks of different composition and structure and under different ambient conditions. This scheme forms the basis of a system of equations for a three-zone model describing desiccation in a moisture-filled rock mass. Results of a comparison between solutions of a simplified system of equations for a quasi-isothermal case with the experimental data are presented; they indicate a satisfactory agreement between the two. Peculiarities of applying this solution in cases where the process is occurring in non-saturated rocks are discussed.

Reports submitted to the 3rd Conference on Geocryology in addition to other publications (Yershov 1979; Yershov et al. 1975, 1978) have summarized and analyzed basic laws governing the process of frost desiccation in unconsolidated rocks. The article attempts to elaborate on the physical concepts in addition to presenting a quantitative theory as a basis for developing a prognostication technique to control the desiccation process at negative temperatures.

PRELIMINARY STATEMENTS AND A PHYSICAL MODEL OF THE PROCESS

Analysis of the physical aspect of the rock desiccation process shows that its intensity is primarily influenced by the capillary, adsorptional, and osmotic relationships between moisture and rock. Consequently, the thermodynamic potential (free enthalpy ϕ) may in this case be represented as the sum of two members, one of which depends on the L potentials, θ_i, i.e. in our case temperature (T), pressure (P_{total}), and the concentration of dissolved substances (Z); the other depends on the (n - L) coordinates of the state χ_j, which defines the system's geometry (the surface of adsorptional interaction Ω_a and the magnitudes of adsorptional surfaces at the interfaces of the solid, liquid, and vaporous phases in the interstitial solution Ω_{23}, Ω_{34}):

$$d\phi(\theta_i,\chi_j) = \sum_{i=1}^{L} \left(\frac{\partial\phi}{\partial\phi_i}\right)_{\substack{\theta_{i\,inv}\\ \chi_{j\,inv}}} \cdot d\theta_i + \sum_{j=L+1}^{n-L} \left(\frac{\partial\phi}{\partial\chi_j}\right)_{\substack{\chi_{j\,inv}\\ \theta_{i\,inv}}} \cdot d\chi; \tag{1}$$

where $\theta_{i\,inv}$, $\chi_{j\,inv}$ indicate that the rest θ_i and χ_j do not change (invariants), here θ_i = P_{total}, T, Z; and χ_j = Ωa, Ω_{23}, Ω_{34}.

The values of the partial derivatives of free enthalpy with respect to χ_j are:

$$\frac{\partial\phi}{\partial\Omega_{34}} = \sigma_{34}; \quad \frac{\partial\phi}{\partial\Omega_{23}} = \sigma_{23}; \quad \frac{\partial\phi}{\partial\Omega_a} = E$$

where σ_{34}, σ_{23} are surface tensions at the water-ice and water-gas interfaces and E is the energy of adsorptional interaction.

Here and further we shall be considering the case in which the concentration of dissolved salts in the interstitial moisture is low, therefore the latter may be regarded as a binary solution composed of free water and water bound by surface forces. To describe the phase equilibrium we used relationships elaborated in the theory of solutions. A decrease in the freezing temperature may in this case be regarded as a displacement of the triple point on the phase P-T diagram of the state of the water. Keeping this analogy in mind it appears from this diagram that the desiccation process is dependent on the relationship between the freezing temperature (T_{fr}) and the temperature of the medium (T_{med}) and can occur in three ways: 1) by ice sublimation, $T_{fr} > T_{med}$; 2) by desorption, $T_{fr} < T_{med} < T_o = 273°K$; 3) by both mechanisms occurring simultaneously, $T_{med} < T_{fr} < T_o$ (ice sublimation occurring in pores with the radius $\mu > \mu_{equ}$ and desorption in the pores $\mu > \mu_{equ}$). Depending upon the thermodynamic conditions prevailing in the vapor-gaseous medium and the properties of the dispersive system, these processes can take one of two forms. The first is typical for systems in which the surficial form of phase transition is stable (frontal transition). The other is distinguished by the presence of a phase transition zone (voluminal transition).

Detailed analysis of possible causes for the instability of the surficial form (Komarov 1979) has shown that apparently the formation of a zone is due to capillary effect occurring at relatively high negative temperatures when ultrapores with a radius $r < 10^{-7}$ m are present in the rock. In this case, the difference between surface tensions at developed water-ice and water-air interface areas exerts a significant effect on the process kinetics, resulting in the formation of an evaporation zone. As temperature decreases, the process kinetics is largely influenced by adsorptional surface forces, which results in degradation of the evaporation zone. In the absence of ultrapores, for example in sands, the zone is essentially absent.

The dehydration kinetics will be discussed with respect to a capillary model of the process, which is a further development and generalized form of Krisher's bicapillary model (1961) as applied to the case involving the presence of the solid water-ice phase in the pores. The concepts incorporated in the model are based on the assumption that unfrozen water is present in the rock partly in between the blocks of mineral particles (microaggregates indicated by circles) and ice (blackened areas) and partly in ultra-capillaries and pores permeating the particles themselves (Figure 1a). In the first case the equilibrium is influenced by the interface curvature with respect to ice and vapor, while in the second case it is exposed to the effect of adsorptional forces. Furthermore, the interaggregate moisture (in the capillary of the R_2 radius) is assumed to be mobile with respect to the moisture present inside the aggregates (capillary R_1). Thus, the overall moisture content of unconsolidated rock can be arbitrarily subdivided into three categories (the

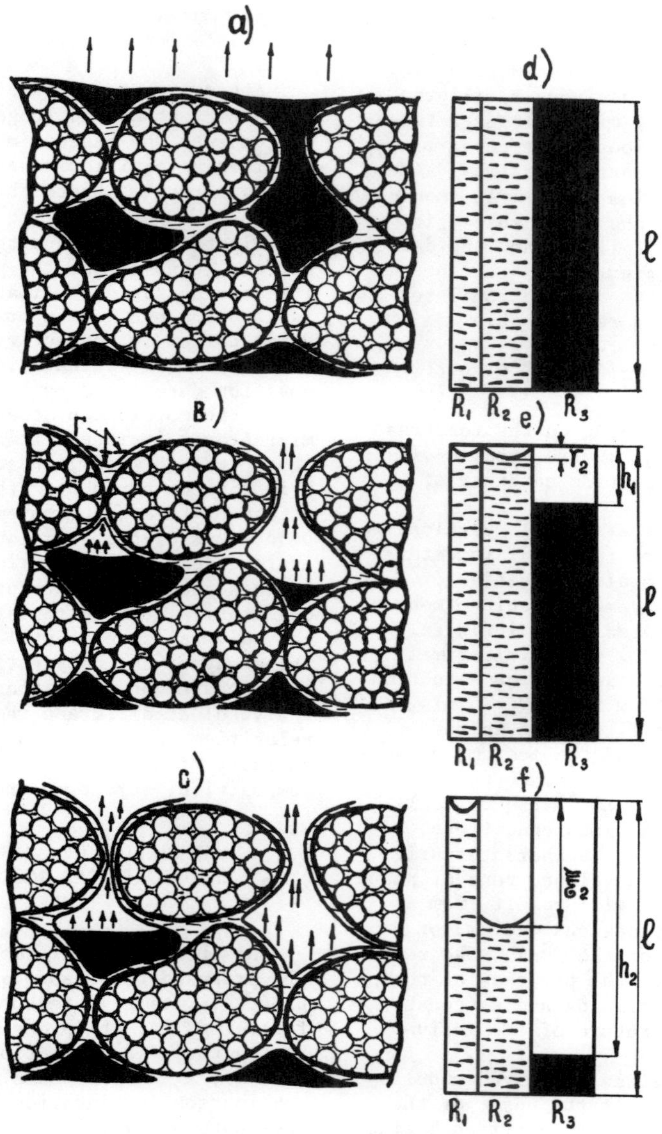

FIGURE 1 Three-capillary model and schematic figure of dehydration process.

3rd category being moisture in the solid state, capillary R_3). Accordingly, to describe the kinetics we may use a diagram of three capillaries of different diameter, linked with one another throughout the length of the wall in the absence of resistance to moisture transfer across these walls (Figure 1d,e,f).

The main features of the physical image resulting from this model are as follows:

At the initial moment of time the level in all the three capillaries is flat (Figure 1a). Once the condition $P/P_s < 1$ is satisfied in the vapor-gaseous medium, the dehydration process begins, its intensity (I_I) at the onset being controlled by the resistance to moisture transfer in the boundary layer of the air medium. Over time, menisci are formed in capillaries R_1 and R_2, while in capillary R_3 the level drops. Simultaneously, the moisture evaporating from the R_2 capillary is compensated for due to ice meltage and subsequent percolation of this moisture through the capillary wall. Until the difference in the capillary pressures, which is determined as $\Delta P = 2\sigma_{23}/\mu$, is sufficient to bring up as much fluid to the height h_1 as will not force down the level in the capillary R_2 but will only increase the curvature of the meniscus r_2 (Figure 1e), the dehydration intensity will be constant (so-called first period). Variation of the curvature radius will continue until a certain maximum value, which represents its equilibrium value. As soon as the meniscus in capillary R_2 has taken shape, it starts to deepen inside the capillary and the desiccation intensity decreases (Figure 1f). At the onset of the second period, the dehydration intensity will be determined by the relationship:

$$I_{II} \cdot (h_2 - \xi_2) = \sigma_{23} \cdot \alpha \cdot \rho_{mv}/\eta_{mv} = B = \text{const}$$

where α is a coefficient depending on the features of the capillary system, η_{mv}, ρ_{mv} are dynamic viscosity and density of the migration category of water.

As seen from the model, the spatial (vertical) length of the evaporation zone $(h_2 - \xi_2)$ is determined by the ratio of withdrawal intensities of replenishment vapors. An extreme situation (zone degeneration) may be interpreted as an infinitely high evaporation rate ($I_{II} \to \infty$) or infinitesimal replenishment ($\eta_{mv} \to \infty$). Such an extreme situation occurs, in particular, for flow rates $V > 5$ m/s, when the evaporation surface lies close to the boundary separating the rock and the medium, as well as in the case where process takes place *in vacuo* (for instance, $P_{total} = 3.10^{-3}$ MPa) (the value I_{II} being large), whereas, in analogous conditions, at $P_{total} = 0.1$ MPa and greater, a characteristic feature is the presence of a developed zone of phase transition. A similar effect will result from a higher molecular mass of the gaseous medium, or higher salinity and density of the rock. The zone is observed to grow also with the progress of the dehydration process (I_{II} - declining and $B = \text{const}$) and with an increasing temperature of the medium (η_{mv} - decreasing).

We shall more clearly define the given model by means of a schematic layout. At the onset of the process, sublimation of ice from the surface begins (Figure 1a). As sublimation continues, the water collar at its surface comes into greater contact with the surrounding air, rather than with ice, while on the opposite side the water collar continues to be in contact with ice (Figure 1b). The equilibrium condition for the water collar meniscus with ice and vapor will be written as $(\psi_{23}) = (\psi_{34})^{\sigma_{23}/\sigma_{34}}$. Since in every case $\psi < 1$ amd $\sigma_{23} > \sigma_{34}$, consequently $\psi_{34} > \psi_{23}$, which stipulates a difference between chemical potentials $\Delta\mu = RT(\ln\psi_{34} - \ln\psi_{23}) > 0$ on the opposite sides of the water collar. This difference in potentials causes melting of the ice localized within the space of the pore, and transition of part of the moisture into the interaggregate water films along which it is being transported to the evaporation zone, which results in "locking" of the evaporating surface in the narrow cross-section of the water collar. When above-ice free porosity occurs, this process can be intensified by vapor transfer, resulting from a succession of evaporation-condensation processes, and by the formation of a low-pressure region in the free volume of the pore. On the other hand, for the glacial lens dividing rock blocks, the surface curvature is unimportant ($\mu \to \infty$) and the sublimation front advances without delay into the rock.

After a certain period of time, when decreasing replenishment can no longer compensate for vapor efflux into the gaseous medium, the water collar acquires an equilibrium shape and the meniscus breaks through. Since the total pressure relaxes at the speed of sound, the opened pore is practically instantly filled with air and the evaporation surface jumps across the pore space (Figure 1c). The ensuing rapid sublimation of the ice remaining in the pore once again brings the moisture of the water collar in contact with the gaseous medium and the evaporation surface is again "stuck" in a stable position determined by the water collar's geometry. Thus, the movement of the evaporation (sublimation) microfronts in interstitial space is discrete in character. However, on account of a specific distribution of pores over the radius, the total effect of these micromovements is continuity and smoothness of the functions $I = I(\tau)$ and $\xi = \xi(\tau)$, as recorded in the experiment.

This proposed three-capillary model provides in qualitative terms a satisfactory explanation for many diverse experimental results obtained during the study of the dehydration process under laboratory conditions (more than 3000 specimens were analyzed) at different temperatures, pressures, velocities, molecular composition, and relative humidity of the medium, in rocks of different composition, structure and properties.

DEHYDRATION OF MOISTURE-SATURATED UNCONSOLIDATED ROCKS AT NEGATIVE TEMPERATURES

The three-capillary model provides a basis for the proposed three-zone quantitative model of the process. Its main features can be reduced to the following. As the process begins three zones appear in the rock: a desorption zone of unfrozen water $(0 - \xi)$; a zone of phase transitions $(\xi - h)$ in which sublimation, evaporation, and melting occur; and a zone not involved in the process (glacial zone) $(h - 1)$. Heat transfer in each of the zones occurs in a purely conductive manner. Moisture is transported towards the vapor-gas interface by various mechanisms: a) in the desorption zone it is primarily by concentrated vapor diffusion (coefficient K_v); b) in the phase transition zone by the above mechanism in combination with migration of unfrozen water and a succession of micro-evaporations-condensations. The total transfer coefficient K_{eff} in this zone may be one or two orders higher than K_v. In the glacial zone the moisture transfer is negligible.

In the proposed model in order to describe moisture transfer we have made use of sorptional kinetics equations that have coefficients that are only slightly dependent on moisture, rather than equations for moisture conduction. This stems from the fact that, on the one hand, the values of moisture concentration in the solid or liquid phase do not represent transfer potentials at negative temperatures and, on the other hand, because the essence of the physical phenomena occurring in the entire region involved in the dehydration process are reflected better according to the concepts expressed in the models. Desiccation occurs in the phase transition zone through the liquid phase by way of desorption of unfrozen water films when there is continuous replenishment due to ice meltage.

The equation describing sorptional energy and kinetics for the identified zones in a unidimensional formulation assumes the form:

$$C_1\gamma_1 \frac{\partial T}{\partial\tau} = \frac{\partial}{\partial x}(\lambda\nabla T) + q_{des}(W)\,\frac{\partial[\gamma_i W_3(\psi,T)]}{\partial\tau}, (0<x<\xi(\tau)) \quad (2)$$

$$\frac{\partial\nu\rho_{52}}{\partial\tau} = \frac{\partial}{\partial x}[K_w(T)\cdot\rho_{52}\cdot\nabla\nu] - \frac{\partial[\gamma_i W_3(\psi,T)]}{\partial\tau}, (0<x<\xi(\tau)) \quad (3)$$

$$C_{\rho eff}(W,T)\,\frac{\partial T}{\partial\tau} = \frac{\partial}{\partial x}[\lambda(W,T)\cdot\nabla T] + q_{con}\,\frac{\partial[\gamma_i W_4(\psi,T)}{\partial\tau},$$

$$(\xi(\tau)<x<h(\tau)) \quad (4)$$

$$\frac{\partial\nu\rho_{52}}{\partial\tau} = \frac{\partial}{\partial x}\,[K_{eff}(W,T)\cdot\rho_{52}\cdot\nabla\nu + \delta\cdot\nabla T] +$$

$$\frac{\partial[\gamma_i W_4(\psi,T)]}{\partial\tau} \quad (\xi(\tau)<x<h(\tau)) \quad (5)$$

$$C_{\rho eff}\cdot\frac{\partial T}{\partial\tau} = \frac{\partial}{\partial x}\,(\lambda\nabla T) \quad (6)$$

The system is closed by the equations of state and bonding energy. The latter with respect to the desiccated zone corresponds to adsorptional equilibrium and with respect to a phase transition zone, to the ice-unfrozen water equilibrium.

$$P = P(\rho,T) \quad (7)$$

$$E = RT\ln\psi = \begin{cases} \psi(W_3,T) & (8) \\ \psi(W_4,T) & (9) \end{cases}$$

Here $C_{\rho eff} = \sum_{i=1}^{i=4} C_i\cdot\rho_i + q_{melt}\,\rho\frac{\partial W_3}{\partial T}$; $\nu = \rho_2/\rho_{52}$; δ is a thermogradient coefficient.

The content of unfrozen water in desorption zones and phase transitions is dependent on the partial pressure of aqueous vapors and temperature. Consequently,

$$\frac{\partial W_3}{\partial\tau} = \left(\frac{\partial W_3}{\partial P}\right)_T \frac{dP}{d\tau} + \left(\frac{\partial W_3}{\partial T}\right)_P \frac{\partial T}{\partial\tau}.$$

To find the derivative $(\partial W_3/\partial P)_T$ it is necessary to have an equation or experimental data for the desorption isotherm. The derivative $(\partial W_3/\partial T)_P$ can be obtained if we have the curve of the phase composition of moisture. For a sufficiently great number of rock types undergoing desiccation at negative temperatures the contraction is insignificant (i.e. $\gamma_1 \neq \gamma_1(W)$) and, consequently, it can be written $\partial(\gamma_1 W)/\partial\tau = \gamma_1(\partial W/\partial\tau)$; otherwise $\partial(\gamma_1 W)/\partial\tau = \gamma_1(W)\partial W/\partial\tau + W\cdot(\partial\gamma/\partial\tau)$.

If the process is accompanied by filtration, i.e. molar gas or water transfer, it is necessary to include convective components of heat and moisture transfer in the left-hand part of equations 2-5. To close the system, equations of continuity and movement must be added.

As far as formulation of the problem for non-moisture-saturated rocks is concerned, it does not appear possible in this case to close the basic equation system unless additional hypotheses are introduced. A promising possibility involves the following assumptions (desiccation of coarse-unconsolidated rock): 1) the shape of ice inclusions in the rock is globular; 2) the radius of the inclusion decreases in the course of the process without changing the shape of the inclusion; 3) the density of the mass sources in the unit volume of rock and the function of their distribution are known (in the case of irregular distribution). To find the moisture-content fields requires solution of the diffusion equations in spherical coordinates in the presence of a mobile phase transition boundary (for unitary ice inclusion) and for diffusion equations in which the source in the right-hand part is written in orthogonal coordinates (for the rock layer as a whole).

SIMPLIFICATION OF INITIAL SYSTEM OF EQUATIONS AND COMPARISON OF DATA

Analysis of the above system of differential equations 2-9 shows that, given the corresponding boundary conditions, even numerically it appears to be extremely difficult to solve this problem. Considerable simplification of the initial system is achieved whenever the process of kinetics is fully determined and controlled, either by heat or moisture transfer. The latter case, as demonstrated by the joint analysis of experimentally determined heat and mass flows, is characteristic of natural conditions during the dehydration process (Yershov et al. 1975). In the absence of an external temperature field (the natural geometric gradient in rock masses is very insignificant, viz. 10^{-4} °C/cm) this process occurs under near isothermal conditions (for a fixed unit of moisture mass the value of the derivative $((dP/dT)_w \to \infty)$). Another extreme case $(dP/dT)_w \to 0$ occurs, for example, under vacuum drying conditions.

The degree of approximation in the two extreme cases can be assessed by considering the heat and moisture balance for an isolated elementary nonshrinkable volume of rock (Komarov 1979). The balance equation assumes the form:

$$\left(\frac{dP}{dT}\right)_w = \left(\frac{K_t}{K_v}\right) \cdot \left(\frac{1-\pi}{\pi}\right) \cdot \left(\frac{q_{con}}{q_{con} + q_{wet(w)}}\right) \cdot A;$$

$$A = RTC_i \gamma_1 / q_{con} \qquad (10)$$

As seen from equation 10, the value of the derivative $(dP/dT)_w$ depends upon: the relative inertia of the temperature and moisture field (K_t/K_v); the relationship $(1-\pi)/\pi$ which relates to the fact that heat is, for the most part, transported through the skeleton of the rock, while moisture is transported through the pore space; and the combined physical properties (A) and the adsorbent concentration $(q_{con}/q_{con} + q_{wet}(W))$. Since the amount of moisture that fills the pores is high, the latter relationship is close to unity, whereas if first monolayers are removed the quantity q_{wet} attains a magnitude $q_{wet} = (1.5 + 2.0)\ q_{con}$.

Quasi-isothermal approximation makes it possible to significantly simplify the initial system by reducing it to equations 3, 5, 7, 8, and 9. The main premises of the solution were: 1) we analysed a moisture-saturated semilimited rock mass; 2) the relaxation time for phase transitions and desorption is much inferior to that of the mass fields; 3) shrinkage is absent.

Since the general form of the function $W = W(P/P_S)$ for either of the zones involved was unknown, our analysis was made with respect to the simplest case of linear dependence $W = W(\psi)$ for the zones in question:

$$W_{des} = W_{mh} - a \cdot \psi_{des} / \gamma_1 \quad (\psi_{med} \leq \psi \leq \psi_{mh});$$

$$W_{Pt} = W_o - B \cdot \psi_{Pt} / \gamma_1 \quad (\psi_{mh} \leq \psi \leq 1.0)$$

where the parameter "a" characterizes the polymolecular adsorption areas (according to BET), and parameter "b" the region of "polycapillary condensation" (ice-unfrozen water equilibrium). The problem was not considered with respect to the monomolecular adsorption region, since the application of quasi-isothermal approximation appeared to be highly problematic. The system (3,5,7,8,9) was solved for the simplest boundary conditions of the 1st, 2nd, and 3rd kind, which are easy to simulate in the experiment, for 4 highly characteristic cases of dehydration (Komarov 1976, 1979).

1. ($a \neq 0$; $b \neq 0$). Dehydration of finely dispersed hydrophilic rocks at negative temperatures ($t > -7$°C) in the 2nd period of the process.

2. ($a = 0$; $b \neq 0$ - degeneration of the desorption zone). Dehydration of hydrophilic rocks in the 1st period, as well as under the condition.

3. ($a \neq 0$; $b = 0$ - degeneration of the phase transition zone). Dehydration of hydrophilic coarse-disperse rock and fine-disperse rocks at a temperature range below -7°C + 10°C.

4. ($a = 0$; $b = 0$ - degeneration of the desorption and phase transition zones). Dehydration of pure sands or artifically hydrophobized rock.

The first two cases were described by sorptional kinetics equations whose mass sources are linearly distributed over the zones subject to the condition of moisture flow conjugation at the zonal boundaries; the third case by an analogous equation in the presence of a mobile boundary (Stefan-type problem); and the fourth case by a diffusion equation in the presence of a mobile boundary (Stefan-type problem).

Experimental verification of the relationships was carried out at three installations whose design experimental procedures and technique for determining moisture-conductive and adsorptional characteristics were described in papers (Yershov 1979, Komarov 1979). The setups made it possible to measure continuously for several months the variability ranges of the medium parameters: $0 > t > -25$°C, $0 < P_{total} < 2.5$ MPa, $0 < V < 12$ m/sec, $0.3 < \psi < 1.0$. Investigations were carried out using a vapor air medium and non-condensing inert gases: helium, argon, nitrogen.

Comparison of the calculated and experimental moisture fields has revealed their analogous character and satisfactory convergence. The degree of error did not exceed, as a rule, 15% for the total moisture balance. This, in our opinion, indicates that the concepts elaborated in the present article may prove helpful in future, more complex studies of the process.

PRINCIPAL CONVENTIONAL SYMBOLS AND INDICES

T; t - temperature, °K, °C; P_{total}, P - total and partial pressure, Pa; x - coordinate, m; τ - time, hour; ρ, γ - density and volume weight, kg/m³; R - gas constant, J/kg·deg; W_o - initial moisture; ψ - relative moisture of the vapor-gaseous medium; V - flow velocity, m/s; μ - radius, m; C - thermal capacity, J/kg·deg; P - porosity; K_t - temperature conductivity coefficient, m²/s; λ - heat conductivity coefficient, W/m·deg; q_{wet}, q_{des}, q_{con}, q_{melt} - differential heats of wetting, condensation, desorption, and melting, J/g. 1 - mineral skeleton; 2 - (av) aqueous vapor; 3 - (uw) - unfrozen water; 4 - (i) ice; 5 - (g) gas; equ - equivalent; p.t. - related to phase transition zone; d - related to desorption zone; med - related to phase; S - parameter values on the phase equilibrium curve; mh - maximal hygroscopy.

REFERENCES

Komarov, I. A., Method of calculating desiccation zone thickness in frozen dispersed rocks composing the walls of underground engineering structures. Materials of the 3rd conference of post-graduates and young scientists: "Geocryology" Series, Moscow State University's

Geological Faculty (manuscript in VINITI, No. 24776). (In Russian)

Komarov, I. A., 1979, Investigation of heat and mass-exchange in dehydration of dispersive systems under below-zero temperatures (in application to rocks): Candidate's dissertatio thesis, LTIKhP, Leningrad. (In Russian)

Krisher, O., 1961, Scientific principles of drying technology: Moscow, Foreign Literature Publishing House. (Russian translation)

Yershov, E. D., 1979, Moisture transfer and cryogenic textures in dispersed rocks: Moscow, Moscow State University Publishers. (In Russian)

Yershov, E. D., Gurov, V. V., Komarov, I. A., and Kuchukov, E. Z., Experimental investigations of heat and moisture transfer in the case of water sublimation-desublimation in finely dispersed rocks, Materials of Third International Conference on Geocryology, v. 1: Ottawa, Canada's National Research Council.

Yershov, E. D., Kuchukov, E. Z., and Komarov, I. A., 1975, Ice sublimation in dispersed rocks: Moscow, Moscow State University Publishers. (In Russian)

THERMAL INTERACTION BETWEEN PIPELINES AND THE ENVIRONMENT

B. L. Krivoshein

Research Institute on Management for
Oil and Gas Pipeline Construction, USSR

Aspects of thermal interactions between large-diameter underground pipelines and the environment are considered. Consideration of this topic is necessary for achieving design solutions under northern conditions. Criteria of choice are formulated. A classification of various methods for pipeline laying is developed for northern regions and appropriate operating conditions are discussed. Numerical simulations have been carried out with the aim of determining the permissible negative temperature of the external surfaces of underground pipelines from an ecological point of view. The influence of water migration on the rate of freezing around the pipe is evaluated. Soil expansion during the transport of chilled gas has also been evaluated. It has been shown that the actual thermodynamic properties of the gas make a significant contribution to the level of gas cooling at compressor stations. A method for determining the optimal thickness of insulation along the length of a gas pipeline is suggested with due consideration of environmental protection. Generalized relationships for heat transfer in underground gas pipelines are given, in the case of both constant and variable temperature (where apparatus for air cooling is used). Results obtained were applied to the design of gas pipelines in the Soviet North.

The experience of operating large-diameter gas pipelines (1420 mm in the USSR North) has shown that the temperature of compressed gas can reach 40-45°C. Therefore, air-cooling apparati (AAS) are presently being installed at compressor stations. In the winter the minimum allowable gas temperature, following AAS, is set based on the freezing stability of the pipe steel, the temperature gradient between compressor stations (CS) and the geocryological conditions along the route (Krivoshein 1982). Pipes currently in use are designed to operate at gas temperatures up to -10°C. A pipeline may undergo annual variations in gas temperature equal to 50-55°C. The allowable temperature gradient for both operation and construction of 1420 mm gas pipelines, at a working pressure of 7.5 MPa at first and second category sites, is 58°C (i.e. it is approximately equal to the temperature gradient under operating conditions). As the range of the gas working pressure increases the safety of pipeline operation is reduced: longitudinal temperature deformations increase and, as a result, there is danger of a loss of stability; under positive ground temperatures corrosion of the pipeline intensifies; mass transfer processes in soil also intensify.

At sites made up of continuous subsidence permafrost it is necessary to ensure a gas transport regime in which the pipe-ground thermal flux approaches zero. This means that the gas temperature should correspond to the ground temperature at a depth of pipeline axis at various times. Such a solution has been adopted for the Urengoi-Pomary-Uzhgorod pipeline, which is subject to permafrost conditions. In order to determine the required level of gas cooling, the temperature changes over distance, the CS and cooling station capacities, etc., it is necessary to take into consideration the pipeline-environment thermal interaction.

STATEMENT OF THE PROBLEM

The thermal interaction between large-diameter pipelines and the ground has a number of features which stem from the actual thermodynamic properties of gas under high pressure and low temperatures, the conjugate nature of gas-soil heat transfer, in addition to the ecological limitations imposed by environmental protection.

Construction and operation conditions for large-diameter pipelines are related to their temperature regime. As diameter and working pressure increase, the stability of buried pipelines has become a crucial factor for reliable gas transport (Batalin 1979). When gas which has been chilled to a negative temperature is transported (chilling can be done either year round or in the winter using AAC), the pipeline's surface also acquires a negative temperature. Therefore, when a pipeline intersects marshy land and water obstacles ice forms on the surface of the pipe, the surrounding ground expands and the pipe heaves upwards towards the surface.

This can result in blockage of ground water run-off and other disturbances in the zone of pipeline-environment interaction. Pipeline icing increases buoyancy, which must be compensated for by additional ballast. Disturbance of the ecological equilibrium might well result in irrevocable damage to the natural landscape and reduction of pipeline reliability. A number of methods of restraint and in some cases elimination of ecological disturbances during construction and operation are already well known.

Experience in designing northern pipelines in the USSR and abroad has shown (Krivoshein et al. 1975) that the problem in question can be solved: by applying methods for laying pipe which eliminate or limit the thaw or freezing depth of the bedrock; by chilling the gas to a level that preserves the frozen bedrock underlying the pipeline during the total period of operation; by combining methods of construction and operation which ensure "thawing-freezing" of the soil in natural ranges; by using local chilling devices.

The prognosis for thermal interaction between the gas transported and the environment plays a significant role in the choice of technical solutions.

We shall first consider the criteria for choosing methods of laying pipe and operating regimes in the North and, secondly, analyze the results of calculations used to preaict thermal and mechanical interaction between the construction process and the environment.

CONSTRUCTION SOLUTIONS WHICH TAKE INTO ACCOUNT ENVIRONMENTAL FACTORS

In above-ground construction of gas pipelines the thermal impact on the ground is eliminated. Limitation of this influence in required ranges may be achieved by a combination of the following: thermal isolation and gas chilling by AAC; gas chilling to annual negative temperatures by AAC and cooling devices, local cooling of soil near the pipeline using cryoanchors, thermopiles, cooling agents, etc. As is well known, the North of Western Siberia is characterized by a considerable variety in soil temperature and humidity and a wide spectrum of cryogenic processes. Therefore, construction techniques and the choice of operating regimes is an involved scientific and engineering problem.

Investigations and design calculations carried out at different institutions, as well as analyses of foreign experience have shown that in northern regions it is necessary to combine various technical solutions depending on the specific geocryological conditions.

CRITERIA FOR GAS PIPELINE CONSTRUCTION AND OPERATING REGIMES

The main problem is either to maintain the prescribed temperature and humidity of the bed or to limit these factors to a range that will ensure stability and integrity of the pipeline, as well as an unviolated environment. The choice of construction method and operational behavior of pipelines in thawed soils depends on the following: subsidence (at $T_N > 0°C$) or heave (at $T_N < 0°C$) of the bedrock; the character of the propagation of ground subsidence or heave in the vicinity of the pipe route; the vertical cryogenic structure of bedrock, soil temperature, depth of the active layer, the location of the ground water horizon and the degree of marshiness of the adjacent territory; the degree to which the pipeline route has been studied taking into account the reliability of prediction for permafrost conditions at the onset of operation; the geography of the district; the presence of the materials needed to ensure construction of the pipeline (cold-resistant steel pipe, supports, etc.) and cost-effective operation (cooling devices, etc.), the character of gas temperature changes along the length of the pipeline and gas temperature changes over time.

The main factors are the following: depth of ground thaw under (or above) a pipe in relationship to the temperature of the pipe's exterior surface (T_N); the magnitude of ground heave, the extent of hydrological disturbance in the vicinity of the route; the temperature of the pipe's exterior surface in marshy districts and water crossings (T_N). Depending on the specific conditions, one of the factors above is the dominant one, and others are subordinate.

In case of poorly timed start-up of the cooling equipment ($T_N > 0°C$), the depth of ground thaw serves as a validating criterion for the method of laying pipe. The depth in question should provide stability of the pipeline, operation in a non-calculated regime (in the absence of AAC), and restoration of the ground temperature regime after start-up of cooling operations.

During transport of AAC chilled gas, all of the main factors must be taken into account, moreover, in the presence of annual negative gas temperatures one must also consider ground heave and the T_N value. As experience in designing northern pipelines has shown, sections of frozen and thawed soils having different properties alternate and this makes different technical solutions necessary.

To select levels for chilling gas, Table 1 presents a classification of pipe-laying techniques. In northern areas design and operating regimes (mainly thermal conditions) are inter-related. Sectors having a high level of ground subsidence and in which thaw below the active layer is not possible (type II) represent the greatest danger. The minimum allowable mean annual temperature of gas in thawing sections (type II) is derived from the formation of frozen patches underneath the base of the active layer (Kudryavtsev et al. 1974).

For sections of continuous (or prevailing) frozen soils (type I and type III), cooling the gas by refrigeration (in the summer) and by AAC (in the winter) to seasonal ground temperatures at the depth of the pipeline's axis is acknowledged as the most rational solution. Results of mathematical simulation performed at a number of institutions (Ivantsov et al. 1977) and foreign experience show that temperature for gas cooling, in the course of the annual cycle, varies from -10°C to -6°C. The institute, GIPROSPETSGAS, has designed a refrigeration station for the Urengoi-Nadym pipeline which will cool the gas to -2°C. The selection of allowable temperatures of a pipe's exterior surface (T_N) affects the thickness of insulation for pipelines designed to operate at negative gas temperatures.

Actual thermodynamic gas properties affect the choice of temperature. For chilled gas pipelines having a negligible gas-soil temperature gradient, the throttle-effect is the determining factor influencing gas temperature changes occurring over distance.

For determining allowable negative temperature, from the point of view of ecology, $(T_N)_{min}$ investigations of thermal interaction between pipelines and soils have been conducted (Kovalkov

TABLE 1 Classification of Construction Methods and Operation Regimes for Northern Gas Pipelines

Route conditions	Methods of laying pipe	Operation regime	Temperature range of a gas pipeline
Continuous frozen soils (type I)	Laying pipe below the active layer	$T_N = T_{gr}(ho)$	$T_g(\tau) = T_f \div T_{inav}$
Thawed soils with islands of frozen soil (type II)	a) Laying pipe in thawing soils	$\frac{\lambda_t}{\lambda_m}\Omega^+ > \lvert\Omega^-\rvert$	$T_g(\tau) = T_w + (10 \div 1) - \Delta T_{dr}$
	Frozen sites, subject to ground subsidence	$h_t(\tau^+) \leqq h_{dop}$	$T_g(\tau) \leqq T_f$
	Ground not subject to subsidence	arbitrary	$T_g(\tau) = T_w + (10 \div 15) - \Delta T_{dr}$
	b) Laying pipe above ground with thermal isolation on ground subject to subsidence	$h_t(\tau^+) \leqq h_{est}(\tau^+)$	$T_g(\tau) = T_w - T_{inav}$
Continuously frozen ground with taliks (type III)	a) Underground pipe laying in frozen soil	$T_N = T_{gr}(ho)$	$T_g = T_{gr}(ho)$
	in thawing soil	$\frac{\lambda_t}{\lambda_m}\Omega^+ > \lvert\Omega^-\rvert$	$T_g = T_{gr}(ho) - \Delta T_{dr}$
	b) Above-ground pipe laying with thermal isolation on thawing heaving soil	$T_N = T_{gr}(ho)$	$T_g \geqq 0°C$

et al. 1977).

Mathematical formulation of the problem for interaction of a chilled gas pipeline ($T_N < 0°C$) includes equations describing external medium and gas with the corresponding boundary conditions that express the boundary thermal balance. Usually the problem is three-dimensional, nonlinear, due to Stephan's conditions (in equations for external medium) and due to terms for gas flow. Because of substantial temperature gradients in longitudinal and transverse directions (to the pipeline axis) the problem may be differentiated as flat for soil around the pipe and one-dimensional for gas flow across the pipeline (Krivoshein 1982, Agapkin et al. 1981).

Within the limitations of traditional assumptions the heat zone of the soil is usually described by a system of equations:

$$\frac{\partial T_i}{\partial \tau} = a_{i,j}\, \Delta T_i \qquad (1)$$

$$(x,y) \in S_i \cap G_j, \qquad i,j = 1.2 \qquad (2)$$

$$\tau = 0: \quad T_2 = T_{est} \qquad (3)$$

$$B_2,\ B_4 \quad \frac{\partial T_i}{\partial x} = 0 \qquad (4)$$

$$B_3 \quad T_i = T_{ns} \qquad (5)$$

$$B_5 \quad \lambda_i \frac{\partial T_i}{\partial r} = \frac{1}{r_{is}}(T_i - T_g) \qquad (6)$$

$$B_6 \quad \begin{cases} \lambda_1 \dfrac{\partial T_1}{\partial n} - \lambda_2 \dfrac{\partial T_2}{\partial n} = Q_f \dfrac{d\xi}{d\tau} \\ T_1 = T_2 = T_f , \end{cases} \qquad (7)$$

where $i = 1,2$: 1 is thawing soil; 2 is frozen soil; B_2, B_3 are, respectively, boundaries along the axis of the pipe and beyond its influence (natural conditions); B_4 is the boundary of soil and the pipeline; B_7 is the boundary of soil zones with different thermo physical properties.

Applying well-known analytical methods of solution (Krivoshein 1982, Agapkin et al. 1981) one encounters insurmountable difficulties. In the USSR Academy of Sciences algorithms have been derived for numerical solution of such problems (Alekseeva et al. 1974), however due to the absence of precise solutions and reliable experimental data it has not been possible to evaluate the accuracy of these numerical calculations. In papers

(Kovalkov and Krivoshein 1977, Kovalkov et al. 1982) a Lukyanov hydrointegrator was used. Non-rectangular blocks were used for the section contiguous to the pipe, while concentrated and hidden thermal capacity was calculated proportionally to the area of blocks.

Initial conditions were obtained by solving for the natural ground field over an 8-12-year period during which boundary conditions simulating thermal influence of the pipeline were not imposed. This experience made it possible to assume that during construction the vegetation cover is disturbed and re-establishes only 5 years following pipeline construction.

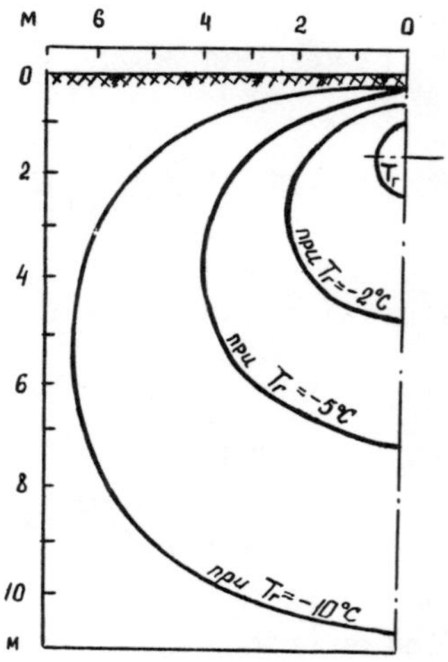

FIGURE 1 Combined graph of the position of freezing front in soil for the zone of laying-up a non-isolated 1420-mm gas pipeline in the Surdut area by the 1st of November for the 9th year of operation and under different gas temperature ($T_Г = T_g$).

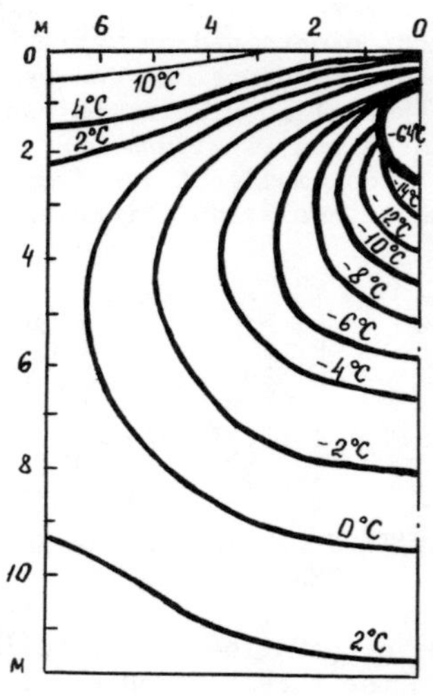

FIGURE 2 Temperature field for soil in the zone of laying-up a heat-isolated 1420-mm gas pipeline in the Surgut area under T_g = -64°C by the 1st of May for the 11th year of operation when vegetation cover is absent.

δ_{is} = 0.16 m; λ_{is} = 0.0405 Wt/m·grad

ANALYSIS OF RESULTS

Some results of multi-variant calculation are shown for non-insulated (Figure 1) and thermal isolated (Figure 2) pipelines. Paper (Kovalkov et al. 1977) suggests a possible criterion for allowable environment disturbance in the zone affected by a chilled gas pipeline.

An impact is considered permissible if, due to the influence of the chilling source (the pipe), the level of ground waters (LGW) in the pipeline construction zone does not exceed the annual LGW in marshy areas of the same territory. A necessary condition for choosing the required temperature $(T_N)_{min}$ is location of the upper freezing boundary in the vicinity of the hydrologically active layer in winter minimal ZGW. The annual average temperature $T_{(N)min}$ = -2°C for the northern zone fields is derived from the above conditions.

Approximate analytical solution for the choice of $(T_N)_{min}$ assumes the following form:

$$\theta = (\bar{h}_t - 1)(1 + 2Bi_p^{-1})^{-1}(2\bar{h} + 2Bi_{is}^{-1} - \bar{h}_t - 1) \qquad (8)$$

where $\theta = \dfrac{T_f - (T_N)_{min}}{T_f - T_p}$; $\bar{h} = \dfrac{h_o - R_1}{h_{dop}}$; $\bar{h}_t = \dfrac{(hsum)max}{h_{dop}}$;

$$Bi_p = \frac{h_{dop}}{\lambda_m r_p} ; \qquad Bi_{is} = \frac{h_{dop}}{\lambda_m r_{is}}$$

These studies (Krivoshein 1982, Krivoshein 1979, and others) are based on a ground thermal transfer equation. The additional thermal input, stemming from ground water filtration flow in the layer between the ground surface and the upper pipe's surface, was not taken into account because reliable experimental data on mass transfer soil characteristics are not available.

The effect of moisture migration on the velocity of soil freezing has been estimated. It has been shown that for actual soils ($L_U \leq 0.1$; $K_O \leq 2$; $\bar{w} < 2$)* under the influence of moisture migration, the depth to which the ground freezes is reduced by not more than 1%.

EVALUATION OF GROUND HEAVING IN THE VICINITY OF A CHILLED GAS PIPELINE

Krivoshein (1982) gives the equation of the velocity of ground freezing, allowing for moisture migration in the direction of the freezing front. Using the relationship given and assuming the value of ground heave to be proportional to the increase in volume by 1.09 times † per unit of area, one obtains:

$$h_{puch} = \gamma\sqrt{\tau^{(-)}}, \qquad (9)$$

where

---

* $L_u = \dfrac{a_m}{a_2}$; $K_o = \dfrac{\rho_s \sigma (w_n - w_o)}{C_m \rho_m |T_p| (1 + \varepsilon)}$; $\bar{w} = \dfrac{w_n}{w_o}$

† 1.09 is the coefficient of water expansion due to freezing.

$$\gamma = \frac{1.09\rho_s}{3(1+\varepsilon)\rho_w}\,[\alpha(w_r - w_o) + \beta(2w_r + w_o)];$$

$$\alpha = \frac{[A/\beta - \beta(2w_n - w_o)]}{w_n - w_o}\,; \quad A = \frac{6\lambda_m |T_p| (1+\varepsilon)}{\rho_s \sigma}\,;$$

$$\beta = \left\{ \frac{[A(w_n + w_o) + 4a_m(w_n - w_o)^2 - (w_n - w_o) \cdot \sqrt{16a_m^2(w_n - w_o)^2 + 8a_m A(w_n + w_o) + A^2}}{6w_o w_n} \right\}^{0.5}$$

For evaluation calculations were performed using the following initial data:

$$a_m = 2 \cdot 10^{-4}\,\mathrm{m^2/t};\ \varepsilon = 0.97;\ w_o = 0.17;\ w_n = 0.27;$$

$$\rho_s = 2830\ \mathrm{kg/m^3}\ ;$$

$$\lambda_m = 1.74\,\frac{\mathrm{Wt}}{\mathrm{m \cdot grad}};\ \sigma = 333.6\,\frac{\mathrm{KD_j}}{\mathrm{kg}}\ ;\ \zeta_w = 1000\,\frac{\mathrm{kg}}{\mathrm{m^3}}\ ;$$

$$T_p = -1;\ -2;\ -5;\ -10°\mathrm{C}.$$

The results of the calculations are given in Figure 3. It is obvious that if $T_N = -2°C$, then the value of ground expansion doesn't exceed 0.3 during the winter. Under the above indicated condition, pipeline stability and normal passage of ground waters are maintained.

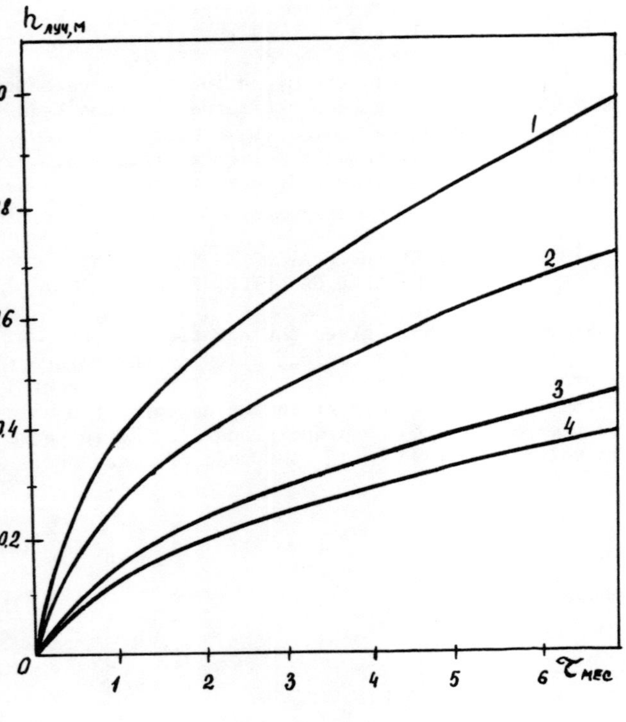

FIGURE 3 Swollen soil-dependence of time under different temperatures: 1) -10°C; 2) -5°C; 3) -2°C; 4) -1°C.

ON THE EFFECT OF REAL GAS PROPERTIES

Krivoshein (1982) reviews the investigations carried out in the USSR on the problem of the thermal interaction of pipelines with environment. On the basis of solutions already obtained, the decrease in gas temperature (due to the effect of Joule-Tompson) for a 1420-mm pipeline with working pressure of 7.5 MPa was evaluated. They have demonstrated that along the section between CS the decrease in temperature amounts to $T = 10°C$. Taking into account the ecological criterion obtained above, the relationship for the level of gas cooling at a CS assumes the form (Krivoshein 1979):

$$T_n = \left[2T_N - aT_{gr}\left\{\frac{1+\exp[-(A_1+a)]}{A_1+a} + \frac{1+\exp[-(A_2+a)L]}{A_2+a}\right\}\right] :$$

$$[\exp(-(A_1+a)L) + \exp(-(A_2+a)L)], \tag{10}$$

where:

$$a = \frac{K\pi D}{GC_p};\quad A_n = D_i(P,T)B(P,T)/P;$$

$$b = \lambda RG^2 z_o(P,T)/2g^2Df^2;\ n = 1,2.$$

DETERMINATION OF OPTIMAL GAS TEMPERATURE PROFILES AND THERMAL INSULATION THICKNESS ALONG THE LENGTH OF A PIPELINE

Since the gas temperature decreases along a pipeline and since it is necessary to provide the condition $T_S = T_N$ at the surface of pipe-soil contact, the thickness of the thermal insulation should increase with distance. Krivoshein (1979) demonstrates a solution obtained by variational calculation for determining optimal gas temperature profiles and thermal insulation thickness. Using computer-assisted numerical calculations it has been shown that optimal profiles for gas temperature and thermal insulation thickness are close to linear:

$$T_{opt} \cong T_n - (T_n - T_k)\,\frac{x}{L} \tag{11}$$

$$\delta_{i,s}(x) \cong \delta_{is}(T_n) + [\delta_{is}(T_n) - \delta_{is}(T_k)]\,\frac{x}{L} \tag{12}$$

CONSIDERATION OF PIPE-GROUND THERMAL TRANSFER

For extensive operation of a pipeline under $T_S = 0°C$ one should use a regression equation, derived from mathematical simulation and field measurements:

$$K_{i\infty} = 0.4 + 2\left(\frac{h_{p2}}{R_1}\right)^{-2}, \tag{13}$$

where

$$h_{p2} = h_o + \frac{\lambda_{g2}}{\alpha_{we}} ; \ K_{i\infty} = \frac{KR_1}{\lambda_{gr}}$$

For chilled gas pipelines data from Figure 4 may be used where

$$K_i = \frac{gc}{\lambda_m(T_{gr} - T_g)} ; \quad c = \sqrt{h_o^2 - R_o^2} .$$

Heat losses of pipelines under transient conditions are defined by the method (Krivoshein et al. 1977). Thermal conditions for gas pipelines equipped with AAC are calculated using the method (Krivoshein et al. 1981) which takes into account the dynamics of ground phase transitions and seasonal variation in gas temperature.

The results described above serve as a foundation for the thermo-physical calculations used in designing Northern gas pipelines in the Soviet Union.

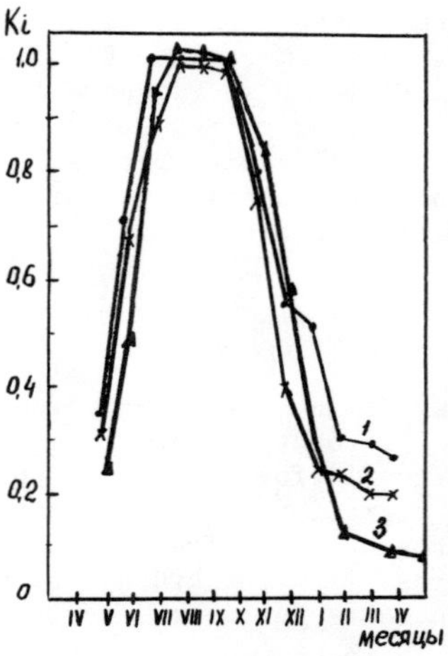

FIGURE 4 Variation of Kirpichev's criterion in the annual cycle for a subground gas pipeline in the Urengoi area under different gas temperatures for the 6th year of operation:
1) T_g = -20°C; 2) -15°C; 3) -10°C

NOMENCLATURE

$a_m, a_{i,j}$ - coefficients of mass and thermal conductivity, respectively; Bi - Bio criterion; C_p, C_M - specific thermal capacity of gas and soil; $D_i(P,T)$ - throttling effect; D,f - diameter and area of pipe section; G - mass flow of gas; h_{all}, h_{nat} - depth of allowable ground thaw and under natural conditions; h_o - depth of pipeline axis; K - coefficient of thermal transfer; K_i, K_o, L_u - criteria of Kirpichev, Kosovich, and Lukov; L - length of pipe; P - pressure; 0_f, σ-latent heat of ice thaw per unit of volume and unit of frozen soil; q - thermal flow from pipeline to soil or in reverse direction; R - gas constant; R_o, R_i - pipe radius; r_{is}, r_v - values of thermal resistance of insulation and vegetative cover, respectively; T_a, T_g, $T_{s(T_i)}$ - temperature of air, gas, and ground, respectively (i = 1 - thawing soil, i = 2 - frozen soil); T_{nt}, T_b - ground temperature at the neutral layer level and at the temperature of beginning of ground moisture freezing; ΔT_{dr} - drop in gas temperature due to throttling effect; x,y - Cartesian coordinates; $Z_o = Z_o(P,T)$ - compressibility factor; α_{we} - thermal transfer coefficient from ground to air; λ - hydraulic resistance coefficient; $\lambda_{gr} = \lambda_{i(i=1,2)}$ - ground thermal transfer coefficient; λ_{is} - insulation thermal transfer coefficient ρ_{sk}, ρ_w - density of soil skeleton and water; w_o, w_{nat} - moisture level of soil at the rolled boundary, and under natural conditions; δ_{is} - thickness of insulation; ε - porosity of soil; τ - time; $\Omega^{+(-)}$ - warming (cooling) impulse ($\Omega^{+(-)} = T_s \tau^{+(-)}$); ξ - coordinate of moving boundary.

INDICES

T, M - thawing and frozen zones; N - outside layer of insulation; 0,1 - internal and external pipe radius; n, k - beginning and end of pipe; natural conditions; in av.an.cond. - mean (average) annual conditions; n - ground surface.

REFERENCES

Agapkin, V. M., Krivoshein, B. L., and Jufin, V. A., 1981, Heat and hydraulic calculations of oil pipelines and product lines: Moscow, Nedra.

Alekseeva, G. V., Birykova, N. J., and Koshelev, A. A., 1974, Development and use of final-difference algorithms for solving the Stephan problem. In Asymptotical methods in the theory of systems, ed. 7: Irkutsk, University.

Batalin, J. P., 1979, Experience of gas pipeline construction Vyngapur-Chelyabinsk. Ed., Informneftegasstroi.

Ivantsov, O. M., Dvoiris, A. D., Krivoshein, B. L. et al., 1977, Construction efficiency for low-temperature gas pipelines—Design and construction of pipelines and gas-oil field facilities: Moscow, NIPIESUoil-gas-stroi.

Kovalkov, V. P., Krivoshein, B. L. et al., 1977, Freezing of underground gas pipe and forecast of disturbance of underground water flow: Stroitelstvo gasoprovodov, no. 7.

Kovalkov, V. P., and Krivoshein, B. L. et al., 1982, Interference of oil and gas facilities with cryogenic rock, Summary of reports (Geocryological approach in regions of Extreme North, Moscow.

Krivoshein, B. L., 1979, Investigation and optimization of heat regimes of gas and oil mains: Izvestia, USSR Academy of Sciences, Energia i transport, no. 4.

Krivoshein, B. L., 1982, Thermo-physical calculations of gas pipelines: Moscow, Nedra.

Krivoshein, B. L., and Agapkin, V. M., 1977, Transient thermal loss of underground pipelines, 1 FJ. vol. 33.

Krivoshein, B. L., Agapkin, V. M., and Dvoiris, A. D., 1975, Construction and operation of pipelines in permafrost: Moscow, VNIIOENG.

Krivoshein, B. L., Kovalkov, V. P., and Sumarokov, V. S., 1981, Temperature range of underground gas pipelines transmitting chilled gas: Stroitelstvo truboprovodov, no. 1.

Kudryavtsev, V. A., Garagulya, L. S., Kondratieva, K. A. et al., 1974, Basic permafrost prognosis for engineering-geological investigations: Moscow, University.

DEFORMATION OF FREEZING, THAWING, AND FROZEN ROCKS

Yu. P. Lebedenko, L. V. Shevchenko, O. A. Kondakova,
Yu. V. Kuleshov, V. D. Yershov, A. V. Brushkov,
and V. S. Petrov

Geology Department, Moscow State University
Moscow, USSR

The paper examines the deformation of earth materials during the processes of heat and mass exchange while undergoing freezing, thawing, and in the frozen state. Study of the deformation of the materials during the processes of freezing and thawing has shown that variously directed and nonuniform strains develop in unconsolidated materials. Shrinkage strains are directed downward due to migration of water toward the freezing front. Heaving strains are directed upward due to the 9% increase in the volume of pore water on freezing and due to the accumulation of segregation ice at the expense of water which has migrated from the unfrozen zone. The accumulation of segregation in enhancing the upward deformation of the earth materials was also found in thawing soils where there was a frozen zone in a gradient temperature field. During a comprehensive study of the deformation of frozen materials in an isothermal temperature field, the samples were tested for uniaxial compression with and without possible lateral expansion. Investigations revealed bi-axial relationships between the pattern of deformation, the final magnitude of deformation, the composition and structure of the materials and the magnitude of the deforming load and rate of deformation.

INTRODUCTION

Deformation of freezing, thawing, and frozen earth materials is an important problem for both general and engineering geocryology, physicochemistry, and frozen soil mechanics. The solution of this problem depends upon thorough investigation of the theoretical basis of the evolution of cryogenic processes (heaving, shrinkage, settling, structurization, and texturization, etc.) and transformations in cryogenic composition, structure, and the properties of frozen soils. It is also important for applied geocryology to provide reliable support for engineering projects in frozen, freezing, and thawing grounds. Thanks to the efforts of many researchers from various countries, considerable theoretical and practical evidence has been accumulated revealing the processes of heaving in frozen ground, of settling in soils, and of mechanically-induced deformation in frozen soils. At the same time, deformation in thawing materials, in which water exchange between the thawing and frozen zones subject to ice segregation occurs, have not been sufficiently studied; likewise the interrelationships between the deformation of frozen soils and water migration, structure, and texture formation within the field of mechanical forces also require further study. Physicochemical factors associated with the deformation of frozen, freezing, and thawing soils must also be examined more thoroughly. This paper deals with these problems.

Experimental and field observations demonstrate that heaving may occur not only during freezing but also when the ground is frozen or in a state of unilateral thaw. Unfrozen water in frozen earth materials within a gradient temperature field migrates towards lower temperatures, resulting in segregation ice and deformation. In thawing grounds, water migration and ice segregation also occur in the frozen zone. The ground surface may be thrust upwards. Thawing ground heave may be, in general, presented as follows:

$$h_{heav} = h_{I_w} + h_{exp} - h_{shr} - h_{set} ,$$

where h_{I_w} represents deformation due to migration ice accumulation in the frozen zone; h_{exp} is deformation of ground surface due to expansion in the thawed zone; h_{shr} is the deformation of physicochemical shrinkage of the thawed zone due to dehydration; h_{set} is thermal settling deformation due to the decreasing ground volume resulting from thawing of pore and segregational ice.

Heaving strain in thawing soils due to the accumulation of migration ice is possible only if there is a temperature gradient in the frozen zone. This occurs, usually, when, due to water migration and the accumulation of segregation ice, a heaving strain develops in the local ground layer, in which case its surface may subside due to strain or settling and shrinkage. The ground surface rises when the expansion strain in the thaw zone occurs simultaneously with the accumulation of migration ice. Ground thaw coupled with positive strain is followed by extreme thermal settling caused by thawing of pore and migration ice. As a result, the ground surface settles. Thus, thaw deformation depends on the sum total of variously directed strains that result in both the fall and rise of the earth materials.

Deformation processes occurring in materials of

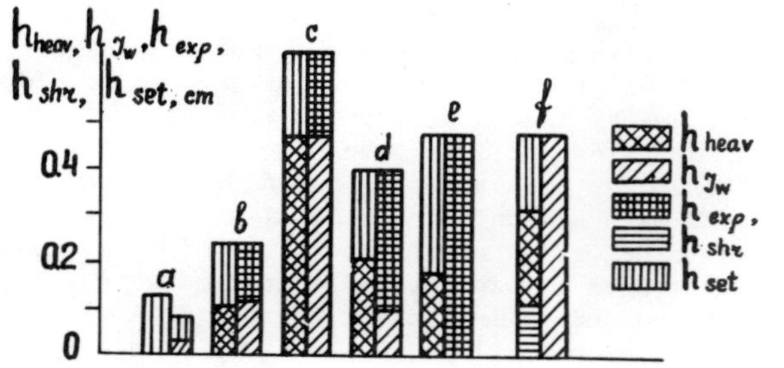

FIGURE 1 Diagram of relationship between strains induced by the accumulation of migration ice (h_{I_W}), expansion (h_{exp}), shrinkage (h_{shr}), settling (h_{set}), and heaving (h_{heav}) in soils of different composition: a) sandy loam; b) loam; c), d), e) kaolinite, polymineral, montmorillonite clays thawing in open systems; f) kaolinite clay thawing in a closed system.

different granulometric and mineral composition and subject to diverse thermal and water conditions during thawing were studied. The results revealed the following: As soil dispersity increases from sands to clays, the flow of migration water and the accumulation of segregation ice in frozen ground increases. In clay of different mineral composition, water migration and the accumulation of segregation ice increase when such soils increase their content of kaolinite minerals (Figure 1). Deformation resulting from the accumulation of segregation ice in the frozen zone of thawing clays and sandy montmorillonite loams comprised 5-10% of the total value of positive strain causing heaving, while in kaolinite clay the deformation due to this process comprised 80-90% of the total value. Conversely, in montmorillonite clays the expansion strain occurs more frequently (80-90%) while in kaolinite clays they constituted about 20% of the total positive deformation. Sandy loams and loams are considerably less subject to expansion: in a thawing loam the magnitude of expansion was half of that in a kaolinite clay, while sandy loam did not reveal any expansion at all (Figure 1).

Our observations indicate that expansion deformation in soils thawed in an open system is always higher than in a closed one. This can be attributed to the fact that when water flows in from the outside, part of the external migration flow is expended on the accumulation of segregation ice in the frozen zone, and part of this water goes to enhance expansion deformation. When thawing occurs within a closed system, the strains in the thawed soil caused by desiccation and shrinkage increase while those due to swelling decrease.

Positive strains in thawing soils which result in heaving usually level off due to developing shrinkage and settling strain. Unlike freezing soils where practically the entire freezing layer undergoes shrinkage, thawing soils shrink only near the border of thawing and their shrinkage lasts only a short time. This phenomenon can be explained by a flooded horizon appearing near the interface border at the expense of the thawed layers of ice. It was found that shrinkage strain during thaw is lower in an open system as compared to thaw in a closed system (Figure 1b,e). Shrinkage strains do not level heaving much; in montmorillonite grounds of highest shrinkage they seldom make up more than 20% of the total value of negative strains.

In freezing soils, unlike thawing ones, the value of heaving strains can in general be expressed as follows:

$$h_{heav} = h_{heav(9\%)} + h_{I_w} - h_{shr}$$

where $h_{heav(9\%)}$ is the value of ground heaving due to the 9% increase in the volume of frozen pore water (mass expansion); h_{I_W} is the value of heaving due to the accumulation of segregation ice; and h_{shr} is the value of physico-chemical shrinkage strains in the thawed ground due to desication.

The calculated value of heaving due to water migration is equal to

$$h_{I_w} = 1.09\ K_{an} \cdot K_w\ \delta_t^{ph}\ \text{grad.}t \cdot \tau\ ,$$

where K_{an} is the coefficient of anisotropy allowing for the inclination angle of ice layers; K_w is the coefficient of water diffusion, m^2/s, and δ^{ph} is the thermogradient coefficient, I/grad C. The heaving value due to the 9% increase of pore water volume during freezing can be expressed as follows:

$$h_{heav(9\%)} = 0.09\ (W_\xi - W_{unfr})\xi$$

where W_ξ is the water content of the ground's mineral layers, which is quantitatively equal to the water content at the phase interface boundary, fractions of unit, and W_{unfr} is the amount of unfrozen water in the soil, fractions of unit.

The shrinkage that takes place in the thawed portion of the soil can be calculated as follows:

$$h_{shr} = K_{shr}\ \beta \cdot \Delta W \cdot \Delta\xi\ ,$$

where K_{shr} is the anisotropic shrinkage coefficient showing the contribution of vertical shrinkage to total volumetric shrinkage; β is the coefficient of relative volumetric shrinkage of the soil; $\Delta W = W - W_{shr}$ is the difference between the initial water content of the soil and the water content at the freezing front; and $\Delta\xi$ is the layer undergoing dehydration, m. In terms of the density distribution of the soil skeleton, the value of shrinkage can be calculated as follows:

$$h_{shr} = K_{shr}\ (I - \gamma_{sk}^{in} / \gamma_{sk}^{f})\ \Delta\xi$$

where γ_{sk}^{in} and γ_{sk}^{f} are, respectively, the initial and final values of the weight density of the soil skeleton before and after freezing. When materials of different composition, structure, and properties freeze under diverse thermal and moisture conditions the total value of heaving depends on the patterns for development of individual deformations induced by frozen pore water, segregation ice, and shrinkage (Figures 2 and 3). On the whole, ground heaving can be expressed as follows:

$$h_{heav} = 0.09\ (W_\xi - W_{unfr})\Delta\xi + 1.09\ K_{an} \cdot K_w\ \delta_t^{ph}\ \text{grad} \cdot t\tau - K_{shr}\ \beta\Delta W\Delta\xi$$

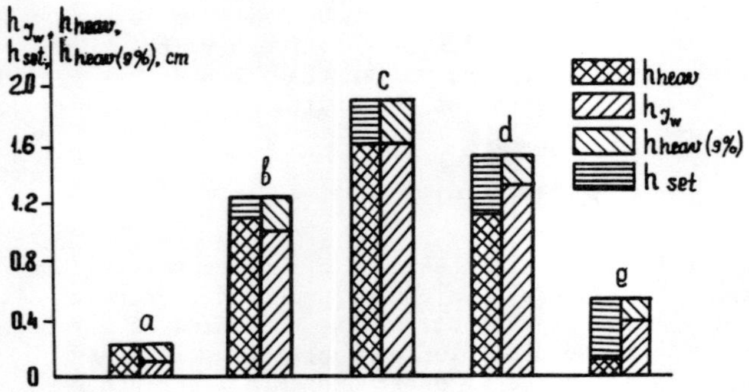

FIGURE 2 Diagram of the relationship between strains in frozen soils of different composition in an open system: a) sandy loam; b) loam, c) kaolinite clay; d) polymineral clay; e) montmorillonite clay.

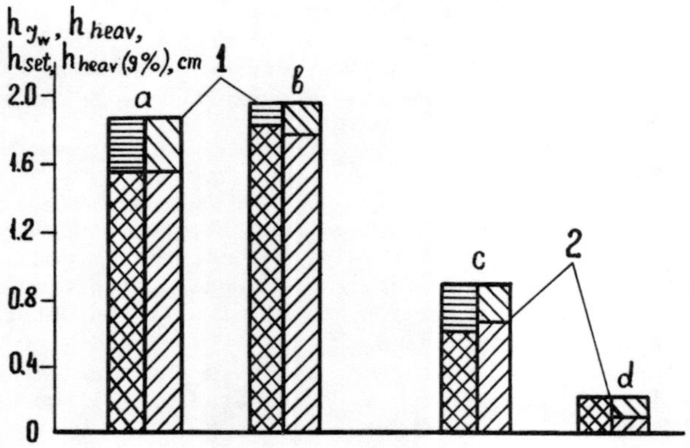

FIGURE 3 Diagram of the relationship between strains in samples of kaolinite clay with different initial density a) and c) compaction load up to 0.1 MPa; b) up to 0.2 MPa; d) up to 0.8 MPa; freezing in an open (1) and in a closed (2) system.

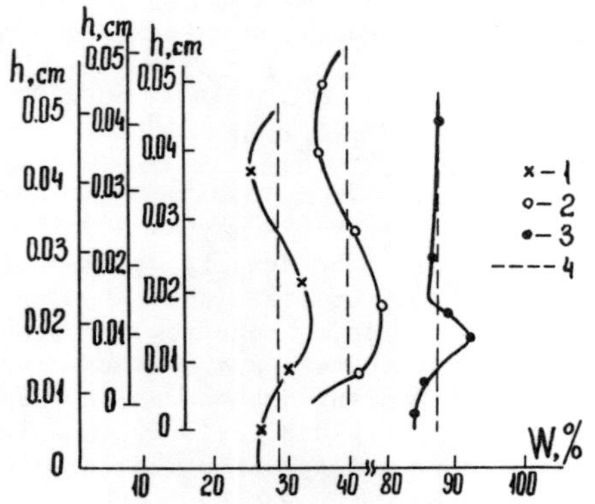

FIGURE 4 Displacement-induced water redistribution in frozen soil samples with different mineral composition: 1 - polymineral clay; 2 - kaolinite clay; 3 - montmorillonite clay; 4 - water content before the experiment; 5 - displacement plane.

This technique for assessing the values of expansion deformation in freezing and thawing earth materials was tested in both laboratory and field studies and yielded satisfactory results.

The dynamics and laws governing the development of stresses in freezing and thawing rocks due to heat and mass exchange, as well as physico-chemical processes occurring during phase transitions, indicate that such strains are interrelated with soil deformation and soil structurization. Greater strains develop in clays, while lower ones are observed in loams and sandy loams. Freezing-induced strains increase in open systems as compared with closed ones, and also under higher external pressures. An increase in both the external load and the integrity of structural bonds in the freezing soils results in reduction of deformations; moreover, when critical load values are attained, deformations fade out to zero. Further increase in pressure results in compaction of freezing and frozen materials.

These studies on ground deformation in a gradient-free thermal field show that an external load on a frozen soil results in migration of unfrozen water and ice segregation, in addition to transformation of the soil micro and macro structure. Water redistribution and micro- and macrostructural transformation were observed during both the slow displacement of frozen soils and compression compaction.

Examination of the moisture distribution curves before and after the experiment (Figure 4) showed that maximum ice accumulation caused by displacement strains in soils of different mineralogical composition occurred in the displacement region: in kaolinite clays (W_{in} = 39%, γ_{sk} = 1280 kg/m^3, in polymineral clay (W_{in} = 29%, γ_{sk} = 1410 kg/m^3, in bentonite clay (W_{in} = 87.5%, γ_{sk} = 800 kg/m^3. (The experiment was carried out at -3°C, τ_{displ} = 0.2 MPa). The thickness of the ice accumulation zone decreased from kaolinite (0.017 m) to polymineral (0.016 m) and bentonite clay (0.0035 m). Maximum desiccation (ΔW_{des}) was found in the upper and lower parts of the samples; it comprised 5.5% in kaolinite, 3.5% in polymineral, and 1.5% in bentonite clay. The desiccation zone also decreased in thickness from kaolinite to polymineral and to bentonite clay, comprising 0.025; 0.017, and 0.0125 m, respectively. Variations in the thickness of the desiccation zone and the extent of desiccation can be attributed to the unequal water conductivity of these clays. The diffusion coefficient in kaolinite clay is the largest, attaining $3.5 \cdot 10^{-9}$ m^2/s, which is why the desiccation zone in kaolinite clay is the greatest. This coefficient is the lowest in bentonite clay ($K_w = 0.5 \cdot 10^{-9}$ m^2/s), which has the thinnest desiccation zone and the least amount of desiccation. The largest migration flows of water are formed in kaolinite clay, namely, $12.6 \cdot 10^{-7}$ kg/m² s; in polymineral clay they are $8.9 \cdot 10^{-7}$ kg/m² s; in bentonite clay they are equal to $1.42 \cdot 10^{-7}$ kg/m² · s. As soil density increases and soil dispersity decreases, the migration flow is reduced.

Microstructure studies of montmorillonite, polymineral, and kaolinite clay samples before and after the experiment show that the shape and size of ice inclusions undergo changes during the process of

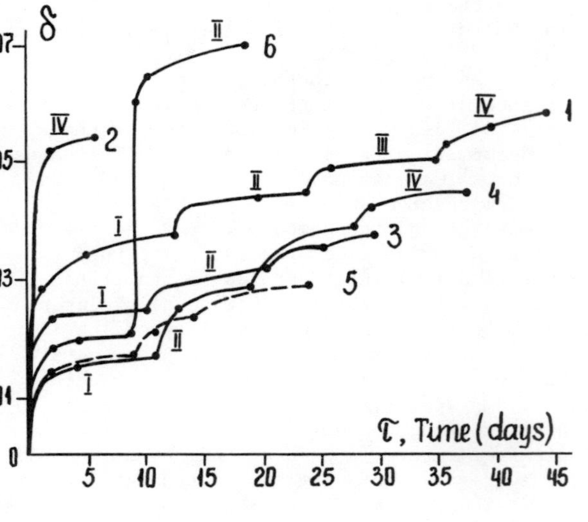

FIGURE 5 Deformation in frozen soils (-1.5°C) under diverse external loads: 1, 2 - montmorillonite clay; 3, 4 - polymineral and kaolinite clays; 5, 6 - sandy loams with different degree of pore saturation with water (q = 1 and q = 0.6); I, II, III, IV - compaction load equal to 0.3 MPa, 0.6 MPa, 0.9 MPa, respectively.

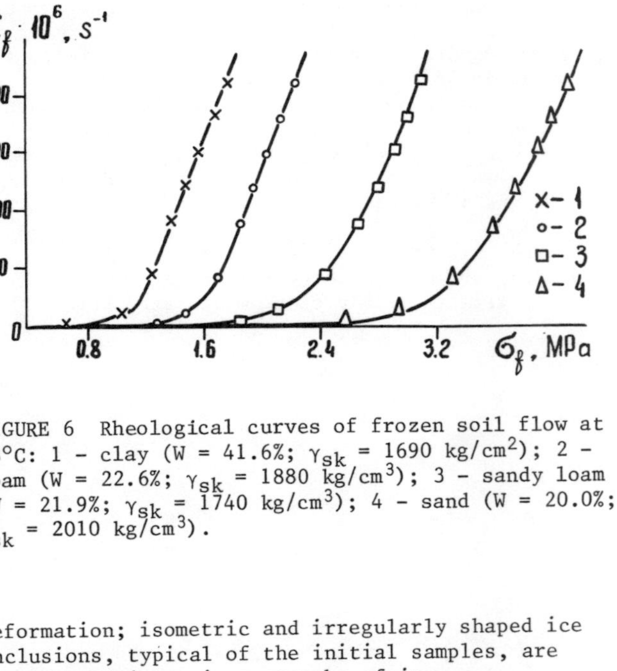

FIGURE 6 Rheological curves of frozen soil flow at -3°C: 1 - clay (W = 41.6%; γ_{sk} = 1690 kg/cm^2); 2 - loam (W = 22.6%; γ_{sk} = 1880 kg/cm^3); 3 - sandy loam (W = 21.9%; γ_{sk} = 1740 kg/cm^3); 4 - sand (W = 20.0%; γ_{sk} = 2010 kg/cm^3).

deformation; isometric and irregularly shaped ice inclusions, typical of the initial samples, are transformed into micro-streaks of ice.

Our investigations have shown that the progression of water transfer and structural transformations in frozen rocks induced by external loads depends on their deformability. Analysis of the deformability of frozen soils under external loads has demonstrated that deformations in less dispersed soils are markedly weaker (Figure 5). The more dispersed a soil is, the longer the deformation lasts. For example, deformation in sandy loams under an external load of 0.3 MPa is completed in 8 days; in the case of clays, according to their dispersity, the deformation process required 10, 11, and 13 days (curves 5, 3, 4, 2). The effect of dispersity on soil deformation is even more manifest when the soil pores are not completely filled in. When the degree to which pores are filled is 0.6 (Figure 5, curve 6) in a polymineral sandy loam, this soil reveals another pattern of deformation that can be explained by the deformation stages of the soil's ice-mineral skeleton and its pore space.

Soil deformability is greatly affected by the mineral composition of soils. In comparing curves 3, 4, 5 one can see that soils with larger contents of flexible laminated particles of hydromica (polymineral clay and sandy loam) undergo greater deformation than kaolinite clay composed of the larger anisodiametric particles. In terms of their relative compressibility, the soils under study can be arranged as follows: montmorillonite clay > polymineral clay > kaolinite clay > polymineral sandy loam.

The investigation of the compression of frozen soils at constant rates of deformation (within the range from $8.5 \cdot 10^{-6}$ to $1.7 \cdot 10^{-3}$ s^{-1}), and with possible lateral expansion, showed that the curves depicting the relationships between flow stress and the deformation rate were similar for frozen clays and sands (Figure 6). Linear approximation of these relationships clearly demonstrates an area of low rates with high plastic viscosity coefficients and an area of high rates with low values of this coefficient. The transition zone is confined to a definite range of deformation rates (from 8.5 to $85 \cdot 10^{-6}$ s^{-1}); it seems to be related to the relaxation nature of the mechanical properties of the pore ice cementing the mineral particles of the soil.

The investigation of the effect of relative strains (E_T) on the moments when the strains become stable has revealed some interesting regularities. In the case of clayey (clay, loam) and sandy (sand, sandy loam) frozen soils, as the deformation rate increases, i.e. as the soil's elasticity response to compression increases, the value of E_T increases. A similar regularity was observed when both the temperature of soils and their number of monovalent cations decrease. The generation of strains in frozen clays and sand varies. In frozen clays the value of E_T at low deformation rates is half of that at high deformation rates. In frozen sands, despite the large strains, this relationship is less evident, and within the range of -1.5°C, the value of E_T is close to unity. This peculiarity may be primarily due to the difference in the compressibility of the mineral skeleton of soils. The compression tests carried out have yielded results that agree with this supposition.

It was also found that at temperatures as low as 0° to -10°C, the flow in the soil samples almost always began when the sample began to increase its diameter. In sandy soils the degree of compression at the time the diameter began to increase (E_d) was lower than that in clayey soils. In the area of high deformation rates, in the case of clays, at temperatures below -1.5°C the values of E_d do not change much (about 2%) (Figure 7). As the temperature decreases, the value of ($E_T - E_d$) increases, which attests to increased volumetric plastic

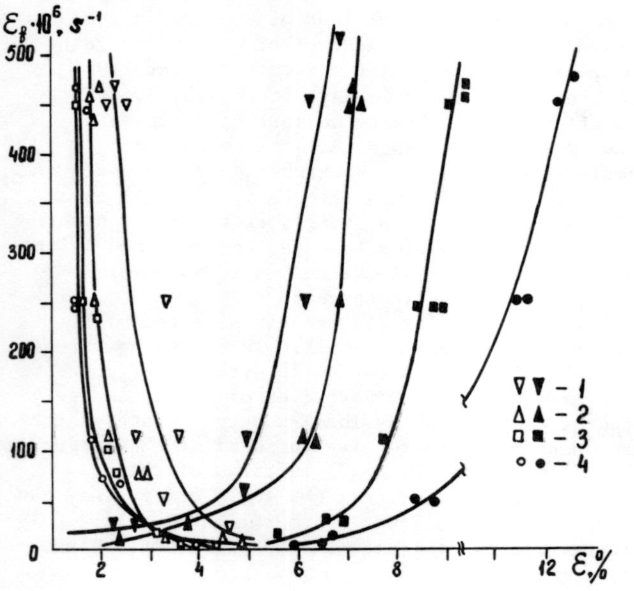

FIGURE 7 Relationship between the degree of compression in clay and deformation rate by the initial moment of lateral expansion (unblackened signs) in samples and by that of flow (blackened signs) at different temperatures. Symbols: 1,2,3, 4 - soil temperature equal to -0.5°, -1.5°; -3° and -10°C, respectively.

strains resulting from the increase in the sample diameter. At the same time, at temperatures above -1.5°C and at deformation rates below $8.5 \cdot 10^{-5}$ s^{-1} the flow of clay for a certain time did not affect the sample diameter, i.e. the process was of the "compression" type.

These regularities and the results of microstructure studies of deformed soils make it possible to identify the main characteristics of the process of accumulation of volumetric plastic strains in frozen fine-grained soils under uniaxial compression. In the first place, two processes occur during volumetric deformation (dilatation): "compressional" constriction and plastic dilatation caused by an increase in the sample diameter. The importance of the first process increases with greater soil dispersity, elevated temperatures, and lower deformation rates. In favorable instances this process may dominate. In the second place, during compression deformation of clayey soils, their mineral and ice components undergo a degree of differentiation (when the initial cryogenic texture is bulky) which depends on the critical compressibility of the soil's mineral skeleton and the volumetric plastic deformability of ice stemming from its relaxational nature. Due to this feature of ice, this process is time-dependent. In the third place, plastic dilatations of soils occurring during the enlargements of their diameter result from interrelated dislocations of consolidated aggregates (of clay) or of separate particles (sand). The mechanical resistance of soil to such displacements (soil viscosity) depends directly on the phase composition, the relationship between the deformation rate, and the time of strain relaxations in the ice. As the dispersity of soils decreases and the deformation rate increases, the local continuity of samples may be interrupted more frequently (the appearance of microcracks).

Therefore, our studies of the mechanism and laws governing deformation in freezing, thawing, and frozen soils of different granulometric and mineral composition, structure and properties, under diverse thermal, water content, mechanical, and dynamic external conditions have demonstrated the following. Deformation of freezing, thawing, and frozen soils is interconnected with processes of heat and mass exchange, of structurization, and texturization. On the one hand, soil deformation can be induced by heat and mass exchange; on the other hand, deformation and stresses themselves induce water migration, and, on the whole, result in heat and mass exchange. In each type of deformation (free or induced), the development of soil deformations derives from their micro and macro structure.

MIGRATION OF ELEMENTS IN WATER IN TAIGA-PERMAFROST LANDSCAPES

V. N. Makarov

Permafrost Institute, Siberian Branch
Academy of Sciences, Yakutsk, USSR

The peculiarities of the migration of elements in water in the main taiga-permafrost landscape types in Northeastern Yakutia are examined. Zonal and azonal classes of geochemical landscapes associated with permafrost and bioclimatic peculiarities are distinguished on the basis of the conditions of migration of elements in water. Differing intensities of the migration of elements in water as between permafrost and non-permafrost zones were demonstrated. Both surpluses and deficits of certain elements were identified and also paragenic associations of elements characteristic of specific geochemical landscapes.

The identification of taiga-permafrost landscapes of the north-eastern part of Yakutia is based on principles developed by Perelman (1966). The largest generally accepted categories of geochemical landscapes are represented by bioclimatic units that define the natural conditions and the character of migration of substances. According to water migration conditions associated with bioclimatic and permafrost factors, zonal classes are identified as: acid (H^+), acid gley (H^+-Fe^{++}), calcium (Ca^{++}) and azonal-sulphate (H^+ - SO_4^{11}). In addition to azonal landscapes of sulphate class occurring in areas of sulphide mineralization, azonal landscapes associated with icings and characterized by contrasting geochemical features can also be distinguished.

Taiga-permafrost landscapes of class (H^+) orthoeluvial on granitoid and effusive formations generally form under conditions of highly unconsolidated relief. These landscapes occur in medium-high and high mountains composed of granitoid and effusive rock. Autonomous landscapes of the acid class occur on a weathering debris crust and on mountain-tundra and mountain-taiga rocky primitive soils which form on them.

The unconsolidated relief and ledge rock govern the poor development of vegetation, low degree of marshiness and intensive water exchange. Typical plant forms are: fragmented moss-lichens, thickets of ledum-cladonium cedar creepers in the lower part of slopes and bush-lichen tundras on northfacing slopes.

The physical weathering that leads to the formation of large-size placer debris and kurum-talus is especially pronounced. Rapid water seepage through the coarse upper debris of the kurum layers provides for downward placer migration of elements which are transported downwards along the section.

We will examine the chemical weathering characteristic for landscapes of this type, using granitoid formations as an example.

Surface weathering leads to the formation of iron-bearing surface-eroded layers of loose granite. The main hypergenic changes of granitoids consists of fragmentation and disintegration of rock, partial destruction of feldspars, transformation in stages of biotite into hydrobiotite, accumulation of iron in the form of films and scums of amorphous and weakly crystalline oxides, and the transport of sodium, calcium, potassium and silica from the eroded layers. The soils are represented by crushed-rock supes having high acidity (pH 3.4 - 4.7) and a rather high organic matter content in the form of coarse humus (up to 7-8%).

The landscapes under consideration are characterized by the propagation of ultra-fresh chloride-hydrocarbonate, mixed in composition, cations of low-acid waters, poor in iron (less than 0.1 mg/ℓ) and organic matter. The chemical composition of surface waters (averaged over 100 samples), taken from 15 granitoid formations, is as follows:

$$0.019\,\frac{HCO_3 73\ Cl\ 27}{Na58\ Ca22\ Mg20}\quad pH\ 6.0\ Eh\ 0.54B$$

Within zones of exo- and endocontacts of granitoid strata, mineralization of natural waters is increased to 30-40 mg/ℓ and sulphate-hydrocarbonate waters appear. Natural waters of landscapes of this type are characterized by a relationship between the ions $HCO_3 > Cl$ and $Na >$ Ca, Mg. As a result of weathering of granitoids and eruptive rock, the composition of solutes (ion sink) sodium, calcium, and magnesium hydrocarbonates predominate.

In order to quantitatively estimate the water migration of elements for the landscapes under consideration we shall determine the value of ion sink for which qualitative characteristics of the migration capacity of elements may be obtained by computing the water migration coefficients.

According to our calculations, the annual surface run-off of solutes from the granitoid strata varies between 1 and 4 t/km^2/year and averages 2.2 t/km^2. Dividing the value of ion run-off by the average weight density of the region's granites, 2.60, we obtain a value for the chemical denudation (0.85 mk/year), which is 4-5 times less than that for granitoid rock in moderate climatic zones (Strakhov, 1965).

The values for ion run-off that we obtained are much lower than those hitherto available for granitoids in the north-eastern part of the USSR, 16-17 t/km² /year (Walpeter 1972). This, to our knowledge, has been overestimated by a factor of 3 because the investigator did not take into account atmospheric fallout of solutes (Janda 1971). The rock masses, in which the ion run-off has been determined, lie along littoral zones of the Sea of Okhotsk where the atmospheric fallout rate of salts is 10-12 t/km² /year (Climate Atlas of the USSR, 1960).

We have computed coefficients (K_x) for migration in water using the formula reported by Perelman (1966):

$$K_x = \frac{m_x \cdot 100}{a \cdot n_x},$$

where m_x is the volume of the element in the waters, g/ℓ; a is water mineralization, g/ℓ; and n_x is the volume of the element in rock, %.

The values m_x and a were derived analytically, while n_x represent the amount (in percentages) of acid rock in the lithosphere (Vinogradov 1962).

The rows representing the migration of elements in water are listed in Table 1.

TABLE 1 Mobility of Elements in Water of Granitoid Strata

Migration rate	Value of coefficient of migration in water, K_x				
	100	10	1	0.1	0.01
Very high	C,Cl, Hg,As, Bi,Tl, Ag,Sb				
High		Sn,B,Mg, Ni,F			
Medium			Zn,Pb,Cu, V,Na,Y,Ca, Cr,Be,Mo, Mn		
Weak				Ga,Ba, Si,Co	
Very weak					Ti,Al, Fe

As compared to the rows compiled by Perelman (1966) for the weathered crust of silicate rock in a temperate zone, the zone of granitoid hypergenesis, which is situated in the subarctic climatic region, tends towards inertness of some of the macrocomponents, namely sodium, calcium and silicon. The abovementioned values for ion run-off also lend support to a reduction in the amount of chemical weathering of granitoids under conditions of hypergenesis.

Figure 1 shows a value for the migration of elements in water, derived using K_x, for natural landscapes in different climatic zones. Within the range of the migration distribution of chemical elements having varying degrees of contrast in

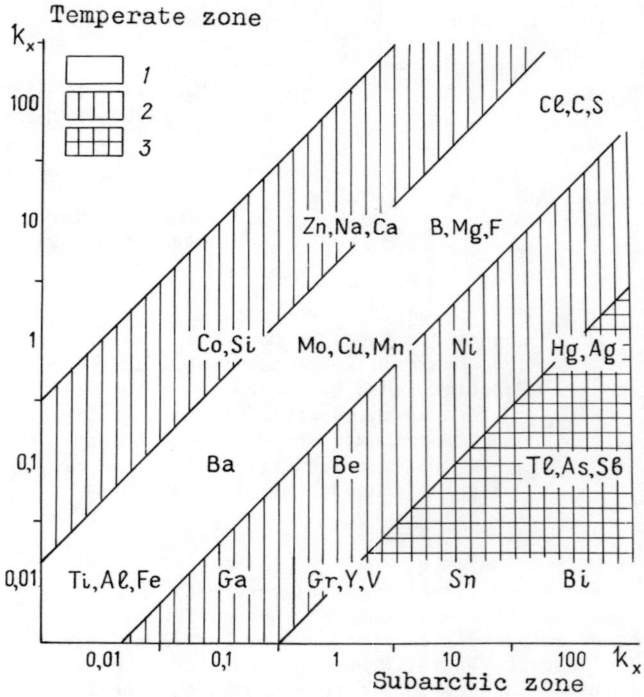

FIGURE 1 Contrast range of element migration in granitoid landscapes of different climatic zones. Fields of distribution of chemical elements:
1 - non-contrast migration (K_k = 1);
2 - contrast migration (K_k = 2 - 10);
3 - high-contrast migration (K_k = more than 10).

climatic zones, groups of elements having similar mobility can be distinguished, i.e. unique geochemical associations. Migration contrast can be compared using K_k, the coefficient for the range of contrast equal to

$$K_k = \frac{K_x \text{ of granitoids in subarctic zone}}{K_x \text{ of granitoids in temperate zone}}$$

Maximum migration contrast (K_k is larger than 10) is typical for Sn, As, Ti, B, Ag, Sb, Hg. Contrast migration (K_k = 2-10) is observed in Ni, Pb, Cr, V, Y, and Be.

Nearly all the elements refer to very strong (Hg, As, Bi, Ti, Ag, Sb) and strong migrators (Sn, B, Ni, F) in granitoids of north-eastern Yakutia (cf. Table 1).

The set of elements that possess high mobility and maximum contrast for migration in water can be regarded as a unique paragenic landscape association.

Taiga-permafrost landscapes of acid class (H^+), orthoeluvial in terrigenous stratified formations, are widely distributed throughout regions having low and medium height mountains. Autonomous acid class terrains are distributed in weathered debris crust and the mountain-taiga and northern taiga permafrost soils which have formed on them. Watershed areas are composed of dense sandstone rock and saddlebacks occur adjacent to stratified shale outcrops. Sandstone rock is characterized by coarse shales and on slopes produce typical rocky talus made up of large-clod materials. Destruction of clayey and aleurite shales leads to the formation of deluvial gravels.

The thickness of the seasonal thaw layer on south-facing slopes is 1.2-1.4 m. On north-facing slopes it is practically absent, with frozen rocks or ice occurring directly below the moss cover.

Mountain-taiga suglinok soils have an acid reaction (pH 3-4), a high concentration of exchange bases within the upper humus layer and considerable (up to 25%) organic matter content in the form of coarse humus.

Landscapes of this type are characterized by the occurrence of ultra-fresh hydrocarbonate waters with mixed composition of cations. The average composition of natural waters in taiga-permafrost terrain (over 4,000 samples) is:

$$0.028 \frac{HCO_3 80\ Cl\ 18\ SO_4 2}{Na37\ Ca33\ Mg30} \text{ pH 6.3 Eh 0.48B}$$

As compared to taiga-permafrost landscapes occurring in volcanic bedrock, the mineralization of natural waters increases in this case by a factor of 1.5-2.0, while the anion-cation ratio remains the same.

The composition of the ion run-off is predominantly calcium, magnesium and sodium hydrocarbonates. The ion run-off value is, on the average, estimated at 4.25 t/km²/year; chemical denudation, at 1.6 per year (with an average weight density of stratified rocks in the Verkhoyansk complex of 2.65); this comprises about one-fifth of the mechanical denudation value which is typical of terrigenous rocks in the Verkhoyansk complex.

The rows for the migration of elements in water for stratified rocks of the Verkhoyansk complex are presented in Table 2.

TABLE 2 Mobility of Elements in the Water of Terrigenous Stratified Formations

Migration rate	Coefficient values for migration in water, K_x				
	100	10	1	0.1	0.01
Very high	C,Cl,S, F,B,Ag, Na,Au,Hg				
High		W,Be,As, Mg,Ca,Tr, Sr,Ba,Mo, Mn			
Medium			Sn,Sb,Ti, Pb,Cr,Cu, Zr,V,Ti, Ni		
Weak				Zn,Fe	
Very weak					Co,Ga, Al,Si

A maximum range of contrast for migration in water in taiga-permafrost landscapes of stratified terrigenous formations, as compared to those of a temperate climatic zone, is characteristic for gold, mercury, tungsten, and beryllium (a sharp increase of migration activity, K_k more than 100) and for zinc, silicon, and cobalt (substantial decrease of migration mobility, K_k less than 0.01, cf. Figure 2).

Taiga-permafrost landscapes of the calcium class (Ca^{++}) on carbonate stratified formations. These landscapes provide more favorable conditions for the biological cycle, producing larger amounts of biotic material; there are more grasses and bushes in the plant cover, the amount of humus increases in the soil, its peat content decreases and carbonate gleization develops. The waters are more mineralized and contain more calcium and less acid organic matter and the pH reaction approaches the neutral one.

$$M0.150 \frac{HCO_3 97\ \ Cl\ \ 2\ \ SO_4 1}{Ca50\ Mg30\ Na20} \text{ pH 6.8 Eh 0.27 B}$$

There is a substantial increase in the ion run-off rate, amounting to 28 t/km², or nearly twice the mechanical one (Seimchan River, location Chapaevo).

In a weak alkaline medium many metals (copper, lead, nickel, titanium) exhibit low migration capacity and are transported from soils in small amounts (Table 3).

The low migration capacity of most of the elements in the landscapes has been confirmed by the evaluation of "soil mobility" made using the value of eluvial-accumulative coefficient (Mikhailova 1964) which indicates very low mobility for molybdenum, zinc, lead, copper, etc. (Table 4).

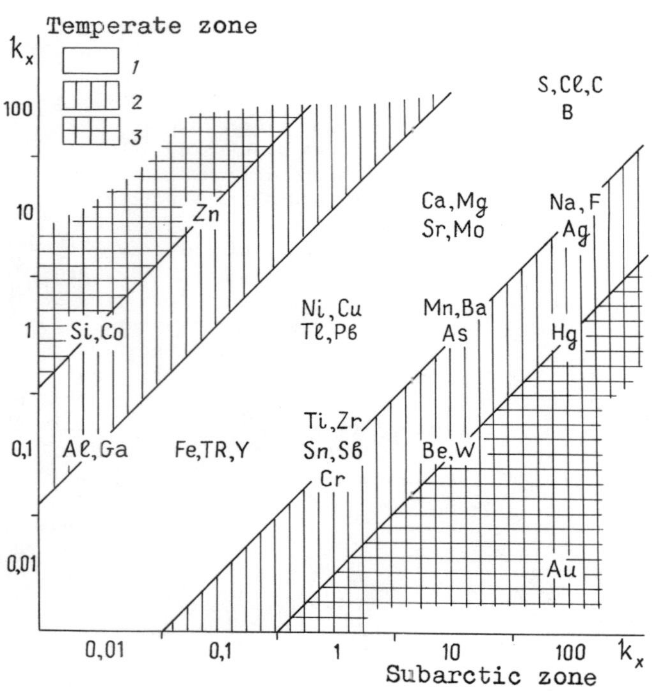

FIGURE 2. Contrast range of element migration in terrigenous landscapes of different climatic zones. Legend same as in Figure 1.

TABLE 3 Mobility of Elements in the Water of Carbonate Stratified Formations

Migration rate	Coefficient values for migration in water, K_x				
	100	10	1	0.1	0.01
Very high	C,S,Ag, Cl				
High		Cd,Hg			
Medium			Ca,Mg,Mn, Na,Ti,Sn		
Weak				Cu,Ni,Mo, Cr,Pb,Zn, Ge,B,V,Ga	
Very weak					Ti

TABLE 4 Soil Mobility of Elements in Carbonate Formations

Intensity index of element migration ($K_{el.-acc.}$)			
Highly mobile, below 0.35	Mobile 0.35-0.7	Weakly mobile, 0.7-1.4	Inert, above 1.4
Ge, B, Mn	Co, Sn, Cr, Ni	Cu	Y, Ga, V, Pb, Zn, Ti, Mo

The saturation of the soils of taiga-permafrost carbonate landscapes of the calcium class with microelements has a favorable impact on soil fertility. During geochemical exploration in terrain such as those under discussion, it is advisable to employ lithochemical methods because in the soils which overlie deposits contrasting residual haloes of most of the ores arise.

Taiga-permafrost landscapes of the acid gleyey ($H^+ - Fe^{2+}$) class, accumulative superaquifer are for the most part characteristic of low-lying areas, composed of loose, stratified, terrigenous, cainozoic accumulations and peat-bog soils forming on them. The latter develop underneath the plant cover which consists of carex-veinik and bush-ledum-kassandra-carex-moss-bog species. An especially severe soil climate forms underneath the moss cushion and the peat horizon, which subsequently hinders warming of mineral soil in the summer. This is because peat-bog soils in most cases have an acidic reaction within the peat layer (water pH 5.2 - 5.8) that is close to the neutral one occurring in the underlying mineral layer of soil (6.4 - 6.7 pH). Within the peat layer the organic matter content, in terms of humus, varies from 26 to 70%, while that of gross phosphoric acid ranges from 0.26 to 0.46%. The sum of exchange bases varies from 30 to 100 mg - equ per 100 g of dry substance, of which 75-93% is Ca^{2+} and 7-25% Mg^{2+}.

This type of landscape is widely distributed in river valleys and lowlands. In pH and the degree of mineralization, natural waters approach waters of acid class. The abundance of organic matter (peat, residuals of vegetation) produce a regeneration medium. Gleization processes, well developed in the soils of such landscapes, are responsible for the typomorphic nature of the Fe^{++} ion.

All the landscapes of this group are characterized by seams of bedrock underlying a cover of friable cainozoic formations. The influx of elements originates mainly from contiguous landscapes. There is practically no removal of elements because the processes are dominated by accumulation (except for alluvial landscapes). Bedrock, in places where the thickness of the cainozoic formation is small due to the formation of saline bulbs on top of the ore bodies, is another possible source of element accretion.

Plain-type landscapes, widespread occurrence of hover rocks and geocryological and hydrogeological conditions make it possible to consider these landscapes as superaquatic. The processes of mechanical migration of substances everywhere, except for river-beds and floodlands, are greatly weakened. Physical and chemical and biogenic migration is very pronounced. The rate of water migration of microelements is increased (Table 5).

TABLE 5 Basement Levels of Elements in Accumulative Superaquatic Landscape Water

Migration rate	Value of coefficient of water migration, K_x				
	100	10	1	0.1	0.01
Very high	Hg				
High		Mo,Be,As, Ni,Sn,Sc, Zn,B,Nb, Ag,Bi			
Medium			Mn,Y,Pb, Cu,Cr,Ba, V,Ti,Co, Ga		
Weak					
Very weak					

The presence of weak-acid, ultrafresh, chloride-hydrocarbonate water with mixed cations is typical of these landscapes. The composition of natural waters of the Verkhne-Adychanskaya depression is typical (average over 19 samples):

$$M0.023 \frac{HCO_3 74 \ Cl \ 24 \ SO_4 2}{Na43 \ Mg35 \ Ca22} \ pH \ 6.1 \ Eh \ 0.46B$$

The ions characteristically have the following ratio: $HCO_3 > Cl > SO_4$ and $Na^+ > Mg^{++} > Ca^{++}$.

In peatland waters, the mineralization sometimes increases substantially (up to 100 mg/ℓ), while the acidity is decreased due to the presence of humic acids (pH 5.2 or less).

This type of landscape is littoral and is characterized by several other hydrochemical properties. The proximity of a sea, resulting in greater input of salts with atmospheric precipitation, is responsible for the occurrence of chloride-hydrocarbonate, predominantly sodium surface and ground waters with a relatively high mineralization and a weak acid or near-neutral pH. These are exemplified by spring water and shallow lakes in the eastern Primorye lowlands, in the Kolyma River basin (the average for 52 samples from an area of about 2,000 km²):

$$M0.070 \frac{HCO_3 51 \ Cl \ 49}{Na 71 \ Ca 22 \ Mg 7} \ pH \ 6.7 \ Eh \ 0.5B$$

In the surface water composition, the chlorine-ion content averages 25 mg/ℓ (12-50 mg/ℓ), whereas in landscapes of this class occurring further inland, the background content of chlorine-ion is 4 mg/ℓ. The relatively high concentrations of chlorine-ion accounts for the decrease in the biogeochemical activity of silver, lead and copper (Table 6).

Table 7 gives the geochemical characterization of the landscapes under study, the geochemical specialization of a given landscape being defined on the basis of paragenic associations, i.e. groups of elements that are characterized by a high contrast range of migration in the subarctic zone, in comparison to the migration mobility in the temperate climatic zone. "Abundant" elements are those with K_k in excess of 10 while deficient elements include those with K_k below 0.1. The geochemical specialization of the main types of landscape is well defined.

The absence of a number of elements is determined by the very small contribution of underground waters in the surface run-off, by low mineralization of atmospheric precipitation and by weak development of the technogenic processes.

The behavior of concentrating elements, characterized by a high migration contrast range, that form paragenic associations, for example, gold, mercury, antimony and tungsten in landscapes of terrigenous mesozoic formations, is determined mainly by the metallogenic properties of the geological formations, the high regional percentages of the lithosphere.

The paragenic associations of elements isolated here are typical of certain landscapes and indicate not only geochemical but also ore specialization. The abundance or shortage of a number of elements determines the balneological or economic value of a landscape.

TABLE 6 Biochemical Absorption of Microelements in Accumulative Superaquatic Landscapes

Region	Coefficient of biological absorption		
	10	10-1	1-0.1
Verkhne-Adychanskaya depression	Cd,Ag,As	Mo,B,Zn,Pb, Cu,Ni,Sn,Cr, Co	Li,V,Ga,Ti, Y,Mn,Sc
Primorye depression	As,Mo	Mn,Zn,Co,Nb, Ag,Cr,B	Ni,Li,Sn, Y,Pb,Ge,V, Ga

TABLE 7 Geochemical Characteristics of the Main Types of Landscapes of North-East Yakutia

Type of landscape	Mineralization (g/ℓ)	Ion run-off (t/km² /year)	Class of migration	Geochemical specialization: Abundant elements	Deficient elements
Terrigenous MZ	0.028	4.2	H^+	Au,Be,W,Hg, Na,F,Cu,Ag,Mn, Ba,As,Sb,Sn	Si,Zn,Co,Al
Granitoid MZ	0.019	2.2	H^+	Bi,Sn,Ti,As, Hg,Ag,Ni,Be,Cr	Na,Ca,Si,A, Zn,Co
Carbonate PZ	0.150	28.0	Ca^{++}	Cd,Mn,Ti,Ag, Hg,Sn	Pb,Zn,Ni,Ti, V,Ga,B,Mo
Terrigenous KZ	0.023 (0.070)	7.0	H^+-Fe^{++}	Mo,Be,Ni,Si, Ti,Co,Ga,Sn	Ag,Pb,Cu

REFERENCES

Climate Atlas of the USSR, 1960, Leningrad, Gidrometeoizdat.

Janda, R. J., 1971, An evaluation of procedures used in computing chemical denudation rates. Bull. Geol. Soc. Amer., 82, No. 1, 67-79.

Mikhailova, R. P., 1964, Some relationships of microelement abundance in soils of the taiga-zone of the Urals in relation to geochemical exploration. In Soil-geographic and landscape-geochemical research. Moscow, Izd-vo MGU, p. 113-120.

Perelman, A. I., 1966, Landscape geochemistry. Moscow, Vysshaya shkola, 392 p.

Strakhov, N. M., 1965, Foundations of the theory of the lithogenesis. Moscow, Izd. AN SSSR, P. I, 212 p; P. II, 574 p.

Vinogradov, A. P., 1962, Average abundances of chemical elements in main types of eruptive rocks of the Earth's crust. Geokhimiya, No. 7, p. 641-664.

Walpeter, A. P., 1972, The character of present-day processes of weathering in the north-eastern part of the USSR. In Current problems of the gold geology in the north-eastern part of the USSR. Trudy SVKNII, Magadan, v. 44, p. 157-173.

INVESTIGATION OF THE DEFORMATION OF CLAYEY SOILS RESULTING FROM FROST HEAVING AND THAWING IN FOUNDATIONS DUE TO LOADING

M. A. Malyshev, V. V. Fursov, and M. V. Baluyura

Institute of Civil Engineering
Tomsk, USSR

The relationships between the deformation of clayey soils beneath loaded foundations and the number of freeze-thaw cycles, soil composition, pressure, depth of the foundation footing, and ground water levels with respect to the depth of freezing were demonstrated. When the pressure on the foundations increases, heaving begins at the maximum depth of freezing penetration beneath the foundation footing. Intensity of heaving decreases. An increase in the depth of the foundation within the freezing zone reduces the impact of heaving at the ground water level with respect to the depth of freezing. Loaded clayey foundation materials whose settlement had stabilized prior to freezing experienced additional settlement after further cyclical freeze-thaw in the materials. Additional compaction of the soils, as a result of repeated freeze-thaw in the loaded foundation materials, leads to reduction in the deformation.

The aim of the investigation was to study clayey soil deformation under the influence of seasonal freezing and thawing. To that end deformations in loaded foundation beds of settlement plates and laboratory tests of soil-bearing capacity and deformation under the influence of freezing and thawing have been investigated.

The construction site with foundations is located in Tomsk. The soils represented are alluvial loams and sandy loams with soft plastic and fluid consistency. These loams are dusty and carbonaceous, 70-95% of the light fraction consists of quartz and feldspar; in the colloid-dispersive fraction, kaolinite prevails. For the freezing period the ground water level was at 1.8-2.8 m. The soils were soft, water-saturated, and strongly heaving. The depth of seasonal freezing at the regularly freed of snow site during the investigation time in 1978-82 was 1.8-2.2 m.

The area of the settlement plates was 10,000 cm² (Figure 1), their depth was 1.0 and 1.5 m, and the pressure on the soil was 0, 0.1, 0.2, and 0.3 MPa. Four of the foundations, constructed in 1978, had a depth of 1 m and the pressure of (MPa) : F_1 = 0.05; F_2 = 0.1; F_3 = 0.2; F_4 = 0.3. In 1979 six more foundations were put up, two of which had the depth of 1 m and the pressure of: F_5 = 0; F_6 = 0.3; and four foundations at the depth of 1.5 m had the pressure of: F_7 = 0; F_8 = 0.1; F_9 = 0.2; F_{10} = 0.3.

The field test procedure is described below (Malyshev and Fursov, 1982).

Before setting a foundation, a hole 50 mm in diameter and 3.0 m deep was bored below the foundation footing. According to the selected type of soil, the degree of moisture and plasticity were determined. The surface of the foundation beds was leveled. Foundations were set on a 1-cm-thick layer of cement and sand mortar. The side surface of the settlement plate and the metal frame were protected against freezing together with the soil. To measure vertical displacement Maximov's deformation meters and clock-type indicators were used. In experiments the leveling by anchors and marks on foundations was made. The depth of soil freezing was determined by Ratomsky frostmeters, and the soil temperature by soil thermometers. Up to the given pressure, foundations were loaded in 0.2 MPa increments until settlement had stabilized. For the entire freeze-thaw season observations of temperature, depth of freezing, soil deformations,

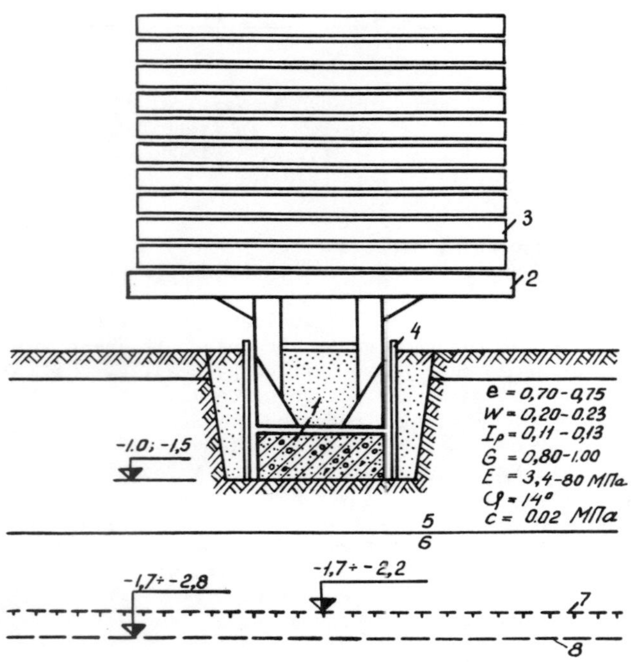

FIGURE 1 Schematic drawing of the settlement plate 1- model of the settlement plate 1.0 × 1.0 × 0.4 m; 2- frame for load; 3- load; 4- film, covered with solidol and a board box; 5- loam I_L = 0.4-0.7; 6- loam I_L = 0.75-1.0; 7- depth of freezing; 8- ground water level.

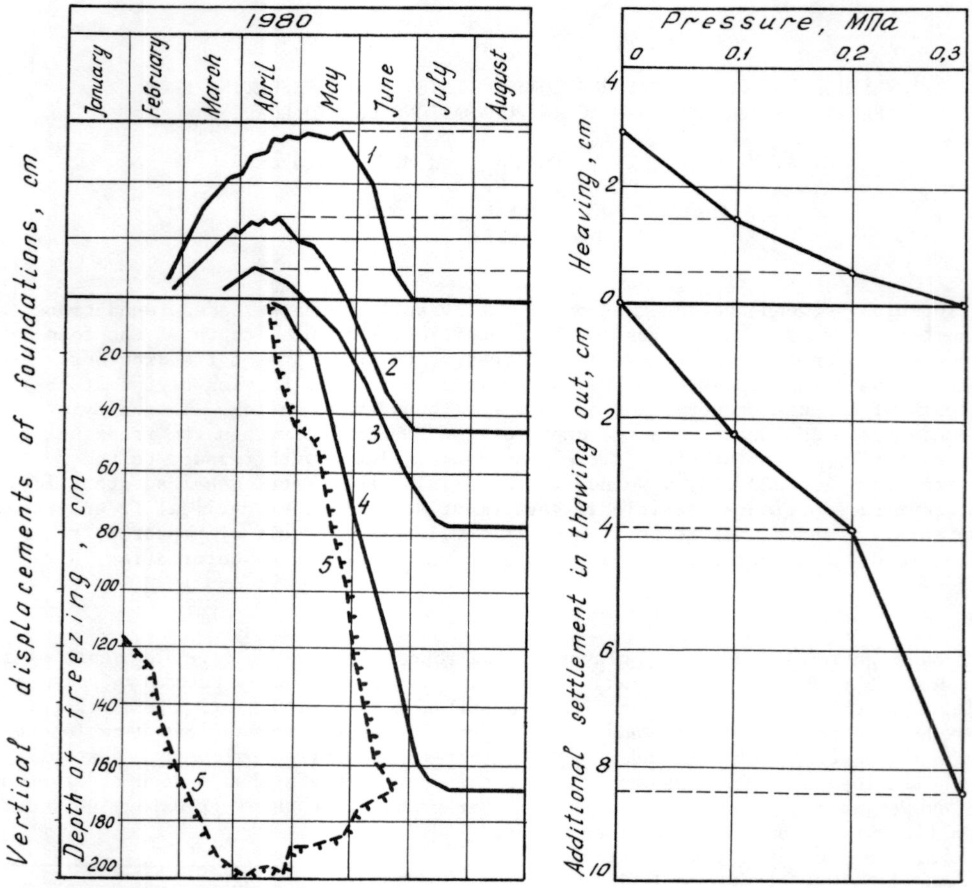

FIGURE 2 Vertical displacement of settlement plates at a depth of 1.5 m in freezing and thawing of foundation beds: 1,2,3,4 - foundations F_7, F_8, F_9, and F_{10}; 5 - depth of freezing.

and foundation displacement were made. At the beginning and at the end of winter the level of ground water was measured.

Observations in 1978-82 indicated that foundations set at the same depth but under different soil pressure were characterized by heaving, the beginning of which did not coincide with the soil freezing of foundation beds and was different in value. See also Shvets and Mel'nikov (1969), Kiselev (1971), and Karlov (1979). Foundations with higher pressure on the soil had a thicker layer of frozen soil below the foundation footing, so that foundations did not heave, and the total value of heaving was decreased. For instance, when the settlement plate pressure increased from 0 to 0.3 MPa, the maximum value of heaving of foundations set at a depth of 1.5 m decreased from 2.8 to 0 cm (Figure 2). Of foundations with the same soil pressure, but set at different depths (1.0 and 1.5 m), the greater capacity of the frozen soil below the footing, before heaving began, was inherent to foundations that were deeper. These foundations had a ground water level at the end of winter that was approximately 40-80 cm lower than the depth of freezing. At higher ground water levels, with the depth of 1.7-1.9 m from the surface and with freezing to 1.9 m in 1979-80, foundation F_6, set at a depth of 1.0 m with a footing pressure of 0.3 MPa, heaved 7.4 cm. The heaving began when the first below foundation footing reached 40 cm and sharply accelerated during further freezing up to the ground water level. Under the same conditions, foundation F_9, set at a depth of 1.5 m and under the lower pressure of 0.2 MPa, heaved only 0.5 cm. The foundation displacement curves are presented in Figure 3. The greater heaving of foundation F_6 as compared with foundation F_9 is accounted for by the increase of water migration through the freezing soil, while the pressure influence is sharply decreased because of the redistribution in the frozen soil under the foundation footing. Thus, the level of ground water being in the frost zone, the influence of pressure on the decrease of heaving is more effectively manifested in the high-temperature lower part of the frozen layer. According to observations on the displacement of foundations with different pressures it can be stated that, at the beginning of heaving, the normal pressure under the foundation footing resulting from freezing of 1 cm soil for foundations set

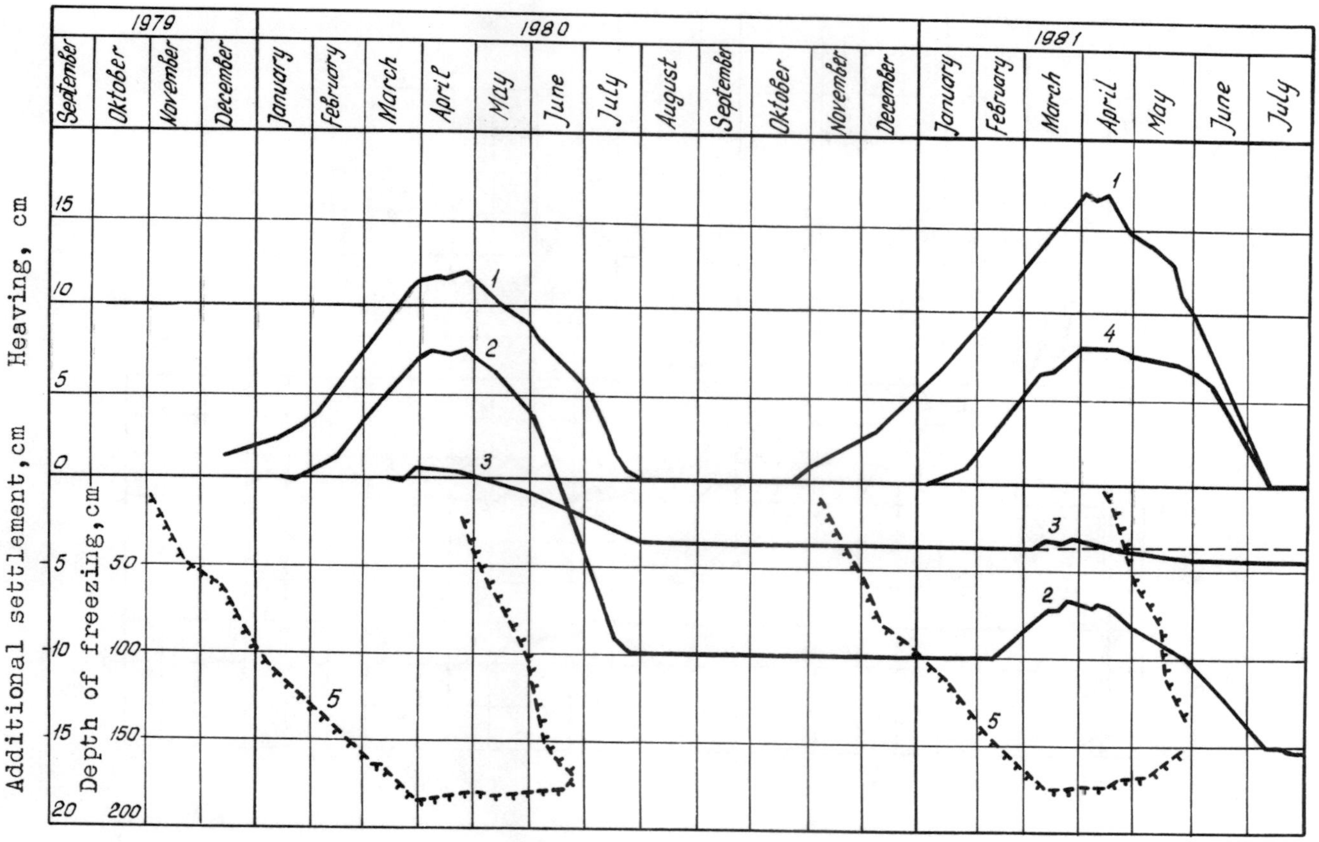

FIGURE 3 Vertical displacement of settlement plates and marks - heavemeters in freezing and thawing of foundation beds' soil.
1 - surface mark; 2 - foundation F_6; 3 - foundation F_9; 4 - mark at the depth 1.0 m; 5 - depth of freezing.

at the depth of 1 m, is approximately equal to 0.01 to 0.007 MPa/cm, and for foundations at 1.5 m it is not more than 0.005 MPa/cm. These values were recorded when the level of frozen soil under the footing was approximately 0.4-0.8 m. The influence of pressure on deformations of loaded foundation beds during the freeze-thaw process can be clearly seen in the displacement curve of foundation F_6 under a pressure of 0.3 MPa (Figure 3) set at the depth of 1.0 m and marks located at different depths close to foundation. Four stages can be established:

1. No foundation heave takes place when the soil is frozen to a depth of 1.4 m (40 cm below the foundation footing) and the temperature at the bottom of the footing is -2°C. In such a case the rise of the mark at the level of foundation footing is 3 cm, and at the level of soil surface is 10 cm.

2. Foundation heaving occurs during freezing from a depth of 1.4 m down to 1.75 m, with the temperature at the footing being -5°C. For this period the foundation heaving reached 3.2 cm, mark -7.8 and 15.8 cm.

3. Foundation heaving comes to an end, while foundation settlement and heaving of marks is still going on. The soil temperature at the depth of freezing levels off. The foundation settlement reaches 0.7 cm, the rise of marks reached the level of 8.1 and 16.8 cm. The depth of freezing does not increase. In thawing the layer between the surface and the foundation footing, the foundation settlement was 3.2 cm, that is it was equal to the value of frost heaving, the settlement of marks was 0.8 and 4.0 cm.

4. Thawing at the depth of freezing below the footing initiates foundation settlement. The foundation settlement stops a month later after the soil is completely thawed and the temperature rises to +3-5°C. The additional foundation settlement was 5.1 cm, the value of marks settlement was equal to the value of their heaving and had no additional settlements.

Examination of displacement stages of the loaded foundations made it possible to note some features in the nature of deformations observed in all the experiments. For instance, under increased pressure the first stage becomes longer, at the second stage the heave intensity becomes slower, and at the third and fourth stages settlements increase. All the loaded foundations in the last stage had additional settlement, which was greater for more heavily loaded and less deep foundations, but it was especially characteristic of foundations with 0.3 MPa pressure on the foundation bed (Figure 3). With decreased pressure and increased depth of

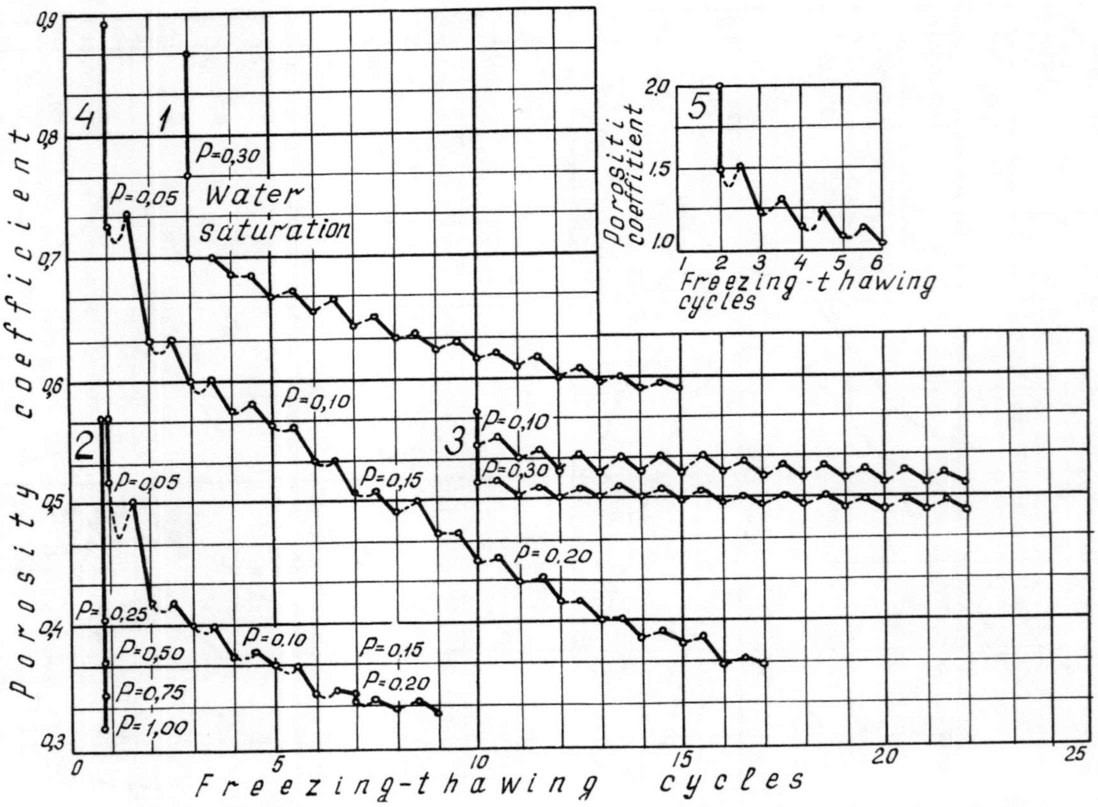

FIGURE 4 Deformation of the soil samples under repeated freezing-thawing conditions.
1,2,3,4,5 - soil samples
--------- heaving; __________ thawing settlement

Table 1

No.	Type of soil	Granulometric composition of particle, %					Void Ratio (e)	Plastic Limit (W_P)	Liquid Limit (W_L)	Liquidity Index (I_L)
		0.25-0.10	0.10-0.05	0.05-0.01	0.01-0.005	<0.005				
1	Loam	5.6	36.5	45.9	3.7	8.3	0.86	0.15	0.25	0.55
2	Loam	50.9	15.6	8.5	12.2	12.8	0.61	0.12	0.22	1.00
3	Loam	28.2	10.2	40.1	10.0	11.5	0.57	0.18	0.28	0.00
4	Clay	4.9	11.2	36.8	43.9	3.9	0.89	0.13	0.37	0.85
5	Clayey fracture	-	-	-	-	100.0	1.94	0.41	0.86	0.66

foundation, additional settlements decrease and, in the case of unloaded foundations, were not observed.

Soft soils under positive pressure during the first year of freezing and thawing give only settlement. The four years of observations show that for the first winter season, additional settlements are greatest and they subsequently decay as the freeze-thaw cycle is repeated.

Additional compression of clayey soils under the influence of freezing and thawing was caused by the complex influences acting inside the soil forces, causing dislocation of mineral particles and film water, formation of cryogenic textures, decrease of soil-bearing capacity and increase of deformation ability in thawing out.

To estimate the influence of repeated freezing and thawing on the deformation and on the bearing capacity of loaded foundation beds of clayey soil, laboratory tests have been carried out (Malyshev, 1969; Fursov, 1979).

In consolidometers made of organic glass, ne samples with different types of soil, density, and moisture were compressed under pressures of 0.1-0.3 MPa, until absolute stabilization was achieved. Then they were frozen and thawed under the same loading. In some of the samples after a series of freeze-thaw cycles the loading was reduced or totally removed. While examining water-saturated soils there was water in the lower porous disk of the apparatus. In examining the soils not saturated with water the upper disk was covered with paraffin to protect the soil from drying. Freezing was carried out for 48 hours at temperatures from -5°C to -20°C, and thawing took place at +18°C. The cycles were repeated from 6 to 30 times.

The tests, given in Figure 4 and Table 1 for typical types of soils yielded the following results. Clayey soils were compressed to a considerably greater extent by cyclic freezing and thawing under loading than without freezing. The most intensive compression developed during the first cycles. As the compression increased and the freeze-thaw cycles were repeated, soil deformation decreased, and the frost heaving and settlement during thawing levelled out. The ultimate density depended on the loading, the degree of moisture, and the type of soil. In the clayey, water-saturated soil with the prevalence of clayey and dusty particles, 0.2-0.3 MPa pressure density corresponded to the equation:

$$e = \frac{w_p G_s}{S}$$

where e = void ratio
w_p = plastic limit, %
G_s = specific gravity of solids
S = saturation, %

The ultimate loam density depended on the loading and optimal compressibility of the sandy and large dusty particles constituting them. Unloading after compression and the repetition of freeze-thaw cycles increased heaving and decreased density.

Analysis of the results of tests on different types of soils under repeated freezing and thawing conditions provides approximate deformation dependancies of loaded foundation beds on the physical characteristics of the soil.

Experiments on resistance to the shift of loams and sandy loams before freezing and immediately after thawing, run in the regime of the quick unconsolidated-undrainaged shift, showed the decrease of their soil-bearing capacity. Figures of soil-bearing capacity of dusty loams of native structure ($e = 0.7$-0.9; $I_p = 0.1$-0.14; $I_L = 0.1$-0.8) were reduced from 10 to 80%, of sandy loams to 10%. The increase of density and sandy particles content, the consistency and the degree of moisture reduction ends in the decrease of freezing and thawing influence on the soil-bearing capacity.

The results of the investigation allow the depth of foundation footing to be chosen based on the foundation bed calculations according to ultimate states.

REFERENCES

Fursov, V. V., 1979, Issledovanije vlijanija promorazhivanija i ottaivanija gruntov na ikh fiziko-mekhanicheskije svojistva.- Issledovanija po stroitel'nym konstruktsijam i fundamentam. Izdatel'stvo Tomskogo universiteta, Tomsk, p. 214-217.

Karlov, V. D., 1979, Frost heave of unsaturated loamy soil under field conditions. Engineering Geology, vol. 13, no. 1-4, p. 41-52.

Kiselev, M. F., 1971, Meroprijatija protiv deformatsiji zdaniji i sooruzheniji ot dejstvija moroznogo vypuchivanija fundamentov. Moscow, Strojiizdat, p. 47-54.

Malyshev, M. A., 1969, Deformatsija glinistykh gruntov pri promerzanii i ottaivanii.- Inzhenernaja geologija, osnovanija i fundamenty. Sbornik nauchnykh trudov Tomskogo inzhenernostroitel'nogo instituta, Tom XIY, Tomsk, p. 58-62.

Malyshev, M. A. and Fursov, V. V., 1982, Vlijanije sezonnogo promerzanija i ottaivanija glinistykh gruntov na rabotu osnovaniji i fundamentov.- Osnovanija, fundamenty i mekhanika gruntov, no. 3, p. 16-18.

Shvets, V. B. and Mel'nikov, B. N., 1969, Ustoichivost' melkozaglublennykh fundamentov v uslovijakh glubokogo sezonnogo promerzanija.- Osnovanija, fundamenty i mekhanika gruntov, no. 2, p. 13-18.

ARTIFICIAL ICE MASSES IN ARCTIC SEAS

P. I. Melnikov, K. F. Voitkovsky, R. M. Kamensky, and I. P. Konstantinov

Permafrost Institute, Siberian Branch
Academy of Sciences, Yakutsk, USSR

Natural climatic conditions in the arctic seas create preconditions for extensive use of ice and frozen ground as construction materials for creating artificial islands and structural foundations in areas of shelf. Experiments were carried out off the Kaya Sea coast into the formation of ice masses both by freezing layers of sea water and by sprinkling. The rate of ice formation by means of repeated flooding was up to 12 cm daily. The rate of ice formation by the sprinkling method was about 0.7 cm per hour at an air temperature of -15°C, and 1.1 cm per hour at -30°C. In the laboratory, the rate of freezing of layers of water at -20°C was from 0.2 to 0.55 cm per hour depending on the wind velocity. Ice formed by the sprinkling method was more porous and saline than ice formed by the freezing of layers of water or with natural sea ice, and its compressive strength was only half that of natural sea ice.

The Arctic Sea shelf near the coast of the Soviet Union occupies more than a million square kilometres and is rich in various mineral and natural resources (Slevich 1977), particularly oil and gas.

Construction of artificial islands and foundations for oil and gas prospecting and extraction is one of the most important problems related to the development of the shelf. The movement of thick ice in the arctic seas precludes the use of traditional elevated marine platforms and other artificial structural foundations.

Conditions on the Arctic shelf require specific designs and means of constructing artificial bases that take into account climatic and ice conditions. The Arctic shelf is known to have negative mean annual air temperatures and permafrost along the coast, as well as a thick (1-2 m) ice cover that exists for more than nine months of the year. Considerable movement of the ice cover occurs in most regions. The action of winds and tides, together with temperature fluctuations, cause ice hummocking. As a result of ice movement and hummocking, the ice cover exerts considerable pressure on artificial structures and islands.

Natural and climatic conditions of the arctic seas create preconditions for the wide use of ice and frozen ground as building materials for the construction of artificial islands in the shelf's shallow waters. Solution of the given problem requires the following measures: Control of ice production and ground freezing; the rational use of ice and frozen ground in engineering; and effective prevention of ice massifs and thawing of frozen ground.

The Soviet Union has had experience in constructing artificial ice massifs from fresh water and in using ice as a building material (Bubyr 1965, Krylov 1951, Schelokov 1982). Investigations have been carried out to study ice-production processes and to determine ways to amplify them (Gordeichik and Sosnovskiy 1981, Gordeichik et al. 1980, Sosnovskiy 1982). The construction of artificial ice massifs from sea water has not yet been demonstrated. In this connection it became necessary to study layered ice freezing from sea water.

Experiments on layered freezing of water were conducted early in 1981 during the construction of a temporary experimental ice island in the Kara Sea (Voltkovskiy and Kamenskiy 1981). At a distance of 300 m from the coast, the snow was removed from a natural ice site 50 m in diameter. The water was pumped on to the ice's surface by means of an engine-driven pump, the water layer being about 50 mm deep. After this layer froze and cooled down, water was poured successively and ice layers from 15 to 60 mm thick were produced. At the same time, observations were carried out aimed at studying the process of ice production, air and ice temperature regimes, and the structure and physical-mechanical properties of artificial ice.

It was found that monolithic ice formed under conditions in which the next layer was frozen only after cooling the previously produced ice layer down to -10°C. If this condition was not met, caverns with unfrozen brine remained in the body of the ice.

During the ice massif production abnormally high air temperatures and two long-term warmings occurred. Frequent strong winds and snowstorms presented difficulties for the production of ice in layers and sometimes the rate of ice-formation was slowed down due to snow accumulation on the surface. The highest rate of ice production was 12 cm daily at an air temperature of less than -20°C. The daily mean ice production rate with consideration of unfavourable weather conditions was 5 cm.

By the end of the tests on layered ice production from sea water, the total thickness of the ice massif was 4 m. It was submerged to sea bottom and rose 1.2-2.0 m above the surface of the water. There were no measures to protect the ice massif from solar radiation and sea waves. Systematic observations of the ice massif's temperature regime and melting were carried out prior to the onset of ice movement. After the surrounding ice cover had been destroyed and the ice had been carried out into the open sea, melting of the ice's surface and sides intensified. By the end of July, due to the decrease in the ice massif's thickness, it came to the water surface and broke up during a storm.

Studies of layered ice production from sea water under laboratory conditions were undertaken, in addition to in-situ ice production. To do this, a wind tunnel was designed in the underground refrigerating chamber of the Permafrost Institute of the Siberian Branch of the USSR Academy of Sciences. Inside the wind tunnel there was a special $30 \times 8 \times 4$ cm thermoinsulated pan, equipped with sensors for measuring water and ice temperature. The air temperature was constant during all the tests in the chamber, the temperature ranging from -12 to -30°C. The air speed in the tube above the pan was held at from 1 to 15 m/s.

The pan was filled with sea water at 0°C and placed in the working part of the wind tunnel, then the ventilation system was switched on. Readings of the temperature sensors indicated the time for water freezing and ice cooling down to the control temperature of -10°C.

The thickness of the layer of frozen water in the pan was recorded from 5 to 40 mm. To determine the relationship between the rate of ice production and the air temperature and wind velocity, a series of tests related to freezing a 25 mm layer of water were undertaken (Table 1). In addition, experiments have been carried out on freezing water layers of different thicknesses at constant temperature and wind speed. On the basis of these tests, it was found that the time t_δ required for a given water layer to freeze and for the ice to cool down to -10°C is in approximate linear dependence on this layer thickness δ (within 5 and 30 mm) and can be expressed by the empirical formula

$$t_\delta = \frac{\delta}{25} t_{25}$$

where δ is the thickness of the freezing water layer in mm, t_{25} is the time required to freeze a water layer 25 mm thick and to cool it down to -10°C.

The time it took to freeze a water layer 40 mm thick was $t_\delta = 1.7\ t_{25}$. This means that by freezing a water layer with a thickness greater than 30 mm, the linear dependence changes towards the increase of freezing time, when compared to the above-mentioned formula. When the layer of freezing water is thin, the rate of layered ice production demonstrates negligible dependence on the thickness of the freezing water layers if they do not exceed 30 mm. This can be explained by the fact that the intensity of the heat emanating from the surface of the freezing layer by means of turbulent heat exchange as the next water layer begins to freeze is presumably determined by temperature and wind velocity; only when the ice thickness exceeds 30 mm does the reduction of the intensity of the heat being given off become pronounced as well as usual decrease in water freezing rate when ice thickness increases.

From the aforementioned considerations, the actual rate of layer-wise production process can be estimated based on the test results of freezing 25 mm thick layers of sea water (Table 1). The dependence of the rate of production upon the air temperature at different wind speeds is shown in Figure 1. As indicated in the plot, the rate of layered ice production, taking into account cooling down to -10°C, increases substantially when the air temperature decreases to -20°C. Lowering the air temperature further does relatively little to increase the ice production rate. The effect produced by the wind velocity is more obvious. With an increase of wind velocity from 1 to 5 m/sec, the rate of ice production increased 1.3-1.4 times and twice as much at a wind velocity of 15 m/sec.

TABLE 1. Duration (hr) of production and cooling of sea water ice to a temperature of -10°C (thickness of water layer is 25 mm)

Air temperature (°C)	Wind speed (m/s)				
	1	2	5	10	15
-12	14.5	13.8	11.1	8.3	6.5
-14	12.7	11.8	9.8	7.4	5.7
-16	11.8	11.2	9.1	6.8	5.3
-18	11.2	10.3	8.3	6.2	4.9
-20	10.7	9.8	7.9	6.0	4.6
-25	9.8	9.0	7.3	5.5	4.2
-30	9.1	8.4	6.8	5.1	4.0

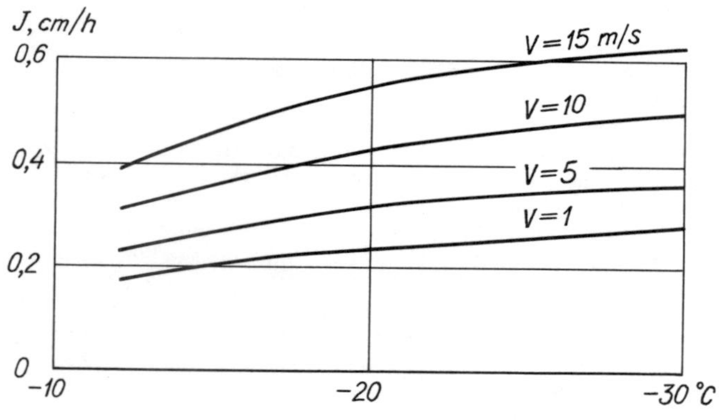

FIGURE 1. The water layer freezing rate (J) vs air temperature and wind speed (V).

The strength of the layered ice depends to a great extent on its salt content. To define this dependence more accurately a series of ice stress tests were conducted. Ice samples 73 mm in diameter and 130 mm high were produced by freezing layers of water having a given salinity. For this purpose sea water was diluted with fresh water. The tests were carried out at four temperature points. The rate of load application in all the trials was equal to 2 MPa/min. The results of these experiments are presented in Table 2.

A pronounced increase in the ultimate strength of the ice is observed at an ice temperature of -4.5°C, as its salinity decreases. This dependence becomes more complicated at lower temperatures. A slight increase in ice salinity results in increased strength as compared to ice produced from fresh water, whereas further increase in ice salinity leads to decreased strength. In all cases the strength of ice increased as the temperature was

TABLE 2 Maximum resistance of ice produced in layers to uniaxial compression (MPa)

Ice salinity (0/00)	Strength of ice at temperature(°C)			
	-4.5	-7.0	-14.6	-27.5
30	0.8	1.2	2.5	6.3
22	1.1	1.6	3.4	7.4
17.4	1.5	2.6	4.1	7.5
9.2	2.4	3.3	4.8	7.6
0	2.6	2.9	3.7	5.0

TABLE 4 Physical characteristics of ice

Kind of ice	Density (g/cm^3)	Porosity (%)	Salinity (0/00)
Natural sea ice	0.90	3.5	4.5
Manufactured ice, produced in layers	0.87	5.7	9.9
Manufactured ice, produced by showering	0.84	9.0	16.5

TABLE 3 Ice production of sea water by the shower method

Avg. air temp. (°C)	Pre-experimental temp. of ice surface (°C)	Wind velocity (m)	Shower duration (hr)	Thickness of produced ice layer (cm)	Intensity of ice production (cm/hr)	Time to cool ice layer to -10°C (hr)
-18	-16.5	12	2.9	6.0	2.1	5.4
-18	-12.4	11	2.4	6.0	2.5	-
-21	-19.4	0	2.0	5.0	2.5	-
-23	-21.4	1	2.25	3.0	1.3	-
-19	-18.2	3	4.25	2.0	0.5	0.5

lowered. However, the higher the salinity, the greater its relative strength.

The rate of layered ice production from sea water is relatively low, and this limits the thickness of the ice massif created in the course of one winter. It is known that the rate of ice production when the shower method is used to build ice masses is faster than layered ice method by a factor of four (Krylov 1951). By spraying sea water, during which its drops are partially frozen in the air prior to their contact with the surface of the ice mass, the process of ice formation can be accelerated to a great extent.

Experimental ice production from sea water using the shower method was carried out in December 1981 in the same area where the experiments on layered ice production had been conducted. Five series of experiments were carried out (Table 3) in which the water expenditure varied from 2.5 to 9.5 ℓ/sec at spurt rotation speed ranging from 0.35 to 2.4 revolutions per minute. When there was no wind, the radius of watering ranged from 23 to 35 m. The rate of shower intensity varied from 0.15 to 0.29 mm/min. The rate of ice production depended on the pre-sprinkling air temperature, the wind velocity and temperature of the ice surface. The rotation speed of the stream of water and shower intensity had considerable impact on the rate of ice production. The intensity of ice production increases initially as the amount of water being poured on the ice surface is increased but then the rate of ice formation decreases. The density of the ice manufactured by the shower method was less than that of the layered ice production (Table 4). This ice was also characterized by its high content of salts and spherical air blobs; it is a fine-grained ice having chaotic structure.

The temperature of the underlying ice increased during ice production by the shower method. This indicates that to create a monolithic ice mass it is necessary to limit the shower time and to interrupt the process periodically, so that the new ice layer can cool. The higher the intensity of sprinkling and the thicker the manufactured layer of ice, the longer the required cooling time. Based on experiments and theoretical calculations it was found that the optimum thickness of a single ice layer ranges from 5 to 40 mm. In field conditions at weak wind the actual rate of sea water ice production (J) by the shower method attains, depending on the air temperature (t_a), the following values:

at t_a = -15°C J = 0.7 cm/hr
t_a = -20°C J = 0.9 cm/hr
t_a = -25°C J = 1.0 cm/hr
t_a = -30°C J = 1.1 cm/hr

The amount of ice produced can be somewhat increased when there is a wind, although in this case it is difficult to ensure uniform ice production. It is possible to increase the rate by decreasing the size of the drops and by increasing the cooling and partial freezing time. This can be brought about by raising the position of the stream of water and by expanding the dispersion area. It can also lead, however, to a substantial increase in the porosity and salinity of the ice being produced.

To compare the mechanical properties of the ice obtained by both the shower and layered methods with those of natural ice, impression resistance tests using a ball punch were conducted on corresponding samples of ice. Changes in the resistance

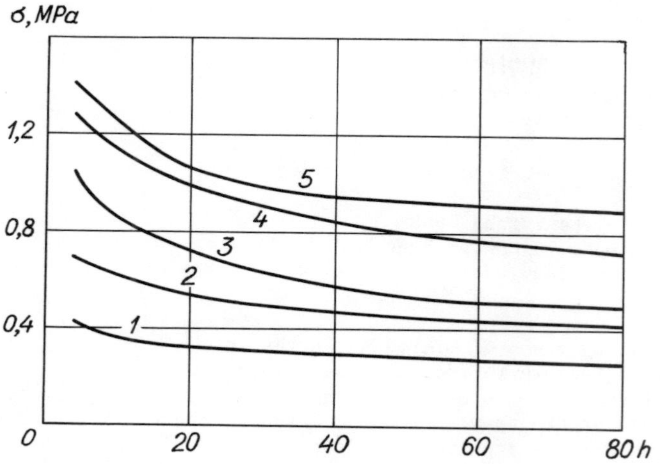

FIGURE 2. Strength of ice (σ, MPa) to ball stamp press in. 1 - ice, manufactured by sprinkling; ice temperature θ_i = 6.4°C; 2 - the same ice, θ_i = -11.8°C; 3 - ice, produced layer-by-layer, θ_i = -6.4°C; 4 - the same ice, θ_i = -11.8°C; 5 - natural sea ice, θ_i = -11.8°C.

of ice to the pressure exerted on it by a 20 mm diameter punch having a 4.65 kg applied load weight are presented in Figure 2.

Strength of the ice produced by the shower method turned out to be about half that of natural sea ice and 1.6-1.7 times less than that of layered ice. The relatively *low strength of the ice* obtained by the sprinkling method is due to higher porosity and salinity of the ice.

Because of the higher porosity and salinity of the manufactured ice, especially at temperatures higher than -7°C, its strength is less than that of natural sea ice. Partial removal of brines during ice production could be a means of increasing ice strength. As sea water freezes, ice crystals with low saline content begin to form. While the crystals grow and their temperature decreases, the inter-crystal pores become smaller, the salts are partially captured by the growing crystals, and concentration of brine in the pores increases. The pore brines are able to filtrate through ice. This property can be used to reduce the salinity of manufactured ice by draining the brine off into bore-wells or holes and then pumping it out of an ice mass. Drainage pipes from various systems can be used to increase the intensity of brine filtration.

The strength of the manufactured ice mass can be substantially increased by maintaining within it a lower temperature by means of artificial freezing as well.

When constructing artificial bases in the arctic seas it is advisable to use ice combined with ground and other materials. Designs that allow for reliable construction of ice and frozen ground masses and that ensure thermal and mechanical stability are also quite promising. The theoretical preconditions for the design of similar artificial bases already exist. A USA patent for the construction of artificial ice islands in the open sea is also known (US Patent, 1973). There are, however, numerous unclarified points and problems related to construction technology that require additional study. Studies aimed at enhancing the ice formation process and at increasing ice mass strength; improvement in the methods used to freeze the grounds saturated with sea water in the foundation of ice masses; the elaboration of methods to protect ice masses and frozen grounds from thawing, wave destruction and ice movement, can be referred to as urgent scientific and technological problems.

REFERENCES

Bubyr, A. A., 1965, Water-development projects of ice and how to improve their construction: Trudy Koordinatsionnykh Soveshchaniy po gidrotekhnike, v. 23, Energiya, Leningrad, p. 47-59.

Gordeichik, A. V., and Sosnovskiy, A. V., 1981, Experiments on water spray cone ice production: Materialy glaciologicheskikh issledovaniy, v. 43, Moscow, p. 131-135.

Gordeichik, A. V., Khodakov, V. G., and Charushnikov, Yu. A., 1980, Some effects of manufactured ice production by sprinkling: Materialy glaciologicheskikh issledovaniy, v. 38, Moscow, p. 254-260.

Krylov, M. M., 1951, Isothermal ice stores: AN SSSR, Moscow, 88 p.

Slevich, S. B., 1977, Shelf. Development and exploitation: Leningrad, Gidrometeoizdat, 236 p.

Sosnovskiy, A. V., 1982, Application of water spray cone method for intensification of layer-by-layer ice production: Materialy glaciologicheskikh issledovaniy, v. 44, Moscow, p. 79-86.

Schelokov, V. K., 1967, Ice stores: Moscow, Nauka, 119 p.

US Patent, 1973-3, August 7, 412. Method of forming and maintaining offshore ice structure. J. L. Fitch, and L. G. Jones.

Voitkovskiy, K. F., and Kamenskiy, R. M., 1981, The problems of engineering glaciology, related to development of the Arctic Shelf: Materialy glaciologicheskikh issledovaniy, v. 41, Moscow, p. 46-49.

APPLICATION OF MICROWAVE ENERGY FOR ACCELERATING EXCAVATION IN FROZEN SOIL

Yu. M. Misnyck[1], O. B. Shonin,[1] N. I. Ryabets[2], and V. S. Petrov[2]

[1]Leningrad Mining Institute, Leningrad, USSR
[2]North Mining Engineering Institute, Siberian Branch
Academy of Sciences, Yakutsk, USSR

The paper examines the heating of frozen ground by means of microwave energy in order to determine the most rational thawing regimes where frozen masses are being prepared for excavation by formation of local thawed zones or by thawing of a slope in layers. The derived equation of the advance of the thawing front created by the absorption of microwave energy by the rocks as well as experimental data revealed that the main critical parameters for thawing frozen rocks by microwave energy are the depth of penetration of the electromagnetic field, the area of formation of a thawed zone at the surface through which the electro- and thermo-physical properties of the rock exert an influence on the thawing, and the frequency and strength of the microwave energy flow. It was shown that when operating under optimal conditions, typical depths of microwave thawing, depending on the frequency and type of rock, are 0.1-0.6 m with a consumption of microwave energy of 20-30 kWh/m^3.

INTRODUCTION

Unconsolidated frozen soils, particularly those containing coarse grained materials, are very difficult to disintegrate because of their great strength and abrasiveness. Therefore in most cases, the excavation of frozen soil can be achieved only after their strength has been reduced. One method of preparing massive permafrost for excavation is by thawing. Traditional heating, however, is not sufficient for this purpose: it is time-consuming due to low frozen soil heat conductivity and has negative effects caused by the heaters on the thermal regime of the excavation site. This explains the interest in the use of microwaves as an efficient thawing method. Such energy treatment of materials is already widely used in numerous fields of science and technology for the enhancement of various processes.

The use of microwave energy in the new area of permafrost thawing is based on the following features of the electromagnetic field interaction with frozen soils (Misnyck 1982):

- inner massive heating due to the direct transformation of microwave energy into heat as a result of its absorption during polarization of the frozen soil in the electromagnetic field;
- the limited impact of heat conductivity on the formation of a temperature field, as compared to the action of internal heat sources caused by the absorption of electromagnetic energy;
- technologically acceptable, microwave dependant, thawing depths of 0.1-0.6 m attained in the course of several minutes of microwave action, depending on frequency and flux density;
- capability of highly efficient, non-contact transmission of microwave energy into the soil;

These features together with the capacity for structural and electromagnetic compatibility of microwave technology with existing mining equipment indicate cases where microwave energy would be effective. Examples are mining deep gravels, geological exploration, driving of prospect pits, opening of damaged zones of underground networks, etc.

In all of the cases in which microwave technology is to be used one must consider the optimum values of such characteristics as the depth of the thawed zone, the time of electromagnetic action and the energy efficiency. This paper addresses the results of experimental and theoretical study of these characteristics.

THEORETICAL DEFINITION OF THE PATTERNS OF PHASE FRONT MOVEMENT

Figure 1 shows a microwave device for thawing frozen soil consisting of a generator operating at one of the frequencies (2375, 915, 430 MHz) designated for commercial application, a feeder, and a microwave irradiator which distributes the heat required in a stope. The mass permafrost to be excavated is prepared by forming local thawed zones in the mass in the form of thawed strips or by layered thawing of the entire stope.

Without going into the details of any particular type of microwave irradiator, we shall examine the heating of a permafrost mass where a plane electromagnetic wave strikes the surface in a normal fashion and where the energy's flux density S depends only on coordinate Z, counted at mass interior from its surface. This approximation is assumed because the absorption of electromagnetic energy takes place in the near-irradiator region in which the energy has not yet dissipated.

Since we are interested primarily in phase front movements, and to a lesser degree in temperature distribution, we may limit the examination of Stephan's problem to the derivation of a rule for phase front movement based on the following considerations.

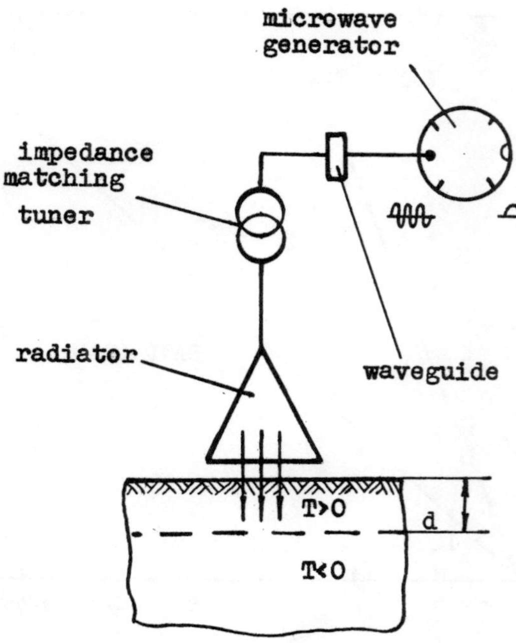

FIGURE 1 Diagram of microwave device for frozen soil thawing.

Electrical properties of frozen and thawed soils differ enough to represent temperature-dependent changes of complex permittivity $\varepsilon = \varepsilon' - i\varepsilon''$ by step function at T = 0, averaging the value of ε separately for temperature zones above and below zero. Typical values of the dielectric constant ε' and the dielectric loss factor ε'', based on 2375 MHz frequency measurements (Shonin and Sokolova 1981), are $\varepsilon_f' = (3\text{-}7)$, $\varepsilon_f'' = (0{,}1\text{-}3)$ and $\varepsilon_t' = (10\text{-}25)$, $\varepsilon_t'' = (2\text{-}7)$, depending on soil type. Subscripts f and t refer to frozen and thawed states of soil, respectively.

If a step approximation of $\varepsilon(T)$ is adopted, frozen soil occupying a half-space $Z \geqq 0$, may be considered in electromagnetic terms as a layered lossy dielectric medium, in which the boundary between electrically contrasted media coincides with a moving phase front.

As the period of oscillations is obviously much smaller than the time needed for thawed layer depth d to change markedly, the electromagnetic field distribution in the ground may be defined assuming the value of d to be constant. These assumptions allow solution of the intensity of heat sources per unit of volume q = -div S, where S is power flux density at observed point in frozen ground, using well known one-dimensional wave equation solution for stratified medium. Omitting intermediate transformations, we get

$$q(z) = 2S_o/h_f \cdot N(d) \exp(-2z/h_f), \quad d\ z \leqq \infty \qquad (1)$$

where S_o is incident energy flux density, N(d) is penetration coefficient, taking into account the energy reflected from ground surface and energy absorbed in thawed layer. This coefficient is a complicated function of thawed layer depth d, complex permittivities ε_f and ε_t, and wave length λ_o. Note, that $N(0) = 2\sqrt{\varepsilon_f'} / (i + \sqrt{\varepsilon_f'})$ and $N(d \to \infty) =$ 0 is the depth of electromagnetic field penetration determined from relationship $E(0)/E(h) = e$, where E is electrical field intensity

$$h = \lambda_o\sqrt{\varepsilon'}/\pi\varepsilon'' \qquad (2)$$

Experiments show that much less time is required for permafrost thawing by microwaves than by heat conductivity. Therefore we can ignore heat conductivity and derive the phase front movement equation assuming evident condition—energy absorbed at given point Z by the time t of phase front approach to this point to be equal $Q = c_v|T_i| + q_p$, where c_v, T_i, and q_p are respectively heat capacity per unit volume, initial temperature, and phase transition energy per unit volume. This brings about the following equation (Nekrasov and Rickenglass 1973).

$$\tau[\exp(2d/h_f) - 1] = \int_\tau^t N(d)\, dt \qquad (3)$$

where τ is the time of surface layer thawing

$$\tau = Qh_f/S_oN(0) \qquad (4)$$

After integro-differential transformations and a number of algebraic manipulations the desired function d(t) can be found as (Rickenglass and Shonin 1980).

$$d = 0.5h_t \ln[1 + \beta(t/\tau - 1)] \qquad (5)$$

where $\beta = h_f/h_t$. This expression helps to determine the specific microwave energy consumption as a transmitted energy flux density per depth of thawed zone

$$W(t) = St/0.5h_t \ln[1 + \beta(t/\tau - 1)] \qquad (6)$$

Analysis of equation (6) allows examination of microwave thawing with minimal energy consumption. If S is given, the parameters of such a regime may be derived from condition $\partial w/\partial t = 0$, which results in the following equation

$$\ln[1 + \beta(t_{opt}/\tau - 1)] = \frac{\beta t_{opt}/\tau}{1 + \beta(t_{opt}/\tau - 1)} \qquad (7)$$

Next, the value of d_{opt} is easily defined from (5), where $t = t_{opt}$.

The expressions derived above show that introducing such characteristics of thawing as h and τ is better considered as space and time scales of the process studied, respectively. This dramatically simplifies digesting phase front movement histories, and facilitates understanding the way in which microwave parameters S_o and λ_o (just as permafrost electrical and thermal properties), depending in turn on soil types, moisture content, soil density, soil grain size etc., influence permafrost thawing.

The developed theoretical model of permafrost heating promoted further experimental investigation of frozen soil thawing under microwave action.

EXPERIMENTAL DETERMINATION OF MICROWAVE PARAMETERS FOR THAWING FROZEN SOIL

The initial experiments aimed at study of thawed zone formation depths employed blocks of artificially frozen sand-clay, placed in a refrigerator chamber at temperatures of -10°C and -30°C. There were two soil types under study: sand-loam (moisture content ω = 10%, soil density γ = 1.55 g/cm³, initial temperature T_i = -30°C) and sand (ω = 5%, γ = 1.4 g/cm³, T_i = -10°C and -30°C).

The radiator utilized had flanges, open ends, a rectangular waveguide with a cross sectional area of 9×4.5 cm^2 and equipped with a dielectric quarter wave length plug. The waveguide output was controlled by means of a dissipative adjustable attenuator, while the power radiated was indicated by thermistors, installed in the ports of a bidirectional coupler. Power flow density radiated was 110, 40 and 16 W/cm² at 2375 MHz frequency. Frozen soil block dimensions were chosen (0.3 × 0.3 × 0.4 m³) to simulate a semi-infinite dielectric half space so far as propagation of electromagnetic waves was concerned.

The time of surface layer thawing was determined as a time when surface temperature of the block attained zero. Thawed zone depth was measured along the waveguide axis where energy flow density was maximum and equal to $2P/\sigma$, where P is power transmitted into the frozen soil block and σ is the cross sectional area of waveguide.

When experimental verification of the theory was sought, analysis of the τ vs Q/S experimental data linear function approximation, as seen from Figure 2, proved to be valid. It should be noted that proportionality constants of both lines appeared to be equal to h_f, as was predicted by expression (4).

The relation between the depth d of thawed zone, obtained for different values of power flux density S, and normalized time of microwave action t/τ is shown in Figure 3. Inspection of the plotted data indicates that, if the normalized variable t/τ is used, the influence of S on d in this case vanishes. This makes the experimental data approximately the same as the theoretical curve (5). Comparison of theory and experiment revealed some scatter in the data of d vs t/τ when h_f = 19.5 cm and h_t = 6.4 cm. Thus, the results justify the theoretical model of microwave heating of frozen soils developed above. In spite of a number of assumptions, this model represents microwave thawing histories correctly and may be recommended for approximate calculations of the process parameters.

The investigations of microwave thawing of permafrost employed experimental microwave devices at 2375 MHz and 915 MHz frequencies, providing an energy flux density at the i radiator center of $4 \cdot 10$ and $23 \cdot 10^4$ W/m² respectively. In the former case, a pyramidal horn with cross sectional area of aperture σ = 0.75 × 0.38 m^2 was used, while in the latter E-sectorial horn with σ = 0.5 × 0.22 m^2 was chosen.

There were two soil types covered. The first one heated at 2375 MHz was coarse-grained soil-pebbles cemented by clay 50-60% filling (T_i = -6°C, ω = 9-10%, γ = 1.9 tn/m^3, C_v = 0.7-0.8 MJ/m³ · °C) and the second one heated at 915 MHz was represented by the sand of active layer in

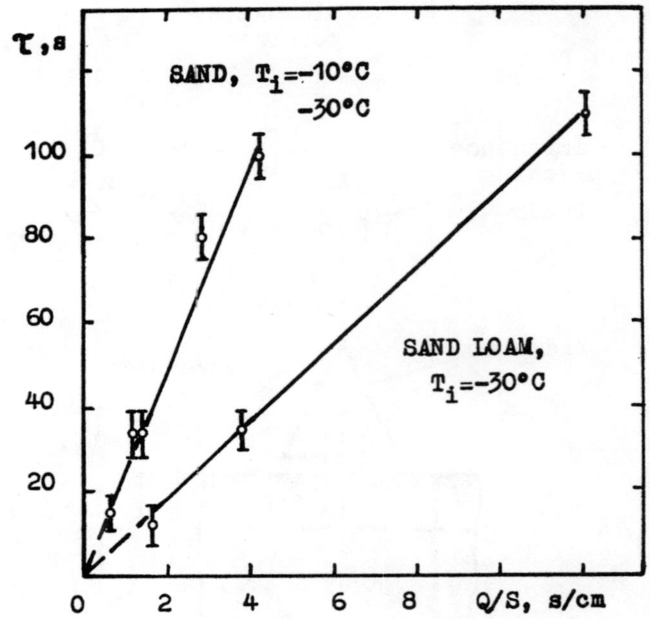

FIGURE 2 Influence of microwave energy flux density S and energy of phase transition Q on the time τ of surface layer thawing.

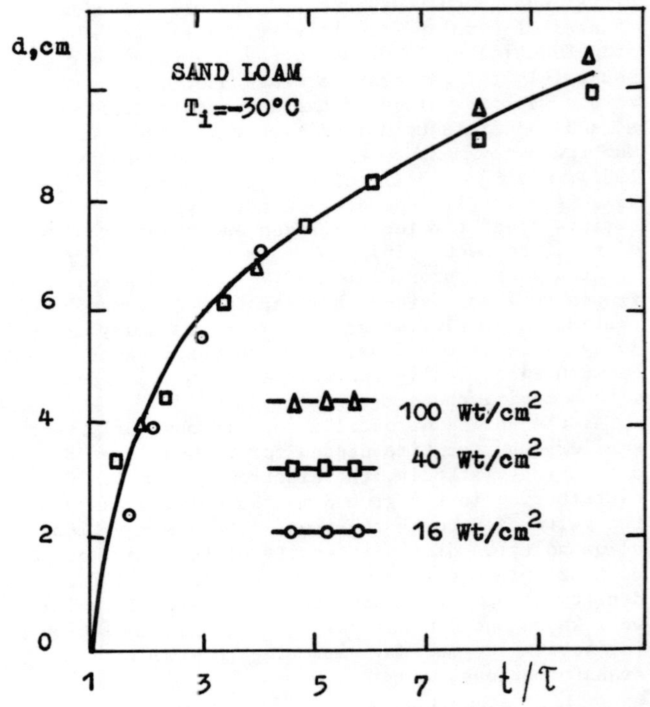

FIGURE 3 Dependence of thawed zone depth d on normalized time t/τ of microwave action.

Yakutsk region (T_i = -7°C, ω = 12-14%, γ = 1.6 tn/m³, C_v = 1.2 MJ/m³ · °C).

Figure 4 illustrating the dynamics of permafrost thawing shows that both curves change most rapidly from τ (40s - curve 1 and 50s - curve 2) until t = (4-5)τ. During microwave action the depth of the thawed zone reaches 0.28-0.32 m at 915 MHz and

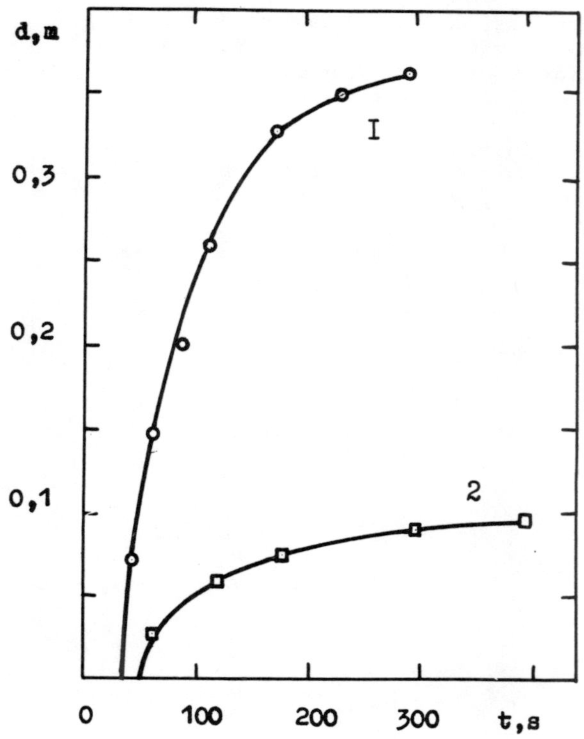

FIGURE 4 Dynamics of permafrost thawing by microwaves at 915 MHz (curve 1, sand) and 2375 MHz (curve 2, coarse grained soil).

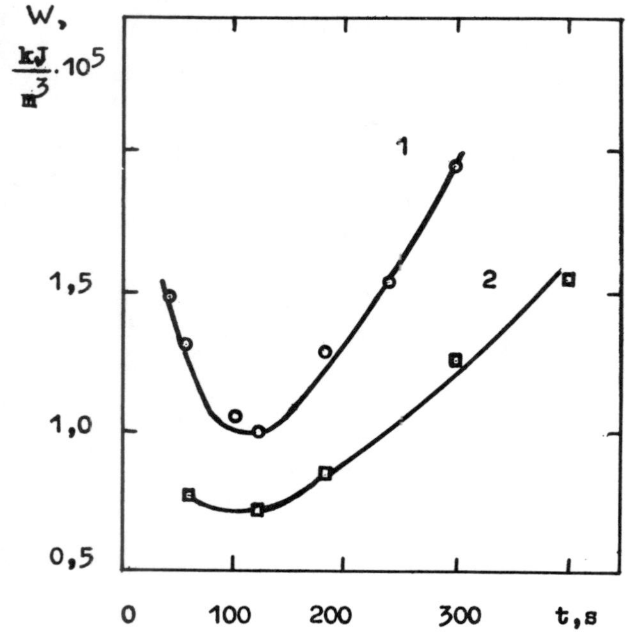

FIGURE 5 Variations in microwave power consumption W with time of electromagnetic action: 1 - sand, heated at 915 MHz, 2 - coarse grained soil, heated at 2375 MHz.

0.07-0.08 m at 2375 MHz. Further microwave treatment of the massive permafrost slows down the rate of phase front movement. This is caused by attenuation of the electromagnetic wave propagating through the thawed layer.

Figure 5 indicates that during thawing, microwave energy consumption per unit of volume W changes so that the curves W(t) have minimums. This fact is of practical interest because it promotes thawing with maximum productivity. The experimental data indicate that such a regime for heating frozen sand at 915 MHz is realized when d_{opt} = 0.27-0.28 m, St_{opt} = (2.5-2.8) $\cdot 10^4$ kJ/m² and W = 10^5 kJ/m³ or 30 Wh/m³.

If inner heat sources are distributed uniformly in the ground, the energy consumption for its thawing is 0.75 $\cdot 10^5$ kJ/m³. Comparing this value to that obtained above, we see that the coefficient of microwave power utilization is rather high and equal to 75%.

Similar parameters of coarse grained soil thawing at 2375 MHz may be defined as d_{opt} = 0.06-0.07 m, St_{opt} = (0.5-0.55) $\cdot 10^4$ kJ/m² and W = 0.7 $\cdot 10^5$ kJ/m³ or 20 kWh/m³.

The energy consumption listed above was determined taking into account radiated microwave power only. If the generation losses of 30-40% are considered the actual value of this parameter increases 1.4-1.7 times.

The curves of Figure 5 indicate that microwave energy consumption decreases when moisture content or, what amounts to the same thing, phase transition energy diminishes. That is why it is especially promising to use microwave thawing of permafrost in coarse grained frozen soils.

According to equations (2) and (5) as well as experimental data (Figure 4), the depth of thawing increases when frequency decreases. Therefore, the estimation of parameters of thawing at 430 MHz is especially interesting. New values of d_3 may be defined based upon experimental data d_1 and d_2 obtained at 2375 and 915 MHz respectively, if the electrical properties of soils are assumed to be frequency independent, and it is required that t/τ be constant at every frequency covered. Using expressions (2), (4), and (5) this produces

$$d_m = d_n\lambda_m/\lambda_n, \qquad t_m = t_n\lambda_m S_n/\lambda_n S_m \tag{8}$$

where subscripts m, n = 1, 2, 3 refer to wavelengths 12.6, 33, and 70 cm, which correspond to frequencies 2375, 915 and 430 MHz respectively. Formula (8) gives values of optimum depths of thawing at 430 MHz as follows: 0.55-0.6 m for frozen sand and 0.35-0.40 m for coarse-grained soil. The depth limits of thawing restricted by slowing down of phase front movement may be estimated as 0.6-0.7 and 0.40-0.45 m respectively. Permafrost thawing to greater depths should be performed layer-by-layer.

Note that, to secure a permanent rate of thawing independent of frequencies used, the characteristics of microwave power action should satisfy the following condition $\lambda_m S_n/\lambda_n S_m$ = const. Energy consumption in this case, as it is seen from equation (6), does not change.

In conclusion, these results may serve as a basis for the rational use of microwave energy in various technical applications related to permafrost.

REFERENCES

Misnyck, Yu. M., 1982, Philosophy of frozen soil slackening by the action of microwave energy: Leningrad University Press.

Nekrasov, L. B., Rockenglass, L. E., 1972, Microwave energy reflection from semi-infinite lossy dielectric medium in presence phase transition in it: Journal of Technical Physics, v. 7, p. 1339-1342.

Rickenglass, L. E., and Shonin, O. B., 1980, Approximate calculation of the parameters of frozen soil thawing by microwaves: Technical Institute Transactions on Mining, v. 4, p. 43-46.

Shonin, O. B., and Sokolova, N. V., 1981, Super high frequency dielectric properties of frozen soils: Physical Processes in Mining, v. 9, p. 29-35.

CRYOGEOTHERMAL PROBLEMS IN THE STUDY OF THE ARCTIC OCEAN

Ya. V. Neizvestnov, V.A. Soloviev, and G.D. Ginsburg

All-Union Institute of Mineral Resources of the World Ocean
Leningrad, USSR

One of the most notable geological-geophysical features of the Arctic Ocean is the presence in any sea-bed section of a cryolithozone, involving frozen sediments and ground ice. The stages of its evolution in shelf areas correspond to cyclic regressions and transgressions in the Polar Arctic Basin. This involves an oscillation from unfrozen submarine sediments through a newly formed submarine frozen zone to a frozen terrestrial zone, and then back through a relic submarine frozen zone to an unfrozen submarine zone. Sedimentation on the sea bed on the shelf is closely related to the evolution of the cryolithozone. Widespread occurrence of ice-rich lacustrine-alluvial deposits on the Arctic shelf in the Late Pleistocene is inferred on the basis of peculiarities of sedimentogenesis in the course of thermal abrasion of the sediments during the transgressive phase of development of the Arctic Basin. Submarine lithogenesis is accompanied by unique changes in the fluid phase, viz., salinization of pore water and the formation of gas hydrates. It is possible to outline possible gas hydrate-bearing areas on the basis of cryogeothermal reconstructions. Submarine permafrost also dictates a number of peculiarities in the geophysical fields, especially the mobility of the thermal fields; this is extremely critical in the case of formation and degradation of the frozen and/or gas hydrate-bearing strata.

The cryolithozone is the most distinctive and characteristic geologic-geophysical feature of the Arctic Ocean area. This paper deals mainly with cryogeothermal aspects related to paleo-geography, polar lithogenesis, and changes in the fluid phase of rocks and bottom sediments, in particular, gas hydrate formation, peculiarity of geophysical field is discussed as well.

A cryolithozone, submarine inclusive, may be determined as a part of the lithosphere within the negative temperature belt. Physically, it may be frozen (ice-bearing, in particular ice-bonded) and unfrozen (saturated by negative temperature saline waters-cryopegs). To distinguish types of submarine cryogeothermal environments it is reasonable to take into account a thermic field pattern, which can be stationary and nonstationary. The nonstationary state is typical of the submarine frozen cryolithozone (frozen zone, for short) both for the aggradation during the formation and the degradation of relic frozen ground.

Conditions favorable for deep freezing and frozen zone formation on land in the northern polar area seem to have existed during the entire Late Cenozoic since the Eopleistocene. Within the present shelf area, freezing could take place only during subaerial stages, though it may start even under subaqueous conditions when the water depth does not exceed the ice thickness. A transition of the frozen zone, formed under subaerial conditions, into the subaqueous position is accompanied by abrasion and thermal abrasion of the frozen ground. Thus the evolution of the shelf cryolithozone is subject both to subaerial and subaqueous conditions and by the change of environments.

Two final stages—Late Pleistocene (pre-Holocene) regressions and the last Late Pleistocene-Holocene transgression—affected most of the present submarine frozen zone. The first stage favored deep freezing of much of the shelf, at the present approximately 110 m isobath. In the eastern Eurasian shelf, over a vast area adjacent to the New Siberian Islands, off the southern Laptev Sea, and the southwestern East Siberian Sea, there was a concurrent freezing sedimentation and the formation of thick syngenetic polygonal wedge ice. The present subaqueous frozen rocks outside the littoral zone are mainly relics of that time.

During the last transgression, which began 18-19,000 yr B.P., the frozen zone was not completely degraded and it fell within the subaqueous environment. This stage was marked by thermal abrasion that resulted in the reworking of older rocks that were different in genesis and age. The heavily iced syngenetically frozen deposits from the Upper Pleistocene were then thermally abraded for the entire thickness. Thermal abrasion was strongest on the eastern Eurasian Arctic (Are, 1980) and Beaufort Sea shelves.

The above two stages of cryolithozone evolution are based on numerous datings of the sea level in various tectonically stable regions of the world ocean over the last 30,000 yr (Curray, 1961, and others) including the Laptev Sea Shelf (Holmes and Creager, 1974). The average curve plotted using these data (Figure 1) is free of a tectonic constituent and hence reflects actual eustatic changes (the nature of which can be hydrocratic, particularly glacioeustatic and geocratic). The eustatic changes were a global factor that determined the cyclic development of the shelf cryolithozone. In certain cases the change of subaqueous and subaerial regimes is independent of eustatic changes, though against their background. Local coastline displacements now taking place in some areas off the Arctic Ocean are mainly locally controlled and

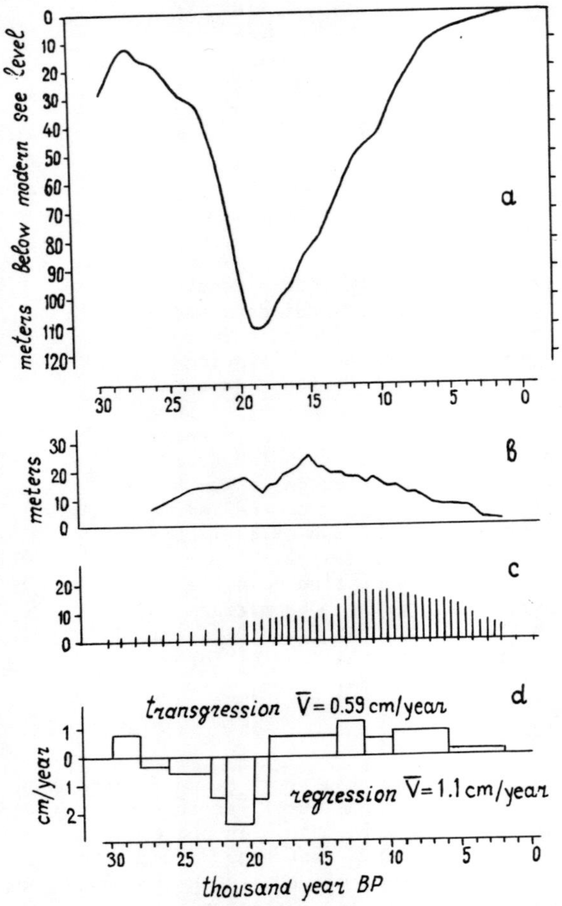

FIGURE 1 Averaged curve of eustatic sea level fluctuations during the past 30,000 years. a) changes in level; b) deviation from the mean; c) number of individual curves used; d) rate of eustatic change in sea level.

are due to recent tectonic and, in particular, glacioisostatic movements of the Earth's crust. In such cases regime changes are oscillatory in nature, like the eustatic sea-level changes.

Thus, cyclic regressions and transgressions in the northern Polar Basin during the Late Pleistocene-Holocene were due to eustatic changes in the shelf cryolithozone of the World Ocean. During the regressive stage (28,000 to 18,500 B.P. for the Arctic shelf as a whole), the cryolithozone successively evolved from a submarine unfrozen zone through a submarine newly frozen zone to a continental frozen zone. The transgressive stage (18,500 B.P. to the present) shows the evolution of the cryolithozone from a continental frozen zone through a relic submarine frozen zone to a submarine unfrozen cryolithozone. A similar sequence may be established for any particular region where the coastline migration and related environmental changes were due to recent tectonic movements. The relationship between a stage of cryolithozone development and transgressive-regressive cycles may be used to describe both direct (extent of permafrost zone assessed using paleogeographic data) and inverse geocryology problems.

The authors have solved one such problem. An area of probable lacustro-alluvial Late Pleistocene deposits with high ice content and thick syngenetic polygonal wedge ice was mapped on the recent shelf (Figure 2). On land these deposits are known as "edoma"; they are common on the maritime lowlands of Yakutia and the New Siberian Islands. Such deposits with high ice content seem to be intensely thermally abraded during marine transgression. If a normal marine basin is defined by a stastistical relationship between its depth, the distance from the coast, and the particle size distribution of the bottom sediments (Richter, 1965), then in this case it does not exist. The reconstructions are based on these assumptions (Soloviev, 1978).

The coefficient of relative clay content γ

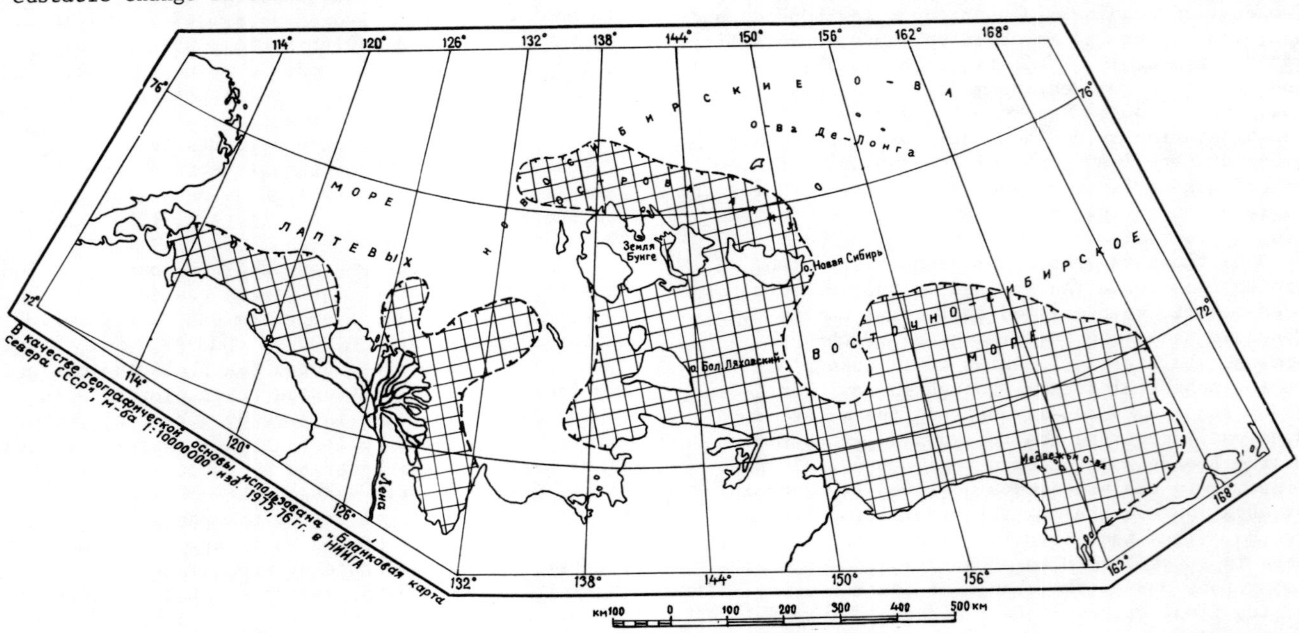

FIGURE 2 Position of Late-Pleistocene coastline on the Laptev and East Siberian shelves. Hatched area represents land area within which a complex of ice-rich deposits developed.

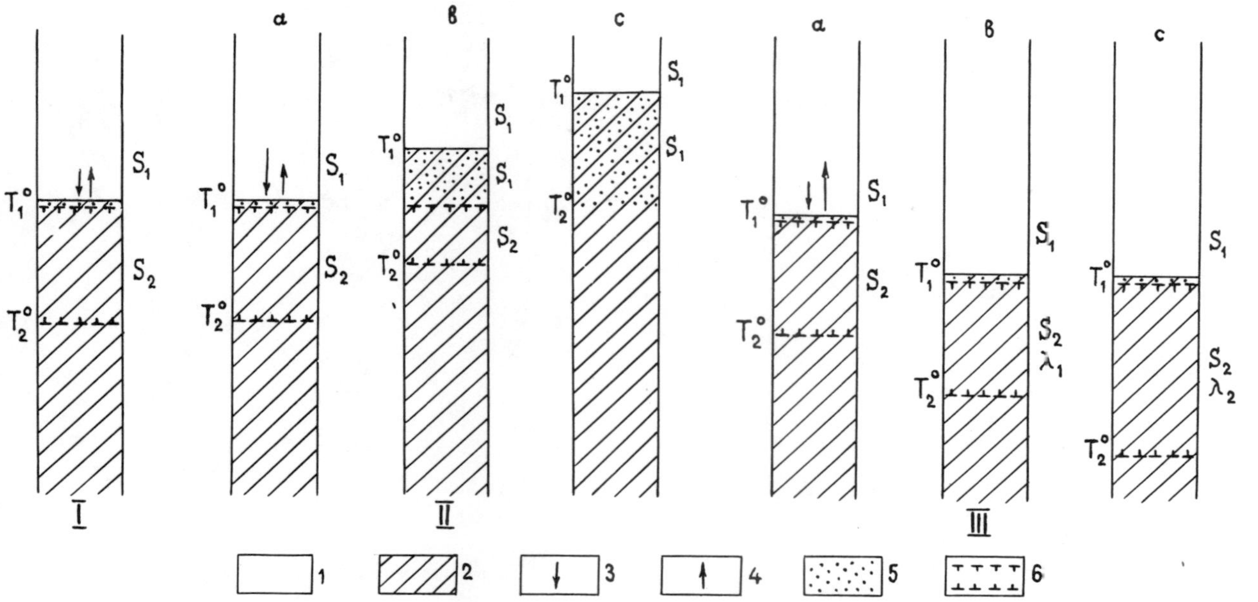

FIGURE 3 Some variants (I-III as explained in the text) in the development of relict submarine permafrost depending upon hydrodynamic conditions. 1 - sea water; 2 - unconsolidated deposits; 3 - vector of sediment accumulation; 4 - vector of erosion; 5 - unfrozen sediments; 6 - boundary of the frozen zone. S_1 and S_2 - salinities of sea water and pore water respectively ($S_1 > S_2$); T_1^0 and T_2^0 - temperatures of phase change in sea water and pore water respectively; λ - thermal conductivity of deposits ($\lambda < \lambda_2$).

(Richter, 1965) was used as an index of the particle size distribution of the bottom sediments. On the Laptev and East Siberian Sea shelves more than 1200 determinations were made for γ and H (water depth), resulting in two sets of data that are reliably geographically linked. One set of data is characterized by a statistically significant (at 0.05 significance level) γ-to-H relation and is restricted to shelf areas where Late Pleistocene-Holocene transgression has not been followed by some specific processes. There is no γ-to-H relation at the same significance level for the other set of data. In our opinion, this is due to the fact that, at the final stage of transgression, intense thermal abrasion strongly affects "edoma"-type deposits with high ice content (volumetric ice content of 80-90%).

A sequence about 40 m thick was reworked by thermal abrasion to form a section of Holocene sediments several meters thick; they are nonuniformly distributed over the leveled sea floor. Mean values for the coefficient of relative clay content of these sediments and of Upper Pleistocene ice-containing deposits from the New Siberian Islands at the 0.01 significance level do not differ, supporting their common genesis.

The data for the southeastern Beaufort Sea obtained from 244 stations (Pelletier, 1975) reveals two similar sets of data. The absence of the γ-to-H relation shows, in particular, in a large area north of Tuktoyaktuk Peninsula, stretching west-east approximately for 250 km and 100 km south-north. It is considered a Pleistocene land that was transformed into subaqueous state during the Holocene transgression. This conclusion is supported by the presence of numerous submarine pingo-like-features (O'Connor, 1980) whose extent is confined to the region discussed.

A Late Pleistocene paleoland recognized on the Laptev, East Siberian, and Beaufort Sea shelves may be important to forecast the extent of a submarine relic frozen zone. It is there that frozen rock may occur; this is confirmed by drilling data for offshore areas of the New Siberian Islands and the Beaufort Sea shelf. Recently vanished due to thermal abrasion, Semenov, Vasiliev, Diomid, and other islands (Gakkel, 1958; Grigorov, 1946) also occupied the above areas of the East Arctic shelf of the USSR. Though it is quite improbable, some thin relics of frozen rocks may occur within the area delimited by the 110-m isobath.

The data available for the Beaufort (Sellman and Chamberlain, 1979), Laptev, and East Siberian Sea shelves suggest that there is no regular relationship between the depth to the top of subaqueous frozen rocks and the distance from the coast. The development of a relic submarine frozen zone and in particular the depth of occurrence seem to be controlled mainly by the hydrodynamic environment. Figure 3 shows some evolutionary patterns of a submarine frozen zone in shelf areas with negative bottom temperature.

When the thickness of a relic frozen zone is determined by the mean annual bottom temperature and the sub-bottom geothermal gradient while bottom sediment erosion is compensated for by sedimentation, we have a situation that corresponds to a quasi-stationary condition. In this case thawing from below ceases, but the upper part of a frozen sequence may deteriorate (thaw) for some depth owing to marine water salt diffusion. In such a state, neither temperature nor hydrodynamic conditions change (Figure 3-I). The top of frozen rocks may lie just near the bottom, as in the Dmitriy Laptev Strait and the Ebelyakh Bay (southeastern Laptev Sea) areas.

On the shelf, where sedimentation is intense (Figure 3-II) a stratum saturated by saline water is formed above a frozen zone. The negative temperature area shifts up the section with deposition and the frozen rocks thaw from below (Figure 3-IIb). This may well be the case for Gedenstrom Bay (New Siberian Islands) where at a water depth of 4 m the top of the frozen rock is exposed at a depth of 9 to 24 m. The process discussed can result in a completed degradation of the frozen zone and the formation of an unfrozen cryolithozone (Figure 3-IIc), such as in the Sannikov Strait (New Siberian Islands).

Continuous erosion of the bottom underlain by frozen rocks (Figure 3-III) can also occur. While its thickness remains constant, the top of the frozen zone moves down the section (Figure 3-IIIb). If the thermal conductivity of the underlying deposits is higher than that of the original (overlying) frozen rocks, then the thickness of the submarine frozen zone may increase (Figure 3-IIIc). This may occur in areas of present submarine uplifts at a water depth of 100 m, which were situated on land at the end of the Late Pleistocene due to the eustatic drop in the sea level. In areas of recent subsidence an intense sedimentation and, under certain conditions, complete degradation of a submarine relic frozen zone may occur.

The submarine cryolithogenesis is followed by a peculiar change of the fluid phase, both pore waters and gases.

The evolution of pore water similar in composition to marine water starts on the seafloor by shallowing to a depth of 2 to 5 m when fast ice is on the bottom most of the year and at temperatures on the freezing front below -1 or -2°C. Under lagoonal conditions, freezing may begin with the presence of a water layer under the ice. In this case, freezing increases marine water salinity to 60-80 g/kg and more, and the temperature of freezing brine drops to -4 or -5°C.

The freezing of deposits saturated by salt water is accompanied by the crystallization of ice and calcite. The average content of calcite in ice formed by freezing of normal marine water amounts to 0.09 g/kg (Gitterman, 1937). At a freezing temperature lower than -7.8 to -8.2°C, mirabilite begins to crystallize. The content of mirabilite in ice under these conditions may reach 6.3 g/kg. At a freezing temperature below -23°C, hydrohalite crystals are liberated from the liquid phase; their content in ice-cement may exceed 20-30 g/kg.

The water that saturates rocks is separated into solid ice, ice left in the frozen zone, and freezing brines. The total salt content of the brines at equilibrium with frozen rocks gradually changes from 35-40 g/kg at -2°C to 115 g/kg at -7.6°C, and reaches 230 g/kg at -22.5°C (Gitterman, 1937).

A composition of relic freezing brines and minerals may be used to infer freezing paleotemperatures. For instance, at the transgressive stage of the Polar Basin development, when a frozen zone degrades, the ice thaws, and the salts that were

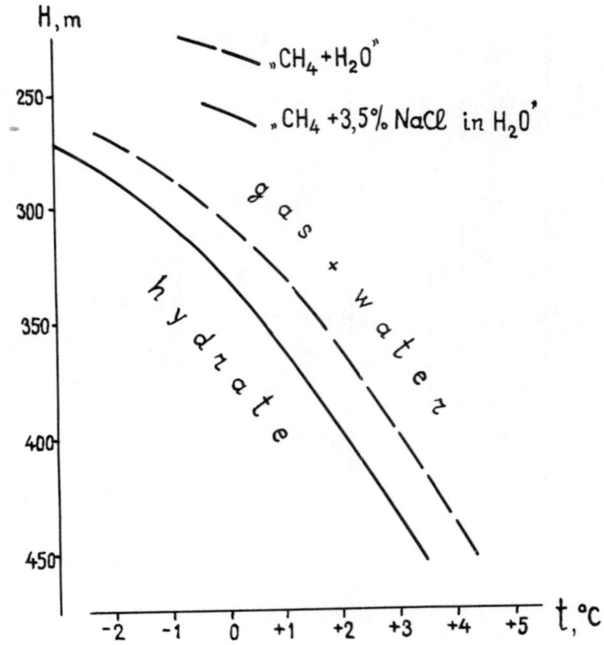

FIGURE 4 Conditions for the existence of methane hydrates at the bottom of the arctic seas. N = sea depth; t = sea bed temperature.

crystallized in freezing, being the most resistant, may be preserved in unfrozen rocks.

The formation of gas-hydrate bearing rocks and sediments holds a unique position in submarine cryolithogenesis. It may be stated that the process of gas hydrate formation in the earth's interior along with PT-conditions is determined by sufficient amounts of gas and water, their composition and genesis, mineral composition, the structure and texture of rocks and sediments, the presence of organic matter, and a mechanism that occurs when gas and water molecules are brought into contact. The influence of the above factors has not been studied in detail, though it is quite evident that they impede gas hydrate formation in the rocks (Ginsburg, 1969).

To formulate a problem to assess the possible existence of gas hydrates in offshore arctic seas it is reasonable to use an equilibrium curve for the methane + 3.5% sodium salt solution in PT-coordinates. One should take into account that when true PT conditions comply with the field of methane hydrate stability in the above system, hydrates might not exist at all.

Calculations show that areas of probable methane hydrate stability in the Arctic are situated where the total thickness of the cryolithozone and the seawater layer exceed 250-300 m (Figure 4). Hence, a discontinuous distribution of these areas may be inferred for the Arctic shelf.

The cryolithozone (primarily the frozen zone) defines some peculiarities of the geophysical fields of arctic offshore areas, i.e. abnormally high velocity for small depths and abnormally low velocity for the electric conductivity of sub-bottom deposits; probable gravity anomalies related to ice content; and the generation of natural electric fields due to phase transitions at the frozen rock/salt water interface.

The similar physical properties of gas-hydrate bearing and frozen deposits result in unambiguous interpretation of geophysical data. In debatable cases, the absence of frozen ground at a water depth of more than 100-110 m and of hydrate-bearing deposits when the net sequence of the cryolithozone and water body are at less than 250-300 m are helpful for interpretation.

The most important geophysical feature of the Arctic Ocean is a clearcut instability of the thermal field. This is most evident in shelf areas where a frozen zone is present. As noted above, the present 100-110 m isobath outlines areas where a frozen zone was widely (almost ubiquitously) developed during the Late Pleistocene. Heat flow measurements obtained for the bottom sediments in such regions cannot be used to describe geothermal conditions at depth. In this case measurements should be made in boreholes whose depths well exceed the maximum paleothickness of the frozen zone. The paleothickness may be within a 500-m range depending on the water depth and the distance from the coast. Such specific recommendations cannot be given at the present state of knowledge for the shelf areas beyond the 110 m isobath.

The thermal non-stationarity of deep-sea rocks is inferred from the thermal structure of the water body. It is characterized by negative-temperature bottom water (ocean cryopegs) separated from cold surface water by a layer of relatively warm (positive temperature inclusive) Atlantic waters. The ocean cryopegs are underlain by the unfrozen submarine cryolithozone. There are two alternative viewpoints on the nature of the ocean bottom cryopegs. This water body is considered to be either a relic of the glacial epoch (Neizvestnov, 1970) or a result of present-day shallow water discharge. If the former case, the bottom waters and underlying sediments are now being warmed, and if the latter, they are being cooled. Hence, the representativeness of heat flow measurements in Arctic Ocean bottom sediments is to be judged with due regard for heat balance.

When deep-water heat flow measurements in the inner ocean area, including the continental slopes, are analyzed, one should take into account the oceanic gas-hydrate formation, i.e. the latent heat of a gas $\rightleftarrows$ hydrate phase transition may become an additional and important fact resulting in a non-stationary heat field. In addition, the thermal conductivity of gas-hydrate bearing deposits differs from that of water-saturated sediments. The possible presence of gas hydrates should be taken into consideration, however, not only in studies of the Arctic Ocean but of the other oceans as well.

The cryogeothermal study of the Arctic Ocean is a separate scientific problem. At the same time, the authors have tried to show that these studies are closely connected with those of the Late Cenozoic paleogeography and the present thermal state of the Arctic Ocean. Most of the problems have not yet been fully solved. Problems to be solved in the near future are: assessment of the extent of the submarine relic frozen zone on the arctic shelf, mapping it, and determining to what degree the sub-bottom deposit heat field is non-stationary in various parts of the Arctic Ocean.

REFERENCES

Are, F. E., 1980, Thermal abrasion of shores: Moskva, Nauka (in Russian).

Curray, J. K., 1961, Late Quaternary sea level: a discussion: Geological Society of America, Bulletin, 72, p. 1707-1712.

Gakkel, Ya. Ya., 1958, Distribution of Semenovsky Island, in Problemy Arktiki, vyp. 4: Leningrad, Morskoi Transport, p. 95-98 (in Russian).

Ginsburg, G. D., 1969, On the formation of crystal-hydrates of natural gases in the earth's interiors, in Hydrogeologia Yeniseyskogo Severa: Leningrad, NIIGA, p. 109-128 (in Russian).

Gitterman, K. E., 1937, Thermal analysis of marine water, in Trudy solyanoi laboratorii AN SSSR, vyp. 15, ch. I: Moskva, AN SSSR, p. 5-32 (in Russian).

Grigorov, I. P., 1946, Vanishing islands: Priroda, N 10, p. 65-68 (in Russian).

Holmes, M. L., and Creager, J. S., 1974, Holocene history of the Laptev Sea continental shelf, in Marine geology and oceanography of the Arctic seas: New York, Springer, p. 211-229.

Neizvestnov, Ya. V., Tolstichin, N. I., and Tomirdiaro, S. V., 1970, Negative temperature waters of the Earth, in Tezisy dokladov Vsesoyuznogo soveshchania po merzlotovedeniyu 1970 goda: Moskva, MGU, p. 78-79 (in Russian).

O'Connor, A. D., 1980, Study of pingo-like-features detected in Beaufort Sea: Sidney, B.C., Canada.

Pelletier, B. R., 1975, Sediment dispersal in the southern Beaufort sea, in Technical Report No. 25a: Victoria, B.C., Canada.

Richter, V. G., 1965, Methods used in studies of neo- and recent tectonics of sea and ocean shelf zones: Moskva, Nedra (in Russian).

Sellman, P. V., and Chamberlain, E. J., 1979, Permafrost beneath the Beaufort Sea: near Prudhoe Bay, Alaska, in Proceedings of the 11th Annual Offshore Technology Conference, Houston, Texas, v. 3: Dallas, Texas, p. 1481-1494.

Soloviev, V. A., 1978, Laptev and East Siberian Sea shelves in Late Pleistocene (Mapping technique of Late Pleistocene coastal line), in Theoreticheskie i methodologicheskie osnovy kompleksnogo izuchenia i osvoenia shelfov: Leningrad, VGO SSSR, p. 127-129 (in Russian).

INVESTIGATIONS INTO THE COMPACTION OF THAWING EARTH MATERIALS

G. M. Pakhomova

Design and Research Institute for Industrial Enterprises
Eastern Baikal Division, Yakutsk, USSR

The author has developed a method for solving one- and two-dimensional problems concerning the compaction of thawing earth materials. The method was used to investigate the impact of the thawing regime, the depth of the layer being thawed, and its compaction properties on trends in the process of compaction of the thawing material. The results achieved are presented as nomograms which allow one to predict the degree of compaction of various types of thawing and recently thawed materials at any point beneath the foundations of buildings or structures at a given point in time. They may be used to resolve specific practical problems in location and design.

Field data on buildings and structures on frozen soil show that their stability and strength are significantly influenced by the character of thaw settlement (consolidation) over time.

In this connection, a thorough understanding of the settlement of thawing soil over time is essential to determining design requirements that ensure the operational reliability of buildings and structures built according to the second principle on foundations in permafrost.

Analysis of the research available on thawing soil strain (Zhukov, 1958; Vyalov, 1981) shows that this is a complicated process dependent on a number of factors, such as soil type, soil consolidation properties, thawing conditions, and thickness of the thawing layer.

The influence of the factors is difficult to quantify due in part to the labour-consuming character, high cost, and long duration of the experiments and in part to the mathematical difficulties (Tsytovich, 1966).

The results of research on thawing soil consolidation by hydroanalogy made by this author under Prof. S. S. Vyalov are presented in this paper.

The hydroanalogy method allows a wide range of calculations, to account for the nonlinearity of the physical parameters and to get a visual picture of the consolidation process according to the factors outlined above.

The pattern of the compaction (consolidation) of thawing soil over time is defined by the laws of thawing and compaction (Vyalov, 1981; Malyshev, 1966). There is a comprehensive analogy between these processes and the process of water movement in the hydrointegrator (Lukyanov, 1947) because they are mathematically described by the same type of equations.

$$e\frac{\partial T}{\partial \tau} = \frac{\partial}{\partial x}\left(a\frac{\partial}{\partial x}\right) + \frac{\partial}{\partial y}\left(b\frac{\partial}{\partial y}\right) + \frac{\partial}{\partial z}\left(c\frac{\partial}{\partial z}\right) \quad (1)$$

where T is a function; x,y, z, and τ are independent variables (coordinates, time); and e, a, b, and c are physical characteristics of the medium.

To solve the problem of thawing soil compaction over time, it is necessary to solve two successive tasks:

- the change of depth of thawing over time (the thermal physical task),
- the pore pressure change over time (the consolidative task.

The result of the first task, whose method of solution is described by Golovko (1958), Porkhaev (1970), and others, is the boundary condition for the solution of the second task. To solve the compaction problem, the boundary conditions of the region under investigation are defined taking into account the results of the thermal physical task solution. The region under investigation is divided into blocks: the blocks are smallest in the region of the most intensive compaction (near to the filtering boundary with z=0 coordinate). Block volumes and resistivities between their centers have been calculated by the formula:

$$C_{i,k} = a_{o_{i,k}} \cdot V_{i,k} = a_{o_{i,k}} \cdot \ell_{i,k} \cdot F_{i,k} \quad (2)$$

$$R_{i,k} = [(\ell_i/2k_i) + (\ell_k/2k_k)](\ell/F_k) \quad (3)$$

where $C_{i,k}$ - block volume i, k;
$a_{o_{i,k}}$ - compressibility coefficient of the blocks of the region i, k under investigations;
$F_{i,k}$ - block cross-section;
$R_{i,k}$ - resistivity between centres of the blocks;
ℓ_i, ℓ_k - block thickness;
k_i, k_k - block filtration coefficient.

Then the scaling correlation between the block volumes and the cross-section of the hydrointegrator pipes M_C, and between the block centre resistivities and the hydrogenerator hydraulic resistivities M_R are determined.

$$M_C = C/w \quad (4)$$

$$M_R = R/\rho \quad (5)$$

where M_c - the scale volume;
C - the block volume;

w - volume of the hydrointegrator;
M_R - the scale of resistivities;
R - the resistivity between block centres;
ρ - hydraulic resistivity between hydrointegrator pipes;

Time scale M_τ and height scale M_h have been computed on the basis of the scale correlations:

$$M_\tau = M_c\ M_R \tag{6}$$

$$M_h = U/h_r \tag{7}$$

where U is the pore pressure in the blocks of the region under investigation; and
h_r is the water level in the hydrointegrator piesometric tubes.

The working scheme of the hydromodel is composed and solved on the hydrointegrator using devices that provide the necessary boundary conditions, which in turn provide the only solution.

When the assembly is finished and the necessary boundary conditions have been established, computation can begin. The hydromodel's elements (hydrovolume w, hydraulic resistivities) are incorporated into the operation in time increments determined by the thermal task solution. Before each successive block connection, the hydrointegrator is switched off and water level readings in the piezometric pipes are noted. At the moment of complete pore pressure dissipation, when the water level in the piezometric tubes is equal or near to zero, the hydrointegrator is switched off. The complete dissipation criterion is that the water level change in the piezometers over a simulated 10-year time period does not exceed 0.1 cm with the time scale taken into account.

The computation results are processed by (Lukyanov, 1974):

$$S^p_\tau = \sum_o^n (h_r - h_{r,\tau})\ell_{i,k}\ a_{o_{i,k}} \tag{8}$$

where: S^p_τ - thawing soil compaction settlement at a given point in time;
h_r, $h_{r,\tau}$ - water level in the hydrointegrator piezometric pipes, initial and at a given point in time;
$\ell_{i,k}$; $a_{o_{i,k}}$ - same as in equations 2 and 3.

Thawing soil consolidation has been investigated using the hydrointegrator, which might be applied to the next three cases:

Case 1 - permafrost thawing during construction;
Case 2 - permafrost thawing at the full depth of the presumed thaw zone, when the thawing soil layer is underlain by low compressed bedrock.
Case 3 - partial permafrost thawing before beginning of construction with subsequent thawing assumed during the period of exploitation.

For all this, the series of tasks with changing parameters of the factors outlined above has been solved in order to define the consolidation process.

On the basis of the interpretation of the solution, the relationship of consolidation extent (S^p_τ/S^p) of thawing soil to the rate of its thawing and compaction (β/α) (Figure 1) has been obtained. Visual support of this proposition is given in Tsytovich et al. (1966).

When the process is one-dimensional, the change over time in the depth of thaw and the thawing soil compaction S^p_τ might be determined by equations 9 and 10:

$$h_\tau = \alpha\sqrt{\tau} \tag{9}$$

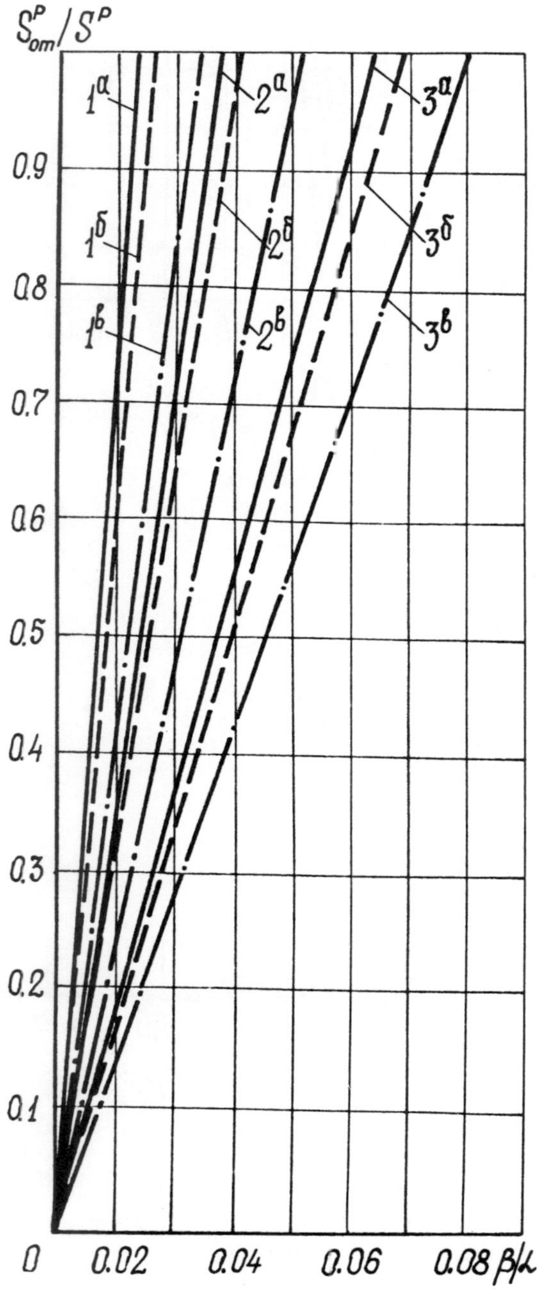

FIGURE 1 Dependence S^p_τ/S^p from β/α. For thawing soil by different values P; h 1, 2, 3 - P = (0.05 + P_o); (0.2 + P_o) MPa; P_o - dead weight of the thawing soil; P = 0.05; 0.1; 0.2 MPa - compression pressure; a, b, c, - $h_{\tau=\infty}$ = 6, 10, 20 m.

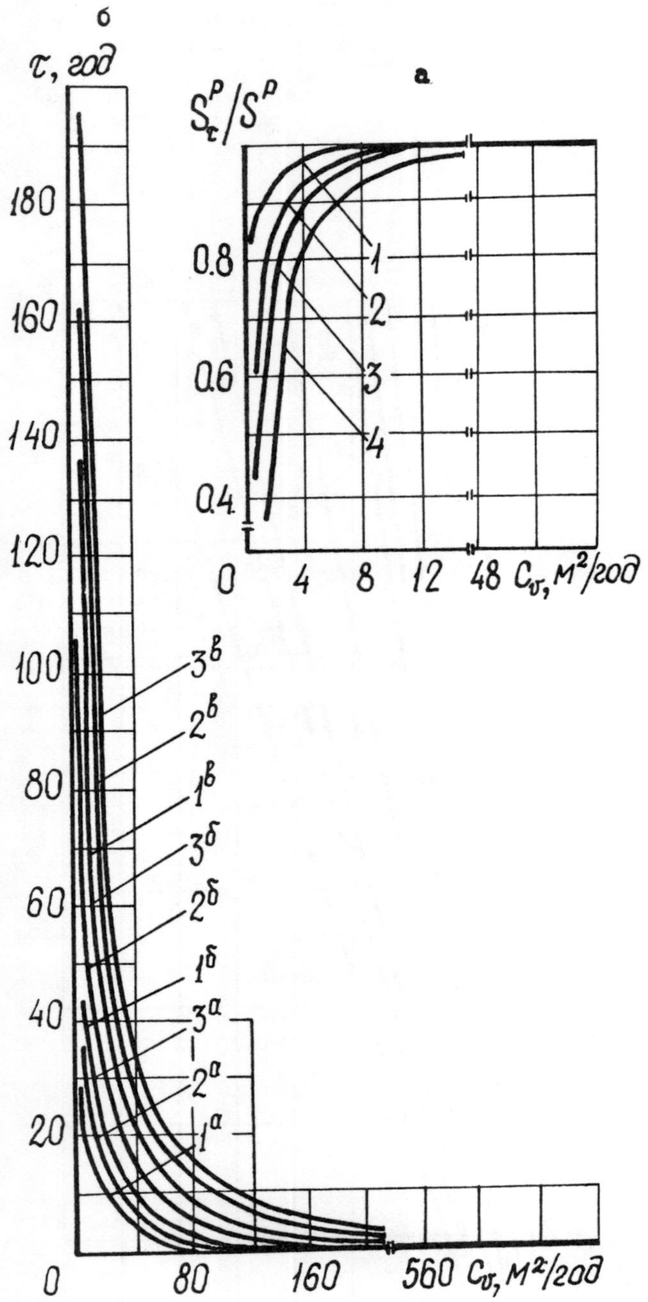

FIGURE 2 a) Relationship S^p_τ/S^p from C_v 1, 2, 3, 4 - h = 2, 6, 10, 20 m; b) Dependence of the settlement stabilization time upon preliminary thawed soil - C_v; 1, 2, 3 - P = 0.05; 0.1; 0.2 MPa; a, b, c - $h_{\tau=\infty}$ = 6, 10, 20 m.

$$S^p_\tau = \beta\sqrt{\tau} \quad (10)$$

where h_τ - depth of thawing at a given point in time;
α - parameter of thaw rate/thermal coefficient;
S^p_τ - the same as in eq. 8;
β - parameter of compaction rate;
τ - time;

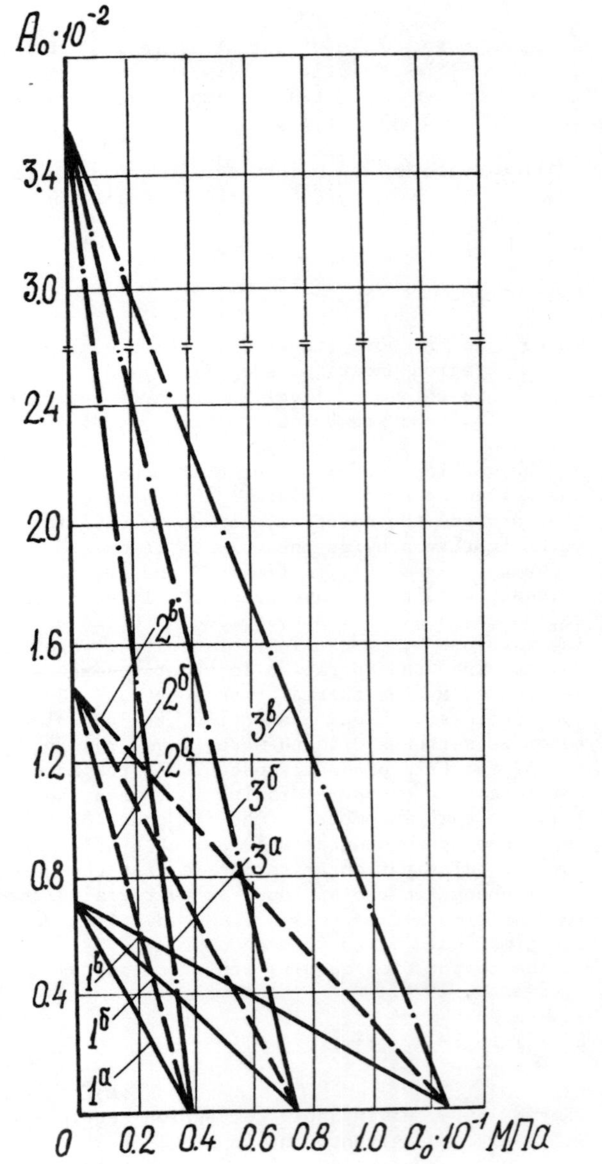

FIGURE 3 Nomogram to establish optimal depth of thawing before the beginning of construction. a, b, c - h = (0.05; 0.8) $h_{\tau=\infty}$; 1, 2, 3 - P = 0.05; 0.1; 0.2 MPa.

Dividing both parts of eq. 10 into ($S^p = a_o h_{\tau=\infty} P$) one obtains an analytical expression for the evaluation of the degree of thawing soil consolidation at the moment of thaw completion:

$$S^p_\tau/S^p = \beta\sqrt{\tau}\ /\ a_o Ph \quad (11)$$

where S^p - stabilized settlement of compaction at the moment $\tau=\infty$;
h - depth of thawing at the moment of time $\tau=\infty$
a_o - the same as in eqs. 2, 3, and 8;
P - pressure under which soil is thawing.

It has been stated that when $(\beta/a_o p\alpha) \geqq 1$ (Golovko, 1958), soil compaction has been completed during the process of thawing, but when $\beta/a_o p\alpha < 1$ (Porkhaev, 1970), thawing soil compaction has continued to develop after thawing completion.

Data computations on hydrointegrator have identified the dependencies of the consolidation of the preliminary thawed soil layer 50 years after construction upon the pressure value, layer thickness, and the soil's consolidation characteristics (Figure 2a,b).

It is evident that the consolidation time varies over a wide range (from a month to hundreds and thousands of years), and at small values of C_V, the degree of soil consolidation over a period of 50 years is 30-50% of the stabilized settlement S^p. Not only the finite stabilized settlement of the foundation bed, but also the settlement corresponding to moment τ can be taken into account. For the computed curve of consolidation of the preliminary thawed layer, dependence $S^p_\tau = \beta\sqrt{\tau}$ is valid within the limits of $S^p_\tau/S^p = 0.8\text{-}0.9$.

The nomogram has been compiled on the basis of the task solutions. It is used to select the optimal depth of the preliminary thawed layer according to the value of its strain characteristics (A_o, a_o) (Figure 3). The nomogram is compiled for the maximum depth of thawing (h = 20 m) under mean permissible settlement (S = 0.15 m) for

a) $h/h_{\tau=\infty} = 0$; b) $h/h_{\tau=\infty} = 0.5$

c) $h/h_{\tau=\infty} = 0.8$

The nomogram allows the depth of the preliminarily thawed layer h relative to the depth of the thaw zone $h_{\tau=\infty}$ to be established without additional computations. The hydroanalogy method gives a precise solution.

The investigations lead to the following conclusions:

1. The hydroanalogy method gives a satisfactory solution.
2. Developed techniques for calculating settlement and the solutions obtained may be used to solve practical problems of construction on permafrost.

REFERENCES

Bondarev, P. D., 1957, Strain of building in Vorcuta region, their reasons and methods of their prevention: Moscow, USSR AS Publishing House, 98 p.

Feldman, G. M., 1966, Solution of one-dimensional problem in thawing soils consolidation accounting the variable permeability and compressibility, in Proceedings of the XIII All-Union Meeting on Geology: v. 5, Yakutsk, p. 185-191.

Golovko, M. D., 1958, Computation method of bowl of thawing in the structures bed built on permafrost: Moscow, TsNIIS, report 141, p. 52.

Gorelic, L. B., and Tsybin, A. M., 1979, To the consolidation theory of thawing soils, in Collection of scientific papers VNIIG, v. 134, p. 119-127.

Lukyanov, V. S., 1947, Hydraulic analogy as the new means of technical problems investigations, Thesis presented by the doctor technical science (typeprinting): USSR, Lenin State Library, p. 177.

Malyshev, M. K., 1966, Computation of foundation settlement built on thawing soils, in Beds, foundations and soil mechanics, no. 4, p. 20-24.

Nixon, J. F. and Morgenshtern, N. P., 1973, Practical development of the consolidation theory for thawing soils, in Proceedings of the II International Conference on Permafrost: Moscow, Nauka, p. 112.

Porkhaev, G. V., 1970, Thermal interaction of buildings and structures with permafrost: Moscow, Science, p. 208.

SNPP-18-76, 1976, Beds and foundation of buildings and structures built on permafrost. Designing specifications: Moscow, Stroyizdat, p. 46.

Tsytovich, N. A., 1973, Mechanics of frozen soils: Moscow, High school, 446 p.

Tsytovich, N. A., Grigorieva, V. G., and Zaretsky, Yu. K., 1966, Investigations of the thawing soil consolidation, in Beds and foundations: Moscow, Stroyizdat, v. 56/NIIOSP, p. 92-112.

Vyalov, S. S. and Pakhomova, G. M., 1981, Test investigation of thawing soils using field hot punches, in Thawing soils used as beds of structures: Moscow, Science, p. 47-55.

Zaretsky, Yu. K., 1968, To the computation of thawing soil settlement, in Beds, foundations and soil mechanics, no. 3, p. 3-6.

Zhukov, V. F., 1958, Preliminary thawing of permafrost for construction: Moscow, USSR AS Publishing House, 148 p.

THE THERMAL REGIME OF THERMOKARST LAKES IN CENTRAL YAKUTIA

A. V. Pavlov and F. E. Are

Permafrost Institute, Siberian Branch
Academy of Sciences, Yakutsk, USSR

Annual values of the components of the heat balance of the surfaces of thermokarst lakes and of land areas in the taiga zone of Eastern Siberia were obtained. The radiation balance and the effective radiation were higher in the case of the lake surface than from a meadow surface; the evaporation from the water surface, totalling 500 m, was twice as high as the precipitation and 1.8-2.4 times higher than from the soil surface in meadows. In contrast to the situation on land, turbulent heat exchange over the lakes was positive in winter and negative in summer. All components of the heat balance increased with lake depth. The mean annual temperature on the lake surface was 4-6°C higher than that on land. Freeze-up occurred at a water temperature of 1.5-4°C. In springtime under-ice warming of the water resulted in temperatures of 10°C and melting of the ice from below amounted to 35-45% of the total melting. In summertime the surface water warms up to 25-30°C. Mean annual temperatures of the lake bed are positive throughout the entire water area. The maximum warming effect on the bottom materials is produced by a water layer 1-1.5 m in depth. In the central areas of large lakes the annual heat absorption by the bottom is equal to the heat loss but exceeds it as one approaches the shore.

Investigations of the thermal regime of lakes, needed in the study of thermokarst processes, were placed on a broad footing by Soviet permafrost researchers in the 1960's (Dostovalov, Kudryavtsev 1967; Gavrilova 1969, 1973; Tomirdiano 1965, 1972; Are and Tolstyakov 1969; Are 1972; Are 1974), but the observations gathered were not comprehensive enough and usually did not cover the annual cycle of variation of the parameters under study. An experimental study of all components of the thermal balance within an annual cycle was first carried out under the leadership of these authors during 1973-1977 on Syrdakh Lake, a large thermmokarst lake measuring 2×1 km and up to 12 m deep. In addition to studying the thermal regime of the lake itself, great attention was given to actinometric and meteorological observations because the energy circulation within lakes is mainly due to heat exchange through the surface and is therefore closely related to atmospheric processes. The observations were made at the center of the lake and at varied distances from its shore as well as on land in a larch-type forest and in open meadow areas. The methodology of these observations has been described in several publications (Are 1974; Pavlov 1975, 1979; Pavlov and Prokopiev 1978; Tishin 1980; Pavlov and Tishin 1981).

The observational data acquired on Syrdakh Lake permit the generalization of the available data set to the thermal regime of thermokarst lakes of Central Yakutia. These lakes occur within hollow-shaped depressions with steep sides and comparatively flat bottoms. In some cases the hollows are as deep as 40 m but their depth mostly varies within 5-7 m. They are normally surrounded by woodland with trees from 5 to 10 m tall. Depending on the stage of development, the lakes occupy a larger or a smaller part of the hollow bottom. They are mostly oval-shaped and measure up to several kilometers across but most measure from several tens to several hundreds of meters. The depth of some lakes reaches 10-15 m but most do not exceed 3 m. There are numerous lakes only 0.3-0.5 m deep with an extensive water surface. They usually have a flat bottom covered with silt.

The climate of Central Yakutia is strongly continental. It is notable for a rapid change of seasons, large diurnal oscillations of climatic elements, and a hot summer with high insolation and gentle breezes. The ice-free period lasts 4-4.5 months.

Central Yakutia is situated within the continuous permafrost zone with a maximum permafrost depth of 450 m. Weak winds, shores overgrown with forest and steep and relatively lofty shore ledges greatly diminish frictional stirring of lake water. The high insolation combined with low transparency of the water and high summer air temperatures produce a strong heating of surface water layers. The permafrost lowers the water temperature from below. These factors render water masses highly stable during the summer and are responsible for a variety of interesting features of the lacustrine thermal regime.

The albedo (A) of lakes up to 2.5 m deep varies from June to September within 0.08 and 0.24 at mean values of 0.11-0.14. The albedo of Lake Syrdakh with a mean depth of 4.5 m and more transparent water averages 0.08 during the summer. During the winter months the albedo ratio of lakes and land varies from 0.8 to 1. A maximum albedo of 0.6-0.8 is observed on lakes in December-January and it is decreased to 0.18-0.25 at the time of snow cover thawing.

Since during an ice-free period the albedo of lakes is less than that of meadow areas, the net radiation balance $R = Q(1-A) - I_{ef}$ of an open water surface is larger. In the equation Q is total radiation, I_{ef} - effective radiation equal to the difference in radiation between the ground surface I_s and the atmosphere I_A. On the average for the summer net radiation balance for lakes is by a factor of 1.21-1.34 larger than for meadow areas (Pavlov 1979). In the winter period the values of R are nearly equal and have a negative sign.

The effective radiation from a lake surface (cf. Table 1) is 1.4-1.7 times larger than that of a meadow surface, both in winter and in summer. This is accounted for by the higher temperature of lake surface and by a corresponding increase of I_s.

The evaporation E from the water surface of lakes is in magnitude close to the theoretical maximum value. Maximum occurs in July: 8-11 mm/day on Lake Syrdakh. The July total of evaporation is nearly one third of summer total which for shallow lakes makes up 400-450 mm. The normal evaperoles from the water surface of Lake Syrdakh reaches 500 mm which is approximately twice the precipitation and 1.8-2.4 times the soil evaporation in meadow areas. Evaporation is reduced in shoreline wooded areas. Here about two thirds of the annual total of precipitation are expended on drainage. Evaporation from snow cover on lakes amounts only to about 10-15 mm (Are 1978), or less than 3% of the annual total. Heat expenditure for evaporation from lakes is about 80% of R or half the Q for a warm period and 90% of R per year.

TABLE 1 Totals of thermal balance components of Syrdakh Lake averaged over Sept. 1974-Aug. 1976 (10·MJ/m²)

Components	Period		
	τ_s	τ_w	τ_y
Q	245	113	358
S	21	70	91
I_{ef}	74	55	129
R	150	-12	138
P	-13	26	13
E	119	3	122
B	43	-43	0
B_{TS}	1	2	3
B_{TI}	27	-27	0
B_W	10	-10	0
B_B	6	-6	0

Notes: τ_s - period with air temperature $t_A > 0°C$, τ_w - period with $t_A < 0°C$, τ_y - year, S - reflected solar radiation.

Turbulent heat exchange P on lakes differs from the land where it is positive in summer and negative in winter. On lakes, on the contrary, it is usually positive in winter and negative in spring and partially or even completely negative in summer. For example, in 1977 on Lake Syrdakh P was negative from April to August. Its monthly means for that period varied from -6 to -42 W/m² (Pavlov and Tishin 1981). The negative values of P in spring and at the beginning of summer are explained by the fact that the lakes become completely ice-free 1 month after the establishment of an above-freezing daily mean air temperature. Since the air temperature is several degrees higher than that of a thawing ice surface, the latter receives a considerable input of turbulent heat. For the same reason the P would also be negative on a summer day.

For seasonally freezing lakes, thermal flow through the surface (B) can be determined as the sum of heat expenditures for changing the heat content of the water mass (B_W), for thawing of ice (B_{TI}), and for heat exchange with bottom deposits (B_B). For an intermittently established regime the above components are recurrent within an annual cycle because the process of lake warming is replaced by cooling (cf. Table 1). The heat expenditure for thawing of ice is so large that it is comparable with a limit change of heat content of water masses. Conversely, the heat expenditure for thawing of snow B_{TS} constitutes a small fraction of B_W.

The heat flow across the lake surface becomes positive toward the end of March and again changes sign 4-4.5 months later. Thermal input into soil in meadow areas becomes negative 1 month later (cf. Figure 1).

All the components of B increase with increasing depth of lakes. On Lake Syrdakh when the depth increases from 1.5 to 6 m the heat accumulated by water masses increases three times. For the lake on the whole it is 7-8 times larger than that accumulated by soil on land. For a period with above-zero air temperatures the B/Q ratio for Lake

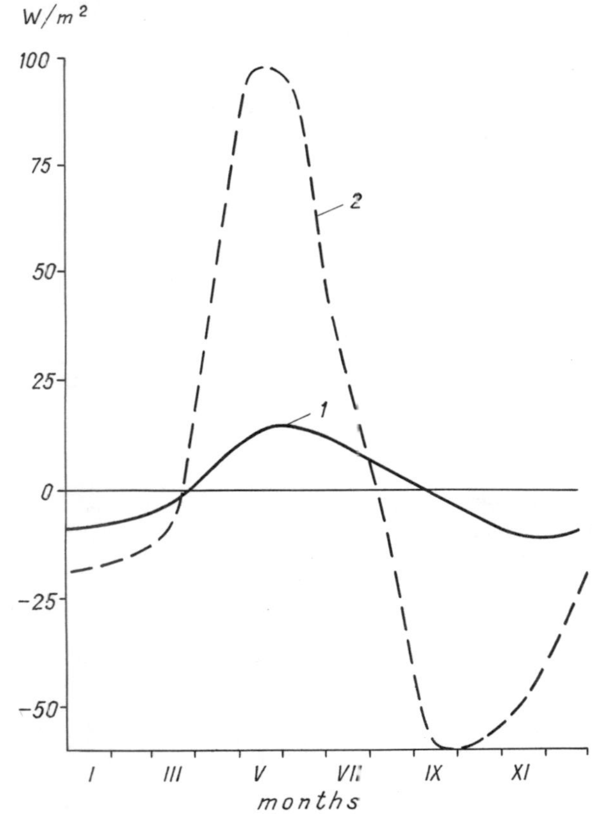

FIGURE 1. Annual variation of monthly mean values of heat flow across the surface of Lake Syrdakh (2) and through soil in a meadow area (1).

Syrdakh amounts to nearly 0.12 while the B/R ratio is 0.21. The mean annual temperature of the surface of lakes is by 4-6°C higher than that on land because of large heat amounts accumulated by water masses.

The temperature regime of water masses has been studied in detail on small and large lakes (Are 1974, Tishin 1978). Usually the lakes become icebound in the first half of October, this occurring 1.5-2 weeks later on larger lakes in comparison with smaller ones, at a comparatively high temperature of water of 1.5-2°C in small shallow lakes and of 3-4°C for large deep lakes. The temperature distribution with depth is nearly isothermal with gradients of 0.2-0.3°C/m. Only near the water surface and bottom do gradients become steep.

In the winter period lake waters cool due to atmospheric cooling. Heat flow from the bottom does not produce any increase of water temperature after the lakes become icebound, which is accounted for by a relatively high temperature of the water before the formation of ice cover and by a decrease of the amount of heat flow due to the cooling effect of a frozen mass. There are no observed diurnal variations of temperature below ice. Once the lake becomes ice covered, the increased vertical temperature gradients of water within the near-surface layer disappears rapidly but remains unchanged within the near-bottom layer throughout the entire winter, reaching 2-3°C/m toward the end of March. Within the main mass of water the temperature distribution in depth is close to isothermal. In shallow lakes the water cools during the wintertime to the freezing point and the water in deep lakes cools to 2-3°C.

The spring warming of the water starts at the end of April after the daily mean air temperatures have traversed 0°C, that is long before ice break-up on lakes. Initially the warming is produced by thermal flow from the bottom. The first manifestations of the thermal effect of solar radiation penetrating the ice appear in mid-May after the snow cover disappears completely and the ice thickness is decreased to 1 m. Subsequently, the water temperature below the ice increases rapidly due to the penetrating radiation and can be as high as 10°C by the beginning of the ice cover break-up.

During the period of under-ice warming, the water temperature distribution depends on the size of lakes, their depth, and water transparency. In lakes that are less than 1.5 m deep, the penetrating radiation warms not only the water but also the bottom sediments. For instance, within a lake 1.35 m deep, at the water temperature of 8.5°C, the bottom temperature was measured to be 6°C. Within small turbid lakes 2-3 m deep the warming penetrates to 1.5 m while deeper layers remain nearly isothermal prior to break-up. Within large and deep lakes with a transparent water the warming penetrates to greater depths. For instance, warming penetrates to a depth of 6 m in Lake Syrdakh. In this case the temperature gradient gradually decreases with depth. The period of warming below ice is notable for daily fluctuations of water temperature of 1°C to a depth of 1 m. The radiation warming of the water below ice initiates intensive thawing of ice from below, which constitutes 35-45% of the total thawing for the time before the start of ice cover movements.

The summer temperature regime of water is established after the lakes become ice-free in the first half of June. Smaller lakes open 10-15 days earlier than larger ones. After breakup there is some decrease of temperature within the upper one-meter layer of water but thereafter a rapid warming starts and even within a few days the temperature of surface layers on large lakes usually reaches 10-15°C and on small lakes 15-20°C. In this case on lakes deeper than 1.5-2 m a normal three-layer temperature distribution in depth is established. The thickness of epilimnion is 1-1.5 m on smaller lakes and 2-3 m on larger ones; that of metalimnion, respectively, ranges from 0.5 to 4 m. The vertical temperature gradients in the smaller lake metalimnion are 20-25°C/m while those for larger lakes are 2-3°C/m. The surface water layers are warmed up to maximum temperatures of 25-30°C. In this case the epilimnion decreases in thickness while on lakes up to 2 m deep the temperature distribution sometimes becomes rectilinear throughout the entire depth, particularly during the daytime. The vertical temperature gradients reach 8°C/m. Diurnal fluctuations of temperature within the epilimnion reach 3-4°C and on occasions 6°C and those within the metalimnion do not exceed 1°C.

The summertime cooling of water masses starts in the first half of August. In this case the epilimnion increases in thickness and a icothermal condition is eatablished throughout the entire depth at a temperature of about 10°C in the second half of September even in the deepest lakes. Subsequently, in the period prior to freezing the water temperature decreases and the character of the temperature gradient remains unaltered.

There are no substantial horizontal temperature gradients of water for most of the year. Only at the onset of the summer warming in June and in the first half of July in shallows less than 1 m deep is the water temperature higher than that within the same layer of a deep part of lakes.

An important parameter needed in calculations of the thawing of permafrost below water reservoirs is the temperature of the bottom surface. It is largely determined by the water temperature and depends on the depth of a lake. The observations on 12 lakes have been used to infer a dependence of the mean annual temperature of the bottom surface on the depth for smaller shallow lakes and for Lake Syrdakh, shown in Figure 2. The plots in Figure 2 show that the maximum warming effect on underlying ground is produced by a water layer of 1 m thickness for smaller lakes and of 1.5 m for larger ones. A further increase of the water layer of smaller lakes leads to an abrupt decrease of bottom temperature. For larger lakes this effect is much weaker. The maximum temperature of the bottom surface of smaller lakes is about 6°C and that of larger ones is 7°C. It is found that under climatic conditions of Central Yakutia the mean annual temperatures of the bottom surface are above zero throughout the entire water area of lakes up to the water edge.

The heat exchange of water masses with the bottom is characterized by a heat flow B_B, which varies over the annual cycle both in value and in sign. In the part of the water area of lakes that do not freeze up to the bottom, B_B becomes positive in the second half of May and then reverses sign in

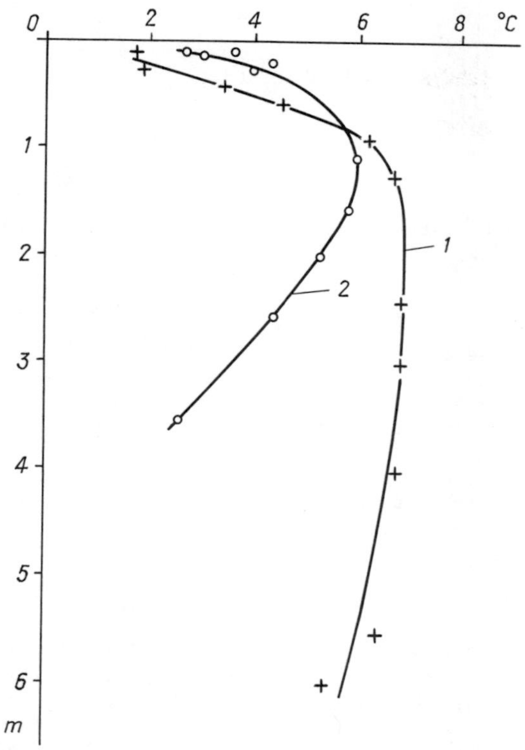

FIGURE 2. Mean annual temperature of the bottom surface versus depth: (1) Lake Syrdakh (Pavlov and Tishin 1981), (2) small lakes (Are 1972).

the first half of September. The maximum amount of heat of about 30 MJ/(month·m²) is received by the bottom in July and the maximum loss of heat of 17-20 MJ/(month m²) takes place in September or in October. The maximum heat absorption by the bottom on Lake Syrdakh is observed at a depth of about 1.6 m. The dependence of this quantity on the predominant depth of lakes is illustrated by figures listed in Table 2, obtained by Gavrilova (1973), Are (1974), and Pavlov (1979).

TABLE 2 Heat absorption of bottom of lakes for a season

Lake	Predominant depth (m)	B_B (MJ/m²)
Syrdakh	4.5	58
Prokhladnoye, Tyungyulyu	2.0-2.5	63-66
Kradenoye	0.2-0.5	138

For smaller lakes with closed taliks below lakes and for larger lakes near the shore the annual heat input into the bottom exceeds substantially the heat loss. Irreversible heat absorption of the bottom is spent for maintaining the talik below a lake within permafrost. For larger lakes with an open talik below a lake at a considerable distance from the shore the permafrost effect is missing and the annual heat absorption of the bottom is equal to the losses (cf. Table 1).

The above relationships of energy exchange in atmosphere-water medium-underlying ground systems hold true not only for thermokarst lakes of Central Yakutia but also for the whole taiga zone of East Siberia.

REFERENCES

Are, A. L., 1978, Snow cover of Central Yakutia, some features of its radiative and hydrothermal regime, in Heat exchange in permafrost landscapes: Yakutsk, IM SO AN SSSR, p. 30-42.

Are, F. E., 1972, Das Wärmeregime der Seen Zentral-Jakutiens: Verhandlungen der Internationalen Vereinigung für Limnologie, 18, S. 574-579.

Are, F. E., 1974, Thermal regime of shallow lakes of the taiga zone of East Siberia (in the example of Central Yakutia), in Lakes of the permafrost zone of Siberia: Novosibirsk: Nauka, p. 98-116.

Are, F. E., and Tolstyakov, D. N., 1969, On penetration of solar radiation into water: Meteorologiya i gidrologiya, No. 6, p. 58-64.

Dostovalov, B. N., and Kudryavtsev, V. A., 1967, General cryology: Moscow, Izd. Mosk. un-ta, 403 p.

Gavrilova, M. K., 1969, Microclimatic and thermal regime of Lake Tyungyulyu, in Questions of Yakutia geography: Yakutsk, Kn. izd-vo, v. 5, p. 57-72.

Gavrilova, M. K., 1973, Climate of Central Yakutia: Yakutsk, Kn. izd-vo, 119 p.

Pavlov, A. V., 1975, Heat exchange of soil with atmosphere at northern and moderate latitudes of the USSR: Yakutsk, Kn. izd-vo, 302 p.

Pavlov, A. V., 1979, Thermal physics of landscapes: Novosibirsk, Nauka, 285 p.

Pavlov, A. V., and Prokopiev, A. N., 1978, Heat exchange of soil with atmosphere in conditions of a larch forest in Central Yakutia, in Heat exchange in permafrost landscapes: Yakutsk, IM SO AN SSSR, p. 3-29.

Pavlov, A. V., and Tishin, M. I., 1981, Thermal balance of a large lake and of the area adjacent to it in Central Yakutia, in Structure and thermal regime of frozen earth materials: Novosibirsk, Nauka, p. 53-63.

Tishin, M. I., 1978, Temperature regime of water masses of Lake Syrdakh, in Geothermophysical research in Siberia: Novosibirsk, Nauka, p. 58-66.

Tishin, M. I., 1980, Thermal regime of the bottom of a large thermokarst lake in Central Yakutia, in Permafrost exploration in development areas of the USSR: Novosibirsk, Nauka, p. 40-47.

Tomirdiano, S. V., 1965, The physics of lacustrine thermokarst in polar plains and Antarctic Region and cryogenic transformation of earth materials: Kolyma, no. 7, p. 30-34; no. 8, p. 36-40.

Tomirdiano, S. V., 1972, Permafrost and exploitation of highlands and plains: Magadanskoye kn. izd-vo, 174 p.

PREDICTION OF CONSTRUCTION CHARACTERISTICS OF SLUICED MATERIALS USED IN FOUNDATIONS UNDER PERMAFROST CONDITIONS AND IN SEVERE CLIMATES

V. L. Poleshuk

Design and Research Institute for Industrial Enterprises
Eastern Baikal Division, Yakutsk, USSR

During engineering preparation of an area for construction in permafrost areas by means of hydraulic action it is essential to predict the thermal interaction of the sluiced and natural materials and their structural properties. The investigation revealed that the use of sluiced and underlying materials as foundations can be handled in three different ways: the sluiced material forms the levelled building surface; a combination of sluiced materials and the underlying materials is used for the building foundation; no load is exerted directly on the underlying material and only the sluiced material forms the foundation. With these schemes in mind, requirements in terms of the construction properties of materials are examined and principles for using them as foundations are formulated.

Building in permafrost areas requires special engineering preparation of the territory that takes into account the protection of the environment. Construction codes permit the use of permafrost ground as a building foundation in accordance with two conditions: in a frozen state (condition 1) and in a thawed or state of thawing during operations (condition 2). When either condition mentioned above exists, the engineering preparation of the site must provide for the conservation of the calculated temperature regime of the foundation soil and for the preservation of the natural environmental conditions. In order to drain storm-water and other kinds of surface water from the construction site and to preserve the vegetative and soil covers, the engineering preparation of a site should be carried out by using fills during vertical levelling and grading. As a rule, shear cuts are not permitted. Bedding fills are used in the construction of road embankments, accesses, building sites for one or more buildings.

During the engineering preparation of a construction site for large industrial enterprises and large-scale developments (towns, cities, individual housing), it becomes necessary to construct different types of embankments of considerable size.

From a technical-economic point of view, the optimum method for the use of fill in construction must be selected during the engineering preparation of large areas. Both foreign and domestic practices in temperate climate zones widely employ the "hydromechanical" method in earth works related to municipal construction. This method is cheaper and less labor-intensive in comparison with conventional "dry" technology. It does not require the construction of quarries and roads, the extensive use of trucks and machinery, or the transport and placement of soil into the fill. The net cost of the soil extracted and placed by the hydro-mechanical method is 1.5-3 times less, while the labor required is 15 times less than in the conventional "dry" method. The high degree of efficiency of fill construction by hydromechanical means has promoted the widespread use of sluiced materials for the engineering preparation of sites in different cities of the Soviet Union including Moscow, Leningrad, Kiev, Gomel, Gorky, Omsk, and Archangel. In recent years, the areas in which sluiced materials are being used has expanded to regions of Western, Central, and Eastern Siberia and to the east of BAM.

Several years ago in permafrost regions several embankments were constructed for different purposes using sluiced materials. For instance, in 1960-62 this method was used to form the site of the present river port in Yakutsk and to construct a dike to protect the port area. To the north of Krasnoyarsk, Yakutsk and Magadan, a number of small tailing storage dams for ore enrichment plants was built using the sluice method. The extraction of bottom sand and gravel deposits in river beds and other water basins, where they are in a thawed state all year round and easily developed using hydromechanical means, has been made possible by the use of sluiced materials under such extreme conditions.

These first attempts to use sluiced materials in permafrost revealed a number of specific characteristics of these materials which form when they are used as foundations. These features hinder the direct application of experience in building on sluiced materials in temperate climate zones. The main feature is the thermal interaction between fill and the underlying perennially frozen natural ground and a tendency for the development of cryogenic processes. This interaction determines the formation of the structural properties of sluiced materials and their suitability as foundations.

It became necessary to carry out special investigations on the properties of sluiced materials in permafrost. These investigations were conducted by the Yakutsk Division of the "Zabaikalpromstroyiniiproekt" Research Institute of the Ministry of Eastern Construction at a special experimental site in Yakutsk. The site is situated on the Lena River floodplain on the bank of its channel which floods every year. The alluvial deposits mainly consist of fine-grained sand 14-20 mm thick underlain by sandstone and aleurolite bedrock. The sand stratum includes silty sediments, peat and, in some places, layers of dusty, sandy loam and sandy clay from 0.3 to 3.0 m thick. The floodplain is characterized by typical meadow vegetation; the average thickness of the soil's vegetative layer is 10-20 cm. The depth of seasonal thaw varies from 0.9 to 2.6 m depending upon soil type and surface cover. The entire area of the floodplain is characterized by flat topography. The annual fluctuation of soil temperature varies from -0.3 to -1.9°C. Moisture content of the alluvial deposits usually is in the range of 20-30% and ice content is 14-22%. The distribution of ice along the section is relatively uniform, and the soils are characterized by massive cryogenic texture. Information on the permafrost thickness in the floodplain is contradictory but at the site it exceeds 100 m. This was demonstrated by drilling a special well: the lower limit of the frozen zone was not established. Experimental fill, having a total volume of 300,000 m^3 and varying in thickness from 1 to 6 m, was constructed by the sluice method on a 12-hectare test site.

The processes of heat exchange were studied. The foundation was tested by static loading and the construction techniques were examined. The rate of soil drainage was determined. The technical-economic efficiency of the use of sluiced materials for the construction in Yakutsk was evaluated.

The investigations outlined above demonstrate the feasibility of developing the floodplain adjacent to the city for large-scale construction using the sluice method. An area of about 800 hectares has to be developed at an average depth of a 6-7 m sluiced embankment. Therefore, Yakutsk will be the first building site where construction on sluiced materials in permafrost is to be performed on a large scale. In the absence of analogous construction in permafrost, the decision was made to begin the floodplain development with the construction of three experimental buildings on sluiced materials. They represent 5-storey residential buildings which are being constructed by different methods on different types of foundations (pile, columnar and foundations-enveloped). Examination of these types of foundation and the techniques used to build them on beds composed of sluiced materials should provide the optimal solutions for their future application in large-scale development.

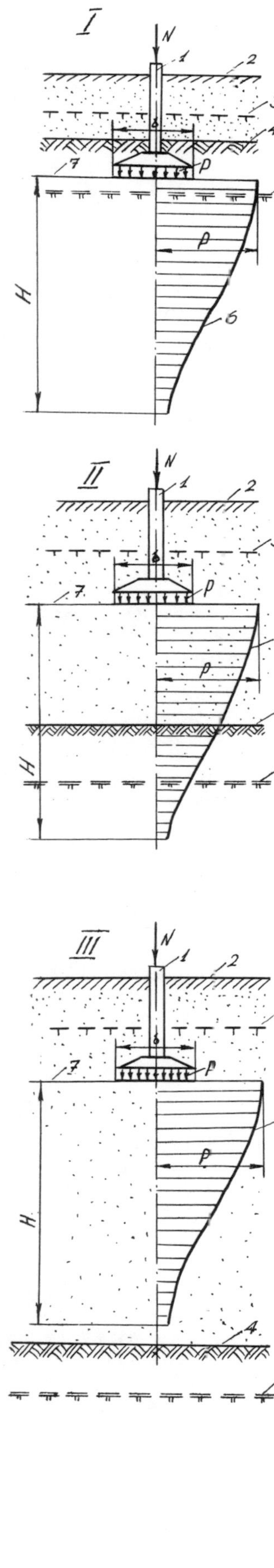

FIGURE 1 Schemes for using fill soil as a bed: (I) Foundation does not influence fill; (II) Fill is influenced by the foundation together with underlying layer; (III) Foundation loads are totally transferred to the fill: δ = width of foundation bottom; H = computable compressed layer; N = loading on foundation; p = pressure on the soil under foundation bottom. 1. foundation; 2. planning mark; 3. limit of seasonal freezing-thawing in the sluiced layer; 4. surface of natural soil; 5. limit of seasonal freezing-thawing in the natural soil before sluice; 6. plot of distribution of compressed pressure; 7. foundation bottom mark.

Depending upon the vertical levelling done for the engineering preparation of the site and the use of sluiced materials as the foundation bed, three scenarios can be distinguished (Figure 1):

I. The foundation does not exert load on the embankment fill; it functions as the levelling layer.

II. The foundations of structures exert loads on the embankment fill and its natural underlying layer.

III. The embankment fill accepts all of the loads exerted on it by the foundations and structures. The underlying layers are not affected by the foundation loads.

In the first scheme the use of a fill embankment, irrespective of the use of foundations, requires that there be no evidence of heave in the sluice; gravel, sand or a combination of both are used.

It is necessary to provide high load bearing capacity of the natural ground. It is compressed under the sluiced layer and frozen in the compressed state, which improves its structural properties. As for Yakutsk, when scheme I is used, freezing takes place over the course of one winter if the fill height does not exceed 2 m.

If foundation works for the account of soil freezing along the lateral surface, it is recommended to use fine-grained soils. Their highest load bearing capacity will be reached under conditions of total moisture capacity and maximum density. Maximum density is provided by the special selection of fractions. The best degree of pore filling in the mixture of two fractions is obtained when the small fraction weight is 30% (Bannik 1976). Filter particles must not be frost-heaved and have a dimension greater than 0.1 mm (Dalmatov and Lastochkin 1978).

When scheme II of sluiced fill use has been chosen, it is necessary that maximum load bearing capacity of both the sluice materials and the underlying ground be maintained. Under sluicing this soil thaws, but under the weight of the sluiced materials it is compacted and a redistribution of moisture takes place. It is recommended to take measures for draining the thawed layer in order to prevent frost heave. Experiments show that sandy soil compresses during the thawing period. When compression is completed they might be used as beds. The calculation of strain deformation must be done according to norms established for two-layered foundation beds. In the case of Yakutsk, the degree of settling under working loads does not exceed allowable standards. Strain deformation of the underlying soil can be reduced when the warming effect of sluicing decreases. Depth of thawing can be diminished by depositing sluiced materials on the frozen ground in the spring, increasing the sluicing intensity and by thermal insulation of the natural surface. In some cases, it appears to be feasible to obtain maximum depth of thawing of natural soils for their melioration in the sluicing process. The technical and engineering techniques that allow for the control of this process within definite limits in a prescribed direction are well known.

When scheme III for the use of sluiced materials is adopted, the thickness of working layer exceeds the computed thickness of the linearly deforming layer. Therefore, the total depth of the sluiced embankment will be 10 m or more in the case of big skeleton soils.

The deformation requirements for the structural properties of the sluiced and underlying soils are the same as for scheme III. If underground ice, ice-rich soils, frozen peat and other types of soil, which undergo considerable settling during thawing, are found in the underlying natural layer sluicing must be done with an eye toward restriction of its thermal effect on the underlying soil. The possibility that the underlying soil could thaw during the construction and operation of buildings must be completely eliminated. The third scheme expands the limits of application of the second principle of soil use as a foundation bed, because the depth of the layer of sluiced materials promotes the formation of an extensive, positive temperature thaw zone. Under natural conditions it can take more than 15 years for this layer to freeze. Measures designed to accelerate this freezing process are frequently ill-advised due to technical and economic considerations.

The choice of scheme depends upon the thickness of the fill and the characteristics of the natural topography of the area being developed. Therefore, in the engineering preparation for future construction on the Yakutsk Lena River floodplain, all three schemes could be adopted.

When the fill embankments are constructed by the sluice technique in permafrost and severe climatic conditions, the problem of choosing the correct scheme for the use of sluiced materials as a foundation bed is of particular importance. This problem has evoked considerable controversy among specialists. The problem, first of all, relates to acceleration of the time required to freeze the sluiced materials, while increasing the thickness of this layer. In addition, the engineers are interested in obtaining a frozen foundation bed as fast as possible. Periods of freezing and stabilization of the temperature depend on the intensity of sluicing, the season when the work is being performed, moisture content of the alluvial fill and also upon the design of building foundations. In order to use permafrost as a foundation bed (in thawed or frozen state) it is recommended to go with the natural development of cryogenic processes in the new fills. Technological and engineering solutions must facilitate intensification of these processes and under no circumstances impede or oppose them. This makes it possible to provide the necessary reliability of the beds and stability of building and structures.

Insofar as temperature regime of the fill soil affects the condition of the ground water, it determines to a great extent the soil's construction properties. In connection with this, the prediction of fill thermal conditions is particularly important for different technological schemes of the use of sluiced materials, thickness of the alluvial layer and conditions of thermal exchange between the alluvial layer and the underlying soil in the bottom of fill and the atmosphere. For the conditions in Yakutsk, the problem of such prediction was solved for the first time using hydro-integration and computer technology at the Permafrost Institute of the USSR Academy of Sciences (Konstantinov and Votyakova 1981).

These calculations proved that there is a tendency for permafrost to form in the alluvial embankment fill under the heat exchange conditions in Yakutsk. However the freezing time indicated by these predictions was considerably longer than for the results of natural experiments.

Based on the results mentioned above and new experimental data from a test polygon and an experimental construction site, the scientists of the Yakut research division have now developed more precise methods of calculating alluvial soil temperature conditions using a computer. The calculations make it possible to get accurate data on fill freezing periods as a function of its thickness, time and sluice intensity, soil composition and moisture distribution taking into account the effect of the inhomogeneity of thermophysical properties on fill depth. An algorithm for the numerical solution of this problem has been developed.

Mathematical models make it possible to select any principle for the use of fill soil as a foundation bed. It should be remembered that formulation of the principles for definition of the conditions for the use of sluiced materials changes in comparison with standards available. This makes it necessary to take into account thermophysical processes in alluvial soils.

Thus, the formulation of the first principle's definition under conditions for the use of sluiced materials might be written as follows:

- alluvial and natural underlying soils of bed are used in a frozen and freezing state (freezing may occur during construction and operation of a building or structure).

For the second principle:

- alluvial and underlying soils are used in a thawed state (thawing of underlying soils at a calculated depth is assumed during the deposition of sluiced materials, building or use of a structure). In this case, steps must be taken to prevent subsequent soil freezing.

In the process of sluicing big skeleton soils and when schemes II and III are used it is advisable to use foundations with developed surface of support, which make greatest use of the resistance of alluvial soil to normal pressure.

When principle I is employed the freezing periods can be considerably (2-3 times) shortened when such measures have been undertaken as construction of ventilated sub-basements, forced cooling of the soil by blowing cold air through the cavities of foundations and engineering lines, the use of automatic closed convective or condensation cooling system, cleaning of snow cover, etc.

The optimum use of sluiced materials, intensification of natural cryogenic processes as a function of nature (climatic and frost-soil conditions), and specialized designs for buildings and structural foundations make it possible to manage the construction properties of foundation beds on sluiced materials both during construction and during the use of buildings and structures.

REFERENCES

Bannik, G. J., 1976, Technical melioration of soils: High School, Kiev.

Dalmatov, B. J., and Lastorkin, V. S., 1978, Gaspipe construction in heaved soils: Nedra.

Konstantinov, J. R., Makhotko, S. P., and Li, G. E., 1981, Temperature conditions prediction of alluvial fill and its bed, in Engineering investigations of frozen soils: Novosibirsk, Nauka, Siberian Division.

Votyakova, N. J., 1981, Computation of sluice soil and its bed freezing, in Composition and thermal conditions of frozen soils: Novosibirsk, Nauka.

ANALOG METHODS FOR DETERMINING LONG-TERM DEFORMABILITY OF PERMAFROST MATERIALS

L. T. Roman

Design and Research Institute for Industrial Enterprises
Eastern Baikal Division, Yakutsk, USSR

The paper demonstrates the possible use of temperature-and-stress-time analog methods for predicting long-term deformation of frozen earth materials. A solution was achieved which allows one to process the data from ground testing using a spherical punch by analog methods.

To compute foundations composed of elastic frozen soil it is necessary to predict settlement for a period comparable to the lifetime of the building and construction. The reliability of the prediction depends on the choice of the optimal mathematical model. It should be capable of revealing the strain character during the primary phase of loading and of giving an exact quantitative description of the deformation over a time of several orders of magnitude greater than the experimental one.

Studies of the rheological properties of frozen soil (Vyalov, 1978) have shown that strain is significantly influenced by temperature. As temperature lowers, creep of soil decreases, other things being equal. In its turn, the effect of temperature on the strain value is interconnected and interequivalent with the effect of time: under stress that is an equal ratio of either instantaneous or long-term strength over a longer time period in low-temperature soil the same strain value appears as in high-temperature soil. Therefore if the influence of temperature and time on soil deformation is determined using the results of short-term specimens in high-temperature tests under loading, long-term soil deformation at low temperatures might be predicted. An analogous relationship has been observed between the effect of stress and time on deformability. It suggests that temperature and stress-time analogy techniques may be used to evaluate deformability of frozen soil. These techniques intensify the strain by one of the affecting factors (e.g., temperature increase) in order to observe the strain under the increased rate and to extrapolate the results of short-term experiments for longer time periods.

To make this extrapolation it is necessary to determine the time reduction ratio (a). Its basis is the correlation of compaction intensity as a function of time and temperature (Urzhumtsev, 1975):

$$g(t',\theta) = (\rho_o\theta_o/\rho\theta)g(t,\theta) \quad (1)$$

where $g(t,\theta)$ - the function of the compaction intensity;
ρ,ρ_o - densities under temperatures θ,θ_o respectively;
t - time;
t' - t/a_θ;
a_θ - temperature-time reduction ratio.

The equation is complicated to use. In practice, the reduction coefficient is determined according to the dependence of the specimen's compliance upon the effect of the time of loading under different temperatures. Compliance is the relationship of relative strain to stress (σ):

$$J = \delta/\sigma \quad (2)$$

Methods of strain computation based on the uniaxial tension and compression (Urzhumtsev, 1975) and on spherical punch pressing (Takahashi, 1964) have been the test techniques widely used in frozen soil mechanics. They simplify the use of analog methods for frozen soil strain prediction and permit the combination of the specimen strength and strain tests.

Using the analog method it is necessary to establish the character of the thermorheological behaviour of the samples under loading. In the theoretical prerequisites of the temperature-time analogy two types of such behaviour are considered (Urzhumtsev, 1975).

1. Under temperature change, compliance depends only upon the displacement of the macromolecules. Isothermal curves of viscous-elastic compliance obtained for different temperatures are similar. The reduction ratio a_θ, which permits the generalized curve under comparative experimental conditions to be plotted, is the function of the single argument of temperature.

Geometrically, the sense of this proposition is: under temperature change the compliance curves J vs ln t are barely displaced along the time axis and the parallelism of their displacement is not disturbed. This significantly simplifies the determination of the reduction ratio a_θ because temperature change stipulates only the horizontal time shift of the compliance.

This type of strain is the characteristic feature of thermorheologically simple materials.

2. When strain is complicated by the effect of structure transformations, the relationship J vs ln t also becomes complicated. A vertical shift as well as the horizontal one has been marked. The

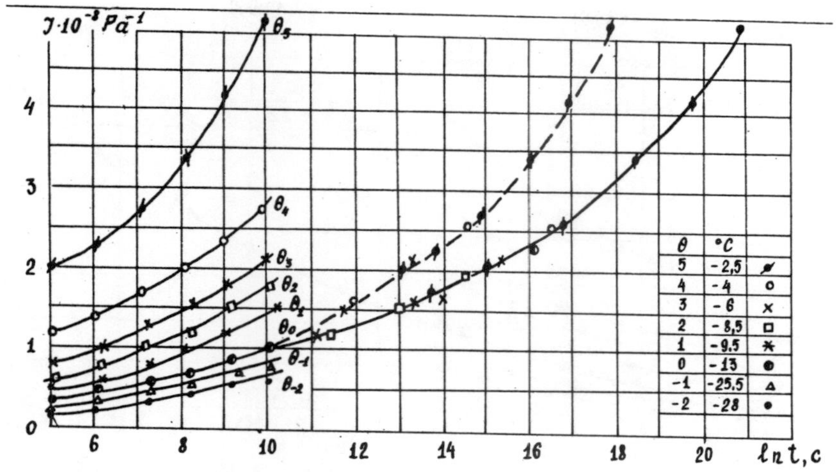

FIGURE 1 Compliance of the log peat ($R = 20\%$, $\gamma_o = 1.03\ g/cm^3$, $\gamma_r = 1.5\ g/cm$, $W = 4.95$) according to the spherical punch test.
1 - generalized curve plotted as for thermorheologically simple materials.
2 - generalized curve plotted as for thermorheologically complex materials.

reduction ratio depends on both the period of strain and the temperature. These are referred to as thermorheologically complex materials.

Computation of the reduction ratio for thermorheologically complex bodies is a complicated problem, so, when the effect of vertical shift is insignificant, it may be ignored. In this case one may plot the generalized curve as for thermorheologically simple materials. The temperature-time analogy does not permit accurate long-term strain prediction when the vertical shift factor cannot be determined.

To apply the analog method to frozen soil it is necessary to understand whether the frozen soil is a simple or a complex material.

The results of frozen soil tests with a 22-mm diameter spherical punch (Roman, 1982) have been analyzed. Under all temperatures the spherical punch was loaded so that its settlement was similar at any given moment in time. It makes identical the stress influence upon strain, and revealed only the effect of temperature. The specimens were made from an upper layer of bog peat. The degree of peat decomposition was from 18 to 20%. To provide homogeneity, the air-dried soil was sieved. The diameter of the sieve was 4 mm. The specimen was then saturated with water and kept in the sieve for 48 hours covered with a filter for free drainage. The soil mass prepared by this technique was used to prepare metaform blocks 100 × 10 × 100 mm. Compaction was by layered tamping and freezing in the underground laboratory at 4.2°C. The specimens were then kept at the test temperature. The experiments had been carried out in the chambers of an underground laboratory and in a heat-proof room where temperature conditions were provided by the effects of outside air temperature in winter. The experimental data for analysis were obtained under isothermal conditions. The physical properties of the soil tested were as follows: soil density, 1.03; soil skeleton density, 0.172; particle density, 1.5; moisture, 4.95; and initial freezing temperature of soil water, -0.08°C. The unfrozen water content, determined by calorimetric method, is shown in Table 1.

TABLE 1 Unfrozen water content (W_H unit fractions) in peat test specimens vs. temperature

θ°C	-2.5	-4	-6	-8.5	-9.5	-13	-25.5	-28
W_H	2	1.3	0.9	0.7	0.6	0.55	0.51	0.5

The most appropriate method must be chosen to determine the relationship between the settlement soil strain modulus load and the punch dimensions for frozen soils to calculate the strain value using the results of the spherical punch test.

Experimental analysis showed that the modulus of strain of frozen soil may be computed from the rigid spherical punch pressing into the elastic half-space (Bezukhov, 1953). In this case, as well as for thawed clayey soil, the radius of the punch print as the initial parameter in the solution must be equal to the radius of the truncated segment of the punch being submerged in the soil.

Since the compliance (J) is the reciprocal magnitude of the strain module using the above solution it can be determined by the formula:

$$J = \frac{4S^{3/2}(d-S)^{\frac{1}{2}}}{3(1-\mu^2)P} \tag{3}$$

where S is the punch settlement (cm);
μ is the Poisson ratio; and
P is the load on the punch.

In Figure 1, the compliance of the tested specimens in conditions J vs ln t calculated by eq. 3 is represented.

The generalized curve has been plotted for -13°C for both thermorheologically simple materials (curve

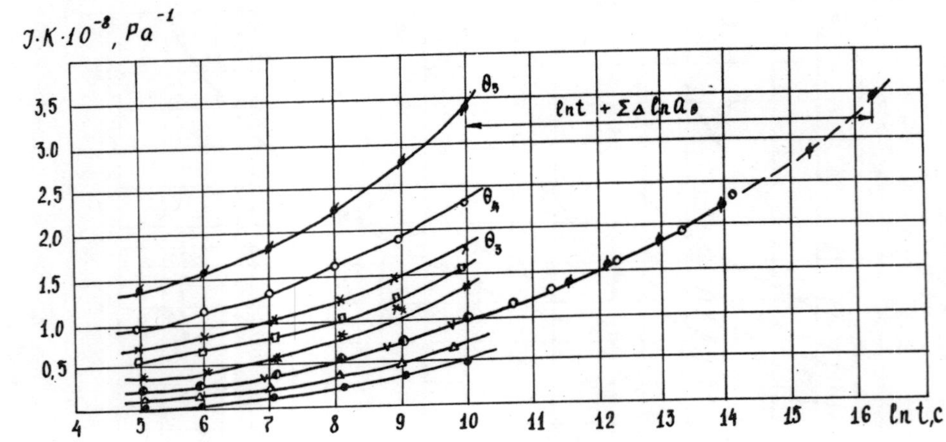

FIGURE 2 Compliance with K-parameters being taken into account and generalized curve plotted as for thermorheologically simple materials.

1) and thermorheologically complex ones (curve 2). The procedure for plotting the generalized curves by both methods is detailed in Roman (1982). To evaluate the degree of reliability of any method, i.e. to decide whether the frozen soil is a thermorheologically simple or complex material, long-term, 30-day tests were conducted. The conditions of the base experiment were preserved. The results show that frozen soils should be referred to as thermorheologically complex materials: the test points are on curve 2.

Factors that disturb the frozen soils' continum and homogeneity are the main reasons for stipulating the vertical shift of the compliance curves. To plot the generalized curve, results of experiments carried out under a wide range of temperatures were used, so the influence of unfrozen water content will be different in either test.

The volume of soil occupied by unfrozen water and gases as well as the volume occupied by ice and soil particles can be determined. The relative content of ice and particles (K) is easily expressed as a proportion of the soil volume:

$$K = \gamma_{ck}[1/\gamma_r + (W_c - W_H)/\gamma_\wedge] \qquad (4)$$

where γ_{ck} - density of the soil skeleton;
γ_r - soil particle density;
W_c - total moisture;
W_H - unfrozen water content
$\gamma_\wedge$ - ice density.

To take the influence of the gas content and unfrozen water on compliance into account, the stress should be referred not to the geometric square of the specimen but to the square occupied by soil particles and ice. An approximation is acceptable if it is proportional to the relative content of soil and ice (K). Therefore, the compliance value related to the source of soil particles and ice will be equal to:

$$J_k = \delta k/\sigma = JK \qquad (5)$$

For the experimental data given in Table 1, K-values calculated with eq. 4 are represented in Table 2.

TABLE 2 K-values for specimens tested

θ°C	-2.5	-4	-6	-8.5	-9.5	-13	-25.5	-28
K	0.625	0.746	0.815	0.85	0.867	0.876	0.883	0.884

Plots JK vs ln t and the generalized curve plotted for thermorheologically simple materials are shown in Figure 2. It can be seen that the K parameter allows the influence on the compliance of the structure modifications connected with moisture phase transition to be reduced, and simplifies long-term strain prediction.

Interpretation of the experimental data on the J vs. ln t coordinates shows that the family of compliance curves has converged at the pole. The abscissa of the pole was approximately equal to the time log of the atoms' free oscillation (10-13) and the ordinate was approximately equal to the value of the instantaneous-elastic compliance. To calculate the pole ordinate for the specimen tested, the module of instantaneous elasticity E_y was obtained with an ultrasonic technique (Urghumtsev and Maksimov, 1975). Instantaneous-elastic compliance taken as a reciprocal value E_y was from 1.1^{-6} to 2.1^{-6} MPa^{-1}. It is one to two orders of magnitude less than the compliance value computed on the total strain of the specimens (elastic and residual). It makes it possible to determine the pole ordinate, which is equal to 0.

The availability of the pole suggests the plot technique of curve construction. The rays crossing the compliance curves are drawn. The coordinates of the points of the rays' concurrence and either curve have been determined. The relationship JK vs. ln t from $\theta_t - \theta_o$ under temperature-time analogy and $\sigma_i - \sigma_o$ under stress-time analogy have been plotted. (σ_o, θ_o are stress and temperature correspondingly, and the base curve has been plotted for them). Extrapolating these relationships to the ordinate axis we determine the JK and ln t values corresponding to $\sigma_i - \sigma_o = 0$ or $\theta_i - \theta_o = 0$; i.e. we have the coordinates of the points of concurrence of the secant rays with the base curve. Using those coordinates, the base curve for a period of

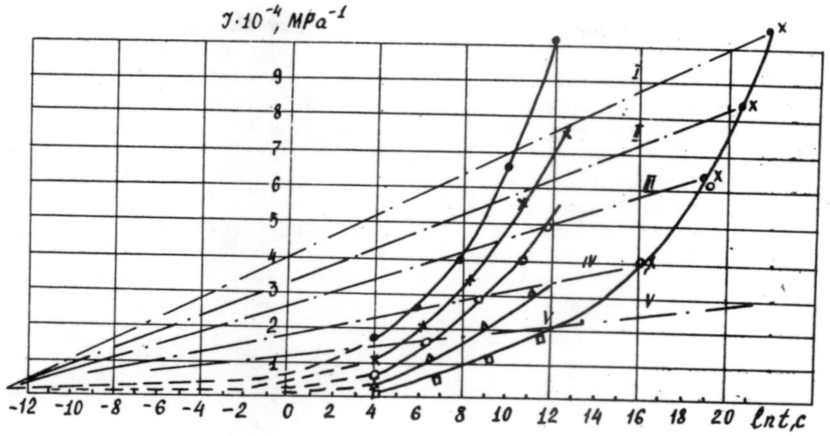

FIGURE 3 Plotting method of generalized curve construction using the log peat experimental results by the stress-time analogy method (θ = 4.5°C).

time exceeding the experimental one by several orders of magnitude was plotted. Figure 3 is an example of generalized curve plotting based on the results of frozen peat tests using the stress-time analogy. In practice, the generalized curve coincides with the curve obtained by the conventional method, i.e. a rigid horizontal shift to the compliance curves to the right of the reduction ratio value.

The secant ray technique proposed in this paper is less labour-consuming and it permits the period of strain prediction to be increased.

A series of experiments carried out to plot the generalized curve allows long-term strains to be predicted under other than base stress. Interpretation of the experimental data shows that the plotting relationships lnJK vs $\sigma_i - \sigma_o$ determined for either fixed time form a family of parallel lines (Figure 4A). Therefore it is easy to get the orbit line lnJK - $\sigma_i - \sigma_o$ for the time being equal to each value within a predicted period including the maximum one (t_{max}). For this purpose it is necessary to find lnJK value on the abscissa of the lnJK vs $\sigma_i - \sigma_o$ plot taken from the base curve for t_{max}. Across this point a line is drawn parallel to the experimental ones. This line is the compliance relationship with stress for time period t_{max}.

For the temperature-time analogy method the lnJK vs ln t relationships prove to be linear. Their extrapolation to the given time period allows the long-term compliance under all experimental values of t to be determined (Figure 4b).

The reliability of the results of frozen soil long-term strain prediction by the analogy method has been verified in two ways:

1. Comparison of the results of the strain prediction based on the creep equations suggested by

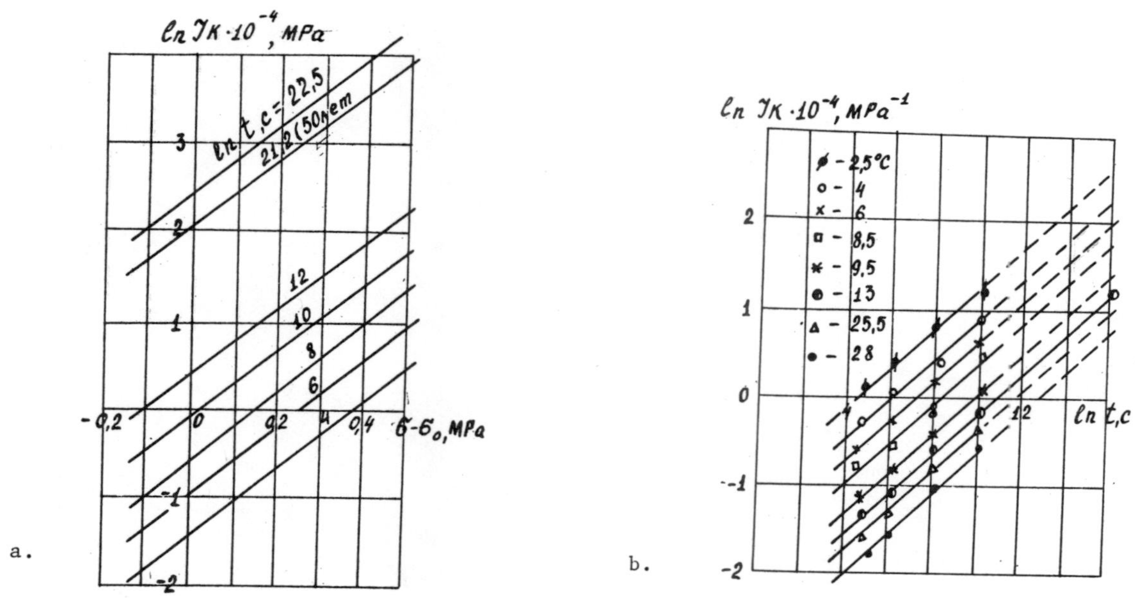

FIGURE 4 ln JK relationships: a) ln JK - $\sigma_i - \sigma_o$ plotted on test data given in Figure 3. b) ln JK - ln t plotted on test data given in Figure 2.

Vyalov (1978) and those based on the time-stress analogy method.

2. Comparison of the punch experimental settlement computed for the case when the strain module had been determined by the analogy method. The same data were used as in the first case.

The equation was written as a power function, accounting for the similarity of the creep curves. A comparison of the results is given in Table 3.

TABLE 3 Significance of bog peat compliance tested at 4.5°C under pressure of 0.2 MPa for 908 hr.

Technique	$JK \cdot 10^{-4} MPa^{-1}$
Using unconfined compressive strength	1.45
Creep equation	1.68
Stress-time analogy	1.39

The punch test was used to test the second method of comparison of the settlements. In the underground laboratory (θ = -4°C), a cylindrical hole 30 cm in diameter and 17 cm deep was drilled in a sand layer of natural composition. It was filled with peat moss under layered tamping.

After freezing and temperature stabilization a 17-cm-diameter punch was installed. A jack provided incremental loading. The process of settlement over time and its dependence upon stress has been fixed. The frozen peat compliance at the same temperature was also determined by the stress-time analogy. The generalized curve could be plotted and the relative settlement determined for these specimens. At the end of the stabilization period, the settlement was determined by the layer-adding technique. Comparison of the experimental and the computed settlement values show a good agreement for the first step of loading. For the last steps, the predicted values were somewhat less than the experimental ones. This was caused by ignoring the fact that the loading of previous steps occurred over longer periods of time. In conclusion it should be pointed out that settlement prediction is most reliable by the analogy method. It accounts for the character of the soil rheological properties, in particular the nonlinearity of the strain dependence upon stress.

REFERENCES

Bezukhov, N. J., 1953, Basis of the theory on elasticity, plasticity and bearing capacity of frozen soils: Moscow, USSR Academy of Science, p. 188.

Roman, L. T., 1982, Prediction of creep of frozen peat by the temperature-time analogy method, in Soil rheology and engineering geocryology Publication.

Takahashi, M., Shan, M., Taylor, R., and Tobolyky, A., 1964, Master curves for some amorphous polymers. J. Appl 1. Polymer. Sci., v. 8, no. 4, p. 1549-1564.

Urzhumtsev, Ju. C. and Maksimov, P. D., 1975, Prediction of polymer material strain ability: Riga, "Zinatne", p. 416.

Vyalov, S. S., 1959, Rheologic ability and bearing capacity of frozen soils: Moscow, USSR Academy of Science, p. 188.

Vyalov, S. S., 1978, Rheological basis of soil mechanics: Moscow, High school, p. 447.

RESISTIVITY LOGGING OF FROZEN ROCKS

A. M. Snegirev

Permafrost Institute, Siberian Branch
Academy of Sciences, Yakutsk, USSR

The article examines the present state of some aspects of resistivity logging as a means of studying the cryolithozone, and the results of investigations in the past few years. The efficacy of resistivity logging is determined by the differentiation of the electrical properties of frozen ground, by the accuracy of measurement and by the technical conditions of drilling. The potential of the major methods is being fully realized in holes drilled without using fluid solutions as long as the values for electrical resistivity of the frozen rocks do not exceed 800-1000 ohm-meters. An increase in the range of measurements turned out to be possible only by using measuring equipment with a high input resistance and of new technical elements that substantially improve the accuracy and operational characteristics of resistivity logging equipment where parameters of constant and low frequency electric fields were being measured in dry holes. The investigations carried out and the observational error achieved allow one to apply resistivity logging studies of low-gradient processes occurring within the permafrost zone, to improve the methods and technology of studying geoelectrical cross sections, and to study variability in the physical and chemical properties of permafrost in relation to the effects produced by external physical factors.

Resistivity logging as a method of obtaining information about frozen rocks is based upon the assumption that the geocryological section can be represented as spatially distributed electrical parameters of the medium of study (for example, resistivity, electrochemical activity, dielectric permittivity), varying in time and depending on composition, ice content, concentration of mineral salts in ice and pore moisture, temperature, and other characteristics of rocks. In principle, finding correlations between these parameters will contribute to a mass determination of the geotechnical characteristics of permafrost that are needed by building workers and specialists studying heat and mass exchange in permafrost and the deep freezing of the Earth's crust, as well as by geologists and geophysicists for the purposes of reliable interpretation of field data.

Study of resistivity logging with regard to permafrost is justified given the rich background of experience associated with the application of resistivity logging in prospecting for oil, coal, and ore in areas where no permafrost is present.

In a more specific context—that is, the determination of analytical and experimentally statistical relationships between the recorded components of an electric field and the properties of the medium being studied—this problem becomes extremely complicated, not only because of the special features of permafrost but also because of the metrological characteristics of measuring instrumentation and equipment, and peculiarities of the state of frozen rocks in the vicinity of a borehole, which are in many respects determined by the drilling method.

This last factor is known to play an important role in selecting electrical methods for frozen rock studies in boreholes and should be taken into account in field data interpretation.

In this connection we now consider, in the first line, the conditions of conducting resistivity logging in boreholes that are drilled using a fluid, whether the fluid remains in the borehole shaft or is removed after drilling operations are completed, or that are drilled using compressed air.

The interpretation of resistivity logging diagrams and electrical field parameter measurements in boreholes filled with drilling fluid is marked by the appearance of an unfrozen layer in the vicinity of the borehole shaft by a high-gradient temperature field perpendicular to the borehole axis, and by the properties of the drilling fluid. The unfrozen layer produced by the drilling surrounds the borehole shaft (on occasion its thickness reaches 1-2 m (Maramzin 1963) and allows only lithological boundaries within the cross section of the borehole to be revealed through resistivity logging (for example, Musin and Sedov 1977), and to a considerable extent it hinders or even rules out the possibility of studying the electrical properties of the permafrost levels.

In the presence of such a layer and temperature gradient normal to the borehole axis, large resistivity installations have to be employed to reveal the cross section and to investigate only relatively thick frozen rock benches. The use of drilling fluid in drilling operations plays an important role in assessing the thickness of the permafrost zone using resistivity logging diagrams of the natural electrical field. The experience accumulated in the Soviet Union (Volodko 1976, and others) indicates that the accuracy of such determinations is fairly high.

In boreholes where the drilling fluid is

removed after drilling is completed, there also arises a relatively thick layer of modified rocks. The processes of thawing at the time of drilling and of refreezing in the vicinity of the borehole shaft after drilling operations are completed lead to irreversible changes in the rock properties. In particular, once drilling is terminated, it is possible to apply methods using direct and alternating current in order to subdivide the cross section lithologically, but after the temperature regime has recovered it is difficult to resolve this task. In a number of cases the high electrical resistivity of frozen rocks makes it impossible to apply frequency methods in investigations (for example, the induction method) and postulates special requirements to the measuring technique for electrical conductivity, natural electrical field, and polarizability (for example, ensuring low contact resistance at the electrode-rock contact, the use of instrumentation with high input resistance, and others).

Similar measuring conditions are created within boreholes that are drilled through permafrost while purging the face with cooled compressed air. However, experience shows that these boreholes are more suitable for studying the geoelectrical cross section of permafrost because the maximum thickness of rocks warmed by the drilling does not exceed 0.2-0.3 m, and the natural temperature regime recovers in the course of a few days (Balobaev et al. 1977).

For several reasons, notably the lack of water in the winter time and the difficulties of preparing and using unfrozen drilling fluids, drilling operations using compressed air is widely used in the North for drilling test and prospecting holes. The volume of such drilling for the Yakutian Republic alone exceeds many hundreds of running meters. At the same time, techniques for resistivity logging of these boreholes, judging by the available publications (e.g., Bakulin 1973, Akimov 1973, Hunter 1975, Jackson 1956, Avetikyan and Dorofeev 1970, Bakulin 1967, and others) have not yet been adequately developed.

Most of the publications include an analysis of the results of investigations, but technical and methodical questions are discussed to a lesser extent, although logically it is they that should be emphasized in connection with the specifics of the investigations being conducted. Poor information about the results of the investigations and imperfect resistivity logging techniques for frozen rocks in dry boreholes seem to be the reasons why, in the practice of engineering-geological and geocryological work, the resistivity logging method is being used to study only an insignificant number of boreholes. In addition, these boreholes are most frequently filled with a specially prepared fluid and observations are made with techniques developed in oil and ore geophysics using standard measuring instruments. In this case the rocks near the borehole shaft, particularly dispersive ones, become mechanically not solid and therefore there are frequent cases of a tack of electrical probes. It is understandable that the effectiveness of such research is on the whole low not only because of the need for performing a number of additional labor-consuming operations but also in connection with time limitation of such observations and therefore of operations in each borehole.

All of this requires further solution of the theoretical, methodological, and technical problems of resistivity logging of dry boreholes because they provide the best conditions for the study in the permafrost zone.

Theoretically, due to the wide range of electrical conductivity values of frozen rocks it is practically infeasible in the investigations to give preference to resistivity logging methods using either direct or alternating electromagnetic fields. The primary application in the Soviet Union and abroad of resistivity logging using direct current could be due to the relative simplicity of the technical and methodical capabilities of this kind of research, but analysis of the published data and of results of special research shows this explanation is premature.

Several problems face resistivity logging, namely the identification of particular features of a geoelectrical cross section (determining electrical parameters of individual levels, their time variation and space variation, etc.), the refinement of the structure of a geocryological cross section (ice content distribution in the cross section and within the lithological-stratigraphical levels, the thickness of particular rock benches, etc.), identification of the structure of natural and artificial electrical fields in the vicinity of the borehole and within the area of study, and so forth. These problems will be successfully resolved through the use of appropriate measuring installations and a sufficiently high accuracy of measurements of electrical field parameters is ensured.

The former (except for the solution of special problems such as detection of ice in the borehole cross section, evaluation of the amount of unfrozen water, etc.) does not require special developments and is satisfied comparatively readily on the basis of an extensive scope of theoretical and methodological investigations in the field of oil, coal, and ore resistivity logging.

The latter requirement however can only be met after the main factors that distort the results of measurement of electrical parameters are identified and removed. We now consider these factors in the context of resistivity logging using direct current.

The contact resistance electrode-frozen rock (ρ_{el}) predetermines the choice of a generating device and measuring instruments for the resistivity logging installation. For a dry borehole drilled in frozen rocks, ρ_{el} is of particular importance because the known methods for decreasing and ensuring the stability of this value (moistening, increase of the contact area, enhancement of clamping of the electrode to the hole wall, etc.) are not always feasible. For example, increasing the electrode's area of contact with the hole wall decreases ρ_{el}, limits the minimum dimension of resistivity logging installations and necessitates complex calculations of the geometrical coefficient of the probe (Oparin 1978, and others). In this case, the value ρ_{el} nevertheless remains dependent on the electrical resistivity of rock (ρ_n).

Tests of various borehole electrode designs (brush, spring-type, and others) have demonstrated that a maximum value of ρ_{el} is reached as high as 500 Kohm; therefore, to obtain a relative error of measurement δ = 1-20%, the input resistance of the measuring device should be of at least 50 Mohm. It can be seen that operating instrumentation with such input resistance is possible only when definite microclimatic conditions are provided in the field laboratory, which is not always feasible.

In one of the latest papers (Snegirev et al. 1981) it is proposed to dampen the surface of the contact electrode-rock with an electrolyte contained in a nonpolarizing electrode. Automatic control of the electrolyte flow foreseen in the design ensures a value of $\rho_{el} \leq 100$ Kohm.

In addition, stable electrode potential (no more than 3-5 mV) makes it possible to study in a dry borehole the natural electrical field and the induced electrochemical activity of rocks.

One other property of the measurements may include current leakage from the current line into the measuring line through the surface of the resistivity logging cable.

In general, the influence of current leakage in resistivity logging installations upon the results of the investigations is known, as are the steps to be taken to avoid leakage (Geophysicist's Manual 1961). Their observance in investigating boreholes filled with drilling fluid normally yields stable, high-precision results. For dry boreholes the fulfillment of these conditions is necessary but insufficient.

As a result of friction of the electrodes on the borehole walls, the detaching ice particulates touching the relatively warm cable thaw to produce a wet film on its surface between the probe electrodes. Such a film may also result from electrolyte leakage from the non-polarizing electrode or from moisture that penetrates into the borehole from a seasonally unfrozen layer, as well as from other causes. The electrical conductivity of the rising layer is unstable and varies due to variations in its thickness, salt concentration, etc. Depending on voltage, the electrical resistivity of the rock, and the electrode earthing resistance, the amount of current leaked may be substantial. It is easy to show that with increasing rock resistivity and electrode earthing resistance the relative fraction of leaked current on the cable surface between the electrodes increases, other conditions being equal; therefore the measurement error will increase. (According to calculations and experimental observations the relative error reaches 100% and more.)

This fact increases the existing requirements for insulation resistance among all structural elements of the resistivity logging installation (not less than 50 Mohm) and requires the development of special methods for eliminating the current leakage between borehole electrodes that occur over the surface of the cable. One of these devices, proposed in Snegirev et al. (1981), breaks the current-conducting film by hollow packings, made of dielectric hydraulic insulating material shaped like cones bell-mouthed downwards and mechanically supported by the cable between the electrodes.

The evaluation of the influence of complicating factors, such as the diameter of the borehole wall, the Earth's surface, and the altered rock zone, upon other kinds of measurement results can in principle be carried out analytically and via modeling.

While excluding the possibility of using them, nevertheless in practical work it is advisable to apply such resistivity logging installations, whose readings would be independent of these factors. In particular, probes that are more than 5-6 times the diameter of the borehole satisfy quite reasonably the solving of most of the problems; in this case the borehole diameter can be omitted in interpreting the observations. Efficient is also the application in probe designs of special devices that preserve the separation of clamping borehole electrodes (for example, Borisenko et al. 1977, and others).

Figure 1 exemplifies some results of resistivity logging, as obtained within one of the regions of Yakutiya.

The scientific objective included an assessment of the abilities of resistivity logging installations to distinguish between various frozen lithological-stratigraphical levels. To aid comparison, a gradient probe and middle and vertical gradient installations were employed.

The last two modifications were used to preclude shielding effects in the results of the investigation. For that purpose, a system of fixed feed electrodes installed on the Earth's surface or within the borehole was used to create a d.c. vertical electrical field.

The measuring system of all installations (Figure 1e) incorporated nonpolarizing electrodes, a resistivity logging cable, and a millivoltmeter with input resistance of more than 50 Mohm. The electrode potential drop and natural electrical field potential were compensated for at every point before cutting in. The measuring step was 0.5 m.

The set of all elements of the system and allowance for special features of frozen rock resistivity logging described above provided at least 3% error of measurement.

The geocryological cross section depicted in Figure 1 is representative of typical rocks of the region, exhibiting in the vicinity of the borehole a uniform thickness, horizontal bedding, and different electrical conductivity. The behavior of the ρ_k plots within each lithological difference is due both to the peculiarities of the formation of the geocryological cross section and to the properties of the measuring installation being used.

The curve ρ_k on the interval of rocks of trap formation (0-40 m) has a maximum in connection with the different degree of metamorphism of the central and peripheral parts, while the peaked appearance of the ρ_k plot obtained with a gradient probe is accounted for by current shielding effects produced by rock cracks. Given fixed feed electrodes (Figure 1b, 1c), the curve displays a smooth behavior.

We have similar plots on the borehole interval between 95 and 150 m, composed of benches of dolomites, marls, and limestones, which have different electrical resistivities. By reducing the effects of shielding, the positions of separate benches of seams can be reliably determined.

The middle part of the cross section made up of sand from the Permian age is characterized by smooth ρ_k plots. Despite the different conditions of electrical field excitation, the ρ_k values do not practically differ from each other, so the sand cross section, especially its upper part (50-70 m), can be regarded as uniform and isotropic.

For a specific conduction of work this interval can be selected as a reference in determining transverse resistances (Alpin 1978).

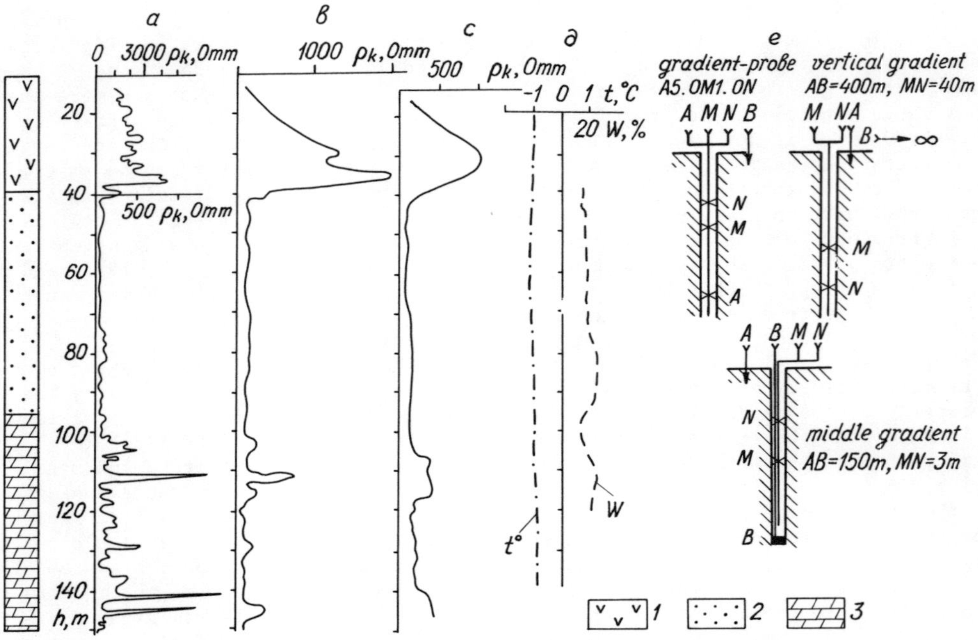

FIGURE 1 Results of resistivity logging.
1) crack dolerites; 2) sand; 3) alteration of marls, limestones, dolomites; a) ρ_k curve, gradient probe; b) $\rho_k$curve, middle gradient installation; c)ρ_k curve, vertical gradient installation; d) temperature t^oC and weighting humidity W % of frozen rocks; e) layouts of measuring resistivity logging installations.

Comparison of the plots in Figure 1b and 1c suggests that, in principle, the use of a middle gradient installation makes it possible to determine the transverse resistance of seams that form the middle part of the cross section because within the borehole it helps create a vertical electrical field, ensures the measurement of fairly high values of electrical potentials, and removes the influence of shielding effects on the measurement results. Nevertheless, operations with this installation are complicated by the labor-consuming operations needed to provide reliable grounding of the feed electrode in the dry borehole shaft and by the presence within the borehole of a feed cable, which displaces the measuring dipole. Therefore the vertical gradient installation for a study of shallow dry boreholes provides a number of advantages as compared to the known point resistivity logging technique KC.

The essence of these merits lies in the possibility of obtaining plain-shaped plots of observations, uncomplicated by shielding effects, that permit solution of a major logging problem, namely the determination of the parameters of rock seams that have different composition and properties. In addition, it is also possible to run continuous resistivity logging of rocks within a dry borehole. No attempts have been made to address this problem through the use of three-electrode probes because of the extreme difficulty of producing a current stabilizer that operates where there is sharp variation of load. With the vertical gradient installation, the need for such a stabilizer no longer arises, while the variation of the installation coefficient as the measuring dipole moves along the borehole shaft can be compensated for.

The example just considered reflects an insignificant fraction of work on the effectiveness of resistivity logging methods. (See Akimov 1973, Bakulin 1967, Snegirev et al. 1980, 1981, and others).

In summarizing, the following has to be emphasized. It is advisable to conduct studies of geoelectrical cross sections of frozen rocks with resistivity logging within boreholes driven without the application of drilling fluids. The technical and methodological problems can be overcome if the factors that influence the accuracy of observations and distort logging plots are taken into consideration and eliminated.

The investigations carried out permit measurement of electrical field parameters within a dry borehole with a relative error of 1-3%; which makes it possible to utilize logging installations in experiments and observations of weak-gradient processes occurring within permafrost.

The peculiarities of resistivity logging of rocks outlined above do not cover the diversity of factors that researchers may encounter, but we feel they are crucial. The degree of reliability in determining the electrical parameters of the cross section using resistivity logging data will largely determine the successful application of electrical methods in areas of development of permafrost. Therefore there is a need for further development of the techniques for borehole investigations. An in-depth study and accumulated experience will make it possible, taking into account the state-of-the-art of the measuring technique, to start implementation of the most advantageous continuous resistivity logging of frozen rocks in dry boreholes.

REFERENCES

Akimov, A. T., 1973, Logging in shallow dry boreholes aimed at studying geotechnical and geodynamical characteristics of permafrost, II International Conference on Permafrost: Yakutsk, Yakut Publishing House, p. 18-25.

Alpin, L. M., 1978, Determination of transverse resistivity of strata from measurements in boreholes: Izvestiya VUZ, Geology and prospecting, no. 4, p. 52.

Avetikyan, Yu. A., and Dorofeev, E. A., 1970, High-quality logging of boreholes for determining geoelectrical properties of rocks, Engineering investigations in civil construction: Synopses 4(7), Ser. II, Moscow, p. 61-65.

Bakulin, Yu. A., 1967, An instrument for resistivity logging of dry boreholes and results of its application in geocryological investigations: Komi, p. 213-219.

Bakulin, Yu. A., 1973, On the use of resistivity logging of dry boreholes in determination of composition and some properties of frozen rocks: Proceedings of the VNII for hydrogeology and engineering geology, v. 62, p. 59-65.

Balobaev, V. T., et al., 1977, Manual on thermistor calibration and their use in geothermal measurements: Yakutsk, p. 28.

Borisenko, Yu. N., et al., 1977, Design of the probe for the apparent resistivity method, in Geophysical instruments, v. 62, Leningrad, "Nedra," p. 108-111.

Geophysicists Manual, 1961, v. II, Moscow, p. 157-158.

Hunter, J. A., Borehole geophysical methods in permafrost: Geological Survey of Canada, Paper 75-31, p. 67.

Jackson. Equipment and method of borehole logging, US patent no. 2755450. 17.07.56.

Maramzin, A. V., 1963, Experience of drilling geological prospecting boreholes in frozen rocks: Moscow, Gosgeoltekhizdat, p. 21-26.

Miller, A. V., 1970, The technique of resistivity logging of dry boreholes, in Proceedings of the 2nd Republican scientific-theoretical conference of young geologists of the Kazakh SSR: Ust-Kamenogorsk, p. 181-182.

Musin, A. M., and Sedov, B. M., 1977, Some peculiarities of borehole resistivity logging under conditions of permafrost: Abstracts of papers and communications "Methods of engineering-geological investigations and mapping of the permafrost region," v. 2, Yakutsk, p. 30-32.

Oparin, V. N., 1978, Depth range of borehole electrometry, Journal Physico-technical problems of mineral resources exploitation: Novosibirsk, Nauka, v. 4, p. 84-87.

Snegirev, A. M. et al., 1980, Some peculiarities of resistivity logging (apparent resistivity) of frozen rocks in dry boreholes: Izvestiya VUZ, Geology and prospecting, no. 6, p. 76.

Snegirev, A. M. et al., A probe for resistivity logging of boreholes, Patent certificate no. 802891. 7.2.81, Bulletin no. 15.

Snegirev, A. M. et al., A non-polarizing electrode. Patent certificate no. 894655. 30.12.81, Bulletin no. 46.

Snegirev, A. M. et al., 1981, Geoelectrical cross section of frozen rocks of the Viluy River watershed area, in Field and experimental investigations of permafrost: Yakutsk, p. 7-16.

Volodko, B. V., 1976, The use of the AR method in determining the thickness of frozen terrigenous banks, in Regional and thermal-physics investigations of frozen rock in Siberia: Yakutsk, p. 212-214.

STRESS-STRAIN CONDITION AND THE ASSESSMENT OF SLOPE STABILITY IN AREAS OF COMPLEX GEOLOGICAL STRUCTURE UNDER VARIOUS TEMPERATURE REGIMES

S. B. Ukhov, E. F. Gulko, V. P. Merzlyakov, and P. B. Kotov

Institute of Civil Engineering
Moscow, USSR

Calculations of the stress-strain condition and the stability of rock masses in northern regions are complicated by the effects of temperature and by associated physical and mechanical processes. In terms of thermodynamics these processes are controlled by the stress tensor, temperature, concentrations of individual components and phases, crack density tensor, and by irreversible internal parameters. Equations of the condition of the rock, including the above-mentioned variables, complete the system of equations of equilibrium expressing geometric relationships and heat-mass exchange. It is possible to simplify the proposed scheme to the level of engineering possibilities by taking into account adequate representation of results. In this case a set of engineering investigations essential for assessing the stability of the sides of reservoirs under different regimes would include the following: 1) On the basis of data derived from engineering exploration the real structure is represented in the form of engineering and geological models and reflecting the peculiarities of its composition, structure and condition; 2) the laws governing the behavior of rocks under loads in frozen, thawed, and thawing conditions are determined under both field and laboratory conditions. On this basis the physical and mechanical properties of the rocks of separate zones and the elements of the actual slope are determined. Geomechanical models of the rock mass are then built on the basis of the data obtained; 3) items 1) and 2) allow one to compile calculations of the stress-strain condition of the slope under varying conditions; 4) on the basis of the data derived from calculations of the stress-strain condition minimum stability safety factors may be determined in each element of the layout network, potential slip planes may be identified and integral safety factors of slope stability may be calculated.

Analysis of the stress-strain condition and the stability of rock masses in northern regions includes the physical and mechanical processes connected with temperature. For a general case, thermal strains of frozen and thawing rock are controlled by the behavior of skeleton and ice crystal lattice, phase transitions, structural transformations, water and air compressibility, chemical reactions, and migration and filtration of moisture. In this connection, the classical concept of thermal strains

$$\varepsilon = \alpha\Delta\theta \tag{1}$$

should be more accurate. Here ε is the relative strain, α is the thermal expansion factor, and $\Delta\theta$ is the temperature change.

In frozen masses of complicated geological composition, several nonhomogeneous types can be distinguished. The nonhomogeneity of properties is conditioned by the nonhomogeneity of the mass composition, its structure, and state (Ukhov 1975). Nonhomogeneity caused by phase transitions (frozen and thawing zones) is clearly seen within a homogeneous rock mass. It is necessary to apply thermodynamics to describe the processes in such a medium (Merzljakov 1980).

To describe a thermodynamic system one should choose the so-called external parameters characterizing conversible processes. First, there are the stress tensor components σ_{ij} (or strains ε_{ij}) and temperature θ. The influence of the intensive phase transition zone is taken into account by adding concentrations of components and phases m_1, variable in this process, to the mentioned variables. The degree of fissuring described by crack density tensor β_{ij} (Vakulenko 1974) also influences the mechanical properties of rock to a considerable extent. Thus, thermodynamic functions, namely full thermodynamic potential ϕ, are the functions of four variables

$$\phi = \phi(\sigma_{ij}, \theta, m_1, \beta_{ij}), \tag{2}$$

and a potential change in a reversible process may be expressed by

$$d\phi = \left(\frac{\partial\phi}{\partial\sigma_{ij}} + \frac{\partial\phi}{\partial\beta_{kl}} \cdot \frac{\partial\beta_{kl}}{\partial\sigma_{ij}}\right) d\sigma_{ij} + \left(\frac{\partial\phi}{\partial\theta} + \frac{\partial\phi}{\partial m_1} \cdot \frac{\partial m_1}{\partial\theta}\right) d\theta . \tag{3}$$

Also taking into consideration that in the reversible process,

$$d\phi = \mu_1 dm_1 + S d\theta - \varepsilon_{ij} d\sigma_{ij} \tag{4}$$

we get

$$\varepsilon_{ij} = -\frac{\partial\phi}{\partial\sigma_{ij}} - \frac{\partial\phi}{\partial\beta_{kl}} \cdot \frac{\partial\beta_{kl}}{\partial\sigma_{ij}} \tag{5}$$

where μ_1 is the chemical potential of a component or phase 1 and S is the entropy of the system. Here and further, summing up is implied by repeated indices. Tensor ε_{ij}, in general, is a tensor of finite strains. It should be noted that eq. 4, generally speaking, makes it possible to take into account mass exchange processes by introducing of the source of the K-component:

$$dm_k = G_k dt \tag{6}$$

where G_k is weight flow per time unit. By disintegrating a thermodynamic potential (eq. 2) into Taylor's series regardless of initial stressed state and differentiation in accordance with eq. 5 we get

$$\varepsilon_{ij} = C_{ijkl}\sigma_{kl} + C_{ijklpq}\sigma_{kl}\sigma_{pq} + \ldots + \alpha_{ij}\theta +$$

$$+ \alpha_{ijkl}\sigma_{kl}\theta + \ldots \tag{7}$$

Here, the constant coefficients are respectively equal to the following:

$$C_{ijkl} = [\partial^2\phi/\partial\sigma_{ij}\partial\sigma_{kl} + (\partial^2\phi/\partial\sigma_{ij}\partial\beta_{pq})(\partial\beta_{pq}/\partial\sigma_{kl})]_o,$$

$$2C_{ijklpq} = [\partial^3\phi/\partial\sigma_{ij}\partial\sigma_{kl}\partial\sigma_{pq} - (\partial^3\phi/\partial\sigma_{ij}\partial\sigma_{kl}\partial\beta_{st})$$

$$(\partial\beta_{st}/\partial\sigma_{pq})]_o,$$

. . . .

$$\alpha_{ij} = [\partial^2\phi/\partial\sigma_{ij}\partial\theta - (\partial^2\phi/\partial\sigma_{ij}\partial m_p)(\partial m_p/\partial\theta)]_o,$$

. . . .

As a rule, not every elastic constant C_{ijkl} is independent. It is possible to demonstrate that, for instance, for fissured rock the number of independent constants does not exceed six. Instead of eq. 7, it is possible to use the equivalent relationship

$$d\varepsilon_{ij} = C_{ijkl}d\sigma_{kl} + \alpha_{ij}d\theta \tag{8}$$

where coefficients C_{ijkl} and α_{ij} are functions of variables σ_{kl}, θ, β_{kl}, and m_p, which are determined experimentally.

Equations of state 7 or 8 complete the equilibrium equation system, geometric relationships, and equations of heat-mass exchange. As has been shown by Porkhaev (1970), mass exchange processes may be neglected in most cases while calculating thermal field in mass rock. Therefore further we shall consider that heat movement inside frozen and melted zones and displacement of a "demarcation" line are carried on only by conductivity heat exchange. Stephen's problem solution may be obtained according to enthalpy techniques, which take into account the intensive phase transition zone. One of the versions may be found in Plotnikov (1978). The variable surface of frozen water in such a case is a known temperature function (Ershov and Akimov 1979). Common calculation of the stress-strain state and filtration is one of the indirect ways of using eq. 6.

Describing the mechanical behavior of rock in terms of non-linear equations 7 or 8 causes a nonlinear equation system of the stress-strain state calculation problem. The solution of the problem is possible only by means of numerical methods according to the iteration scheme, provided actual nonhomogeneous and boundary outlines are taken into account. Let us explain the scheme by a particular example. In the case of an isotropic nonlinear elastic medium, eq. 7 is equivalent to the relationships:

$$I_\sigma = (3\lambda - 2\mu)(I_\varepsilon - \alpha\theta) \tag{9}$$

$$II_\sigma = 2\mu II_\varepsilon \tag{10}$$

where

$$I_x = x_{kk}/3,$$

$$II_x = \sqrt{(x_{11}-x_{22})^2 + (x_{11}-x_{33})^2 + (x_{22}-x_{33})^2}$$

$$+ 6(x_{12}^2 + x_{13}^2 + x_{23}^2).$$

Here, λ and μ are elastic characteristics in terms of Lame factor, concentrations m_1 are considered to be temperature functions, and their change are evaluated by α, the coefficient, thus,

$$\alpha = \alpha' + \delta_1 \cdot \partial m_1/\partial\theta$$

(Merzljakov 1980). To determine the parameters of relationships 9 and 10 I_ε, II_ε, I_σ, II_σ, μ_ε, θ it appears to be necessary to add two dependencies that are determined experimentally:

$$\psi_i(I_\varepsilon, II_\varepsilon, I_\sigma, II_\sigma, \mu_\varepsilon, \theta) = 0, \quad i = 1,2 \tag{11}$$

where μ_ε is the Lode parameter. It is assumed that a stress tensor and a strain tensor are similar (μ_ε, μ_σ); temperature changes during the experiment and parameter α may be determined by the supplementary experiments. Dependencies 11 also include strength conditions. Then, using equilibrium equations in displacements

$$(\mu \cdot u_{i,j})_{,i} + (\mu \cdot u_j)_{,i} + 2(\lambda \cdot u_{kk})_{,j} + (\gamma\theta)_{,j}$$

$$+ x_j = 0 \tag{12}$$

where i,j = 1,2,3. We can draw up an iteration solution of nonlinear elastic problem, in the same way as was done by Gulko (1979), but taking into account temperature changes. Equation 12 makes it possible to determine displacements with the given elastic constants λ and μ, thus determining I_ε, II_ε, and μ_1. I_σ and II_σ may be determined from eq. 11, and more accurate values of λ and μ from eqs. 9 and 10. The process is repeated. It should be noted that eqs. 9 and 10 express the so-called deformation theory, while relationship 8, generally speaking, is non-holonomic.

To describe irreversible processes in the system, it is necessary to consider particularly chosen internal parameters along with the external parameters of a medium (Sedov and Eglit 1962). The so-called "non-compensated" heat dq′ is added to the right side of eq. 4. As the simpliest hypothesis describing creep, it is possible to assume

$$dq' = \kappa(dx/dt)dx, \ \kappa>0 \qquad (13)$$

where κ is scalar functions of external parameters and internal parameter. Non-holonomic relations 8 under condition 13 turn into the following:

$$d\varepsilon_{ij} = C_{ijkl}d\sigma_{kl} + \alpha_{ij}d\theta + A_{ij}dx \ldots \qquad (14)$$

where coefficients C_{ijkl}, α_{ij}, and A_{ij} are external parameters, functions, and x. If necessary, due to the experimental data, hypothesis 13 can be altered and internal parameters can be added.

The full equation system described above, even with the stated limitations, cannot be completed because of the great expense of the research under complicated geological conditions. By simplifying the scheme to the capacity of an engineer and to considerable reliability of the results obtained, acceptable practical solutions of the given problem are possible.

The suggestions made above concerning mechanical behavior of frozen-thawing rock is the basis for the solution of a stress-strain state problem. The problem of evaluating slope stability is closely connected with it, and the ultimate stable state of the rock is considered to be the final state. In this case, the stability safety factor (K_s) of a slope may be defined as the ratio of critical parameters of the problem (failure load, critical embedding of slopes, etc.) to the actual spatial parameters. However, this method was not a success in practice since it does not have enough practical grounds. In this connection, in design practice traditional techniques are used based on the definition of (K_s) as the ratio of forces keeping a creeping fill to forces shifting it along the given surface of sliding. Experimental data of K_s relevant to this technique have been stored which assures slope stability. However, traditional techniques have a significant shortcoming which is to the following effect: confining and shearing forces are determined rather approximately and without any connection with the solution of a stress-strain state boundary-value problem. Boundary condition influence and changes in maintenance period, force, and temperature influence cannot be accurately taken into consideration.

The following method of evaluating slope stability of complicated geological composition, developing upon suggestions of Ukhov (1975), may be considered a rather important specification of traditional techniques: 1) On the basis of the design data and the geological and geomechanical research of an object, the geomechanical model and design scheme are worked out. They reflect its structural peculiarities, composition, state, and other factors responsible for its physical and mechanical behavior. 2) Stress components are calculated for the boundary area in question by the solution of some physical problems (e.g., thermo-technical) determining a state of the object and a stress-strain state problem. 3) Strength safety factors (K_{ss}) are considered to be the ratio of ultimate values of shear stresses (the latter is determined by the condition of strength, e.g. Coulomb's equation) to the maximum acting shear stresses, the stress field being known. 4) By analyzing the minimum K_{ss} values and the stress-strain state pattern, the area that is evidently responsible for the formation of the dangerous collapse surface is revealed and serial of potentially collapse surfaces is chosen. 5) For the series mentioned above, integral total stability safety factors (K_s) are defined as the ratio of epure square of ultimate shear stresses at the dangerous surface to epure square of acting shear stresses along the same surface. The design surface is chosen from the possibly dangerous surfaces with minimum K_s value.

Below, as an example, the results are given of the slope evaluation of the K_s of the complicated geological composition of a northern water-storage reservoir. The slope in question consists of 12 types of rocks whose disposition and the location of the supposed great wreck are shown in Figure 1. Due to the lack of experimental data, the sequence of quasi-stationary (temperature and mechanical) states is under consideration to make eq. 7 more concrete. The calculations are carried out on the basis of the thermo-physical and mechanical properties of rock in frozen, thawing and thaw states. The design versions and calculation results of K_s are given in Table 1.

In different periods (pre-construction, construction and maintenance of a dam) the rock's physical state and its mechanical properties may be quite different. To compile the calculation schemes, 26 quasihomogeneous rock areas were considered. The permafrost and thawed areas are demarcated in Figure 2-6 by a dotted line. A boundary of a slope melting area at filling of a water storage reservoir (Figures 5 and 6) is meant for 50-year maintenance.

In deciding upon the design (calculation) schemes and the numerical solution of the problem, the choice of the dimensions of a calculation area is important. The dimensions were derived from the test calculations (other conditions being equal). Preliminary calculations and analysis carried out in this way demonstrated that marking the dimensions of a calculated area by Version 1 (Figure 2) caused erroneous results for K_s, which is 10-14% by Version 2 (Figure 3-6) and to the error not exceeding 1-2% in calculation versions (the area of Version 2 is increased by 35% in comparison with Version 1). For example, the value of K_s for a slope in the preconstruction period is 1.09 and 0.95, according to Versions 1 and 2, respectively. As shown, the results obtained are considerably different. In the first case, the slope is stable, in the second it is not. Thus, when applying numerical methods of evaluating slope stability, one should be very accurate in outlining the calculation area.

The following conditions are assumed as the calculation area boundaries: at the lateral boundaries, horizontal displacements are equal to zero;

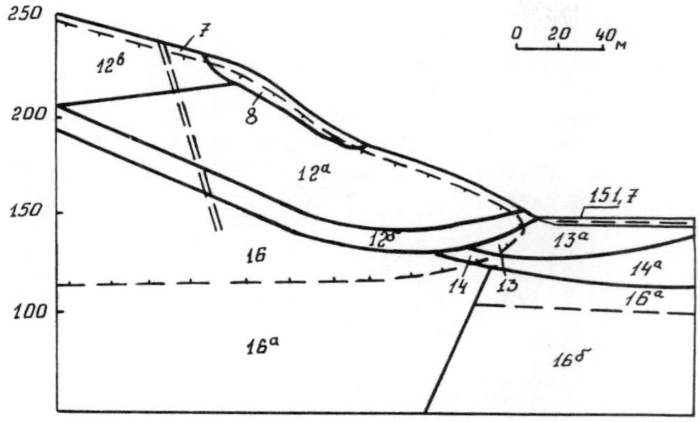

FIGURE 1 Geological scheme of a left bank and a river bed

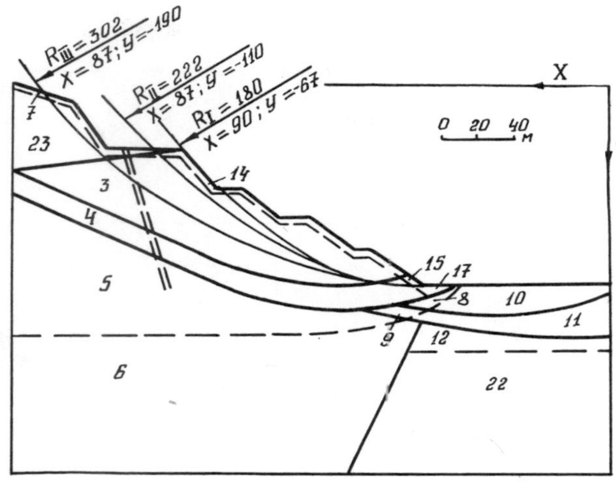

FIGURE 4 Calculation area schematic in construction period (Section I-I)

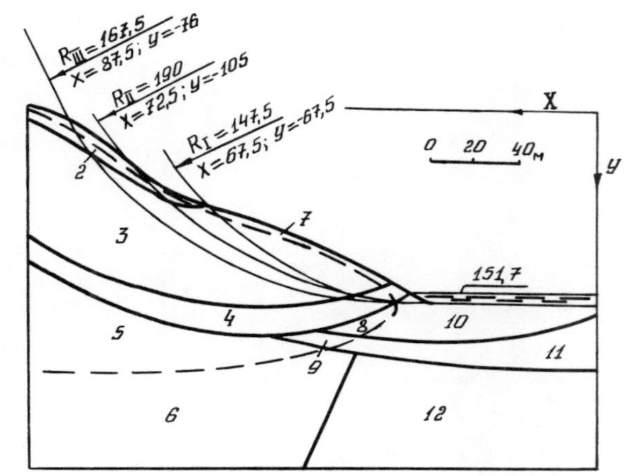

FIGURE 2 Calculation area schematic in preconstruction period (Version 1)

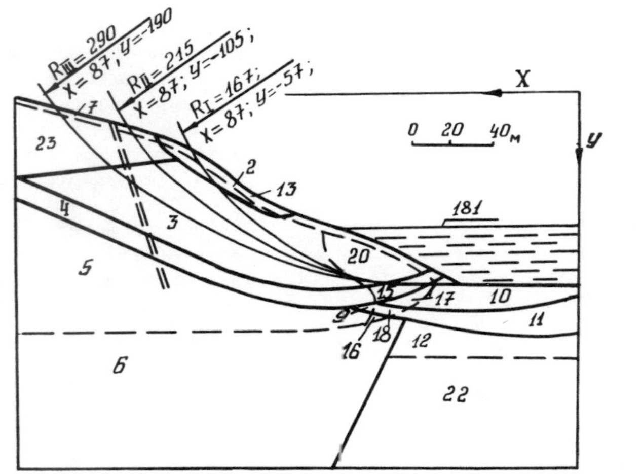

FIGURE 5 Calculation area schematic in maintenance period (cross-section is higher than dam)

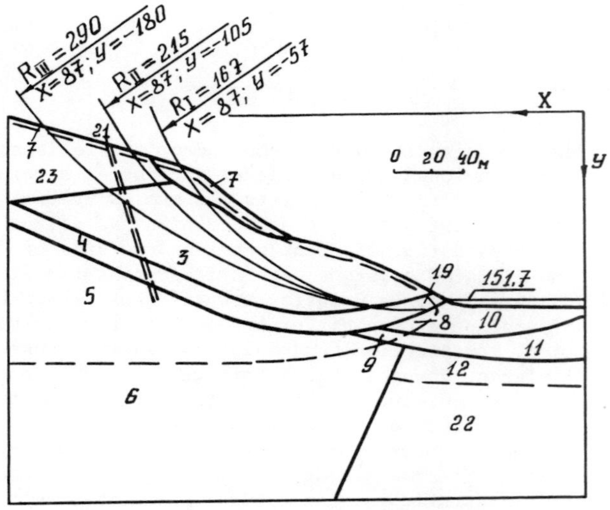

FIGURE 3 Calculation area schematic in preconstruction period (Version 2)

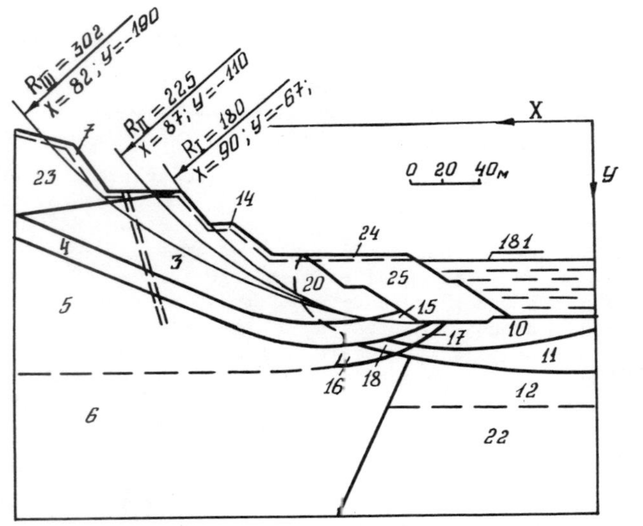

FIGURE 6 Calculation area schematic in maintenance period (Section I-I)

TABLE 1 Calculation Results of Minimum Stability Safety Factor (K_s)

Calculation number	Description	K_s values	
		Without cracks	With a crack filled with breccia
1	Preconstruction period (natural slope)	1.35	1.18
2	Construction period (design out, "discharge" stress considered)	1.26	1.09
3	Similar to No. 2 but without "discharge" stress	1.57	1.39
4	Maintenance period, natural, melt slope at TWL	1.46	1.38
5	Similar to No. 4, at instant discharge of NTWL up to tail-water level and non-stabilized depression curve, stress discharge involved	1.28	1.20
6	Similar to No. 5, but at stabilized depression curve	1.23	1.13
7	Maintenance period, cross-section I-I (within a dam and upstream), "discharge" stress considered at construction period cut	1.42	1.32
8	Similar to No. 7, regardless "discharge" stresses	1.73	1.60
9	Similar to No. 7, but at instant discharge NTWL up to TWL and non-stabilized depression curve, "discharge" stress involved	1.33	1.21
10	Similar to No. 9 but at stabilized depression curve	1.36	1.27

at the lower boundaries vertical displacements are equal to zero; and at the upper boundaries external loads are zero. During the construction and maintenance periods, in shear areas external forces are assumed to be equal to discharge forces. Hydrostatic water pressure upon frozen mass outline, in accordance with breaking of the area into finite elements, is applied in terms of horizontal and vertical components of discretion forces. After the normal top water level is removed, the "discharge" forces are applied to the areas of frozen mass outline free of hydrostatic pressure.

For every design, K_s has been calculated for the probable sliding surfaces disposed in the area of minimum values of local strength safety factors. The typical surfaces (I, II, III) corresponding to minimum values of K_s are presented in Figure 2-6. In the calculated versions with a supposed crack, dangerous surfaces are composed of two regions: the first includes a crack plane, the second a formation surface III. Table 1 lists minimum K_s values. In the versions without cracks, minimum K_s values correspond (in different versions) to surfaces II and III, for versions with cracks—to surfaces IV. Comparison of calculation results shows that the K_s of a slope during different periods (pre-construction, construction, and maintenance) undergoes different changes. Rock thickness during construction period causes considerable change of a stress-strain state and K_s value decrease.

When a water storage reservoir is filled, slope stability increases due to the horizontal component of hydrostatic water pressure. At maximum water discharge and a stabilized depression curve, K_s tends to minimum values. The latter is conditioned by the size of the thawed slope area and by the difference in mechanical rock properties in frozen, thawing, and thawed states. In a dam cross-section an extra load from upstream fill increases slope stability in all calculated versions.

Thus, calculation result analysis makes it possible to draw the following conclusions:

1. The differences in rock's strain and strength properties in the frozen, thawed, and thawing states changes considerably the stress-strain state and the stability of a slope.

2. Slope stability is conditioned by the construction period and by the maintenance regime of the reservoir. The design value of K_s is influenced by several variable force factors dependent upon a) construction cutting of rock thickness, b) changing hydrostatic head, c) unit weight change of a thawed part of a hillside along with a change of a depression curve disposition, and d) the dimensions and configuration of the thawing area of a hillside outline.

3. The suggested method of evaluating slope stability, based on numerical calculations of the stress-strain state, is efficient for a great extent actual non-homogeneity of slopes (initial and changing with the maintenance of the water power project, variable force, thermal effects, and other factors.

REFERENCES

Ershov, E. D., and Akimov, U. P., 1979, Phase moisture composition in frozen rock: Moscow, Moscow University.

Gulko, E. F., 1979, Forecast of stress-strain state of soil fill in its spatial action: Candidate thesis, manuscript, Moscow.

Merzljakov, V. P., 1980, Application of the basic thermodynamic relations in description of soil strains, Proceedings of the 3rd All-Union Symposium on soil rheology: Erevan, p. 273-278.

Plotnikov, A. A., 1978, Temperature regime calculation of permafrost footings, Energeticheskoye stroitelstvo, no. 8, p. 70-73.

Porkhaev, G. V., 1970, Heat interaction of buildings and structures with permafrost soils: Moscow, Nauka.

Sedov, L. I., Eglit, M. E., 1962, Non-holonomic modelling of continuous media by finite strain application and some physical and chemical effects: Doklady Academii Nauk USSR, v. 142, p. 54-57.

Ukhov, S. B., 1975, Rock footings of water power structures: Moscow, Energia.

Vakulenko, A. A., 1974, Interaction of micro and macro models of an elastic-plastic body, Elastic and Plastic Research, vol. 10: Leningrad: Leningrad University, p. 3-37.

A NEW TECHNIQUE FOR DETERMINING THE STATIC FATIGUE LIMIT OF FROZEN GROUND

V. V. Vrachev, I. N. Ivaschenko, and E. P. Shusherina

Geology Department, Moscow State University
Moscow, USSR

Existing techniques for determining the static fatigue limit (R_ν) of the frozen ground are time-consuming and labor-intensive (Vyalov et al. 1966, 1978; Tsytovich 1973). The proposed new method can achieve an accelerated determination of the static fatigue limit by means of brief tests by gradually increasing loading on sublimated samples of frozen ground. The physical bases of the method, which has been verified experimentally, are as follows: 1) the static fatigue limit of frozen clay materials depends upon the structural bonding of the mineral framework rather than on bonding by ice-cementing; 2) the structure and texture of the mineral framework do not vary considerably during sublimation of the icecrete until the ground moisture equals the content of unfrozen water under the same thermodynamic conditions; and 3) the static fatigue limit of sublimated ground approximates in magnitude its theoretical instantaneous strength R_o. To determine R_ν by the proposed method within limits of accuracy which are acceptable for practical purposes, it is recommended that one select for analysis samples of frozen clay material with a massive or micro-veined cryogenic texture within the temperature range of -3 to 20°C.

INTRODUCTION

The mechanical properties of frozen ground stem mainly from porous moisture crystallization and icecrete bonding (Tsytovich 1973, Vyalov 1978). Two types of bonding can be distinguished: intracrystalline bonding within the polycrystalic icecrete and ice-cementing between the ice and mineral particles. Both types are of chemical (hydrogenic) nature, but can vary considerably in their energy depending upon the chemical affinity between the surface contacts and the presence of an unfrozen water film between them. The rheological processes of creep, relaxation, and decrease of strength in time are associated with the presence of ice in the frozen ground.

The strength and deformability of frozen clay ground can also be determined by structural bonding of the mineral framework caused by molecular and ion electrostatic interaction of particles and their cementation by various cryohydrates. Elimination of ice by means of sublimation allows evaluation of the individual role of structural bonding and ice-cementing in the formation of mechanical properties of frozen ground.

The present paper considers the use of ice sublimation for the evaluation of the static fatigue limit R_ν, which is a basic engineering parameter used in the design of structures with foundations in frozen ground (Vyalov et al. 1962).

R_ν is used in different design formulae to predict the origin and development of cryogenic processes (Grechischev et al. 1966). Existing methods to determine parameters of strength and deformability require strict observation of the temperature regime during the test because of the instability of the mechanical properties of the frozen ground. The allowable temperature deviations in the field of intensive phase transformation should not exceed ±0.1°C. In addition, to protect samples of frozen ground from erosion it is necessary to keep them hermetically sealed in specialized cases. These factors complicated the preparation and performance of extensive tests; moreover, slight deviations from these requirements causes significant scattering of the experimental data. The most reliable values of R_ν can be obtained when a number of identical ground samples are tested by loads that are different in strength, but constant over time. The first sample was tested by rapid loading and its conditional instantaneous strength R_o was then determined. The rest of the samples (no less than 6) were tested for creep effect at various loads less than R_o. Each sample was pushed to failure and the corresponding time was recorded. The resulting static fatigue curve approximates, in particular, the experimentally verified expression:

$$R(t) = \frac{\beta}{\ln\frac{t}{T}} \qquad (1)$$

where β, T are parameters of the static fatigue curve and t is the time to destruction. Assuming t to be equal to 100 years, we obtain the static fatigue value, R_ν.

The proposed method of accelerated estimation of R_ν was at the Cryology Division of the Geological Department, Moscow State University. It is based on the study of the process of ice sublimation and its specific effect upon the composition, structure, and mechanical properties of frozen ground.

Ice sublimation is a phase transformation of the first type, and is a direct transformation of water molecules from solid to gas phase (Lykov 1968).

Taking into consideration experimental data that indicate the existence of a thin "quasiliquid" layer of bound water on the surface of the ice crystals, this sublimation is, in fact, evaporation of unfrozen water and simultaneous enrichment of its reserves due to melting from the lower ice layer brought about by the thermodynamic balance between the solid and liquid phases (Ershov et al. 1975).

METHOD OF TEST PREPARATION AND PERFORMANCE

The research performed included the study of the composition and structure of the frozen and sublimated ground by electron and optical microscopy, and the determination of strength and deformability parameters with consideration of the time factor. Samples of Paleogenic clay (mPg_2^{kv}) having natural and disturbed structure and a moisture content (W), on the order of about 32-34%, and a framework density (γ_{ck}) of 1.45 g/cm³ were analyzed. More precise definition of the physico-mechanical properties of the clay have been studied in a paper by Ershov et al. (1975). Cylindrical samples of ground (d = 3.8 cm, h = 8 cm) were frozen at a temperature of -30°C to form massive cryogenic texture. The experimental subject was selected due to the fact that sandy types of disperse grounds (sands, fine sandy loam) with massive cryogenic texture and clays with veined cryogenic texture did not fit the purpose of research because during frost desiccation they broke down into their constituent elements or units (Zhestkova et al. 1972).

Ice has been sublimated at -10°C in the special sublimators in which air at an optimum velocity of 10-20 m/sec. was blown over and around it. During ice evaporation the changes in the sample were periodically monitored by the formula

$$P_H - P_K = V \cdot \gamma_{ck} (W_H - W_{H.B.}) \quad (2)$$

where P_H is initial sample mass (in grams); P_K is final mass (in grams), V is sample volume (in cm³); γ_{ck} is density of ground framework (in g/cm³); W_H is initial sample moisture (in %); and $W_{H.B.}$ is unfrozen water content (in %).

Desiccation of the samples stopped when moisture of clay W became equivalent to the unfrozen water content for the given temperature of -10°C ($W_{H.B.}$ = 7.5%).

The moisture limit $W_{H.B.}$ is of special consideration; below this limit volume densification of the ground occurs, resulting in a considerable increase in strength. After ice sublimation is completed, the samples were sealed hermetically and kept at a given temperature regime for more regular distribution of the moisture within the sample. Homogeneity of samples was verified by nondestructive monitoring of the velocity of acoustical wave propagations in orthogonal directions (Zykov et al. 1974).

The microstructure of frozen and sublimated clay samples has been studied using a light polarization microscope, MIN-8 (Vrachev and Rogov 1975). Simultaneously, the micro-aggregate composition of the described samples was determined by Kachinsky's method (1957). Creep tests have been performed at -10°C under uniaxial compression by various acting stresses constant in time (Vyalov et al. 1966). All creep tests were repeated three times.

DISCUSSION OF RESULTS

Results of the study of the microstructure of the frozen clay ground by electron microscope Tesla B5-242 (Rogov and Zabolotskaya 1978) show that the mineral framework is initially represented by separate elementary particles of various size (from clay to sands) and by their aggregates. Aggregates vary in shape and size from 0.1 to 2 mm. Icecrete was not detected even at 5000X magnification. Therefore, we can suppose that structural bonding in the aggregates is determined mainly by the molecular and ion electrostatic interaction of mineral particles. Ice inclusions should be considered typically cement-like, since they form as the result of water freezing in the ground pores with negligible displacement. As a rule, isometric icecrete inclusions consist of one or, rarely, several crystals with chaotic orientation of optical axes. The ice inclusion-ground aggregate boundary is indistinct, which can be explained by the existence of an unfrozen water film in the gap between them. The type of bonding between the crystals within the ice inclusions can be regarded as intercrystalline, and between the ice and mineral aggregates as ice-cementation. Intra-aggregate ground porosity (n), in particular, is another basic element of microstructure. The optical method used to study the surface of a cross-section of clay samples in reflected light revealed the distribution of pores over 0.02 mm. To estimate the effect of ice sublimation upon a given component in the microstructure of frozen ground we derived individual indices of its porosity. General porosity n_Σ is determined by

$$n_\Sigma = \frac{\Delta - \gamma_{ck}}{\Delta} \cdot 100 \quad (3)$$

where Δ is density of mineral mass of the ground (2.72 g/cm³) and γ_{ck} is density of solid in g/cm³. Porosity formed in 0.02 mm voids has been calculated as a percentage ratio between the density of these pores and the total area of the analyzed sample in its plane section. Pores of this size are the most sensitive to any volume changes of the ground.

Table 1 illustrates data on micro-aggregate composition and porosity of frozen and sublimated clay samples.

Analysis of the structure of the mineral framework in the sublimated clay samples shows that elimination of ice leads to formation of post-cryogenic (inherited) structure and texture in the ground. There is no significant change in the geometry and average size of the aggregates, of their mutual alignment and the degree to which they fill in the structural space. The latter agrees well with results (Lykov 1968) on the study of the effect of vacuum-sublimation drying on the structure of capillary porous bodies. Lykov noted that "this drying technique keeps the pores of the material uncompressed and the structure almost unchanged. Slight shrinkage does occur during elimination of adsorptive bound water." All this allows the sub-

TABLE 1 Composition and Structural Properties of Clay Samples

Clay	Micro-aggregate composition: Size of particles (mm) / Content of particles (%)			γ_{ck} (g/cm^3)	n(%)	
	Sand >0.05	Silt 0.005–0.05	Clay <0.005		n_Σ	n > 0.02 mm
Frozen	46.2/33.2	44.6/37.3	9.2/29.5	1.45/1.43	46.5/47.4	–/9.9
Sublimated	48.2/30.0	43.8/38.4	8.0/31.6	1.47/1.46	46.0/46.3	–/8.1

Note: Numerator stands for indices of natural composition of ground, denominator is for destructed ground.

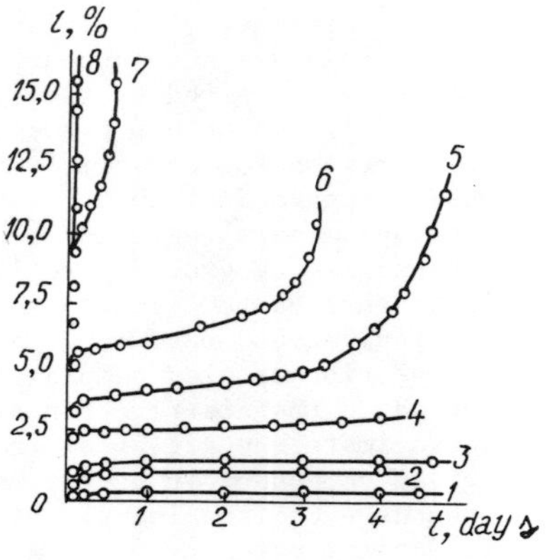

FIGURE 1 Creep curves for frozen clay of natural composition; stresses equal to: 1 - 0.47; 2 - 1.24; 3 - 1.49; 4 - 2.10; 5 - 2.48; 6 - 2.64; 7 - 2.79; 8 - 2.94 MPa.

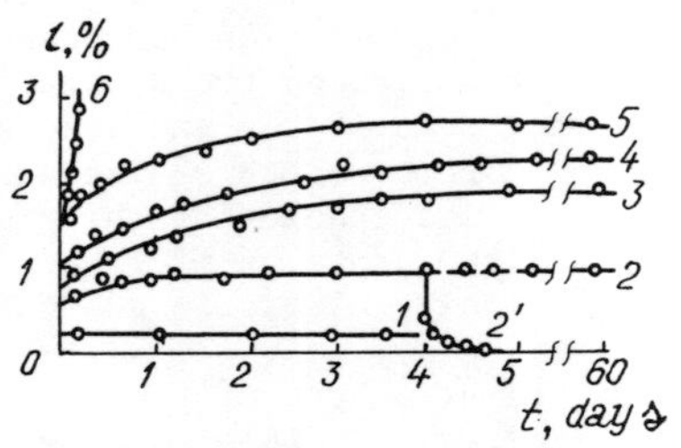

FIGURE 2 Creep curves for sublimated clay of natural composition; stresses equal to: 1 - 0.1; 2 - 0.5; 3 - 1.27; 4 - 1.58; 5 - 1.7; 6 - 1.78 MPa.

limated ground to be viewed as an adequate model of the mineral framework structure of the frozen ground. The strength of the clay material is known to be determined not by the strength of their structural elements (i.e. micro-aggregates), but by the bonding forces between them (Osipov 1979). As has already been noted, icecrete and structural bonding can be distinguished in the frozen ground. Formation of icecrete bonding in the freezing ground with complete water saturation occurs throughout the intraphase surface of particles, while formation of structural bonding occurs only at sites of greatest contact. Resistance to the breakdown of such a porous system as the mineral framework of the ground depends upon the bonding strength between micro-aggregates at contacts and upon spatial distribution of these contacts. As far as the difference in the reaction of structural and icecrete bonding to external mechanical influences is concerned we shall consider the behavior of frozen and sublimated clay samples under conditions of prolonged loading. The origin and behavior of rheological processes in the frozen ground stems mainly from ice inclusions, i.e. icecrete and ice bands in their composition. Practically any load can cause plastic flows in them and a re-orientation of their ice crystals, in addition to viscous films of unfrozen water (Vyalov 1978, Tsytovich 1973). The process of deformation of the non-sublimated clay samples over time is typical of frozen ground (Figure 1). The creep curves show three distinct stages when stresses exceed static fatigue limit ($\sigma > R_\nu$): 1) non-stable creep with a gradually decreasing rate of deformation; 2) stable creep with an approximately constant rate of deformation (i.e. plastic viscous flow); and 3) progressing flow with an increasing rate of deformation.

In the case of natural ground the last stage ends in breakdown of the sample with the formation of visible fractures. Samples of irregular composition demonstrate plastic deformation with the following barrel-line formation. The stable creep stage dominates over time. The second and the third stages are absent in the presence of small stresses ($\sigma < R_\nu$), and the deformation process occurs as attenuated creep with a gradually decreasing rate of deformation.

Ice elimination due to frost drying of the frozen ground sharply changes the character of the deformation and the destruction of the samples (Figure 2). Even for stresses R_o from 0.85 to 0.9, deformation of the sublimated samples occurs during creep reduction when the test time exceeds 6 months. When this limit was exceeded, a typically fragile destruction of the ground occurred 1 to 4 hours

TABLE 2 The Effect of Ice Sublimation on Physico-Mechanical Properties of Clay (θ = 10°C)

State of the ground	W (%)	γ_{ck} (g/cm^3)	R_ν (MPa)	R_ν (MPa)
Frozen	$\frac{32\pm2.0}{34\pm2.0}$	$\frac{1.45\pm0.04}{1.44\pm0.03}$	$\frac{3.10\pm0.3}{3.90\pm0.2}$	$\frac{1.74\pm0.25}{1.50\pm0.25}$
Sublimated	$\frac{9.7\pm0.5}{10.2\pm0.5}$	$\frac{1.45\pm0.01}{1.45\pm0.01}$	$\frac{2.00\pm0.1}{1.35\pm0.1}$	$\frac{1.73\pm0.05}{1.27\pm0.05}$

Note: Numerator stands for properties of undisturbed ground, denominator stands for disturbed ground.

after the test began. Deformations developing in the initial period are caused by elastic compression of the ground framework and by a predominantly reversible relative shift of mineral particles along viscous films of unfrozen water. General creep deformation of the frozen ground is 15-20% depending on the effective stress and is mainly plastic (reverse). That of sublimated samples is reduced to 3-5% with domination of the elastic component.

When ground freezes its strength is sharply increased due to crystallization of porous water and formation of ice-cementing. This phenomenon is widely used in civil engineering, for instance, during tunneling or trenching in weak and water-saturated ground by means of artificial freezing (Vyalov et al. 1962). Ice-cementing depends on the amount of ice and the ground temperature; it is less stable than structural bonding. Ground loading causes a shift in the balance between the ice and unfrozen water due to a concentration of stresses at particle contacts. Under the influence of the gradient that arises, the unfrozen water, which is replenished by melting ice, shifts to the less tense zone where it refreezes. Simultaneously, the ice recrystallizes and a basal plane reorientation of crystals occurs parallel to the shift forces. These processes cause a decrease in the frozen ground's strength when the loading is prolonged. The load for clay samples R_o is determined by (1) and equals 0.45 to 0.5 (Table 2).

Sublimated clay samples, as compared to frozen samples, do not demonstrate considerable decrease in strength over time; their static fatigue resistance to compression is 0.85 to 0.9 of the conditional instantaneous strength. R_ν of the frozen ground and R_o of the sublimated ground show considerable similarity. Such coincidence seems to be regular if we assume R_ν as maximal stress at which no deformation of the ground occurs during plastic-viscous flow or breakdown. A similar conclusion about the physical properties of the R_ν of capillary porous bodies was reached by Bykovsky (1954) when he analyzed the after-effects of deformation of wood.

If we treat the frozen ground as an elastic-plastic-viscous medium pierced by an elastic spatial grid, the system could be destroyed when the stress limit is surpassed; as a result, further increase of deformations in the frozen ground is no longer inhibited by its mineral framework. As far as interground bonding is concerned, this stress limit corresponds to the value of structural bonding of the mineral framework, which can be determined by short-term tests for the strength of the sublimated ground samples.

If we compare the proposed and existing methods for determining R_ν, we note that the first has obvious merits. In particular, it reduces test time since it allows us to determine the R_ν of the frozen ground by conducting short-term tests of their sublimated analogues. Besides, sublimated samples are less sensitive to temperature variations, especially beyond the field of intensive phase transformations (below -5°C). And finally, the accuracy of the test results for sublimated samples is higher (±4%) than for frozen samples (±12 to 15%).

At the same time, it should be noted that the proposed method can be used only to determine the R_ν of clay ground in massive cryogenic structures. The possibility of using this method at temperatures below -10°C is also debateable because we have little knowledge about the rheology of low-temperature frozen ground.

CONCLUSIONS

1. To determine the R_ν of the frozen clay ground of a massive cryogenic structure, the sublimated analogues can be used, because the structure of the mineral framework is not disturbed during the process of frost drying until $W_{H.B.}$, and its static fatigue limit corresponds to the conditional instantaneous strength of the sublimated ground.

2. The proposed method for determining the R_ν of frozen ground has certain merits as compared to existing methods. It allows a reduction in test time, is not as sensitive to temperature variations (±3°C), and its results have high accuracy.

3. Ice sublimation allows for more detailed investigation into the nature of the strength and deformability of frozen ground. In particular, it allows us to determine the individual roles of structural bonding and ice-cementing in the formation of their mechanical properties.

REFERENCES

Bykovsky, V. N., 1954, Definition of static fatigue limit of the wood by the after-effects deformation: Zhurnal tekhnicheskoi fiziki, v. 24, is. 9, p. 1631-1635.

Ershov, E. D., Ivaschenko, I. N., et al., 1972, Effect of ice sublimation on physico-mechanical properties of frozen rocks, in Permafrost research, is. 13, Izd. MGU, p. 172-176.

Ershov, E. D., Kuchukov, E. Z., Komarov, I. N., 1975, Ice sublimation in dispersed rocks: Izd. MGU, Moscow, p. 27-34.

Grechischev, S. E., Chistotinov, L. V., Shur, Yu. L., 1980, Cryogenic physical-geological processses and their prediction: Moscow, Nedra, p. 243-246.

Kachinskii, N. A., 1957, Mechanical and micro-aggregate composition of the soil and methods of its investigation: Izd. AN SSSR, Moscow, p. 34-36.

Osipov, V. I., 1979, Nature of strength and deformability properties of clay rocks: Izd. MGU. Moscow, p. 139.

Rogov, V. V., Zabolotskaya, M. I., 1978, Fine structure of frozen rocks with icecrete, in Problems of cryolithology, is. 7: Izd. MGU. Moscow, p. 169-171.

Tsytovich, N. A., 1973, Mechanics of frozen ground: Vyshaya shkola, Moscow, p. 126, 127.

Vrachev, V. V., Rogov, V. V., 1975, Frost drying as a technique to investigate the strength nature of the frozen clay ground, in Permafrost research, is. 16: Izd. MGU, p. 230-235.

Vyalov, S. S., 1978, Rheological basis for mechanics of the ground: Moscow, Vyshaya shkola, p. 138, 263.

Vyalov, S. S., Gorodetskii, S. E., et al., 1966, Parameterization method for creep, static fatigue limit and compression of the frozen ground: Izd. AN SSSR, Moscow, p. 41-65.

Vyalov, S. S., Gmoshinskii, V. G., et al., 1962, Strength and creep of the frozen ground and calculations for ice-soil barrier: Izd. AN SSSR, Moscow, p. 144-158.

Zhestkova, T. N., Ivaschenko, I. N., et al., 1972, On the dependence of mechanical properties of the frozen ground and their cryogene texture, in permafrost research, is. 13, Izd. MGU, p. 166-172.

Zykov, A. V., 1968, Theory of drying, Energy, p. 74.

Zykov, Yu. D., Baulin, Yu. I., Vrachev, V. V., 1972, Some data on elastic moduli variations in the frozen clay ground in the result of their frost drying, in Permafrost research, is. 15, MGU, p. 248-250.

PERMAFROST BENEATH THE ARCTIC SEAS

L. A. Zhigarev

Geography Department
Moscow State University, Moscow, USSR

Frozen seabed materials are widespread beneath the Arctic seas, thus forming a submarine cryolithozone. The major part of this submarine cryolithozone consists of mineralized materials containing water supercooled below 0°C. Perennially frozen rocks occur as a continuous belt along both continental and island coasts, and also as isolated masses on the open shelf within the limits of the Zyryan-Kargin lacustro-alluvial plain. The conditions of formation and survival of the frozen sediments beneath the seabed in the arctic seas are discussed.

Cryogenic rocks, which are widespread beneath the Arctic Ocean and its marginal seas and which are one of the most striking phenomena of the Arctic, constitute a vast subaqueous cryolithozone. The study of cryogenic rocks in the seabed is of both theoretical and practical significance. Both Soviet and foreign experience in the economic development of the Arctic shelf show that during design and construction in offshore zones, lack of knowledge about and insufficient account of, or disregard of, the characteristic features of development of the subaqueous cryolithozone and its cryogenic rock constituent can cause no less grave consequences than on dry land.

Changes in cryogenic conditions in the offshore zone of the Arctic seas take place with hydrotechnical construction, prospecting, and especially the exploitation of mineral and oil-gas deposits. An essential change in the cryogenic-geological conditions is to be expected in estuarine offshore regions as a result of inter-basin diversion of northern rivers, as well as in shoals of bays and straits as pipeline construction progresses.

The first data on the occurrence of frozen sediment in the Arctic seabed were obtained in 1739 during the Great Northern Expedition, when Kh. P. Laptev, commander of the naval detachment, observed sea ice adfreezing to bottom deposits at the northeastern coast of the Taimyr Peninsula. He suggested that sea ice in the western part of the Taimyr also adfreezed to the bottom sediments. Such a phenomenon was recorded by U. R. Parry in 1819-1820 in Viscount Melville Bay (the eastern part of the Beaufort Sea).

Later, data on the subaqueous cryolithozone resulted from the expeditions of Nordensheld, Toll, and Sverdrup, as well as some Soviet expeditions for the exploration and exploitation of the Northern Shipping Route. It is to be noted that a particular contribution to the study of the special features of cryogenic rock distribution in the seabed was made by the Russian polar expedition headed by E. V. Toll.

In the USSR, systematic studies of the subaqueous cryolithozone with the use of boring have been carried out since the 1930s in connection with port surveys and prospecting for mineral deposits. Abroad, investigations into cryogenic rocks, based on drilling, were carried out in 1932 by McLachlan in the northern part of the Atlantic Ocean on the western coast of the Hudson Bay. Systematic studies of the subaqueous cryolithozone in the American-Canadian North were commenced only in the late '60s in connection with the discovery of oil and gas deposits in the Beaufort Sea.

These studies have revealed patterns of distribution of cryogenic rocks in the Arctic seas, as well as special features of the subaqueous cryolithozone formation. They allowed evaluation of permafrost conditions within a large part of the water area in the Arctic basin. Such evaluations were made by Baranov (1958), Grigoryev (1962), Kudryavtsev and Romanovsky (1975), Are (1976), and Danilov and Zhigarev (1977). Recent research into the subaqueous cryolithozone makes it possible to essentially clarify the cryogenic conditions in the Arctic seas.

But is is to be recalled, first of all, that cryogenic rocks are divided (Danilov and Zhigarev, 1977) into recent and relict by age, perennial and seasonal by existence period, and into frozen (negative temperature, containing ice and unfrozen water), frosty (negative temperature, containing no ice and practically no water), and supercooled (negative temperature, containing unfrozen water but no ice) by physical state.

The subaqueous cryolithozone develops due to thermodynamic, chemical, and hydrodynamic interaction between the atmosphere, the hydrosphere, and the lithosphere. The basic external agents that determine freezing or thawing of rocks in the sea are those of absorbed radiation, turbulent and convective circulation of the water mass, and advection of heat by currents, which is determined to a considerable degree by the sea depth, the bottom relief, and the ice cover. The main internal agents are the lithological composition of rocks, humidity, mineralization of deposits, and pressure of the overlying water mass and sediments. The integral expression of the external agents is the temperature of the bottom sea-water layer, and that of the internal agents is the temperature of the pore water freezing in the bottom sediments.

As has been established earlier (Molochushkin and Gavrilyev, 1970; Zhigarev and Plakht, 1974), the mean annual temperature of the bottom water layer is almost identical to that of the deposits at the base of the layer of zero annual amplitudes. Hence the correlation between the temperatures that determine the external and internal agents indicates directly the possibility of freezing or thawing of the deposits, i.e. the formation of recent (newly formed) frozen rock series, conditions of conservation of relict perennially frozen rocks, and distribution of seasonally or perennially supercooled deposits. Only frosty rocks are characterized solely by the mean annual temperature of the bottom sea-water layer.

Existing perennially frozen rocks are found to occur in all recently emerged localities as well as in the water areas of the seas, but at depths not exceeding the maximum thickness of the fast ice. Bottom sediments in the coast-ice areas freeze deep due to the high heat conductivity of ice and the insignificant thickness of the snow cover. The annual repetition of deep freezing results in new growths of permafrost. With the sea depth exceeding the mentioned value, recent perennially frozen rocks do not form even near the shore. In all the other areas, the recent frozen rocks encircle the entire offshore zone at the continental and island coasts with a belt of varying width. The width of this belt is greatest in the Laptev Sea (up to 35 km) and in the western part of the East Siberian Sea (up to 30-40 km). The temperature of the recent perennially frozen rocks at the depth of the annual variation base was -11°C in one of the sections of the Laptev Sea, and their thickness for 35 years of freezing approximated 60 m, as was estimated in the East Siberian Sea.

The problem of occurrence of relict permafrost in Arctic seas is extremely complicated; it can be solved only on the basis of research into the paleogeographical conditions of the Late Pleistocene period. These conditions comprise the development of transgressions and regressions of the Arctic seas and the sea level during their maximum phases, the relief of the shelf and possible formation of an ice sheet on it, the thickness of the perennially frozen rocks prior to their becoming subaqueous, the interior heat flow, as well as the above-mentioned external and internal agents of freezing and thawing.

Relict perennially frozen rocks have undoubtedly survived in some comparatively recently abraded areas in the Laptev Sea and in the western part of the East Siberian Sea within the bounds of the lacustro-alluvial plain of the Zyryan-Kargin age. The boundary of this plain probably ran at a depth of about 30 m. This is demonstrated by numerous sand fields and banks that are residual outcrops of the former continental plain and that do not exceed that depth in the Arctic seas, and also by the fact that there is little or no subaerial relief over the entire area of the sand fields and banks.

During the period of development of transgressions of the Arctic seas, as is known, there took place thermoabrasional destruction of the shores built of ice-rich deposits of the Zyryan-Kargin lacustro-alluvial plain. The deposits thawed in the destroyed shore sections as they were submerging. It has been noted (Zhigarev, 1981) that the most intensive thawing of frozen sediments takes place at sea depths ranging between 2-3 m and 5-7 m. The mean annual temperatures of the bottom water layer within this depth range are positive and can reach relatively high values. The sea takes hundreds, and possibly thousands, of years to transgress through this depth range. Therefore, extremely intensive degradation of the frozen rocks from top and from bottom develops during this time period. The continuing rise of the sea level and bottom sediment washout increase the depth, lower the mean annual temperatures of the bottom water layer, and hence stop the thawing of frozen rock from above and retard the thawing rate from the bottom. Occurrences of relict permafrost are distinguished on the basis of an analysis of sea depths, bottom relief, and the lithological composition of the bottom sediments. Relict frozen rock series have survived in areas of the completely thermoabraded Semenov, Vasilyev, Figurin, and Diomid islands.

The survival of relict permafrost is also possible in regions with intensive present-day accumulations of marine sediments where the sea depth has decreased to a value equal to the maximum thickness of the fast ice. Thus, for example, highly ice-rich rocks were recorded at a depth of from 86 m to the borehole face at 100 m on an inlet offshore bar in the Laptev Sea, with its depth here being 1.9-2 to 2.5 m (Zhigarev, 1981). Relict frozen rock masses can be supposed to occur in regions of extensive sand field distribution over shoals of the seas. It can be assumed, according to the mentioned characteristics, that relict frozen rock series are most widespread in the Laptev Sea, but they also occur in the western part of the East Siberian Sea and more rarely in the Kara Sea. Their thickness can reach hundreds of meters in freshly abraded areas, but it probably does not exceed tens of meters in regions of old thermo-abrasion within the Zyryan-Kargin lacustro-alluvial plain.

The recent and relict perennially frozen rocks near continental and island coasts form original "visors" (overhangs) directed towards the sea where they wedge out. Grigoryev (1966) was the first to record such forms of deposition of perennially frozen rocks. In some gulfs of the Laptev and East Siberian Seas, we happened to observe double "visors", with the upper and lower ones being characteristic of the recent and relict frozen rocks respectively.

As is known, the zone of the Arctic seas is a mosaic structure consisting of a number of large upheaved or lowered blocks divided by folded formations or by wide ruptured belts. The upheaved blocks are built of sedimentary rocks of the Silurian, Devonian, or Carbonic periods. In the eastern sector of the Arctic, these are overlain by the Cretaceous effusive rocks and by a thin cover of loose deposits, but this cover is locally washed away completely, and the effusive rocks, under negative temperatures, are overlain by frost rocks. The latter are widespread in the Kara, Laptev, and East Siberian seas.

The lowered blocks in the western sector of the Arctic are built of a series of poorly consolidated or unconsolidated sediments with a thickness of up to several kilometers of the Triassic-Quaternary or Jurassic-Quaternary age, and those in the eastern sector are built of a series of Cretaceous-Cenozoic deposits. The occurrence of relict frozen rocks is theoretically possible in lowered block areas under negative temperatures. However, pore water of loose deposits in the offshore zone of the Arctic seas, as seen from the studies carried out by the MSU Problem Laboratory for Development of the North and other organizations, is characterized by extremely high mineralization of chloride-sodium, sulfate-sodium, or chloride-calcium types, reaching 50 to 160 g/kg and penetrating right down to the underlying bedrock. Under the conditions of such mineralization and of relatively high negative temperatures of the bottom deposits (-0.4° to -1.8°C), loose rocks supercooled below 0°C are most widespread in the Arctic seas. Perennially supercooled rocks occur in all the Arctic seas, as well as in the abyssal part of the Arctic Ocean. These are absent only in areas where the warm Atlantic and Pacific waters affect the bottom, and in the zones of influence of the major Siberian rivers, where the mean annual temperatures of the submarine deposits are always positive.

The recent and relict perennially frozen and perennially supercooled rocks recorded in the Arctic seas are overlain by seasonally frozen and seasonally supercooled rocks in the offshore zone. Seasonally thawed rocks occur in the emerged areas of the newly formed frozen rock masses. Seasonally supercooled rocks are rather widely spread in the Arctic seas. According to estimations, these are deposited as a layer up to 4 m thick from zero sea level to depths of 16-18 m, 14-16 m, and 20-22 m in the Kara, Laptev, and East Siberian Seas, respectively. Within the above depths, these rocks are absent only in areas of localization of fast-ice masses (the Northern Land, Yana, and New Siberian massifs).

Seasonally frozen rocks in areas with rapidly growing depths extend into the sea for tens or hundreds of meters, while on shoals with depths not exceeding the fast-ice thickness they extend for kilometers or even tens of kilometers. The maximum thickness of seasonally frozen rocks at zero sea level reaches 2 m. The thickness invariably decreases towards the water. Such rocks, as well as the newly formed frozen rock masses discussed above, occur at sea depths not exceeding 2.5-3 m. Sections of submarine deposits washed by fresh river underchannel flows in estuaries and estuarine offshore zones are an exception to this rule. Seasonally frozen rocks in this case can form at greater depths as well. In the autumn-winter period, due to the reduction of the warm flows of the river waters, sea waters supercooled below 0°C intrude along the deepest sections of the rivers. Because the sea waters have negative temperatures, while the pore water of the poorly permeable underchannel deposits is fresh, freezing of the latter takes place. The extension of the halocline in the estuaries of individual rivers can reach hundreds of kilometers, and most deep-water and poorly permeable sediments over this entire distance are seasonally frozen. The depth of their freezing, according to estimations, reaches 1 m.

A unique characteristic of freezing bottom sediments is observed in regions of stamukha occurrences. Stamukha is an ice hummock (or grounded ice) of dynamic origin that forms as a result of the wind effect and the development of sea-ice compression pressure forces on shoals, banks, and reefs, with sea depths not exceeding 20 m. Usually, floating hummocky ice penetrating shallows adfreezes to the bottom sediments, becomes immobile, and turns into stamukhas. The latter have a shape of gently sloping eminences with slope steepness not greater than 30°. In some parts of the East Siberian Sea, an average of one stamukha per km^2 can be observed (Gorbunov et al., 1977). Their lengths, widths, and heights vary here from 50 to 4750 m, 20 to 3500 m, and 2-4 to 20 m, respectively. Occasionally some stamukhas standing near one another form chains or ramparts and ridges stretched for several tens of kilometers. No one has studied the temperature regime of stamukhas. There are, however, some temperature data on a perennial ice mass from the section. Cherepanov (Peschansky, 1963), for example, studied the temperature regime in an 8-m ice mass at the SP-6 Station (North Pole drifting station). The temperature at its base varied during the year from -2.4°C (June, 1957) to -3.8°C (April, 1958).

A rather intensive destruction of stamukhas takes place in summer in the ice-free water area due to the effects of heat and wave agents. The greater the size of the stamukhas, the slower their destruction. After the stamukhas are destroyed completely, the bottom sediments in places of their bedding can be found to be nonsaline or even frozen. It is probably just such sediments that were observed by Molochushkin (1973) in the Dm. Laptev Strait of the East Siberian Sea. The very wide distribution of stamukhas in this region is corroborated by observations carried out during navigation in various years (Khmyznikov, 1937). The existence of sediments in a frozen state is determined by the time period of existence of the stamukhas themselves. It can be either seasonal or of 2- to 3-years existence. The areas of such freezing, like the areas of stamukha occurrences, have no permanent character. The depth of freezing of the bottom sediments in these areas apparently does not exceed several tens of centimeters. Hence it follows that small sporadic islands of the original seasonally frozen rocks, or pereletoks of frozen ground, occur in the bottoms of the Arctic seas.

Seasonally thawing rocks ("osushki", beaches, spits, barriers, bars, etc.) are widespread on the emerged surfaces. Due to the high mineralization of the pore water, these rocks thaw considerably earlier and freeze later than the nonsaline deposits. The thickness of seasonally thawed deposits in the region of zero sea level is equal to the thickness of the layer of seasonal freezing.

Thus, the Arctic seas are characterized by the wide distribution of various cryogenic rocks that create a very complicated cryolithozone structure. This factor should be given proper attention and must be taken into account in any kind of economic development of the vast Polar territory.

REFERENCES

Are, F. E., 1976, On the subaqueous cryolithozone of the Arctic Ocean, in Regional and thermophysical studies of frozen rocks in Siberia,

Yakutsk, Knizhnoe Izdatelstvo, p. 3-26.

Baranov, I. Ya., 1958, Some regularities in the development of perennially frozen rock series and in seasonal freezing of soils, in Transactions of the Academy of Sciences of the U.S.S.R., Series of Geography, no. 2, p. 22-35.

Danilov, I. D., and Zhigarev, L. A., 1977, Cryogenic rocks of the Arctic shelf, in Frozen rocks and the snow cover, Moscow, Nauka, p. 17-26.

Gorbunov, Yu. A., Losev, S. M., and Timokhov, L. A., 1977, Ice drift with stamukhas present, in The hydrological regime and dynamics of the Arctic Ocean, Leningrad, Gidrometeoizdat, p. 79-94.

Grigoryev, N. F., 1966, The perennially frozen rocks of the maritime zone of Yakutia, Moscow, Nauka, 123 p.

Grigoryev, N. F., 1962, The role of cryogenic factors in the formation of the sea coasts in Yakutia, in Perennially frozen rocks and the accompanying phenomena in the territory of the Yakutsk A.S.S.R., Moscow, Academy of Sciences of the U.S.S.R. Publishers.

Khmyznikov, P. K., 1937, A description of the navigations of vessels in the Laptev Sea and in the western part of the East Siberian Sea, Leningrad, Glavsevmorput Publishers, 180 p.

Kudryavtsev, V. A., and Romanovsky, N. N., 1975, The cryolithozone of the shelf of the Polar basin, in Problems of the geology of the shelf, Moscow, Nauka, p. 60-66.

Molochushkin, E. N., 1973, The effect of thermoabrasion on the temperature of perennially frozen rocks in the offshore zone of the Laptev Sea, in The Second International Conference on Permafrost, Papers and Communications, v. 2, Regional Geocryology, Yakutsk, Knizhnoe Izdatelstvo, p. 52-58.

Molochushkin, E. N., and Gavrilyev, R. I., 1970, The structure, phase composition, and thermal regime of the rocks constituting the bottom of the offshore zone in the Laptev Sea, in The Arctic Ocean and its coasts in the Cenozoic, Leningrad, Gidrometeoizdat, p. 503-508.

Peschansky, I. S., 1963, Ice science and ice engineering, Leningrad, Morskoi Transport, 345 p.

Zhigarev, L. A., 1981, The development of the cryolithozone of the East Siberian and Chukchi Seas in the Pleistocene and Holocene, in The history of development of perennially frozen rocks in Eurasia, Moscow, Nauka, p. 181-191.

Zhigarev, L. A., and Plakht, I. R., 1974, Special features of the structure, distribution, and formation of the subaqueous cryogenic rock series, in Problems of Cryolithology, v. 4, Moscow, Moscow University Press, p. 115-124.

Problems of Geocryology

ПРОБЛЕМЫ ГЕОКРИОЛОГИИ

P.I. Mel'nikov, Editor-in-Chief
Nauka, Moscow, 1983

PREFACE

The present collection contains papers by leading Soviet experts in the field of geocryology, prepared for the Fourth International Conference on Permafrost, held in Fairbanks, Alaska, July 1983. The collection contains the results of research conducted between 1978 and 1982 in various areas of permafrost study. Problems of theoretical, experimental, and engineering geocryology are examined, and principles of new scientific trends are formulated.

This collection combines the results of research carried out on general geocryology, cryolithology, the thermal properties of permafrost, rheology, mechanics and engineering geocryology, as well as on the principles of environmental protection and the recultivation of landscapes that have been altered by man and that are found in the vicinity of permafrost.

The combination of leading experts in different areas of geocryology, who describe important problems for future research, is of great interest within the framework of a single collection.

The collection opens with a paper by Mel'nikov et al. that, based on a geohistorical analysis of the conditions of development of cryogenic strata during the Quaternary period, presents zonal and regional principles of the cryogenic transformation of rocks and groundwater in various hydrogeological structures within the USSR. Kudryavtsev's paper presents an analysis of the thermal bases of geocryology and describes ways to improve geocryological surveys and to obtain raw data for forecasting changes in permafrost conditions and for considering cryolithogenesis as a zonal type of lithogenesis. Specific zonal features are distinguished in the structure of epicryogenic rock strata, and the spatial correlation between types of cryogenesis and the conditions of unidirectional tectonic development is noted.

A number of the papers present results of research in the area of soil mechanics and the construction of foundations. Tsytovich et al. present a unique method for forecasting the thermal stress-strain states of soil structures that is based on the theories of thermoelasticity and thermomechanics. The paper by Vyalov et al. presents new methods for calculating pile foundations on the basis of two groups of limiting conditions that take into consideration the rheological characteristics of frozen soils. A new approach to the thermal physics of the foundations of buildings and structures is described by Khrustalev.

The collection includes papers that are devoted to the study of cryogenic structures and the structure and characteristics of freezing, frozen, and thawing soils. There are papers describing research into the dynamics of the cryolithozone as it is affected by changes in climate and by human activity, the role of thermal abrasion in the development of the cryolithozone of the Eurasian Arctic shelf during the late glacial period, and other problems. Current views on the history of the development of permafrost in the USSR are given by Baulin et al. Gorbunov presents correlated results of studies on permafrost in the mountainous regions of Central Asia.

Individual problems of environmental protection and the recultivation of disturbed landscapes are examined in the papers by El'chaninov and Grigoryev.

The materials in this collection shed light on various problems and trends in geocryology and, in some measure, sum up the work that Soviet geocryologists have conducted in recent years.

ABSTRACTS

New trends in ground water study in cryogenic regions. Mel'nikov, P.I., Romanovskii, N.N., and Fotiev, S.M., p. 7-21.

Based on a geohistorical analysis of the developmental conditions of cryogenic strata during the Quaternary period, the zonal and regional characteristics of the cryogenic transformation of rock strata and ground waters in various hydrogeological structures in the USSR are examined and indicated on a map. The characteristics of the cryogenic transformation of the conditions of water exchange in different geocryological zones and hydrogeological structures are evaluated. Particular attention is given to the results of new studies of open underwater taliks and icing, which are indicators of ground water discharge foci. The importance of studying the composition and mineralization of cryometamorphosed fresh and salt waters is discussed.

Initial assumptions of the thermophysical (geophysical) bases of geocryology.
Kudryavtsev, V.A., p. 21-27.

A broad geophysical approach, which unites the geostructural, geographic, geohistorical, and thermophysical approaches, is substantiated for the study of the frozen zone of the Earth's crust. It involves the use of a large variety of methods to study the qualitative and quantitative relationships between the parameters of seasonally and perennially frozen strata, as well as the principles of formation of the composition and the cryogenic structure of permafrost, cryogenic processes and phenomena with the constituents of the geological and geographical environment, examined within a geohistorical framework. The permafrost survey is the basic method of obtaining the initial information.

Geocryological conditions of Yakutiya in relation to the construction of trunk pipelines. Mel'nikov, P.I., and Grave, N.A., p. 28-35.

The paper presents a brief physico-geographical description of Yakutiya. Geocryological regions are described, and the territory is schematically divided into regions according to the resistance of the surface to technological activities. Data are presented of observations of the development of surface damage along the route of a buried pipeline.

Cryolithogenesis as a zonal type of lithogenesis. Popov, A.I., p. 35-43.

The general position of cryolithogenesis within the lithogenesis system is determined. Cryolithogenesis is examined as a phenomenon of diagenesis and hypergenesis (cryodiagenesis and cryohypergenesis) that is characteristic of the thermodynamic situation of the zones of constant cooling of the Earth. Specific zonal features are distinguished in the structure of epicryogenic rock strata that are characteristic of northern Eurasia (as an example of the zonal nature of cryolithogenesis). The spatial relationships of types of cryolithogenesis to the conditions of equidirectional tectonic development are noted.

Predicting the thermal stress-strain condition of earthworks by the finite element method. Tsytovich, N.A., Kronik, Ya.A., Gavrilov, A.N., and Demin, I.I., p. 44-56.

How to formulate and obtain an approximate solution to the problem of predicting the thermal stress-strain state of earthworks that are constructed under severe climatic conditions is examined. A unified method is developed for predicting the thermal and stress-strain states of earthworks on the basis of the theories of thermoelasticity and thermomechanics in the formulation of a disconnected plane problem. As an example, the thermal stress-strain state of a test embankment constructed of macroelastic soil is calculated by the finite element method with the aid of a computer. The results of the computations are compared with two years of field observations of the test embankment.

Settlement and bearing capacity of pile foundations on permafrost. Vyalov, S.S., Slepak, M.E., and Tikhomirov, S.M., p. 57-67.

Methods are presented for designing pile foundations on the basis of two groups of limiting conditions. In contrast to traditional methods, the methods described take into account the rheological characteristics of frozen soils and make it possible to carry out pile computations under both constant and variable external loads and temperatures. Examples are presented to illustrate the new methods.

Foundations on permafrost that thaws while the building or structure is in use. Kolesov, A.A., p. 68-75.

Effective methods are examined for the construction of foundations on permafrost for industrial and civil uses. Recommendations are given for the design of foundations, how to lay them, and how to prepare the ground prior to construction.

Ventilated solid foundations on fill. Kutvitskaya, N.B., and Dashkov, A.G., p. 76-82.

Design solutions are examined for solid foundations of the slab, strip, and column types, which combine bearing and cooling functions. On the basis of experimental and theoretical research, a solution is given to the problem of the interaction between buildings on ventilated solid foundations and the permafrost. Examples are presented of design solutions and of actual construction.

Calculating the thawing depth of multilayer coverings on permafrost, including a highly efficient layer of thermal insulation. Ivanov, V.N., Merzlyakov, V.P., and Plotnikov, A.A., p. 82-89.

An analytical method is proposed for calculating Stefan-type thermal conductivity in multilayer constructions that include an efficient layer of thermal insulation as used in permafrost coverings. The solution is reduced to a series of recurrence formulas that are used to write a computer program in Fortran IV. The results of the calculations are compared with field observations.

Fundamentals of the thermorheology of cryogenic soils. Grechishchev, S.E., p. 90-100.

The thermorheology of cryogenic soils (TCS) is defined. A brief overview is presented of certain scientific trends in the physics and mechanics of cryogenic soils from the point of view of the applicability of the obtained results to the development of the principles of TCS. The thermodynamic and mechanical conditions at the phase boundaries in waterlogged soils are presented, as is a theoretical model of cryogenic heaving. Several known cryogenic soil phenomena are explained on the basis of the proposed theory.

Shear resistance of permafrost during thawing. Shusherina, E.P., Maksimyak, R.V., and Martynova, G.P., 100-108.

Results of experimental studies of the shear resistance of permafrost during thawing are examined. The research was carried out on undisturbed natural soils of various origins, characterized by different structural and textural features, moisture contents, and compositions of the skeleton. The shear resistance of thawed soil at the thawing boundary as a function of these factors is presented, as is the relation of the size of the normal load and the angle of inclination of the ice layers to the shear plane.

The flow of frozen soils on rock slopes. Sadovskii, A.V., and Bondarenko, G.I., p. 108-113.

The mechanism of the shear deformation of frozen soil during flow over a rock slope was studied. The shear testing of samples of soil that are adfrozen to rock slopes and observations of the deformation of frozen soils during dumping on rock slopes established the existence of two flow zones in soils that are flowing down a rock channel: a contact layer and a layer that is located higher in the soil. The rate of flow at the soil surface is equal to the sum of the rates of flow of these layers. The rates of flow that are calculated using the suggested formulas are in good agreement with observed rates of creep of frozen soil dumps.

Physicochemical aspects of the creation of man-made ice structures using seawater. Savel'ev, B.A., Gagarin, V.E., Zykov, Yu.D., Latalin, D.A., Rozhdestvenskii, N.Yu., and Chervinskaya, O.P., p. 113-118.

The creation of man-made ice structures is examined from three points of view: the quality of the artificial ice, the adfreezing of the structure to the bottom of the reservoir, and the durability of the structure. On the basis of experiments that used structural analysis and geophysical methods, the advantage is shown of creating ice structures and of securing them to the bottom by freezing ice within a volume of water. The question of the thermal regime of a man-made ice structure in shallow water is examined.

Elastic and electrical properties of ice-rich soils and ice. Zykov, Yu.D., Rozhdestvenskii, N.Yu., and Chervinskaya, O.P., p. 118-127.

To successfully use geophysical methods to study ice-rich soils and ice, it is necessary to study their properties, beginning with their elastic and electrical properties. Experimental data are examined on specific electrical resistances and speeds of elastic waves in ice of varying composition and structure, as well as the dependence of these parameters, as measured on model samples with layered cryogenic structure, on the ice content. Significant anisotropy was observed, and its dependence on various factors was studied. The results are compared with calculations. The possibility of evaluating the properties of ice and ice-rich soils by electrical and acoustic measurements is demonstrated.

Thermophysical problems of permafrost engineering. Khrustalev, L.N., p. 127-136.

Practical problems that currently confront researchers in the construction of foundations on permafrost are examined. The author discusses the problems of thermal interaction between erected structures and permafrost, foundation cooling using natural cold with the aid of various types of self-regulating cooling installations, and evaluating foundation reliability to arrive at a scientifically valid way of choosing the method for building a foundation. The most effective approaches and methods of solving these problems are indicated.

Significance of soil complex changes on the permafrost temperature regime. Shur, Yu.L., Shvetsov, P.F., Slavin-Borovskii, V.B., Moskalenko, N.G., and Malevskii-Malevich, S.P., p. 136-143.

The effects of weathering, soil formation, and vegetation on the temperature regime of permafrost are examined. Changes in mean annual temperature are evaluated in connection with a directed change in the properties of the seasonally thawing layer. Temporary variability of the components of the temperature balance is studied.

Formation of cryogenic structures during epigenetic and syngenetic freezing of dispersed soils. Ershov, E.D., p. 143-152.

The mechanism of the formation, development, and transformation of various types of cryogenic structures is examined on the basis of heat-mass exchange and physicochemical and physicomechanical processes that occur in freezing and frozen soils. The formation of migration ice and segregated ice layers in permafrost, which have frozen epigenetically and syngenetically, is examined.

Investigation of the main factors and the mechanism of the cryogenic transformation of minerals. Konishchev, V.N., Kolesnikov, S.F., and Rogov, V.V., p. 152-157.

A theoretical model of the cryogenic stability of particles of various sizes of the main soil-forming minerals is examined on the basis of notions about the protective role of unfrozen water. The cryogenic breakdown of quartz, feldspars, hornblendes, pyroxenes, and ores is modeled in the laboratory. Tables are given of the cryogenic stability of minerals at various temperatures and moisture contents.

Permafrost facies structure of modern alluvial deposits of plain rivers in the eastern Subarctic. Rosenbaum, G.E., p. 158-161.

The facies structure of modern alluvial deposits of plain rivers of the eastern Subarctic is examined. Riparian and inner-floodplain facies are distinguished in the structure of the river plain alluvium. Subfacies are distinguished within each facies, depending on the nature of the development of the polygonal relief and the ice veins in different sections of the plain surface. Superimposed river plains are distinguished in whose cross sections the facies of the inner floodplain is underlain not by channel alluvium but rather by deposits of an alas complex.

Cryolithogenic deposits and their permafrost-formation complexes. Katasonov, E.M., p. 162-169.

Deposits that form near permafrost are examined. Depending on the conditions of accumulation and freezing, they are classified as permafrost formation complexes - strata that are distinguished by the types of facies and their content and distribution of ground ice. Special attention is devoted to ice complexes. Standard sections are presented.

Cryolithogenic consequences of non-unidimensional freezing of soil. Zhestkova, T.N., p. 169-177.

The freezing of soils under two- and three-dimensional conditions was investigated. The relationship between the cryogenic structure and the thermal field of the frozen soil is established. The existence of horizontal heaving, which may attain significant proportions under non-unidimensional soil freezing conditions, is proven.

Paleogeographic reconstructions of the Pleistocene based on the thickness and structure of permafrost. Baulin, V.V., and Chekhovskii, A.L., p. 177-184.

Problems of paleoclimatic reconstructions based on the analysis of the distribution of thick permafrost are examined. It is concluded that the thickness and structure of permafrost can serve as indicators of changing heat exchange conditions at the Earth's surface.

Cryolithozone dynamics as related to climatic changes and anthropogenic influences. Balobaev, V.T., and Pavlov, A.V., p. 184-194.

An analysis is presented of climatic and geocryological changes effected by the natural global cycle (which has a period of several thousand years) and influenced by anthropogenic changes in the composition of the atmosphere, for the next century, and, as affected by local technology, in the short term. An attempt is made to evaluate the quantitative limits of possible transformations of the cryolithozone, its thermal regime, thickness, and local development.

The role of thermal abrasion in the development of the cryolithozone of the Arctic Shelf of Eurasia during the late- and post-glacial period. Are, F.E., p. 195-201.

Based on studies of the deterioration of the shores of arctic seas and on paleogeographic concepts, the role of thermal abrasion in the development of the subaqueous cryolithozone of the Arctic Shelf of Eurasia during the late- and post-glacial period is presented. In the Soviet Arctic, six regions are distinguished that differ with regard to the development of thermal abrasion. It is shown that in the East Siberian region the rise in the sea level during the post-glacial transgression was responsible for a rapid rate of change in the position of the shoreline. During this process thermal abrasion broke up the upper layers of the continental glacial deposits, which then came to be in a subaqueous position.

Methodological principles of the evaluation, prognosis, and mapping of natural complexes with varying resistances to thermal subsidence. Zhigarev, L.A., and Parmuzina, O. Yu., p. 201-206.

The principles of evaluating and forecasting thermal subsidence in natural complexes are described. Methods are proposed for mapping such complexes, which have varying resistance to thermal subsidence.

The effects of mining on the environment and methods for alleviating them. El'chaninov, E.A., p. 207-214.

Mining for minerals causes the removal from the natural environment and the dumping into the environment of a series of components. An evaluation is presented of the subsystems of the technological cycle and the degree to which they affect the environment. Information is given on developed technological procedures that contribute to the elimination or attenuation of the harmful effects of mining on the environment.

The Central Asia permafrost region. Gorbunov, A.P., p. 214-222.

Data about the cryolithozone and the cryogenic relief of the mountains of Central Asia are correlated. The extent of the area of permafrost in the region is estimated to be 2.09 million km^2, the volume of the cryolithozone is 160,000 km^3 and the volume of ground ice is 9,000 km^3. The upper permafrost boundary in the northern part of the region is at 2200 m above sea level, and in the south at 5000 m above sea level.

History of the development of permafrost in the USSR. Baulin, V.V., Velichko, A.A., and Danilova, N.S., p. 222-229.

A brief paleogeographical scheme is proposed for the formation of permafrost throughout the entire territory of the Soviet Union for the main stage of the Quaternary. The Upper Pleistocene and the Holocene, which are subdivided into stages of multi-directional permafrost dynamics, are covered in the greatest detail. All of the reconstructions are made on the basis of an analysis of the cryogenic structure of the permafrost, of wedge ice, and of associated pseudomorphosis, as well as the thickness of the permafrost. Special features of the development of permafrost in large regions of the USSR, such as the European plain and western Siberia, are examined.

The use of cryochemical data in the study of the origin of ground ice deposits. Anisimova, N.P., and Kritzuk, L.N., p. 230-239.

The physicochemical processes that determine the chemical composition of ground ice deposits that were formed in various ways are examined. The comprehensive study of the unique deposit of ground ice in the valley of the Se-Yakha River on the Yamal Peninsula is presented as an example. It is concluded that it originated underground, and that the injection mechanism of formation played a role.

Suprapermafrost water of the cryolithozone: Its classification and description. Shepelev, V.V., p. 239-244.

The characteristics of the suprapermafrost waters of the cryolithozone and schemes for their classification are examined. A brief description is given of the individual subtypes, classes, subclasses, and varieties suprapermafrost water, distinguished according to their relationships to the various cryogenic confining strata, the degree of freezing, their conditions of occurrence, hydraulic properties, and temperature. The great importance is pointed out of suprapermafrost water in the development of many permafrost processes and phenomena, as well as in changes in the permafrost-hydrogeological circumstances that are due to man's activities.

Thermodynamic criteria of the stability of the soil complex in the Subarctic. Shvetsov, P.F., p. 245-247.

The expenditure of energy for cryogenic processes is evaluated on the basis of heat storage data. The thermodynamic and mechanical stability of subarctic soil complexes is examined.

The cryolithosphere of Mars and cryogenic phenomena on its surface. Kuzmin, R.O., p. 247-255.

Structural features of the planetary permafrost mantle of Mars' crust (cryolithosphere) and the possible distribution within it of stable phases of water and carbon dioxide are examined. Data are presented about the determination of the depth at which soil strata that contain ice occur in the cryolithosphere, as are results of the geomorphological study of various permafrost indicators in the relief of Mars.

Autoregeneration (optimization) of disturbed surface conditions in northern West Siberia. Grigoryev, N.F., p. 256-263.

Examples are presented of relatively rapid (3-4 years) arrest of thermokarst during its initial stages, when it is caused by the disturbance of the vegetative cover of tundra on ice-rich soil. The summer thawing layer is observed to dry out gradually after the removal of the vegetative cover, which helps to improve the stability of the surface.

Permafrost surveys and forecasting of changes in permafrost conditions during economic development of permafrost regions. Kudryavtsev, V.A., p. 264-269.

The existing requirements for permafrost surveys, as stated in Departmental and All-Union standards and technical regulations for surveys, are inadequate for producing scientifically substantiated forecasts of permafrost conditions during the construction and use of structures in the permafrost zone. Recommendations are presented for conducting permafrost surveys in such a way as to make it possible to produce a geocryological forecast, and measures are proposed that will ensure the stability of structures and of the environment.

Other Contributions

WATER FLOWS INDUCED BY THERMAL GRADIENTS

D. Blanchard and M. Frémond

Laboratoire Central des Ponts et Chaussées
58 Blvd. LeFebvre, 75732 Paris, France

Water flows are important in soil freezing. At the frost line (0°C isotherm) which separates the frozen soil from the unfrozen soil a large depression, the cryogenic suction, draws in a large amount of water, which freezes. We describe a macroscopic model of thermal gradients and water flows, including the cryogenic suction. The model is based on the deformability of the soil considered as a porous medium and the existence of unfrozen water in the frozen soil. It uses the conservative laws (Fourier's law, Darcy's law, unfrozen water content versus temperature in the frozen soil). Its outputs are the temperature and hydraulic potential functions of space and time. A two-dimensional computer program is available.

THE NATURE AND DISTRIBUTION OF SHALLOW SUBSEA ACOUSTIC PERMAFROST ON THE CANADIAN BEAUFORT CONTINENTAL SHELF

S.M. Blasco

Geological Survey of Canada
Bedford Institute of Oceanography, P.O. Box 1006
Dartmouth, Nova Scotia, Canada B24 4A2

M.J. O'Connor

M.J. O'Connor and Associates Limited
101-4712, 13th Street
Calgary, Alberta, Canada T2E 6P1

The correlation of high resolution analog reflection seismic profiles with shallow and deep borehole data has led to the identification of several types of shallow acoustic permafrost (APF) to depths of 100 m below seabed, including discontinuous hummocky APF, continuous APF, stratigraphically controlled APF, APF lensing, and possibly massive APF associated with offshore pingo-like features. The occurrence of shallow APF has been mapped over the shelf from the coast to the 70-m isobath and from Baillie Island to the eastern edge of the Mackenzie Trough. On a regional scale, shallow APF is laterally and vertically variable and discontinuous. East of the 135° meridian, the distribution is less well defined because of the marginal nature of the ice bonding.

The distribution is closely related to the surficial geology. The shelf is overlain with a thin veneer of fine-grained marine deposits (unit A) that overlie or grade downward into a transgressive sequence of interbedded fine- to medium-grained sediments (unit B). These two units unconformably overlie older pre-Holocene sediments (unit C) of possible reworked fluvial-deltaic origin. The APF is primarily confined to unit C (although some ice content has been observed in the overlying units). Depth to the APF may vary from 0 m to 50 m below the unconformity. In some areas, the top of the APF is found to mirror the unconformity surface.

FROST-HEAVED STRUCTURES IN BEDROCK: A VALUABLE PERMAFROST INDICATOR

J.C. Dionne

Département de Géographie, Université Laval
Québec City, Québec, Canada G1K 7P4

Frost-heaved structures in bedrock are periglacial features produced by the vertical displacement of bedrock fragments. Blocks, frost-wedged from bare bedrock along joints, are raised above the general surface by frost processes. They are common and exclusive to permafrost regions, and are considered by French as features representative of very active periglacial regions.

Although they were mentioned long ago in the literature by Tyrrell, they have been described and studied in some detail only recently by Bournérias, Payette, Dyke, and Dionne. Among the various features reported, isolated heaved blocks (monoliths) and conical mounds are the most common. Their existence, to our knowledge, is restricted to the Northern Hemisphere: Canada, Greenland and Spitsbergen. Numerous sites occur in subarctic Québec, between latitudes 51°45' and 60°20'N, and in subarctic Canada between latitudes 58°30' and 66°10'N.

They develop in various lithologies: basalt, diabase, andesite, gabbro, granite, gneiss, crystalline schists, and occasionally in sedimentary rocks such as sandstone and quartzite. However,

This section contains non-Soviet abstracts and papers that were not available at the time of printing of either the Abstract and Program volume or the first Proceedings volume.

most features are in volcanic, igneous and metamorphic rock with well-developed systems of joints.

Single heaved blocks from 50 to 200 cm long are often raised as much as 150 cm above the surrounding surface. Conical mounds made of several blocks (up to 15 fragments) are commonly 3 to 10 m in diameter, and a central depression (crater) characterizes a few structures. Most heaved blocks exhibit weathered and lichen-covered sides except at their bases where freshly exposed surfaces indicate recent heaving. As already mentioned, these features result from frost action, particularly frost wedging and heaving due to pressure of ice in vertical, oblique and horizontal joints.

Frost-heaved features in bedrock are all located in the permafrost zone. The southernmost occurrence is at the summit of the Groulx Mountains, Québec (51°45'N), near the southern limit of the discontinuous permafrost zone. The mean annual air temperature in the area considered ranges from -4° to -9°C. Although snow depth is not known for each locality, most sites consist of bare, well-exposed bedrock surfaces which are wind-blown during the winter. Consequently, a thin snow cover is common to the environment.

Dating heaved structures is a difficult problem. However, recent heaving is usually evidenced by lichen-free sides at the base of heaved blocks. The occurrence of several structures in a locality indicates primarily that the conditions required for their development do or did exist; the structures may have developed in a relatively short time (a few decades) as well as over many centuries. Thus, further studies are needed to document the time required for the development of heaved features in bedrock and to determine their age in a given locality.

Based on their geographical distribution, on climatic data, and on the occurrence of permafrost at almost every site, it is suggested that heaved features in bedrock could be considered a valuable indicator of permafrost. If found outside the permafrost regions today, relict heaved features will indicate former permafrost conditions.

ROCK GLACIERS OF THE EAST CENTRAL BROOKS RANGE, ALASKA

J.M. Ellis

Gulf Oil Exploration and Production Co.
P.O. Box 37048
Houston, Texas 77236, USA

P.E. Calkin

Dept. of Geological Sciences, State University of New York, Buffalo, NY 14226, USA

M.P. Bruen

Harza Engineering Company, 203 W. 15th Avenue
Anchorage, Alaska 99501, USA

Rock glaciers here were largely formed by increased mass wasting associated with valley deglaciation in latest Pleistocene time; most have upper surfaces which first stabilized by the early Holocene. They range from 3-km-long, tongue-shaped lobes that are cored by glacial ice and headed by reactivated (Neoglacial) cirque glaciers and late Holocene moraines to smaller lobate masses 15 to 150 m in length lying beyond talus cones. They appear best developed below resistant ridges with large exposures of shale, siltstone, and phyllite; their talus enhances insulation of subsurface ice. All rock glaciers occur within the zone of continuous permafrost at altitudes from 900 to 2,000 m. About 75% of the rock glaciers and almost all of the active forms occur north of the Continental Divide, demonstrating the climatic significance of this east-west trending divide. Active tongue-shaped rock glaciers have a strong N to NE orientation pattern. However, lobate forms face all directions, demonstrating insensitivity to insolation and perhaps a lower ice content relative to debris. Both forms occur 50 to 100 m higher on south-facing slopes than on northerly slopes. No correlation was found between solar energy received at a rock glacier and the rock glacier's present state of activity. Little correlation also exists between the activity state and a) bedrock, b) overall slope, c) orientation, or d) latitude and altitude north of the Continental Divide. Whether or not a rock glacier is active appears to depend most upon the availability of debris from headward cliffs.

ACTIVE LAYER STUDIES IN NORTHERN QUÉBEC: TOWARDS A PREDICTIVE MODEL

J.T. Gray and J. Poitevin

Département de Géographie, Université de Montréal
Montréal, Québec, Canada H3C 3J7

J. Pilon

Earth Physics Branch
Department of Energy, Mines, and Resources,
Ottawa, Ontario, Canada K1A 043

Field studies recently conducted on the west coast of Ungava Bay, in the southern part of the continuous permafrost zone, have permitted the establishment of the pattern of active-layer development for six different geomorphological and vegetational terrain types. All study sites were situated on horizontal surfaces free from local topographic influences in order to permit the explanation of thaw progression in terms of the earth materials, moisture content, and vegetation cover that characterize each terrain type. Thermal properties for organic and mineral soil and for bedrock were calculated from field data on texture, bulk density, and moisture content of the earth materials at the sites, and were used to explain thaw-layer development for each terrain type. A linear relationship was established for each group of sites between the logarithmic transformation of thaw penetration and an atmospheric thawing index obtained through interpolations of air temperature data from the two closest meteorological stations --Kwjjuaq (Fort Chimo) and Koartaq (Cape Hope's

Advance). This relationship was then used in a predictive sense to establish, first, the depth of the active layer at the end of the thaw season for each terrain type, and second, variations in active-layer depths in the region over a 20-year time period. Probable spatial variations of active layers for extreme climatic variations in the Ungava Peninsula are also briefly presented.

THE INFLUENCE OF CLAY-SIZED PARTICLES ON SEISMIC VELOCITY FOR CANADIAN ARCTIC PERMAFROST*

M.S. King

Lawrence Berkeley Laboratory and
University of California
Berkeley, California 94720, USA

Seismic wave velocities have been measured on 37 unconsolidated permafrost samples as a function of temperature in the range -16° to +5°C. The samples, taken from a number of locations in the Canadian Arctic Islands, the Beaufort Sea and the Mackenzie River, were tightly sealed immediately upon recovery in several layers of polyethylene film and maintained in their frozen state during storage, specimen preparation, and until they were tested under controlled environmental conditions. During testing, the specimens were subjected to a constant hydrostatic confining stress of 0.35 MPa (50 psi) under drained conditions. At no stage was a deviatoric stress applied to the permafrost specimens. The fraction of clay-sized particles in the test specimens varied from almost zero to approximately 65%. At temperatures below -2°C the compressional-wave velocity was observed to be a strong function of the fraction of clay-sized particles, but only a weak function of porosity. At temperatures above 0°C the compressional-wave velocity was observed to be a function only of porosity, with virtually no dependence upon the fraction of clay-sized particles. Calculation of the fractional ice content of the permafrost pore space from the Kuster and Toksöz theory showed that for a given fraction of clay-sized particles the ice content increases with an increase in porosity. It is concluded that the compressionalwave velocity for unconsolidated permafrost from the Canadian Arctic is a function of the water-filled porosity, irrespective of the original porosity, clay content, or temperature.

PERIGLACIAL DUNE FORMATION AND NIVEO-EOLIAN FEATURES AS EXEMPLIFIED IN THE KOBUK SAND DUNES, ALASKA

E.A. Koster

Physical Geography and Soil Science Laboratory
University of Amsterdam
Amsterdam, The Netherlands

J.P. Galloway

U.S. Geological Survey, 345 Middlefield Road
Menlo Park, California 94025, USA

Active dune building in the Nogahabara Sand Dunes (approximately 44 km^2) in the Koyukuk Flats and in the Great and Little Kobuk Sand Dunes (approximately 64 km^2) in the central Kobuk River valley is being studied to compare it with paleodune building in the well-known Weichselian inland dune and cover sand regions of Western Europe. The active dune fields in the Kobuk River valley consist primarily of large parabolic and transverse ridges, indicating primary wind directions from the east and southeast. During summer months secondary slip faces on the dunes develop due to westerly winds. Wavy transverse ridges have been deformed into barchan-like dunes. Radiocarbon dates (approximately 24,000 B.P.) of sandy peat and organic mud layers underneath the dune sands indicate a late-Wisconsin age.

Distinct stabilized, canoe-shaped dunes resemble the "Cree Lake-type dune ridges" which occur in northern Saskatchewan. They have been formed by southeasterly paleowinds of uniform direction by elongation of primary parabolic dunes. Recent analogues are found opposite the mouth of the Akillik and Hunt Rivers.

In early summer annual denivation features, several meters high, develop in the steep westward-facing slip faces of the large transverse dunes. Sand layers incorporated in the snow indicate niveo-eolian deposition during winter, related to a discontinuous snow cover. Hummocky topography with a completely cracked, moist sand surface rapidly develops. During the course of summer these features gradually disappear and a smooth slip face is reestablished.

THAW PENETRATION AND THAW SETTLEMENT STUDIES ASSOCIATED WITH THE KIRUNA TO NARVIK ROAD, SWEDEN

S. Knutsson

University of Luleå, S-951 87 Luleå, Sweden

A permafrost study has been undertaken on the recently built road between the towns of Kiruna, Sweden, and Narvik, Norway. The road crosses an area with discontinuous permafrost (68°15'N) where an environmental engineering study was initiated during autumn of 1979.

Eight holes were drilled to maximum depths of 11 m. Four of these were located outside the road in the undisturbed permafrost and four within the road in a cross section perpendicular to the road line.

Temperature gauges were installed in the drill holes at vertical distances of 0.5 m. Measurements of the temperatures are taken approximately once a month, and the measurements from three years of data-logging have been analyzed.

The discontinuous permafrost is warm, -1°C at 6-8 m depth, and the depth of seasonal thaw pene-

*Full paper subsequently published in Canadian Journal of Earth Sciences, vol. 21, p. 19-24, 1984.

tration (active layer) in the undisturbed areas adjacent to the road is on the order of 0.5-1.0 m. Beneath the road the seasonal thaw penetrates about 2.5 m. The difference is explained by the dark colour of the bitumen top cover and of the relatively coarse road material, with its relatively high thermal conductivity during summertime. The permanently frozen soil materials range from rock to ice-rich, fine-grained silts with a top layer of peat.

The core from the undisturbed area is geologically described and the amount of ice and thickness of ice lenses have been measured by means of X-ray technique. This technique seems to be a useful, easy and non-destructive tool for ice-content measurments in cores of frozen soil.

In addition to the temperature measurements, settlement of the road has been observed as well. This shows gradually increasing settlements, which are due to the thawing of the underlying permafrost. The observed settlements are compared to those predicted. The predictions are based on thaw-consolidation tests performed in the laboratory, and these are described as well. Reasonable agreement between the observed and predicted settlements is obtained.

THE EFFECTS OF COAL DUST ON SURFACE ALBEDO AND THAW DEPTH IN NORTHERN ALASKA

J.S. Makihara

(formerly) Geography Department
Pennsylvania State University
State College, Pennsylvania 16801, USA

The effects of coal dust on surface albedo and thaw depth in arctic Alaska were studied via field data acquisition and analysis and computer simulation modeling. Field research was conducted during summer 1981 in the vicinity of Prudhoe Bay. Coal dust reduced albedo by 25 to 90 percent. Apart from a higher rate of thaw exhibited on coal-dusted plots compared to control plots in July, the short-term effects of coal dust were not apparently significant in the field data due to the short period and spatial extent of observation, and natural variability in thaw and soil thermal regimes. In determining long-term effects, computer simulations of one-dimensional heat conduction revealed deeper maximum thaws up to 52.4 cm, 20 percent over the undisturbed condition, and especially a higher thaw rate in the first part of the summer on disturbed plots as compared to control plots. Given comparable parameters, winter-only impact cases produced greater thaw depths than summer-only impact cases. Dusting in winter produces premature spring meltout and consequently deeper thaw; any potential coal mining activities should be restricted in April and May. The model assumes that dust does not affect vegetation mortality. In actuality, dust can reasonably be assumed to damage vegetation with subsequent enhanced thermal impact. Future research should take into account the effect of vegetation mortality on surface albedo.

PALEO-CLIMATIC RECONSTRUCTION FROM MEASURED TEMPERATURES IN BOREHOLES IN NORTHERN QUEBEC

J.A. Pilon, A.S. Judge and A.E. Taylor

Earth Physics Branch
Department of Energy, Mines and Resources
Ottawa, Ontario, Canada K1A 0E4

J.T. Gray and J. Poitevin

Département de Géographie, Université de Montréal
Québec City, Québec, Canada H3C 3J7

B. Lauriol

University of Ottawa
Ottawa, Ontario, Canada K1A 6N5

The analysis of deep borehole temperature gradients can assist in the reconstruction of former climates at the ground surface. In this report, emphasis is on the modelling process itself, but some results are presented in the context of the authors' geothermal studies in northern Quebec. The modelling process begins with the acquisition of temperature profiles from one or several boreholes once the disturbance due to drilling has dissipated. The drill-hole temperature profile integrates the geothermal heat flux from deep in the crust and a number of largely near-surface modifying effects. The latter, which create departures from the constancy of the heat flow profile with depth, reflect the role of 1) varying thermal conductivity in an anisotropic medium, 2) spatial topographic and microclimatic irregularities around the collar of the borehole, 3) water movement in fractures and joints, and 4) temporal fluctuations in ground surface temperatures. Through the modelling process, the influence of the first two groups of factors on the temperature profiles is determined, and secondary profiles are derived from which irregularities results from these influences have been removed. Where characteristic responses indicating water flow are present, the data are usually not suitable for the climatic analysis. In suitable data sets the final step then takes available information on the dates and direction of climatic change at the surface and generates the amplitude of such changes backwards in time. Since the mathematical solutions are non-unique it is necessary to use independently derived information on dates and amplitudes and to use geothermal profiles to determine the unknown parameter.

Other applications of the technique are to determine the development or degradational history of permafrost, whether it is contemporary or relic in nature, and to determine approximate ice-base temperatures beneath the Wisconsin ice sheets, provided the drill holes are sufficiently deep. Such modelling of geothermal data carried out at two sites in northern Quebec indicates major differences in the permafrost and glacial history between the central highlands of Ungava and a lowland region approximately 500 km southeast.

EXCAVATION RESISTANCE OF FROZEN SOILS UNDER VIBRATING CUTTINGS

Shingi Takasugi

Technical Research Center, Komatsu Ltd.
Kanagawa, Japan

A. Phukan

University of Alaska, 3221 Providence Avenue
Anchorage, Alaska 99504, USA

This paper presents dynamic laboratory experimental studies of frozen sands and silts at temperatures varying from -2°C to -11°C, using various cutting velocities from 2 mm per second to 18 mm per second. The 6-Hz vibration model used was varied from 1.3 mm amplitude to 4 mm amplitude. Theoretical analyses of the cutting resistance of frozen soils under vibration cutting are formulated to explain the failure characteristics obtained during the experimental work. Energy ratio versus the number of contacts per inch of cutting (Nc) plots show a similar tendency for both silty and sandy frozen soil samples; that is, as the energy ratio increases, Nc increases. The contact length between the soils and the tool in the vibration cycle is inversely proportional to Nc. At high Nc, the contact length is too small to cause soil failure and thereby increases the vibration energy. Relationships between the energy ratio and Nc at various amplitudes, for both soil types, show a greater energy requirement at 4 mm amplitude of vibration than at 1.3 mm amplitude, in spite of the identical draft energy reduction at both vibration amplitudes. It is possible that by inducing vibration of the cutting tool, a reduction in draft energy can be obtained and thereby the excavation process can be efficiently executed in frozen ground.

SHORELINE REGRESSION: ITS EFFECT ON PERMAFROST AND THE GEOTHERMAL REGIME, CANADIAN ARCTIC ARCHIPELAGO

A.E. Taylor and A.S. Judge

Earth Physics Branch
Department of Energy, Mines and Resources
Ottawa, Ontario, Canada K1A 043

D. Desrochers
Department of Geography, University of Ottawa
Ottawa, Ontario, Canada K1A 6N5

In the Queen Elizabeth Islands of the Canadian Arctic Archipelago, the late Quaternary event with the most profound influence on ground temperatures is shoreline regression accompanying postglacial isostatic uplift. The coastal margins of many islands have experienced hundreds to several thousand years of submergence since 8,000 years ago. The effect on the geothermal regime is far from subtle because of the large contrast between arctic air temperatures and sea temperatures. Permafrost thicknesses measured today reflect this surface temperature history. Two deep wells 1 km apart on Cameron Island have measured permafrost thicknesses of 726 and 660 m; the geothermal analysis attributes the difference entirely to the first, and higher, site having emerged from the sea around 7,000 year B. P., about 2,000 years before the second. Inland sites on the Sabine Peninsula of Melville Island may have, in a similar lithology, permafrost twice as thick as coastal sites. The geothermal analysis explains this variation in terms of a simple sea regression model derived from emergence curves published for the region.

A COMPARISON OF SUCCESSIONAL SEQUENCES FOLLOWING FIRE ON PERMAFROST-DOMINATED AND PERMAFROST-FREE SITES IN INTERIOR ALASKA

K. Van Cleve

Forest Soils Laboratory, University of Alaska,
Fairbanks, Alaska 99701, USA

L.A. Viereck

Institute of Northern Forestry
Fairbanks, Alaska 99701, USA

The structure and function of upland interior Alaskan forest ecosystems has been examined across two secondary successional sequences. One, the most common in interior Alaska, follows fire in black spruce stands on permafrost sites. The other, less common sequence follows fire on warmer, generally south aspect sites and passes through a shrub and hardwood stage to white spruce. On black spruce sites, the thickness of the forest floor is a key factor responsible for maintaining moist soil, low soil temperatures and permafrost. Soil degree-days range from 1250 to 1000 degree-days 1 to 5 years following fire. In mature black spruce temperature ranges from 500 to 800 degree-days. The active layer thickness in later stages of this succession is usually from 40 to 60 cm. Following fire it increases for several years and may reach 1 m. In the south aspect white spruce sequence, soil temperature approaches 1400 degree-days in heavily burned forest in contrast with 9m0 to 1000 degree-days in mature white spruce. On the black spruce successional sequences, low soil temperature results in the lowest rates of soil biological activity, nutrient cycling, and, in turn, the lowest tree productivity. Net annual above-ground tree production ranges from about 7 metric tons per hectare on the warmer, south aspect hardwood-white spruce sequence to less than 1 metric ton per hectare on the black spruce sequence.

SOME CONSIDERATIONS REGARDING THE DESIGN OF TWO-PHASE LIQUID/VAPOR CONVECTION TYPE PASSIVE REFRIGERATION SYSTEMS

E. Yarmak, Jr. and E.L. Long

Arctic Foundations, Inc., 5621 Arctic Boulevard
Anchorage, Alaska 99502, USA

Two-phase liquid/vapor convection type passive refrigeration units are widely used for permafrost foundations and earth stabilization. Aspects that are of primary importance to the designer include determination of an effective freezing radius for a unit, calculation of a thermal resistance for the soil surrounding the embedded portion of the unit, and estimation of an appropriate thermal resistance for the condenser portion of the unit. Design charts and formulas are presented to estimate these values and demonstrate the sensitivity of varying key parameters such as soil type, wind velocity, condenser surface finish and spacing. Computations for two different design applications are detailed.

THE STABILIZATION OF THE NANISIVIK CONCENTRATOR FOUNDATION, NANISIVIK MINES LTD.

D. Williams

Strathcona Mineral Services Limited
401-44 Victoria Street, Toronto
Ontario, Canada M5C 142

Nanisivik Mines Ltd. operates a 2,000-ton/day underground zinc/lead mine in the Canadian Arctic on the north end of Baffin Island at 73°N in an area of thick permafrost. The concentrator building houses the powerplant, workshops, warehouse and offices. The grinding mills and generators produce heavy foundation loads. A suitable construction site was selected following drilling and surface inspection. All bore holes within the site area showed a gently dipping dolomite as bedrock. The ice content was negligible, being present only in small vugs or as thin coatings along bedding planes.

The steel frame building was erected during 1975 and in the following year minor cracks were noticed in the concrete block partition walls. A survey of the steel column bases was carried out to provide reference elevations for future observations. Accelerated movement was observed during 1979, with elevation changes on the order of 50 mm being recorded over the course of the year.

In order to determine the nature and extent of this problem a number of holes were drilled with a diamond drill through the floor of the building. These holes indicated an ice lens under the building at an average depth of 10 m which varied in thickness from 1 to 1.5 m. In some areas the ice had completely melted, leaving a water-filled void. The overlying crown pillar of rock on which the building had been constructed was subsiding into this cavity.

Several support methods were tried which involved placing tailings, cemented tailings and grout into the existing void and into voids which were deliberately created, in an attempt to arrest movement. When none of these methods guaranteed a positive solution to the problem the decision was made to tunnel in under the building and to provide positive support in the form of concrete pillars. Stress meters have been placed in selected pillars. The crown pillar has been monitored by regular level surveys on the building column bases and by the installation of wire extensometers and thermistors into the rock.

This paper deals with the investigations carried out, the remedial attempts, the installation of ground movement monitoring instruments, temperature measurements, the mining problems in ice, the pouring and setting of concrete in freezing conditions, a study of the heat transfers involved, and a review of the successful solution of a unique problem.

CRYOPEDIMENTS IN THE BIGHORN CANYON AREA, SOUTHCENTRAL MONTANA

Gerald E. Nelson

Department of Geology, Casper College/University of Wyoming at Casper
Casper, Wyoming 82601, USA

At least four levels of gently sloping erosional surfaces, representing at least four periods of erosion, are present in the Bighorn Canyon area of southcentral Montana. The truncated bedrock surfaces are irregular in detail, and are overlain with a thin (1 to 3 m) veneer of diamicton. The diamicton is very poorly sorted, and contains erratic blocks up to several metres in diameter. Most of the fine grained portion of the diamicton originated from formations beneath the veneer. Formerly much more active frost wedging and frost churning are indicated by the wide size range and angularity of the erratic blocks, fragmentation and disaggregation of subjacent formations, and the thorough mixing of these materials. Frost wedging of bedrock is thought to have been the main process of pediment erosion; no evidence or prior planation was found. The movement of the diamicton is interpreted to have been mainly by solifluction, as indicated by the extensive, sheet-like geometry of the deposits, preferred downslope block orientations, the presence of diamicton lobes, and the transportation of huge blocks several kilometers on slopes between 2° and 11°. The major conclusion drawn from data and interpretations presented in this paper is that these pediments are cryopediments, developed through the action of intensive frost processes in a periglacial environment.

INTRODUCTION

Numerous studies have documented the expansion of the periglacial zone that occurred during the Pleistocene. Various features are interpreted to have formed under periglacial conditions, but, with the exception of inactive sand wedges and ice-wedge casts (Black, 1969), the recognition of many features is difficult and may be subject to multiple interpretations (French, 1976, Chapter 12). Ice-wedge casts and inactive sand wedges have been recognized and studied in the high basins and plains of Wyoming (Mears, 1981, and Nelson, 1980) and Montana (Schafer, 1949). It would seem highly likely that additional relict periglacial features could be found in this region, but, to date, little work has been done on the problem.

Deposits of blocky, poorly sorted diamicton, veneering pediments, are present in many of the high valleys and basins of Wyoming and Montana (e.g. Bailey et al., 1968, and Nelson, 1983, p. 129). The pediments and their associated veneers appear to be similar to cryopediments (Czudek and Demek, 1970). One such case, in the Bighorn Canyon National Recreation Area in southcentral Montana, was studied in detail (Nelson, 1983). The present paper summarizes the results of that study.

DESCRIPTIONS OF THE PEDIMENTS AND THEIR DIAMICTON VENEERS

General Characteristics

At least four levels of gently inclined pediment systems, each consisting of several individual surfaces at or near the same elevation above local base levels, are present in the transition zone between the Pryor Mountain front and the Bighorn River (Figure 1). The erosional surfaces truncate gently-dipping Mesozoic sandstones, siltstones, and shales. They are covered with a thin (1 to 3 m) veneer of poorly sorted, blocky diamicton. Downslope profiles of the surface are smooth, hyperbolic curves, steeper in their upper portions and of such shallow concavity in their lower reaches that they are nearly rectilinear. In plain view, the pediments are very nearly rectilinear. Average size, slope, and elevation of each pediment system are presented in Table 1.

TABLE 1. Summary of Size, Slope, and Elevation of Pediment Systems.

	Ave.Area km^2	Ave.Length km	Ave.Width km	Ave.Slope Range, Deg.	Ave.Elev Range, Metres
Qp 1 16 Examples	1.03	1.26	1.35	5-11	1330-1490
Qp 2 24 Examples	.25	.76	.39	4-8	1380-1500
Qp 3 18 Examples	1.14	1.04	.55	2-3	1425-1520
Qp 4 9 Examples	.20	.76	.55	3-6	1500-1580

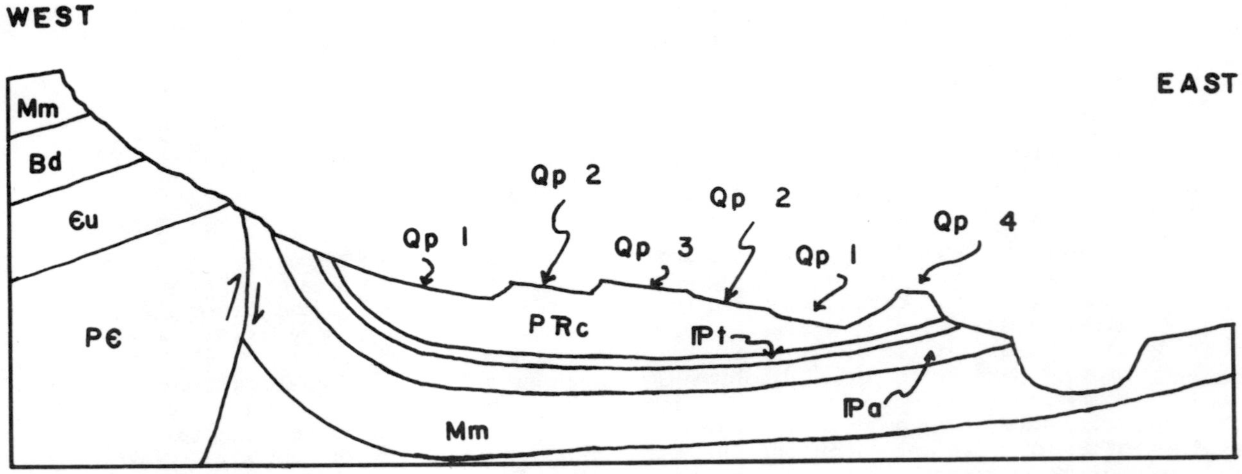

Figure 1. Generalized geological cross-section between the Pryor Mountains on the west and the Bighorn River on the East, showing the area of pediment development. The pediments shown on this cross-section are diagramatic, to show the relationships between the various pediment systems.

Most pediments join steep upper slopes with relatively sharp angles of inflection. This angle coincides with changes in bedrock geology only along the mountain front fault. Others do not join steep upper slopes at all; instead they occur on top of isolated, butte-like landforms. The lower ends of the lowest surfaces grade to present-day streams or, in a few instances, to ledges of resistant rock. The lower edges of pediments in the higher systems project as planes into space over the surrounding lowlands. The diamicton veneers of all pediments contain erratic blocks that could only have originated at outcrops along the front of the Pryor Mountains. This indicates that the isolated pediment remnants were, at one time, connected to a source of erratic material, either at the mountain front or a higher pediment. Isolated surfaces are often surrounded or nearly surrounded by pediments of lower systems. Such crosscutting relationships allow classification of the systems into a relative age sequence, with the highest being the oldest, and on down to the lowest which is the youngest.

Two examples of pediments were selected for detailed study of the diamicton veneer and the truncated bedrock surfaces, one each from the lower two systems. Gully erosion and construction of a highway through the study area provided good cross-sections of these pediments and their veneers.

Pediment Veneer

Natural and man-made exposures indicate a generally uniform thickness for the veneer, from 1 to 3 metres, with about 2 metres being the most common (Figure 2). The veneer consists of one, rarely two or three, layers of poorly sorted material. Individual layers consist of chaotic mixture of various grain sizes, with a tendency for larger than average blocks to occur at or near the top of the layer (inverse grading).

Figure 2. View of the pediment veneer in a gully. The contact between bedrock and diamicton is approximately at the feet of the people. Note the large angular blocks and the bedrock ledge near the bottom of the photograph.

Blocks in the diamicton reflect the composition of formations exposed in source areas directly upslope from the pediments. They have low sphericity and range between bladed and elongate in shape. Estimates of roundness, using a comparison chart, range from very angular to subrounded. It was impractical or impossible to measure a large enough number of blocks to obtain an average block size. An estimate of maximum block size was made at sample sites on the two pediments selected for study (sample sites are the same as those shown in Figures 4 and 5). These estimates consist of the average nominal diameter (the diameter of a sphere with the same volume as the block) of the 10 largest blocks found within a 100 metre radius of each sample site. Average nominal diameters of the largest blocks range between 40 and 250 cm for pediment 1, and 40 to 350 for pediment two.

Based on sand-silt-clay ratios of samples taken at the same sample sites, the matrix of the diamicton is a silty sand; most of the sand is in the very fine to fine-sand size range. Histograms of average grain-size distributions of the same samples show that the matrix is texturally and compositionally bimodal (Figure 3). The coarse-grained mode is composed chiefly of erratic material from Pryor Mountain source areas, mainly carbonate rocks and chert; this type of material decreases in abundance with decreasing fraction size until, in the fine-grained mode, most of the material is composed of disaggregated grains of the subjacent formation (Permo-Triassic Chugwater sandstone and siltstone).

Block Fabric

Because of the obvious indication of block transportation, the orientation of the long axis of 100 blocks was determined at each sample site on both pediments. Fabric diagrams (Figures 4 and 5) are generally composed of two well-defined modes; one oriented downslope and a second, generally weaker, parallel with the contours. The strength of the downslope mode increases with increased distance from source areas. In the lower reaches of the pediments, the cross-slope mode is as strong or stronger than the downslope mode.

Diamicton Lobes

The apparent smoothness of the land surface when viewed from a distance is, upon close inspection, broken by lobate accumulations of blocky diamicton (Figure 6). The lobes are elongate downslope, have arcuate, convex fronts, and merge into the general slope or are covered over by additional lobes in the upslope direction. Very large blocks are often found at lobe fronts, packed tightly together. The lowest two pediment systems have the clearest and greatest number of lobes, while the upper two have at best only vague, arcuate concentrations of deeply implanted blocks.

Veneer Bedrock Interface

The eroded bedrock surface is the actual pediment; the zone containing the uppermost portion of the bedrock, the pediment, and the lowermost portion of the diamicton is herein named the veneer-bedrock interface. In all exposures, the interface is a relatively thin zone, ranging from a few centimetres to a few tens of centimetres thick

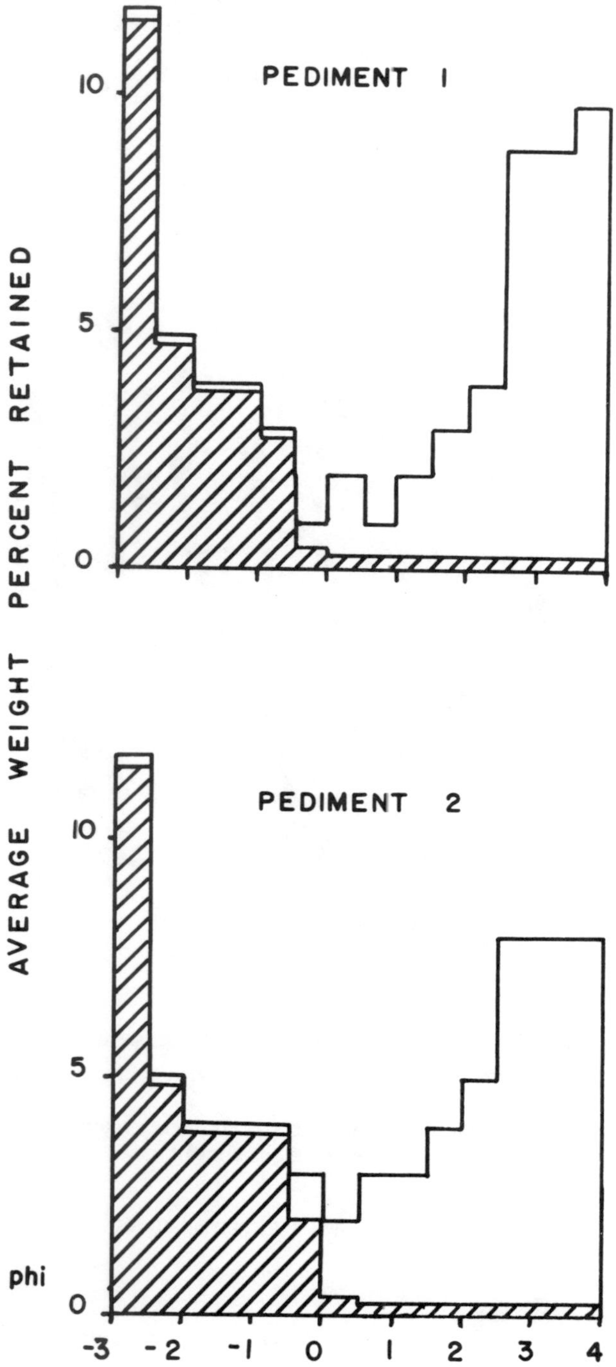

Figure 3. Average grain-size distributions of diamicton matrix. Diagonal shading represents erratic material, mainly carbonate rock fragments, and the clear area represents material derived from subjacent formations.

(figure 7). Many interfingerings between broken and disturbed bedrock and the overlying diamicton are present. Below the interface, primary bedding is preserved and the bedrock is in place although cementation is lacking and they are unconsolidated. Between these fragments, the material is a fine-grained mixture of disaggregated grains from subjacent formations plus erratic carbonate clasts. The fragmented bedrock zone grades upwards over a very short distance into typical diamicton, containing fewer local clasts and more erratic clasts.

Figure 4. Fabric diagrams of the long axes of 100 blocks at each sample site, pediment one.

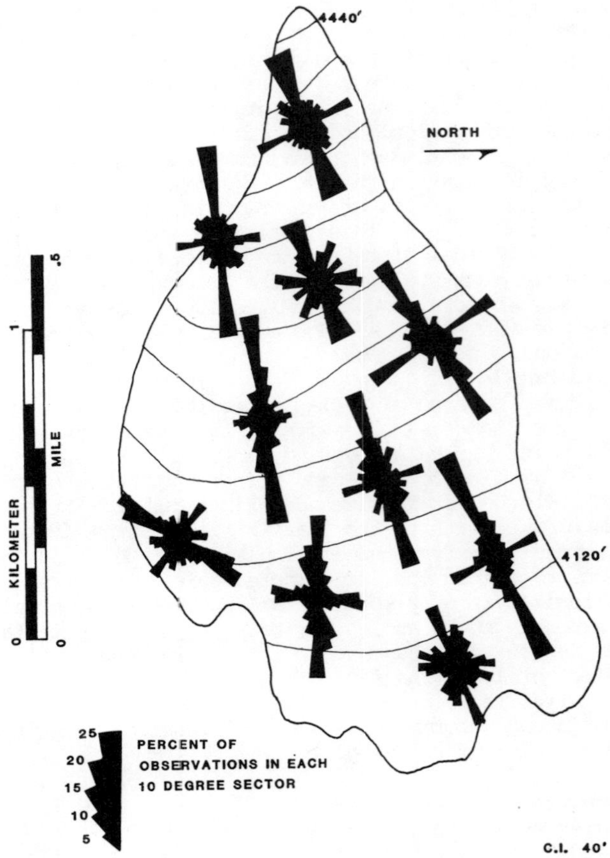

Figure 5. Fabric diagrams of the long axes of 100 blocks at each sample site, pediment two.

DISCUSSION AND INTERPRETATIONS

Introduction

Implicit in the following discussion is the assumption that the diamicton deposits are not now forming or moving and that pediment erosion is no longer taking place. Evidence for lack of diamicton movement consists of fence lines and roadcuts that remain undisturbed, blocks whose exposed surfaces are covered with lichens and whose buried surfaces are covered with caliche, pine trees several inches in diameter growing in the diamicton, undisturbed stone circles or tipi rings and stone cairns used by prehistoric people, and development of soil horizons within the diamicton.

Origin and Movement of Diamicton

Analyses of the samples prove that the diamicton is texturally and compositionally bimodal. The only possible sources for the generally coarse-grained components are outcrops along the front of the Pryor Mountains. The main source for the bulk of the fine-grained components is bedrock subjacent to the diamicton deposits. The angularity and wide size range of all these components suggest that they were produced by frost wedging (see, for example, Washburn, 1980, p. 76-78 and p. 222-223). Gradual, continual mixing of material from the two source areas is indicated by the diffuse contact between the veneer and the underlying bedrock. Mixing could have occurred during diamicton movement or by frost churning or both.

The type of slope movements reported in the literature which produce deposits with characteristics similar to those described in this

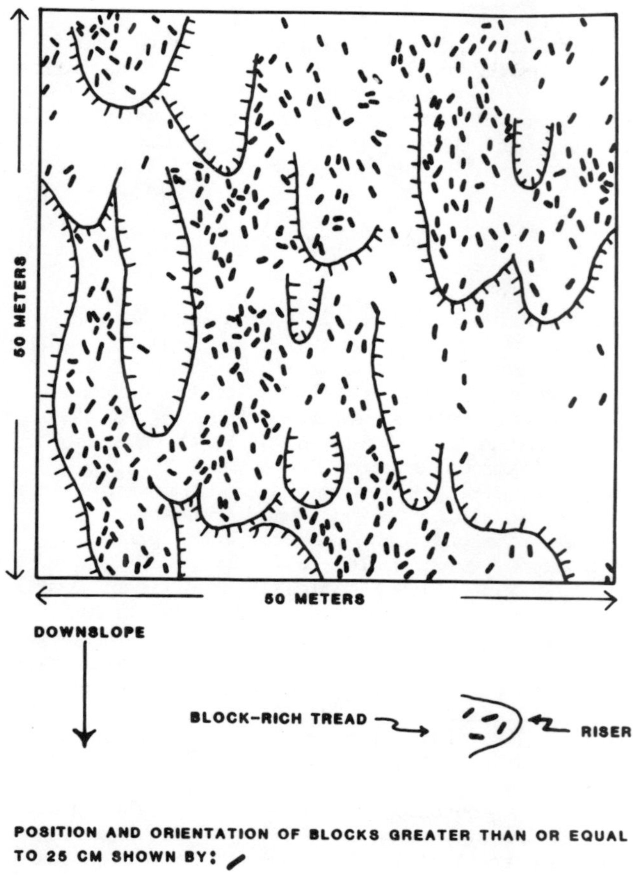

Figure 6. Brunton and tape sketch map of a 50 by 50 metre site on pediment one. The lobes illustrated here are typical of those found on the lower two pediment systems.

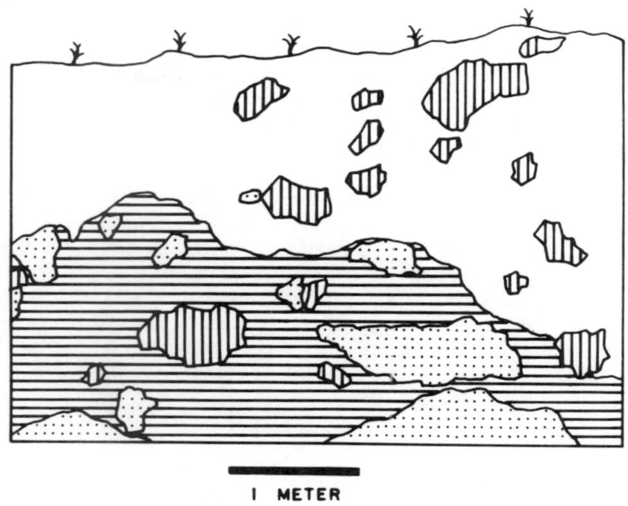

Figure 7. Camera lucida drawing of the bedrock-veneer interface, pediment two. The stippled pattern is bedrock (Permo-Triassic Chugwater formation in this case) which is in-place at the bottom of the drawing and as separate fragments elsewhere. The vertical stripe pattern represents erratic carbonate clasts, the horizontal stripes represent diamicton rich is disaggregated grains from the subjacent formation, and the clear area represents typical diamicton with fewer local clasts and more erratic clasts. Disaggregated grains of the local bedrock are mixed thoroughly into the diamicton even near the top of the deposit.

paper are solifluction and frost creep (for example, see summary in Washburn, 1973, p. 189). Solifluction is indicated by: 1) obvious movement of a silty diamicton, containing huge erratic blocks, on very gentle slopes; 2) lobes of diamicton; 3) preferred downslope orientations of long axes of blocks; and 4) the uniformly thin, sheet-like geometry of the deposits. In addition to the above, intensive frost creep is indicated by the concentrations of blocks found at lobe margins (Benedict, 1970).

It is possible that the flow component of diamicton movement was gelifluction (movement over a frozen substrate) rather than solifluction (substrate not necessarily frozen). The accumulated evidence of formerly more intensive frost action suggests that at least a seasonal frost table existed in the area throughout more of the year than is presently the case, but positive proof of permafrost is lacking. Slopewash and stream action must also be considered as probable adjunct processes of erosion and transportation, but direct evidence of these processes is lacking.

Erosion of Pediments

Development of lower, younger surfaces in the study area was at the expense of higher, older surfaces, as indicated by the isolated, butte-like pediment remnants surrounded or nearly surrounded by lower pediments. The fact that higher surfaces have retained their pediment form suggests that backwearing of steep slopes at the upper edges of the pediments is dominant over downwearing. However, some regrading must take place as the pediment extends itself upslope and gently sloping portions replace steeper portions.

Solifluction (or gelifluction) movement by itself appears to have little erosive ability. This is demonstrated by vertical velocity profiles showing movement concentrated in the upper few decimetres, and attenuating to zero at some shallow depth (Rudberg, 1964, and Williams, 1966), and by numerous cases of solifluction lobes overriding organic material (e.g. Benedict, 1970). The angularity and wide size range of fragments of the eroded bedrock in the diamicton suggests that frost wedging was responsible for pediment erosion, with the material thus produced moved downslope mainly by solifluction. No indications were found during the course of this study which would suggest that the bedrock surface had been eroded by, for example, stream action, prior to diamicton movement. Furthermore, the diffuse veneer-bedrock interface indicates that diamicton movement and bedrock planation were contemporaneous.

The lowest pediments in the Bighorn Canyon area are between the foot-portions of higher slopes and local base levels. No similar features have developed on the front or the crest of the Pryor Mountain Range. In this geomorphic position, and with evidence suggesting their formation by periglacial processes, the pediments in the study area may more properly be termed cryopediments (Czudek and Demek, 1970).

Age of Bighorn Canyon Pediments

Relative age relationships among the four levels of pediments have been established on the basis of crosscutting relationships. The relative age sequence is supported by differences in microrelief of blocks and lobes; upper pediment systems are smoother and differ in surface texture from each other and from the lower two systems. The highest pediment system is the oldest, and each successively lower system is younger. No evidence for the absolute ages of the pediments is available. They are, in general, Pleistocene in age because: 1) similar features are not now developing in the area; 2) the youngest system grades to levels dated archaeologically at least as old as 8,000 years (Husted, 1969); and 3) the accumulated evidence presented in this paper strongly indicates a relationship between pediment erosion and diamicton movement, and cold climatic conditions.

Climatic Considerations

Inactive sand wedges and ice-wedge casts have been found in numerous scattered localities throughout the state of Wyoming by Mears (1981) and Nelson (1980). Fossil periglacial features have been found at least as low as 1,500 masl in the northern Bighorn Basin, 35 kilometres south of the present study area. Although similar features have not yet been found in the Bighorn Canyon area, it seems likely that this area (1,300 to 1,600 masl) also experienced a permafrost climate during Pleistocene cold periods. Late Pleistocene vertebrate material from Natural Trap Case, three kilometres southeast of the present study area, include remains of tundra dwellers such as the collard lemming *Dicrostonyx* sp. (Gilbert et al., 1978). Although a spatial relationship between permafrost features and tundra dwellers, and pediments in the Bighorn Canyon does not prove a temporal relationship, it does support the interpretations based on geomorphology alone. At the very least, the area had a climate rigorous enough to have supported more intense frost wedging, frost churning, and solifluction than is presently the case. The cryopediments in this area indicate that the effects of Pleistocene periglacial climates in the Middle Rocky Mountains are more widespread than previously demonstrated.

Acknowledgements

Partial funding for this project was provided by grants from Sigma Xi and a Dissertation Fellowship from the University of Kansas. I am very grateful for this aid, and the support of the faculty in the department of geology at the University of Kansas.

REFERENCES CITED

Bailey, R.L., Harp, E.L., Jasmin, P.H., Lahren, L.H., Lavin, S.J., Mackin, R.L., Meuller, R.G., Schneider, G.B., Stradley, A.C., Todd, S.G., and Montagne, J., 1968, Pediments of the northeast slope of Canyon Mountain, Livingston, Park Co., Montana: Montana State Univ., Earth Science Miscellaneous Research Series, no. 1, 17 p.

Benedict, J.B., 1970, Downslope soil movement in a Colorado alpine region: Rates, processes, and climatic significance: Arctic and Alpine Research, v. 2, p. 165-226.

Black, R.F., 1969, Climatically significant fossil periglacial phenomena in northcentral United States: Biuletyn Peryglacjalny, v. 20, p 225-38.

Czudek, T., and Demek, J., 1970, Pleistocene cryopedimentation in Czechoslovakia: Acta Geographica Lodziensia, v. 24, p. 101-108.

French, H.M., 1976, The Periglacial Environment, Longman Group Ltd., London, 309 p.

Gilbert, B.M., Martin, L.D., and Chomko, S.A., 1978, Paleontology and paleoecology of Natural Trap Cave, Wyoming: 20,000-10,000 BP: American Quaternary Association, Abstracts 5th Biennial Meeting, Alberta, Canada, p. 203.

Husted, W.M., 1969, Bighorn Canyon archaeology: Smithsonian Institute River Basin Surveys, Publications in Salvage Archaeology, no. 12, 138 p.

Mears, B., Jr., 1981, Periglacial wedges and the late Pleistocene environment of Wyoming's intermontane basins: Quaternary Research, v. 15, p. 171-198.

Nelson, G.E., 1980, Inactive sand wedges at Casper Wyoming: Indicators of the former presence of permafrost: Geological Society of America, Abstracts with programs, v. 12, no. 6, p. 299.

1983, Origin and paleoenvironmental significance of pediments in the Bighorn Canyon area, southcentral Montana: Unpublished PhD manuscript, The University of Kansas, Lawrence, Kansas, 159 p.

Schafer, J.P., 1949, Some periglacial features in central Montana: Journal of Geol., v57, p154-74.

Rudberg, S., 1964, Slow mass movement processes and slope development in the Norra Storfjall area, southern Swedish Lappland: Zeitschriftfur Geomorphologie N.F., Supplement 5, p. 192-203.

Washburn, A.L. 1973, Periglacial processes and environments: St. Martins Press, New York, 320 p

1980, Geocryology: a survey of periglacial processes and environments: John Wiley and Sons, New York, 406 p.

Williams, P.J., 1966, Downslope soil movement at a subarctic location with regard to variations with depth: Canadian Geotechnical Journal, v. 3, p. 191-203.

ELECTRICAL THAWING OF FROZEN SOILS

Alfreds R. Jumikis

Department of Civil and Environmental Engineering
Rutgers University-The State University of New Jersey
Piscataway, New Jersey 08854, USA

In foundation engineering activities in the Arctic, permafrost is a dominant factor the builder must cope with. To facilitate excavation and foundation work in frozen soil, the latter can be loosened up, among other methods, by thawing electrically. Electrical thawing is a convenient and expedient method for application to small, dispersed structures, especially in built-up areas during cold seasons of the year. This article presents a method of calculation for electric thawing of frozen soils. The thawing system is a hexagonal electrode grid. It consists of six, vertically embedded, electrodes around one central grounded electrode. Single-phase, three-wire alternating current for thawing was used in this calculation.

INTRODUCTION

The principle of radial, electrical, thawing of a frozen soil by means of vertical electrodes may be described as follows. A soil, through which passes an electric current, heats up due to resistive dissipation, that is, the applied electrical energy is transformed into an equivalent thermal energy.

The electrode-soil-electrothermal system under discussion is a vertical cylindrical system where an a-c current emanating from the vertical, central electrode flows radially into the soil. There are also currents flowing between adjacent electrodes on the periphery of the hexagonal grid.

The region of influence of thawing of a uniform soil around one electrode is assumed to be a vertical annulus of soil whose inner radius is r_e (the outer radius of the electrode), and outer radius is r, see Figure 1. Because of the electrical resistance of the entire body of the frozen soil, Joules' heat is generated "en masse" within the soil around the electrode.

Electrodes

For direct, radial, electrical, thawing of frozen soils, solid, steel electrodes, ~25.4 mm to ~50.8 mm in diameter, may be used. Because lateral and downward drainage of the melt water of ice is blocked by the still frozen permafrost around the electrodes, the thawed soil becomes more or less water saturated. To cope with a muddy site condition, the melt water can be removed and discharged if perforated metal pipe electrodes, instead of solid steel rod electrodes, are used.

The electrodes may be installed in the frozen soil in various electrode grid patterns at equal spacing of approximately s = 2-3 m, center to center, in a straight row. The spacing of rows is ~2.5 m to ~4.0 m. The array of electrodes in an electrical thawing system may also be in staggered rows (equilateral triangular pattern or in a checkerboard arrangement). Figure 1 shows a hexagonally arranged electrode grid pattern.

For safe operation of an electrical thawing facility, the electrical thawing installation must be grounded. The site must be fenced in, and provided with warning signs. Also, for reasons of safety, direct electrical defrosting (with bare electrodes) of frozen soils containing metallic ores should be categorically prohibited because of stray electric currents in the ground.

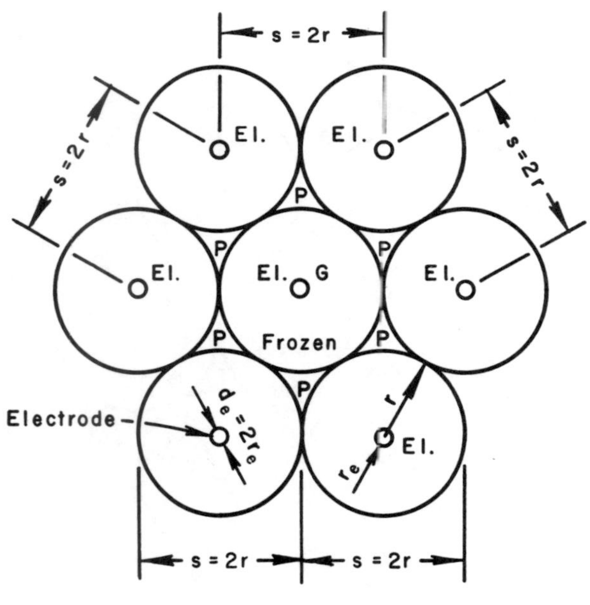

FIGURE 1 Hexagonally arranged electrode grid pattern. Horizontal cross-section through electrodes and frozen soil.

El. = electrode.
G = grounded tap electrode
r_e = $d_e/2$ = radius of electrode.
r = outer radius of frozen soil cylinder.
s = 2r = spacing of electrodes.
P = pocket.

THAWING

Generally, the process of thawing of frozen soils may be visualized to take place in three major steps:

1. heating of the frozen soil from its initial temperature T_{in} to the temperature of the melting point T_m of the soil ice;
2. isothermal melting of soil ice at constant melting temperature T_m.

   The latent heat of melting is supplied by heating the soil electrically and the ice in the frozen soil changes its phase from solid to liquid. A pure ice transforms to water at a theoretically constant melting temperature of $T_m = 0°C$. At high mineralization of ice, its melting begins at less than 0°C. The change in the phase ice to water is accompanied by a simultaneous decrease in the resistivity of the soil, and by changes in its thermal conductivity and heat capacity. And,
3. after melting of the ice, the thawed soil continues to be warmed up to some final, positive, design temperature T_{fin}.

The accumulated heat in the thawed soil distributes slowly beyond the designed thaw boundaries, and practically increases the design-volume of the thawed soil. This heat might help to thaw the "triangular" spaces or "pockets" between the merged, already thawed soil cylinders.

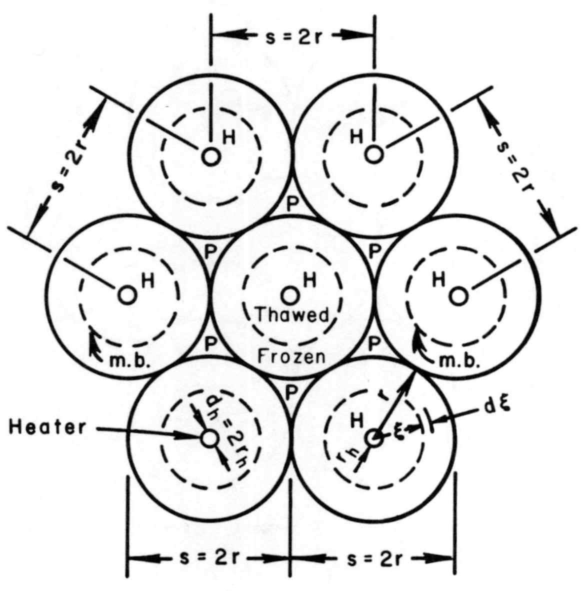

FIGURE 2 Hexagonally arranged vertical heaters. Horizontal cross-section through heaters and thawed and frozen zones of soil.

H = heater.
r_h = $d_h/2$ = outer radius of heater.
ξ = radius of thawed zone.
r = outer radius of frozen zone.
s = 2r = spacing of heaters.
m.b. = moving isothermal thaw-frost boundary.
P = pocket.

In Figure 2, the cylindrical surface whose radius is ξ, is the interface between the thawed and frozen zones of the soil. Here the absorption of the latent heat of melting of the ice occurs. This isothermal interface is also known as the moving isothermal boundary between the thawed and frozen zones of the soil. On this isothermal boundary, changes in phase and thermal properties bring about a discontinuity in soil temperature distribution, and in heat transfer. Because of the discontinuity, as well as because of the nonlinearity that results upon introducing the latent heat equation into the system of Fourier's heat transfer equations for frozen and unfrozen states, there is no general, exact, closed-form solution for changes in phase in cylindrical coordinates available. Further discussion on thawing of this type is beyond the scope of this article.

As mentioned already, on passing of electrical current through the cylindrical body of the frozen soil around the electrode the soil heats up en masse because of the electrical resistance of the soil. The frozen soil changes temperature, and eventually the ice of the soil melts. Because there are no thermal gradients in this thawing soil system, there is no heat flow. Consequently, there is no isothermal phase change boundary present upon change of ice to water.

The interwoven complex phenomena of electricity and electromagnetism in frozen and thawing soils, the complex nature of the multicomponent electrothermal system of soils, and the multitude of soil physical and electrical properties indicate that thermal and electrical calculations during electrical thawing of frozen soils are very difficult. Therefore, to make such a problem accessible to reasonable calculations, the thawing problem must be idealized by making simplifications and assumptions. Simplifications facilitate a practical approach to dealing with the electrical thawing problem in frozen soils. The solution to such a simplified problem can be acceptable in practice, but, the solution to the problem of artificial thawing of frozen soils electrically in a cylindrical system can be treated only approximately.

Assumptions made for electrical thawing of frozen soils are:

1. the soil is homogeneous, uniformly frozen, permafrost;
2. the cryogenic texture of the frozen soil is horizontally layered soil platelets (scales) (Jumikis, 1978[1]; 1978[2]);
3. the frozen soil is fully saturated with water;
4. there are no heat convection processes;
5. to avoid undue upward flow of heat from the thawed soil to the atmosphere, and to protect the already thawed soil from refreezing during a cold season, the ground surface at the site is covered with an appropriate thermal insulation (as is commonly practiced in other methods of thawing of frozen soils);
6. the upward flow of the terrestrial heat from the interior of the earth toward the underside of the permafrost, being relatively small, is ignored;
7. the heat loss from the thawed soil to the frozen subsoil (or to the unfrozen subsoil,

if applicable), as is usually done, may be included as a percentage of the overall energy loss of the thawing system;

8. the initial temperature T_{in} of the uniformly frozen soil is constant.

ELECTRICAL RELATIONS

There are various electrode grid configurations possible. Figures 1, 3 and 4 show a plan of a hexagonal grid or group of six vertically embedded electrodes. They are arranged radially about one central, grounded electrode in a triangular pattern. Each electrode interacts with adjacent electrodes spaced from one another at s = 2r (Figures 2 and 3). It is assumed that in such an electrode grid all electrical paths between the electrodes have equal electrical resistance R because of the hexagonal geometry of the grid (Figure 3), and because of the assumed homogeneity of the soil. However, the homogeneity may vary more or less with the local conditions at the site, especially as the thawing of the soil progresses. Notice in Figure 3 that in each of the six inner or radial resistance paths the voltage is U, and that the power in the six outer or circumferential paths will be four times the power in the six inner or radial paths.

The grounded center tap is connected by an insulated zero wire (the third wire) to the single phase alternating current three-wire electrical power line.

The progress of thawing may be monitored by temperature measurements of the soil in temperature-measuring boreholes. Also, soil cores can be extracted to check whether all ice inclusions in the soil are melted.

Soil Resistance

The electrical resistance R between a pair of parallel cylindrical electrodes in an infinite, homogeneous medium, assuming the spacing s of the electrodes is large compared to their diameter, is given by Della Torre and Longo (1969) as

$$R = (\rho/2\cdot\pi\cdot H_{el})\cdot \ell n(s/r_e), \quad (1)$$

where R = soil resistance, ohm
ρ = resistivity of soil, in ohm meters
H_{el} = embedded length of an electrode, in meters
s = spacing of electrodes, in meters
r_e = radius of electrode, in meters

The electrical resistivity of soils depends on mineralogical composition, porosity, moisture content, amount of electrolytes in the soil, cryogenic texture, ice content, temperature, and possibly some other factors.

According to Rennie, Reid, and Henderson (1978), the range of resistivities of an unfrozen low-plastic clay is 25-45 Ωm, that of a frozen low-plastic clay is 40-80 Ωm, that of a high-plastic silt in an unfrozen condition is 40-60 Ωm, and that of a frozen high-plastic silt is 60-110 Ωm.

Power

The power loss in an electrical resistance traversed by an alternating current is expressed by the well-known formula

$$P = U^2/R, \quad (2)$$

where P = the power lost in heating the material, watts
U = voltage, volts

Thus the power P_r in each of the six radial electrical resistance paths is:

$$P_r = U^2/R, \quad (3)$$

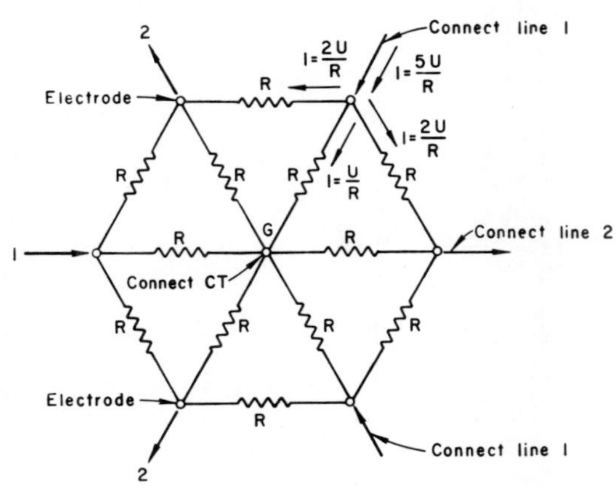

FIGURE 3 Radial and peripheral electrical resistance paths in soil (after personal communication with Prof. C. V. Longo).
G = grounded central tap electrode.
I = current.
R = electrical resistance of soil between each pair of electrodes.
U = voltage.
CT = central tap.

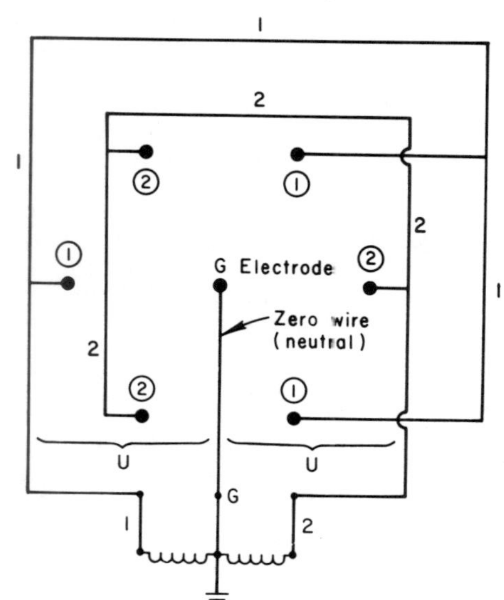

FIGURE 4 Connection of electrode grid to a single-phase source of electric power.
G = grounded central tap electrode.
U = voltage.

because the voltage is U for these paths. The power P_c in each of the six circumferential paths is

$$P_c = 4 \cdot U^2/R, \tag{4}$$

because the voltage is 2U for each circumferential path. The power P_T in the entire grid of all electrical paths is

$$P_T = 6 \cdot P_r + 6 \cdot P_c = 30(U^2/R) \tag{5}$$

The total power P_T in the entire grid heats up the frozen soil, as well as provides for the latent heat of melting for the soil ice upon its changing state from solid to liquid.

THERMAL CALCULATIONS

In designing artificial defrosting facilities for frozen soils, one needs to know the amount of thermal energy Q (electrical energy $E = P_{nec} \cdot t$), necessary, and the total duration of time t necessary for thawing the given volume V of the frozen soil.

The thermal calculations of the necessary amount of heat Q to be added to the frozen soil for its thawing are usually made for each of the soil components - solids, ice contents, unfrozen water - separately (Jumikis, 1977). The air component is usually disregarded in these calculations.

Example

Given $V = 110\ m^3$ of a low-plastic clay to be thawed electrically.

The average initial temperature of this frozen soil is: $T_{in} = -6°C$.

The average final temperature of the thawed soil is: $T_{fin} = +2°C$.

The resistivities of this soil are given as: $\rho_{unfrozen} = 25\ \Omega m$ and $\Omega_{fr} = 40\ \Omega m$.

The unit weight of the soil is: $\gamma = 1950\ kg/m^3$.

Its saturation moisture content by dry weight is given as w = 30%.

The dry unit weight of the soil is: $\gamma_d = \gamma/(1+w) = 1500\ kg/m^3$.

Moisture content at plastic limit of the soil: $w_{P.L.} = 14\%$.

The plasticity index of this soil is: $w_{P.I.} = 11\%$.

Mass heat capacity of soil mineral particles or skeleton: $c_{md} = 0.20$ kcal/kg·°C.

Mass heat capacity of ice: $c_{mi} = 0.50$ kcal/kg·°C.

Mass heat capacity of water: $c_{mw} = 1.00$ kcal/kg·°C.

Latent heat of melting: L = 80 kcal/kg.

Heat losses from ground surface of the thawing soil to the atmosphere: assessed at ≈ 10% of the energy requirements.

25.4 mm-diameter (= d_e) steel electrodes are spaced s = 4.00 m apart.

The thawing depth H should be 1.5 m.

The effective depth of the embedded electrode is $H_{el} = 1.00$ m.

The amount of unfrozen water in frozen soil is given by

$$w_{uw} = k\, w_{P.L.} = (0.55)(0.14) = 0.077\ (=7.7\%), \tag{6}$$

where k = a coefficient.
Then the ice content is

$$w_i = w - w_{uw} = 30\% - 7.7\% = 22.3\% \tag{7}$$

By Eq. (1), the electrical resistances of the soil under consideration are:

for unfrozen condition: $R_{th} = 23$ ohm

for frozen condition: $R_{fr} = 36.6$ ohm

The volumetric heat capacities of the frozen and thawed soil systems are given by the usual relations:

$$c_{vfr} = \gamma_d\left[c_{md} + (c_{mi})(w_i) + (c_{mw})(w_{uw})\right] \tag{8}$$

$$c_{vth} = \gamma_d\left[c_{md} + (c_{mw}) \cdot (w)\right] \tag{9}$$

The total energy required is:

$$Q = Q_1 + Q_2 + Q_3 + Q_\ell \tag{10}$$

where Q_1 - energy needed to increase soil temperature from T_{in} to T_m

Q_2 - latent heat of ice melting

Q_3 - energy to increase soil temperature from T_m to T_{fin}

Q_ℓ - heat losses ≈ $(0.1)(Q_1+Q_2+Q_3)$

Thus

$$Q = (1.1)V\left[c_{vfr}(T_m - T_{in}) + L\, w_i\, \gamma_d + c_{vth}(T_{fin} - T_m)\right] \tag{11}$$

Then $Q_1 = 447.3$ kwh

$Q_2 = 3423.4$ kwh

$Q_3 = 191.8$ kwh

$Q_\ell = (0.1)(Q_1+Q_2+Q_3) = (0.1)(4062.5) = 406.3$ kwh

$Q = E = 4468.8$ kwh.

Assuming that the thawing project is planned to be completed in t = 60 days, the required grid power P_T is:

$$P_T = E(kwh)/t(h) = 4469/1440 = 3.1\ kw. \tag{12}$$

By Eq. (5), the necessary voltage U_{nec} for this thawing work is

$$U_{nec} = \left[(R_{fr}/30) \cdot P_T\right]^{0.5} = 61.5 \approx 65\ v \tag{13}$$

line-to-grounded center tap, or $2U_{nec} \approx 130$ v line-to-line.

If U is taken as 65 v then the power dissipations in the frozen and thawed soils are

$P_{fr} = 3.46$ kw

$P_{th} = 5.51$ kw.

Thawing Time

1. The time t_{th} in hours necessary for heating up the frozen soil from its initial negative temperature $T_{in} < 0°C$ to the melting temperature $T_m \approx 0°C$ of the ice in the soil is

$$t_{th} = Q_1/P_{fr} = 129.3\ h = 5.4\ days.$$

2. Duration of time t_m for isothermal melting of ice at $T_m \approx 0°C$:

$$t_m = Q_2/P_{fr} = 988.6 \text{ h} = 41.2 \text{ days.}$$

3. Duration of time t_{fin} for final heating up the soil from $T_m = 0°C$ till final temperature $T_{fin} = +2°C$ of the soil:

$$t_{fin} = Q_3/P_{th} = 34.8 \text{ h} = 1.5 \text{ days.}$$

4. Total approximate thawing time:

$$t_{total} = t_{th} + t_m + t_{fin} = 48.1 \text{ days.}$$

The required grid power is:
3.46 kw for frozen soil, and
5.51 kw for thawed soil.
The necessary voltage is:
$U_{nec} = 65$ volts
The total thawing time is:
$t_{total} = 1155$ hours < 1440 hours
The electric power and its voltage can do the thawing job in 1155 hours.

CONCLUSIONS

Under permafrost conditions, direct, radial, electrical thawing can be pursued year-round and at any time if electrical energy - from stationary electric plants or mobile generators - is readily available at the site. This method would speed up defrosting time considerably as compared to hydraulic thawing, because the electric defrosting is independent of climate, and temperature of water and steam used in the other methods of defrosting of soils. The radial method of electrical thawing is good not merely for year-round thawing of frozen soils but it is also convenient for protecting the already defrosted soils against their refreezing.

The main disadvantages of the direct, radial thawing of frozen soils electrically are:

a) a more complex thawing operation than that when air, water, solar energy, or steam is used as a heat source for thawing;
b) the relatively high cost of operation of an electric thawing installation; and
c) the necessity to adhere to strict safety regulations in operating an alternating current electrical thawing installation with bare electrodes.

In the future, in the economical development of cold regions, artificial, radial thawing of frozen soils electrically will probably be increasingly used.

The natural thawing process of frozen soils by means of radiation heat or solar radiation is not too dependable for continuous, uninterrupted thawing work, especially in cold climates. The hydraulic methods of thawing of frozen soils such as by means of river water, warm water, and/or steam, take a lot of time and involve great, unproductive heat losses.

Today, calculations of thawing frozen soil cylindrical systems electrically are atill performed based on incomplete soil electro-thermal data.

Although electrical properties of permafrost soils have been studied for their own purpose and/or interest by Arcone and Delaney (1981); Olhoeft (1978); Rennie, Reid and Henderson (1978), and others, there still exists a great need for systematically integrated knowledge about electrical and thermo-physical properties of freezing, frozen and thawing soils, and the functional relationships of the various factors involved, with particular reference to radial thawing of permafrost. Of particular need in calculations are usable values of resistances and resistivities for unfrozen and frozen soils as functions of soil geotechnical properties, and electrode grid configurations.

ACKNOWLEDGMENTS

The author expresses his cordial thanks to Professor Emeritus C. V. Longo of Rutgers University, Department of Electrical Engineering, for his assistance on electrical circuits and power calculations, and detailed review of the draft of this article. The writer is grateful to Dr. V. J. Lunardini, M.E., of the U.S. Army Cold Regions Research and Engineering Laboratory at Hanover, New Hampshire, for his valuable suggestions and final editing of the author's article. Also, the author expresses his sincere appreciation to Dr. E. H. Dill, Dean, College of Engineering, and Dr. E. G. Nawy, Chairman, Department of Civil and Environmental, both of Rutgers, The State University of New Jersey, for sponsoring tangibly the preparation of this article.

REFERENCES

Arcone, S. A., and Delaney, A. J., 1981, VHF Electrical properties near Point Barrow, Alaska: Hanover, N.H., U.S. Army Corps of Engineers, Cold Regions Research and Engineering Laboratory, CRREL Report 81-13, 24 pages.

Della Torre, E., and Longo, C. V., 1969, The electromagnetic field: Boston, Mass., Allyn and Bacon, p. 192-198.

Jumikis, A. R., 1977, Thermal geotechnics: New Brunswick, N.J., Rutgers University Press, 375 pages.

Jumikis, A. R., 1978[1], Cryogenic texture and strength aspects of artificially frozen soils, in Proceedings of the First International Symposium on Ground Freezing, Ruhr University, Bochum, Germany, v. 1, p. 75-85.

Jumikis, A. R., 1978[2], Some aspects of artificial thawing of frozen soils, in Engineering Geology, Amsterdam, Holland, Elsevier Scientific Publishing Company, v. 13, 1979, p. 287-297.

Olhoeft, G. R., 1978, Electrical properties of permafrost, in Proceedings of the Third International Conference on Permafrost, Ottawa, National Research Council of Canada, v. 1, p. 127-131.

Rennie, J. A., Reid, D. E., and Henderson, J. D., 1978, Permafrost in the southern fringe of the discontinuous permafrost zone, Fort Simpson, N.W.T., in Proceedings of the Third International Conference on Permafrost, Ottawa, National Research Council of Canada, v. 1, p. 439-444.

Closing Plenary Session
Friday, July 22, 1983

TROY L. PÉWÉ - Good afternoon, ladies and gentlemen. Welcome to the final formal session of the Fourth International Conference on Permafrost. We have been meeting here in Fairbanks for the past five days to listen to excellent papers on all aspects of permafrost. Now we are gathered for a few final items of business, a few closing remarks, and presentations. I would like to thank all of you on behalf of the Organizing Committee for attending the conference. I am sure you all agree that we had five days of very fruitful discussion with friends from many countries involved in permafrost research, and I know we look forward to continuing these contacts in the days ahead. I would first like to call on Dr. A. Lincoln Washburn to read the necrology.

Closing session of the Fourth International Conference on Permafrost, University of Alaska, Fairbanks, Alaska, July 22, 1983. Left to right: Kaare Flaate, Delegate from Norway; Li Yusheng, Delegate from the People's Republic of China; Jay Barton, President of the University of Alaska Statewide System; A. Lincoln Washburn, USA; Michael Hammond, interpreter; P.I. Melnikov, Delegate from the USSR.Standing: Troy L. Péwé, Chairman of the U.S. Organizing Committee for the conference; Governor of Alaska, Bill Sheffield; Shi Yafeng, Delegate from the People's Republic of China; Hugh French, Delegate from Canada; Jerry Brown, Delegate from the U.S.

A.L. WASHBURN - Ladies and gentlemen, it is important for the Fourth International Conference on Permafrost to honor our colleagues who have died since the last Congress. They include:

Ivan Baranov - USSR
Robert F. Black* - USA
Sietse Bylsma - The Netherlands
Roger J.E. Brown - Canada
Don Gill - Canada
Robert Gilpin - Canada
Reuben Kachadoorian - USA
Vladimir Kudryavtsev - USSR
Daisuke Kuroiwa* - Japan
Ralph R. Migliaccio - USA.
Edward Patten - USA
Val Poppe - Canada
Georgi Porchaev - USSR
E.F. "Eb" Rice - USA
January Slupik - Poland
Sigurdur Thorarinsson - Iceland
Kenneth B. Woods - USA
Kenneth A. Linell* - USA

To commemorate these colleagues, and any others whose passing remains to be reported, let us rise for a moment of silence in their honor.

Thank you.

TROY L. PÉWÉ - Alaska is the only state in the Union with much permafrost. In fact, 85% of this state is directly or indirectly affected by permafrost, and it affects almost all of the taxpayers, one way or another, in the State of Alaska. This conference is directly related to this state, even more than others. It was our pleasure last night, and it is as well today, to be joined at this conference by Governor Bill Sheffield. I would like to now present the Governor of Alaska, Bill Sheffield.

BILL SHEFFIELD - Thank you very much, Dr. Péwé. It is an honor and a pleasure to again welcome this distinguished group of scientists and researchers to Alaska. I had the opportunity to enjoy dinner with many of you last evening. From all over the world you have traveled to this conference, and for that I not only welcome you, I want to take this opportunity to also thank you. I welcome you because we in Alaska are proud of our state in a world too often prone to the ills of mankind such as air and water pollution. For example, we in Alaska are proud that we still have a relatively clean environment. And even though Alaskans endure long and cold winters, a high cost of living, and summers that are far too short, a clean and spectacular environment makes the living here very enjoyable. So welcome to our state.

I also want to thank you for coming here, because the work that you do is very important to my state and nation. Indeed, it is important to the entire world because you know better than I that so many of the resources we need for the survival of the human race are in the Arctic and Subarctic. The arctic regions of our state now produce most of the revenue that we enjoy as the 49th state of the Union. If you haven't already, many of you will have the chance to see the Trans-Alaska oil pipeline. That pipeline and the road that parallels it are good examples of why the work you are doing in promoting the understanding of the arctic environment is so important. The oil pipeline was built with private financing and government supervision. In that regard, it represents the best our nation can offer the world: The initiative and vision of individuals working toward a specific goal like building a pipeline, and the cooperation and broader concerns of a government working for the common good, like protecting the environment.

This kind of approach to development of the Arctic and Subarctic will be even more important to the world in the future. In our state alone, we have millions of tons of coal and hard rock minerals that some day will have to be brought to market. To do the job right, we will require all the knowledge you can give us on the complex phenomenon called permafrost. This is one reason why the State of Alaska has joined the National Academy of Sciences in helping to sponsor this international conference. We know that the work you are doing and sharing with your colleagues is important to our development as a state. As Governor, I also know that the work you are doing is important to the people who have called the Arctic their home for thousands of years. The Inupiat Eskimos depend upon all kinds of mammals, waterfowl, and fish to keep their culture alive as well as to pass that culture on to future generations. And make no mistake about it, the Inupiat are very determined that their culture not become extinct. As a matter of fact, the Northwest Territories of Canada will play host next week to the Annual Circumpolar Conference, where the residents of all but one arctic region will gather to renew their natural ties and fashion their common goals. That conference in Frobisher Bay is an important one for the future of our state, and the arctic region of Alaska.

So, too, is this conference, because the engineering and other studies that are a part of your work as scientists will play a central role in future development. And unless that development goes ahead on a sound environmental basis, the future of the Arctic and its people will be threatened. I see no reasons why the future should be threatened for the arctic regions of the world. With the talent that is assembled in this room and the determination of government and industry leaders to make sound practices and sound projects happen, we can have rational economic development in the Arctic without sacrificing either the past or the future.

To do this will require a coordinated approach among scientists, politicians, and laymen. As an American, I am sad to say that my country is the only polar nation without a coordinated effort to conduct the necessary scientific research in the Arctic. But I am glad to report that we may finally be resolving the problem. As you may know, the Alaskan delegation to the U.S. Senate and House of Representatives has sponsored bills to create an Arctic Science Policy Council and an Arctic Research Commission. We hope these bills will pass very soon, and that with them will come a commitment of money to bring the arctic sci-

*Died since the conference.

ences the coordination and prestige that I know they deserve. The State of Alaska has endorsed the concept of both the House and Senate legislation, and I would encourage those of you here today to let our Congress know how you feel, if you haven't already.

As Governor of Alaska, I know how my constituents feel about arctic research, because in the last year the privately financed University of Alaska Foundation has raised more than a quarter of a million dollars for arctic research. For that, we have many people to thank, including Paul Gavora, Tom Miklautsch and William R. Wood, all of Fairbanks, and Attorney John Hughes and Bill Scott of Anchorage, who spearheaded the fund raising drive. That kind of financial commitment from individual Alaskans is positive proof that we support knowing more about the arctic regions. This research can benefit all of us, now and in the years ahead. And you know, when you consider the wealth of our state's resources, our geographic location, and the potential markets we could serve on the Pacific Rim, the importance of the research you will be doing becomes all that more apparent. We need your knowledge because we are a maturing frontier state, and in many ways we are like an emerging nation that is only vaguely aware of both its potential and the problems that lie ahead. People like you will help us overcome the problems--people who know that research and scientific investigation have made the world a better place to live and can make it even better still in the future.

It is an exciting time to be in Alaska, and I am sure you will understand how fortunate I feel being governor at this point in our state's development. So, I again welcome you to Alaska, and I hope you share some of that same excitement as you travel throughout our state. Thank you so very much.

TROY L. PÉWÉ - Ten years ago in Yakutsk, Siberia, discussion concerning the possibility of an International Permafrost Association occurred. It was a short discussion, but we realized such an organization was needed. Five years ago at Edmonton, our Canadian colleagues agreed to seriously look into the formation of such a group, and here to report the latest events is Dr. Hugh French, head of the Canadian delegation.

HUGH FRENCH - Mr. Chairman, ladies and gentlemen: An informal meeting was held yesterday among the official delegates from Canada, The People's Republic of China, the United States of America, and the Soviet Union. I am very pleased to announce that it was unanimously agreed to form an International Permafrost Association. The four countries just mentioned will be founding members of this association, and we encourage other nations to join it. Those who join within the next year will also become founding members.

The primary aim of the association will be to provide for the on-going organization and coordination of International Permafrost Conferences. This will be done through the Adhering National Bodies. An International Secretariat will be established in Canada at the University of British Columbia. Professor J. Ross Mackay has agreed to serve as Secretary-General.

At the meeting yesterday, Academician P.I. Melnikov of the Soviet Union accepted the position of President of the new association, and Professor Troy L. Péwé of the United States accepted the position of Vice-President.

Details concerning national and individual memberships can be obtained from the Constitution, an outline of which will be available at the rear of the hall after this meeting [text follows]. I am sure that you will all join with me in wishing

Participants in founding meeting of the International Permafrost Association. Front row: Louis DeGoes, Li Yusheng, Shi Yafeng, J. Ross Mackay, P.I. Melnikov, T.L. Péwé, G.H. Johnston, N.A. Grave. Back row: Zhao Yuehai, Cheng Guodong, Hugh French, Jerry Brown. Photograph by J.A. Heginbottom, who was also a participant in the founding meeting.

Founding officers of the newly formed International Permafrost Association: Kaare Flaate, Norway, 2nd Vice-President; Academician P.I. Melnikov, USSR, President; Troy L. Péwé, USA, Vice-President; J. Ross Mackay, Canada, Secretary-General. University of Alaska, Fairbanks campus. (Photograph No. PK 25499 by T.L. Péwé, July - 1983.)

every success to the new association, which, I should emphasize, as Professor Péwé has just done, is the culmination of several years of discussion. I believe that it is both timely and appropriate, particularly given the success of this and previous international conferences on permafrost.

Therefore, it remains for me to congratulate Academician Melnikov and Professor Péwé upon their appointment of office, and to thank Professor Mackay for accepting the position of Secretary-General. I am confident that under their very capable leadership and guidance, the future of the association and the continuing development of permafrost science and engineering are internationally assured. Thank you very much.

TROY L. PÉWÉ - Thank you, Dr. French. We will now have a few closing remarks from the leaders of the national delegations. We will hear first from Academician Melnikov of the USSR.

P.I. MELNIKOV - Mr. Chairman, ladies and gentlemen: I participated in this fourth conference on permafrost and in the third, and I was also one of the organizers for the second permafrost conference. I am a witness to an increase in the number of scientists who are participating in permafrost-related research and also engineers working in this area. We are indeed very pleased by this because this will aid us in the very important task ahead of us in developing enormous territories which contain some of our most valuable resources. I could mention, for example, that in Yakutia, which contains almost 90% of all the diamond resources in our country, there is considerable diamond extraction activity. It also contains large deposits of gold and other valuable minerals.

As you probably know, approximately one-half of all the gas and oil is being extracted from western Siberia. Also, the Far North contains enormous coal reserves. The hydroelectric power stations being constructed there are also extremely productive. As already mentioned, the development of these northern territories requires specialized techniques and methods. We have already, at the present time, learned how to construct the equipment used to develop these territories. We are now faced with the further problem of making the equipment less expensive.

Again, I repeat that I am very pleased to see that the number of people working in this discipline has increased. We think that this is going to further the development of the entire science.

I would like to especially thank the organizing committee for all of the efforts that it has undertaken in order to organize and arrange for the Fourth International Conference on Permafrost. To all of you personally, I would like to extend my best wishes for successful scientific activities and for success in your personal lives. Thank you for your attention.

TROY L. PÉWÉ - Thank you, Academician Melnikov. Now I would like to call on Professor Shi Yafeng, representing the People's Republic of China.

SHI YAFENG - Mr. Chairman, ladies and gentlemen: First of all, I would like to congratulate you on the great success of this international conference and on the founding of the International Permafrost Association. Congratulations to Academician Melnikov, the first President of the association, to Professor Troy L. Péwé, the Vice-President of the association, and to Professor J. Ross Mackay, Secretary-General of the association.

In this conference, the Chinese delegation presented their papers, exchanged scientific ideas on the study of permafrost, and learned a great deal. During this conference, the Chinese delegates have been warmly hosted by the U.S. Organizing Committee, the University of Alaska, the City of Fairbanks, the State of Alaska, and especially the people of the United States and Alaska. We feel the friendship among the scientists and the peoples. Thanks for all the help given to the conference from all aspects.

In recent years, permafrost research has developed rapidly and much has been achieved. There still exist many problems to be dealt with, and more new ones will develop by the time the old ones have been solved. Thus, we have a great deal of work to do in the days ahead. In China, there is an old story. It tells us that an old man wished to move two mountains in front of his house. Therefore, he dug at the mountains every day without stopping. When he died, his sons carried on. And when those sons died, his grandsons took over the job. Finally, the two great mountains were moved away. I am fully convinced that in the next interval, with the cooperation of all the scientists of permafrost in all the countries, we will surely solve the problems of permafrost. Finally, I sincerely wish all the delegates, all the members of this conference, health and happiness. Thank you very much.

TROY L. PÉWÉ - Next, we will hear from Dr. Jerry Brown, representing the United States.

JERRY BROWN - On behalf of our Committee on Permafrost, I would like to thank you again for attending the conference. I believe the count, as of this morning, was that more than 950 individuals have participated in various events of the conference, and that does not include the tremendous support staffs that have put the whole conference together here in Fairbanks. As you know, the published record of the conference is already quite substantial. I would like to remind you that the papers will be printed and distributed in a large volume before the end of the year. So you have a great deal of reading ahead of you.

Some 140 people will be leaving tomorrow on field trips throughout Alaska and into the Canadian Northwest. I wish you good weather and interesting experiences. We are all looking forward to seeing you at the fifth conference, at which time some of us will have an opportunity to relax. I would like to thank you again. And to all the visitors: have a safe and enjoyable journey home. Thank you.

TROY L. PÉWÉ - Thank you, Jerry. And now for Dr. Hugh French of Canada.

HUGH FRENCH - Mr. Chairman, ladies and gentlemen: I speak on behalf of all those Canadians who have participated in this conference. I would like to say that we have found it extremely beneficial to our own understanding of permafrost conditions. We have seen a very smooth and highly organized conference. We appreciate fully the magnitude of the logistics behind all of this, and we want to congratulate you upon the successful completion of this tremendous job. Mention was also made in the opening remarks of this conference on Monday of Canadian participation in the review process and organization of the conference. Let me say, Mr. Chairman, on behalf of all those Canadians, that it was a pleasure to be so involved. Canada and the United States have many common interests in the North, and permafrost is central to many of them. We look forward to future cooperation and scientific exchange in the future. We also look forward to maintaining our contacts with our friends in the Soviet Union. Permafrost and drilling technology are but two of the number of topics that have recently been agreed upon as being areas of mutual technological and scientific exchange between our two countries. Finally, it has been a pleasure to discuss permafrost problems with our colleagues from other countries. We are particularly impressed with the growth of permafrost studies in China, and we look forward to hearing more about these in the future. Therefore, Mr. Chairman, on behalf of all the Canadian delegates, we thank you for organizing and holding this Fourth International Conference. Thank you.

TROY L. PÉWÉ - At this time, I would like to introduce Dr. Kaare Flaate. Dr. Flaate is Chairman of the Permafrost Committee of the Royal Norwegian Council for Scientific and Industrial Research that is based in Oslo. Dr. Flaate.

KAARE FLAATE - Ladies and gentlemen: I am here in order to present an invitation. But I think that I will first start by making a confession. It was only after I arrived here Sunday afternoon that I appreciated all the work that went into organizing this conference. And that so affected me that I considered turning around and going home. Well, there are always simple reasons that keep you in line, and I tell you I have two good reasons for not leaving. First, I had an APEX airline ticket, and I could not return immediately. And secondly, I was what you would call "dead tired." And I am glad I stayed. It has been a pleasure to participate in the conference. I know I gained professionally as well as socially.

We would like to invite you to Norway for the next conference. I will give you two good reasons. One, somebody has to arrange the conference, and we feel that we also have obligations for international cooperation. We must do our share of the work. So much for the unselfish part. Second, we believe that the whole country has much to gain by having the conference. We will gain a lot professionally and we will have the pleasure of having visitors. So that was the selfish side.

Let me finish this brief talk by presenting the official invitation for you.

INVITATION

The Norwegian Committee on Permafrost
is pleased to act as host for
The Fifth International Conference on Permafrost
in June 1988
at the Norwegian Institute of Technology
in Trondheim, Norway

The Norwegian Committee on Permafrost was appointed by the Royal Norwegian Council for Scientific and Industrial Research with the aim of promoting permafrost research and technology. The Norwegian Institute of Technology is well equipped for and has long experience in handling conferences. The academic community at Trondheim has a professional background in permafrost problems. The city of Trondheim has a population of about 130,000 and was founded by the Vikings in 997. The city has many interesting sights and is a natural starting point for excursions related to permafrost phenomena.

TROY L. PÉWÉ - Academician Melnikov, new President of the International Permafrost Association, will now accept the invitation.

P.I. MELNIKOV - Ladies and gentlemen: I think at this time we should all extend our gratitude to the Government of Norway for the opportunity to conduct the Fifth International Conference on Permafrost in Norway. I think that in connection with this it would be a very good idea to propose that a second vice-president of the newly formed International Permafrost Association be Professor Kaare Flaate, the representative of Norway. I think that all of you will join us in supporting the proposal to hold the next international conference in Norway. We support this proposal.

Again, permit me at this time to express my heartfelt gratitude to all of you for the honor you have accorded me, to be the first president of the newly founded International Permafrost Association. I have already dedicated my entire life to permafrost studies. When I was still only a student in the Leningrad Mining Institute, I spent three years conducting research on the proposed Baikal-Amur Railroad, whose construction is presently being completed. Upon completion of my studies, I was then assigned to activities in the Academy of Sciences of the Soviet Union. The Academy of Sciences in turn sent me to work in the far northern regions as the director of a scientific research institute. I spent three years there in the Indigirka region, and then I spent the next 23 years working in Yakutia, conducting scientific research and also as the director of the Yakutsk Permafrost Institute. At the same time, for the last 13 years, I have been a representative of the Academy of Sciences of the Soviet Union in Moscow. The Scientific Cryology Council, whose activities I participate in, in Moscow, is responsible for organizing and coordinating all the research projects carried out by both the academic community and the industries throughout the entire Soviet Union. I anticipate that the new activities that I will have to be carrying out as the president of the newly founded International Permafrost Association will be quite difficult, since it is always difficult when a new association or organization is founded. But again, I will exert all possible energies and talent to see that the work carried out by our association will further all of the research in permafrost. I would like to thank you once again.

TROY L. PÉWÉ - Thank you, President Melnikov: I am sure that all of you noticed the very large banner proclaiming the Fourth International Conference on Permafrost, all four conferences, on the wall in the Great Hall during the past week. I have now taken down this banner. In addition to the conferences being proclaimed in three languages, there are the logos of the past four conferences. I think you remember this banner. It is now my duty to pass this banner on to Kaare Flaate, who will be in charge of it until the next conference. Perhaps he will add another official language, and indeed another logo. Dr. Flaate, please accept this banner.

You will be happy to know that we are nearing the end of our plenary sessions. However, before we have our final statements, I have a pleasant duty. We are fortunate to have in the audience the oldest active investigator of permafrost in Alaska and probably North America. Here today to listen to our deliberations is Professor Earl H. Pilgrim, first professor of Mining Engineering and Metallurgy of the University of Alaska. Professor Pilgrim was one of the six original faculty of the Alaska Agricultural College and School of Mines (now the University of Alaska) when it opened its doors on September 18, 1922. Since 1926, he has been actively mining in perennially frozen gravel and bedrock and now operates an antimony mine in the Kantishna District near Denali National Park in the Alaska Range. Now in his nineties, Professor Pilgrim has been actively working with permafrost for 60 years. It is my pleasure to introduce Professor Earl H. Pilgrim of Stampede Creek. Professor Pilgrim, would you please stand and be honored.

I would like to call on Dr. Jay Barton, President of the University of Alaska, and a generous host of the conference for a closing statement.

JAY BARTON - Dr. Péwé, it has been a great honor and pleasure to have the Fourth International Conference on Permafrost here on the campus of the University of Alaska at Fairbanks. We hope you all will return. Thank you.

TROY L. PÉWÉ - We have now reached the end of the formal program of the Fourth International Conference on Permafrost, and it is a pleasure to note that this has been the largest international conference on permafrost to date. I have been informed that it was the largest and most complex conference ever held at the University of Alaska in Fairbanks.

As Chairman of the U.S. Organizing Committee for the Fourth International Conference on Permafrost, I would like to thank you for participating in the conference and wish you well on your return journeys. I now declare the Fourth International Conference on Permafrost closed.

INTERNATIONAL PERMAFROST ASSOCIATION
BASIC PRINCIPLES OF THE CONSTITUTION
(July 1983)

The organizations referred to in the Constitution and By-laws are defined as follows:

(1) Council is the governing body of the Association.

(2) Adhering National Body is a representative body, such as a national committee, designated by the appropriate authority to represent in the Council of the Association the interests in permafrost of scientists and engineers of a country.

OBJECTIVE

To foster the dissemination of knowledge concerning permafrost and promote cooperation among persons and national or international organizations engaged in scientific investigations or engineering work on permafrost.

MEMBERSHIP

Membership in the Association is through Adhering National Bodies. There shall be only one Adhering National Body per country.

In countries where no Adhering National Body exists, an individual may apply directly to the Association to take part in Association activities.

OFFICERS OF THE ASSOCIATION

The officers of the International Permafrost Association are:

(i) President
(ii) Vice Presidents
(iii) Secretary General

The officers of the Association shall serve from the end of one International Conference on Permafrost to the end of the next Conference.

COUNCIL

The Council shall consist of the President and two representatives from each Adhering National Body.

The Council shall determine, with the advice of the Secretary General, the annual subscription fee to the Association.

There will not be any fees unless approved by the Council.

INTERNATIONAL CONFERENCE ON PERMAFROST

The organization and financing arrangements of an International Conference are the responsibility of the Adhering National Body of the host country.

AMENDMENTS TO CONSTITUTION AND BY-LAWS

Amendments to the Constitution and By-Laws may be proposed by an Adhering National Body.

Appendix A: Field Trips

Field trips were a major part of the Fourth International Conference on Permafrost, and almost everyone took advantage of either the extended trips, which took place immediately before or after the conference, or the local geological and engineering permafrost field trips, which took place during the conference.

The local field trips permitted hundreds of participants to observe permafrost and permafrost-induced construction problems as well as the relationship of permafrost to vegetation in the Fairbanks area.

The five pre- and post-conference trips had a total of 250 participants. A list of these trips is as follows (the names of participants on each trip are indicated in Appendix D): A-1, Alaska Railroad and Denali National Park; A-2, Fairbanks to Prudhoe Bay via the Elliott and Dalton Highways; B-3, Northern Yukon Territory and the Mackenzie River Delta; B-4, Fairbanks to Anchorage via the Richardson and Glenn Highways; and B-6, Prudhoe Bay and the Colville River Delta. Guidebooks for these trips were available to all field trip participants (see accompanying list).

The carefully planned and executed field trips and publications will be long remembered and were appreciated by all the participants of the conference.

Oscar J. Ferrians, Jr., Chairman
Field Trips Committee

Available Guidebooks

Published by Alaska Division of Geological and Geophysical Surveys, 794 University Avenue, Basement, Fairbanks, Alaska 99701

Guidebook 1: Richardson and Glenn Highways, Alaska. T.L. Péwé and R.D. Reger (eds.), 1983, 263 p., $7.50.

Guidebook 2: Colville River Delta, Alaska. H.J. Walker (ed.), 1983, 34 p., $2.00.

Guidebook 3: Northern Yukon Territory and Mackenzie Delta, Canada. H.M. French and J.A. Heginbottom (eds.), 1983, 183 p., $8.50.

Guidebook 4: Elliott and Dalton Highways, Fox to Prudhoe Bay, Alaska. J. Brown and R.A. Kreig (eds.), 1983, 230 p., $7.50.

Guidebook 5: Prudhoe Bay, Alaska. S.E. Rawlinson (ed.), 177 p., $6.00.

Guidebook 6: The Alaskan Railroad Between Anchorage and Fairbanks. T.C. Fuglestad (ed.), approx. 130 p.

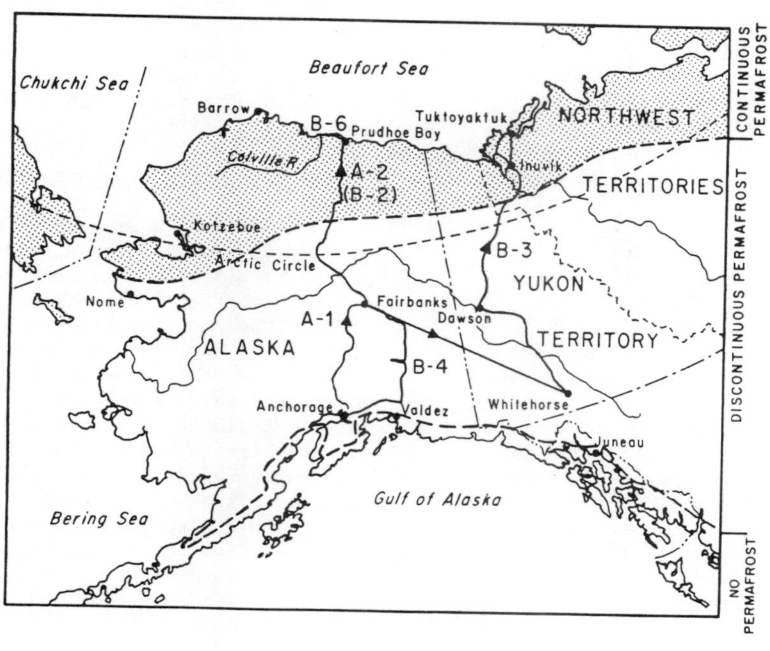

Local Field Trips

Introduction

One of the main reasons for holding the Fourth International Conference on Permafrost in Fairbanks was that permafrost and its effect on man's activities in the North are well illustrated locally. It is possible to see many permafrost phenomena and many permafrost-related construction problems within a few miles of the University of Alaska (Fairbanks) campus.

Therefore, among the highlights of the conference were the field trips in the immediate Fairbanks area to examine engineering and geological aspects of permafrost. During the conference, three afternoons (Tuesday, Thursday and Friday) and two evenings (Monday and Wednesday) were set aside for these trips. No technical sessions were held during those times. The trips were provided at no extra cost to the registrants, and published permafrost information was issued to all of the participants.

There were seven tours. Two different tours covered the geological and vegetation aspects of permafrost and its relation to construction, and two covered the engineering aspects of permafrost. A special tour was scheduled to a permafrost tunnel, and one to a frost-heave facility. In addition to these bus trips, there was a self-guided walking tour (with brochure) to examine permafrost features (thermokarst mounds) on the University of Alaska campus.

It is estimated that more than 500 registrants participated in these local field trips and received published and unpublished scientific and engineering reports. Only draft copies were completed of the guidebooks to the permafrost and Quaternary geology of the Fairbanks area, and the permafrost and engineering features. Therefore, all participants were provided with the following publications: 1) Péwé, T.L., 1958, Geology of the Fairbanks (D-2) Quadrangle, U.S. Geological Survey, Geological Quadrangle Map GQ-110, scale 1:63,360, 1 sheet; 2) Péwé, T.L., Ferrians, O.J., Jr., Karlstrom, T.N.V., Nichols, D.R., 1965, Guidebook for field conference F central and south-central Alaska International Association for Quaternary Research, 7th Congress, Fairbanks, 1965: Lincoln, Nebraska, Academy of Science, 142 p. (reprint 197, College, Alaska, Division of Geological and Geophysical Surveys); and 3) Péwé, T.L., 1982, Geologic hazards of the Fairbanks area, Alaska: Alaska Division of Geological and Geophysical Surveys Special Report 15, 109 p. A brief description of the field trips is outlined in this volume of the proceedings.

Acknowledgments

The two field trips on the geological-construction aspects were organized and conducted by Richard D. Reger, Alaska Division of Geological and Geophysical Surveys, and Troy L. Péwé, Arizona State University. They were very ably assisted during the trip by the following members of the Alaska Division of Geological and Geophysical Surveys staff: 1) Bus leaders: Rodney A. Combellick, Jeffrey T. Kline, and Stuart E. Rawlinson, and 2) Assistant bus leaders: Jeffrey H. Ewing, Duncan R. Hickmott, Terry M. Owen, David A. Vogel, and Christopher F. Waythomas.

Field trips emphasizing the engineering aspects of permafrost were organized and carried out by A.J. Alter, Civil Engineering Consultant, and F.L. Bennett, University of Alaska. They were ably assisted at the many stops in the field by local members of the American Society of Civil Engineers.

In addition, the organizers of the local field trips appreciate the cooperation of Dan Egan of the Alaskan Gold Company in Fairbanks; Alice Ebenal, vice president of E.V.E. Co., Inc.; officials of the Northwest Alaska Pipeline Company in charge of the frost-heave facility in Fairbanks; and many others for enabling the local field trips to be a success.

Special acknowledgment is made of the Cold Regions Research and Engineering Laboratory, especially Paul V. Sellmann, John Craig and Bruce F. Brockett, for organizing and conducting the many trips that allowed hundreds of participants to examine the tunnel cut into perennially frozen, organic-rich silt near Fox, north of Fairbanks.

Permafrost in the Fairbanks Area

The Fairbanks area is in the discontinuous-permafrost zone of Alaska, and perennially frozen ground is widespread, except beneath hilltops and moderate to steep south-facing slopes. Since 1903, fires and disturbance of vegetation by land development have increased the depth to permafrost by 25 to 40 ft in many places, although much of the ground refroze after the surface was revegetated. Sediments beneath the floodplain are perennially frozen as deep as 265 ft. Permafrost frequently occurs as multiple layers of varying thickness, and in many areas is not present. Unfrozen areas occur beneath existing or recently abandoned river channels, sloughs, and lakes. Elsewhere, layers of frozen sand and silt are intercalated with unfrozen beds of gravel. Depth to

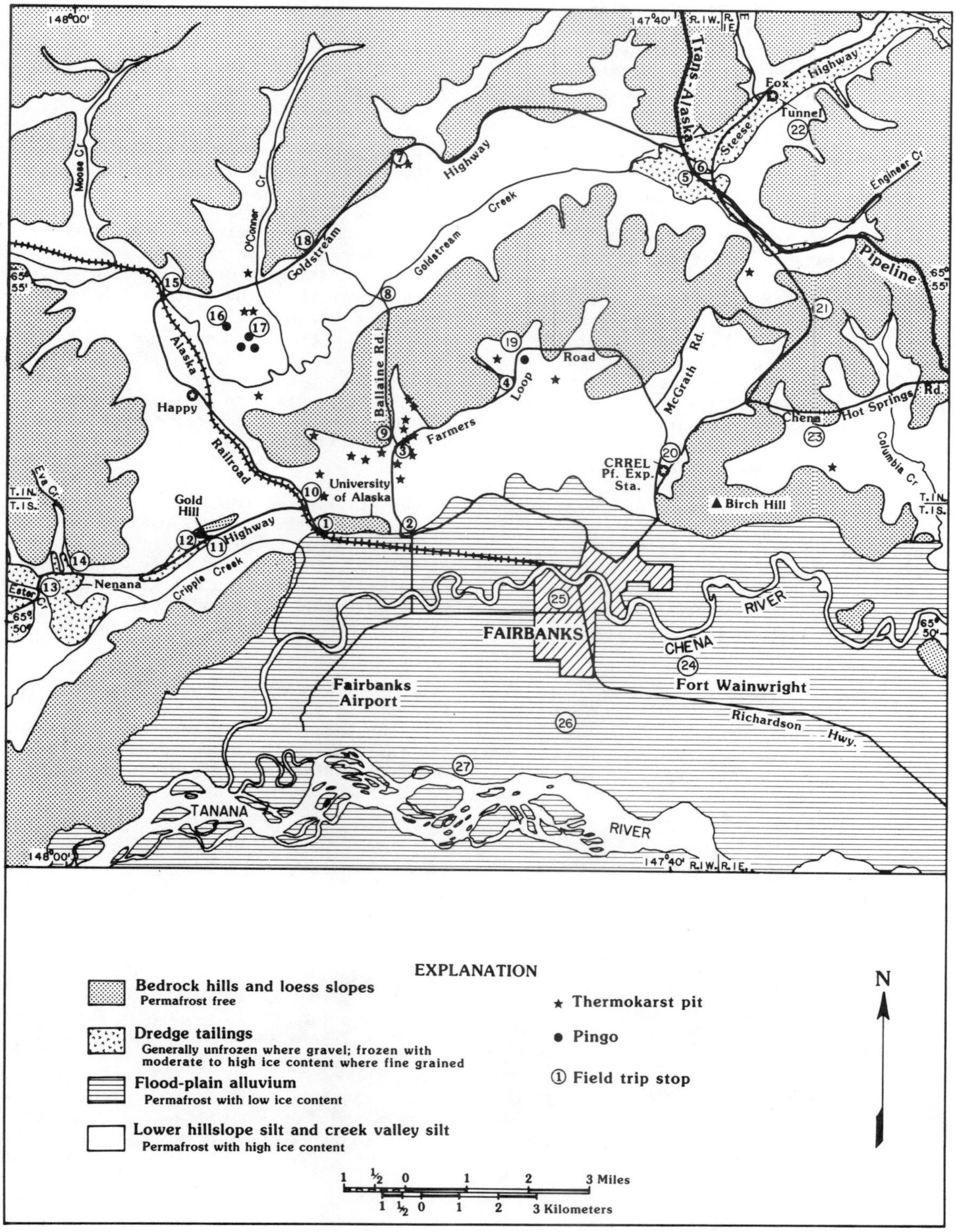

Generalized permafrost map of the Fairbanks area, Alaska, showing field trip stops. Modification of map prepared by T.L. Péwé.

permafrost varies from 2 to 3 ft in undisturbed areas and is more than 4 ft on the slip-off slopes of rivers. Ice in perennially frozen sediments beneath the floodplain consists of granules that cement grains. Large ice masses are absent.

Permafrost in the retransported valley-bottom silt in creek valleys and on lower hillslopes is thickest at lower elevations--up to at least 360 ft near the floodplain of the Tanana-Chena Rivers. Permafrost in these sediments contains large masses of clear ice in the form of horizontal to vertical sheets, wedges, and saucer- and irregular-shaped masses. Ice masses are foliated (ice wedges) and range from less than 1 ft to more than 15 ft in length. Much of the ice is arranged in a polygonal or honeycomb network that encloses silt polygons 10 to 40 ft in diameter. Perennially frozen ground does not exist beneath steep south-facing slopes, but extends nearly to the summit of north-facing slopes.

The temperature of permafrost in the Fairbanks area at a depth below the level of seasonal temperature fluctuations (30 to 50 ft) is about 31°F. If the vegetation cover is removed, the relatively warm, temperature-sensitive permafrost thaws (degrades). Thawing of permafrost in floodplain alluvium with low ice content results in little or no subsidence of the ground, but in creek-valley bottoms on lower slopes, thawing of ice-rich retransported loess (muck) results in considerable differential subsidence of the ground surface.

Although much permafrost in the area is probably relict (from colder Wisconsin-age conditions), it also forms today under favorable circumstances. Pre-Wisconsin permafrost probably disappeared during the widespread thawing in Sangamon time.

Description of Numbered Field Trip Stops*

Stop 1 - WEST END OF COLLEGE HILL

College Hill is a low bedrock hill covered with up to 80 ft of loess. The top and south sides are permafrost free and have long been the sites of agricultural fields and buildings of the University of Alaska. From the top, a fine view to the south shows the wide Tanana River floodplain with the glacier-sculptured Alaska Range and Mt. McKinley in the far distance.

Stop 2 - FARMERS LOOP "SINKHOLE"

Almost all of Farmers Loop Road crosses ice-rich permafrost and continuing maintenance is necessary to ensure a passable surface. In 1962, Farmers Loop Road was rerouted across ice-rich peat and organic silt between College Road and the east entrance to the University of Alaska. This one-third-mile rerouting cost $170,000 (Woodrow Johannson, Alaska Department of Transportation and Public Facilities, written comm., October 3, 1977). Shortly after the subgrade was paved, differential settlement occurred because of thawing of the underlying ground ice and ice-rich peat and subsequent compacting of the peat. In 1977, leveling of the road shoulders cost $5,000, and similar annual maintenance, including leveling with asphalt, has been necessary since 1974.

The area has been called the Farmers Loop sinkhole because of the tremendous amount of subsidence during the past few years and continued filling with gravel and asphalt. A thickness of up to 5 ft of pavement used as patching may be present in limited areas.

In 1983, the area was painted white to promote reflection of solar radiation. In the winter of 1982-83, thermal heat tubes were installed at the edge of the road on the west side in an attempt to freeze the ground beneath the highway and prevent further thawing and subsidence.

Stop 3 - THERMOKARST TOPOGRAPHY, GOLF COURSE, KFAR TRANSMITTER BUILDING

The perennially frozen retransported loess at the Fairbanks golf course and at the KFAR transmitter building contains large ice wedges in a

Stop 1. Richard D. Reger lecturing to group.

Stop 2. Note in the distance, tilting of the light standards and signs and telephone poles along both sides of the road which indicates thawing of the ice-rich permafrost with associated subsidence of the ground. (Photograph 25443 by T.L. Péwé, May 15, 1983.)

*Location of stops indicated on accompanying generalized permafrost map of the local Fairbanks area.

Stop 3. Troy Péwé lecturing to group at KFAR transmitter station. The station had subsided 2 to 3 ft by 1948; by 1983 it had subsided 4 to 5 ft. The ground surface is now near the base of the windows.

Stop 4. House built in 1975 on ice-rich permafrost. (Photograph 3997 by T.L. Péwé, 1977.)

polygonal network. As the ground thaws, thermokarst pits and mounds form. Pits are steep-walled, 5 to 20 ft deep, and 3 to 30 ft across. Broad mounds 50 to 100 ft in diameter and 2 to 4 ft high have formed and are slowly growing higher as the ice continues to melt. They are best displayed in the golf course, which was established in 1946 and is perhaps not only the world's farthest north golf course, but the only one in which thermokarst mounds and pits as natural hazards cause serious difficulty.

The radio transmitter building and tower were built in 1939. The building is of reinforced concrete, and, although the underlying ground has settled considerably due to melting of ground ice, the foundation has not been seriously deformed. Ground subsidence became evident when the well in the building began to "rise." The casing of the well is frozen to 120 ft in permafrost, but the pump, which is set on top of the casing, continues to "rise" above the floor. When it was 1 to 2 ft above the floor, the well casing was cut off and reset at floor level. Contrary to popular opinion, the casing was not rising out of the ground; rather, the ground and building were sinking around the stable casing. The building has not been used as a transmitter station for several years, and stands partially subsided (3 to 5 ft) in the middle of the golf course surrounded by well-developed thermokarst pits and mounds.

Stop 4 - HOUSES DEFORMED BY UNDERLYING PERMAFROST

Houses built on the south-facing hillside at this field trip stop on Farmers Loop Road are underlain with unfrozen ground and have excellent foundations. Houses on the gentle lower slopes or flatlands here are underlain by ice-rich, perennially frozen silt. Some homes with heated basements have subsided in a disastrous way. For example, the foundation of one house was built in 1974 and the house (see photograph) was built in 1975. It was occupied for one year and vacated during the summer of 1977. By this time, evidence was abundant that the underlying ice-rich permafrost was thawing. With the shifting of the house, vertical drain pipes dropped off, two of the basement doors and the front door were out of plumb, and the large concrete block front-door step had subsided. By 1983, the house had further subsided and was even more severely distorted. The large concrete block steps at the front door had completely disappeared because of subsidence, and the level of the ground was up to the base of the front door. The basement room next to the garage was severely cracked and the garage doors were no longer functional. The house was not occupied in 1983, except for a caretaker.

Stop 5 - TRANS-ALASKA PIPELINE SYSTEM

The Trans-Alaska Pipeline System is the largest completed, privately funded construction effort in history. It was built in 1974-1977 and is designed to transport up to 2 million barrels of crude oil per day. The cost of constructing the pipeline was about $8 billion dollars. About $1 billion of that amount was necessary to learn about, combat, accommodate and otherwise work with the perennially and seasonally frozen ground, which emphasizes the impact frozen ground had on the cost of this major construction project.

The pipeline transports warm crude oil, at temperatures up to 145°F, for about 800 miles from the North Slope of Alaska to the ice-free port of Valdez. Originally, the Trans-Alaska Pipeline System was to be buried along most of the route. The oil temperature was initially estimated at 158° to 176°F during full production. Obviously, such an installation would thaw the surrounding permafrost, but, because such an enterprise had never before been undertaken, there was scant knowledge of the serious problems that would be created by a warm-oil pipeline in frozen ground.

Stop 5. Elevated section of Trans-Alaska Pipeline at Engineer Creek, 5 miles north of Fairbanks. The pipe is insulated with 4 inches of resin-impregnated fiberglass jacketed by galvanized steel. (Photograph 3986 by T.L. Péwé, July 17, 1977.)

Stop 6. Local field trip participants examining huge ice wedges in perennially frozen retransported loess of Wisconsin age near Fairbanks, Alaska.

Thawing of the widespread ice-rich permafrost by a buried warm-oil line could cause liquification and loss of bearing strength, thus producing soil flow and differential settlement of the line. The greatest differential settlement could occur in areas of ice wedges. About half of the pipeline was built above the ground because of the presence of ice-rich permafrost.

Although the elevated pipeline successfully discharges its heat into the air and does not directly affect the underlying permafrost, other frozen ground problems must be considered. The above-ground pipe is placed on a cross beam installed between steel vertical-support members (pilings) placed in the ground. The 120,000 vertical-support members used along the pipeline route are 18-in.-diameter steel pipes that are subject to frost heaving. To eliminate frost heaving, each steel pipe is frozen firmly into the permafrost using a thermal device that is installed in the pipe. The device consists of metal tubes filled with hydrous ammonia that becomes a gas in winter and rises to the top of the tubes. In the cold atmosphere it liquefies, running down the pipe and thereby chilling the ground whenever the ground temperature exceeds the air temperature. The devices are non-mechanical and self-operating. Aluminum fins on top of the steel piles permit rapid dispersal of heat. Both above-ground and below-ground construction modes for the pipelines are used in the Fairbanks area.

Stop 6 - PLACER GOLD MINE IN PERENNIALLY FROZEN GROUND

Placer gold lying at the base of perennially frozen stream creek gravels was discovered in 1902 about 16 miles north of Fairbanks. Within a few years the region became one of the greatest gold-producing regions in Alaska. Early placer mining used underground methods, but by the early 1920's most of the richer deposits were exhausted. In the middle 20's a revival of mining in the Fairbanks gold region was stimulated by the initiation of large-scale operations with huge gold dredges and hydraulic stripping of the overlying ice-rich frozen silt. Placer gold mining in the area was essentially terminated in 1964 with the closing down of the last dredge. However, small-scale mines still operate.

Gold in the areas generally lies on bedrock and is overlain by 3 to 10 ft of perennially frozen gravel. Overlying this is 100 to 300 ft of perennially frozen, ice-rich retransported loess. This barren silt must be removed and the gravel thawed before the gold can be extracted. Until today, the removal of the overlying frozen retransported silt was done by washing away the ground with water under pressure through giant hydraulic nozzles.

At the E.V.E. Co. Mine at Stop 6, the 45 ft of overlying perennially frozen, ice-rich silt was removed during the winter using huge mechanical tractors (bulldozers), and frozen silt blocks were pushed to one side. The frozen gravel, 25 ft thick, was then removed in the summertime with mechanical equipment. The gold-rich gravel is processed through a screen and sluice box and the gold recovered.

The removal of 300,000 yd^3 of perennially frozen ice-rich silt created excellent exposures of late Pleistocene and Holocene frozen silt deposits with large, complex ice wedges. The large ice wedges are exposed in the Goldstream Formation, which is Wisconsin in age. Overlying the Goldstream Formation is the 5-ft-thick Holocene retransported silt with no ice wedges.

Stop 7 - THERMOKARST MOUNDS AND PITS

Mounds as much as 8 ft high and 40 ft in diameter and isolated thermokarst pits occur in the cultivated field on the downhill side of the road. Removal of natural vegetation caused thawing of permafrost and melting of ice wedges.

Stop 8 - SEASONAL STREAM ICINGS (AUFEIS)

Where Goldstream Creek crosses Ballaine Road, icings are formed and are sites of study by scientists from the University of Alaska. Aufeis begins to accumulate at Goldstream Creek in early November and reaches a maximum thickness of about 4 or 5 ft by March or April; it is completely melted by the middle of May. The stream freezes to the bottom in the winter, causing water to overflow and accumulate as thin layers of ice. Aufeis plugs the stream channel during spring break-up, and melt waters are forced onto surrounding floodplain surfaces. This type of flooding is unique to subarctic and arctic regions.

Stop 9 - BUILDINGS DISTORTED BY GROUND SUBSIDENCE

Some of the homes built in this subdivision are sited on perennially frozen ice-rich retransported silt. As the ground thaws from the heat of the houses, the buildings become distorted and eventually have to be removed.

Stop 10 - THERMOKARST PITS AND MOUNDS

The best-developed mounds (with the best-documented history in the Fairbanks area) are in a field on the gentle, north-facing slope at the University of Alaska experimental farm. The smooth field was cleared of black spruce in 1908; by 1922, pronounced individual and connected depressions that interfered with the operation of farm machinery had formed. The field was then seeded to pasture. By 1938, the mounds were 3 to 8 ft high and 20 to 30 ft in diameter. In November 1938, a bulldozer removed the upper part of every hummock and filled each pit until the land surface was approximately uniform. The smooth surface existed for nearly a year; by July 1939--nearly a year later--irregularities in the smooth surface began to form. During succeeding years, thermokarst mounds formed as the ground surface subsided over melting ice. By 1947, mounds in the area that had been graded in 1938 were as large and as high as those in the field that were not graded. The maximum height was 8 ft. Comparisons of 1938 and 1948 aerial photographs reveal a similar size and shape for the mounds. A soil-auger probe on July 14, 1948, revealed no ice or frozen ground 9 ft below the surface in one of the trenches. Paper birch, willow, alder and various forbs now grow where black spruce once stood.

Stops 11-12 - GOLD HILL

An exposure 5 ft long and 0.8 ft wide on Gold Hill was created by placer gold mining operations in 1949-53. Here is exposed a rather complete stratigraphic section of Quaternary deposits. Thawing and slumping have now destroyed the details, but the 202-ft-thick section of Gold Hill Loess is still present, as are the brown gravels of Pliocene-Pleistocene age in the tailing piles. The top of Gold Hill is free of permafrost, but permafrost occurs on the lower slope and in the valley bottom of Cripple Creek. The organic-rich Goldstream Formation with ice wedges is exposed on the south side of the mining exposure. On the

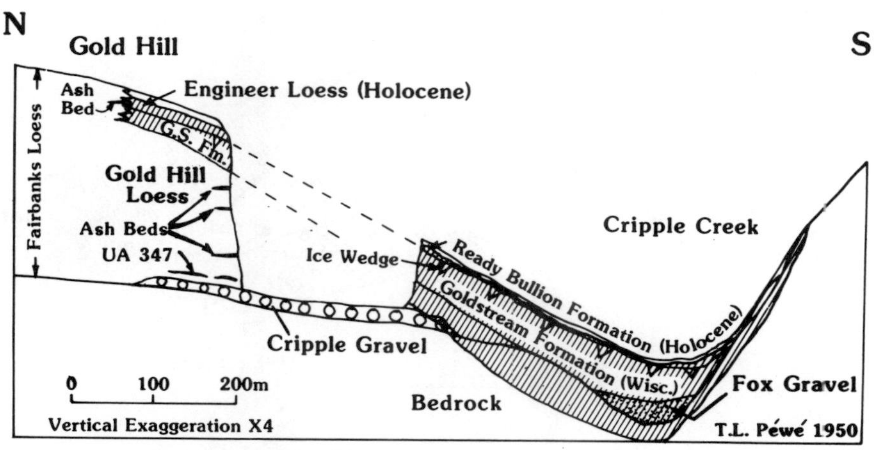

Stops 11 and 12. Stratigraphic section of the Gold Hill-Cripple Creek area 5 miles west of Fairbanks, Alaska, showing late Cenozoic formations and distribution of modern permafrost as exposed in placer gold mining excavations. Ash bed UA 347 is approximately 500,000 years old. Diagonally lined area represents existing permafrost.

north side there exists a 200-ft-high cliff of loess with only a small amount of permafrost near the top. Various volcanic ash layers have been identified, and the one at the base of the exposure (see illustration) is about 500,000 years old. The type locality of the Gold Hill Loess is the north wall of the Gold Hill mining cut.

Stop 13 - ESTER-CRIPPLE CREEKS MINING AREA

The Ester-Cripple Creeks mining area has been extensively mined for placer gold for about the last 80 years. Two ages of gold exist, one associated with the older brownish gravel (Cripple Gravel), which may be Pliocene-Pleistocene in age, and a younger placer, which is associated with the greyish gravel (Fox Gravel) and is early Pleistocene in age. These gravels are overlain by thick deposits of loess of Illinoian and Wisconsin age.

Today, permafrost in the area is restricted to the valley bottoms of the creeks and is up to 100 ft thick. This frozen ground formed in Wisconsin time.

Stop 14 - EVA CREEK

This placer gold mining excavation exposes excellent examples of perennially frozen Pleistocene gravel and retransported loess with large ice wedges.

Stop 15 - BRIDGE OVER GOLDSTREAM CREEK ON GOLDSTREAM HIGHWAY

Goldstream Creek is entrenched 12 ft into the retransported loess, and permafrost is at a depth of 27 ft below the bottom of the stream. On the sides of the creek, permafrost is from 1 to 10 ft deep. Pilings at the south end of the bridge annually settled at a rate of about 2 inches per year for several years after the bridge was built in 1965. This movement was arrested each winter and frost heaving raised the pilings up to half an inch. By 1974, total pier settlement reached 6 inches and the bridge deck had to be raised.

Stops 16-17 - OPEN-SYSTEM PINGOS

Open-system pingos are low, generally forested, ice-cored, perennial mounds, and several occur in the Fairbanks area. They develop where cold, subpermafrost or intrapermafrost water slowly flows through an artesian system and finds its way near the surface where it is blocked and freezes. Hydrostatic stress and the force of ice crystallization are relieved by doming of the near-surface material, which is silt in the Fairbanks area. These mounds may grow to a height of 25 ft and then rupture, with subsequent melting of the pingo ice and collapse of the mound. A large collapsed pingo with a lake about 30 ft across is exposed at Stop 16, and a small pingo with a white spruce-aspen growth at least 100 years old is formed at Stop 17.

Stop 18 - THERMOKARST MOUNDS AND PITS

On the downhill side of the Goldstream Highway, natural vegetation covering ice-rich permafrost has been removed, with subsequent formation of thermokarst topography.

Stop 19 - SOLAR-ASSISTED CULVERT-THAWING DEVICE

Highway culverts in the Subarctic and Arctic are very susceptible to filling by ice in winter. Gradual ice accumulation causes damming of stream and spring waters, resulting in flooding and highway icings. Traditional methods for keeping the culvert open so that water does not back up and

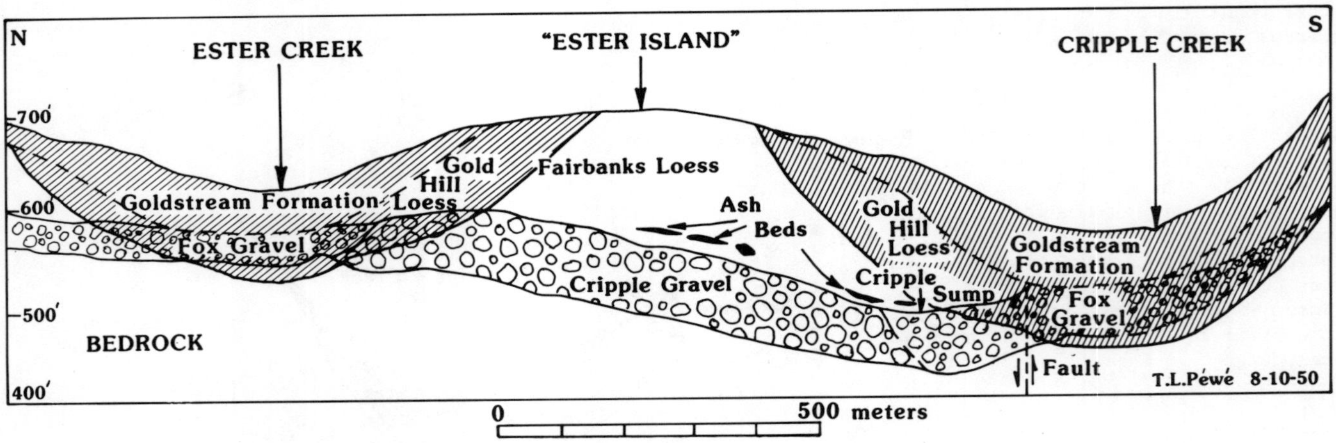

Stop 13. Cross section 1 mile west of the confluence of Ester and Cripple Creeks 10 miles west of Fairbanks, Alaska, showing probable distribution of late Cenozoic sediments prior to their partial removal by gold mining operations. The lower part of the Fairbanks Loess, where it is overlain by the Goldstream Formation, is termed Gold Hill Loess. Holocene Engineer Loess and the Ready Bullion Formation, as well as ice wedges in the Goldstream Formation, are omitted for simplicity. The diagonally lined areas represent distribution of modern permafrost. Permafrost data, surface topography, top of gravel and bedrock, and the fault are based on data from U.S. Smelting, Refining and Mining Company, Fairbanks, Alaska.

form ice on the roadway have involved 1) pumping hot water into the culvert; 2) using a steam generator to force steam through the culvert; 3) using electrical heat tapes installed in the pipe; and 4) melting the ice in the pipes with an oil-fired barrel stove.

To reduce costs for thawing of ice and culverts, a maintenance-free, solar-assisted culvert thawing device has been designed, constructed, and installed at this site. This system circulates fluid in a closed-loop circuit through flat solar-collector plates. Preliminary tests indicate that the device is feasible and is cheaper than other methods of thawing culverts.

Stop 20 - PERMAFROST EXPERIMENT STATION

The Alaska Field Station of the Cold Regions Research and Engineering Laboratory (CRREL) on Farmers Loop Road was established by the Corps of Engineers in 1946 to obtain information related to problems of design and construction in permafrost regions. The station is built on retransported, perennially frozen, ice-rich silt overlying creek gravel. Permafrost is 126 ft thick but does not extend to bedrock. Groundwater flowing from the permafrost-free slope on Birch Hill lies between permafrost and bedrock and is under artesian pressure. Drilling in 1946 produced a flowing artesian well that is now capped. This was the first major permafrost research field station in the United States and was active from 1946 to 1978. More recently, private companies and a state agency have utilized the area for short-term permafrost studies.

The experimental area was used to verify conclusions reached by theoretical and or laboratory research, to observe results of construction methods on permafrost, and to investigate some current empirical designs and methods of construction. Some of the studies included the testing of types of pilings and piling-installation methods for structural support, testing anti-frost-heaving devices, insulation of pavements, rate of permafrost thawing under natural vegetation cover, cleared areas, and artificial highway and airfield test sections.

Stop 21 - RELOCATION OF THE STEESE HIGHWAY

The relocation of the Steese Highway in the late 1970's placed the road over perennially frozen, retransported silt with large ice masses. An opportunity was provided to conduct a geophysical survey to locate ice masses prior to excavation when they were exposed. Much of the perennially frozen, ice-rich silt is now removed; however, some still remains under the road, and thawing is occurring.

Stop 22 - PERMAFROST TUNNEL

The only research tunnel in permafrost in the western world was built by the Cold Regions Research and Engineering Laboratory (CRREL) near Fox in the early and mid-1960's. It enters a near-vertical silt scarp left at the edge of the valley by gold-mining operations. Tunneling was performed by a prototype continuous mining machine (Alkirk Cycle Miner) that had never before been used in permafrost. When operating properly, the machine could cut as much as 7 ft of tunnel in an hour. The 360-ft-long tunnel is now maintained by CRREL and the University of Alaska for permafrost studies and demonstration purposes.

The tunnel is cut in retransported loess of the Goldstream Formation. It exposes ice wedges as well as mammal bones of Wisconsin age. The tunnel is 5 to 7 ft high and the permafrost has a temperature of about 28°-30°F. The tunnel is now chilled in winter by circulating cold outside air and in summer by mechanical refrigeration, especially near the opening. An inclined drift from the main tunnel extends downward through poorly sorted stream gravel into disintegrated schist bedrock. This 200-ft-long inclined drift reaches a depth about 20 ft below the level of the main tunnel.

The warm permafrost is subject to slow, continuous deformation, which has been extensively studied in the tunnel. Deformation rates range from 0.17 to 0.24 in. per week vertically and from 0.07 to 0.029 in. per week horizontally, with the highest rates in both directions occurring at the rear of the main tunnel. However, sublimation of pore ice may offset closure from deformation in the main tunnel. Creep of permafrost is very noticeable in the room at the end of the inclined drift, where measurement poles have been twisted and deformed.

Stop 23 - FROST-HEAVE TEST FACILITY

Enormous reserves of natural gas are now known in arctic Alaska. It is proposed to transport this gas through western Canada to the contiguous United States or to southern Alaska by pipeline. Because oil or natural gas warmer than 32°F transported through a buried pipeline thaws the perennially frozen, ice-rich ground, it is planned to transport the gas by refrigerating it to a temperature lower than 32°F. A chilled gas line may become a problem in areas of discontinuous permafrost and seasonally frozen ground. In these areas, the unfrozen ground will freeze around the cold pipeline, forming ice segregations that could seriously distort the pipe. The amount of frost heave depends upon the type of soil and the moisture available. If the refrozen ice-rich ground thaws, there will also be loss of soil-bearing strength, resulting in soil instability.

To understand fully the nature of the freezing of fine-grained moist sediment in the vicinity of the chilled gas pipeline, the Northwest Alaskan Pipeline and the Foothills Pipelines Companies established a frost-heave test facility on Chena Hot Springs Road near Fairbanks in August 1978, and data collection began in March 1979. This test facility is on perennially frozen, retransported silt with considerable ground ice. The active layer is 1 to 4 ft thick and the silt is very susceptible to frost action if water is available. At the test facility, various measures for controlling or minimizing frost heave are undertaken. Two of the most effective basic control measures involve various designs of pipe insulation and selected materials that surround the pipe. There are ten buried sections of 48-in.-diameter pipe. Each section provides testing for different design combinations.

Stop 24 - UNDERGROUND UTILIDORS

In areas of rigorous climate, it is common to put water, sewer, steam, telephone, and other service lines in a subsurface structure that connects all buildings and central power plants. In the Arctic and Subarctic, this utilidor system is extensively used to protect water and sewer lines from freezing in seasonally or perennially frozen ground.

An extensive utilidor system was first used in central Alaska during World War II construction of military bases. It is also used at the University of Alaska as well as several cities and villages. Because considerable heat emanates from utilidor systems, surrounding permafrost may thaw. Utilidors in Fairbanks are in perennially frozen gravel with low ice content and have functioned well. However, utilidors installed in frozen ground with high ice content have slumped, sagged, and ruptured as the ground thawed.

Stop 25 - FAIRBANKS WATER SYSTEM

Prior to 1953, the citizens of the City of Fairbanks were largely dependent for their water supply on shallow to mid-depth wells into the gravel of the floodplain. Water from these private wells was heavily mineralized or organic rich. A privately owned steam plant and small water system served a very limited downtown zone in the city. With rapid growth, it was necessary to establish a large-scale municipal facility to provide electricity, telephone, and water services. Water supplies in cities in the Arctic and Subarctic generally require the use of heated utilidors, which are very expensive. Low winter temperatures and permafrost present special problems for a community water system.

In 1953, a special type of water system was designed and installed in the City of Fairbanks, and it has been successful for the last 30 years. The concept of this system has been copied throughout the North. It consists of a recirculating (looped) water-distribution system in which water, warmed by heat at the power plant, constantly moves from the street to the house connection and back.

Stop 26 - DISPOSAL OF SOLID WASTE IN THE SUBARCTIC

Disposal of solid wastes is more difficult in the Arctic and Subarctic than in temperate areas, in part because of the presence of seasonally and perennially frozen ground. Frozen ground complicates landfill site excavation, decreases the availability of unfrozen cover material, prevents or reduces the natural processes of decay and absorption into the soil system, and restricts movement of ground water. Because of frozen ground and for other reasons, much of the solid waste in the Fairbanks area is baled before being placed in the landfill or transported for salvage. Volume reduction by baling and landfill disposal of compacted material has several advantages: all types of waste can be handled; it is possible to salvage some material prior to baling; volume of waste is reduced; and the life of the landfill area is doubled.

Stop 27 - FAIRBANKS WASTEWATER TREATMENT FACILITY

The Fairbanks Wastewater Treatment Facility is a regional facility that was completed in 1976. It includes a central headworks for reception of wastes, biological treatment units serviced by an oxygen plant, clarifiers with provision for sludge return, chlorine chamber, outfall for effluent discharge into the Tanana River, and an aerobic sludge digester and facilities for handling and disposal of solid waste. The treatment facility is built on perennially frozen floodplain gravel with low ice content.

Stop 24. Concrete utilidor under construction at Fort Wainwright near Fairbanks. This area is on the Chena River floodplain and is underlain by perennially frozen sand and gravel with low ice content. (Photograph 139 by T.L. Péwé, July 16, 1947.)

SELF-GUIDED TOUR TO THE THERMOKARST MOUNDS IN THE BOREAL ARBORETUM

In this area of second-growth trees and shrubs you can see a network of low mounds separated by trenches. These were formed by subsidence of the land following the melting of ice-wedge polygons.

The black spruce forest was cleared for cultivation during the earliest stages in the development of the Agricultural Experiment Station in 1908, only four years after the initial settlement of Fairbanks. By 1922 when the University first opened its doors to students, pronounced mounds and depressions had formed, finally precluding the use of farm machinery.

In 1938, at a time when Otto Geist was starting to assemble collections that stimulated the creation of the University of Alaska Museum, the mounds were bulldozed to form a smooth surface, but a year later the mounds re-formed as melting of the ice and consequent subsidence continued. Studies in 1948 revealed that finally there was no ice or frozen ground to a depth of 9 ft below the surface of the trenches.

The mounds are roughly circular, reflecting the modified outline of the original ice-wedge network, 10 to 50 ft in diameter and 1 to 8 ft in height. Paper birch, willow, alder, and various forbs now grow where black spruce forest once stood. See Stop 10 discussion for additional details.

Low-angle, oblique aerial view of thermokarst mounds in an abandoned agricultural field at the U.S. Department of Agriculture Experimental Farm at the University of Alaska, Fairbanks. These mounds are located on the north-facing side of College Hill and are 10 to 30 feet in diameter. (Photograph by R.F. Black and T.L. Péwé, September 10, 1948).

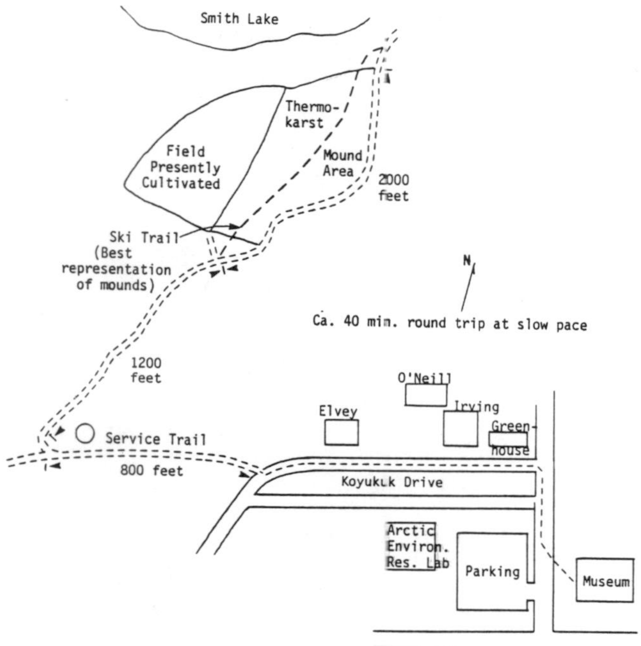

Extended Field Trips

FIELD TRIP A-1: ALASKA RAILROAD/DENALI NATIONAL PARK, JULY 14-16, 1983

The participants were transported by a chartered train with a specially prepared open gondola car, accessible at any time, that provided excellent visibility for observation and photography.

July 14: Alaska Railroad Station, Anchorage, to Denali Park Station

- Upper Knik Arm (photo stop)
- Test section of concrete ties
- Dramatic view of Mount Denali and other nearby peaks (photo and lunch stops)
- Hurricane Gulch Bridge
- Large beaver dams and ponds on both sides of tracks (photo stop)
- Denali Park Station
- McKinley Chalets (dinner and overnight)

July 15: Denali Park Area

- Six-hour bus tour of part of Denali National Park (a.m.).
- Tour of Usibelli Coal Mine (p.m.). Return to Chalets (dinner and overnight)

July 16: Denali Park Station to Fairbanks

- Healy Canyon (Nenana River Gorge): large landslides and related maintenance problems
- Town of Nenana on the Tanana River
- Goldstream Valley, underlain by ice-rich permafrost (lunch stop)
- Insulated track section
- Siding from which to view fully automat-tamping machine used to raise and align track
- Fairbanks

Field Guides

- Francis C. Weeks, Alaska Railroad
- T.C. Fuglestad, Alaska Railroad
- Ted B. Trueblood, Alaska Railroad
- Oscar J. Ferrians, Jr., U.S. Geological Survey
- Joe Usibelli, Usibelli Coal Mine

The cooperation and logistical support of the Alaska Railroad and its staff are greatly appreciated.

Train crossing active segment of Moody Landslide. Piles along shoulder were installed to help control downslope movement of the roadbed.

Members of Railway Delegation from People's Republic of China, inspecting the Alaska Railroad, central Alaska. They are discussing permafrost problems involved with railway construction with the former Chief Engineer of the Alaska Railroad. (Photograph No. PK25552 by T.L. Péwé, July 7, 1983.)

FIELD TRIP A-2: FAIRBANKS TO PRUDHOE BAY, JULY 12-16, 1983

July 12: Fairbanks to Yukon River

- Fox: permafrost tunnel and surroundings
- Washington Creek overview
- Wickersham Dome overview (lunch)
- Livengood area
- Hess Creek
- Yukon River Campground (dinner and overnight)
- BLM briefing on transportation corridor and archeology

July 13: Yukon River to Coldfoot

- Highway maintenance camp
- Ray River overview
- No Name Creek
- Finger Mountain uplands (lunch)
- Old Man runway
- Arctic Circle (photo stop)
- Gobbler's Knob overview (comfort station)
- South Fork Koyukuk (photo stop)
- Rosie Creek
- Marion Creek Campground (dinner and overnight)

July 14: Coldfoot to Toolik Lake

- Sukakpak Mountain: ice-cored mounds and Koyukuk River (comfort station)
- Scenic views of Brooks Range to Atigun Pass (lunch)
- Glacier hike (optional)
- Atigun Valley
- Pump Station 4 overview
- Toolik Lake (dinner and overnight)

July 15: Toolik Lake to Deadhorse

- Tour of Toolik Lake research area
- Slope Mountain: ice-cored mounds
- Happy Valley cut and erosion (lunch)
- Franklin Bluffs
- Coastal plain
- Deadhorse
- Kodiak oilfield camp (dinner and overnight)

July 16: Deadhorse to Fairbanks

- Bus tour of Prudhoe Bay (east)
- SOHIO camp (lunch)
- Bus tour to Kuparuk River oilfield
- ARCO and SOHIO camps (dinner)
- Depart for Fairbanks by plane

Field Guides

Fox Tunnel

- P.V. Sellmann, CRREL, Geology
- John Craig, CRREL, Paleobiology
- Bruce E. Brockett, CRREL, Drilling

Elliott Highway to Yukon River

- Leslie Viereck, Forest Service, Plant Ecology
- Hal Livingston, Alaska Department of Transportation and Public Facilities, Engineering Geology
- Glenn Johns, Formerly Federal Highway Administration, Historical and Environmental Engineering

Yukon River

- David Wickstrom, BLM, Management of Transportation Corridor
- John Cook, BLM, Archeology

Yukon River - Prudhoe Bay

- Jerry Brown, CRREL, Permafrost, Environment
- Ray Kreig, Kreig & Associates, Terrain Analysis
- Mike Metz, GeoTec Services, Geotechnical, Pipeline
- K.R. Everett, Ohio State University, Soils, Landforms
- Pat Webber, University of Colorado, Plant Ecology
- Palmer Bailey, Military Academy, Periglacial Landforms
- Len Gaydos, USGS, Landsat, Landcover Classification
- Richard Haugen, CRREL, Climate

Atigun Pass

- Parker Calkin, State University of New York, Glacial Geology
- Leah Haworth, State University of New York, Glacial Geology

Prudhoe Bay

- Tony Kinderknecht, SOHIO, Oil Field Activities
- Sue Degler, SOHIO, Environmental Protection
- Caesar Barrera, ARCO, Public Affairs
- K.R. Everett, Ohio State University, Soils, Geomorphology
- Patrick Webber, University of Colorado, Plant Ecology

The special logistics required to accomplish the field trip were supported by funding from the Alyeska Pipeline Service Company and the University of Alaska Foundation (derived from a variety of conference sponsors). The actual logistics were provided through the University of Alaska's Logistic Center under David Witt's supervision and with the volunteer help of the Explorer Post 47-Search and Rescue (Fairbanks). Pam Tinsley of the Logistics Center and Dianne Nelson of CRREL arranged many of the final details for the trip.

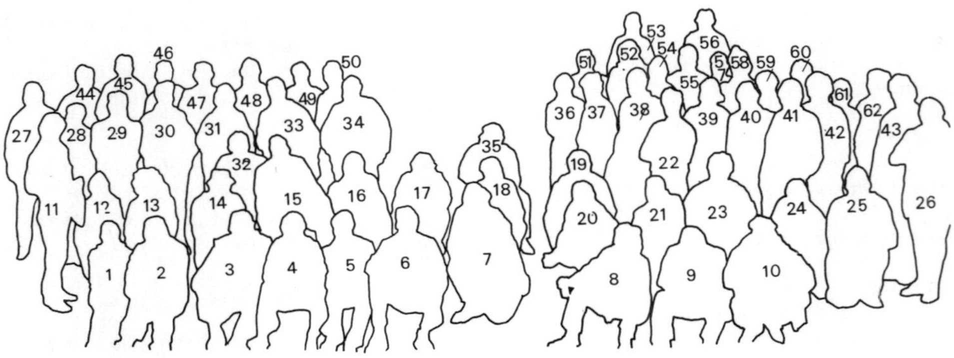

Participants in A-2 Field Trip

1. Celia Brown
2. Glenn Johns
3. Louis DeGoes
4. Jerry Brown
5. Cheng Guodong
6. Lorenz King
7. G. Michael Clark
8. Baerbel K. Lucchitta
9. Shi Yafeng
10. Cui Zhijiu
11. Ursee Burghardt
12. Kaye MacInnes
13. Lori Greenstein
14. Richard Haugen
15. Arturo Corte
16. Jaime Aguirre-Puente
17. H. Makela
18. Philip Tilley
19. Noel Potter
20. Edward Maltby
21. Guy Larocque
22. Michael Brown
23. Patrick Webber
24. Juerg Suter
25. Vernon Rampton
26. Arno Semmel
27. Michael Metz
28. Xu Bomeng
29. Thomas Osterkamp
30. Wang Liang
31. Patricia McCormick
32. Palmer Bailey
33. Alfred Jahn
34. Henry Holt
35. Steve Blasco
36. Linda Witt
37. Boris Zamsky
38. Esa Eranti
39. Thora Thorhallsdottier
40. Oskar Burghardt
41. Max Maisch
42. Kaye Everett
43. Colin Crampton
44. Liu Hongxu
45. Zhu Qiang
46. Wilbur Haas
47. Brainerd Mears
48. Leonard Gaydos
49. Ernest Muller
50. Jim Hamilton
51. Ray Herman
52. Ray Kreig
53. Hal Livingston
54. Terrence Hughes
55. Martin Gamper
56. Jim Scranton
57. Horst Strunk
58. William Barr
59. Donald Coulter
60. Dionyz Kruger
61. Barbara Gamper
62. Morris Engelke

Not pictured: Kathleen Ehlig, Dianne Nelson

FIELD TRIP B-3: DAWSON CITY TO TUKTOYAKTUK ALONG THE DEMPSTER HIGHWAY, JULY 23-30, 1983

July 23: Fairbanks to Dawson City via two DC-3 aircraft

Themes: Permafrost and vegetation relationships in discontinuous permafrost; Quaternary geology and history; Klondike goldfields and modern placer mining; Dawson City
Peat plateau and fen complex
Too Much Gold Creek: permafrost-vegetation relations, and effects of aspect
Hunker Creek, Mayes Claim: permafrost exposure revealing ice wedges and ice bodies within organic-rich "muck"; placer mining techniques
Bear Creek: open system pingo
Lower Klondike Valley: old tailing, Yukon Ditch, historical aspects
Dawson City (dinner and overnight)

July 24: Sixtymile Highway: Dawson City to the Yukon-Alaska international boundary

Themes: Regional geomorphology of the Klondike district; periglacial phenomena and tundra on the Yukon Plateau
Midnight Dome, Dawson City: panoramic view
Yukon River ferry crossing, Dawson City
Km 54: tors
Km 103-106: Yukon-Alaska boundary, cryoplanation terraces and patterned ground, solifluction and tundra
Lunch in field
Return: castellated tors
Dawson City overlook (photo stop)
Yukon River ferry crossing, Dawson City (dinner and overnight)

July 25: Dempster Highway: Dawson City to Eagle Plain Lodge

Themes: Glaciated and unglaciated terrain; the Ogilvie Mountains; rock glaciers; seasonal frost mounds; periglacial phenomena (tors, slopes)
Rock Glacier, Ogilvie Mountains
Tombstone Mountain overlook: regional geomorphology (photo stop)
North Fork Pass, Ogilvie Mountains: seasonal frost mounds
Dempster Highway near Chapman Lake: man-induced thermokarst
Lunch in field
Engineer Creek and Col: springs, hydrology, slopes
Castles' Hill, Ogilvie River Valley: springs, tors, slopes
Seven Mile Hill, Dempster Highway: Ogilvie River and Wernecke Mountains (photo stop)
Eagle Plain Lodge (dinner and overnight)

The North Fork Pass, southern Ogilvie Mountains. Seasonal frost mounds under study, July 25, 1983.

July 26: Dempster Highway: Eagle Plain Lodge to Inuvik

Themes: Eagle River Bridge, engineering and permafrost considerations; the Arctic Circle; patterned ground (mud boils); the Richardson Mountains, alpine tundra and periglacial phenomena; hydrology and road construction, culverts, icings
Eagle River Bridge, Dempster Highway: permafrost engineering and monitoring
The Arctic Circle: mud boils and patterned ground
Icing locations, Dempster Highway
Rat Pass, Richardson Mountains: unglaciated terrain, tundra conditions, solifluction, blockstreams and patterned ground, pediments
Lunch in field
Eastern Foothills, Richardson Mountains: view of forest fire south of Fort MacPherson
Peel River ferry crossing
Mackenzie River ferry crossing (Arctic Red)
Inuvik (dinner and overnight)

July 27: Inuvik

Themes: Soils and vegetation in continuous permafrost; non-sorted circles; organic terrain; Mackenzie Delta ecosystems; construction and engineering in Inuvik and adjacent area
Campbell Lake lookout: regional setting
Km 711, Dempster Highway: earth hummocks and vegetation
Km 737, Dempster Highway: insulated road test site, permafrost engineering and monitoring
Km 749, Dempster Highway: snow road test site, vegetation response
Km 754, Dempster Highway: Gaynor Lake, impact of forest fire in 1968 upon active layer, vegetation and terrain
Inuvik (lunch)
CFS Inuvik: mud hummocks, firebreaks and vegetation

Soil and vegetation studies at Bombardier Channel in the Mackenzie Delta near Inuvik, July 27, 1983.

Oil pipeline test pad
Boat excursion to Bombardier Channel: Mackenzie Delta ecosystems, soils, vegetation
Inuvik (dinner and overnight)

July 28: Inuvik to Tuktoyaktuk

Themes: a.m., Continuation of previous day; p.m., Mackenzie Delta and Tuktoyaktuk: Permafrost terrain, polygons, ice wedges, pingos from the air
Inuvik Scientific Resource Centre, Department of Indian Affairs and Northern Development, Government of Canada: tour of facilities and introduction by Scientific Manager, D.A. Sherstone
Inuvik (lunch)
Air Transect: Inuvik to Tuktoyaktuk
Polar Continental Shelf Project Base Facility, Tuktoyaktuk, Department of Energy, Mines and Resources, Government of Canada: dinner and welcome, Director, PCSP, G.D. Hobson
Tuktoyaktuk Village ice cellars: ground ice and Quaternary history

July 29: Tuktoyaktuk and the Pleistocene Delta

Themes: Pingos, massive ground ice, ice wedges, tundra terrain, field experiments
Ibyuk pingo
Peninsula Point: massive ground ice exposures, ice wedges
Lunch in field
Illisarvik drained lake experimental site, Richards Islands: field experiments in permafrost

July 30: Tuktoyaktuk to Inuvik and departure for Edmonton

Field Guides

Fairbanks to Dawson City

H.M. French, University of Ottawa
S.A. Harris, University of Calgary
R.O. van Everdingen, Environment Canada

Sixtymile Highway

H.M. French
S.A. Harris

Dawson City to Eagle Plain Lodge

S.A. Harris
R.O. van Everdingen
W.H. Pollard, University of Ottawa

Eagle Plain Lodge to Arctic Red

S.A. Harris
H.M. French
R.O. van Everdingen

Arctic Red to Inuvik

J.A. Heginbottom, Geological Survey of Canada
C. Tarnocai, Agriculture Canada

Inuvik to Tuktoyaktuk

J.A. Heginbottom
C. Tarnocai
H.M. French
J.R. Mackay, University of British Columbia

Participants in B-3 Field Trip

1. Herbert Liedtke
2. Roger Langohr
3. Stuart Harris
4. Baerbel Lucchitta
5. Jonas Åkerman
6. Wayne Pollard
7. Jean-Claude Dionne
8. J.A. Heginbottom
9. Ming-Ko Woo
10. John Vitek
11. Charles Tarnocai
12. Wayne Rouse
13. Rendel Williams
14. Philip Tilley
15. Irene Heyse
16. Hugh French
17. Paul Haesaerts
18. William Krantz
19. Kevin Gleason
20. Jeff Vandenberghe
21. Robert Gunn
22. Margaret Kiely
23. Matti Seppala
24. Colin Thorn
25. James Willis
26. Johannes Karte
27. Francesco Dramis
28. Lori Greenstein
29. Charles Harris
30. Eduard Koster
31. William Barr
32. Ray Kreig
33. Else Kolstrup
34. Harald Svensson
35. John Kiely
36. Ross Mackay
37. Hans Kerschner
38. Horst Hagedorn

Not pictured: Esa Eranti, John Flory, Peter Hale, Michael Metz, James Neuenschwander, R.O. Van Everdingen, David Harry, Edwin Clarke

FIELD TRIP B-4: FAIRBANKS TO ANCHORAGE, RICHARDSON AND GLENN HIGHWAYS, JULY 23-27, 1983

July 23: Fairbanks to Delta Junction

Harding Lake: investigation of lake on north side of the Tanana River held in by high level terraces
Birch Lake: shallow late-Quaternary lake in a re-entrant of the Yukon-Tanana Upland dammed on the west side by sediments of the Tanana River
Tanana River Overlook: view of the braided glacial river
South wall of road cut at top of hill at Mile 292.6: examination of ice-wedge casts and ventifacts
Shaw Creek Bluff: panorama of the Alaska Range and the broad Tanana River Valley; terminal moraines of the Delta and Donnelly advances are seen in the distance
Shaw Creek Road: road cut exposes weathered bedrock, ventifacts, ice-wedge casts, dune sand and loess
Tanana River Bridge: confluence of the Tanana River and the braided Delta River and the crossing of the Tanana River by the Trans-Alaska Pipeline System
Jack Warren Road: well-developed ventifacts on the outwash fan of the Delta advance

July 24: Delta Junction to Paxson

FAA Station: overlook of the Alaska Range
Delta moraines
Examination of the delta moraine and outwash of Donnelly age
Polygonal ground and ice-wedge casts
Donnelly till
North-central Alaska Range: crossing of the Richardson Highway by the Trans-Alaska Pipeline System
Examination of Holocene moraines of Black Rapids Glacier: alternate methods for dating, including radiocarbon, stratigraphy, lichenometry and tree ring counting

Participants of B-4 field trip at the trans-Alaska pipeline crossing the active Denali fault, central Alaska Range, one of the largest crustal breaks in Alaska. Average rate of displacement along the Denali fault is 0.1 to 3.5 cm per year; 5 to 60 m of right-lateral movement and 6 to 10 m of vertical movement have occurred during the past 10,000 years. The pipeline is so constructed here that it can slide laterally as much as 6 m on Teflon "shoes" to accommodate lateral displacement along this right-lateral, strike-slip fault. It can also accommodate 1.8 m vertical displacement. The 1.2-m-diameter warm-oil pipeline was completed in 1977 and no serious maintenance problems have developed along the 1,285-km route over permafrost terrain. (Photograph No. PK25557 by T.L. Péwé, July 24, 1983.)

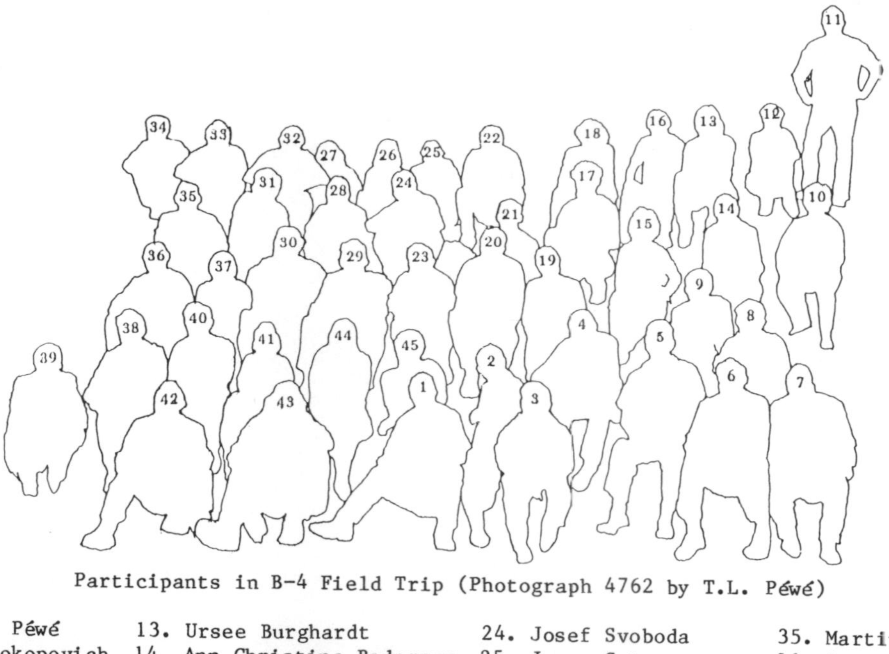

Participants in B-4 Field Trip (Photograph 4762 by T.L. Péwé)

1. Troy L. Péwé
2. N.P. Prokopovich
3. Cui Zhijiu
4. Andre Pancza
5. Jim Bales
6. Arturo Corte
7. Shi Yafeng
8. Qui Guoqing
9. Cheng Guodong
10. Zhou Youwu
11. Richard Reger
12. Barbara Gamper
13. Ursee Burghardt
14. Ann Christina Bedegrew
15. Kathleen Ehlig
16. Oskar Burghardt
17. Dave Vogel
18. Heinz-Peter Jons
19. Peggy Smith
20. Henry Holt
21. Takeei Koizumi
22. Max Maisch
23. Eric Karlstrom
24. Josef Svoboda
25. Juerg Suter
26. Dixie Brown
27. Lisa Clay
28. George Stephens
29. Brainerd Mears, Jr.
30. Noel Potter, Jr.
31. John Bell
32. Gary Clay
33. Thomas Baker
34. Liu Hongxu
35. Martin Gamper
36. James Walters
37. Peter Bayliss
38. Bjorne Bederson
39. Gerhard Sontag
40. Kaare Flaate
41. Julie Brigham
42. Alfred Jahn
43. Donald Hyers
44. Peggy Winslow
45. Keith Scoular

Crossing of the Denali fault by the Trans-Alaska Pipeline System
Rainbow Mountain
Active rock glacier
Gulkana Glacier view at Richardson Monument
Summit Lake

July 25: Paxson to Mile 50 on the Denali Highway and return

Alaska Range panorama
Road cut in small esker
Whistler Ridge: cryoplanation terrace site
Palsas
Road cut through an esker

July 26: Paxson to Glennallen on the Richardson Highway

Frost-rived granite blocks from rigorous Wisconsin periglacial environment (near site of Meier Road House)
View of Wrangell Mountains
Simpson Hill road cut and Copper River bluff
Permafrost problems on man-made structures in Glennallen
Tolsona No. 1: mud volcano

July 27: Glennallen to Anchorage on the Glenn Highway

Overlook of Matanuska Glacier
Matanuska Glacier: examination of terminal ice thrusting
Upper Matanuska Valley (photo stop)
Typical cross section through a crevasse-fill ridge
View of western Chugach Mountains in the Twin Peaks area
Landslide in Anchorage created by 1964 Great Alaska Earthquake; Government Hill School

Field Guides (and Lecturers*)

Troy L. Péwé, Arizona State University
Richard D. Reger, Alaska Division of Geological and Geophysical Surveys, Fairbanks
Daniel E. Lawson, Cold Regions Research and Engineering Laboratory*
Randall G. Updike, Alaska Division of Geological and Geophysical Surveys, Anchorage*

The following individuals provided special assistance at Big Delta: Glen Chowning, Superintendent of Public Schools, furnished overnight accommodations in the school gymnasium; and R.L. Dinger of the Cold Regions Test Center furnished cots and blankets and assisted in other local arrangements.

FIELD TRIP B-6: PRUDHOE BAY AND ARCTIC COASTAL PLAIN, JULY 23-31, 1983

July 23: Fairbanks-Prudhoe Bay

Fairbanks to Deadhorse (commercial flight)
Kodiak Camp
SOHIO Base Camp (lunch)
Putuligayuk River
Weather Pingo
Putuligayuk River reservoir and gravel pit
Putuligayuk River archeology site
West Dock
Kodiak Camp, Nabors Drilling (dinner and overnight)

July 24 and 25:

Abandoned road near Deadhorse Airport
Sagavanirktok River bridge; thaw gully and ice wedges
Sagavanirktok River floodplain
ARCO Prudhoe Bay operations center
Flow Station 2
Sand dunes between Drill Site 4 and East Dock
Flaxman boulders in lake bed northeast of East Dock
East Dock: traverse north along beach
Pump Station 1
Prudhoe Monument
Drill Site 14
Mount Prudhoe
Africa Lake
Kuparuk River, east side and bridge
Thaw stream
C-60 Pingo
Kuparuk Oil Field; thaw gully and Kuparuk C gravel pit
Oliktok Dock
Kodiak Camp (overnight)

July 26: Colville River Boat Trip, Nechelik Channel (fly to and from)

Nuiqsut and Nuiqsut Airfield
Putu Channel
Tapped lake, eroding pingo, and thaw lakes
Peat banks, ice wedges, and polygons
Gubik Formation
Tapped lake and lake fill
Slough entrance, willow vegetation, and turf-house site
Kodiak Camp (overnight)

July 26: Coastal-Plain Overflight and Barrow

North Slope Borough
Barrow utilidor
Gravel pit
Archeology site
Naval Arctic Research Laboratory
Barrow spit
Anchorage (via commercial flight)

Participants examining an ice wedge and peat along the western bluff of the Sagavnirktok River on the third day of the field trip. (Photograph by Duncan Hickmott.)

Participants probing for permafrost table in sand-dune field near east dock at Prudhoe Bay.

Participants examining beaded drainage along the Kuparuk route near Kuparuk oil field on the second day of the field trip. (Photograph by Duncan Hickmott.)

Field Guides (and Lecturers*):

Stuart E. Rawlinson, Alaska Division of Geological and Geophysical Surveys, Leader
David M. Hopkins, U.S. Geological Survey
Donald A. Walker, Institute of Arctic and Alpine Research*
John Harper, Woodward-Clyde Consultants (presently Dobrosky Seatech Ltd.)*
Beth Knol, Shell Oil Company*
H. Jess Walker, Louisiana State University, Colville River Guidebook Author
Duncan Hickmott and Terry Owen (DGGS), Field trip assistants

ARCO Alaska, Inc., and Sohio Alaska Petroleum Company arranged logistics and ground transportation, and representatives from these companies accompanied the field trip. Alyeska Pipeline Service Company arranged tours of Pump Station 1. Shell Oil Company was enthusiastic and helpful with a proposed visit to an artificial-island drill site. Kodiak-Nabors, a subsidiary of Anglo-Energy, provided accommodations at their facility at Prudhoe Bay. The North Slope Borough provided assistance with field-trip activities at Barrow, Alaska.

Participants examine well head at Drill Site 7.

Appendix B: Formal Program

Monday, July 18, 1983

8:30 - 10:00 a.m.

OPENING PLENARY SESSION

10:30 a.m. - noon

Panel Session: PIPELINES IN NORTHERN REGIONS

O.J. Ferrians, Jr. (Chairman), U.S. Geological Survey, USA
H.O. Jahns, Exxon Production Research Co., USA
E.R. Johnson, Alyeska Pipeline Service Co., USA
A.C. Mathews, Office of the Federal Inspector, ANGTS, USA
M.C. Metz, GeoTec Services, USA

1:30 - 3:30 p.m.

INVITED CHINESE SESSION

Chair: T.L. Péwé, Arizona State University, USA
J. Brown, Cold Regions Research and Engineering Laboratory, USA

Shi Yafeng and Cheng Guodong. A brief introduction to permafrost research in China.
Li Yusheng, Wang Zhugui, Dai Jingbo, Cui Chenghan, He Changgen and Zhao Yunlong. Permafrost studies and railroad construction in permafrost areas in China.
Zhou Youwu and Guo Dongxin. Some features of permafrost in China.

THERMAL ENGINEERING DESIGN

Chair: V. Lunardini, Cold Regions Research and Engineering Laboratory, USA
J.C. Harle, Alyeska, USA

Jahns, H.O. and C.E. Heuer. Frost heave mitigation and permafrost protection for a buried chilled-gas pipeline.
Walker, D.B.L., D.W. Hayley and A.C. Palmer. The influence of subsea permafrost on offshore pipeline design.
Cronin, J.E. Design and performance of a liquid natural convection subgrade cooling system for construction on ice-rich permafrost.
Zirjacks, W.L. and C.T. Hwang. Underground utilidors at Barrow, Alaska: A two-year history.
Reid, R.L. and A.L. Evans. Investigation of the air convection pile as a permafrost protection device.

PIPELINES

Chair: U. Luscher, Woodward-Clyde, USA
A.I. Gritsenko, Research Institute of Gas Industry, USSR

Thomas, H.P. and J.E. Ferrell. Thermokarst features associated with buried sections of the Trans-Alaska Pipeline.
Stanley, J.M. and J.E. Cronin. Investigations and implications of subsurface conditions beneath the Trans-Alaska Pipeline in Atigun Pass.
Hanna, A.J., R.J. Saunders, G.N. Lem and L. Carlson. Alaska Highway Gas Pipeline Project (Yukon section): Thaw settlement design approach.
Carlson, L. and D. Butterwick. Testing pipelining techniques in warm permafrost.
Heuer, C.E., J.B. Caldwell and B. Zamsky. Design of buried seafloor pipe lines for permafrost thaw settlement.
Mitchell, D.E., S.W. Laut, N.K. Pui and D.D. Curtis. Well casing strains due to permafrost thaw subsidence in the Canadian Beaufort Sea.

ICE AND SOIL WEDGES

Chair: M. Seppälä, University of Helsinki, Finland
J.R. Mackay, University of British Columbia, Canada

Carter, L.D. Fossil sand wedges on the Alaskan Arctic Coastal Plain and their paleoenvironmental significance.
Black, R.F. Three superposed systems of ice wedges at McLeod Point, northern Alaska, may span most of the Wisconsinan stage and Holocene.
Lawson, D.E. Ground ice in perennially frozen sediments, northern Alaska.

Jahn, A. Soil wedges on Spitzbergen.
Ferrians, O.J. Pingos on the Arctic Coastal Plain, northeast Alaska.

4:00 - 6:00 p.m.

THERMAL ANALYSIS

Chair: L. E. Goodrich, National Research Council, Canada
C. E. Heuer, Exxon Production Research Co., USA

Odom, W.B. Practical applications of underslab ventilation system: Prudhoe Bay case study.
Hildebrand, E.E. Thaw settlement and ground temperature model for highway design in permafrost areas.
Hromadka, T.V. II, G.L. Guymon and R.L. Berg. Comparison of two-dimensional domain and boundary integral geothermal models with embankment freeze-thaw field data.
Zhao Yunlong and Wang Jianfu. Calculation of thawed depth beneath heated buildings in permafrost regions.
Lunardini, V.J. Thawing beneath insulated structures on permafrost.

MECHANICS OF FROZEN SOIL

Chair: R. Dunning, Sohio, USA
T. S. Vinson, Oregon State University, USA

Baker, T.H.W. and G.H. Johnston. Unconfined compression tests on anisotropic frozen soils from Thompson, Manitoba.
Mahar, L.J., R. Wilson and T.S. Vinson. Physical and numerical modelling of uniaxial freezing of a saline gravel.
Ladanyi, B. and H. Eckardt. Dilatometer testing in thick cylinders of frozen sand.

PLEISTOCENE PERMAFROST CONDITIONS

Chair: A. Jahn, University of Warsaw, Poland
L. Christensen, University of Aarhus, Denmark

Haesaerts, P. Stratigraphic distribution of periglacial features indicative of permafrost in the Upper Pleistocene loesses of Belgium.
Landohr, R. The extension of permafrost in Western Europe in the period between 18,000 and 10,000 y. B.P. (Tardiglacial): Information from soil studies.
Svensson, H. Ventifacts as paleo-wind indicators in a former periglacial area of southern Sweden.
Kolstrup, E. Cover sands in southern Jutland (Denmark).
Vandenberghe, J. Ice wedge casts and involutions as permafrost indicators and their stratigraphic position in the Weichselian.
Lu Guowei, Guo Dongxin and Dai Jingbo. Basic characteristics of permafrost in northwest China.

REMOTE SENSING AND PLANETARY PERMAFROST

Chair: J.A. Heginbottom, Geological Survey of Canada
R.A. Kreig, R.A. Kreig & Associates, USA

Clifford, S.M. Ground ice in the equatorial region of Mars: A fossil remnant of an ancient climate or a replenished steady-state inventory?
Morrissey, L.A. The utility of remotely sensed data for permafrost studies.
Gurney, J.R., J.P. Ormsby and D.K. Hall. A comparison of remotely sensed surface temperature and biomass estimates for aiding evapotranspiration determination in central Alaska.

Tuesday, July 19, 1983

8:30 - 10:30 a.m.

Panel Session: ENVIRONMENTAL PROTECTION OF PERMAFROST TERRAIN

J.E. Hemming (Chairman), Dames & Moore, USA
M.C. Brewer, U.S. Geological Survey, USA
H.M. French, University of Ottawa, Canada
N.A. Grave, Permafrost Institute, Yakutsk, USSR
J. Tileston, Bureau of Land Management, USA
P.J. Webber, University of Colorado, USA

10:30 a.m. - 12:30 p.m.

THERMODYNAMICS AND TRANSPORT PHENOMENA

Chair: J.L. Oliphant, Cold Regions Research and Engineering Laboratory, USA
A.V. Sadovskiy, Foundations and Underground Constructions, USSR

Horiguchi, K. and R.D. Miller. Hydraulic conductivity functions of frozen materials.
Yoneyama, K., T. Ishizaki and N. Nishio. Water redistribution measurements in partially frozen soil by X-ray technique.
Kay, B.D. and P.H. Groenevelt. The redistribution of solutes in freezing soil: Exclusion of solutes.
Aguirre-Puente, J. and J. Gruson. Measurements of permeabilities of frozen soils.
McGaw, R.W., R.L. Berg and J.W. Ingersoll. An investigation of transient processes in an advancing zone of freezing.

MECHANICS OF FROZEN SOIL

Chair: S.R. Stearns, ASCE, Dartmouth College, USA
R.E. Smith, ARCO Oil & Gas, USA

Nelson, R.A., U. Luscher, J.W. Rooney and A.A. Stramler. Thaw strain data and thaw settlement predictions for Alaskan soils.

Zhu Yuanlin and D.L. Carbee. Creep behavior of frozen silt under constant uniaxial stress.
Vinson, T.S., C.R. Wilson and P. Bolander. Dynamic properties of naturally frozen silt.
Chamberlain, E.J. Frost heave of saline soils.
Corapcioglu, M.Y. A mathematical model for the permafrost thaw consolidation.

MOUNTAIN AND PLATEAU PERMAFROST

Chair: Shi Yafeng, Institute of Glaciology and Cryopedology, PRC
S.A. Harris, University of Calgary, Canada

Cheng Guodong. Vertical and horizontal zonation of high-altitude permafrost.
King, L. High mountain permafrost in Scandinavia.
Qiu Guoqing, Huang Yizhi and Li Zuofu. Alpine permafrost in Tian Shan, China.
Haeberli, W. Permafrost-glacier relationships in the Swiss Alps, today and in the past.
Greenstein, L.A. An investigation of mid-latitude alpine permafrost on Niwot Ridge, Colorado Rocky Mountains, USA.

EFFECTS OF MAN-MADE DISTURBANCES

Chair: L. Rey, Comité Arctique International, Switzerland
C.W. Slaughter, U.S. Forest Service, USA

Kershaw, G.P. Some abiotic consequences of the CANOL Crude Oil Pipeline Project; 35 years after abandonment.
Klinger, L.F., D.A. Walker and P.J. Webber. The effects of gravel roads on Alaska Arctic Coastal Plain tundra.
Komarkova, V. Recovery of plant communities and summer thaw at the 1949 Fish Creek Test Well 1, arctic Alaska.
Thorhallsdottir, T.E. The ecology of permafrost areas in central Iceland and the potential effects of impoundment.
Linkins, A.E. and N. Fetcher Effects of surface-applied Prudhoe Bay crude oil on vegetation and soil processes in tussock tundra.
Collins, C.M. Long-term active layer effects of crude oil spilled in interior Alaska.

PLANETARY PERMAFROST

Chair: D.M. Anderson, State University of New York, USA
M.C. Malin, Arizona State University, USA

Fanale, F.P. and R.N. Clark. Solar system ices and Mars permafrost.
Carr, M.H. The geology of Mars.
Lucchitta, B.K. Permafrost on Mars.
Nummedal, D. Permafrost on Mars: Distribution, formation and geological role.

Wednesday, July 20, 1983

8:30 - 10-30 a.m.

Panel Session: DEEP FOUNDATIONS AND EMBANKMENTS

N.R. Morgenstern (Chairman), University of Alberta, Canada
Ding Chingkang, Northwest Institute, PRC
D.C. Esch, Alaska Department of Transportation and Public Facilities, USA
B. Ladanyi, Ecole Polytechnique, Montréal, Canada
F.H. Sayles, CRREL (formerly Office of the Federal Inspector, ANGTS), USA

10:30 a.m. - 12:30 p.m.

ROADS AND RAILWAYS (THERMAL ASPECTS)

Chair: G.H. Johnston, National Research Council, Canada
O. Gregersen, Norwegian Geotechnical Institute, Norway

Zarling, J.P., B. Connor and D.J. Goering. Air duct systems for roadway stabilization over permafrost areas.
Johnston, G.H. Performance of an insulated roadway on permafrost, Inuvik, N.W.T.
McHattie, R. and D.C. Esch. Benefits of a peat underlay used in road construction on permafrost.
Goodrich, L.E. Thermal performance of a section of the Mackenzie Highway.
Esch, D.C. Evaluation of experimental design features for roadway construction over permafrost.
Zhang Shixiang and Zhu Qiang. A study of the calculation of frost heaving.

FOUNDATIONS

Chair: R.W. Fadum, North Carolina State University, USA
J.W. Rooney, R&M Consultants, USA

Andersland, O.B. and M.R. Alwahhab. Lug behavior for model steel piles in frozen sand.
Penner, E. and L.E. Goodrich. Adfreezing pressure on steel pipe piles, Thompson, Manitoba.
Dufour, S., D.C. Sego and N.R. Morgenstern. Vibratory pile driving in frozen sand.
DiPasquale, L., S. Gerlek and A. Phukan. Design and construction of pile foundations in the Yukon-Kuskokwim Delta, Alaska.
Nottingham, D. and A.B. Christopherson. Driven piles in permafrost: State of the art.
Manikian, V. Pile driving and load tests in permafrost for the Kuparuk pipeline system.

FROST MOUNDS AND OTHER PERIGLACIAL PHENOMENA

Chair: H. Svensson, University of Copenhagen, Denmark
R.F Black, University of Connecticut, USA

Mackay, J.R. Pingo growth and sub-pingo water lenses, western Arctic coast, Canada.

Seppala, M. Seasonal thawings of palsas in Finnish Lapland.

Pollard, W.H. and H.M. French. Seasonal frost mound occurrence, North Fork Pass, Ogilvie Mountains, northern Yukon, Canada.

Corte, A.E. Geocryogenic morphology at Seymour Island, Antarctica: A progress report.

Walton, D.W.H. and T.D. Heilbronn. Periglacial activity on the subantarctic island of South Georgia.

EFFECTS OF MAN-MADE AND NATURAL DISTURBANCES

Chair: T.F. Albert, North Slope Borough, USA
P. Duffy, Environment Canada

Ebersole, J.J. and P.J. Webber. Biological decomposition and plant succession following disturbance on the Arctic Coastal Plain, Alaska.

Johnson, A.W. and B.J. Neiland. An analysis of plant succession on frost scars, 1961-1980.

Maltby, E. and C.J. Legg. Revegetation of fossil patterned ground exposed by severe fire on the North York moors.

Racine, C.H., W.A. Patterson III and J.G. Dennis. Permafrost thaw associated with tundra fires in northwest Alaska.

Johnson, L. and L. Viereck. Recovery and active layer changes following a tundra fire in northwestern Alaska.

Shaver, G.R., G.L. Gartner, F.S. Chapin III and A.E. Linkins. Revegetation of arctic disturbed sites by native tundra plants.

Panel Session: CLIMATE CHANGE AND GEOTHERMAL REGIME

J. Pilon (Chairman), Department of Energy, Mines and Resources, Canada

Cheng Guodong, Institute of Glaciology and Cryopedology, PRC

J. Gray, University of Montreal, Canada

T.E. Osterkamp, University of Alaska, USA

M. Smith, Carleton University, Canada

FOUNDATIONS

Chair: O.B. Andersland, Michigan State University, USA
V. Manikian, ARCO Alaska, Inc., USA

Nixon, J.F. Geothermal design of insulated foundations for thaw prevention.

Wojcek, A., P.M. Jarrett and A. Beatty. Cold-mix asphalt stabilization in cold regions.

Luscher, U., W.T. Black and J.F. McPhail. Results of load tests on temperature-controlled piles in permafrost.

Gregersen, O., A. Phukan and T. Johansen. Engineering properties and foundation design alternatives in marine Svea clay, Svalbard.

Liu Hungxu. Calculation of frost heaving force in seasonally frozen soil.

Cui Chenghan and Zhou Kaijiong. Experimental study of the frost heave reaction.

GROUND ICE AND SOLIFLUCTION

Chair: G. Gryc, U.S. Geological Survey, USA
J.H. Akerman, University of Lund, Sweden

Rampton, V.N., J.R. Ellwood and R.D. Thomas. Distribution and geology of ground ice along the Yukon portion of the Alaska Highway Gas Pipeline north of Kluane Lake.

Harry, D.G. and H.M. French. The orientation and evolution of thaw lakes, southwest Banks Island, Canadian Arctic.

Reanier, R.E. and F.C. Ugolini. Gelifluction deposits as sources of paleoenvironmental information.

Gamper, M. Controls and rates of movement of solifluction lobes in the eastern Swiss Alps.

Ellwood, J. and J.F. Nixon. Observations of soil and ground ice in pipeline trench excavations in the south Yukon.

WATERSHED STUDIES IN PERMAFROST REGIONS

Chair: R.F. Carlson, University of Alaska, USA
B.E. Ryden, University of Uppsala, Sweden

Slaughter, C.W., J.W. Hilgert and E.H. Culp. Summer streamflow and sediment yield from discontinuous permafrost headwaters catchments.

Drage, B., J.R. Gilman, D. Hoch and L. Griffiths. Hydrology of North Slope coastal plain streams.

Woo, M., P. Marsh and P. Steer. Basin water balance in a continuous permafrost environment.

Flügel, W.A. Summer water balance of a high Arctic catchment area with underlying permafrost in Oobloyah Valley, N. Ellesmere Island, N.W.T., Canada.

PERMAFROST AND CLIMATE

Chair: S.A. Bowling, University of Alaska, USA
Xu Xiaozu, Institute of Glaciology and Cryopedology, PRC

Rouse, W.R. Active layer energy exchange in wet and dry tundra of the Hudson Bay lowlands.

Smith, M.W. and D.W. Riseborough. Permafrost sensitivity to climatic change.

Harris, S.A. Comparison of the climatic and geomorphic methods of predicting permafrost distribution in western Yukon Territory.

Nelson, F. and S.I. Outcalt. A frost index number for spatial predictions of ground-frost zones.

Xu Xiaozu and Wang Jiacheng. A preliminary study on the distribution of frozen ground in China.

4:00 - 6:00 p.m.

INVITED SOVIET SESSION

Chair: J.R. Kiely, Bechtel Power Corporation, USA
O.J. Ferrians, U.S. Geological Survey, USA

Melnikov, P.I. Major trends in the development of Soviet permafrost research.
Gritsenko, A. and Yu.F. Makoyon. Drilling and operation of gas wells in the presence of natural gas hydrates.
Zakharov, Yu.F., L.D. Kosuklim and Ye.M. Naviosky. The impact of installation for the extraction and transport of gas on permafrost conditions in western Sibera.
Vyalov, S.S. Engineering geocryology in the USSR: A review.
Sadovskiy, A.V. The construction of deep pile foundations in permafrost in the USSR (a summary).
Grave, N.A. Cryogenic processes associated with developments in the permafrost zone.

Thursday, July 21, 1983

8:30 - 10:00 a.m.

Panel Session: FROST HEAVE AND ICE SEGREGATION

E. Penner (Chairman), National Research Council, Canada
R.L. Berg, Cold Regions Research and Engineering Laboratory, USA
Chen Xiaobai, Institute of Glaciology and Cryopedology, PRC
R.D. Miller, Cornell University, USA
P.J. Williams, Carleton University, Canada

10:30 a.m. - 12:30 p.m.

FROST HEAVE

Chair: E.J. Chamberlain, Cold Regions Research and Engineering Laboratory, USA

Holden, J.T. Approximate solutions for Miller's theory of secondary heave.
Lovell, C.W. Frost susceptibility of soils.
Akagawa, S. Relation between frost heave and specimen length.
Konrad, J.M. and N.R. Morgenstern. Frost susceptibility of soils in terms of their segregation potential.
Rieke, R.D., T.S. Vinson and D.W. Mageau. The role of specific surface area and related index properties in the frost heave susceptibility of soils.
Chen Xiaobai, Wang Yaqing and Jiang Ping. Influence of penetration rate, surcharge stress and groundwater table in frost heave.

EMBANKMENTS, ROADS AND RAILWAYS

Chair: C.W. Lovell, Purdue University, USA
K. Flaate, Norwegian Road Research Laboratory, Norway

Bell, J.R., T. Allen and T.S. Vinson. Properties of geotextiles in cold regions applications.
Hayley, D.W., W.D. Roggensack, W.E. Jubien and P.V. Johnson. Stabilization of sinkholes on the Hudson Bay railway.
LaVielle, C.C., S.C. Gladden and A.R. Zeman. Nuiqsut Airport dredge project.
Tart, R.G., Jr. Winter constructed gravel islands.
Wang Liang, Xu Bomeng* and Wu Zhijin. Properties of frozen and thawed soil and earth dam construction in winter.

PATTERNED GROUND AND ROCK STREAMS

Chair: A.E. Corte, Instituto Argentino de Nevologia y Glaciologia, Argentina
Qiu Guoqing, Institute of Glaciology and Cryopedology, PRC

Walters, J.C. Sorted patterned ground in ponds and lakes of the High Valley-Tangle Lakes region, central Alaska.
Muir, M.P. The role of pre-existing, corrugated topography in the development of stone stripes. the Kunlun Shan, China.
Hagedorn, H. Periglacial phenomena in arid regions of Iran.
Cui Zhijui. An investigation of rock glaciers in
Sadovsky, A.V. and G.I. Bondarenko. Creep of frozen soils on rock slopes.

WATERSHED STUDIES IN PERMAFROST REGIONS

Chair: M.K. Woo, McMaster University, Canada
R.O. van Everdingen, Environment Canada

Ashton, W.S. and R.F. Carlson. Predicting fish passage design discharges for Alaska.
Lewkowicz, A.G. Erosion by overland flow, Central Banks Island, western Canadian Arctic.
Chacho, E.E. and S.R. Bredthauer. Runoff from a small subarctic watershed, Alaska.

PERMAFROST GEOPHYSICS

Chair: K. Kawasaki, University of Alaska, USA
G.D. Hobson, Polar Continental Shelf Project, Canada

Collett, T.S. Detection and evaluation of natural gas hydrates from well logs, Prudhoe Bay, Alaska.
Kay, A.E., A.M. Allison, W.J. Botha and W.J. Scott. Continuous geophysical investigation for mapping permafrost distribution, Mackenzie Valley, N.W.T., Canada.
Sinha, A.K. and L.E. Stephens. Deep electromagnetic sounding over the permafrost terrain in the Mackenzie Delta, N.W.T., Canada.
Ehrenbard, R.L., P. Hoekstra and G. Rozenberg.

*Name omitted in Abstract Volume, p. 247.

Transient electromagnetic soundings for permafrost mapping.

Oliphant, J.L., A.R. Tice and Y. Nakano. Water migration due to a temperature gradient in frozen soil.

Pearson, C., J. Murphy, P. Halleck, R. Hermes, and M. Mathews. Sonic resistivity measurements on Berea sandstone containing tetrahydrofuran hydrates: A possible analogue to natural gas hydrate deposits.

Evening: Banquet

Governor Bill Sheffield: Welcoming remarks

S. Russell Stearns, President, American Society of Civil Engineers: Keynote Speaker.

Friday, July 22, 1983

8:30 - 10:30 a.m.

Panel Session: SUBSEA PERMAFROST

D.M. Hopkins and P.V. Sellmann (Chairmen), U.S. Geological Survey, and Cold Regions Research and Engineering Laboratory, USA

S.M. Blasco, Geological Survey of Canada

D.W. Hayley, EBA, Canada

J.A.M. Hunter, Geological Survey of Canada

H.O. Jahns, Exxon Production Research Co., USA

10:30 a.m. - 12:30 p.m.

EXCAVATIONS, MINING AND MUNICIPAL FACILITIES

Chair: R.D. Abbott, Shannon & Wilson, USA

Rooney, J.W. and J.H. Wellman. Fairbanks wastewater treatment facility: Geotechnical considerations.

Simpson, J.K. and P.M. Jarrett. Explosive excavation of frozen soils.

Weerdenburg, P.C. and N.R. Morgenstern. Underground cavities in ice-rich frozen ground.

Sun Yuliang. Deformability of canals due to freezing and thawing.

Ryan, W.L. Design considerations for large storage tanks in permafrost areas.

Williams, D. The stabilization of the Nanisivik concentrator foundation.

COLD CLIMATE ROCK WEATHERING

Chair: H. Hagedorn, Federal Republic of Germany
S. Kinosita, Hokkaido University, Japan

Akerman, J. Notes on chemical weathering, Kapp Liné, Spitsbergen.

Hallet, B. The breakdown of rock due to freezing: A theoretical model.

Hyers, A. Some spatial aspects of climate-dependent mechanical weathering in a high altitude environment.

Ray, R.J., W.B. Krantz, T.N. Caine and R.D. Gunn. A mathematical model for patterned ground: Sorted polygons and stripes, and underwater polygons.

Vitek, J.D. Stone polygons: Observations of surficial activity.

Fukuda, M. The pore-water pressure profile on porous rocks during freezing.

GROUNDWATER IN PERMAFROST

Chair: Zhu Qiang, Water Power Research Institute, PRC

Kane, D.L. and J. Stein. Field evidence of groundwater recharge in interior Alaska.

Michel, F.A. Isotope variations in permafrost waters along the Dempster Highway pipeline corridor.

Wright, R.K. Relationships between runoff generation and active layer development near Schefferville, Quebec.

Price, J.S. The effect of hydrology on ground freezing in a watershed with organic terrain.

SUBSEA PERMAFROST

Chair: G. Shearer, Minerals Management Service, USA
J.A. Hunter, Geological Survey of Canada

Morack, J.L., H.A. MacAuley and J.A. Hunter. Geophysical measurements of the sub-bottom permafrost in the Canadian Beaufort Sea.

Neave, K.G. and P.V. Sellmann. Seismic velocities and subsea permafrost in the Beaufort Sea, Alaska.

Walker, H.J. Erosion in a permafrost-dominated delta.

Swift, D.W., W.D. Harrison and T.E. Osterkamp. Heat and salt transport processes in thawing subsea permafrost at Prudhoe Bay, Alaska.

Melnikov, P.I., K.F. Voitkovsky, R.M. Kamensky, I.P. Konstantinov. Artificial ice masses in arctic seas.

Paterson, D.E. and M.W. Smith, Measurement of unfrozen water content in saline permafrost using time domain reflectometry.

ECOLOGY OF NATURAL SYSTEMS

Chair: H.W. Gabriel, Bureau of Land Management, USA
K. MacInnes, Department of Indian Affairs, Canada

Viereck, L.A. and D.J. Lev. Long-term use of frost tubes to monitor the annual freeze-thaw cycle in the active layer.

Walker, D.A. A hierarchical tundra vegetation classification especially designed for mapping in northern Alaska.

Schell, D.M. and P.J. Zieman. Accumulation of peat carbon in the Alaska Arctic Coastal Plain and its role in biological productivity.

Koizumi, T. Alpine plant community complex in permafrost areas of the Daisetsu Mountains, central Hokkaido, Japan.

Murray, D.F., B.M. Murray, B.A. Yurtsev and R. Howenstein. Biogeographic significance of steppe vegetation in subarctic Alaska.

Van Cleve, K. and L.A. Viereck. A comparison of successional sequences following fire on permafrost-dominated and permafrost-free sites in interior Alaska.

2:00 p.m.

CLOSING PLENARY SESSION

Poster Sessions

Tuesday, July 19, 1983

THERMAL ENGINEERING DESIGN AND THERMAL ANALYSIS

Blanchard, D. and M. Frémond. Computing the cryogenic suction in soils.

Burgess, M., G. Lemaire and A. Dupas. Soil freezing around a buried pipeline: Design of an experiment in a controlled environment facility.

Kinney, T.C., B.W. Santana, D.M. Hawkins, E.L. Long and E. Yarmak, Jr. Foundation stabilization of central gas injection facilities, Prudhoe Bay, Alaska *

Phillips, W.F. Applications of the fast Fourier transform to cold regions engineering.*

Wexler, R.L. Diurnal freeze-thaw frequencies in the high latitudes: A climatological guide.*

Yarmak, E., Jr. and E.L. Long. Some considerations regarding the design of two phase liquid/vapor convection type passive refrigeration systems.

SITE AND TERRAIN ANALYSIS AND PIPELINES

Brockett, B. and D. Lawson. Shallow drilling in permafrost, northern Alaska.

Gill, J.D. and M.P. Bruen. Permafrost conditions at the Watana Dam site.*

Vita, C.L. Thaw plug stability and thaw settlement evaluation for arctic transportation routes: A probabilistic approach.*

FROST MOUNDS, GROUND ICE, AND PATTERNED GROUND

Bailey, P.K. Periglacial geomorphology in the Kokrine-Hodzana Highlands of Alaska.*

Bowling, S.A. Paleoclimate inferences from permafrost features: A meteorological point of view.

Brown, J., F. Nelson, B. Brockett, S.I. Outcalt and K.R. Everett. Observations of ice-cored mounds at Sukakpak Mountain, south central Brooks Range, Alaska.*

Burn, C.R. and M.W. Smith. Thermokarst development: Some studies from the Yukon Territory, Canada.

Dionne, J.C. Frost-heaved bedrock features: A valuable permafrost indicator.

Douglas, G.R., J.P. McGreevy and W.B. Whalley. Rock weathering by frost shattering processes.*

Ellis, J.M., P.E. Calkin and M.J. Bruen. Rock glaciers of the east central Brooks Range, Alaska.

Gregory, E.C. and C.W. Stubbs. A high-resolution technique for measuring motion within the active layer.*

Harris, C. Vesicles in thin sections of periglacial soils from north and south Norway.*

Hughes, T.J. Downslope creep of unstable frozen ground.

Kinosita, S. Analysis of boring cores taken from the uppermost layer in a tundra area.

Mitter, P. Frost features in the karst regions of the west Carpathian Mountains.*

Kreig, R.A. and A.R. Zieman. Ground ice exposed in the Barrow Utilidor.

Nelson, G.E. Cryopediments in the Bighorn Canyon area, southcentral Montana.

Priesnitz, K. and E. Schunke. Pedimentation and fluvial dissection in permafrost regions with special reference in northwestern Canada.*

Souchez, R.A. and R.D. Lorrian. σ_D and $\sigma^{18}O$ composition of successively formed ice layers: Implications in permafrost studies.

Taylor, R.B. Thaw processes in coarse sediment beaches, Somerset and Bylot Islands, N.W.T.

Whalley, W.B. Rock glaciers: Permafrost features or glacial relics?*

PERMAFROST GEOPHYSICS

Kawasaki, K. and T.E. Osterkamp. Application of electromagnetic induction measurements to permafrost terrain.

Kiefte, H., M.J. Clouter and B.L. Whiffen. Acoustic velocities in structure I and II hydrates by Brillouin spectroscopy.

King, M.S. The influence of clay sized particles on seismic velocity for Canadian Arctic permafrost.

Parameswaran, V.R. and J.R. Mackay. Field measurements of electrical freezing potentials in permafrost areas.*

Walker, G.G., K. Kawasaki and T.E. Osterkamp. Use of D.C. resistivity measurement for determining the thickness of thin permafrost.

Wednesday, July 20, 1983

FROST HEAVE

Coulter, D.M. Predicting heave and settlement in discontinuous permafrost.

Guymon, G.L., R.L. Berg and T.V. Hromadka II. Field tests of a frost heave model.*

Jones, R.H. and K.J. Lomas. The frost susceptibility of granular materials.*

Kettle, R.J. and E.Y. McCabe. Mechanical stabilization and frost susceptibility.

Mageau, D.W. and M.B. Sherman. Frost cell design and operation.*

McCabe, E.Y. and R.J. Kettle. The influence of surcharge loads on frost susceptibility.*

Wood, J.A. and P.J. Williams. Stresses and moisture movements induced by thermal gradients in frozen soils.

* Paper published in first Proceedings volume.

MECHANICS OF FROZEN SOIL, MUNICIPAL AND EMBANKMENTS

Jones, S.J. and V.R. Parameswaran. Deformation behaviour of frozen and sand-ice materials under triaxial compression.*

Keyser, J.H. and M.A. Laforte. Road construction through Palsa fields.

Knutsson, S. Thaw penetration and thaw settlement studies associated with the Kiruna to Narvik road, Sweden.

Mahoney, J.P. and T.S. Vinson. A mechanistic approach to pavement design in cold regions.*

Man, Chi-Sing. Ultimate long-term strength of frozen soil as the phase boundary of a viscoelastic solid-fluid transition.*

Phukan, A. Long-term creep deformation of roadway embankment on ice-rich permafrost.*

Retamal, E. Some experimental results on snow compaction.*

Retherford, R.W. Power lines in the Arctic and Subarctic: Experience in Alaska.*

Schindler, J.F. Solid waste disposal, National Petroleum Reserve in Alaska.*

PERMAFROST MAPS, PERMAFROST IN CHINA, PLEISTOCENE PERMAFROST

Bell, J.W. and T.L. Péwé. Mapping of permafrost in the Fairbanks area, Alaska, for urban-planning purposes.

Christensen, L. The recognition and interpretation of in situ and remoulded till deposits in western Jylland, Denmark.*

Craig, J.L., T.D. Hamilton and P.V. Sellmann. Paleoenvironmental studies in the CRREL permafrost tunnel.

Granberg, H.B., J.E. Lewis, T.R. Moore, P. Steer, and R.K. Wright. A report on a quarter century of permafrost research at Schefferville.

Heginbottom, J.A. Problems in the cartography of ground ice: A pilot project of northwestern Canada.*

Heyse, I. Phenomena of fossil permafrost in sandy deposits of the last glacial in Belgium.

Karlstrom, E.T. Periglacial features and paleoenvironment in the Flagstaff region, northern Arizona.

Kerschner, H. Late glacial paleotemperatures and paleoprecipitation as derived from permafrost: Glacier relationships in the Tyrolean Alps, Austria.*

Kuhle, M. Permafrost and periglacial indicators on the Tibetan Plateau, from the Himalayan Mountains in the south to the Quilian Shan in the north (28-40°N).

Liedtke, H. Periglacial slopewash and sedimentation in northwestern Germany during the Würm (Weichsel) glaciation.*

Lusch, D.P. Late Wisconsinan patterned ground in the Saginaw lowland of Michigan.

Shi Yafeng and Mi Desheng. A comprehensive map of snow, ice and frozen ground in China (1:4,000,000).*

Strunk, H. Pleistocene diapiric upturnings of lignites and clayey sediments as periglacial phenomena in central Europe.*

Wayne, W.J. Paleoclimatic inferences from fossil permafrost indicators in alpine regions.*

WATERSHED STUDIES AND GROUNDWATER IN PERMAFROST

Buska, J.A. Tanana River monitoring and research program.

Corbin, S.W. and C.S. Benson. The thermal regime of a small Alaskan stream in permafrost terrain.*

Dean, K.G. Stream icing zones in Alaska.

Hilgert, J.W. and C.W. Slaughter. Physical, biological, chemical and hydrologic characteristics of a subarctic stream system.

Runchal, A.K. PORFREEZ: A general purpose ground water flow, heat and mass transport model with freezing, thawing and surface water interaction.

Slaughter, C.W. Snow and ice in Caribou-Poker Creeks research watershed, central Alaska.

Stoner, M.G., F.C. Ugolini and D.J. Marrett. Moisture and temperature changes in the active layer of arctic Alaska.*

van Everdingen, R.O. and H.D. Allen. Ground movements and dendrogeomorphology in a small icing area on the Alaska Highway, Yukon, Canada.*

Thursday, July 21, 1983

THERMODYNAMICS AND TRANSPORT PHENOMENA AND RAILWAYS (THERMAL ASPECTS)

Berg, R.L. and D.C. Esch. Effect of color and texture on the surface temperature of asphalt concrete pavement.*

Gori, F.A. A theoretical model for predicting the effective thermal conductivity of unsaturated frozen soils.*

Gosink, T.A. Gas permeability in and emissions from permafrost.

Nakano, Y., A. Tice, J. Oliphant and T.F. Jenkins. Soil-water diffusivity of unsaturated soils at subzero temperatures.*

Riseborough, D.W., M.W. Smith and D.H. Helliwell. Determination of the thermal properties of frozen soils.*

FOUNDATIONS AND EXCAVATIONS

Mindich, A. Behavior of concrete strip foundations on thawed layer of clayey permafrost.*

Takasugi, S. and A. Phukan. Excavation resistance of frozen soils under vibrating cuttings.*

Keusen, H.R. and W. Haeberli. Site investigation and foundation design aspects of cable car construction in alpine permafrost at the Chli Matterhorn, Wallis, Swiss Alps.*

PERMAFROST AND CLIMATE

Brigham, J.K. and G.H. Miller. Paleotemperature estimates of the Alaskan Arctic Coastal Plain during the last 125,000 years.*

Chatwin, S.C. Holocene temperatures in the Upper Mackenzie Valley, determined by oxygen isotope analysis of peat cellulose.*

Gosink, J.P., T.E. Osterkamp and K. Kawasaki. Evidence for recent permafrost warming in Alaska.

Gray, J.T., J. Pilon and J. Poitevin. Active layer studies in northern Quebec: Towards a predictive model.

Harrison, W.D. and S.A. Bowling. Paleoclimate implications for arctic Alaska of the permafrost at Prudhoe Bay.

Haugen, R.K., S.I. Outcalt and J.C. Harle. Relationships between estimated mean annual air and permafrost temperatures in north-central Alaska.*

Odasz, A.M. Climatic and edaphic influences on tree limit, upper Alatna River drainage, central Brooks Range, Alaska.

Prokopovich, N.P. Paleoperiglacial climate and hydrocompaction in California, USA.

Ryden, B.E. and L. Kostov. Thawing and freezing in tundra soils.

Taylor, A., A. Judge and D. Desrochers. Shoreline recession: Its effect on permafrost and the geothermal regime, Canadian Arctic Archipelago.*

Pilon, J., A. Taylor, A. Judge and J.T. Gray. Paleoclimatic reconstruction from borehole temperatures.

SUBSEA PERMAFROST

Blasco, S.M. and M.J. O'Connor. The nature and distribution of shallow subsea acoustic permafrost on the Canadian Beaufort Continental Shelf.

MacAuley, H.A., J.A. Hunter, S.E. Pullan, J.L. Morack, R.M. Gagne and R.A. Burns. Evidence for seasonal variation of nearshore subseabottom permafrost temperatures in the Canadian Beaufort Sea.

Osterkamp, T.E. and W.D. Harrison. Thermal regime of subsea permafrost.

Rogers, J.C. and J.L. Morack. Permafrost studies along the northern Alaska coast using geophysical techniques.

PLANETARY PERMAFROST AND REMOTE SENSING

Boothroyd, J.C. and B.S. Timson. The Sagavanirktok and adjacent river systems, eastern North Slope Alaska: An analog for the ancient fluvial terrains on Mars.*

Gaydos, L. and R.W. Witmer. Mapping of arctic land cover utilizing Landsat digital data.*

Jöns, H.P. Permafrost-related types of large-scale dissection and deformation of Martian landscape.*

Malin, M.C. and D.B. Eppler. Observations of Martian fretted terrain.*

Mellor, J.C. Use of seasonal window for radar and other image acquisition and arctic lake region management.*

PERMAFROST TERRAIN AND ENVIRONMENTAL PROTECTION

Everett, K.R. An 11-year record of active layer thickness, Arctic Gas Test Facility, Prudhoe Bay, Alaska.

Freedman, B., J. Svoboda, C. Labine, M. Muc, G. Henry, M. Nams, J. Stewart and E. Woodley. Physical and ecological characteristics at Alexandra Fjord, a high arctic oasis on Ellesmere Island, Canada.*

Gartner, B.L. Germination characteristics of arctic plants.*

Greene, D.F. Permafrost, fire and the regeneration of white spruce at treeline near Inuvik, N.W.T., Canada.*

Hall, D.K., P.J. Camillo, R.J. Gurney and J.P. Ormsby. Quantitative analysis of the recovery of burned areas in Alaska using remote sensing.

Hettinger, L.R. Vegetation mapping for habitat and wetland assessment along proposed mine transportation routes, northwestern Alaska.

Knapman, L. Reclamation of fire control lines on permafrost sites in interior Alaska.

MacInnes, K.L. Land use controls and environmental protection on permafrost terrain in N.W.T.

Marion, G.M., P.C. Miller and C.H. Black. Cycling of N-15 through tussock tundra ecosystems.

Newbury, R.W. and G.K. McCullough. Shoreline erosion and re-stabilization in a permafrost-affected impoundment.*

Senyk, J.P. and E.T. Oswald. Ecological relationships within the discontinuous permafrost zone of southern Yukon Territory.*

EXHIBITORS

Alaska Department of Transportation & Public Facilities
Alaska Division of Geological and Geophysical Surveys
American Society of Civil Engineers
Arctic Environmental Information and Data Center, University of Alaska
Arctic Foundations
Associated Pile and Fitting Corporation
Cold Regions Research and Engineering Laboratory
Dames and Moore
ERTEC, Inc.
Geologial Survey of Canada
Geophysical Institute, University of Alaska
Harza Engineering
Institute of Arctic and Alpine Research, University of Colorado
Institute of Water Resources, University of Alaska
Magnetics of Alaska
Mobile Augers and Research Ltd.
National Aeronautics and Space Administration
National Research Council, Canada
R.A. Kreig & Associates, Inc.
3M Company
Topp Engineering
United States Geological Survey
World Data Center A for Glaciology

Appendix C: Committees and Reviewers

U.S. ORGANIZING COMMITTEE

T.L. Péwé* (Chairman), Arizona State University
G.E. Weller* (Vice-Chairman), University of Alaska
A.J. Alter, Technical Council on Cold Regions Engineering, ASCE
J. Barton, University of Alaska
J. Brown*, Cold Regions Research and Engineering Laboratory
O.J. Ferrians, Jr., U.S. Geological Survey
H.O. Jahns, Exxon Production Research Company
J.R. Kiely, Bechtel Corporation
A.H. Lachenbruch, U.S. Geological Survey
R.D. Miller, Cornell University
R.D. Reger, Alaska Division of Geological and Geophysical Surveys
A.L. Washburn, University of Washington
J.H. Zumberge, University of Southern California
L. DeGoes, FICOP Secretary

* Member Executive Committee

PAPER REVIEW AND PUBLICATIONS COMMITTEE

R.D. Miller (Chairman), Cornell University
D.M. Anderson, State University of New York, Buffalo
R.F. Black, University of Connecticut
J.M. Boyce, NASA
R.F. Carlson, University of Alaska
H.M. French, University of Ottawa
J. Dennis, U.S. Park Service
L.E. Goodrich, National Research Council of Canada
W.D. Harrison, University of Alaska
H.O. Jahns, Exxon Production Research Company
C.W. Lovell, Purdue University (ASCE)
V. Lunardini, Cold Regions Research and Engineering Laboratory (ASME)
L.A. Viereck, U.S. Forest Service, Alaska
T.S. Vinson, Oregon State University
J.R. Williams, U.S. Geological Survey, Menlo Park
L. DeGoes, Secretary

TECHNICAL PROGRAM COMMITTEE

W.D. Harrison (Chairman), University of Alaska
F.L. Bennett, University of Alaska
J. Brown, Cold Regions Research and Engineering Laboratory
O.J. Ferrians, Jr., U.S. Geological Survey
D.L. Kane, University of Alaska
U. Luscher, Woodward-Clyde Consultants
R.D. Miller, Cornell University
T.E. Osterkamp, University of Alaska
A. Phukan, University of Alaska, Anchorage

FIELD TRIPS COMMITTEE

O.J. Ferrians, Jr. (Chairman), U.S. Geological Survey
A.J. Alter, Technical Council on Cold Regions Engineering (ASCE)
F.L. Bennett, University of Alaska
J. Brown, Cold Regions Research and Engineering Laboratory
H.M. French, University of Ottawa
G.E. Johns, City of Juneau
T.L. Péwé, Arizona State University
S.E. Rawlinson, Alaska Division of Geological and Geophysical Surveys
R.D. Reger, Alaska Division of Geological and Geophysical Surveys
L. Sweet, Alaska Department of Transportation and Public Facilities
H.J. Walker, Louisiana State University
F.W. Weeks, The Alaska Railroad, Federal Railroad Administration

FINANCE AND ADVISORY COMMITTEE

J. Barton, University of Alaska
D.R. Brown, University of Alaska
J.R. Kiely, Bechtel Corporation
T.L. Péwé, Arizona State University
G.E. Weller, University of Alaska

LOCAL ARRANGEMENTS COMMITTEE - FAIRBANKS

W.M. Sackinger (Chairman), University of Alaska
G.E. Weller, University of Alaska
R.D. Abbott, Shannon and Wilson
L. Bennett, University of Alaska
D.R. Brown, University of Alaska Foundation
K. Coffer, University of Alaska
J. Cronin, Shannon and Wilson
D.C. Esch, Alaska Department of Transportation and Public Facilities
W.D. Harrison, University of Alaska
J. Hickey, University of Alaska
K. Kawasaki, University of Alaska
J. Miller, University of Alaska

D.M. Murray, University of Alaska
R.D. Reger, Alaska Division of Geological and Geophysical Surveys
P. Sackinger, Fairbanks, Alaska
L.H. Shapiro, University of Alaska
C.W. Slaughter, U.S. Forest Service
L. Sweet, Alaska Department of Transportation and Public Facilities
L.A. Viereck, U.S. Forest Service

LOCAL ARRANGEMENTS - ANCHORAGE

W.C. Hanson (Chairman), Dames & Moore
M.C. Brewer, U.S. Geological Survey
G. Dickason, University of Alaska
R. Dragnich, Exxon
F. Fisher, Alyeska Pipeline Service Company
P.S. Ford, Harza Engineering Company
J.E. Hemming, Dames & Moore
D. Hickok, University of Alaska
L.E. Johnson, Arctic Slope Consulting Engineers
K. Krause, Alaska Division of Geological and Geophysical Surveys
J. LaBelle, University of Alaska
R.I. Lewellen, Lewellen Arctic Research
W. Parker, Parker Associates, Inc.
L.D. Perrigo, Battelle Alaska Operations
J.F. Schindler, Husky Oil NPRA Operations
C. Sloan, U.S. Geological Survey
J.B. States, Battelle Alaska Operations
T. Swearingen, Chugach Engineers
L.S. Underwood, University of Alaska
World Affairs Council

REVIEWERS OF CONTRIBUTED PAPERS

Civil Engineering
B.D. Alkire, Michigan Technological University, Houghton, Michigan
J.D. Allen, ARCO Oil & Gas Company, Dallas, Texas
A.J. Alter, American Society of Civil Engineers, Juneau, Alaska
O.B. Andersland, Michigan State University, East Lansing, Michigan
T.H.W. Baker, National Research Council of Canada, Ottawa, Canada
R.L. Berg, CRREL, Hanover, New Hampshire
D.E. Bruggers, Harding-Lawson, Anchorage, Alaska
A.L. Burwash, A.L. Burwash Consulting Limited, Calgary, Canada
E.J. Chamberlain, CRREL, Hanover, New Hampshire
R.W. Christensen, Harding-Lawson, Anchorage, Alaska
L.K. Clark, Salem, Oregon
B. Conner, ADOTPF, Fairbanks, Alaska
F.E. Crory, CRREL, Hanover, New Hampshire
R.L. Crosby, Bishop, California
D.C. Esch, ADOTPF, Fairbanks, Alaska
R.E. Fadum, North Carolina State University, Raleigh, North Carolina
K. Flaate, Norwegian Public Roads Administration, Molde, Norway
W.R. Galatiuk, Calgary, Canada
G.L. Guymon, University of California, Irvine, California
W.M. Haas, Michigan Technological University, Houghton, Michigan
B. Hallet, University of Washington, Seattle, Washington
H.W. Hunt, Associated Pile & Fitting Corporation, Clifton, New Jersey
L.E. Janson, VBB AB, Stockholm, Sweden
P.M. Jarrett, Royal Military College, Kingston, Canada
H.L. Jessberger, Ruhr-University, Bochum, W. Germany
N.I. Johansen, University of Alaska, Fairbanks, Alaska
E.G. Johnson, ADOTPF, Anchorage, Alaska
R.J. Kettle, University of Aston, Birmingham, United Kingdom
T.G. Krzewinski, Dames & Moore, Anchorage, Alaska
B. Ladanyi, Ecole Polytechnique, Montreal, Canada
K.A. Linell, Hanover, New Hampshire
U. Luscher, Woodward-Clyde Consultants, Walnut Creek, California
L.J. Mahar, ERTEC Western Inc., Long Beach, California
A. Mahmood, McClelland-EBA, Anchorage, Alaska
J.P. Mahoney, University of Washington, Seattle, Washington
R.W. McGaw, CRREL, Hanover, New Hampshire
D.I. Norum, University of Saskatchewan, Saskatoon, Canada
M. O'Rourke, RPI, Troy, New York
E. Penner, National Research Council of Canada, Ottawa, Canada
A. Phukan, University of Alaska, Anchorage, Alaska
F. Pita, Bellevue, Washington
B.H. Reid, Environmental Protection Agency, Corvallis, Oregon
R.G. Rein, Jr., Bellevue, Washington
W.L. Ryan, Ott Engineering, Anchorage, Alaska
F.J. Sanger, Orange City, Florida
B.W. Santana, ARCO Alaska, Inc., Anchorage, Alaska
F.H. Sayles, CRREL, Hanover, New Hampshire
D.C. Sego, University of Alberta, Edmonton, Canada
R.G. Tart, Woodward-Clyde Consultants, Anchorage, Alaska
H.P. Thomas, Woodward-Clyde Consultants, Anchorage, Alaska
T.S. Vinson, Oregon State University, Corvallis, Oregon
J.J. Walsh, Billings, Montana
P.J. Williams, Carleton University, Ottawa, Canada
A.F. Wuori, CRREL, Hanover, New Hampshire
R.N. Yong, McGill University, Montreal, Canada
W.L. Zirjacks, Frank Moolin & Associates, Inc., Anchorage, Alaska

Mechanical Engineering (Heat Transfer)
G. Abele, CRREL, Hanover, New Hampshire
D.M. Anderson, State University of New York, Buffalo, New York
O.A. Ayorinde, Exxon Production Research Company, Houston, Texas
H. Baker, National Research Council of Canada, Ottawa, Canada
R.L. Berg, CRREL, Hanover, New Hampshire
M. Burgess, Energy, Mines and Resources, Ottawa, Canada
L.E. Carlson, Foothills Pipeline Limited, Calgary, Canada
E.J. Chamberlain, CRREL, Hanover, New Hampshire

E.J. Couch, Mobil Research & Development Corporation, Dallas, Texas
D.M. Coulter, Royal Military College, Kingston, Canada
F.E. Crory, CRREL, Hanover, New Hampshire
J.L. Davis, Earthtech Research Limited, Baltimore, Maryland
B. Dempsey, University of Illinois, Urbana, Illinois
L. Domaschuck, University of Manitoba, Winnipeg, Canada
S. Dufour, Geocon, Inc., Calgary, Canada
J. Ellwood, Foothills Pipeline Limited, Calgary, Canada
D. Elrick, Department of Land Resources Science, Univ. of Guelph, Canada
T. Ersoy, Foothills Pipeline Limited, Calgary, Canada
D.C. Esch, ADOTPF, Fairbanks, Alaska
O. Farouki, Queen's University, Belfast, Northern Ireland
M.G. Ferrick, CRREL, Hanover, New Hampshire
J.W. Galate, Enertech Engineering and Research, Houston, Texas
P. Gaskin, Queens University, Kingston, Canada
L.E. Gold, National Research Council of Canada, Ottawa, Canada
L. Goodrich, National Research Council of Canada, Ottawa, Canada
P.H. Groenevelt, University of Guelph, Guelph, Canada
G.L. Guymon, University of California, Irvine, California
D. Harland, Dames & Moore, Golden, Colorado
R.K. Haugen, CRREL, Hanover, New Hampshire
F.D. Haynes, CRREL, Hanover, New Hampshire
L. Hayes, University of Texas, Austin, Texas
H.N. Haynoe, Land Resource Research Inst., Central Exp. Farm, Agriculture Canada, Ottawa, Canada
J.A. Heginbottom, Geological Survey of Canada, Ottawa, Canada
C.E. Heuer, Exxon Production Research Company, Houston, Texas
S.W. Hopke, Exxon Research and Engineering, Florham Park, New York
C.T. Hwang, EBA Engineering Consultants Limited, Edmonton, Canada
Y.W. Jame, Agriculture Canada Research Station, Swift Current, Canada
T.C. Johnson, CRREL, Hanover, New Hampshire
G.H. Johnston, National Research Council of Canada, Ottawa, Canada
A. Judge, Energy, Mines and Resources, Ottawa, Canada
B.D. Kay, University of Guelph, Guelph, Canada
M. King, University of California, Berkeley, California
J.M. Konrad, National Research Council of Canada, Ottawa, Canada
J.F. Lea, Amoco Production Company, Tulsa, Oklahoma
Y. Lee, University of Ottawa, Ottawa, Canada
S. Lin, Sir George Williams Campus, Montreal, Canada
E.F. Lobacz, Hanover, New Hampshire
E. Long, Arctic Foundations, Inc., Anchorage, Alaska
V.J. Lunardini, CRREL, Hanover, New Hampshire
L.J. Mahar, Earth Technology Corporation, Long Beach, California
D.W. Margeu, R&M Consultants, Inc., Fairbanks, Alaska
R.W. McGaw, CRREL, Hanover, New Hampshire
E.C. McRoberts, Hardy Associates Limited, Calgary, Canada
R.D. Miller, Cornell University, Ithaca, New York
N.R. Morgenstern, Edmonton, Canada
Y. Nakano, CRREL, Hanover, New Hampshire
J.F. Nixon, Hardy Associates Limited, Calgary, Canada
D.I. Norum, University of Saskatchewan, Saskatoon, Canada
J.L. Oliphant, CRREL, Hanover, New Hampshire
K. O'Neill, CRREL, Hanover, New Hampshire
P.H. Oosthuizen, Queens University, Kingston, Canada
T.E. Osterkamp, University of Alaska, Fairbanks, Alaska
E. Penner, National Research Council of Canada, Ottawa, Canada
G. Phetteplace, CRREL, Hanover, New Hampshire
A. Phukan, University of Alaska, Anchorage, Alaska
W.F. Quinn, CRREL, Hanover, New Hampshire
R.L. Reid, University of Texas, El Paso, Texas
W.R. Rouse, McMaster University, Hamilton, Canada
D. Shields, University of Manitoba, Winnipeg, Canada
W. Slusarchuk, R.M. Hardy & Associates Limited, Calgary, Canada
M.W. Smith, Carleton University, Ottawa, Canada
J.E. Sunderland, University of Massachusetts, Amherst, Massachusetts
O. Svec, National Research Council of Canada, Ottawa, Canada
S. Takagi, CRREL, Hanover, New Hampshire
A.R. Tice, CRREL, Hanover, New Hampshire
G.C. Topp, Land Resource Research Inst., Central Exp. Farm, Agriculture Canada, Ottawa, Canada
P.J. Williams, Carleton University, Ottawa, Canada
Y.C. Yen, CRREL, Hanover, New Hampshire
R. Yong, McGill University, Montreal, Canada
J.P. Zarling, University of Alaska, Fairbanks, Alaska
Zhu Yuanlin, CRREL, Hanover, New Hampshire and Institute of Glaciology and Cryopedology, PRC

Geophysical, Geothermal, and Subsea

S.A. Arcone, CRREL, Hanover, New Hampshire
S.A. Bowling, University of Alaska, Fairbanks, Alaska
G.W. Brass, University of Miami, Miami, Florida
R.F. Carlson, University of Alaska, Fairbanks, Alaska
A.J. Delaney, CRREL, Hanover, New Hampshire
C. Economides, University of Alaska, Fairbanks, Alaska
W.D. Harrison, University of Alaska, Fairbanks, Alaska
R.K. Haugen, CRREL, Hanover, New Hampshire
J.D. Hays, Lamont Geological Observatory, Palisades, New York
C.E. Heuer, Exxon Production Research, Houston, Texas
P. Hoekstra, Geo-Physi-Con, Inc., Lakewood, Colorado
D.M. Hopkins, U.S. Geological Survey, Menlo Park, California
W.A. Hoyer, Exxon Production Research, Houston, Texas

J.A. Hunter, Geological Survey of Canada, Ottawa, Canada
A.S. Judge, Energy, Mines and Resources, Ottawa, Canada
G.V. Keller, Colorado School of Mines, Golden, Colorado
A.H. Lachenbruch, U.S. Geological Survey, Menlo Park, California
R.I. Lewellen, Palmer, Alaska
A. Long, University of Arizona, Tucson, Arizona
U. Luscher, Woodward-Clyde Consultants, San Francisco, California
M.N. Nabighian, Newmont Overseas Explorer Limited, Tucson, Arizona
G. Neave, Northern Seismic Analysis, Echo Bay, Canada
G. Olhoeft, U.S. Geological Survey, Denver, Colorado
T.E. Osterkamp, University of Alaska, Fairbanks, Alaska
S.I. Outcalt, University of Michigan, Ann Arbor, Michigan
J.C. Rogers, Michigan Technological University, Houghton, Michigan
W.L. Sackinger, University of Alaska, Fairbanks, Alaska
P.V. Sellmann, CRREL, Hanover, New Hampshire
C.W. Slaughter, USDA Forest Service, Fairbanks, Alaska
M.W. Smith, Carleton University, Ottawa, Canada
A.E. Taylor, Energy, Mines and Resources, Ottawa, Canada
C.L. Vita, R&M Consultants, Inc., Irvine, California

Planetary and Remote Sensing

D.M. Anderson, State University of New York, Buffalo, New York
R. Arvidson, Washington University, St. Louis, Missouri
V. Baker, University of Texas, Austin, Texas
D. Birnie, Dartmouth College, Hanover, New Hampshire
J. Boyce, NASA, Washington, D.C.
M. Darly, Mobil Research & Development, Dallas, Texas
K. Dean, University of Alaska, Fairbanks, Alaska
F. Fanale, University of Hawaii, Honolulu, Hawaii
E. Gibson, Johnson Space Center, Houston, Texas
M. Grolier, U.S. Geological Survey, Flagstaff, Arizona
D. Hall, Goddard Space Flight Center, Greenbelt, Maryland
E. Kasischbe, ERIM, Ann Arbor, Michigan
P. Kiels, Bureau of Land Management, Anchorage, Alaska
K. Kitcho, Woodward-Clyde Consultants, San Francisco, California
H. Lang, Jet Propulsion Laboratory, Pasadena, California
B. Lucchitta, U.S. Geological Survey, Flagstaff, Arizona
M.C. Malin, Arizona State University, Tempe, Arizona
P. Mouginis-Mark, University of Hawaii, Honolulu, Hawaii
D. Nummedal, Louisiana State University, Baton Rouge, Louisiana
P. Patton, Wesleyan University, Middletown, Connecticut
A. Peterfraund, Brown University, Providence, Rhode Island
T.L. Péwé, Arizona State University, Tempe, Arizona
J. Pollack, Ames Research Center, Moffett Field, California
L.A. Rossbacker, Whittier College, Whittier, California
S. Saunders, Jet Propulsion Laboratory, Pasadena, California
T. Schmugge, Goddard Space Flight Center, Greenbelt, Maryland
J. Underwood, Kansas State University, Manhattan, Kansas
G.E. Weller, Geophysical Institute, University of Alaska, Fairbanks, Alaska

Periglacial and Permafrost

D.M. Barnett, Indian and Northern Affairs, Ottawa, Canada
R.F. Black, University of Connecticut, Storrs, Connecticut
N. Caine, University of Colorado, Boulder, Colorado
D.L. Carter, U.S. Geological Survey, Anchorage, Alaska
E.J. Chamberlain, CRREL, Hanover, New Hampshire
M.G. Clark, University of Tennessee, Knoxville, Tennessee
A. Dyke, Geological Survey of Canada, Ottawa, Canada
K.R. Everett, The Ohio State University, Columbus, Ohio
B. Fahey, University of Guelph, Guelph, Canada
O.J. Ferrians, Jr., U.S. Geological Survey, Menlo Park, California
D.C. Ford, McMaster University, Hamilton, Canada
W. Gabriel, Bureau of Land Management, Anchorage, Alaska
L. Goodrich, National Research Council of Canada, Ottawa, Canada
T.D. Hamilton, U.S. Geological Survey, Anchorage, Alaska
S.A. Harris, University of Calgary, Calgary, Canada
D.G. Harry, University of Western Ontario, London, Canada
J.A. Heginbottom, Geological Survey of Canada, Ottawa, Canada
D.A. Hodgson, Geological Survey of Canada, Ottawa, Canada
J. Hunter, Geological Survey of Canada, Ottawa, Canada
A.D. Hyers, University of North Carolina, Wilmington, North Carolina
P.G. Johnson, University of Ottawa, Ottawa, Canada
R.H. King, University of Western Ontario, Canada
R.M. Koerner, Polar Continental Shelf Project, Ottawa, Canada
B. Ladanyi, Ecole Polytechnique, Montreal, Canada
D.E. Lawson, CRREL, Hanover, New Hampshire
A.G. Lewkowicz, University of Toronto, Mississauga, Canada
B. Luckman, University of Western Ontario, London, Canada
V.J. Lunardini, CRREL, Hanover, New Hampshire

F. Michel, Carleton University, Ottawa, Canada
G. Miller, University of Colorado, Boulder, Colorado
F. Nelson, University of Michigan, Ann Arbor, Michigan
W. Nickling, University of Guelph, Guelph, Canada
S.I. Outcalt, University of Michigan, Ann Arbor, Michigan
T.L. Péwé, Arizona State University, Tempe, Arizona
W.H. Pollard, University of Ottawa, Canada
L. Price, Portland State University, Portland, Oregon
V.N. Rampton, Terrain Analysis & Mapping Services Limited, Carp, Canada
S. Rawlinson, ADGGS, Fairbanks, Alaska
R.D. Reger, ADGGS, Fairbanks, Alaska
D. Sharpe, Geological Survey of Canada, Ottawa, Canada
C. Tarnocai, Land Resource Research Institute, Ottawa, Canada
C.E. Thorn, University of Illinois, Urbana, Illinois
F. Ugolini, University of Washington, Seattle, Washington
J.S. Vincent, Geological Survey of Canada, Ottawa, Canada
J.D. Vitek, Oklahoma State University, Stillwater, Oklahoma
A.L. Washburn, University of Washington, Seattle, Washington
S. Watts, Geological Survey of Canada, Ottawa, Canada
W.J. Wayne, University of Nebraska, Lincoln, Nebraska
S. White, The Ohio State University, Columbus, Ohio
S. Zoltai, Environment Canada, Edmonton, Canada

Hydrology

I. Barnes, U.S. Geological Survey, Menlo Park, California
S. Bredthauer, R&M Consultants, Anchorage, Alaska
K.L. Carey, CRREL, Hanover, New Hampshire
R.F. Carlson, University of Alaska, Fairbanks, Alaska
E.F. Chacho, CRREL, Fairbanks, Alaska
T.B. Coplen, U.S. Geological Survey, Reston, Virginia
J. Downey, U.S. Geological Survey, Fairbanks, Alaska
B.T. Drage, Nottingham, Peratrovich and Drage, Anchorage, Alaska
P. Emery, U.S. Geological Survey, Anchorage, Alaska
R.P. Emmanuel, U.S. Geological Survey, Anchorage, Alaska
G.L. Guymon, University of California, Irvine, California
S. Jones, U.S. Geological Survey, Anchorage, Alaska
J. Knott, U.S. Geological Survey, Anchorage, Alaska
R. Lamke, U.S. Geological Survey, Anchorage, Alaska
D.E. Lawson, CRREL, Hanover, New Hampshire
F.A. Michel, Carleton University, Ottawa, Canada
B. Parks, U.S. Geological Survey, Fairbanks, Alaska
C.W. Slaughter, Institute of Northern Forestry, Fairbanks, Alaska
J. Stein, University of Alaska, Fairbanks, Alaska
C.E. Sloan, U.S. Geological Survey, Anchorage, Alaska
D. Trabant, U.S. Geological Survey, Fairbanks, Alaska
J.R. Williams, U.S. Geological Survey, Menlo Park, California
M.K. Woo, McMaster University, Hamilton, Canada
C. Zenone, U.S. Geological Survey, Anchorage, Alaska

Environmental

D. Billings, Duke University, Durham, North Carolina
L.C. Bliss, University of Washington, Seattle, Washington
F.S. Chapin, University of Alaska, Fairbanks, Alaska
R. Densmore, University of Alaska, Fairbanks, Alaska
J.V. Drew, University of Alaska, Fairbanks, Alaska
C.T. Dyrness, Institute of Northern Forestry, Fairbanks, Alaska
R.O. van Everdingen, Environment Canada, Calgary, Canada
K.R. Everett, The Ohio State University, Columbus, Ohio
W. Gabriel, Bureau of Land Management, Anchorage, Alaska
L.W. Gatto, CRREL, Hanover, New Hampshire
G. Guymon, University of California, Irvine, California
L.R. Hettinger, Dames and Moore, Golden, Colorado
M.L. Hienselman, St. Paul, Minnesota
A.W. Johnson, San Diego State University, San Diego, California
L.A. Johnson, CRREL, Fairbanks, Alaska
P.L. Johnson, Lamar University, Beaumont, Texas
G. Juday, Institute of Northern Forestry, Fairbanks, Alaska
G.P. Kershaw, University of Alberta, Edmonton, Canada
A.E. Linkins, Virginia Polytechnic Institute, Blacksburg, Virginia
D.K. MacKay, National Hydrology Research Institute, Ottawa, Canada
J. Major, University of California, Davis, California
J.V. Matthews, Geological Survey of Canada, Ottawa, Canada
J.D. McKendrick, Agricultural Research Center, Palmer, Alaska
W. Mitchell, Agricultural Research Center, Palmer, Alaska
D.M. Murray, University of Alaska, Fairbanks, Alaska
S.I. Outcalt, University of Michigan, Ann Arbor, Michigan
W.A. Patterson III, University of Massachusetts, Amherst, Massachusetts
C. Racine, Johnson State College, Johnson, Vermont
W.R. Rouse, McMaster University, Hamilton, Canada
G.R. Shaver, Marine Biological Laboratory, Woods Hole, Massachusetts
C.W. Slaughter, Institute of Northern Forestry, Fairbanks, Alaska
B. Sveinbjornsson, University of Alaska, Anchorage, Alaska

J. Svoboda, University of Toronto, Mississauga, Canada
L.L. Tieszen, Augustana College, Sioux Falls, South Dakota
K. Van Cleve, University of Alaska, Fairbanks, Alaska
D.A. Walker, University of Colorado, Boulder, Colorado
P.J. Webber, University of Colorado, Boulder, Colorado
R. Wein, University of New Brunswick, Fredericton, Canada
G. Weller, University of Alaska, Fairbanks, Alaska
B. Willard, Boulder, Colorado
M. Wilson, University of Calgary, Calgary, Canada
M.K. Woo, McMaster University, Hamilton, Canada
J. Yarie, University of Alaska, Fairbanks, Alaska
S. Young, Center for Northern Studies, Wolcott, Vermont
J. Zasada, Institute of Northern Forestry, Fairbanks, Alaska

Appendix D: Participation

Approximately 900 individuals from 25 countries participated in the conference. Statistics on participation are provided in Tables 1 and 2. Names and addresses of all participants are given in this Appendix with an indication of field trips in which each participated.

Table 1. Registration and National Representation.

Registration		Participation by Country			
Full	627	United States	595	Norway	3
Spouses	80	Canada	122	Netherlands	3
Students	72	People's Republic of China	20	Belgium	3
Daily	72	Federal Republic of Germany	18	France	2
		United Kingdom	16	Czechoslovakia	1
Total	851	Switzerland	8	Argentina	1
		USSR	6	Chile	1
		Finland	5	Poland	1
		Japan	5	Iceland	1
(does not include press,		Sweden	5	Australia	1
dignitaries and support		Denmark	5	Austria	1
staffs)		Italy	4	South Africa	1

Table 2. North American Participation.

Federal Agencies

	Alaska	Non-Alaska	Canada	Total
U.S. Departments				
Interior				
BLM	22	0		
USGS	7	9		
MMS	2	0		
NPS	2	0		
BIA, FWS, other	2	1		
Agriculture				
Forest Service	4	0		
SCS	7	0		
Transportation				
FHWA	2	1		
Alaska Railroad	2	0		
Defense				
CRREL	5	16		
Other Army	4	3		
Air Force	1	1		
Energy (incl. National labs)	0	5		
Office of Federal Inspector	3	2		
NASA	0	2		
NSF	0	1		
U.S. Total	63	41		104
Canadian Federal Agencies			23	23

Table 2 (cont'd). North American Participation.

	Alaska	Non-Alaska	Canada	Total
State, Local and Provincial Governments				
State of Alaska				
Transportation	43			
Environmental Conservation	5			
Natural Resources - DGGS	7			
OCS	1			
State Government	6			
Local Government	5			67
Canadian Provincial Government			2	2
Universities				
United States	51	62		113
Canada			37	37
Students	21	22	15	58
Private and Others				
Consulting and Industry	116	76	25	217
Spouses	22	35	12	69
Others	8	11	8	27

Registrants

Rohn D. Abbot
Shannon & Wilson, Inc.
P.O. Box 843
Fairbanks, AK 99707

Jaime Aguirre-Puente (A-2)
17 Avenue d'Alembert
g2160 Antony
FRANCE

Kristina Ahlnaes
Geophysical Institute
University of Alaska
Fairbanks, AK 99701

Satoshi Akagawa (B-6)
4-17 Etchujima, 3-Chome
Koto-Ku, Tokyo 135
JAPAN

Jonas Akerman (B-3)
Dept.of Physical Geography
Solvevgatan 13 S-223 62
Lund,
SWEDEN

Tom Albert
North Slope Borough
Box 69
Barrow, AK 99723

James W. Aldrich
Hydro-Con
P.O. Box 60007
Fairbanks, AK 99707

Vera Alexander
Institute of Marine Science
University of Alaska
Fairbanks, AK 99701

Muhammad Amzad Ali
Gulf Oil Corporation
P.O. Box 36506
11000 S. Wilcrest Dr.
Houston, TX 77236

Bernard D. Alkire
Civil Engineering Dept.
Michigan Tech. Univ.
Houghton, MI 49931

Harriet D. Allen
Dept. of Geography
Downing Place
Cambridge, CB2 3EN
ENGLAND

Del Allison
100 Dunbar Ave
Fairbanks, AK 99701

Amos J. and Catherine Alter
P.O. Box 304
Juneau, AK 99802

Jalmar V. Alto
Alyeska Pipeline Service Co.
1835 S. Bragaw Street
Anchorage, AK 99512

Orlando B. Andersland (A-1)
Dept. of Civil Engineering
Michigan State University
East Lansing, MI 48824

Duwayne M. and June Anderson
FNSM Dean's Office
State University of NY/Buffalo
Buffalo, NY 14260

Martha Andrews
Inst. of Arctic Alpine Research
University of Colorado
Campus Box 450
Boulder, CO 80309

Steven A. Arcone
CRREL
72 Lyme Road
Hanover, NH 03755

William Ashton
Box 81372
College, AK 99708

Andrew Assur
CRREL
72 Lyme Road
Hanover, NH 03755

O.J. Astarita
Alaska Int'l. Construction
P.O. Box 1410
Fairbanks, AK 99707

Carl Baek-Madsen (A-1, B-6)
Danish Geotechnical Inst.
1 Maglebiergvei, DK-2880
Lyngby
DENMARK

Dhari Bahiat
Conoco
2661 Bluestem
Ponca City, OK 74601

Palmer Bailey (A-2)
25 3rd Infantry Road
Ft. Leavenworth, KS 66027

Grant C. Baker
P.O. Box 82448
Fairbanks, AK 99708

Thomas Baker (B-4)
Geophysical Institute
University of Alaska
Fairbanks, AK 99701

Richard and Carolyn Baldwin
ARCO Oil and Gas Co.
8881 Pioneer Drive
Anchorage, AK 99504

James T. Bales (A-1, B-4)
Sergent, Hauskins & Beckwith
8643 E. Buena Terra
Scottsdale, AZ 85253

S. Banerjee
University of Washington
132 E. Moore Hall FX-10
Seattle, WA 98195

Mary Bannister
Quadra Engineering
401 E. Fireweed Ln.
Anchorage, AK 99503

Martin Barnett (A-1)
Dept. of Indian & Northern
Affairs
Ottawa K1A OH4
CANADA

William Barr (A-2, B-3)
Dept. of Geography
University of Saskatchewan
Saskatoon, Saskatchewan S7N OWO
CANADA

Jay Barton
The Barton Group
P.O. Box 36360
Tucson, AZ 85740

Conrad Baumgartner
Dept. Renewable Resources
Gov't of Yukon
Box 2703
Whitehorse, Yukon Territory
CANADA

Peter Bayliss (A-1, B-4)
Geology & Geophysics Dept.
University of Calgary
Calgary, Alberta T2N 1N4
CANADA

Anthony Beaty
Royal Military College
Kingston, Ontario
KFM555
CANADA

Gerard Beaudoin
AMOCO Production Co.
AMOCO Bldg.
Denver, CO 80202

Robert C. Beck (A-1)
Bechtel Civil and Mineral Inc.
P.O. Box 3965
San Francisco, CA 94119

Anne-Christina Bedegrew (B-4)
4850 156th Avenue, N.E.
Apartment 20
Redmond, WA 98052

Bjorn Bederson (B-4)
Fylkeshuset
6400 Molde
NORWAY

Earl H. Beistline
Mining Consultant
P.O. Box 80148
Fairbanks, AK 99708

J. R. Bell
Civil Engr. Dept.
Oregon State University
Corvallis, OR 97331

John W. Bell (B-4)
Nevada Bureau of Mines-Geology
University of Nevada
Reno, NV 89557

John Bendz
Quadra Engineering
401 E. Fireweed Ave.
Anchorage, AK 99503

William G. Benjey
Minerals Management Service
P.O. Box 101159
Anchorage, AK 99510

Lawrence Bennett
University of Alaska
Dept. of Engineering Management
133 Duckering
Fairbanks, AK 99701

Carl S. Benson
Geophysical Institute
University of Alaska
Fairbanks, AK 99701

Richard L. Berg
CRREL
72 Lyme Road
Hanover, NH 03755

Michael B. and Sidsel Bergmann
1221 Shore Drive
Anchorage, AK 99515

John Best
ARCO Alaska
P.O. Box 360
Anchorage, AK 99510

C. Birch
Alaska Dept. of Transportation
and Public Facilities
2301 Peger Road
Fairbanks, AK 99701

Charles Bisbee
Bechtel Petroleum
P.O. Box 3456
San Francisco, CA 94119

Roy and Pauline Bjornson
Petro-Canada
4008 - 5th Ave., S.W.
Calgary, Alberta T3C 0C1
CANADA

Charles H. Black
Systems Ecology Research Group
San Diego State University
San Diego, CA 92182

Robert F. Black
65 Sawmill Brook Lane, Rt. 3
Willimantic, CN 06226

Patrick Black
Cornell University
Box 58
Bradfield Hall
Ithaca, NY 14853

R. David Black
Ott Water Eng., Inc.
4790 Business Park Blvd.
Bldg. D, Suite 1
Anchorage, AK 99503

William Black
Office of Federal Inspector
32 Miner's Trail
Irvine, CA 92714

Steve M. Blasco (A-2, B-6)
Geological Survey Canada
Bedford Institute
Dartmouth, Nova Scotia B2Y 4A2
CANADA

David L. and Rosie Boggs
DP Engineering
6010 Acheson Lane
Anchorage, AK 99504

Charles W. Bolen
P.O. Box 1178
Flagstaff, AZ 86002

James F. Boltz
North Pole Refining
P.O. Box 56399
North Pole, AK 99705

Thomas Bond
Standard Oil Co.
4440 Warrensville Ctr. Rd.
Cleveland, OH 44128

Richard Bonwell
Fairbanks North Star Borough
P.O. Box 1267
Fairbanks, AK 99707

Earl J. Boone
Bureau of Land Management
701 C Street
Anchorage, AK 99513

Jon C. Boothroyd
Dept. of Geology
University of Rhode Island
Kingston, RI 02881

Bryan E. Borjesson
Consulting Engineering
S.R. 40455
Fairbanks, AK 99701

William J. Botha
Hardy Associates Ltd.
211 - 18th St SE
Calgary, Alberta T2E 6J5
CANADA

Richard Bouts
Bureau of Land Management
701 C Street
Anchorage, AK 99513

Peter Bowers
Bureau of Land Management
701 C Street
Anchorage, AK 99513

Sue Ann Bowling
Geophysical Institute
University of Alaska
Fairbanks, AK 99701

Gregg Bradley
Alaska Dept. of Transportation
and Public Facilities
Pouch 6900
Anchorage, AK 99502

Herbert K. Brasseur
Bureau of Land Management
701 C Street
Anchorage, AK 99513

D. Brauer
Alaska Dept. of Transportation
and Public Facilities
P.O. Box 60107
Fairbanks, AK 99701

Phillip F. Brease
Bureau of Land Management
P.O. Box 1150
Fairbanks, AK 99707

Bjorn Bredesen
Ingenior F. Selmer
Box 6035
Oslo 6
NORWAY

Lloyd R. Breslau
CRREL
72 Lyme Rd.
Hanover, NH 03755

Kathy Brewer
University of Alaska
Box 82282
Fairbanks, AK 99708

Max C. Brewer
U.S. Geological Survey
4200 Providence Drive
Anchorage, AK 99504

Julie K. Brigham (B-4)
Inst. of Arctic Alpine Research
Box 450
University of Colorado
Boulder, CO 80309

Charles W. and Bernice Britzius
Twin City Testing
19350 Park Ave.
Deephaven, NM 55391

Michael I. Brock
Alyeska Pipeline Service Co.
1835 S. Bragaw Street
Anchorage, AK 99504

Bruce E. Brockett (A-2, B-2)
CRREL
72 Lyme Rd.
Hanover, NH 03755

Dixie Brown (B-4)
University of Alaska Foundation
590 University Avenue
Fairbanks, AK 99701

Jerry, Celia and Mike Brown (A-2, B-2)
CRREL
72 Lyme Rd.
Hanover, NH 03755

Leslie G. Browne (A-1)
Morrison-Forester
1 Market Plaza
San Francisco, CA 94111

Michael P. Bruen
Harza-Ebasco J.V.
8740 Hartzell Road
Anchorage, AK 99507

Rudolph Buchholz
Ring Street 6
D-7778 Markdorf
WEST GERMANY

John Bugbee
Peratrovich & Nottingham Inc.
P.O. Box 6105
Anchorage, AK 99502

Margaret Burgess
Earth Physics Branch
1 Observatory Cres.
Ottawa, K1A 0Y3
CANADA

Oskar and Ursee Burghardt (A-2, B-4)
Nordrhein-Westfalen
De-Greiffstr 195
D-4150 Krefeld
WEST GERMANY

Christopher Burn
Dept. of Geology
Carleton University
Ottawa K1S 4B6
CANADA

James S. Buska
CRREL
Building 4070
Ft. Wainwright, AK 99703

Charles Butler
Int'l Engineering Co., Inc.
6740 Lunar Drive
Anchorage, AK 99504

Edward D. Butterwick
Foothills Pipe Lines
1600 - 205 5th Avenue, SW
Calgary, Alberta T2P 2V7
CANADA

Nel Caine
Inst. Arctic Alpine Research
Box 450
University of Colorado
Boulder, CO 80309

James B. Caldwell
EXXON Production Research Co.
Box 2189
Houston, TX 77001

Kerry J. Campbell
McClelland Engineers
5450 Ralston St.
Ventura, CA 93003

Axel R. Carlson
SR Box 30173
Fairbanks, AK 99701

Lorne E. Carlson
Foothills Pipe Lines
1600 - 205 5th Avenue SW
Calgary, Alberta T2P 2V7
CANADA

Robert Carlson
Institute of Water Resources
University of Alaska
Fairbanks, AK 99701

Michael H. Carr (A-1)
U.S. Geological Survey
345 Middlefield Road
Menlo Park, CA 94025

David L. and Bettie Carter
U.S. Geological Survey
4200 University Drive
Anchorage, AK 99508

L. R. Case
Alaska Int'l Construction
P.O. Box 1410
Fairbanks, AK 99707

Daniel A. Casey
Alaska Dept. of Transportation
and Public Facilities
Pouch Z
Juneau, AK 99811

Edward F. Chacho
CRREL
Building 4070
Ft. Wainwright, AK 99703

Edwin Chamberlain
CRREL
72 Lyme Road
Hanover, NH 03755

Charles A. and Lynell Champion
Northern Technical Service
750 W. 2nd Ave., Suite 100
Anchorage, AK 99501

Earl Chandler
US Army Corps of Engineers
Pouch 898
Anchorage, AK 99506

Chen Guojing (A-1, B-2)
Road Research Division
Ministry of Communications
Beijing
PEOPLE'S REPUBLIC OF CHINA

Chen Xiaobai (A-1, B-6)
Lanzhou Institute of Glaciology
and Cryopedology
Academia Sinica
Lanzhou
PEOPLE'S REPUBLIC OF CHINA

Cheng Guodong (A-2, B-4)
Lanzhou Institute of Glaciology
and Cryopedology
Academia Sinica
Lanzhou
PEOPLE'S REPUBLIC OF CHINA

Edward J. Christensen
Sohio Petroleum Co.
100 Pine St. M/S 111
San Francisco, CA 94111

Leif Christensen (B-6)
Geologisk Inst.
Aarhus University
8000 Aarhus C
DENMARK

J. M. Clark
Alyeska Pipeline Service Co.
1835 S. Bragaw St.
Anchorage, AK 99512

Mark H. Clark
Soil Conservation Service
P.O. Box F
Palmer, AK 99645

Michael G. Clark (A-2)
Dept. of Geological Science
University of Tennessee
Knoxville, TN 37996

Edwin S. Clarke (B-3)
Clarke Engineering
P.O. Box 60007
Fairbanks, AK 99706

Stanton L. Clarke
Woodward Clyde Consultants
SRA Box 1160 B
Anchorage, AK 99502

Gary and Lisa Clay (A-1, B-4)
SR 2, Box 219
Cave Creek, AZ 85331

Stephen M. Clifford
Dept. of Physics & Astronomy
University of Massachusetts
Amherst, MA 01003

John Cobb
Gulf Canada Resources
P.O. Box 310
Calgary, Alberta T2P 2H7
CANADA

Nicholas Coetzee
Engineering Dept.
University of Alaska
Fairbanks, AK 99701

Timothy S. Collett
Geology-Geophysics Dept.
University of Alaska
Fairbanks, AK 99701

Charles M. Collins
CRREL
Building 4070
Ft. Wainwright, AK 99703

Jim Collins
Marathon Oil Co.
P.O. Box 3128
Houston, TX 77253

John Collins
Alaska Dept. of Environmental
Conservation
437 E Street
Anchorage, AK 99501

Rod Combellick
P.O. Box 82422
Fairbanks, AK 99708

Billy Connor
Alaska Dept. of Transportation
and Public Facilities
2301 Peger Road
Fairbanks, AK 99701

John S. Conover
Century Engineering, Inc.
500 L Street, Suite 200
Anchorage, AK 99501

Kelly S. Copley
Cameron Iron Works
P.O. Box 1212
Houston, TX 77251

M. Yavuz Corapcioglu (A-1)
Dept. of Civil Engineering
University of Delaware
Newark, DE 11971

Samuel W. Corbin
1428 N. Lincoln Street
Spokane, WA 99201

Arturo E. Corte (A-2, B-4)
Ianigla
5500 Mendoza
Casilla De Correo 330
ARGENTINA

Donald M. Coulter (A-2)
Dept. of Mechanical Engineering
Royal Military College
Kingston, Ontario
CANADA

D. Cowee
Alaska Dept. of Transportation
and Public Facilities
P.O. Box 60107
Fairbanks, AK 99701

Robert Crackwell
Bureau of Land Management
701 C Street
Anchorage, AK 99513

Charles and Bettie Craddock
Harza Engineering Co.
201 E. 16th Ave., Apt. 303
Anchorage, AK 99501

John Craig
CRREL
Bldg 4070
Ft. Wainwright, AK 99703

Colin Crampton (A-2, B-6)
Dept.of Geology
Simon Fraser University
Burnaby, B.C. V5A 1S6
CANADA

John E. Cronin
Shannon & Wilson, Inc.
P.O. Box 843
Fairbanks, AK 99707

Kevin M. Cronin
AMOCO Production Co.
1670 Broadway
Denver, CO 80202

Cui Chenghen (A-1, B-2)
Third Survey and Design Inst.
Ministry of Railways
Tianjing
PEOPLE'S REPUBLIC OF CHINA

Cui Zhijiu (A-2, B-4)
Dept. of Geography
Peking University
Beijing
PEOPLE'S REPUBLIC OF CHINA

Donald Currie
Mobile Augers Research. Ltd.
5736 103 A Street
Edmonton, Alberta T6H 2J5
CANADA

Scott Dallimore
Geography Dept.
Carleton University
Ottawa
CANADA

Douglas H. Dasher
Alaska Dept. of Environmental
Conservation
P.O. Box 1601
Fairbanks, AK 99701

Arlene M. Davis
R. A. Kreig & Assoc., Inc.
1503 West 33rd Avenue
Anchorage, AK 99503

Louis DeGoes (A-2)
Consultant
4727 38th Street N
Arlington, VA 22207

Ken Dean
Geophysical Institute
University of Alaska
Fairbanks, AK 99701

Thomas Dean
Bureau of Land Management
P.O. Box 1150
Fairbanks, AK 99707

Larry L. Dearborn
Alaska Division of Geological
and Geophysical Surveys
P.O. Box 77216
Anchorage, AK 99501

Sue Degler
Sohio Alaska Petroleum Co.
Pouch 6-612
Anchorage, AK 99502

Allan J. Delaney
CRREL
72 Lyme Road
Hanover, NH 03755

James Deninger
Bureau of Land Management
701 C Street
Anchorage, AK 99513

Ding Jinghang (A-1, B-2)
Northwest Research Institute
Ministry of Railways
Lanzhou
PEOPLE'S REPUBLIC OF CHINA

Jean-Claude Dionne (B-3)
Dept. of Geography
Université Laval
Quebec, GLK 7P4
CANADA

Leroy D. DiPasquale
Alaska Area Native Health Service
P.O. Box 7-741
Anchorage, AK 99510

Len Domaschuk
Civil Engineering Dept.
University of Manitoba
Winnipeg, Manitoba R3T 2N2
CANADA

Ingrid Donnerstag
Quadra Engineering
401 E. Fireweed Lane
Anchorage, AK 99503

Brent Drage
Peratrovich & Nottingham Inc.
1506 W. 36th Ave.
Anchorage, AK 99503

Francesco Dramis (B-3)
Universita Di Camerino
Viale Beeti 1
Camerino 62032
ITALY

J. P. DuPlessis (A-1, B-6)
Dept. of Applied Math
University of Stellenbosch
7600 Stellenbosch
REPUBLIC OF SOUTH AFRICA

Patrick Duffy
Federal Environ. Assessment Office
Fontaine Building
Ottawa, Ontario K1A 0N3
CANADA

Alan Dufour
Carleton University
Ottawa, K1S 5B6
CANADA

Sylvain Dufour
26-7200 Huntercrest Rd. NW
Calgary, Alberta T2K 5S4
CANADA

R. Dunning
SOHIO
Pouch 6-612
Anchorage, AK 99502

Kenneth Dunton
Institute of Water Resources
University of Alaska
Fairbanks, AK 99701

Kurt S. Dzinich
Senate Advisory Council
Pouch V
Juneau, AK 99811

Jim Ebersole
Inst. of Arctic Alpine Research
Box 450
University of Colorado
Boulder, CO 80309

William F. Ediger
Trans Canada Pipeline
P.O. Box 54
Commerce Court West
Toronto, M5L 1CZ
CANADA

Christopher M. and Sharon Effgen
Alaska Home and Energy Evaluators
4101 Arkansas
Anchorage, AK 99503

Catherine Egan
Inst. of Water Resources
University of Alaska
Fairbanks, AK 99701

Kurt Z. Egelhofer
7 Green Street
Lebanon, NH 03766

Kathleen A. Ehlig (A-2, B-4)
1560 Via Del Rey
S. Pasadena, CA 91030

James C. Eldridge
Bureau of Land Management
P.O. Box 1150
Fairbanks, AK 99707

Earl P. Ellis
Office of Federal Inspector
690 W Lake Ridge Dr.
Eagle River, AK 99577

John R. Ellwood
NOVA, An Alberta Corporation
P.O. Box 2535, Station 'M'
Calgary, Alberta T2P 2N6
CANADA

Morris J. Engelke (A-2)
Bureau of Land Management
701 C. Street
Anchorage, AK 99513

Warren L. and Mary Ann Enyeart
Box 170 Crestview Lane
Eagle River, AK 99577

Esa Eranti (A-2, B-3)
FINN-STROI OY
Melkonkatu 18
00210 Helsinki 21
FINLAND

David C. Esch
Alaska Dept. of Transportation
and Public Facilities
2301 Peger Road
Fairbanks, AK 99701

K. R. and Lynn Everett
Agronomy Department
Ohio State University
Columbus, OH 43210

R. W. Fadum
North Carolina State University
408 Mann Hall
Raleigh, NC 27650

Walter Faha
P.O. Box 1178
Flagstaff, AZ 86002

Fraser Fanale
University of Hawaii
2525 Correa Rd.
Honolulu, HI 96825

Philip Faulkner Jr. & Family
Stearns-Roger
3416 W. Akron St.
Denver, CO 80231

John E. and Lorraine Ferrell
Alyeska Pipeline Service Co.
1835 S. Bragaw Street
Anchorage, AK 99512

Oscar and Muriel Ferrians (A-1)
U.S. Geological Survey
345 Middlefield Road
Menlo Park, CA 94025

Kaare Flaate (B-4)
Fylkeshuset
6400 Molde
NORWAY

W. J. Fleihmann
Bechtel Canada Ltd.
4 Forest Lane Way #2101
Willowdale, Ontario M2N 5X8
CANADA

Louis A. Fletcher
Soil Conservation Service
2221 E. Northern Lts. Blvd.
Anchorage, AK 99504

John P. Flory (B-3)
P.O. Box 82462
College, AK 99708

Wolfgang Fluegel (B-6)
Geographisches Institut
Universtät Heidelberg
Ladenburger Str. 80
6900 Heidelberg
WEST GERMANY

Howard Flug
Stearns-Roger
2075 Tenth St.
Boulder, CO 80302

Michael and Katherine Foley (A-1)
Battelle-Pacific Northwest Lab.
Richland, Washington 99352

Perry G. Francis
Bureau of Land Management
701 C Street
Anchorage, AK 99513

R. T. and Joeann Frankian
R. T. Frankian & Assoc.
P.O. Box 7762
Burbank, CA 91510

Craig Freas
Tryck, Nyman & Hayes
740 I Street
Anchorage, AK 99501

Jill A. Fredston
AEIDC-University of Alaska
707 A Street
Anchorage, AK 99501

Kevin Freemen
Rt. 1, Box 407
Vashon Island, WA 98070

Bruce R. Freitag
Alaska Dept. of Transportation
and Public Facilities
P.O. Box 3-1000
Juneau, AK 99802

Michel Fremond
Lab. Ctr. Des Ponts
Et Chaussées
58 Bld. Lefebvre
75732 Paris, Cedex 15
FRANCE

Hugh French (B-3)
Dept. of Geography
University of Ottawa
Ottawa, Ontario K1N 6N5
CANADA

Rowland French
SOHIO
1 Lincoln Center #1200
5400 LBJ Freeway
Dallas, TX 75240

John Fritz
Alaska Dept. of Transportation
and Public Facilities
Pouch 6900
Anchorage, AK 99502

T. C. Fuglestad (A-1)
Alaska Railroad
Pouch 7-2111
Anchorage, AK 99510

Masami Fukuda
Institute of Low Temp. Science
University of Hokkaido
Sapporo
JAPAN

John G. Fyles
Canadian Geological Survey
601 Booth St.
Ottawa, Ontario K1A 0E8
CANADA

William Gabriel
Bureau of Land Management
701 C Street
Anchorage, AK 99513

W. Lloyd Gallagher
4260 Ridge Drive
Pittsburg, CA 94565

Francesco Gallavresi
7 Moscova
Milan
ITALY

Dave Galloway
EXXON Company, U.S.A.
Pouch 6601
Anchorage, AK 99502

John Galloway
U.S. Geological Survey
345 Middlefield Rd.
Menlo Park, CA 94025

Martin and Barbara Gamper (A-2, B-4)
Geophysical Institute
University Postfach 8033
Zurich
SWITZERLAND

G. V. Ganapathy
Transport Canada
P.O. Box 8550
Winnipeg, Manitoba R3C 0P6
CANADA

Harold Garbeil
School of Mineral Industry
University of Alaska
Fairbanks, AK 99701

Barbara Gartner
11820 Buena Vista Drive
Los Altos Hills, CA 94022

Leonard J. Gaydos (A-2)
U.S. Geological Survey
Ames Research Ctr 240-8
Moffett Field, CA 94035

Bernard J. Gayewski
Alaska Dept. of Environmental
Conservation
437 E Street
Anchorage, AK 99501

Tom George
Geophysical Institute
University of Alaska
Fairbanks, AK 99701

Stephen Gerlek
ARCO Alaska Inc.
P.O Box 360
Anchorage, AK 99510

James Gill
ERTEC Northwest Inc.
600 W 41st #203
Anchorage, AK 99503

Harold Gillam
104-2nd Ave.
Fairbanks, AK 99701

Scott C. Gladden
Rittenhouse-Zemen & Assoc.
13837 NE 8th Street
Bellevue, WA 98005

Roy Glass
U.S. Geological Survey
1209 Orca Street
Anchorage, AK 99501

Kevin J. Gleason (B-3)
University of Colorado
Dept. of Engineering
Box 424
Boulder, CO 80309

Robert Gleck
Institute of Water Resources
University of Alaska
Fairbanks, AK 99701

SanJay Godbole
Petroleum Engineering Dept.
University of Alaska
Fairbanks, AK 99701

Douglas J. Goering
Box 81478
Fairbanks, AK 99708

David Good
100 Bradford Street
Ottawa, Ontario K2B 5Y8
CANADA

Laurel Goodrich (B-6)
Division of Building Research
National Research Council Canada
Ottawa, Ontario K1A 0R6
CANADA

Fabio and Roberta Gori
Finica Technica Dip
Energetica Universita
Via Smarta 3
Firenze
ITALY

Joan Gosink
Geophysical Institute
University of Alaska
Fairbanks, AK 99701

Chris A. Graham
Gulf Canada Resources, Inc.
401 9th Ave., SW
Calgary, Alberta T2P 2H7
CANADA

Mike Grahek
Alaska Dept. of Transportation
and Public Facilities
P.O. Box 60107
Fairbanks, AK 99701

Nikolai Grave (B-2)
Permafrost Institute
USSR Academy of Sciences
677010 Yakutsk - 10
USSR

James Gray (B-6)
Dept. de Géographie
University of Montreal
Montreal, Quebec H3C 3J7
CANADA

Edwin T. and Maureen C. Green
Dept. of Physics/Geology
Southwestern State University
Weatherford, OK 73096

David F. Greene
Geography Dept.
University of Calgary
Calgary, Alberta T2N 1N4
CANADA

Lori Greenstein (A-2, B-3)
University of Colorado
Campus Box 260
Boulder, CO 80309

Odd Gregersen
Norwegian Geotech Institute
Box 40 Tasen
Oslo 8
NORWAY

E. Carrington Gregory
Quaternary Research Center
University of WA AK-60
Seattle, WA 98195

George L. Griffin
2866 Marnar Drive
Springfield, OR 97477

Aleksandr I. Gritsenko (B-6)
All Union Natural Gas Scientific
Research Institute
Minneftegazprom
Bol'shaya Scrpukhovskaya Ulitsa 10
Moscow 113093
USSR

Gerhard Groeschel
Bureau of Land Management
701 C Street
Anchorage, AK 99513

Dora L. Gropp
Robert W. Retherford Assoc.
6728 Diamond Blvd.
Anchorage, AK 99502

Lucy Grove
Dept. of Geology
Carleton University
Ottawa, Ontario K1S 5B6
CANADA

Victor Gruol
Geophysical Institute
University of Alaska
Fairbanks, AK 99701

George Gryc (B-2)
U.S. Geological Survey
345 Middlefield Road
Menlo Park, CA 94025

Robert D. Gunn (B-3)
Dept of Chemical Engineering
University of Wyoming
Laramie, WY 82071

Gary L. Guymon
Civil Engineering
University of California
Irvine, CA 92717

Wilbur M. Haas (A-2)
Civil Engineering Dept.
Michigan Tech. University
Houghton, MI 49931

Wilfried Haeberli
Vaw/Agieth Zürich
ETH-Zentrum
8092 Zürich
SWITZERLAND

Paul Haesaerts (B-3)
IRSNB Rue Vautier 31
B-1040 Bruxelles
BELGIUM

Horst Hagedorn (B-3)
Allerseeweg 33
D8706
Hochberg
WEST GERMANY

Nassrin Haji
Tu Clausthal Geo Ins.
Leibnizstr 10/D 3392
Clausthal-Zellerf
WEST GERMANY

Peter Hale (B-3)
Canada Oil and Gas Land
Administration
355 River Road
Ottawa, K1A 0E4
CANADA

Bernard Hallet
Quaternary Research Center
University of Washington
Seattle, WA 98195

John Hallin
Federal Highway Administration
708 SW 3rd Avenue
Portland, OR 97204

Jim J. Hamilton (A-2)
Northern Pipeline Agency
400-4th Avenue SW
Calgary, Alberta T2P 0J4
CANADA

Richard A. Hamilton
Alaska Department of Transportation
and Public Facilities
B-3-1000
Juneau, AK 99811

H. Hammer
Alaska Dept. of Transportation
and Public Facilities
2301 Peger Road
Fairbanks, AK 99701

Theodore Hammer
Dames & Moore
800 Cordova, Suite 100
Anchorage, AK 99501

Mikhail Hammond (B-6)
Interpreter
200 N Adams, Apt. 611
Arlington, VA 22201

Vince Haneman
Dept. of Engineering
University of Alaska
Fairbanks, AK 99701

Peter Hanley
SOHIO Petroleum
Pouch 6-612
Anchorage, AK 99502

Sam A. Hannas
Stearns-Roger
4175 W. Chenango Ave.
Littleton, CO 80123

Susan Hansen
ARCO Oil and Gas Co.
3201 C Street, Suite 560
Anchorage, AK 99503

Paul Hanson
ARCO Oil and Gas Co.
3201 C Street, Suite 560
Anchorage, AK 99503

Warren W. Hanson
Dept. of Civil Engineering
University of Alaska
Fairbanks, AK 99701

Wayne C. Hanson
Dames & Moore
800 Cordova, Suite 100
Anchorage, AK 99501

William R. Hanson
Elmendorf Air Force Base
2704 Orca Place
Anchorage, AK 99501

Peter Hardcastle
P.O. Box 81308
College, AK 99708

Michael Hare
Carleton University
Geography Dept.
Ottawa, K1S 5B6
CANADA

James C. Harle
Alyeska Pipeline Service Co.
1835 S. Bragaw Street
Anchorage, AK 99512

Charles Harris (B-3)
Dept. of Geography
University College
P.O. 78 Cardiff CF1 1X1
UNITED KINGDOM

Stuart A. Harris (B-3)
University of Calgary
6316 Dalton Drive
Calgary, Alberta T3A 1E4
CANADA

Will Harrison
Geophysical Institute
University of Alaska
Fairbanks, AK 99701

David G. Harry (B-3)
Dept.of Geography
University of Western Ontario
London, Ontario N6A 5C2
CANADA

Judith Hartt
Hartt Enterprises
1561/2 Second Avenue
Fairbanks, AK 99701

Richard K. Haugen
CRREL
72 Lyme Road
Hanover, NH 03755

John Hayes
Mobile Augers Research Ltd.
5736 103 A Street
Edmonton, Alberta T6H 2J5
CANADA

Donald W. Hayley
EBA Engineering Consult. Ltd.
14535 118th Avenue
Edmonton, Alberta T5L 2M7
CANADA

Beez Hazen
Northwest Alaska Pipeline
333 Michalsen
Irvine, CA 92717

David and Shirley Heatwole
Anaconda Minerals Co.
2550 Denali Street, Suite 1000
Anchorage, AK 99503

J. A. Heginbottom (B-3)
Geological Survey of Canada
601 Booth Street
Ottawa K1A 0E8
CANADA

Tommy and M. J. Heinrich
Alaska Dept. of Transportation
and Public Facilities
Pouch 6900
Anchorage, AK 99502

Dot Helm
University of Alaska
P.O. Box 3136
Fairbanks, AK 99707

Kenneth T. Hemenway
Alaska Dept. of Transportation
and Public Facilities
Pouch 6900
Anchorage, AK 99502

James E. Hemming
Dames & Moore
800 Cordova, Suite 100
Anchorage, AK 99501

Ray E. Herman (A-2)
Bureau of Indian Affairs
P.O. Box 3-8000
Juneau, AK 99802

Nils Hetland
SF/Braun
505 S. Main
Orange, CA 92668

Loren R. Hettinger (A-1)
Dames & Moore
1626 Cole Blvd
Golden, CO 80401

Christopher E. Heuer (B-6)
EXXON Production Research
P.O. Box 2189
Houston, TX 77001

I. Heyse (A-1, B-3)
Geophysical Institute
University of Ghent
Krijgslaan 281-S8
B-900 Ghent
BELGIUM

Duncan Hickmutt (B-6)
Alaska Division of Geological
and Geophysical Surveys
794 University Avenue, Basement
Fairbanks, AK 99701

David Hickok
AEIDC
University of Alaska
707 A Street
Anchorage, AK 99501

Eldo Hildebrand
Dept of Civil Engineering
University of Waterloo
Waterloo, Ontario N2L 3G1
CANADA

Jerry W. Hilgert
U.S. Forest Service
Institute of Northern Forestry
University of Alaska
Fairbanks, AK 99701

Jim Hillier
Quadra Engineering
401 E. Fireweed Lane
Anchorage, AK 99503

George D. and Arlie Hobson
Polar Shelf Project
880 Wellington Street, Rm 401
Ottawa, K1A 0E4
CANADA

David Hoch
Peratrovich & Nottingham Inc.
1506 W. 36th Avenue
Anchorage, AK 99503

Grant and Elsie Hocking (B-6)
Applied Mechanics, Inc.
3000 Youngfield Street, #285
Lakewood, CO 80215

Pieter Hoekstra
Geo-Physi-Con Inc.
12860 W. Cedar Drive, Suite 201
Lakewood, CO 80228

Philip Hoffman
Geophysical Institute
University of Alaska
Fairbanks, AK 99701

J.E.F. and Jean Ann Hoggan
Hoggan Engineering & Testing Ltd.
8803 50th Avenue
Edmonton, Alberta T6E 5H4
CANADA

Charlotte Hok
Institute of Water Resources
University of Alaska
Fairbanks, AK 99701

John T. Holden (A-1)
Dept. of Theoretical Mechanics
University of Nottingham
Nottingham
ENGLAND

Donald W. Holland
Alaska Dept. of Transportation
and Public Facilities
Pouch 6900
Anchorage, AK 99502

Henry E. Holt (A-2, B-4)
Planetary Geology, EL-4
NASA Headquarters
Washington, D.C. 20546

David M. Hopkins (B-6)
U.S. Geological Survey
345 Middlefield Road
Mail Stop 90-B
Menlo Park, CA 94015

Theodore and Laura Hromadka
24731 Via Alvorado
Mission Viejo, CA 92691

Scott Huang
School of Mineral Industries
University of Alaska
Fairbanks, AK 99701

Terence and Beverly Hughes (A-2)
Dept. of Geological Sciences
University of Maine at Orono
Orono, ME 04469

Jim Hunter
Geological Survey of Canada
601 Booth Street
Ottawa, Ontario K1A 0E8
CANADA

Ronald G. Huntsinger
Bureau of Land Management
701 C Street
Anchorage, AK 99513

Timothy Hushen
Polar Research Board
2101 Constitution Ave. NW
Washington, D.C. 20545

Albert D. and Peggy Hyers (B-4)
Earth Science Dept.
University of North Carolina
Wilmington, NC 28406

Mark Ingrham
Alaska Division of Geological
and Geophysical Surveys
P.O. Box 1721
Eagle River, AK 99577

Ron Innes
Mobile Augers Research Ltd.
5736 103 A Street
Edmonton, Alberta T6H 2J5
CANADA

Ralph and Barbara Isaacs
Northwest Alaska Pipeline
3333 Michelson Drive
Irvine, CA 92717

Edmonton S. Jacobsen
Alaska Dept. of Environmental
Conservation
437 E Street
Anchorage, AK 99501

George Jacobsen
Tower Arctic Ltd., Suite 920,
1350 Sherbrooke Street West
Montreal, Quebec H3G1J1
CANADA

Alfred Jahn (A-2, B-4)
Department of Geography
University of Wroclaw
Pl. Uniwersytecki 1
50-137 Wroclaw
POLAND

Hans and Suse Jahns (A-1, B-6)
EXXON Production Research Co.
P.O. Box 2189
Houston, TX 77001

Ric Janowicz
Dept. of Indian Affairs
200 Range Rd.
Whitehorse, Yukon Y1A 3V1
CANADA

Lars-Eric Janson
VBB Consulting Group
Box 5038
S-102 41 Stockholm
SWEDEN

John Jardine
Arctic Foundations
5 Chester Street
Winnipeg, Manitoba R2L 1W5
CANADA

Peter and Lesley Jarrett
Dept. of Civil Engineering
Royal Military College
Kingston, Ontario K7L 2W3
CANADA

Larry Jaycox
Anaconda Minerals Co.
555 17th Street, Rm 2146
Denver, CO 80202

John D. Jenkins
U.S. Army Corps of Engineers
P.O. Box 2870
Portland, OR 97208

Hans L. Jessberger
University of Bochum
Universitätsstr. 150
463 Bochum
WEST GERMANY

Ann Jochens
Gulf Oil Corporation
P.O. Box 111683
Anchorage, AK 99511

Nils I. Johansen
School of Mineral Industries
University of Alaska
Fairbanks, AK 99701

Glenn E. Johns (A-2)
City of Juneau
P.O. Box 1393
Juneau, AK 99802

Albert W. Johnson
Vice President Academic Affairs
San Diego State University
San Diego, CA 92182

Danny Johnson
Alaska Dept of Transportation
and Public Facilities
2301 Peger Road
Fairbanks, AK 99701

Elden R. Johnson
Alyeska Pipeline Service Co.
1835 S. Bragaw Street
Anchorage, AK 99512

Eric Johnson
Alaska Dept. of Transportation
and Public Facilities
Pouch 6900
Anchorage, AK 99502

Larry A. Johnson
CRREL
83 A Arctic Health Building
Fairbanks, AK 99701

G. H. and Terry Johnston (A-1)
Division of Building Research
National Research Council Canada
Ottawa, K1A 0R6
CANADA

Ronald H. Jones (A-1)
Dept. of Civil Engineering
University of Nottingham
Nottingham NG72RD
ENGLAND

Stephen J. Jones
N.H.R.I. Inland Water
Environment Canada
Ottawa, K1A 0E7
CANADA

Heinz-Peter Jons (B-4)
Geol. Inst. der T.U.
Leibnizstrasse 10
3392 Clausthal-Zellerfeld
WEST GERMANY

Darryl F. Jordan
ARCO Oil and Gas
700 G. St. (P.O. Box 100939)
Anchorage, AK 99501

Harold Jorgenson
2039 New Hampshire Ave. NW
Washington, D.C.

Noel and C. Journeaux
8803 50 Avenue
Edmonton, Alberta T6E 5H4
CANADA

Glenn Juday
U.S. Forest Service
Institute of Northern Forestry
University of Alaska
Fairbanks, AK 99701

James H. Justice
University of Calgary
ES 104/2500 University Dr. NW
Calgary, Alberta T2N 1N4
CANADA

Douglas Kane
Institute of Water Resources
University of Alaska
Fairbanks, AK 99701

Eric T. Karlstrom (B-4)
Dept. of Geology
Northern Arizona University
Box 15106
Flagstaff, AZ 86011

Johannes Karte (B-3)
Deutsche Forschungsgemeinschaft
Kennedyallee 40
D-300 Bonn, 2,
WEST GERMANY

Gary Kast
3701-19 Street NE
Calgary, Alberta T23 6S8
CANADA

Koji and Virginia Kawasaki
Geophysical Institute
University of Alaska
Fairbanks, AK 99701

B. D. Kay
Dept. of Land Resources
University of Guelph
Guelph, Ontario
CANADA

Geoffrey and Bron Kellaway (A-1)
14 Cranedown
Lewes, Sussex BN7 3NA
UNITED KINGDOM

Robert Kenyon
Dept. of Civil Engineering
University of Manitoba
Winnipeg, Manitoba R3T 2N2
CANADA

Hans Kerschner (B-3)
Institute für Geographie
Universtät Innsbruck
A 6020 Innsbruck
AUSTRIA

G. Peter Kershaw
Dept. of Geography
University of Alberta
Edmonton, Alberta T6G 2H4
CANADA

Roger J. Kettle
Dept. of Civil Engineering
U/Aston Gosta Green
Birmingham B4 7ET
ENGLAND

Harry Kiefte
Physics Department
Memorial University
St. John's, Newfoundland A1B3X7
CANADA

John and Margaret Kiely (B-3)
Bechtel Power Corporation
P.O. Box 3965 (B50112)
San Francisco, CA 94119

Lorenz King (A-2)
Geophysical Institute
University of Heidelberg
Otto Hahn Platz 2
D-6900 Heidelberg
WEST GERMANY

Michael S. King
Earth Sciences Division
Lawrence Berkeley Laboratory
Berkeley, CA 94720

Robert E. King
Bureau of Land Management
P.O. Box 1150
Fairbanks, AK 99707

Thomas C. Kinney
Shannon & Wilson, Inc.
P.O. Box 843
Fairbanks, AK 99707

Seiiti Kinosita
Institute Low Temperature Science
Hokkaido University
Sapporo
JAPAN

Bruce H. Kjartanson
Department of Civil Engineering
University of Manitoba
Winnipeg, Manitoba R3T 2N2
CANADA

Len Knapik
418 Hillside Park
51112 RGE Rd. 222
Sherwood, Park, Alberta T8C 1G9
CANADA

Larry Knapman
Bureau of Land Management
P.O. Box 1150
Fairbanks, AK 99707

Sven Knutsson
University of Lulea
Division of Soil Mechanics
S-95187 Lulea
SWEDEN

Daniel J. Koelzer
Alyeska Pipeline Service Co.
1835 S. Bragaw Street
Anchorage, AK 99512

Byron E. Koepke
Soil Conservation Service
P.O. Box F
Palmer, AK 99645

Takeei Koizumi (A-1, B-4)
Dept. of Geography
Tokyo Gakugei University
Koganei City, Tokyo 184
JAPAN

Else Kolstrup (B-3)
Jemtelandsgade 8, 4th
2300 Kobenhavn S
DENMARK

Vera Komarkova
Inst. Arctic and Alpine Research
Campus Box 450
University of Colorado
Boulder, CO 80309

Jean-Marie Konrad
National Research Council Canada
Ottawa, K1A OR6
CANADA

Eduard Koster
Physical Geogr. Lab
University of Amsterdam
Dapperstraat 115
1093-BS Amsterdam
THE NETHERLANDS

Nat W. and Mrs. Krahl
Brown & Root, Inc.
P.O. Box 03-635
Houston, TX 77001

William B. Krantz (B-3)
Department Chemical Engineering
University of Colorado
Boulder, CO 80309

Raymond A. Kreig (A-2, B-3)
R. A. Kreig Assoc.
1503 West 33rd Avenue
Anchorage, AK 99503

Dionyz Kruger (A-2)
Foothills Pipeline Ltd.
1715 110th Ave., SW
Calgary, Alberta T2W OE4
CANADA

Thomas G. Krzewinski
Dames & Moore
800 Cordova, Suite 100
Anchorage, AK 99501

Matthias Kuhle
Universität Göttingen
Goldschmidtstr 5
D-3400 Göttingen
WEST GERMANY

Vladimir P. Kuleshov (B-6)
All Union Natural Gas Scientific
Research Institute
Minneftegazprom
Bol'shava Scrpukhovskaya Ulitsa 10
Moscow 113093
USSR

William F. Kuper
Crowley All Terrain Corporation
201 Danner Ave., Suite 200
Anchorage, AK 99502

Mark LaForte
Applied Research Center
CRCAC
3800 Cremazie St.
Montreal, H2B 1B8
CANADA

Louanne Labelle
Bedford Institute Oceanography
Box 1006 Dartmouth
Nova Scotia B2Y 4A2
CANADA

Branko and Neva Ladanyi
Civil Engineering
Ecole Polytech.
CP6079 St. A
Montreal. Quebec H3C 3A7
CANADA

J.M.C. Lalonde
Dept. of Geography
University of Toronto
Toronto, Ontario M5S 1A1
CANADA

Hans C. Langager
Greenland Technical Organization
Hauser Plads 20
DK 1127 Copenhagen K
DENMARK

Roger Langohr (B-3)
Earth Sci. Inst.
University of Ghent
Krijgslaan 281
B-9000-Ghent
BELGIUM

H. Gordon and Mary Jo Larew
2500 Hillwood Place
Charlottesville, VA 22901

Guy Larocque (A-2)
Société d'Energie
800 E De Maisonneuve
Montreal, H2L 4M8
CANADA

Wayne F. Larson
Northern Latitude Assoc.
S. R. Box 10326
Fairbanks, AK 99701

Steve Laut
Dome Petroleum
224 McQueen Glen Pl.
Calgary, Alberta T2J 4E2
CANADA

Craig Lavielle
Rittenhouse-Zeman & Assoc.
13837 NE 8th Street
Bellevue, WA 98005

Daniel E. Lawson (B-4)
CRREL
72 Lyme Road
Hanover, NH 03755

L.C.M.F Limited
723 W. 6th Avenue
Anchorage, AK 99501

Rainer Lehmann
Bismarckstr 22A
D-6831 Plankstadt
WEST GERMANY

Gilles Lemaire
Geotechnical Dept.
Ecole Polytecnique
Montreal, Quebec H3C 3A7
CANADA

Debbie Lev
University of Washington
6812 21st NE
Seattle, WA 98195

D. Levine
Alaska Dept. of Transportation
and Public Facilities
2301 Peger Road
Fairbanks, AK 99701

Robert Lewellen
Lewellen Arctic Research
P.O. Box 2089
Palmer, AK 99645

Robert F. Lewis
Alaska Dept. of Transportation
and Public Facilities
Pouch 6900
Anchorage, AK 99502

Antoni G. Lewkowicz (A-1)
Dept of Geography
University of Toronto
Toronto, Ontario M5S 1A1
CANADA

Li Yusheng (A-1, B-2)
Academy of Railway Sciences
Ministry of Railways
Beijing
PEOPLE'S REPUBLIC OF CHINA

Roy A. Lidgren
Public Works Canada
201 Ranger Road
Whitehorse, Yukon Territory
Y1A 3A4
CANADA

Herbert Liedtke (A-1, B-3)
Kellermannsweg 1
D4630 Bochum
WEST GERMANY

Arthur E. Linkins
Biology Dept.
VPI & SU
Blackburg, VA 24061

George Linkletter
Woodward-Clyde Consultants
203 N. Golden Circle Drive
Santa Ana, CA 92667

Liu Hongxu (A-2, B-4)
Low Temperature Construction
Science Research Institute
20 Oingbin Road
Harbin
PEOPLE'S REPUBLIC OF CHINA

Hal Livingston (A-2)
Alaska Dept. of Transportation
and Public Facilities
2301 Peger Road
Fairbanks, AK 99701

Thomas W. and Jan Livingston
Livingston Slone, Inc.
3900 Arctic Blvd.
Suite 301
Anchorage, AK 99503

Kevin J. Lomas
Civil Engineering Dept.
University of Technology
Loughborough
ENGLAND

Erwin Long
Arctic Foundation Inc.
5621 Arctic Blvd.
Anchorage, AK 99502

Ed Lopez
KPFF Consultant Engineers
750 W. 2nd Ave., Suite 207
Anchorage, AK 99501

Kenneth Lovelace
Alyeska Pipeline Service Co.
1835 S. Bragaw Street
Anchorage, AK 99512

C. William and Janet Lovell
School of Engineering
Purdue University
W. Lafayette, IN 47907

Douglas L. Lowery
Alaska Dept. of Environmental
Conservation
Pouch 1601
Fairbanks, AK 99701

Phil Lowry
Dept. of Geology
Louisiana State University
Baton Rouge, LA 70803

Lu Guowei (A-1, B-6)
Yakeshi Institute of Forestry
Survey and Design
Inner Mongolia
Yakeshi
PEOPLE'S REPUBLIC OF CHINA

Baerbel Lucchitta (A-2, B-3)
U.S. Geological Survey
2255 N. Gemini Drive
Flagstaff, AZ 86001

Virgil Lunardini
CRREL
72 Lyme Road
Hanover, NH 03755

David P. Lusch
Center for Remote Sensing
Michigan State University
302 Berkey Hall
East Lansing, MI 48824

Ulrich and Joanne Luscher
Woodward-Clyde Consultants
100 Pringle Avenue
Walnut Creek, CA 94596

H. A. MacAulay
Geological Survey of Canada
601 Booth Street
Ottawa K1A OE8
CANADA

Kaye MacInnes (A-2)
Land Resources
Box 1500
Yellowknife, N.W.T. X1A 2R3
CANADA

Stephen F. Mack
Alaska Division of Geological
and Geophysical Surveys
4420 Airport Way
Fairbanks, AK 99701

J. Ross and Mrs. Mackay (B-3)
Dept of Geography
University of British Columbia
Vancouver, B.C. V6T 1W5
CANADA

Daniel and Eileen Mageau
R & M Consultants, Inc.
59 Goldrush Estates
Fairbanks, AK 99701

Larry Mahar
3351 Ocana Avenue
Long Beach, CA 90808

Arshud Mahmood
McClelland-EBA, Inc.
4649 Business Park Blvd.
Anchorage, AK 99503

Joe P. Mahoney
University of Washington
Moore Hall, FX-10
Seattle, WA 98195

Max Maisch (A-2, B-4)
Geophysical Institute
University of Zurich
Postfach ICH-8033 Zurich
SWITZERLAND

Harri Makela (A-2 B-6)
VTT Geo Lab
SF 02150 Espoo 15
FINLAND

Seppo Maki
Viatek Ltd.
P.O. Box 02101 Espoo 10
FINLAND

Michel C. Malin
Dept of Geology
Arizona State University
Tempe, AZ 85287

Edward Maltby (A-2)
Geography Dept.
Exeter University
Exeter, Devon, 3X4 4RJ
ENGLAND

Chi-Sing Man (B-6)
Dept. of Civil Engineering
University of Manitoba
Winnipeg, Manitoba R3T 2N2
CANADA

Victor Manikian
ARCO Alaska Inc.
3317 Wentworth
Anchorage, AK 99504

Herb Mann
Fairbanks North Star Borough
P.O. Box 1267
Fairbanks, AK 99707

David Marrett
M/S AR-10
University of Washington
Seattle, WA 98195

Philip S. Marshall
AEIDC
University of Alaska
707 A Street
Anchorage, AK 99501

Mark Martinson
Alaska Dept. of Transportation
and Public Facilities
P.O. Box 60107
Fairbanks, AK 99701

Nabic Masri
Alaska OCS Reg.
District Minerals Office
411 W. 4th Ave, Suite 2A
Anchorage, AK 99501

Amos C. Mathews
Office of the Federal Inspector
P.O. Box 6619
Anchorage, AK 99502

Mark Mathews
Los Alamos National Lab
1532 42nd Street
Los Alamos, NM 87544

Ronald Matviyak
Mat-Viak Engineering
1311 4th Avenue
Fairbanks, AK 99701

Lawrence Mayo
U.S. Geological Survey
101 12th Avenue
Box 11
Fairbanks, AK 99707

E. Y. McCabe
Dept. of Civil Engineering
University of Aston, Gosta Green
Birmingham B4 7ET
ENGLAND

David McCaleb
Alaska Dept. of Transportation
and Public Facilities
2301 Peger Road
Fairbanks, AK 99701

Helen McCammon
Dept. of Energy
MS E201
Washington, D.C. 20545

Joseph McCloskey
Soil Conservation Service
2221 E. Northern Lights Blvd.
Anchorage, AK 99504

Patricia A. McCormick (A-2)
SR 30505
Fairbanks, AK 99701

John McCoy
Bureau of Land Management
701 C Street
Anchorage, AK 99513

Mike McCrum
Institute of Water Resources
University of Alaska
Fairbanks, AK 99701

Steven McCutcheon
Alaska Pictorial Service
P.O. Box 6144
Anchorage, AK 99502

Patrick McDevitt
Alyeska Pipeline Service Co.
1835 S. Bragaw Street
Anchorage, AK 99512

Jim McDougall
ESSO Research Canada Ltd.
237 Fourth Ave SW
Calgary, Alberta T5J 2R5
CANADA

Bo McFadden
Quadra Engineering
401 E. Fireweed Lane
Anchorage, AK 99503

Terry McFadden
Mechanical Engineering
University of Alaska
Fairbanks, AK 99701

Richard W. McGaw
CRREL
72 Lyme Road
Hanover, NH 03755

Pete McGee
Alaska Dept. of Environmental Conservation
Pouch 1601
Fairbanks, AK 99701

Robert L. McHattie
Alaska Dept. of Transportation and Public Facilities
2301 Peger Road
Fairbanks, AK 99701

Karen McKenna
Dept. of Renewable Resources
Government of Yukon
Box 2703
Whitehorse, Yukon Territory
CANADA

Harlan L. McKim
CRREL
72 Lyme Road
Hanover, NH 03755

William B. McMullen (B-2)
Alaska Dept. of Transportation and Public Facilities
600 University Ave., Suite B
Fairbanks, AK 99701

Donald C. Meares
Bureau of Land Management
P.O. Box 1150
Fairbanks, AK 99707

Brainerd Mears Jr. (A-2, B-4)
Department of Geology
University of Wyoming
Laramie, WY 82071

Jack Mellor
Bureau of Land Management
P.O. Box 1150
Fairbanks, AK 99707

P. I. Melnikov (B-6)
USSR Academy of Sciences
Council on Earth Cryology
11 Fersman St.
Moscow 117312
USSR

Karen S. Metz
4610 Edinburgh
Anchorage, AK 99502

Michael Metz (A-2, B-3)
Geo Tech. Service Inc.
14379 W. Ellsworth Ave
Golden, CO 80401

W. P. Metz
ARCO Alaska, Inc.
4610 Edinburgh
Anchorage, AK 99502

Kevin G. Meyer
Bureau of Land Management
701 C Street
Anchorage, AK 99513

Paul R. Meyer
AEIDC
University of Alaska
707 A Street
Anchorage, A, 99501

Fred Michel
Dept. of Geology
Carleton University
Ottawa, Ontario K1S 5B6
CANADA

Martti Miettinen
Viatek Oy
Pl 4 02101 Espoo 10
FINLAND

Tom and Helen Miklautsch
Miklautsch & Associates
1001 Noble
Fairbanks, AK 99701

Mark Miles
Institute of Water Resources
University of Alaska
Fairbanks, AK 99701

Bob R. Miller
TSA Systems Inc.
4919 N. Broadway
Boulder, CO 80302

Caroline A. Miller
P.O. Box 701
Eagle, River, AK 99577

Duane L. Miller
Duane Miller & Associates
SRA Box 2045B
Anchorage, AK 99507

John Miller
Geophysical Institute
University of Alaska
Fairbanks, AK 99701

P. Miller
Alaska Dept. of Transportation and Public Facilities
P.O. Box 60107
Fairbanks, AK 99701

Robert D. Miller (A-1)
Cornell University
538 Cayuga Heights Road
Ithaca, NY 14850

Robert E. Miller
Civil Engineer
University of Alaska
Anchorage, AK 99508

Wendell P. Miller
Alaska Dept. of Transportation and Public Facilities
5600 B Street, Suite 202
Anchorage, AK 99502

Alexey Mindich
8339 North Christiana
Skokie, IL 60076

Pavol Mitter (B-2)
Skolska 4 03101
Liptovsky Mikulas
CZECHOSLOVAKIA

Joseph P. Moore
Soil Conservation Service
2221 E. Northern Lights Blvd.
Anchorage, AK 99504

Daryl Moorhead
Ecology Program
University of Tennessee
Knoxville, TN 37916

John Morack
Physics Department
University of Alaska
Fairbanks, AK 99701

Norbert Morgenstern
University of Alberta
Dept.of Civil Engineering
Edmonton, Alberta T6G 2G7
CANADA

Drew Morris
Union Oil Co.
P.O. Box 6247
Anchorage, AK 99502

Leslie A. Morrissey
Technicolor Gov't Service Inc.
MS 242-2 Ames Research Center
Moffett Field, CA 94035

Thomas L. Moses
Alaska Dept. of Transportation
and Public Facilities
Pouch 6900
Anchorage, AK 99502

W. Moyle
Alaska Dept. of Transportation
and Public Facilities
2301 Peger Road
Fairbanks, AK 99701

George Mueller
Institute of Water Resources
University of Alaska
Fairbanks, AK 99701

Jeffrey L. Mueller (B-6)
Conoco, Inc.
1000 S. Pine - 111NT
Ponca City, OK 74603

Mark P. Muir
Quaternary Research Center
University of Washington AK-60
Seattle, WA 98195

Ernest and Wanda Muller (A-2)
Syracuse University
204 Heroy Geology Laboratory
Syracuse, NY 13210

Bob Munson
Dept. of Transportation
and Public Facilities
P.O. Box 60107
Fairbanks, AK 99701

Allen W. Murfitt
8050 King Street
Anchorage, AK 99502

Monica Murphy
General Delivery
Alakanuk, AK 99554

David and Barbara Murray
Museum
University of Alaska
Fairbanks, AK 99701

Douglas H. Murray
University of Alaska
P.O. Box 82282
Fairbanks, AK 99708

Ivan and Huong Muzik
Dept. of Civil Engineering
University of Calgary
Calgary, Alberta T2N 1N4
CANADA

John E. and Peggy C. Myrick
Northwest Alaskan Pipeline Co.
3333 Michelson Drive
Irvine, CA 92730

S. Nakagawa (B-6)
NKK America Inc.
444 S. Flower St., Suite 2430
Los Angeles, CA 90017

H. Nakajima (B-6)
NKK America Inc.
444 S. Flower Street, Suite 2430
Los Angeles, CA 90017

Lynette Nakazawa
Bureau of Land Management
701 C Street
Anchorage, AK 99513

Frank P. Narusch
Alaska Dept. of Transporation
and Public Facilities
Pouch 6900
Anchorage, AK 99502

Gerald and Carol Neave
Northern Seismic Analysis
RR #1
Echo Bay, Ontario POS 1CO
CANADA

Bonita Neiland
University of Alaska
P.O. Box 10095
Fairbanks, AK 99701

Fritz Nelson
Dept. of Geological Sciences
University of Michigan
Ann Arbor, MI 48109

Gerald E. Nelson
Department of Geology
Casper College
125 College Drive
Casper, WY 82601

Gordon Nelson
U.S. Geological Survey
1515 E. 13th Ave.
Anchorage, AK 99501

Robert Nesselhauf
1606 Marika Road, #2
Fairbanks, AK 99701

James F. Neuenschwander (B-3)
SOHIO
Midland Building Rm 877HB
Cleveland, OH 44115

Robert J. Neukirchner (B-6)
Geotec Service Inc.
14379 W. Ellsworth Ave.
Golden, CO 80401

Thomas C. Neunaber
Federal Highway Administration
P.O. Box 1648
Juneau, AK 99802

R. W. Newbury
994 Dorchester Ave.
Winnipeg, Manitoba R3MOS1
CANADA

David Nicholls
Soil Test Inc.
2205 Lee Street
Evanston, IL 60202

Bernard Nidowicz
Harding-Lawson Associates
601 East 57th Place
Anchorage, AK 99502

J. F. Nixon
Hardy Associates (1978) Ltd.
221-18th St. SE
Calgary, Alberta T2E 6J5
CANADA

Dennis J. Normile
Franklin & Allen, Inc.
1813 E First Ave., Suite 207
Anchorage, AK 99501

Nikolas Nugent
School of Mineral Industries
University of Alaska
Fairbanks, AK 99701

Dag Nummedal
Dept. of Geology
Louisiana State University
Baton Rouge, LA 70803

David A. Nuss
U.S. Air Force
PSC #4
Elmendorf AFB, AK 99506

Kenneth Nyman
Arco Pipe Line Co.
515 S. Flower St.
Los Angeles, CA 90071

Deane Oberste-Lehn (A-1)
P.O. Box 4706
Stanford, CA 94035

Ann Odasz
University of Colorado
Campus Box B-334
Boulder, CO 80309

William B. Odom
Sohio Alaska Petroleum Co.
Pouch 6-612 TP2-3
Anchorage, AK 99502

Walter C. Oechel
Ecosystems/Biology Department
San Diego State University
San Diego, CA 92182

OK Lumber
P.O. Box 10449
4 Mi. Old Steese Highway
Fairbanks, AK 99701

Richard Oldford
Alaska Dept. of Transportation
and Public Facilities
Pouch 6900
Anchorage, AK 99502

Joseph L. Oliphant
CRREL
72 Lyme Road
Hanover, NH 03755

R. Olmstead
Alaska Dept. of Transportation
and Public Facilities
P.O. Box 60107
Fairbanks, AK 99701

Kim Olszewski
Soil Conservation Service
P.O. Box F
Palmer, AK 99645

James P. Ormsby
NASA
1367 Poplar Hill Drive
Annapolis, MD 21401

Thomas E. Osterkamp (A-2)
Geophysical Institute
University of Alaska
Fairbanks, AK 99701

Ed Oswald
Pacific Forest Research Center
506 W. Burnside Road
Victoria, British Columbia V82 1M5
CANADA

Thomas R. Ottley
Alaska Dept. of Transportation
and Public Facilities
Pouch 6900
Anchorage, AK 99502

Samuel Outcalt
Dept. of Geological Sciences
University of Michigan
1006 C. C. Little Bldg.
Ann Arbor, MI 48109

Terry Owen (B-6)
Alaska Division of Geological
and Geophysical Surveys
794 University Avenue, Basement
Fairbanks, AK 99701

A. C. Palmer
R. J. Brown & Associates
Ryswijk
HOLLAND

Andre Pancza (B-4)
Prominade - 1
2056 Dombresson
SWITZERLAND

Jerry Parker
Sohio
Midland Building, Rm. 877HB
Cleveland, OH 44115

Walter Parker
Parker Associates, Inc.
3724 Campbell Airstrip Rd.
Anchorage, AK 99504

Bruce Parks
U.S. Geological Survey
101 12th Ave., Box 11
Fairbanks, AK 99701

Tony Pascoal
RR #4
Odessa, Ontario KOH 2HO
CANADA

Almo Paukkonen (B-6)
P.O. Box 22
79101 Leppavirta
FINLAND

Pertti Paukku (B-6)
S.A. Tervo OY
Kauppakatu 28
80100 Joensuu 10
FINLAND

Jon Paulsen
3581 Blue Lake Blvd
Blue Lake, CA 95525

Dan R. Pavey
Alaska Dept. of Transportation
and Public Facilities
Pouch 6900
Anchorage, AK 99502

Mike Pavone
Ebasco Service Inc.
400-112th Ave NE
Bellevue, WA 98004

Edward Penner
Division of Building Research
National Research Council Canada
Ottawa, K1A OR6
CANADA

Ed Perfect
Cornell University
#61 Bradfield Hall
Ithaca, NY 14853

Bob Peters
Century Engineering Inc.
500 L Street, Suite 200
Anchorage, AK 99501

John K. Petersen
Geophysical Institute
University of Alaska
Fairbanks, AK 99701

Robert J. Peterson
3M Company
6121 Coda Avenue
Edina, MN 55436

Troy and Mary Jean Pewe (A-1,B-4)
Dept. of Geology
Arizona State University
Tempe, AZ 85287

Walt Philips
Quadra Engineering
401 E. Fireweed Ave.
Anchorage, AK 99503

Janet Phillips
1420 North Nevada
Colorado Springs, CO 80907

Arvind Phukan
School of Engineering
University of Alaska
3221 Providence
Anchorage, AK 99504

Jean Pilon
Earth Physics Branch
1 Observatory Cres
Ottawa, K1A 0Y3
CANADA

Chien Lu Ping
Agricultural Experiment Station
P.O. Box AE
Palmer, AK 99645

David A. Pischer
Dames & Moore
155 NE 100th, Suite 100
Seattle, WA 98125

R. Platzke
Alaska Dept. of Transportation
and Public Facilities
2301 Peger Road
Fairbanks, AK 99701

Ovid A. Plumb
College of Engineering
Washington State University
Pullman, WA 99164

Jean Poitevin
5009 Clanranald, Apt. 37
Notre Dame De Grace
Montreal, Quebec
CANADA

Denise Poley
2519 - 17th Ave NW
Calgary, Alberta T2M 0R8
CANADA

Wayne H. Pollard (B-3)
Dept. of Geography
University of Ottawa
Ottawa, Ontario K1N 6N5
CANADA

Noel Potter, Jr. (A-2, B-4)
Dept. of Geology
Dickinson College
Carlisle, PA 17013

Suzanne Prestrud
Dept. of Geological Sciences
University of Washington, AJ-20
Seattle, WA 98195

Jonathan Price
112 Clarence Ave. North
Saskatoon, Saskatchewan S7N 1H8
CANADA

Dr. K. Priesnitz (B-6)
Dept. of Geology
Goldschmidtstr. 5
D-3400 Gottingen
WEST GERMANY

Nikola P. Prokopovich (A-1, B-4)
2800 Cottage Way
Sacramento, CA 95825

Anne C. Puffer
U.S. Forest Service
P.O. Box 1628
Juneau, AK 99802

Nam K. Pui
4903 Verulum Place, N Street
Calgary, Alberta T3A 0J9
CANADA

Qiu Guoqing (B-4)
Lanzhou Institute of Glaciology
and Cryopedology
Academia Sinica
Lanzhou
PEOPLE'S REPUBLIC OF CHINA

Charles H. Racine
Division of Environmental Studies
Johnson State College
Johnson, VT 05656

Stephen Rae
P.O. Box 80508
Fairbanks, AK 99708

Ernest R. Rahaim
S & S Engineering Inc
7125 Old Seward Highway
Anchorage, AK 99502

Eric Raisters
Sohio Alaska Petroleum
Pouch 6-612
Anchorage, AK 99502

Vernon and Joyce Rampton (A-2)
Terrain An/Mp Service Ltd.
P.O. Box 158
Carp, Ontario K0A 1L0
CANADA

Stuart Rawlinson (B-6)
Alaska Division of Geological
and Geophysical Surveys
794 University Ave., Basement
Fairbanks, AK 99708

Jerome Raychel
U.S. Army Corps of Engineers
Pouch 898
Anchorage, AK 99506

Richard Reanier
College of Forestry Research
University of Washington
Seattle, WA 98195

Michael W. Reed
Alyeska Pipeline Service Co.
1835 S. Bragaw Street
Anchorage, AK 99512

Richard D. Reger (B-4)
Alaska Division of Geological
and Geophysical Surveys
794 University Ave., Basement
Fairbanks, AK 99701

Robert L. Reid
Mechanical and Industrial
Engineering Dept.
University of Texas at El Paso
El Paso, TX 79968

Richard G. Reider
Dept. of Geology
University of Wyoming
Laramie, WY 82071

Stephen Reimer
University of California, Berkeley
3001 Fulton
Berkeley, CA 94705

Robert Rein
3106 128th NE
Bellevue, WA 98005

Eugenio Retamal
Idiem
Universidad De Chile
Casilla 1420
Santiago
CHILE

Robert W. Retherford
International Engineering Co.
P.O. Box 6410
Anchorage, AK 99502

Louis Rey
La Jaquiere
CH 1066 Epalinges
SWITZERLAND

Ronald D. Rhodehamel
U.S. Fish and Wildlife Service
3950 West Minster Way
Anchorage, AK 99504

John Rich
Bechtel Petroleum Inc.
716 Hillside Avenue
Albany, CA 94706

James Richardson
Bureau of Land Management
701 C Street
Anchorage, AK 99513

Ross D. Rieke
3235 West Viewmont Way West
Seattle, WA 98199

Daniel and Rosalind Riseborough
Carleton University
Dept. of Geography
Ottawa, K1S 5B6
CANADA

Bradley Ritchie
National Park Service
2525 Gambel Street
Anchorage, AK 99503

Richard W. Roberts
Mineral Management Service
P.O. Box 1159
Anchorage, AK 99510

Tom Roberts
Institute of Water Resources
314 Duckering Building
University of Alaska
Fairbanks, AK 99701

Scott Robertson
ARCO Alaska Inc.
P.O. Box 100360
Anchorage, AK 99507

Steve Rog
ERTEC Northwest Inc.
600 W 41st Street #203
Anchorage, AK 99503

Kevin E. Rogers
2009 Westside Drive
Cleveland, TN 37311

J. Romersberger
Alaska Dept. of Transportation
and Public Facilities
P.O. Box 60107
Fairbanks, AK 99701

Arthur D. Ronimus
Crews, MacInnes, Hoffman, Vitro
3812 Spenard Road, Suite 100
Anchorage, AK 99503

James W. and Florence Rooney
R & M Consultants
33925 Faerde Bay
Laguna Niguel, CA 92677

W. R. Rouse (B-3)
McMaster University
Dept. of Geography
1280 Main Street W
Hamilton, Ontario L8A 4K1
CANADA

Akshai Runchal
Analytic Computational Research
3106 Inglewood Blvd.
Los Angeles, CA 90066

Steve Ryan
Alaska Dept. of Transportation
and Public Facilities
Pouch 6900
Anchorage, AK 99502

William L Ryan
Ott Water Engineering
4790 Bussiness Park BD/S1
Anchorage, AK 99503

Bengt Erick Ryden
Physical Geography Department
Uppsala University
Box 554
S75122 Uppsala
SWEDEN

William Sackinger
Geophysical Institute
University of Alaska
Fairbanks, AK 99701

A. V. Sadovski
Research Institute of Foundations
and Underground Structures
Gosstroy-Niiosp
2-A Institutskaja 6
Moscow ZH389
USSR

Lyn Sainsbury
4918 Silver Spruce Lane
Evergreen, CO 80439

J. Sallstrom
Alaska Dept. of Transportation
and Public Facilities
P.O. Box 60107
Fairbanks, AK 99701

John Sanden
Legend Drill Service
P.O. Box 65
Galena, AK 99741

Haakan Sandstedt
Swedish St Power Bd.
S-16287 Vaellingby
SWEDEN

Barry W. Santana
ARCO Alaska, Inc.
P.O. Box 100360
Anchorage, AK 99510

Francis H. and Dorothy Sayles
CRREL
72 Lyme Road
Hanover, NH 03755

Greg Scharfen (A-1)
World Data Center/Glaciology
CIRES Campus Box 449
Boulder, CO 80309

Don Schell
Institute of Water Resources
University of Alaska
Fairbanks, AK 99701

Bob Scher
EXXON Co., USA
Pouch 6601
Anchorage, AK 99502

John F. Schindler
Husky Oil NPR
2473 Captain Cook Drive
Anchorage, AK 99503

R. D. Schlecht (A-1)
Phillips Petroleum
8055 E. Tufts Ave Park
Denver, CO 80237

Mick Schlegel
McClelland-EBA, Inc.
4649 Business Park Blvd.
Anchorage, AK 99503

Ruth A. Schmidt (B-6)
1040 C Street
Anchorage, AK 99501

Scott Schmidt
ARCO Alaska
P.O. Box 100360
Anchorage, AK 99503

Lee and Kathleen B. Schoen
Arco Oil and Gas Co.
P.O. Box 100939
Anchorage, AK 99510

Ekkehard Schunke
Dept. of Geography
Universität Göttingen
Goldschmidtstr 5
D-3400 Göttingen
WEST GERMANY

W. Haeberli Schwitte
Schutzenmott 71
8046 Zurich
SWITZERLAND

Peter C. Scorup
P.O. Box 775
Palmer, AK 99645

William N. Scott
Robertson and Associates
2820 C. Street, Suite 2
Anchorage, AK 99503

Keith W. Scoular (B-4)
1328 E. Lemon Street
Phoenix, AZ 85281

James M. Scranton (A-2)
University of Alaska
Fairbanks, AK 99701

D. C. Sego
Dept. of Civil Engineering
University of Alberta
Edmonton, Alberta T6G 2G7
CANADA

Rich Seifert
Institute of Water Resources
University of Alaska
Fairbanks, AK 99701

Bob Seitz
Instrumentation Service
124 2nd Avenue
Fairbanks, AK 99701

Ken Sellick
F.W. Sawatsky Limited
531 Marion Street
Winnipeg, Manitoba R2J 0J9
CANADA

Paul Sellmann
CRREL
72 Lyme Road
Hanover, NH 03755

Arno Semmel (A-2)
Theodor Koerer Str. 6
D-6238 Hofheim/TS
WEST GERMANY

Stephen Seniuk
Nordsen Industries
5 Chester Street
Winnipeg, Manitoba R2L 1W5
CANADA

John Senyk
Pacific Forest Research Center
506 W. Burnside Road
Victoria, B.C. V8Z 1M5
CANADA

Matti Seppala (B-3)
Department of Geography
University of Helsinki
Hallitusk 11-13 SF-00100
Helsinki 10
FINLAND

Charles W. Seslar
Federal Highway Administration
P.O. Box 1042
Juneau, AK 99802

Dick Sew
Gulf Canada Research
P.O. Box 310
Calgary, Alberta T2P 2H7
CANADA

Lewis H. Shapiro
Geophysical Institute
University of Alaska
Fairbanks, AK 99701

Gus R. Shaver
The Ecosystems Center
Marine Biology Laboratory
Woods Hole, MA 02543

Bill Sheffield
Office of the Governor
Pouch A
Juneau, AK 99811

Mark Sherman
SR 20144-X
Fairbanks, AK 99701

Richard G. Sherman (A-1)
Metcalf & Eddy
50 Stanford Street
Boston, MA 02114

Shi Yafeng (A-2, B-4)
Lanzhou Institute of Glaciology
and Cryopedology
Academia Sinica
Lanzhou
PEOPLE'S REPUBLIC OF CHINA

E. M. Shoemaker
Simon Fraser University
Dept. of Mathematics
Burnaby, British Columbia V5A 1S6
CANADA

Harold Shoemaker
U.S. Dept. of Energy
P.O. Box 880
Morgantown, WV 26505

Carl Siebe
Alaska Dept. of Transportation
and Public Facilities
Pouch 6900
Anchorage, AK 99502

Ramesh Singh
Mitra Niketan
Benaras Hindu University
Varanasi - 221005
INDIA

A. K. Sinha
Geological Survey of Canada
601 Booth Street
Ottawa, Ontario K1A 0E8
CANADA

James A. Sisk
Bureau of Land Management
P.O. Box 1150
Fairbanks, AK 99707

Stephen C. Sisk
Alaska Dept. of Transportation
and Public Facilities
2301 Peger Road
Fairbanks, AK 99701

Charles Slaughter
U.S. Forest Service
Institute of Northern Forestry
University of Alaska
Fairbanks, AK 99701

Joe Sledge
8450 Pioneer Drive
Anchorage, AK 99504

Charles and Marilyn Sloan
U.S. Geological Survey
1515 E. 13th Avenue
Anchorage, AK 99501

Madeleine Smith
9 Francis Drive
Laughborough, LE11 0FE
ENGLAND

Michael W. Smith
Dept. of Geography
Carleton University
Ottawa, K1S 5B6
CANADA

Peggy A. Smith (B-4)
U.S. Geological Survey
345 Middlefield Road MS90B
Menlo Park, CA 94025

Robert Smith
ARCO Oil and Gas
P.O. Box 2819
Dallas, TX 75221

Scott Smith
Box 2703
Whitehorse, Yukon Territory
CANADA

Stephen Smith
Dynatech B/D Co.
99 Erie Street
Cambridge, MA 02139

Gerhard Sonntag (A-1, B-4)
Technical Universität Mech C
Arcisstrasse 21
D-8000 Munchen 2
WEST GERMANY

David Stanley
S & S Engineering, Inc.
7127 Old Seward Highway
Anchorage, AK 99502

Michael J. Stanley
Alyeska Pipeline Service Co.
1835 S. Bragaw Street
Anchorage, AK 99512

Alex P. Stansen
Stearns Roger Engineering Corp.
Box 5888
Denver, CO 80217

S. Russell & Lee H. Stearns
Thayer School of Engineering
Dartmouth College
Hanover, NH 03755

Carl H. Steeby
International Engineering Co.
9271 Campbell Ter.
Anchorage, AK 99502

Jean Stein
Institute of Water Resources
University of Alaska
Fairbanks, AK 99701

George C. Stephens (A-1, B-4)
Department of Geology
George Washington University
Washington, D.C. 20550

Eric and Helma Stewart
ARCO Oil and Gas Co.
P.O. Box 6105
Anchorage, AK 99502

Edward Stondall
National Park Service
2525 Gambell Street
Anchorage, AK 99503

Michael G. G. Stoner
M/S AR-10
University of Washington
Seattle, WA 98195

Edward M. Stram
Earl & Wright Cons. Engineers
One Market Plaza
Spear Street Tower
San Francisco, CA 94105

Carl A. and M. Sam Strande
Anglo Alaska Construction
Box 10191 Curry's Corner
Fairbanks, AK 99701

Horst Strunk (A-2)
Geographisches Institut
der Universität
Universitätsstr. 31
D-840 Regensburg
WEST GERMANY

Robert Stuart
Office of the Federal Inspector
1001 Noble Street
Fairbanks, AK 99701

Christopher W. Stubbs
Quaternary Research Center
University of Washington AK-60
Seattle, WA 98195

Edward and Maree Stuber (A-1)
Assoc. Pile and Fit Corp.
1750 Taylor St, Apt. 1102
San Francisco, CA 94133

Juerg Suter (A-2, B-4)
Geophysical Institute
University of Zurich
Postfach 1 CH-8033
Zurich
SWITZERLAND

Harald Svensson (B-3)
Geophysical Institute
University of Copenhagen
Haraldsgade 68
DK 2100 Copenhagen
DENMARK

Joe Svoboda (B-4)
Dept. of Botany
University of Toronto
Erindale Campus
Mississauga, Ontario L5L 1C6
CANADA

Larry Sweet
Alaska Dept. of Transportation
and Public Facilities
2301 Peger Road
Fairbanks, AK 99701

Dan Swift
Geophysical Institute
University of Alaska
Fairbanks, AK 99701

Charles Tarnocai (B-3)
Land Resource Research Inst.
K. W. Neatby Building
Central Experimental Farm
Ottawa, Ontario K1A 0C6
CANADA

Rupert G. Tart
6448 Village Parkway
Anchorage, AK 99504

Alan Taylor
Earth Physics Branch
1 Observatory Cres.
Ottawa, K1A 0Y3
CANADA

R. B. Taylor (B-6)
Bedford Institute Oceanography
Box 1006 Dartmouth
Nova Scotia, B2Y 4AZ
CANADA

Al S. Telford
ESSO Research Canada Ltd.
237 Fourth Ave SW
Calgary, Alberta T5J 2R5
CANADA

Jean and V. Terasmae (A-1)
Brock University
196 Woodside Drive
St. Catherines, Ontario L2T 1X6
CANADA

William R. Theuer
Fairbanks Native Association
301 1st Avenue
Fairbanks, AK 99701

Howard P. and Ingrid Thomas
Woodward-Clyde Consultants
2611 Brittany Drive
Anchorage, AK 99504

Jim Thomas
8101 Fairwood Circle
Anchorage, AK 99502

Dorothy H. Thompson
Alaska Department of Natural
Resources
P.O. Box 80368
Fairbanks, AK 99708

Grant R. and Iris Thompson
Mobil Research & Development Corp.
320 E. Little Creek Rd.
Cedar Hill, TX 75104

Thora E. Thorhallsdottir (A-2)
Institute of Biology
University of Iceland
Grensasvegur 12
Reykjavik
ICELAND

Colin E. Thorn (B-3)
Dept. of Geology
University of Illinois
Urbana, IL 61801

Jules Tileston
Bureau of Land Management
701 C Street
Anchorage, AK 99513

Philip Tilley (A-2, B-3)
Dept. of Geography
University of Sydney
NSW 2006
AUSTRALIA

Richardo A. A. Todeschini
Bechtel Group Inc.
P.O. Box 3965
San Francisco, CA 94119

Sahin Toraman
DOWL Engineers
4040 B Street
Anchorage, AK 99504

Zelma G. Trafton
P.O. Box 1116
Fairbanks, AK 99707

Ted and Gloria Trueblood (A-1)
Alaska Railroad
1621 Wintercrest Drive
Anchorage, AK 99504

Randall G. Updike (B-4)
Alaska Division of Geological
and Geophysical Surveys
3001 Porcupine Drive
Anchorage, AK 99701

B. Urban
Alaska Dept. of Transportation
and Public Facilities
P.O. Box 60107
Fairbanks, AK 99701

Michael E. Utt
Union Oil Co. of California
P.O. Box 76
Brea, CA 92621

Keith Van Cleve
Forest Soils
University of Alaska
Fairbanks, AK 99701

D. A. Van Everdingen
10 Hickory Street, E.
Waterloo, Ontario N2J 4L4
CANADA

Robert A. Van Everdingen (B-3)
Environment Canada, NHRI
101-4616 Valiant Drive NW
Calgary, Alberta T3A 0X9
CANADA

Doug Van Patton
Soil Conservation Service
P.O. Box 415
Homer, AK 99605

J. Vandenberghe (B-3)
P.O. Box 7161
1007 MC Amsterdam
THE NETHERLANDS

H. Veldhuis
Canada-Manitoba Soil Survey
University of Manitoba
Winnipeg, Manitoba R3T 2N2
CANADA

Ralph P. Vendeland (A-1)
EXXON Company USA
P.O. Box 5025
Thousand Oaks, CA 91359

Leslie A. Viereck
U.S. Forest Service
Institute of Northern Forestry
University of Alaska
Fairbanks, AK 99701

Ted S. Vinson
Dept. of Civil Engineering
Oregon State University
Corvallis, OR 97331

Charles L. Vita
R & M Consultants
#380-19700 Fairchild
Irvine, CA 92715

John D. Vitek (B-3)
Dept. of Geology
Oklahoma State University
Stillwater, OK 74078

David Vogel (B-4)
Alaska Division of Geological and Geophysical Surveys
794 University Ave. (Basement)
Fairbanks, AK 99701

David Walker
Sohio Petroleum Co.
100 Pine Street
San Francisco, CA 94111

Donald and Marilyn Walker
Inst. Arctic Alpine Research
University of Colorado
Box 450
Boulder, CO 80309

Gerald Walker
Geophysical Institute
University of Alaska
Fairbanks, AK 99701

H. Jesse Walker (B-6)
Dept. of Geography
Louisiana State University
Baton Rouge, LA 70803

Kerry E. Walker
2731 Delta Circle
Eagle River, AK 99577

Henry Walters
U.S. Army Corps of Engineers
Pouch 898
Anchorage, AK 99506

James C. Walters (B-4)
Dept. of Earth Science
University of Northern Iowa
Cedar Falls, IA 50614

David W. H. Walton
British Antarctic Survey
Madingley Road
Cambridge, CB3 OET
UNITED KINGDOM

Wang Hong (A-1, B-2)
Research Division
First Survey of Des Inst.
Ministry of Railways
Lanzhou
PEOPLE'S REPUBLIC OF CHINA

Wang Liang (A-2)
Water Power Research Institute
74 Beian Road
Changchun, Jilin
PEOPLE'S REPUBLIC OF CHINA

Ronald W. Ward
Sohio Petroleum Co.
5400 LBJ Freeway, Suite 1200/LB25
Dallas, TX 75240

A. Lincoln and Tahoe Washburn
Quaternary Research Center #AK-60
University of Washington
Seattle, WA 98195

T. Watari (B-6)
NKK America Inc.
444 S Flower Street, Suite 2430
Los Angeles, CA 90017

William J. and Naomi L. Wayne
Dept. of Geology
University of Nebraska
Lincoln, NB 68588

Monte Weaver
Alaska Dept. of Transportation and Public Facilities
Pouch 6900
Anchorage, AK 99502

Patrick Webber (A-2)
Inst. Arctic Alpine Research
University of Colorado
Campus Box 450
Boulder, CO 80309

Florence R. Weber
U.S. Geological Survey
P.O. Box 80745
Fairbanks, AK 99708

Francis Weeks (A-1)
Alaska Railroad
3702 Alexander Avenue
Anchorage, AK 99508

P.C. Weerdenburg
EBA Engineering Consultants Ltd.
145535-118th Avenue
Edmonton, Alberta T5L 2M7
CANADA

E. Wellen
Alaska Dept. of Transportation and Public Facilities
P.O. Box 60107
Fairbanks, AK 99701

Paula Wellen
Institute of Water Resources
University of Alaska
Fairbanks, AK 99701

Gunter Weller
Geophysical Institute
University of Alaska
Fairbanks, AK 99701

John Westgate
University of Toronto
Scarborough Campus
1265 Military TR
Ontario M1C 1A4
CANADA

Ruth L. and Raymond Wexler
Engineering Topographic Lab.
9511 Linden Avenue
Bethesda, MD 20814

W. Brian Whalley
Dept. of Geography
Queens University
Belfast BT7 INN
UNITED KINGDOM

John R. Williams
U.S. Geological Survey
345 Middlefield Road
Menlo Park, CA 94025

Peter J. Williams
Dept. of Geography
Carleton University
Ottawa, Ontario K1S 5B6
CANADA

R. B. G. Williams (B-3)
Geography Lab
University of Sussex
Brighton, Sussex
UNITED KINGDOM

James and Sandra Willis (B-3)
Anglo Alaska Construction
1/2 Mile Pebble Drive
Ester, AK 99725

John Willmott
Oxford University
Hertford College
Oxford OX1 3BW
UNITED KINGDOM

Jerry C. Wilson
Dames & Moore
800 Cordova
Anchorage, AK 99501

M. J. Wilson
University of Colorado
115 S 31st
Boulder, CA 80303

Monte D. Wilson
Dept. of Geology and Geophysics
Boise State University
Boise, ID 83725

Peggy Winslow
Earth Science Dept.
University of North Carolina
Wilmington, NC 28406

Ann Wintle
Godwin Laboratory
Cambridge University
Cambridge
ENGLAND

Ming-Ko Woo (B-3)
Geography Department
McMaster University
Hamilton, Ontario L8S 4K1
CANADA

John Wood
Geography Dept.
Carleton University
Ottawa, Ontario K1S 5B6
CANADA

Ming-Chee Wu
University of Alaska
Fairbanks, AK 99701

Tien Hsing and Pei Hsing Wu
Ohio State University
2070 Neil Avenue
Columbus, OH 43210

Albert F. Wuori
CRREL
72 Lyme Road
Hanover, NH 03755

Xia Zhao Jun
Dept. of Hydraulic Engineering
Northeast Agricultural College
Harbin, Heilongjiang
PEOPLE'S REPUBLIC OF CHINA

Xu Bomeng (A-2)
Water Power Research Institute
74 Beian Road
Changchun, Jilin
PEOPLE'S REPUBLIC OF CHINA

Xu Xiaozu
Lanzhou Institute of Glaciology
and Cryopedology
Academia Sinica
Lanzhou
PEOPLE'S REPUBLIC OF CHINA

Edward Yarmak
Arctic Foundation Inc.
5621 Arctic Blvd.
Anchorage, AK 99503

Edmond J. Yaroch
Adh Coat & Sel Div-3M
3M Center, Bulding 209-1C-13
St. Paul, MN 55144

Bing and Ann M. Yen (A-1)
Dept. of Civil Engineering
California State University
Long Beach, CA 90840

Kiyoshi Yoneyama
Tokyo Gas Co., Ltd.
16-25, Shibaura 1-Chome
Minatoku, Tokyo 105
JAPAN

Ki Ho Yoon
Alyeska Pipeline Service Co.
1835 S. Bragaw St.
Anchorage, AK 99512

Yurii F. Zakharov (B-6)
Tyumen Gazprom
U.S.S.R. Ministry of Gas Industry
NIIG Iprogaz, VPO
Tyumen
USSR

Boris Zamsky (A-2)
EXXON Production Research Co.
P.O. Box 2189
Houston, TX 77001

John P. Zarling
Mechanical Engineering
University of Alaska
Fairbanks, AK 99701

Alvin Zeman (A-1)
Rittenhouse-Zeman & Assoc. Inc.
13837 NE 8th Street
Bellevue, WA 98005

John Zeman
Dept. of Geography
University of British Columbia
Vancouver, B.C. V6T 1R7
CANADA

Zhao Yuehai (A-1, B-2)
Foreign Affairs Bureau
Ministry of Railways
Beijing
PEOPLE'S REPUBLIC OF CHINA

Zhao Yunlong (A-1, B-2)
Institute of Railway Sciences
Qiqihar
PEOPLE'S REPUBLIC OF CHINA

Zhou Youwu (A-1, B-4)
Lanzhou Institute of Glaciology
and Cryopedology
Academia Sinica
Lanzhou
PEOPLE'S REPUBLIC OF CHINA

Zhu Qiang (A-2)
Water Con Dept. Gansu Province
QI-LI-HE, Lanzhou
Gansu
PEOPLE'S REPUBLIC OF CHINA

Zhu Yuanlin
Lanzhou Institute of Glaciology
and Cryopedology
Lanzhou
PEOPLE'S REPUBLIC OF CHINA

Gary W. Zielinski
Dept. of Applied Science
Brookhaven National Lab.
Upto, NY 11973

Paul Zieman
Institute of Water Resources
University of Alaska
Fairbanks, AK 99701

Winston Zirjacks
Frank Moolin & Assoc.
2525 C Street
Anchorage, AK 99503

Chris Zuba
University of Calgary
ES104/2500 University Drive NW
Calgary, Alberta T2N 1N4
CANADA

Author Index

Available Publications of International Permafrost Conferences

National Research Council, 1966. Permafrost: International Conference Proceedings. NRC Pub. 1287. Washington, D.C.: National Academy of Sciences. Available on microfilm in The Cold Regions Science and Technology Bibliography #25-3138, U.S. Library of Congress or NTIS.

National Academy of Sciences, 1973. Permafrost: The North American Contribution to the Second International Conference, Yakutsk. Washington, D.C.: National Academy of Sciences. $50.00

National Academy of Sciences, 1978. Permafrost: USSR Contribution to the Second International Conference. Washington, D.C.: National Academy of Sciences. $19.50

National Research Council of Canada, 1978. Proceedings of the Third International Conference on Permafrost, 2 volumes. Ottawa, Ontario: National Research Council of Canada. $35.00

National Research Council of Canada, 1980. Third International Conference on Permafrost, English Translations of the Soviet Papers, 2 parts. Ottawa, Ontario: National Research Council of Canada. $35.00

University of Alaska, 1983. Permafrost: Fourth International Conference. Abstracts and Program. Fairbanks: University of Alaska. $10.00

National Academy of Sciences, 1983. Permafrost: Fourth International Conference, Proceedings. Washington, D.C.: National Academy Press. $65.00

National Academy of Sciences, 1984. Permafrost: Fourth International Conference, Final Proceedings. Washington, D.C.: National Academy Press. $32.50.

Permafrost: A Bibliography, 1978-1982. Glaciological Data Report GD-14, Boulder, World Data Center for Glaciology, 1983, 172 p., $5.00. (University of Colorado, Box 449, Boulder, Colorado 80309)